Third Edition

# Electric Drives

Third Edition

# Electric Drives

Ion Boldea • S.A. Nasar

CRC Press
Taylor & Francis Group
Boca Raton London New York

CRC Press is an imprint of the
Taylor & Francis Group, an **informa** business

CRC Press
Taylor & Francis Group
6000 Broken Sound Parkway NW, Suite 300
Boca Raton, FL 33487-2742



Printed on acid-free paper
Version Date: 20160115

International Standard Book Number-13: 978-1-4987-4820-9 (Hardback)



**Library of Congress Cataloging-in-Publication Data**

Names: Boldea, I. | Nasar, S. A.
Title: Electric drives, / authors, Ion Boldea and S.A. Nasar.
Description: Third edition. | Boca Raton : Taylor & Francis, CRC Press, 2016.
| Includes bibliographical references and index.
Identifiers: LCCN 2016000766 | ISBN 9781498748209 (alk. paper)
Subjects: LCSH: Electric driving.
Classification: LCC TK4058 .B64 2016 | DDC 621.46--dc23
LC record available at http://lccn.loc.gov/2016000766

**Visit the Taylor & Francis Web site at**
**http://www.taylorandfrancis.com**

**and the CRC Press Web site at**
**http://www.crcpress.com**

Printed and bound in the United States of America by Publishers Graphics, LLC on sustainably sourced paper.

# Contents

# *Preface to the Third Edition*

Ten years (the length of the war in Troy!) after the second edition of *Electric Drives* was published in 2006, we have quite a few reasons to come up with a third (revised) edition, given the steady and strong recent progress in the technology related to energy savings and increased industrial productivity.

After carefully weighing what to maintain and what new material to introduce in order to update the contents without confusing the existing reader (the instructor of the previous editions), here is what we decided to offer in the third edition:

- Check again every phrase, equation, number in numerical examples, and reference, and reflect the necessary changes all over the book.
- Add new, recent references related to existing and new subjects of R&D to refresh readers' views on new trends in the field.
- Add new paragraphs/chapters summarizing around 120 new pages related to subjects of recent representative progress, mainly in the technology of sensorless control of a.c. motor drives:

  1.10 Electric Drive Applications

  8.21 Split-Phase Capacitor IM Transients

  8.21.1 Phase Variable Model

  8.21.2 d-q Model

  9.17 Predictive Control

  10.11 Single-Phase PMSM FEM Modeling for Transients

  11.6.2 Wide Speed Range Sensorless Control of PM-RSM via "Active Flux:" A Case Study

  11.7.4 Sensorless Control of RSM: A Case Study

  14.6 Sensorless Control of a D.C.-Excited SM via "Active Flux:" A Case Study

  15.6.2 PMSG Sensorless Control: A Case Study

- Introduce two new/overview chapters—Chapters 16 and 17—related to two important recent R&D trends in electric drive:

  Chapter 16: Scalar V/f and I–f Control of A.C. Motor Drives: An Overview

  Chapter 17: Multiphase Electric Machine Drives: An Overview

- As we added new material, it was decided to retain all solved numerical examples (problems) and the **10 MATLAB®–Simulink programs available online**. To help develop applications, we strongly urge the diligent reader to take up the new complex case studies mentioned above, write their own MATLAB–Simulink codes for them, to check our results, and even implement at least one of them, to grasp deep-rooted knowledge.

- As is easily visible, we decided to keep most of the content of the second edition as it is fundamental and not outdated.
- Again, the third edition of *Electric Drives* has the attributes of a two-semester (one semester on selected chapters) textbook—with almost enough coverage of power electronics included. It also has attributes of an up-to-date monograph in the field.
- Though we tried to keep the "math" under control, we recognize its wide usage and usefulness in presenting ready-to-use quantitative analysis for modeling, designing, and testing modern power electronics electric drives in low- and high-power range.
- We have ourselves taught yearly through the editions of *Electric Drives* and the third edition reflects our teaching experience. As any book reflects the horizon, limits, and experience of the author(s), we humbly await readers' criticism.

Thanks are due to Dr. Ana Popa for computer-editing the changes for the third edition of *Electric Drives* and to the competent and friendly team of CRC Press who edited the book.

**Ion Boldea**
*IEEE Life Fellow*
*University Politechnica and Romanian Academy*
*Timisoara, Romania, 2016*

Additional material is available from the CRC Press Website: http://www.crcpress.com/product/isbn/9781498748209.

# *Preface to the Second Edition*

The electric drive technologies have grown and matured notably in the last 10 years and prompted this second edition of our book.

Our main problem was how to introduce briefly the required new knowledge without confusing the existing users (faculty and engineers in R&D and manufacturing), while providing for usable new knowledge.

This is what we came up with:

- We added with almost all existing chapters recent literature and a few paragraphs, but as much as possible at the end of the chapter.
- We have introduced new paragraphs on a.c. brush series motor and its drive, on capacitor-split inductor motor and its drive, on single-phase PMSM and on single-phase switched reluctance motors and on tooth wound PMSM in their existing chapters, with numerical examples to facilitate a feeling of magnitudes for the new knowledge.
- For the main a.c. drives, with induction and synchronous motors, we have introduced case studies with ample simulation and test results on advanced motion sensorless control with direct torque and flux control (DTFC) with space vector modulation (SVM) to demonstrate very large speed range (about 1000/1 for IM and 200/1 for the synchronous motor) without signal injection for state estimation.
- We added a complete new (last) chapter on "Control of electric generators." Control at constant and variable speed through power electronics is approached, for excited synchronous, PM synchronous, wound rotor induction and switched reluctance generator, with sample results and applications from standard and distributed power systems (with renewable energy conversion) through standby and standalone to vehicular applications (automotive, railroad, marine and aircraft).

We felt that this is a fast growing breed of technology with large worldwide markets and its place is here with electric drives as electric machines are reversible. However, fast, robust active and reactive power control in constant and variable speed generators exhibits notable peculiarities that need separate treatment in contrast to the so-called regenerative braking of electric drives:

- To the 8 Matlab/Simulink close loop power electronics computer programs: 3 for the IM, 3 for the PMSM, one for the d.c. brush motor and one for SRM we added two more:
    - The doubly fed (wound rotor) induction generator control through a rotor connected bidirectional a.c., d.c., a.c. PWM converter that is used for pump storage hydro and wind energy conversion today.
    - The single phase PMSM standard and low switch count power electronics drive for low power low cost applications.

- We decided not to add to existing proposed problems (as is standard in text books) because today's reader is very busy and because, instead he has now 10 Matlab/Simulink computer programs for 10 different close-loop basic and up to date electric drives to exercise with, make changes in order to deepen his expertise in the field on his own.
- We do believe that the book may be used both as an undergraduate course (one part of it) and a graduate course (the other part of it) with the instructor free to choose his chapters and the corresponding paragraphs of interest.
- Additionally, the book has strong attributes of an up to date monograph that should inspire future work for the MS and PhD students, R&D engineers and new faculty.
- The CD interactive version contains the whole text, selected slides and the solution's manual for the instructor and the 10 Matlab/Simulink simulated close-loop power electronics drives.
- The 10 Matlab/Simulink programs are marked in the paper book by the Matlab icon!
- Special thanks are due to our PhD Student Cristian Ilie Pitic who diligently computer edited this second edition.

**Ion Boldea**
*Timisoara, Romania,*

# Preface to the First Edition

Industrial motion, torque, speed, position, control is paramount in raising productivity and quality and in reducing energy and equipment maintenance costs in all industries.

Electric drives share most of industrial motion control applications. Variable speed is inherent in modern electric drives. Such drives contain various high performance motors, power electronic converters and digital control systems. They are provided with input power filters and comply with electomagnetic interference (EMI) standards. The international markets for electric drives are rising at a 10% annual rate and already enjoy a multibillion U.S. dollars (USD) market worldwide.

This book presents a comprehensive view of modern (variable speed) electric drives. It may be considered as both a practical textbook and as an up-to-date monograph. So the readership spectrum is expected to be broad, from senior and graduate students to R&D and plant engineers who are interested in getting in depth and panoramic knowledge on various electric drive solutions in terms of topology, performance, design elements, digital simulation programs and test results, practical issues in industrial drives.

In order to follow the book, prior thorough knowledge of electric machinery or power electronics is not required. But two semesters of electric circuits and one of linear systems are required.

Various brushless motor (of induction, permanent magnet (PM) or reluctance, switched reluctance types) drives as well as d.c. brush motor drives are dedicated special chapters. Separate chapters are devoted to practical issues in pulse width modulation converter drives and to various large power drives.

There are plenty of solved numerical examples and proposed problems throughout the book. Also digital simulation results from 8 Matlab/Simulink programs are included.

## Book Core

The book is structured into 14 chapters. The first chapter deals with energy conversion in electric drives, application range, saved energy pay-back, typical loads, motion/time profiles or multiquadrant operation, etc.

Electric motor types (configurations) for electric drives are described in Chapter 2. Conventional a.c. motors, d.c. brush motors, and power electronic converter dependent motors are all included.

Chapter 3 synthesizes the main power electronic converters for electric drives. Power electronic switches and various diode rectifiers are treated in some detail while phase-controlled rectifiers, d.c. choppers, voltage-source PWM inverters, current-source inverters and cycloconverters are only briefly reviewed.

Topologies, performance and state space equations of d.c. brush motors are treated in Chapter 4.

Chapters 5 and 6 treat, almost exclusively through explained numerical examples, various rectifier and d.c. chopper fed brush motor drives, respectively. A strong feeling of phenomena and magnitudes is obtained this way. A digital simulation program in Matlab/Simulink treats 4 quadrant d.c. chopper brush motor drives.

Chapter 7 deals with closed-loop motion (torque, speed, position) control in electric drives. Cascade motion controllers, state space controllers as well as non-linear controllers such as sliding mode, fuzzy logic and neuro-fuzzy systems are reviewed and exemplified on d.c. chopper brush motor drives.

Chapter 8 treats the construction, performance, characteristics and space-phasor models for induction motors in electric drives.

Chapter 9 deals extensively with most aspects of modern induction motor drives. Various vector control methods, direct torque and flux control (DTFC) with or without motion sensors are treated in detail, including 3 Matlab-Simulink digital simulation programs. Sensorless control via scalar control methods (V/f with slip compensation) is also included.

Chapter 10 presents PM and reluctance (or hybrid) synchronous motors and their characteristics and d-q (space phasor) models for drives.

Chapter 11 is dedicated to vector (d-q) control and direct torque and flux control (DTFC) of PM and reluctance synchronous motors. Besides sinusoidal, trapezoidal control is included. Three Matlab/Simulink programs are provided to explore the performance of such drives during steady state and transients.

Chapter 12 deals with switched reluctance motors and drives. General and high grade (servo) drive solutions are presented both with and without motion sensors. A Matlab/Simulink program deals with digital simulations for both motoring and regenerative braking operations of such drives.

Chapter 13 includes practical issues with PWM converter drives (all drives presented so far). Input (grid) current harmonics, long power cable effects, bearing currents and remedies are presented in considerable detail.

Chapter 14 treats representative large power drives which use GTOs and thyristors. Three level voltage source inverter and cycloconverter excited synchronous motor drives are followed by current source converter synchronous motor drives and finally by sub- and hyper-synchronous four-quadrant induction motor cascade drives.

## Ways to Use the Book

The book may be considered as a two-semester textbook where the first 7 chapters represent the first semester. It may also be used, through selective paragraphs (chapters), as a one-semester textbook.

Finally the advanced reader may study paragraphs or chapters of his own choice, skipping the basics and jumping directly to new, inspiring knowledge of the subject matter.

### The CD Interactive Version

The CD version, also available through CRC Press, contains:

- The whole text (for browsing);
- Problem solutions;
- Selected slides (to extract) for the teacher;

- 8 digital simulation Matlab/Simulink programs which are highly interactive and user-friendly. They refer to closed loop d.c. chopper drives (1), vector control and DTFC (sensorless) induction motor drives (3), PM or reluctance synchronous motor drives with rectangular current and vector control (3), and switched reluctance motor drives (1). These digital simulation programs written in a known programming medium, provided with comprehensive helps, may constitute a special vehicle to an in-depth independent study of most representative modern electric drives.

## Acknowledgment

The authors express their gratitude to Mr. Jánosi Loránd, their Ph.D. student who worked out the 8 digital simulation programs and computer-edited the entire book.

The rather new philosophy of this book of electrical drives, which does not require prior knowledge of electric machines and power electronics, and includes comprehensive digital simulations and an interactive CD version of the book, is bound to produce, besides beneficial consequences, some secondary effects, also.

Consequently, the authors will welcome readers' feedback.

*Timisoara, Romania*
*Lexington, KY, USA*

# 1

# Energy Conversion in Electric Drives

## 1.1 Electric Drive: Definition

An electric drive is an industrial system that performs the conversion of electrical energy to mechanical energy (in motoring) or vice versa (in generator braking) for running various processes such as production plants, transportation of people or goods, home appliances, pumps, air compressors, computer disk drives, robots, music or image players, etc.

Approximately 50% of the electrical energy produced is used in electric drives today.

Electric drives may run at a constant speed (Figure 1.1) or at variable speeds (Figure 1.2).

The constant speed electric drive constitutes an electric (alternating current [a.c.]) motor, mechanical coupling, mechanical load (plant), and an electromechanical (or electronic) start/stop and protection system. Today, approximately 60%–65% of all electric drives still run at a constant speed because there is not much need for speed control except for starting, stopping, and protection.

However, there is a 35%–40% group of applications, with very high annual expansion rate where the torque and speed must be varied to match the mechanical load.

A typical variable speed electric drive (Figure 1.2) constitutes, in addition, a power electronic converter (PEC) to produce energy conservation (in pumps, fans, etc.) through fast, robust, and precise mechanical motion control, as required by the application (machine tools, robots, computer-disk drives, transportation means, etc.)

Even though constant speed electric drives could constitute a subject for a practically oriented book, in this book we will refer only to variable speed electric drives that make use of power electronic converters (PECs). For brevity, however, we use the simple name of electric drives.

## 1.2 Application Range of Electric Drives

A summary of the main industrial applications and power range of electric drives is shown in Figure 1.3.

Singular applications such as pumped-storage hydroelectric plants are now built for unit powers of 100 MW or more. High performance means, in Figure 1.3, wide speed ranges, fast and precise response in speed, or position control.

Traditionally, for variable speed, d.c. (direct current) brush motors have been used for decades [1] but a.c. motors [2,3] have been catching up lately (since 1990) as shown in Figure 1.4. This radical shift is mainly due to the rapid progress in PECs for a.c. motors [4].

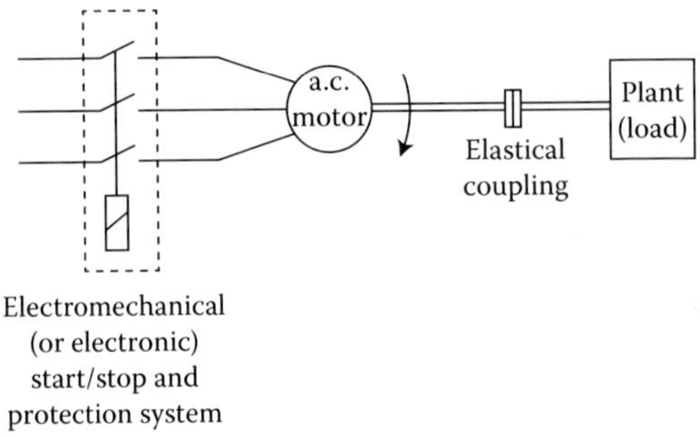


**FIGURE 1.1**
Constant speed electric drive.

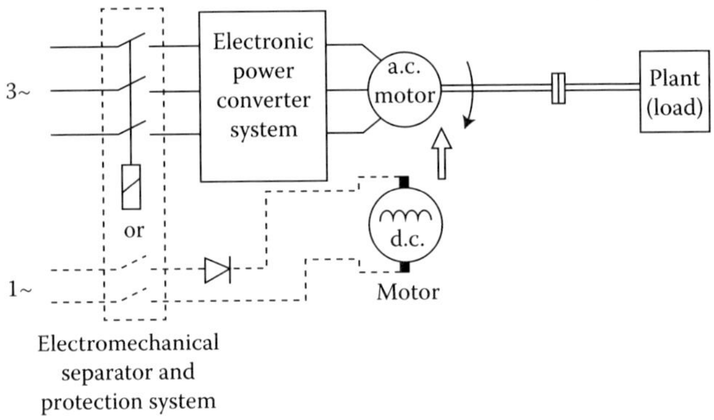


**FIGURE 1.2**
Variable speed electric drive.

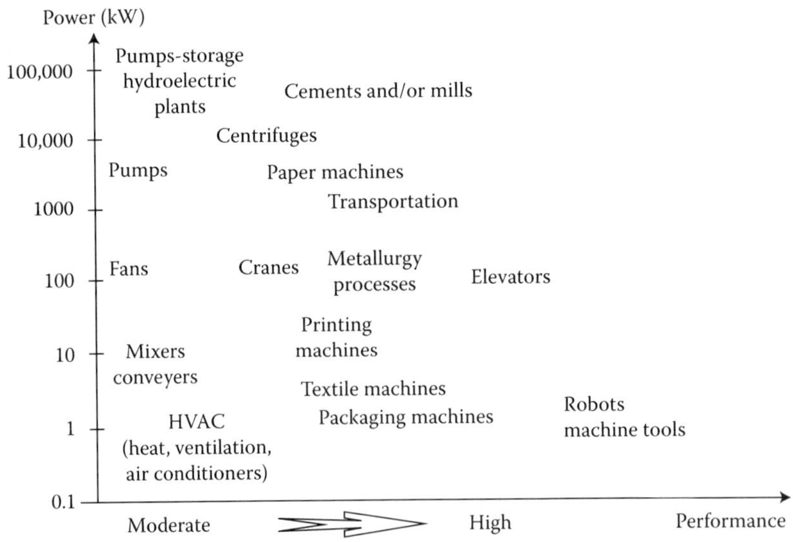


**FIGURE 1.3**
Electric drives—variable speed applications.

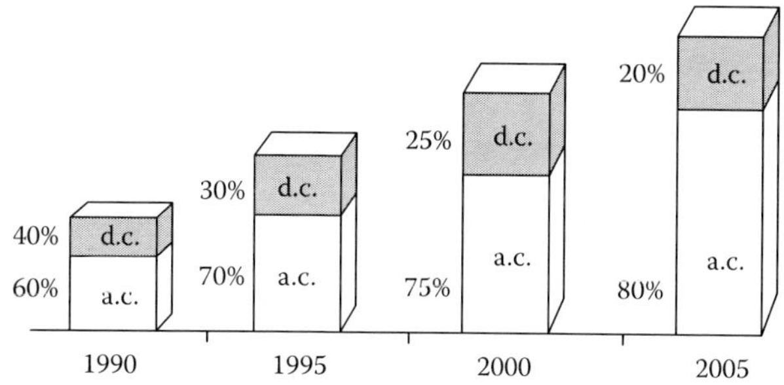


**FIGURE 1.4**
The a.c. versus d.c. electric drives market dynamics.

For variable speed, a.c. motors require both variable voltage amplitude and frequency, while d.c. brush motors require only variable d.c. voltage.

In many cases, a.c. motors are brushless and have a higher torque (power) density (Nm/kg or kW/kg) than d.c. brush motors and lower initial and maintenance costs. Today PECs for both d.c. brush or brushless motors, for reversible speed applications, have comparable prices. The PEC costs are higher than the motor costs, but this ratio (C) decreases as the power and speed levels increase

$$C = \frac{\text{PEC price}}{\text{a.c. motor price}} = 2.5 \div 0.5, \tag{1.1}$$

in general, from the kW and 3 krpm to MW and 15 rpm power/speed range. So what could justify introducing the PEC (Figure 1.2) which is more expensive than the motor?

In most applications, the energy savings pay off the additional investment in the PEC in less than 5 years for applications with energy savings of only 25% for only a 1 to 3 speed variation ratio but 24 h a day operation, above 10 kW rated power level. The larger the power, the lower the revenue period for a given percentage of energy savings.

## 1.3 Energy Savings Pay Off Rapidly

Let us illustrate the energy savings potential pay off for the additional expenses in variable speed drives implied by the presence of the PEC system. Consider a real case when a motor–pump system of 15 kW works 300 days a year, 24 h a day and pumps 1200 m$^3$ of water per day. By on/off and throttling control alone, the system uses 0.36 kW h/m$^3$ of pumped water to keep the pressure rather constant for variable flow rates.

Adding a PEC, in the same conditions, the energy consumption is 0.28 kW h/m$^3$ of pumped water, with a refined pressure control.

Let us consider that the cost of electrical energy is 8 cents/kW h.

The energy savings per year S is

$$S = 1200 \cdot 300(0.36 - 0.28) \cdot \$0.08/\text{year} = \$2304/\text{year} \tag{1.2}$$

Now the conservative cost of a 15-kW PWM-PEC for an induction motor (IM) is less than \$8000. Thus, to a first approximation, only the energy savings pay off the extra investment in less than 4 years.

After this coarse calculation, let us be more realistic and note that the energy costs, interest rates, and inflation rise slowly every year and taxes also lower to some extent the beneficial effect of energy saving through PEC, contributing to an increase in the revenue period.

**EXAMPLE 1.1: THE REVENUE TIME**

Let us consider that S dollars (1.2) have been saved in the first year on energy losses through the introduction of a PEC system, and denote by i the interest rate and by $i_p$ the power cost yearly increase.

Thus the effective interest rate per year $i_E$ is

$$i_E = \frac{1+i}{1+i_p} - 1 \tag{1.3}$$

The net present value (NPV) of losses for an n-year period is

$$NPV = S \cdot \frac{(1+i_E)^n - 1}{i_E \cdot (1+i_E)^n} \tag{1.4}$$

We may now consider the influence of taxes and inflation on these savings:

$$NPV^E = NPV_e + NPV_d \tag{1.5}$$

where $NPV_e$ is the net present value of energy savings and $NPV_d$ is the net present value of depreciation on premium investment (straight line depreciation is assumed). With T being the tax range

$$NPV_e = NPV \cdot (1-T) \tag{1.6}$$

$$NPV_d = \frac{NPV_e}{n} \cdot \frac{[(1+i)^n - 1]}{i \cdot (1+i)^n} \cdot T \tag{1.7}$$

With S = \$2304 as the first year savings, for a period of n = 5 years, with i = 10%, $i_p$ = 5%, and T = 40%, we obtain gradually

$$\text{From Equation 1.3:} \quad i_E = \frac{1+0.1}{1+0.05} - 1 = 0.0476$$

$$\text{From Equation 1.4:} \quad NPV = 2304 \cdot \frac{1.0476^5 - 1}{0.0476 \cdot 1.0476^5} = \$10{,}047.7$$

$$\text{From Equation 1.6:} \quad NPV_e = 10{,}047.7 \cdot (1-0.4) = \$6028.6$$

$$\text{From Equation 1.7:} \quad NPV_d = 6028.6 \cdot \frac{(1+0.1)^5 - 1}{0.1 \cdot (1+0.1)^5} \cdot 0.4 = \$1828.8$$

Finally, the premium investment that can be expanded to achieve \$2034 energy savings in the first year for a period of 5 years is, from Equation 1.5,

$$NPV^{E} = 6028.6 + 1828.8 \approx \$7857$$

This is approximately \$8000 (the costs of the PEC), so more exact calculations have led to an increase of the revenue duration from 4 to 5 years.

Considering that lifetime of PEC drives is more than (10–15) years, the investment pays off well.

## 1.4 Global Energy Savings through PEC Drives

So far the energy savings produced by the PEC in variable speed drives have been calculated for the drive only PEC and motor.

If we consider the costs to produce and transport the saved electric energy to the consumer (drive) location, the rentability of energy savings increases dramatically. Figure 1.5 shows the energy flow from the power plant to the motor–pump system for constant speed operation and 10 kW of useful power, when using throttling to control (reduce) the flow rate.

The total efficiency from the primary energy source is only 11%. This is poor energy utilization. When a PEC is introduced to produce the same useful power, a much better energy utilization ratio is obtained (Figure 1.6).

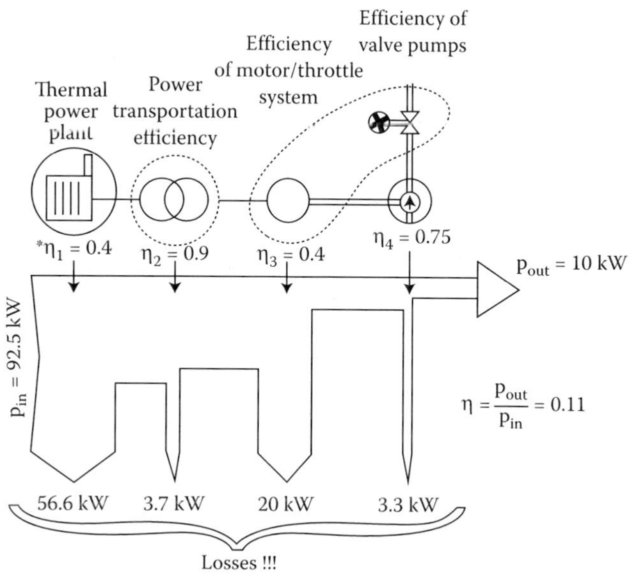


**FIGURE 1.5**
Primary energy consumption for throttle/motor/pump system. (*With recent combined cycle gasturbines $\eta_1 = 0.6$.)

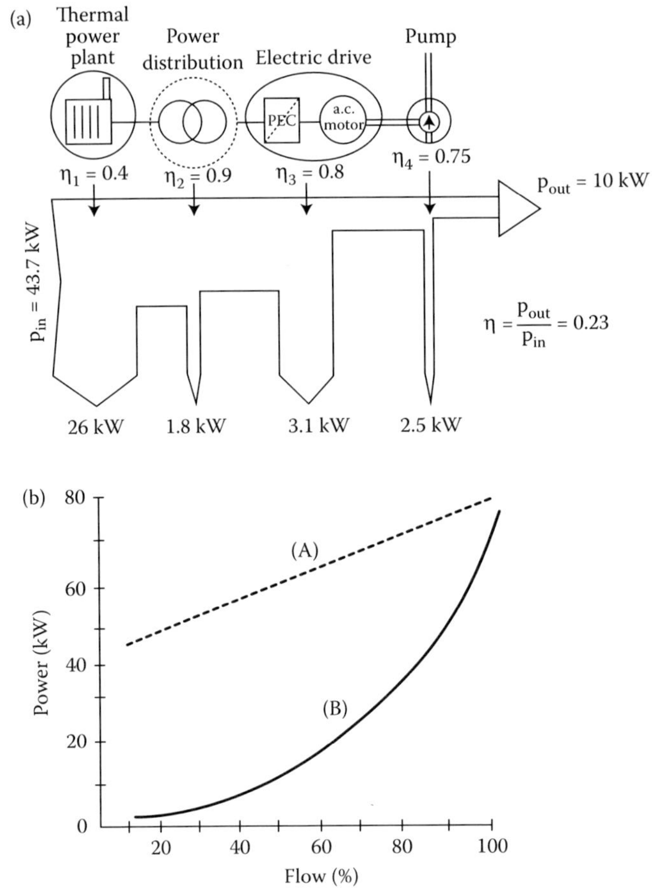


**FIGURE 1.6**
(a,b) Primary energy consumption for PEC/motor/pump systems with variable voltage/frequency 15 kVA power electronics converter.

The energy utilization factor (η) is doubled by the presence of PEC drives for variable speeds. So, to the energy savings value in the drive, we should add the savings along the entire chain of energy conversion and distribution, from the power plants to the consumers.

Variable speed with PEC drives, applied today to 35%–40% of all electric motor power in developed countries is expected to reach 40%–50% by 2020, with an annual extension rate of 7%–8%. If we add to this aspect the beneficial effects of motion control by PEC drives on the quality of products and to manufacturing productivity, we obtain the picture of one of the top technologies of the near and distant future with large and dynamic markets worldwide (already in the range of 10 billion USD in 2015).

## 1.5 Motor/Mechanical Load Match

The role of electric drives is, in fact, to match the electric motor to the mechanical load and to the electrical power grid.

The mechanical load is described by shaft torque/speed or torque/time plus speed/time or position/time relationships.

### 1.5.1 Typical Load Torque Speed Curves

Typical load torque speed curves are shown in Figure 1.7. They give a strong indication of the variety of torque/speed characteristics. Along such curves the mechanical power required from the motor varies with speed.

The base speed (unity speed in Figure 1.7) corresponds to continuous duty rated torque (and power) and rated (maximum) voltage from the PEC:

$$\text{Power} = T_{load} \cdot \Omega_r \tag{1.8}$$

To match the required speed/torque (and power) envelope, the motor and PEC should be carefully chosen or designed. However, the motor to mechanical load match should be provided not only for steady state but also during transients such as drive acceleration, deceleration, or short overload periods. The transients require higher torques, in general below base speed, and both the motor and the PEC have to be able to withstand it.

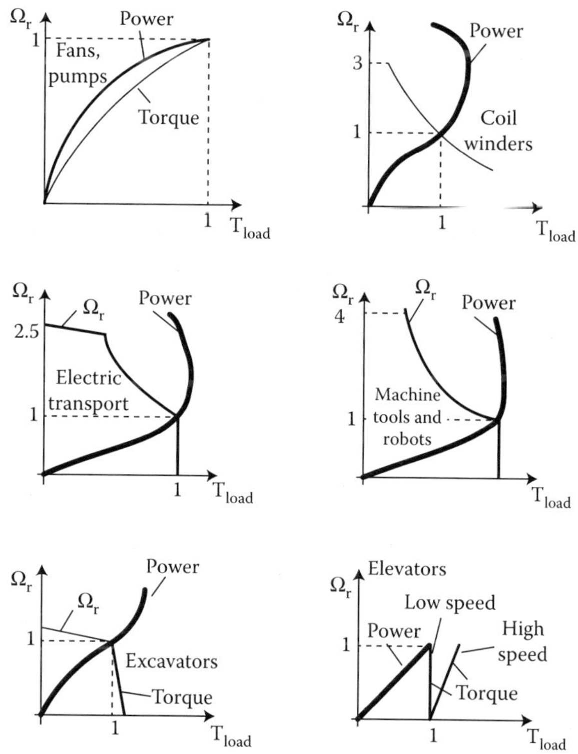


**FIGURE 1.7**
Typical load torque speed, speed/power curves.

## 1.6 Motion/Time Profile Match

For some applications such as robots and electric transportation, for given load inertia and load torque/time, a certain position/time (and) or speed/time profile is required. An example is shown in Figure 1.8. There are acceleration, cruising, and deceleration intervals. They are strong grounds for the design (sizing) of the electric drives. The torque varies with time and so does the motor current (and flux linkage level). The electric, magnetic, and thermal loadings of the motor and the electric and thermal loading of the PEC are definite constraints in a drive specification.

Here is an illustrative example.

### EXAMPLE 1.2: THE DIRECT DRIVE TORQUE/TIME CURVE

A direct drive has to provide a speed/time curve such as in Figure 1.9 against a constant load torque of $T_L = 10$ Nm, for a motor load inertia $J = 0.02$ kg m$^2$.

Neglecting the mechanical losses, let us calculate the motor torque ($T_e$)/time requirements.

The motion equation for a direct drive is

$$T_e(t) = J \cdot \dot{\Omega}_r(t) + T_L(t) \tag{1.9}$$

For the linear speed/time (acceleration–deceleration) zones, the speed derivative is

$$\dot{\Omega}_r = \pm \frac{\Omega_{rmax}}{t_a} = \pm \frac{376.8}{0.2} = \pm 1884 \text{ rad/s}^2 \tag{1.10}$$

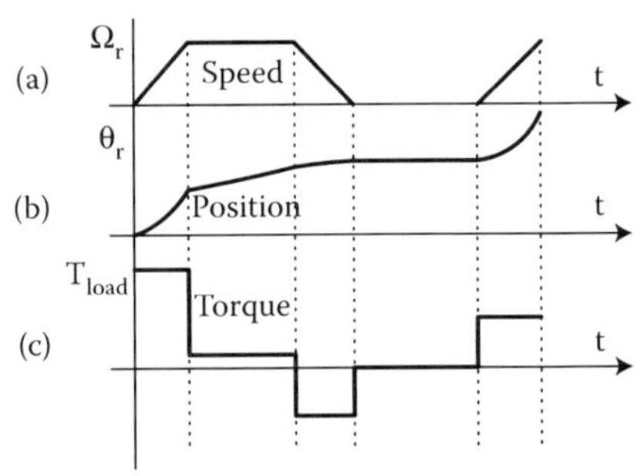


**FIGURE 1.8**
Motion/time profile: (a) speed, (b) position, and (c) the required load torque.

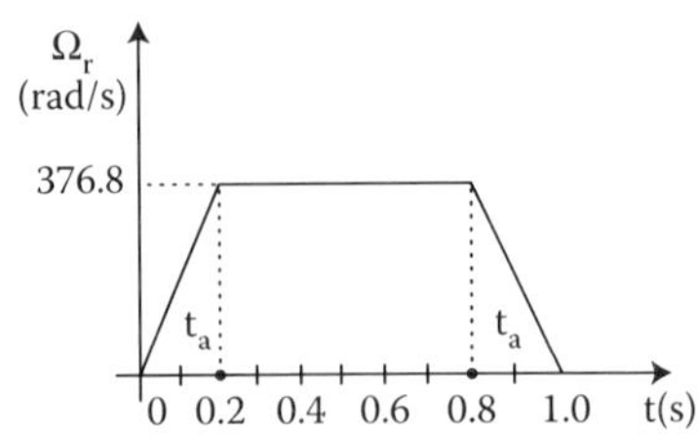


**FIGURE 1.9**
Required speed/time profile.

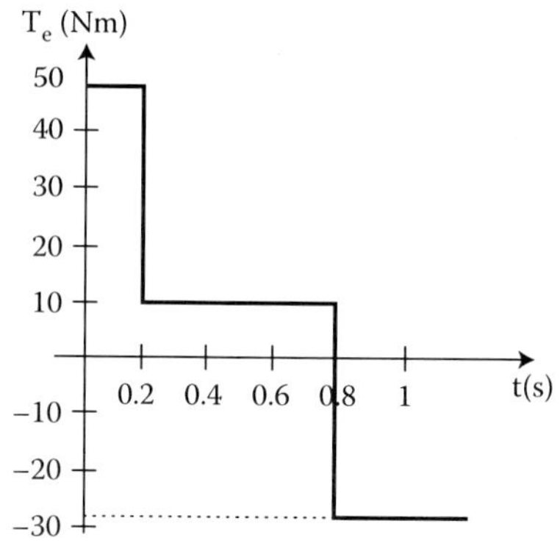


**FIGURE 1.10**
Motor torque/time requirements.

For the constant speed (cruising) zone, $\dot{\Omega}_r = 0.0$.

Consequently, the torque requirements from the motor for the three zones are as follows:

$$T_e = \begin{cases} 1884 \cdot 0.02 + 10 = 37.68 + 10 = 47.68\,\text{N m}; & \text{for } 0 \le t \le 0.2\,\text{s} \\ 0 + 10 = 10\,\text{N m}; & \text{for } 0.2 \le t \le 0.8\,\text{s} \\ -1884 \cdot 0.02 + 10 = -37.68 + 10 = -27.68\,\text{N m}; & \text{for } 0.8 \le t \le 1\,\text{s} \end{cases} \tag{1.11}$$

The motor torque/time requirements of Equation 1.11 are shown in Figure 1.10.

It is known that a theoretical-sudden motor torque is not practical with a limited available source power. However, a quick torque variation is required.

Advanced PEC drives today manage a zero to rated torque ramping in only 1–5 ms. Lower values correspond to low powers (1 kW), while higher values refer to higher powers (hundreds of kW or more). Many electric drives use a mechanical transmission for speed multiplication or for speed reduction by a ratio. The motor and load inertia are "coupled" through mechanical transmission. Let us illustrate this aspect through a numerical example.

### EXAMPLE 1.3: GEARBOX DRIVE TORQUE/TIME CURVE

Let us consider an electric drive for an elevator with the data shown in Figure 1.11.

The motor rated speed $n_n = 1550$ rpm. The efficiency of the gearing system is $\eta = 0.8$.

Let us calculate the total inertia (reduced to motor shaft), torque, and power without and with counterweight.

First, the motor angular speed $\omega_m$ is

$$\omega_m = 2 \cdot \pi \cdot n_n = 2 \cdot \pi \cdot \frac{1550}{60} = 162.22\,\text{rad/s} \tag{1.12}$$

The gear ratios may be defined as speed ratios $\Omega_t/\omega_m$ for $J_4 + J_5$ and $\Omega_d/\omega_m$ for $J_6$ (Figure 1.11).

Consequently, the inertia of all rotating parts $J_r$, reduced to the motor shaft (Figure 1.11), is

$$J_r = J_1 + J_2 + J_3 + (J_4 + J_5) \cdot \frac{\Omega_t^2}{\omega_m^2} + J_6 \cdot \frac{\Omega_d^2}{\omega_m^2}$$

$$= 15 + 8 + 2 + (0.5 + 200) \cdot \left(\frac{2.5}{162.22}\right)^2 + 8 \cdot \left(\frac{7.5}{162.22}\right)^2 = 25.062\,\text{kg m}^2 \tag{1.13}$$

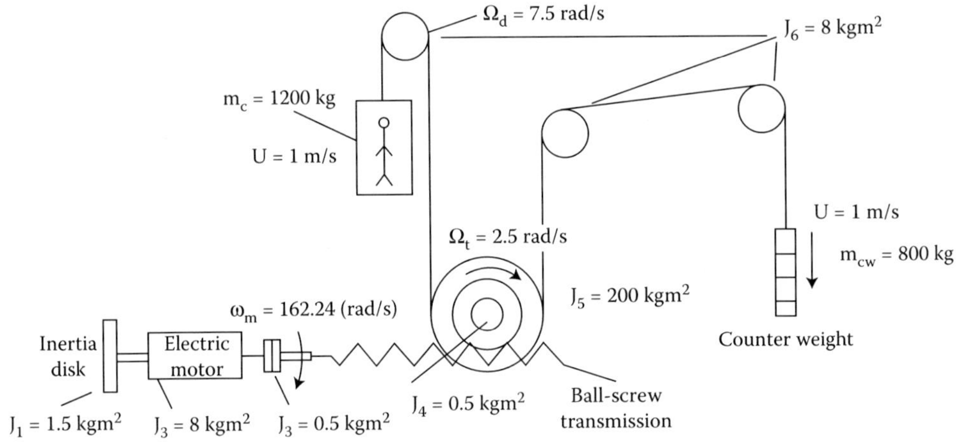


**FIGURE 1.11**
Elevator electric drive with multiple mechanical transmissions and counterweight.

For the cabin and the counterweight, the inertia reduced to motor shaft ($J_e$) is

$$J_e = (m_c + m_{cw}) \cdot \frac{u^2}{\omega_m{}^2} = (1200 + 800) \cdot \frac{1^2}{166.22^2} = 0.07238 \text{ kg m}^2 \tag{1.14}$$

Thus the total inertia $J_t$ is

$$J_t = J_r + J_e = 25.062 + 0.07238 = 25.135 \text{ kg m}^2 \tag{1.15}$$

In the absence of counterweight, the law of energy conservation leads to

$$T_{em} \cdot \omega_m \cdot \eta = m_c \cdot g \cdot u \tag{1.16}$$

Consequently, the motor torque $T_{em}$ yields

$$T_{em} = \frac{1200 \cdot 9.81 \cdot 1}{162.22 \cdot 0.8} = 90.71 \text{ N m} \tag{1.17}$$

The motor electromagnetic power $P_{em}$ is

$$P_{em} = T_{em} \cdot \omega_m = 90.71 \cdot 162.22 = 14{,}715 \text{ W} \tag{1.18}$$

On the other hand, in the presence of a counterweight (1.16), the power balance becomes

$$T'_{em} \cdot \omega_m \cdot \eta = (m_c - m_{cw}) \cdot g \cdot u \tag{1.19}$$

$$T'_{em} = \frac{(1200 - 800) \cdot 9.81 \cdot 1}{162.22 \cdot 0.8} = 30.71 \text{ N m} \tag{1.20}$$

So the motor electromagnetic power $P'_{em}$ is

$$P'_{em} = T'_{em} \cdot \omega_m = 30.71 \cdot 162.22 = 4905\ W \tag{1.21}$$

It should be noted that the counterweight alone produces a 3 to 1 reduction ratios in the motor's electromagnetic power, yielding important energy savings. On top of this, for the acceleration–deceleration periods, the PEC drive adds energy savings and produces soft starts and stops for a good ride quality. To do so in tall buildings, high-speed quality elevators with up to 1/1000 motor speed variation (control) range are required. This is why in Figure 1.3 elevators are enlisted as high-performance drives.

## 1.7 Load Dynamics and Stability

Load dynamics in an electric drive with rigid mechanical coupling between motor and load are described by the equation

$$J_t \cdot \frac{d\Omega_r}{dt} = T_e - T_{friction} - T_{load} \tag{1.22}$$

where $J_t$ is the total inertia of the motor and the load reduced to motor shaft, $T_e$ is the motor electromagnetic torque, $T_{load}$ is the actual load torque, and $T_{friction}$ is the total friction torque of the motor/transmission subsystem. The friction torque $T_{friction}$ has quite a few components:

$$T_{friction} = T_S + T_C + T_V + T_W \tag{1.23}$$

where $T_S$ is the static friction torque (at zero speed), $T_C$ is Coulomb friction torque (constant with speed), $T_V$ is viscous friction torque (proportional to speed), and $T_W$ is windage friction (including the ventilator braking torque, proportional to speed squared):

$$T_V = B' \cdot \Omega_r \tag{1.24}$$

$$T_W = C \cdot \Omega_r^2 \tag{1.25}$$

The friction torque components are shown in Figure 1.12. Only to a first approximation we may write

$$T_{friction} = B \cdot \Omega_r \tag{1.26}$$

In this latter case, Equation 1.22 becomes

$$J \cdot \frac{d\Omega_r}{dt} = T_e - T_{load} - B \cdot \Omega_r \tag{1.27}$$

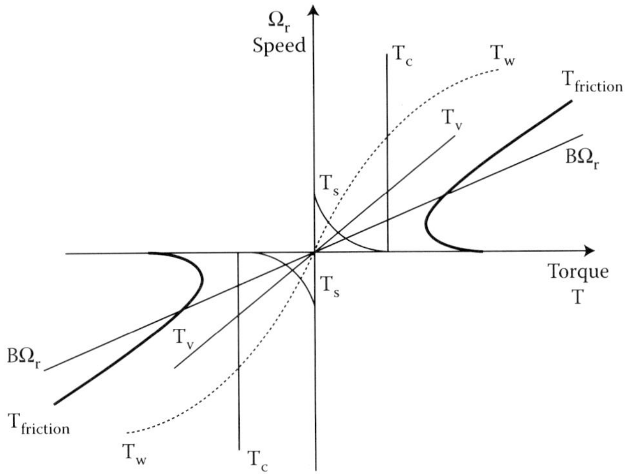


**FIGURE 1.12**
Components of friction torque, $T_{friction}$.

If $T_e$ and $T_{load}$ are constant, the solution of Equation 1.27 is

$$\Omega_r(t) = \Omega_{r\,final} + A \cdot e^{-t/\tau_m} \tag{1.28}$$

where $\Omega_{r\ final}$ is the final, steady-state, speed, $\tau_m = J/B$ is the so-called mechanical time constant, and A is a constant to be determined from initial conditions. Equation 1.28 reflects a stable aperiodic response. As

$$\Omega_r = \frac{d\theta_r}{dt} \tag{1.29}$$

Equation 1.27 becomes

$$J \cdot \frac{d^2\theta_r}{dt^2} + B \cdot \frac{d\theta_r}{dt} - T_e = -T_{load} \tag{1.30}$$

Clearly, for stable operation, the transients in $\theta_r$ must die out. As B is generally small, stability is obtained if the motor torque $T_e$ is of the form

$$T_e \approx -C_e \cdot \frac{d\theta_r}{dt}; \quad C_e > 0 \tag{1.31}$$

or

$$T_e \approx -C_e \cdot \frac{d\theta_r}{dl} - C_i \cdot \theta_r; \quad C_e, C_i > 0 \tag{1.32}$$

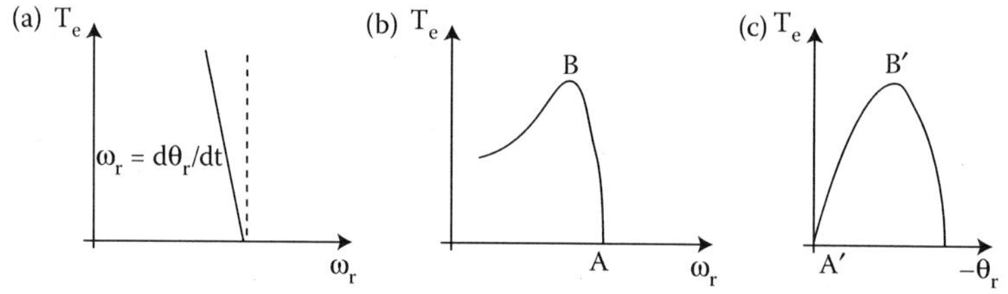


**FIGURE 1.13**
Mechanical characteristics: (a) d.c. brush motor with separate excitation, (b) IM, and (c) SM.

with Equations 1.31 and 1.32, Equation 1.30 becomes

$$J \cdot \frac{d^2\theta_r}{dt^2} + (B + C_e) \cdot \frac{d\theta_r}{dt} + C_i \cdot \theta_r = -T_{load} \tag{1.33}$$

Now it is evident that the $\theta_r$ transients will be stable.

**NOTE:** For the synchronous machine $\theta_r$ is in fact the angle between the e.m.f. and the terminal voltage, which is constant, for steady state.

Let us see if the torque speed curves of basic d.c. brush and a.c. motors satisfy Equation 1.33. Typical torque speed or torque/angle curves for these machines are shown in Figure 1.13.

Condition 1.33 is fulfilled by the d.c. brush motor as

$$-C_e = \frac{dT_e}{d(d\theta_r/dt)} < 0 \tag{1.34}$$

The torque decreases with speed steadily. For the IM (Figure 1.13b) only along the zone AB

$$\frac{dT_e}{d\omega_r} < 0 \tag{1.35}$$

Finally, for the synchronous motor (SM) (Figure 1.13c) only along the zone A′B′

$$-C_i = \frac{dT_e}{d\theta_r} < 0 \tag{1.36}$$

The above discussion serves only to signal the problem of electric drive dynamics and stability. The presence of PECs frees the mechanical characteristics from the forms in Figure 1.13, obtained for constant voltage (and frequency). Consequently, stable response in position, speed, or torque may be artificially obtained through adequate control of voltage and frequency at the motor electrical terminals.

Two simple examples follow.

### EXAMPLE 1.4: D.C. BRUSH MOTOR DRIVE STABILITY

A permanent magnet (PM) d.c. brush motor with the torque speed curve $\Omega_r = 200.0 - 0.1 \cdot T_e$ drives a d.c. generator, which supplies a resistive load such that the

generator torque/speed equation is $\Omega_r = 2T_L$. We calculate the speed and torque for the steady-state point and find out if that point is stable.

**Solution**

Let us first draw the motor and load (generator) torque speed curves in Figure 1.14.

The steady-state point, A, corresponds to constant speed and B = 0 in Equation 1.27. Simply, the motor torque counteracts the generator braking torque

$$T_L = T_e \tag{1.37}$$

Using the two torque speed curves, we find

$$\Omega_{rA} = 200 - 0.1 \cdot \frac{\Omega_{rA}}{2} \tag{1.38}$$

and thus

$$\Omega_{rA} = \frac{200}{1 + 0.1/2} = 190.476 \text{ rad/s} \tag{1.39}$$

and

$$T_{eA} = T_{LA} = \frac{\Omega_{rA}}{2} = \frac{190.476}{2} = 95.238 \text{ Nm} \tag{1.40}$$

The static stability is met if

$$\left(\frac{\partial T_e}{\partial \Omega_r}\right)_A < \left(\frac{\partial T_L}{\partial \Omega_r}\right)_A \tag{1.41}$$

In our case from the two torque speed curves

$$-10 < \frac{1}{2} \tag{1.42}$$

and thus, as expected, point A represents a situation of static equilibrium.

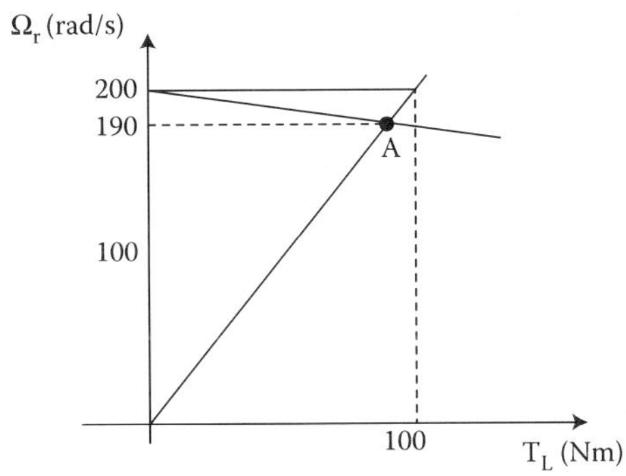


**FIGURE 1.14**
The d.c. brush motor load match.

**EXAMPLE 1.5: IM DRIVE STABILITY**

An IM fed at constant voltage and frequency has the torque/speed dependance given by

$$T_e = \frac{2T_{eK}}{(S/S_K)+(S_K/S)}; \quad S = \frac{\omega_1 - \omega_r}{\omega_1}; \quad S_K = 0.2; \quad T_{eK} = 20\,\text{Nm} \tag{1.43}$$

where S is the slip, $\omega_1$ is the primary frequency, and $\omega_r$ is the rotor electrical speed ($\omega_r = \Omega_r \cdot p$; p is the number of winding pole pairs). This motor drives a d.c. generator with a resistive load, the torque speed curve of which is $T_L = C \cdot \omega_r$. Neglecting the mechanical losses (B = 0), we decide to check the static stability conditions for $T_e = 10$ Nm.

**Solution**

Let us draw the two mechanical characteristics (Figure 1.15).

The slip value for $T_e = 10$ Nm may be found from

$$10 = \frac{2\cdot 20}{(S/0.2)+(0.2/S)} \tag{1.44}$$

The solutions of Equation 1.44 are

$$S_1 = 0.0436; \quad \text{point M in Figure 1.15}$$

$$S_2 = 0.7464; \quad \text{point M' in Figure 1.15} \tag{1.45}$$

The static stability depends on the sign of $\partial T_e/\partial\omega_r$

$$\frac{\partial T_e}{\partial(\omega_r/\omega_1)} = \frac{-\partial T_e}{\partial S} = \frac{2T_e}{((S/S_k)+(S_K/S))^2}\cdot\left(\frac{1}{S_k} - \frac{S_K}{S^2}\right) \tag{1.46}$$

So for

$$S < S_K; \quad \frac{\partial T_e}{\partial(\omega_r/\omega_1)} < 0; \quad \text{stable zone} \tag{1.47}$$

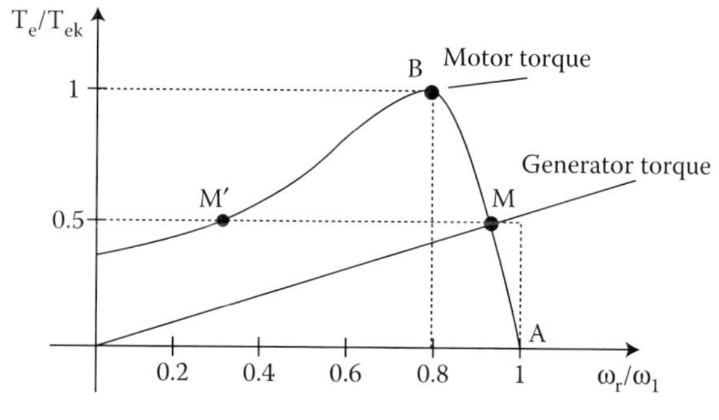


**FIGURE 1.15**
IM/d.c. generator load.

and for

$$S > S_K; \quad \frac{\partial T_e}{\partial(\omega_r/\omega_1)} > 0; \quad \text{unstable zone} \tag{1.48}$$

As at point M $S_1 = 0.0536 < S_K = 0.2$, M corresponds (as expected) to static stability conditions, while at point M′ $S_2 = 0.7464 > S_K = 0.2$; M′ is within the instability zone.

## 1.8 Multiquadrant Operation

An electric drive may be required to provide for forward and reverse motion with rapid (regenerative) braking in both directions. Motor operation implies torque in the direction of motion. In regenerative braking, the torque is opposite to the direction of motion and the electric power flow in the motor is reversed as well (negative). These possibilities are summarized in Table 1.1 and in Figure 1.16.

Positive (+) electric power flow means the electric power drawn from the PEC by the motor, while negative (−) refers to the electric power delivered by the motor (in the generator mode) to the PEC.

The PEC has to be designed to be able to handle this bidirectional power flow. In low- and medium-power PECs (up to hundreds of kW) with slow braking demands, the generated power during the braking periods is interchanged with the strong filter capacitor of PEC or d.c. (dynamic) braking is being used.

For d.c. dynamic braking, the kinetic energy of the motor-load system is converted to heat in the motor rotor. For fast and frequent generator brakings, the PEC has to handle the generated power either by a controlled braking resistor or through *bidirectional* power flow. All these aspects will be discussed in some detail later in subsequent chapters.

For a fast speed response, modern variable speed drives may develop a maximum transient torque up to base speed $\omega_b$ and maximum transient power up to maximum speed, provided that both the motor and the PEC can handle these powers.

A rapid increase in speed from point A to point A′ (Figure 1.16) is managed along the ABCA′ path, which remains in the first quadrant. A slow speed increase from A to A′ is performed along the path AMA′. For a speed reversal (from A to A″), the trajectory goes along the path A′C′D′A″ through the second quadrant (regenerative braking). Such speed transitions indicate the complexity of energy conversion and transfer between the motor, PEC, and power source in a multiquadrant electric drive.

NOTE: So far we discussed only rotary motor electric drives. However, for each rotary motor there is a linear counterpart. Torques and angular speeds will be replaced by thrust

**TABLE 1.1**

For Quadrant Operation Modes

| Mode of Operation | Forward Motoring | Forward Regenerative Braking | Reverse Motoring | Reverse Regenerative Braking |
|---|---|---|---|---|
| Speed ($\omega_r$) | + | + | − | − |
| Torque ($T_e$) | + | − | − | + |
| Electric power flow | + | − | + | − |

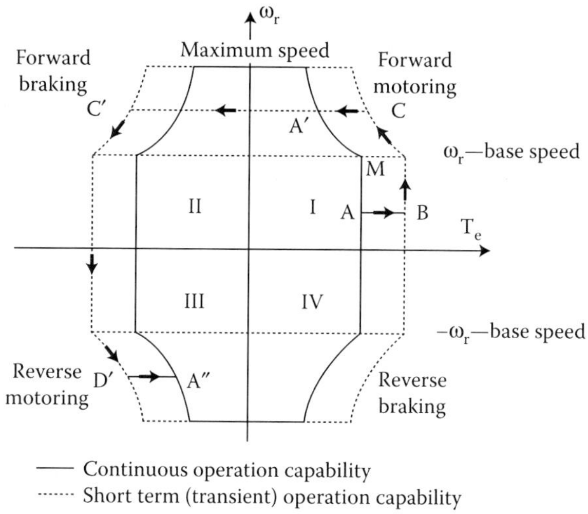


**FIGURE 1.16**
Electric drives with four-quadrant operation.

and linear speed, respectively, in linear motor drives [5]. Linear motors, however, are not treated in this book.

## 1.9 Performance Indexes

The electric drive performance indexes, as defined in what follows, are divided into three main categories: energy conversion indexes, drive response indexes, and costs and weight indexes.

Electric drives perform the conversion of electrical energy to mechanical energy, or vice versa, through the use of magnetic energy as a storage medium. A typical electric drive system contains the following: an electric motor, a static power converter, feedback sensors (or observers), and a digital (or analog or hybrid) motion controller (Figure 1.17).

### 1.9.1 Power Efficiency (for Steady State)

Energy conversion in an electric drive takes place both in the static power converter and in the electric motor. The energy flows from the static power converter to the electric machine in the motor regime (Figure 1.18a), and vice versa for regenerative braking (Figure 1.18b).

Energy conversion is accompanied by losses: conduction (winding), core, and mechanical losses in the motor ($p_{mot}$).

For steady state, the rating of energy conversion from electrical to mechanical or vice versa may be derived through the power efficiency $\eta_p$ defined for the motor ($\eta_{pm}$) or for the drive ($\eta_{pd}$):

$$\eta_{pm} = \frac{P_{out}}{P_{out} + \sum p_{mot}} \quad (1.49)$$

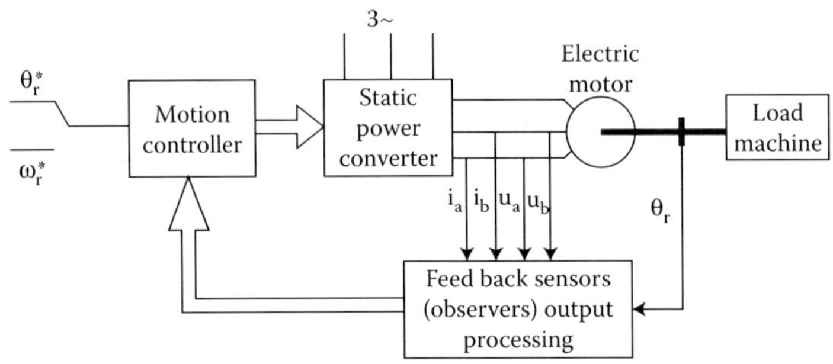


**FIGURE 1.17**
Electric drive basic topology.

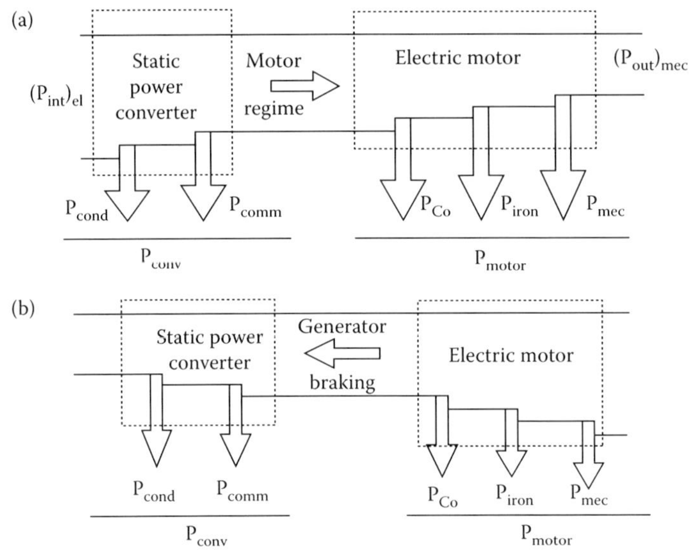


**FIGURE 1.18**
Energy conversion in electric drives: (a) motor regime and (b) generator braking.

$$\eta_{pd} = \frac{P_{out}}{P_{out} + \sum p_{mot} + \sum p_{conv}} \tag{1.50}$$

As speed is variable in electric drives, the power efficiency is relevant for base speed ($\omega_b$), maximum speed ($\omega_{max}$), and rated torque for these two situations.

Base speed $\omega_b$ is the speed for which full voltage is required to produce the peak design continuous/dominant operation torque $T_{ek}$. Above the base speed, the voltage remains (in general) constant and, in a.c. motors, only the frequency increases. In many servo drives, $\omega_b = \omega_{max}$, that is, the peak torque $T_{ek}$ is available up to a maximum speed when the static power converter rating is defined for maximum rather than for base speed.

### 1.9.2 Energy Efficiency (for Mechanical Transients)

In some applications, the variable speed drives undergo very frequent mechanical transients (speed and torque transients), as in hybrid and electric cars.

To assess the energy conversion efficiency for such cases, the energy efficiency $\eta_E$ for the motor ($\eta_{Em}$) and for the drive ($\eta_{Ed}$) are defined:

$$\eta_{Em} = \frac{W_{out}}{W_{out} + W_{motor}} \tag{1.51}$$

$$\eta_{Ed} = \frac{W_{out}}{W_{out} + W_{motor} + W_{conv}} \tag{1.52}$$

where $W_{out}$ is the output (useful) energy, $W_{motor}$ are the motor total energy losses, and $W_{conv}$ are the converter total energy losses.

The energy efficiency may be an important optimization criterion for the design of the drives (motor, converter, and controller) with frequent mechanical transients (robotics, urban transit drives, etc.).

### 1.9.3 Losses/Torque (W/Nm) Ratio

A more flexible energy conversion index is the ratio between the motor losses and the motor torque. The W/Nm ratio may refer to motor winding power losses only. In this case, the W/Nm ratio is most adequate for prolonged low-speed operation, as the core losses at low speed are small.

For high speed, however, all losses in the motor have to be considered in the W/Nm, in order to assess the heating of the motor fairly.

### 1.9.4 RMS kW/kVA Ratio

The RMS kW/kVA ratio for a.c. motors is, in general, computed for the fundamental components and represents, in fact, the displacement power factor (DPF):

$$\text{DPF} = (\text{kW/kVA})_{motor} = \frac{P_{in}}{3V_1I_1} \tag{1.53}$$

For a given torque, the higher the DPF, the smaller the current drawn by the motor for given terminal voltage and frequency. The DPF thus has considerable influence on stator winding losses, and finally on power efficiency.

### 1.9.5 kW/Peak kVA Ratio

As most variable speed servo drives make use of MOSFET or IGBT (insulated gate bipolar transistor), the static power converters are sized according to the peak current and voltage values.

For a.c. motors, at high speeds, in the weakening zone, the motor phase currents depart considerably from sinusoidal waveforms.

The active input power $P_{in}$ is

$$P_{in} = \frac{3}{\pi} V_0 I \frac{1}{K} DPF \tag{1.54}$$

where $V_0$ is the d.c. source voltage at the input of the power converter, I is the peak flat-top phase current, and K is the ratio between the actual peak current and the peak value of the current fundamental (Figure 1.19)—in general, $k \le (1.1\text{–}1.15)$.

$$\text{Peak kW/kV A} = \frac{P_{in}}{S_1} = \frac{3}{\pi} V_0 I \frac{1}{K} \frac{DPF}{6V_0 I} = \frac{DPF}{2\pi K} \tag{1.55}$$

On the other hand, the switched reluctance motor (SRM) (Chapter 12) has unipolar pulse currents with one phase working at a time (Figure 1.20), for three-phase stator winding.

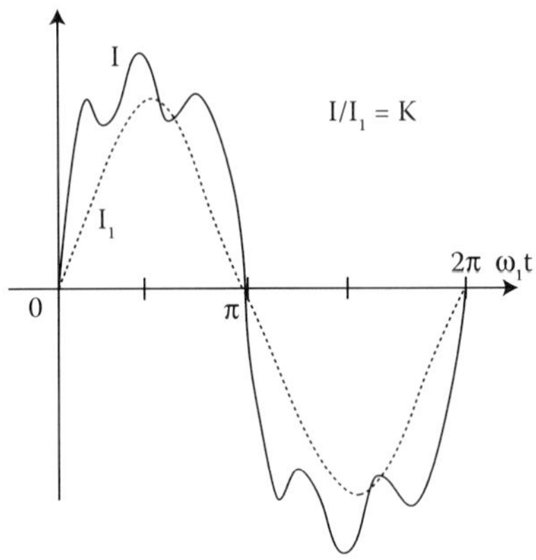


**FIGURE 1.19**
Alternating motor phase current in the field-weakening zone (six-pulse inverter regime).

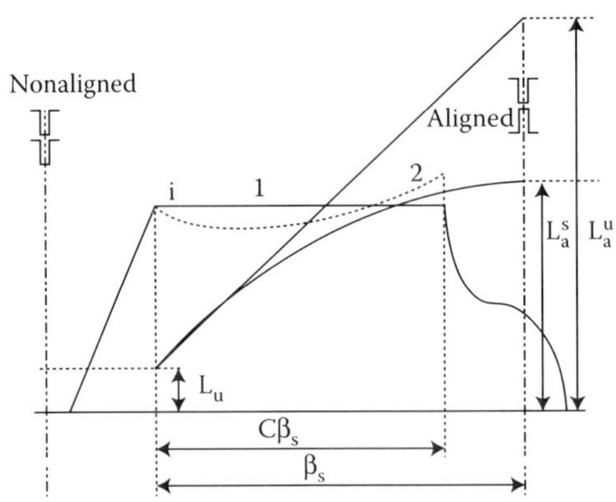


**FIGURE 1.20**
Current waveforms in an SRM: 1—ideal (chopped) current and 2—actual current at full voltage (no chopping).

The same performance index may be defined for the SRM [6] as

$$\text{Peak kW/kVA} = \frac{\beta_s N_r Q}{8\pi} \tag{1.56}$$

where $\beta_s$ is the stator pole ($\beta_s \cong 0.4$ for 6/4 SRMs) and $N_r$ is the number of rotor poles. The performance index Q in Equation 1.56 is [6]

$$Q \approx C\left(2 - \frac{C}{C_s}\right) \tag{1.57}$$

where C is the ratio between the phase turn-on angle and the stator pole angle.

In general, at start (zero speed), C = 1 and decreases with speed ($C \cong 0.65$ at base speed):

$$C_s = \frac{\lambda_u - 1}{\lambda_u \sigma - 1}; \quad \lambda_u = \frac{L_a^u}{L_u} \approx (4 \div 10); \quad \sigma = \frac{L_a^s}{L_a^u} = (0.3 \div 0.4) \tag{1.58}$$

$L_u$, $L_a^u$, and $L_a^s$ are the phase inductance value shown in Figure 1.20.

The peak apparent power $S_1$ is

$$S_1 = 2 \cdot m \cdot V_0 \cdot I \tag{1.59}$$

where m is the number of phases (m = 3, 4 in general), $V_0$ is the d.c. source voltage, and I is the peak phase current value.

In the kW power range, the peak kW/kVA ratio is in the interval 0.55–0.65 for flat-top ideal current, even with a considerable magnetic core saturation level ($\sigma$ = (0.3–0.4)), suggesting a reasonably large peak kVA in the static power converter rating [1].

### 1.9.6 Peak Torque/Inertia Ratio

As already mentioned, most electric drives provide the peak torque $T_{ek}$ only up to base electric speed $\omega_b$. In some servo drives, the base speed is equal to the maximum speed.

The ratio between the peak torque $T_{ek}$ and the rotor inertia J represents the maximum ideal acceleration $a_{max}$ (rad/s$^2$) up to the base electric speed $\omega_b$:

$$a_{max} = \frac{T_{ek}}{J} (\text{rad/s}^2) \tag{1.60}$$

Conversely, the time $t_a$ required to reach the base speed $\omega_b$, under peak torque and no load may be defined as

$$t_a = \frac{\omega_b / p}{a_{max}} = \omega_b \frac{J}{pT_{ek}} (\text{ms}); \quad \text{p is pole pairs} \tag{1.61}$$

Both $a_{max}$ and $t_a$ tend to become catalog data for high-performance variable speed drives.

### 1.9.7 Field Weakening Speed Range ($\omega_{max}/\omega_b$)

Many applications require constant power over a wide range of speeds. For such applications, the motor's full voltage is reached at base speed, while above base speed, the flux is gradually reduced to allow motoring torque.

The ratio between maximum and base speeds ($\omega_{max}/\omega_b$) is a performance index because it shows the capability of the drive to reach high speeds avoiding overheating.

Ratio $\omega_{max}/\omega_b$ of 1.5–2 is characteristic for permanent magnet synchronous motor (PM-SM) and IM. Higher values require special measures and traded converter and motor overrating.

### 1.9.8 Variable Speed Ratio: $\omega_{max}/\omega_{min}$

The speed closed-loop control range, that is, $\omega_{max}/\omega_{min}$, is an indication of the drive capability to accommodate different applications.

Electric drives with $\omega_{max}/\omega_{min} > 200$ are considered to perform wide speed range control and require a position (or at least speed) feedback sensor for precision position (or speed) control.

On the other hand, when $20 < \omega_{max}/\omega_{min} < 200$, the speed range control is medium and the position and speed feedback signals for speed control are obtained through observers.

Speed errors of 2%–3% are allowable for such drives. As the motion (position and speed) sensor costs are important, they should be replaced by motion observers but this is feasible only in applications with medium variable speed range.

For $\omega_{max}/\omega_{min} < 20$ open-loop control with some speed reduction compensation under load is applied with the so-called V/f a.c. drives. V/f a.c. drives are the most common, as for fans and pumps applications they do the job properly at rather low costs (less than 100 USD/kVA).

### 1.9.9 Torque Rise Time ($t_{Tek}$)

In fast response drives, it is important to provide for fast torque variation. Torque transients depend on speed, motor parameters, inertia, converter type, input voltage level, and method of control.

As, inevitably, the torque rise time decreases with speed due to increased back e.m.f.; it seems that the rise time of torque from zero to $T_{ek}$, $t_{Tek}$, with the flux linkage already present in the machine, at zero speed, may constitute a reliable performance index.

In fast response drives, $t_{Tek}$ is in the range of 2–6 ms. For a.c. (brushless) motor drives, such fast torque response is obtained through the field orientation control (FOC) or direct torque control (DTC) of induction, PM-synchronous, and synchronous d.c.-excited or reluctance motors.

### 1.9.10 Torque Ripple Ratio ($\Delta T_e/T_{erated}$)

The pulsations in torque depend on the motor type, parameters, and also on the torque control strategy. They may be assessed through the ratio of torque ripple peak to rated torque ($\Delta T_e/T_{erated}$). The rated torque is defined here as the steady-state continuous torque at base speed. The rated torque pulsations influence the torque, speed and position control precision, vibrations, and noise. Consequently, they should be as low as possible in a trade-off between performance and costs.

### 1.9.11 Thermal Limitation ($\vartheta_{motor}$)

Direct drives for position or wide speed range are in direct mechanical contact with the load (e.g., machine tools). To avoid mechanical deformations in the shaft of the load, in such cases, the temperature of the motor $\vartheta_{motor}$ should be [7–8]

$$\vartheta_{motor} < 20°C + \vartheta_{ambient} \tag{1.62}$$

This is a serious limitation in motor design.

For most drives, the motor temperature is limited by the electric insulation class (B, E, or F).

### 1.9.12 Noise Level, $L_{noise}$ (dB)

The noise radiated by a variable speed drive is due to both the motor and the static power converter. The noise level accepted depends on the application. As an example, for machine tools [2]

$$L_{noise} \approx 70 + 20\log(P_n/P_{n0}); \quad P_{n0} = 1\,kW; \quad P_n = (1 \div 10)\,kW \tag{1.63}$$

where $P_n$ is the rated power of the motor in kW.

### 1.9.13 Motion Control Precision and Robustness

Motion control means torque, speed, and position closed-loop control.

Torque control is the most demanding, as the torque feedback sensor should account for the core losses. In general, motion control precision may be measured as torque, speed, or position error:

$$\Delta T_e; \quad \Delta\omega_r(rpm); \quad \Delta\theta_r(°) \tag{1.64}$$

The torque error is defined in relative values and the speed error in rpm, while the position error is given in degrees.

For speed and position control, the torque loop is present, sometimes as a current limiter.

Robustness of control is defined as the sensitivity of the drive response (in torque, speed, position) with respect to the motor parameters, inertia, and load torque variations. It is the degree of immunity of the drive response to drive parameter detuning.

Between robustness and quickness of speed control there is a conflict in the sense that, for robustness, the available quickness of response has to be partially sacrificed.

Advanced motion controllers such as self-tuning, model reference, variable structure, or fuzzy reasoning controllers are proposed to increase response robustness.

Robustness indexes could be defined as torque error ($\Delta T_e$), for a given parameter detuning $\Delta P_{ar}$ ($\Delta T_e/\Delta P_{ar}$), or speed rise time ratio from zero to base speed for J and 2J, respectively, or for zero speed at rated load torque.

### 1.9.14 Dynamic Stiffness

Dynamic stiffness is the ratio between perturbation torque ($\Delta T_{perturbation}$) and controlled variable error ($\Delta x$) versus frequency of torque perturbation. It is a kind of dynamic robustness to torque perturbations:

$$DS = \frac{\Delta T_{perturbation}}{\Delta x}$$

### 1.9.15 Specific Costs and Weights

The overall costs $C_{total}$ of an electric drive may be considered as a cost criterion:

$$C_{total} = C_{equip} + C_{loss} + C_{maint} \tag{1.65}$$

where $C_{equip}$ is the cost of motor, converter, sensors, and of controller; $C_{loss}$ is the capitalized energy loss cost of the drive over its lifetime; and $C_{maint}$ is the maintenance cost.

The relative importance of $C_{loss}$ increases with drive power (or torque) and should be accounted for in any computation attempt of total costs breakdown.

In the equipment cost $C_{equip}$, the cost of motor, static power converter, sensors (observers), and motion controllers have a relative importance depending notably on the motor power, with the motor cost's importance increasing with the power level. Here, the net present worth cost (including inflation, premium on investments dynamics) for the foreseeable life of the drive is to be considered.

The specific weights are also considered as the performance criteria:

$$\text{Motor specific weight} = \frac{\text{Peak torque}}{\text{Weight}} (\text{Nm/kg}) \tag{1.66}$$

$$\text{Converter specific weight} = \frac{\text{Peak apparent power}}{\text{weight}} (\text{kVA/kg}) \tag{1.67}$$

The first weight criterion (1.66) is a strong comparison criterion between different motors for drives. The second refers, from the same point of view, to the static power converter since the peak apparent power is a design criterion for the MOSFET or IGBT PWM inverters, widely applied to servo drives. For thyristor inverters, the RMS apparent power should be considered, instead of the peak apparent power. Moreover Nm/L and, respectively, kVA/L are important volume criteria for high-energy-density drives.

## 1.10 Electric Drive Applications

Electric drives refer to controlled electric energy (power) flow in electric motors and generators through power electronics, for energy conversion, for distributed power grids (generators), and for motion control to increase productivity and for energy savings by variable speed in variable output processes that require "mechanical work" (motoring).

Though the presence of electric drives—from milliwatt to multimegawatt units—is rather ubiquitous, we single out a few domains where they are very important by beneficial effects in the present and in the future:

- Wind and hydro energy harvesting
- More electric transportation
- Electric drives for high-performance electric power conversion in various industries (from metallurgy, through paper making, chemical cement to clean conveyor technology for making semiconductor wafers, etc.)
- Info gadget drives
- Home appliance (residential) drives

## 1.11 Summary

- Modern electric drives perform electrical to mechanical energy conversion at variable speeds. They contain PECs that modify the voltage (and frequency) with high efficiency [9–14].
- The introduction of PECs in electric drives (35%–40% of all drives in 2015) is justified by energy savings or process control (productivity and quality) performance.
- The d.c. brush motor-PEC drives are now more and more replaced by a.c. motor-PEC drives as the motors are more rugged, less expensive for about the same PEC costs, and of comparative performance, especially for reversible motion control applications.
- Energy savings with PEC drives provide a revenue period of 5 years or less from 10 kW upward, while the global energy savings (from the power plant to the application site) are spectacular.
- Motor torque has to match the load torque during steady state, while stable transients are produced by closed-loop motion control in PEC drives. Load dynamics and stability are priority issues in PEC drives.
- PEC drives provide, in general, motoring in both directions of motion. They are also capable of fast regenerative braking in both directions of motion provided the PEC can retrieve the generated power back to the power source. When this is not possible, for slow dynamics (braking), the energy is dumped into the motor rotor or in an external, controlled, braking resistor attached to the PEC.
- The PECs have a limited voltage ceiling $V_b$ for which the motor produces continuous rated power $P_b$ at the so-called base speed $\omega_b$.
- Most PEC drives can work above $\omega_b$ for constant power $P_b$ (and constant voltage $V_b$) up to $\omega_{max} = (2–4)\ \omega_b$.
- A set of coherent/practical energy conversion drive response quality and cost/weight performance for modern (power electronics) electric drives has been introduced.

## 1.12 Problems

1.1 A PEC drive saves S = \$500 of energy in the first year of operation. For a 4-year (n = 4) money return period, calculate the price of PEC investment if the interest rate i = 8%, the energy costs yearly raise $i_p$ = 4%, and the tax range T = 35%.

1.2 Calculate the global power requirements for 100 kW of useful (mechanical) power of an electric drive at 60% of rated speed if the power plant efficiency is $\eta_1$ = 40%; energy transportation is $\eta_2$ = 90% for two cases:

1. With a conventional drive at $\eta_{34}$ = 60% total efficiency (Figure 1.5).
2. With a PEC drive at $\eta'_{34}$ = 85% total efficiency (Figure 1.5).

1.3 An electric drive uses a mechanical transmission between the motor and load with a reduction ratio a = 1/10. The motor inertia $J_m$ = 0.02 kg m$^2$ while the load-transmission inertia $J_L$ = 2 kg m$^2$. Mechanical losses are neglected and the load torque $T_L$ = 200 Nm. For the speed/time curve in Figure 1.9, calculate the motor torque requirements after reducing the load inertia to motor shaft.

1.4 The base speed and power of a PEC a.c. motor drive are $\omega_b$ = 367 rad/s and $P_b$ = 100 kW. The motor has four poles (2p = 4). Calculate the motor torque $T_{eb}$ at base speed and power and the torque for constant power $P_b$ for $\omega_{max} = 3\omega_b$. What is the torque value and sign for generator braking at $\omega_b$ and delivering $P_b$ with zero motor losses?

## References


1. P. Sen, *Thyristor D.C. Drives*, John Wiley & Sons, New York, 1981.
2. W. Leonhard, *Control of Electric Drives*, Springer-Verlag, Heidelberg, first edition (1985), second edition (1996), third edition (2001).
3. G.K. Dubey, *Power Semiconductor Controlled Drives*, Prentice-Hall, Englewood Cliffs, New Jersey, 1989.
4. N. Mohan, T.M. Undelund, W.P. Robbins, *Power Electronics*, John Wiley & Sons, New York, 1989, 1995.
5. I. Boldea, S.A. Nasar, *Linear Electric Actuators and Generators*, Cambridge University Press, New York, 1997.
6. T.J.E. Miller, Converter volt–ampere requirements of the switched reluctance motor drive, *IEEE Trans.*, vol. IA-21, no. 5, pp. 1136–1144, 1985.
7. W. Weck, G. Ye, Catalog of servodrives for NC machines, *ETZ*, vol. 111, no. 6, pp. 282–285, 1990. [in German]
8. B.T. Boulter, Selecting the right drive, *IEEE IA Appl. Mag.*, vol. 10, no. 4, pp. 16–23, 2004.
9. B.K. Bose, *Modern Power Electronics and Drives*, Prentice-Hall, Upper Saddle River, New Jersey, 2001.
10. R. Krishnan, *Electric Motor Drives*, Prentice-Hall, Upper Saddle River, New Jersey, 2001.
11. H.A. Toliyat, S. Campbell, *DSP-Based Electromechanical Motion Control*, CRC Press, Boca Raton, Florida, 2004.
12. A. Veltman, D.W.J. Pulle, R.W. De Doncker, *Fundamentals of Electric Drives*, Springer-Verlag, Heidelberg, 2007.
13. R.W. De Doncker, D.W.J. Pulle, A. Veltman, *Advanced Electric Drives*, Springer-Verlag, Heidelberg, 2011.
14. D. Schroeder, *Electric Drives—Control of Drive Systems*, second edition, Springer-Verlag, Heidelberg, 2001. [in German.]

# 2

# *Electric Motors for Drives*

In this chapter, we introduce a typical electric drive configuration and various electric motors for electric drives.

## 2.1 Electric Drives: A Typical Configuration

A modern electric drive [1–8], capable of controlled variable speed, is made of some important parts (Figure 2.1) such as

- The electric motor
- The power electronic converter (PEC)
- The electric and motion sensors
- The drive controller
- The command interface

The drive controller may be considered to be made of a motion-speed and/or position-controller, more or less the same for all types of electric drives, and the electric controller, which refers to current and voltage (or flux linkage and torque) control within the PEC.

Electric sensors (observers) refer to voltage, current, flux as measured (or calculated) state variables, while motion sensors (observers) mean position and/or speed and torque, as measured (or calculated) state variables.

The electric controller has, in general, electric sensor (observer) inputs from both the power source and the PEC output. The motion controller handles, in general, only motion sensor (observer) outputs.

On the other hand, the electric controller commands are related to power source side energy conversion performance commands (unity power factor [PF], harmonics elimination), while the motion controller commands relate to motion (speed, position, torque) control commands, handled through an interface from a local digital controller or from a remote process control host computer. As expected, both electric and motion controllers are blended together in the drive controller hard and soft.

By now both the drive controller and the interface are carried out by high-performance digital signal processors (DSPs), which can handle several MFLOP/s, not an excessive feature with modern electric drives.

In this chapter, we will briefly characterize the candidate electric motors used for drives. The various losses occurring in an electric drive (Figure 2.1) will be dealt with in relation to specific motors and PECs in subsequent chapters. The sensors will be described in some detail when applied to various electric drives. The motion controllers, more or less common to all electric drives, will have a separate chapter devoted to them in relation to d.c. brush motor drives, for easy assimilation.

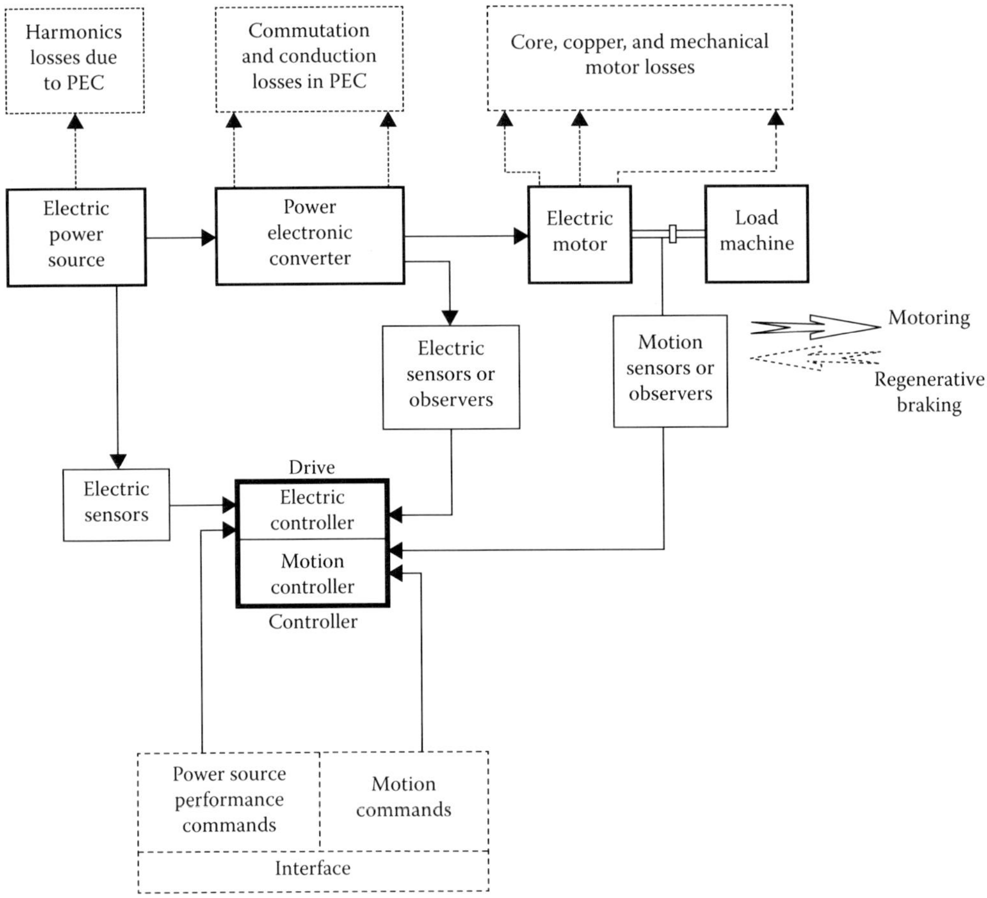


**FIGURE 2.1**
Electric drive basic topology.

Finally, electric controllers tend to be more specific and will be treated during discussions of applications for various electric drives.

A short classification of electric motors for electric drives follows.

## 2.2 Electric Motors for Drives

In essence, all existing types of electric motors may be accompanied by PECs with digital motion controllers, to produce high-performance electric drives. The ideal speed/torque curve for motion control is a straight descending line. This fact and the ease to produce a variable d.c. voltage source for a wide range of speed control had made the d.c. brush motor the favorite variable speed drive up to 1960. Since then, the development of reliable price-competitive PECs with variable a.c. voltage and frequency capability and the linearization of the speed/torque curve through the so-called *vector control* have contributed to the takeover of the variable speed drive applications by a.c. motors. Let us start with the d.c. brush motor.

## 2.3 D.C. Brush Motors

A typical topology of a d.c. brush motor with stator PM (permanent magnet) or d.c. electromagnetic excitation and a rotor comprising the armature winding and the mechanical commutator with brushes is shown in Figure 2.2a. The mechanical commutator is, in fact, an electromechanical d.c.–a.c. bidirectional power flow power converter, as the currents in the rotor armature coils are a.c. while the brush current is d.c.

Figure 2.2c shows an axial airgap PM d.c. brush motor with a printed, winding, ironless-disk rotor, and mechanical commutator with brushes. PM excitation, especially with the nonmagnetic disk rotor, yields extremely low electric time constants L/R (around or less than 1 ms in the sub-kW power range). Thus, a quick response in current (torque) is expected, though the current (torque) harmonics are large if the switching frequency in the PEC is not high enough.

Unfortunately, the mechanical commutator, though not bad in terms of losses and power density, has serious commutation current and speed limits and thus limits the power per unit to 1–2 MW at 1000 rpm and may not be at all acceptable in chemically aggressive or explosion-prone environments.

The PEC, as a d.c. variable voltage source (Figure 2.2b), when capable of four quadrant operation (positive and negative d.c. voltage and d.c. current output), is as complicated and as expensive as the PEC for a.c. motors. Moreover, fast speed reversal is quite a problem.

Still, the d.c. brush motors, especially with PM excitation, are and will still be used for a good period of time, in numerous applications, where 1–2 quadrant operation suffices at low powers and moderate speeds (robotics and car auxiliary drives).

## 2.4 Conventional A.C. Motors

By conventional a.c. motors, we mean traveling field motors—induction, synchronous with electromagnetic excitation, and reluctance synchronous—that may be used either in

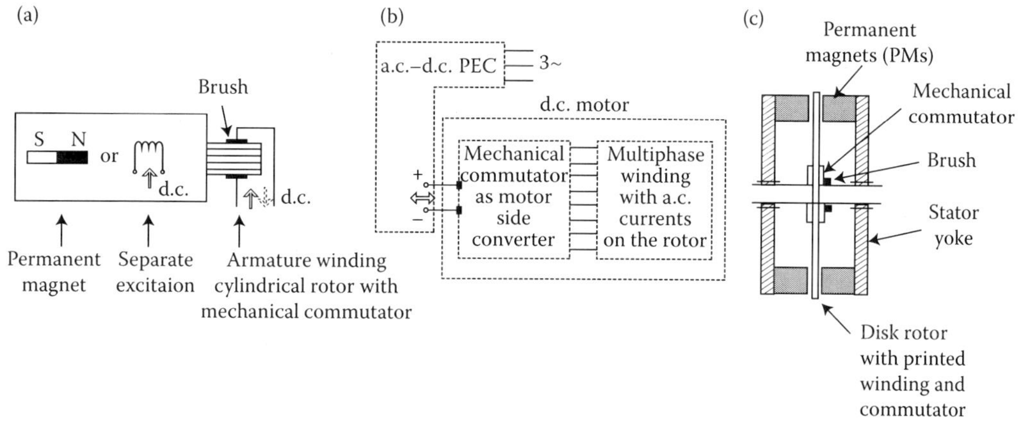


**FIGURE 2.2**
D.c. brush motor: (a) with cylindrical rotor; (b) equivalence of d.c. motor to a.c. motor plus motor side power converter; (c) with PMs and disk rotor.

online start (constant speed) or variable speed (with PEC) applications [9–12]. Consequently, the rotors of all conventional a.c. motors have squirrel cages for asynchronous starting. Among them, the IM (Figure 2.3a) is mostly used for variable speed drives.

A PEC system controls the voltage $V_1$ and frequency $f_1$ in the stator. This is a full-power PEC and is adequate for wide speed control (below 1/2). For high-power applications with limited speed control, a limited-power PEC supplies the wound rotor through the slip rings (Figure 2.3b) with adequate (low) voltage $V_2$ and frequency $f_2$, while the stator is line-fed at constant $V_1$ and $f_1$.

Thus the sizing and costs of PEC wound-rotor IM drives are drastically reduced though, for full torque starting, a special arrangement is required as the PEC cannot handle directly the rather large starting rotor currents and voltages. Still, the slip ring-brush mechanical system poses notable problems of maintenance and may not be used in hostile environments.

The cage-rotor induction motors (IMs) cover wide power ranges from 0.5 kW to 10 MW, while even higher powers (up to 400 MW/unit) may be handled through wound-rotor configurations.

Electromagnetically excited conventional synchronous motors (SMs) have three-phase a.c. windings on the stator and a d.c.-excited cage rotor with salient or nonsalient poles (Figure 2.4).

A conventional SM when used in variable speed drives requires two PECs—one full-power a.c.–a.c. PEC in the stator and a low-power (1%–5%) a.c.–d.c. PEC that supplies, in general through slip rings and brushes, the d.c. excitation windings on the rotor. A coordinated control of the two PECs provides for speed (active power) and reactive power control and for efficient and wide speed range control in high-power applications (MW and tens of MW range).

Why do, above 2 MW, SMs predominate over IMs? This is because the reactive power control through excitation is easily possible with SMs, and thus the requirements in this regard from the full-power PECs are lower. Consequently, the PECs are simpler and less expensive and the overall costs of the drive are lower, though the SM for comparable power and speed is more expensive than the IM. The efficiency of the SM above 2 MW is also, in general, higher than that of the cage-rotor IM.

Conventional reluctance synchronous motors (RSMs) have a cylindrical stator with three a.c. windings and a windingless rotor with a moderate orthogonal axis magnetic saliency of up to 4 (6) to 1. High magnetic saliency is obtained with multiple flux barriers (Figure 2.5).

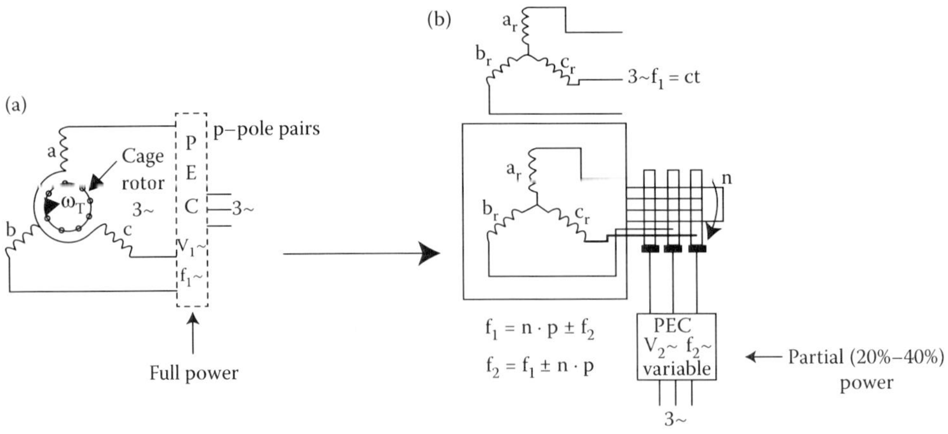


**FIGURE 2.3**
Three-phase IM: (a) with a cage rotor and PEC in the stator; (b) with slip rings wound rotor and PEC on the rotor.

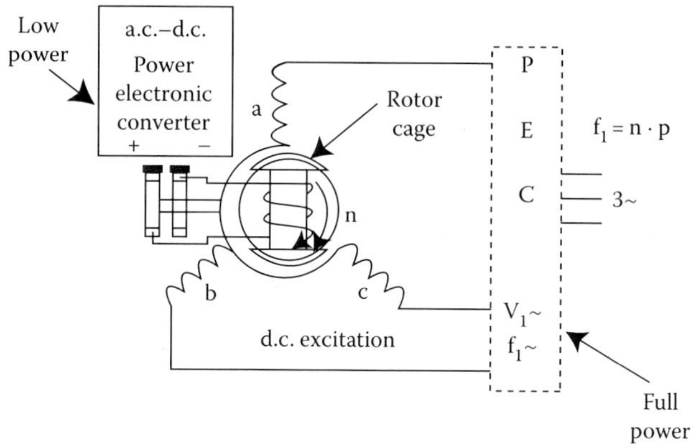


**FIGURE 2.4**
Conventional salient pole rotor SM fed through a full-power PEC in the stator.

A cage on the rotor is sometimes maintained for stability purposes when the motor is fed through a full-power a.c.–a.c. PEC with a rigid relationship between frequency $f_1$ and motor speed n ($n = f_1/p$).

The conventional RSMs are to some extent (up to 100 kW) used in low dynamics variable speed drives with open-loop speed control as the speed does not decrease with load. Consequently, control is simpler than with IMs.

The main drawback of the conventional RSM (reluctance synchronous motor) is the low PF and not high enough torque density (Nm/kg), which leads to higher kVA ratings of the PEC (by 15%–20%).

Besides the conventional a.c. motors (which may also be used for line-start constant speed applications due to the action of the rotor cage), in the last decades, new motor configurations totally PEC-dependent, have been introduced. The main types of PEC-dependent motors are described in Section 2.5.

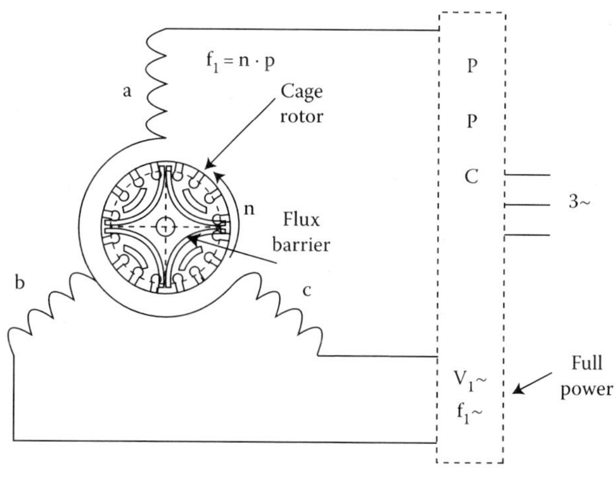


**FIGURE 2.5**
Conventional RSM fed through a full-power PEC in the stator.

## 2.5 PEC-Dependent Motors

Evidently PEC-dependent motors cannot operate without PECs. They are, in general, *multiphase motors* (to limit torque pulsations) and provide for self-starting from any initial position. In essence, this new breed of motors may be fed through unipolar or bipolar multiphase currents intact with the rotor position. Besides, they have singly salient or doubly salient magnetic structures with or without permanent magnets (PMs) on the rotor, or even on the stator.

PEC-dependent motors with PM rotors [13] have evolved from SMs by replacing the d.c. electromagnetic excitation with high-energy PMs (Figure 2.6).

The absence of the rotor cage renders the PM SM fully dependent on PECs as the stator frequency and rotor speed should be in synchronism at all moments.

The stator three-phase windings are concentrated (one coil/pole/phase) when the rectangular bipolar current control (based on inexpensive rotor position sensors) is performed. For a more refined performance (wider speed range [1/1000 range]), sinusoidal current control is performed. A more expensive precise position sensor (or observer) is needed.

For both rectangular and sinusoidal current control, nonoverlap windings have also been introduced recently (from 1 W to tens of kW power levels) to reduce copper losses and frame length and to reduce PM torque at zero current (cogging torque).

The pole magnet and interpole magnet rotors have both advantages and disadvantages as shown in Chapter 11.

The other main category of PEC-dependent motors is the so-called stepper motors. In fact, they are doubly salient multiphase motors with windingless rotor and unipolar position-dependent or position-independent current control through a.c.–d.c. (unipolar current) PECs (Figure 2.7).

It is evident that the switched reluctance motor (SRM) [14]—a commercial name for the power step motor with rotor position controlled slewing (continuous motion)—works on the principle that by sequentially supplying each phase with a current pulse when the inductance of that phase increases will produce positive torque (motoring). Regenerative

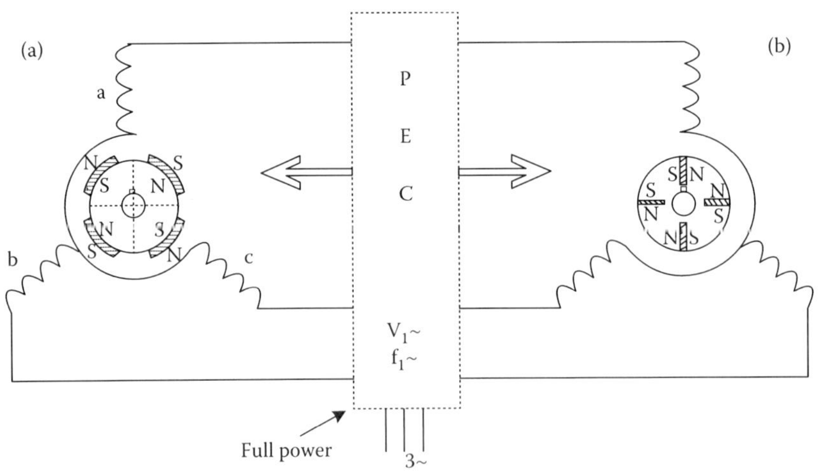


**FIGURE 2.6**
PM SM—singly salient (a) pole magnet rotors; (b) interpole magnet rotors.

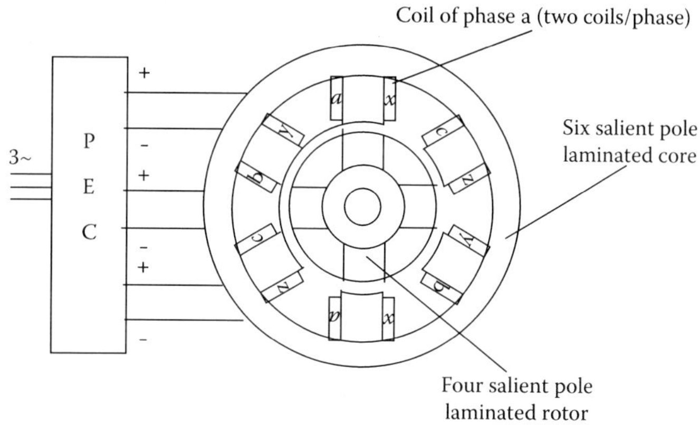


**FIGURE 2.7**
Three-phase switched reluctance 6/4 motor (SRM)—PEC dependent.

braking is obtained if the phase is supplied when the inductance slope is negative. The phases are switched on and off based on position feedback (provided by a position sensor or observer) and thus torque and motion control "without losing steps" is obtained.

The SRM has undergone more than two decades of development and seems ready now for marketing for power from a few watts to megawatt levels, both for low and high dynamics applications. The SRM is, together with PM SM, a competitive motor for high-performance variable speed drives.

The reluctance or hybrid (PM-reluctance) stepper motors resemble the SRM and PM SMs with rectangular current, but they rely on open-loop feed-forward (position independent) control with moderate dynamics and acceptable position precisioning by a notable increase in the number of stator and rotor poles and the number of phases (4–5 in general).

Stepper motors refer to low power positioning applications only. For more details on stepper motors, see Reference 15.

PM SM and SRM drives will be discussed in separate chapters in this book.

Other cageless rotor converter-dependent motors are spin-offs from electromagnetically excited SMs, IMs, RSMs, or doubly salient motors RSM (by adding PMs on the rotor or on the stator).

Despite their severe torque pulsations, single-phase PM synchronous (brushless) motors, as well as single-phase SRM drives, are being gradually introduced in low-power applications (for home appliances or small electric actuators on cars) to cut costs by reducing the number of controlled power electronic switches in the static converter.

They will be discussed later, briefly, in pertinent chapters.

## 2.6 Energy Conversion in Electric Motors/Generators

In all systems, where mass is not destroyed or produced, the principle of energy conservation holds. So energy is not created nor destroyed; it is only converted from one form to another. This principle, together with the Faraday, Ampere, Gauss, Ohm, and Newtonian mechanical laws and theory of electric circuits, govern the electromechanical energy conversion in electric motors–generators.

The electromechanical energy conversion in electric motors–generators implies a few forms of energy:

$$\left\langle \begin{matrix} \text{Energy form} \\ \text{electrical source}(\pm) \end{matrix} \right\rangle = \left\langle \begin{matrix} \text{mechanical} \\ \text{energy}(\pm) \end{matrix} \right\rangle + \left\langle \begin{matrix} \text{stored} \\ \text{magnetic} \\ \text{energy} \\ \text{increase} \end{matrix} \right\rangle + \left\langle \begin{matrix} \text{energy} \\ \text{converted} \\ \text{to heat} \\ \text{(losses)}:(+) \end{matrix} \right\rangle \tag{2.1}$$

The electric energy from the electric source and mechanical energy are considered positive for motoring and negative for generating.

The energy converted into heat has three main causes:

- Electric machine coils (copper) losses: $p_{Co}$
- Mechanical losses (gear friction and windage losses): $p_{mec}$
- Magnetic (hysteresis and eddy current) losses in the magnetic cores of the machine: $p_{iron}$

Including the losses at their location, we obtain

$$\left\langle \begin{matrix} \text{Electric} \\ \text{energy} \\ -p_{Co} \end{matrix} \right\rangle = \left\langle \begin{matrix} \text{magnetic energy} \\ \text{storage increase} \\ +p_{iron} \end{matrix} \right\rangle + \left\langle \begin{matrix} \text{mechanical} \\ \text{energy} \\ +p_{mec} \end{matrix} \right\rangle \tag{2.2}$$

So, in fact, the electrical to mechanical energy conversion in an electric motor implies an intermediate energy form for storage: the magnetic energy, mostly present in the magnetic field of the airgap between the rotor (mover) and the stator (fixed part) of the electric motor (Figure 2.8.).

The net electrical energy increment $dW_e$ may be described in terms of voltage v, current i, and electromotive force (e.m.f.) e:

$$dW_e = (v - Ri)i \cdot dt \tag{2.3}$$

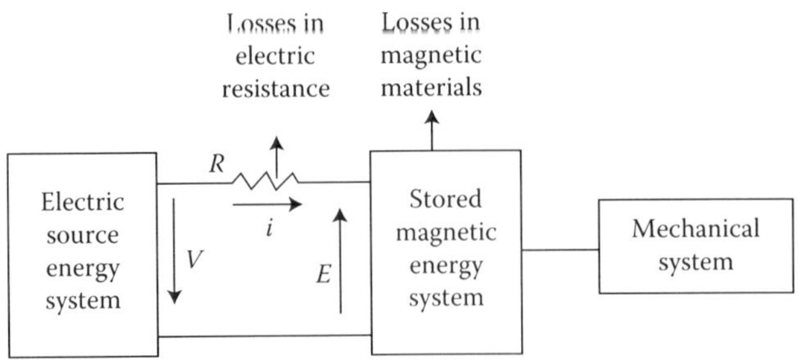


**FIGURE 2.8**
Energy conversion in electric motors/generators.

To realize electrical to mechanical energy conversion, the coupling electromagnetic field in the machine has to produce a reaction to the electric circuit (to electric source voltage v), e:

$$-e = v - Ri; \quad dW_e = -(e)i \cdot dt \tag{2.4}$$

If the electric energy is transmitted to the coupling magnetic field in the machine through more than one electric circuit (phase), then Equation 2.4 contains more terms.

According to Faraday's law, the e.m.f. (e) is

$$e = -\frac{d_s\lambda}{dt} \tag{2.5}$$

$d_s/dt$ is the total time derivative of the flux linkage λ, which may contain two terms: a transformer (pulsational) and a motion-caused term.

According to Equation 2.4

$$dW_e = i \cdot d_s\lambda, \quad dW_{mec} = T_e d\theta_r \tag{2.6}$$

where $dW_{mec}$ is the mechanical energy increment, $T_e$ is the instantaneous electromagnetic torque developed by the electric machine, and $d\theta_r$ is the rotational motion angle increment.

From Equations 2.4 and 2.5 and the energy conservation condition

$$dW_e = id_s\lambda = dW_m + T_e d\theta_r \tag{2.7}$$

where $dW_m$ is the stored magnetic energy increment.

If the magnetic flux linkage is constant ($d_s\lambda = 0$), the energy increment from the electrical energy source is zero and thus the electromagnetic torque $T_e$ is

$$T_e = -\left(\frac{\partial W_m}{\partial \theta_r}\right) \tag{2.8}$$

This situation occurs only when the rotor-stored mechanical energy of the motor is converted into losses in the short-circuited (v = 0) electric generator mode.

The electromagnetic torque is negative and the machine is braked gradually until close to a standstill.

In most applications, the case when there is electric energy transfer from (to) the electric source is more relevant, as it defines the electric motor and electric generator operation modes.

In such cases, Equation 2.7 can be written as

$$dW_m = id_s\lambda - T_e d\theta_r; \quad i = -\frac{\partial W_m}{\partial \lambda} \tag{2.9}$$

with λ and $\theta_r$ being independent variables.

We may choose the current i instead of flux linkage λ as the independent variable. But in that case a new function—magnetic coenergy $W'_m$ —may be defined as

$$W'_m = i\lambda - W_m \tag{2.10}$$

Differentiating this equation, we obtain

$$dW'_m = id\lambda + \lambda di - dW_m \tag{2.11}$$

or, with Equation 2.9

$$dW'_m = id\lambda + \lambda di - id\lambda + T_e d\theta_r \tag{2.12}$$

And finally

$$T_e = \left(\frac{\partial W'_m}{\partial \theta_r}\right)_{i=const}; \quad \lambda = \frac{\partial W'_m}{\partial i} \tag{2.13}$$

$$W_m = \int_0^{\lambda_0} id\lambda; \quad W'_m = \int_0^{i_0} \lambda di \tag{2.14}$$

Equation 2.9 presupposes that the magnetic flux linkage λ varies in time and thus the electric energy exchange with the electric machine $dW_e$ is now nonzero.

In accordance with Equations 2.8 and 2.13, the electromagnetic torque $T_e$ is nonzero only when the magnetic field coenergy varies with respect to rotor position $\theta_r$.

Let us consider a primitive SRM with two stator and two rotor poles (Figure 2.9).

We have, in fact, an inductance L (made of two coils in series or in parallel) with an airgap and a magnetic core. It is evident that when the rotor moves around, the magnetic permeance varies with $\theta_r$.

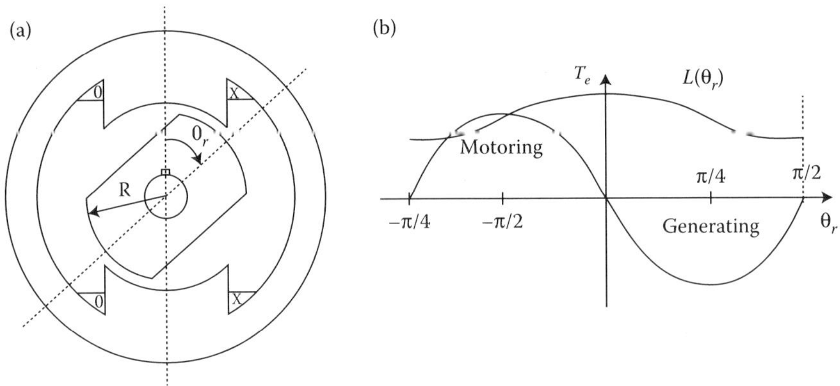


**FIGURE 2.9**
The 2/2 stator/rotor pole primitive SRM: (a) cross section; (b) inductance L and torque $T_e$ versus angle $\theta_r$.

Approximately

$$L(\theta_r) \approx L_0 + L_2 \cos 2\theta_r \tag{2.15}$$

If the magnetic saturation is neglected, $L_0$ and $L_2$ are constant, the magnetic coenergy $W'_m$ of this coil is

$$W'_m = \int_0^{i_0} L(\theta_r) i di = \frac{i_0^2}{2} L(\theta_r) \tag{2.16}$$

The electromagnetic torque $T_e$ is

$$T_e = \left(\frac{\partial W'_m}{\partial \theta_r}\right)_{i_0 = \text{const}} = \frac{i_0^2}{2}\frac{\partial L(\theta_r)}{\partial \theta_r} = -i_0^2 L_2 \sin 2\theta_r = -T_{em} \sin 2\theta_r \tag{2.17}$$

The polarity of the current does not influence the torque polarity. Positive torque means motoring and negative torque implies generating. As the torque has positive and negative values with $\theta_r$ it follows that the 2/2 SRM may not start from any initial rotor position and thus special measures (e.g., a parking PM) are needed to provide self-starting from the designed (parking) position. Multiphase SRMs (switched reluctance motors) may start from any position.

## 2.7 Summary

We summarize the motor candidates for variable speed electric drives in Table 2.1.

There are applications that require linear motion and, provided competitive costs/performance rates are reasonable, linear motors become practical solutions. There is a

**TABLE 2.1**

Electric Motors for Drives

| Motor Type | Singly Salient | Nonsalient | Doubly Salient | Unipolar Current | Bipolar Current | Conventional (Caged Rotor) | Converter Dependent (Cageless Rotor) |
|---|---|---|---|---|---|---|---|
| D.c. brush | X | | | X | | | X[a] |
| Induction | | X | | | X | X | |
| Excited synchronous | X | X | | | X | X | X |
| Reluctance synchronous | X | | | | X | X | X |
| PM synchronous | X | X | X | | X | | X |
| Switched reluctance | | | X | X | | | X |
| Stepper motor | | | X | X | X | | X |

[a] D.c. brush motors may be supplied through a series resistor and in this sense they are conventional (converter independent for variable speed) but this method is energy consuming and, thus, it is hardly used anymore.

linear counterpart for every type of rotary motor. Moreover, their control is similar to that of their rotary counterparts.

Linear motor drives have their separate literature [16,17].

New PMSM(G) configurations with less high-energy PMs (or with low-cost ferrite PMs) on the rotor or stator, in axial or radial airgap topologies have been introduced recently in an effort to reduce initial costs but preserve performance in applications ranging from home appliances to wind (hydro) generators and hybrid electric and electric automobiles [18,19].

## References

1. W. Leonhard, *Control of Electrical Drives*, first and second editions, Springer-Verlag, Berlin, Heidelberg, 1985, 1996.
2. P.C. Sen, *Thyristor D.C. Drives*, John Wiley & Sons, New York, 1981.
3. G.K. Dubey, *Power Semiconductor Controlled Drives*, Prentice-Hall, Englewood Cliffs, New Jersey, 1989.
4. P. Vas, *Vector Control of AC Machines*, Oxford University Press, Oxford, UK, 1990.
5. I. Boldea, S.A. Nasar, *Vector Control of AC Drives*, CRC Press, Boca Raton, Florida, 1992.
6. M.P. Kazmierkovski, H. Tunia, *Automatic Control of Converter-Fed Induction Motor Drives*, Elsevier, Amsterdam, the Netherlands, 1994.
7. A.M. Trzynadlowski, *The Field Orientation Principle in Control of Induction Motors*, Kluwer Academic Press, Norwell, MA, 1994.
8. D.W. Novotny, T.A. Lipo, *Vector Control and Dynamics Motor of AC Drives*, Clarendon Press-Oxford, New York, 1996.
9. S.A. Nasar, I. Boldea, *Electric Machines Steady State*, CRC Press, New York, 1990.
10. S.A. Nasar I. Boldea, *Electric Machine Dynamics and Control*, CRC Press, Boca Raton, Florida, 1992.
11. P. Krause, *Analysis of Electric Machinery*, McGraw-Hill, New York, 1986.
12. I. Boldea, *Reluctance Synchronous Machines & Drives*, Oxford University Press, Oxford, UK, 1996.
13. S.A. Nasar, I. Boldea, L.E. Unnewehr, *Permanent Magnet, Reluctance and Selfsynchronous Motors*, CRC Press, Boca Raton, Florida, 1993.
14. T.J.E. Miller, *Switched Reluctance Motors and Their Control*, Magna Physics Publishing & Clarendon Press, Oxford, 1993.
15. T. Kenyo, *Stepping Motors and Their Microprocessor Controls*, Clarendon Press, Oxford, 1984.
16. I. Boldea, S.A. Nasar, *Linear Electric Actuators and Generators*, Cambridge University Press, New York, 1997.
17. I. Boldea, S.A. Nasar, *Linear Motion Electromagnetic Devices*, Taylor & Francis, New York, 2001.
18. I. Boldea, L. Tutelea, F. Blaabjerg, High power wind generator designs with less or no PMs: An overview, *17th International Conference on Electrical Machines and Systems* (*ICEMS*), 2014, pp. 1–14 (IEEExplore).
19. I. Boldea, L. Tutelea, D. Dorrell, L. Parsa, Automotive electric propulsion systems with less or no PMs: An overview, *IEEE Trans.*, vol. IE-61, no. 99, 2014, pp. 5696–5711.

# 3

# *Power Electronic Converters for Drives*

PECs for drives are now a mature technology with notable and dynamic worldwide markets. Extraordinary research and development efforts are still devoted to the subject today and are likely to continue in the future.

There is a distinct rich literature on the subject [1]. Here, we only summarize the main configurations and the voltage–current waveforms, while more details will be given in subsequent chapters devoted to various motor drives.

PECs may be classified in many ways. Among them, we may include the type of electronic switch used, the steps in power conversion, and current (voltage) output waveforms.

Let us first review the main power electronic switches (PESs).

## 3.1 Power Electronic Switches

PESs used in PECs undergo frequent commutations from a blocked to saturated (open) condition to provide high energy conversion rates (efficiency above 85%–98%) over a wide range of voltages and, when it is the case, frequencies.

The blocked state is characterized by high voltage over the power terminal of the PES when the conducting current is practically zero. The saturated (open) state is defined by the presence of the conducting current while the voltage at the power terminals is low, in the 0.5–2 V range.

PESs may be classified as

1. Uncontrolled
2. Semicontrollable
3. Fully controllable

The *diode* is an uncontrolled PES (Figure 3.1a) whose state of conduction is determined solely by the direction of the current from anode (A) to cathode (K).

The ideal characteristic is shown in Figure 3.1b and shows that the diode conducts only for positive current, when the ideal voltage drop on it $V_{AK}$ is zero. In reality $V_{AK}$ is not zero but small (around 1–2 V).

The diode is present in all PECs either as a rectifier or for protection. The diode does not have a control circuit or a driver for it.

The *thyristor* is a semicontrollable PES. Its state of conduction may be controlled only in the blocked—saturated sequence, when in the command circuit (driver) gate (G)—cathode (K) a positive current is present and if the PES is traveled by a positive current from anode (A) to cathode (K) (Figure 3.2). The saturated state is kept even after the command signal is inhibited until the current in the power circuit (A–K) becomes zero.

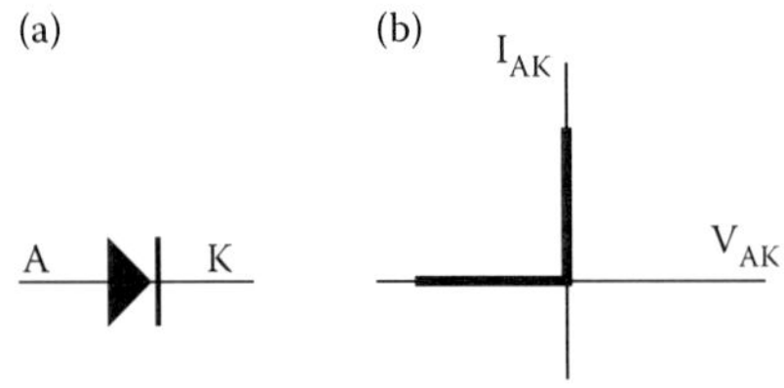


**FIGURE 3.1**
The diode symbol (a) and its ideal characteristic (b).

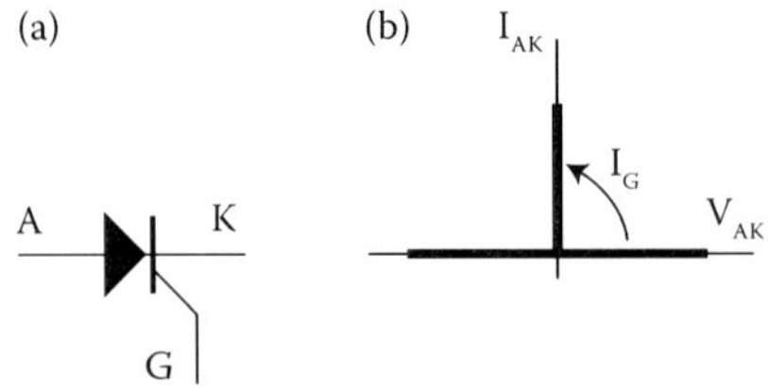


**FIGURE 3.2**
The thyristor symbol (a) and its ideal characteristic (b).

The thyristor is used especially in PECs having an interface with a.c. power grids at high power levels and low commutation frequencies (up to 300 Hz in general).

The *GTO* (gate turn off thyristor) (Figure 3.3) is a fully controllable PES. Its saturation is obtained as for the thyristor but its blocking is accessible when a negative current $i_G$ is applied to the command (driver) circuit.

This PES is used for high-power PECs and for interfacing with a.c. grids. Its switching frequency is higher than that of the thyristor but still below 1 kHz in general.

The *bipolar junction transistor* (BJT) (Figure 3.4) is also a fully controllable PES but at high commutation frequency and low and medium (up to tens of kW) powers. Its command—saturation—is obtained through its base (B) current with respect to the emitter (E) and is maintained only in the presence of the command current. When the command current becomes zero, the BJT gets blocked. The commutation time is lower than for the thyristor and the GTO. The BJT may operate only in the first quadrant (Figure 3.4) as its power circuit collector (C)–emitter (E) may not withstand negative polarization voltages.

The *MOS transistor* is also a fully controllable PES. The MOS transistor—in contrast to other PESs—works with voltage (rather than current) signals in the command (driver) circuit gate (G)–source (S). Consequently, the command power is practically zero, which notably simplifies the command circuitry (Figure 3.5).

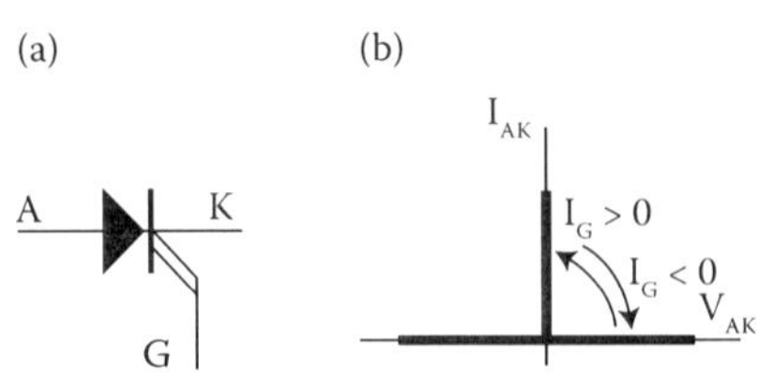


**FIGURE 3.3**
The GTO's symbol (a) and its ideal characteristic (b).

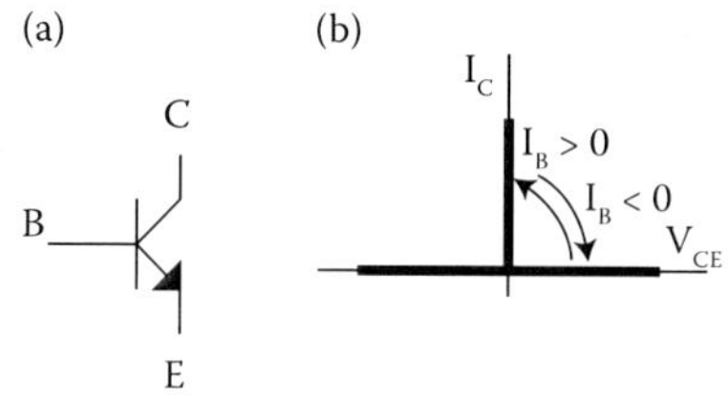


**FIGURE 3.4**
The BJT symbol (a) and its ideal characteristic (b).

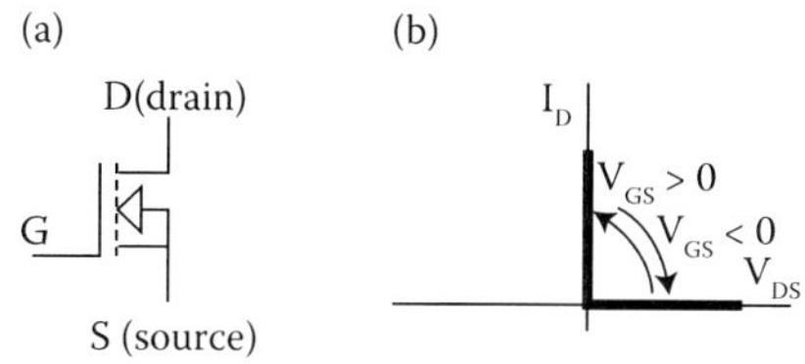


**FIGURE 3.5**
MOS transistor symbol (a) and its ideal characteristic (b).

The MOS transistor has the lowest commutation time, allowing for very high switching frequency (tens of kHz), however, at low and moderate power and voltage levels.

The IGBT (Figure 3.6) possesses the MOS command merits (voltage command signal) and the low commutation time (up to 20 kHz switching frequency) and power circuit assets (low saturation voltage, larger voltage [current] ratings) of the BJT.

In the voltage/current ratings, the switching frequencies are getting higher and higher and their conduction and commutation losses are becoming lower at a high pace. As an example, PECs with IGBTs have reached already 500 kVA per unit in 1996 and, with four units in parallel and simultaneously controlled, 2 MVA.

At the same time, thyristors are less expensive and, in the megawatt (and higher) power range, they are still competitive in many applications and in particular topologies. GTOs (gate turn off thyristors) tend to win rather quickly over the power range from thyristors.

For high power, medium-voltage MOS-controlled thyristors (MCTs) have been recently (1992) introduced. MCTs are turned on and off by a short voltage pulse on the MOS gate with thousands of microcells in parallel on a chip to yield high current levels.

Since 1997, the integrated gate-commutated thyristors (IGCT) are available also for high voltage, high power (4500 V, 300 A per unit). The turn off negative gate current is also 300 A but is very short.

High switching frequency characterizes both MCTs and IGCTs.

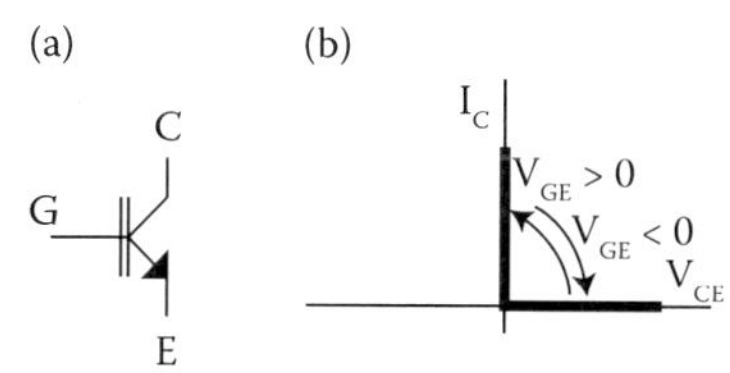


**FIGURE 3.6**
IGBT's symbol (a) and its ideal characteristic (b).

Still in the labs, SiC-based higher-voltage (15 kV, 300 A) medium-current static power switches with a few kHz switching frequency seem likely to revolutionize the medium- and large-power electronics by 2020.

The PECs may be classified many ways. In what follows, we will refer to their input and output voltage/current waveforms and distinguish

- a.c.–d.c. converters (or rectifiers)
- d.c.–d.c. converters (or choppers)
- a.c.–d.c.–a.c. converters (indirect a.c.–a.c. converters)—two stages
- a.c.–a.c. converters (direct a.c.–a.c. converters)

We should note that a.c.–d.c.–a.c. converters contain an a.c.–d.c. source-side converter (rectifier) and a d.c.–a.c. converter called an inverter. These converters are mostly used with a.c. motor drives of all power levels.

For bidirectional power flow, the rectifier should be capable of working as an inverter, while the inverter is in principle capable of working as a rectifier.

Direct (one-stage) a.c.–a.c. converters (cycloconverters) are applied in large-power a.c. drives while the a.c.–d.c.–a.c. converters are used within the whole power range.

A.c.–d.c. converters are either part of a.c.–d.c.–a.c. converters or they are used independently for driving d.c. brush motors.

D.c.–d.c. converters are either connected directly to batteries (d.c. sources in general) or they are fed from uncontrolled (diode) rectifiers for driving d.c. brush motors, in multiphase PEC configurations.

As the diode rectifier is used in many PECs as a line side converter, it will be presented here in some detail. All other PECs will merely be introduced here, for brevity. Further details are given in subsequent chapters on various electric motor drives.

## 3.2 Line Frequency Diode Rectifier for Constant D.C. Output Voltage $V_d$

In most electric drives, the power is provided by the local grid at 60 (or 50) Hz in single-phase or three-phase configurations. Single-phase a.c. is available at low power in general in various buildings, while industrial power grid is three phase. We treat them in sequence.

The output voltage of the diode rectifier should be as ripple free as possible and thus it requires a rather large capacitor filter (Figure 3.7).

Let us consider first a basic rectifier circuit (Figure 3.8) with instantaneous commutation and a line-source inductance $L_s$ providing constant $V_d$ at no load.

The diode starts conducting when $V_s \geq V_d$ at $t_1$. At $t_2$, $V_s = V_d$ but, due to inductance $L_s$, the current goes into the diode until it dies out at $t_3$ such that $A_{on} = A_{off}$. In fact the integral of inductance voltage $V_L$ from $t_1$ to $t_1 + T$ should be zero, that is, the increment of the flux (and the increment of the stored magnetic energy) in the coil per cycle is zero

$$\int_{t_1}^{t_1+T} V_L dt = 0 = A_{on} - A_{off} \tag{3.1}$$

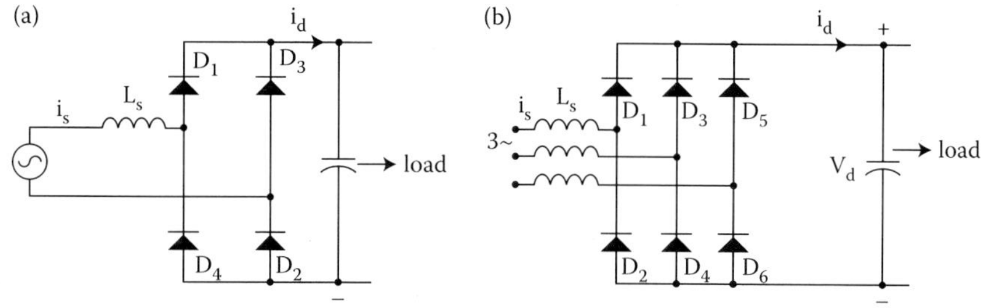


**FIGURE 3.7**
Diode rectifier with output filter capacitor: (a) single phase; (b) three phase.

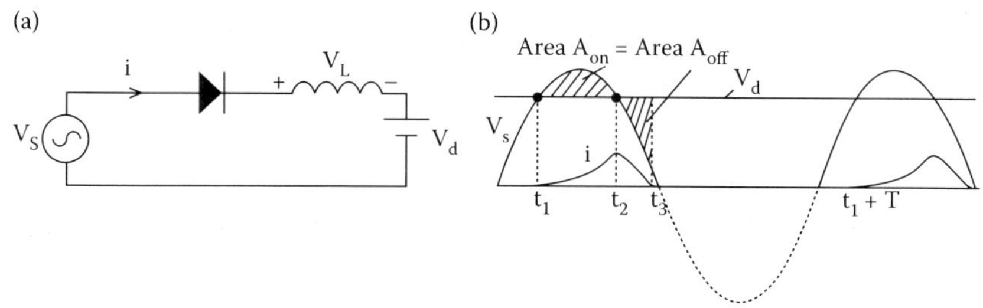


**FIGURE 3.8**
(a) Basic rectifier equivalent circuit and (b) the voltage and current waveforms.

As $V_d$ is rather close to the maximum value $V_s\sqrt{2}$, the current i becomes zero prior to the negative (next) cycle of $V_s$.

Figure 3.8 illustrates only the positive voltage, that is, diodes $D_1$–$D_2$ conducting. For the negative $V_s$, $D_3$–$D_4$ are open and a similar current waveform is added (Figure 3.9).

As long as the current $i_d$ is nonzero

$$V_L = L_s \cdot \frac{di_d}{dt} = \sqrt{2} \cdot V_s \cdot \sin \omega t - V_d \tag{3.2}$$

$$\omega \cdot L_s \cdot i_d(\omega t) = \int_{\omega t_1}^{\omega t} \left(\sqrt{2} \cdot V_s \cdot \sin \omega t - V_d\right) d(\omega t); \quad \theta_{on} < \omega t < \theta_{off} \tag{3.3}$$

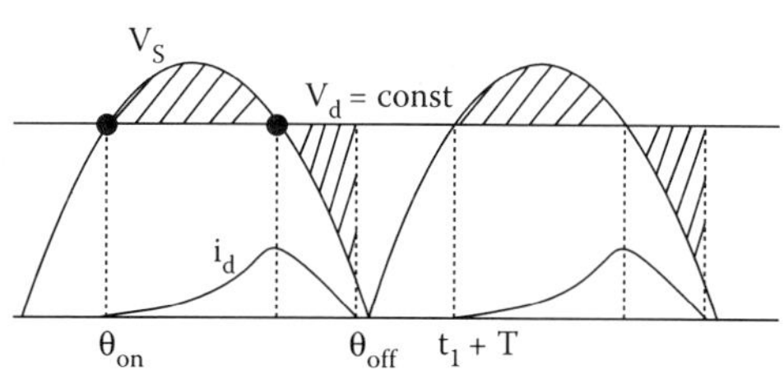


**FIGURE 3.9**
Single-phase rectifier—the waveforms.

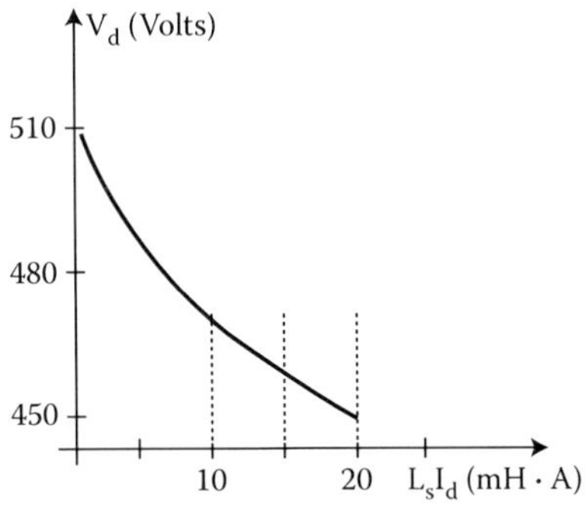


**FIGURE 3.10**
$V_d$ versus $L_sI_d$.

For $\theta_{on}$

$$V_d = \sqrt{2} \cdot V_s \cdot \sin\theta_{on} \tag{3.4}$$

For $\omega t = \theta_{off}$, $i_d(\omega t) = 0$ and from Equation 3.3 we may calculate $\theta_{off}$ as a function of $\theta_{on}$. Finally, the average coil flux linkage $L_sI_d$ is

$$L_s \cdot I_d = \frac{1}{\pi} \cdot \int_{\theta_{on}}^{\theta_{off}} L_s \cdot i_d(\omega t) \cdot d(\omega t) \tag{3.5}$$

For given values of $L_sI_d$, iteratively $\theta_{on}$, $\theta_{off}$ and finally $V_d$ are obtained from Equations 3.3 through 3.5 (Figure 3.10).

## 3.3 Line Current Harmonics with Diode Rectifiers

The line current has the same shape as $i_d$ in Figure 3.9 but with alternate polarities (Figure 3.11). It is now evident that the line (source) current in a rectifier is rich in harmonics. This distortion from the sinusoidal shape can be described by some distortion indexes.

Also, the current fundamental is lagging the source voltage by the DPF angle $\varphi_1$

$$\text{DPF} = \cos\varphi_1 \tag{3.6}$$

The source current rms value is $I_s$. Thus the apparent power magnitude S is

$$S = V_s \cdot I_s \tag{3.7}$$

where $V_s$ is the rms voltage value.

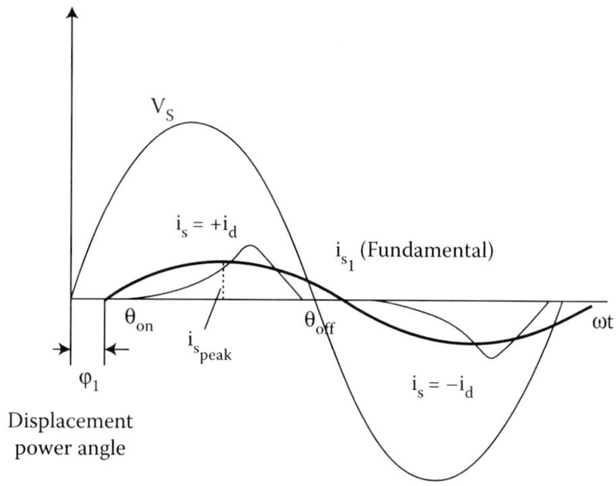


**FIGURE 3.11**
Source current shape.

The power factor

$$PF = \frac{P}{S} \tag{3.8}$$

where

$$P = V_s \cdot I_{s1} \cdot DPF \tag{3.9}$$

So

$$PF = \frac{I_{s1}}{I_s} DPF \tag{3.10}$$

A strong distortion in the line current will reduce the ratio $I_{s1}/I_s$ and thus a small PF is obtained even if DPF is close to unity.

Now

$$I_s = \sqrt{I_{s1}^2 + \sum_{\upsilon=2}^{\infty} I_{s\upsilon}^2} \tag{3.11}$$

The total harmonic current distortion (THD) (%) is

$$THD\% = 100 \frac{I_{dis}}{I_{s1}} \tag{3.12}$$

where

$$I_{dis} = \sqrt{\sum_{\upsilon=2}^{\infty} I_{s\upsilon}^2} \tag{3.13}$$

The peak current $I_{speak}$ is also important to be defined as a relative value constant called the crest factor (CF)

$$CF = \frac{I_{speak}}{I_s} \tag{3.14}$$

or the form factor (FF)

$$FF = \frac{I_s}{I_d} \tag{3.15}$$

It has been shown that the DPF is above 0.9 but the PF is poor if the source inductance $L_s$ is small.

The ratio $V_d/V_s$ only decreases slightly with an $L_sI_d/V_s$ ratio but both the CF and the FF sharply decrease at low (below 0.03) $L_sI_d/V_s$ values. After that, however, they only slightly decrease with $L_sI_d/V_s$.

The distorted waveform of the source current suggests that a.c. filters are required to eliminate the harmonics pollution of the local power grid. The tendency is that every PEC drive which uses a diode rectifier on the source side be provided with a source harmonics filter. More on this aspect will be given in Chapter 13.

**EXAMPLE 3.1**

A single-phase diode rectifier with constant e.m.f. is fed from an a.c. source with the voltage $V_s(t) = V_s\sqrt{2} \cdot \sin\omega t$ ($V_s = 120$ V, $\omega = 367$ rad/s). The discontinuous source current (Figure 3.11) initiates at $\theta_{on} - 60°$ and becomes zero at $\theta_{off} = 150°$. The source inductance is $L_s = 5$ mH. Calculate the d.c. side voltage $V_d$ and the waveform of the source current $i_d(\omega t)$.

According to Figure 3.9 from Equation 3.3, we obtain

$$\int_{\theta_{on}}^{\theta_{off}} \left(\sqrt{2} \cdot V_s \cdot \sin\omega t - V_d\right) \cdot d(\omega t) = 0 \tag{3.16}$$

$$0 = \sqrt{2} \cdot V_s \cdot (\cos\theta_{on} - \cos\theta_{off}) - V_d \cdot (\theta_{off} - \theta_{on}) \tag{3.17}$$

From Equation 3.17

$$V_d = \frac{\sqrt{2} \cdot 120 \cdot (\cos 60 - \cos 150)}{(5/6)\pi - (1/3)\pi} = 147 \text{ V} \tag{3.18}$$

Now from Equation 3.3 again

$$i_d = 0; \quad 0 < \theta < \theta_{on}$$

$$i_d(\omega t) = \frac{1}{\omega L_s} \cdot \int_{\theta_{on}}^{\omega t} \left(\sqrt{2} \cdot V_s \cdot \sin\omega t - V_d\right) \cdot d(\omega t); \quad \text{for } \theta_{on} \le \theta \le \theta_{off} \tag{3.19}$$

$$i_d = 0; \quad \theta_{off} < \theta < 180°$$

Consequently

$$i_d(\omega t) = \frac{\sqrt{2}\cdot V_s\cdot(\cos\theta_{on} - \cos\omega t) - V_d\cdot(\omega t - \theta_{on})}{\omega L_s}$$

$$= \frac{\sqrt{2}\cdot 120\cdot(0.5 - \cos\omega t) - 147\cdot(367t - \pi/3)}{367\cdot 5\cdot 10^{-3}}$$

$$\text{for } 60° < \theta < 150° \tag{3.20}$$

Though not convenient to use, Equations 3.19 and 3.20 allow for the computation of $I_s$ (rms), peak current $I_{speak}$, fundamental $I_1$, TDH% (3.12), CF (3.14), and average d.c. output current $I_d$.

## 3.4 Current Commutation with $I_d$ = ct and $L_S \neq 0$

For the constant d.c. current $I_d$ = ct (Figure 3.12), ideally ($L_s = 0$), the source current, will change stepwise from $-I_d$ to $I_d$ at $\omega t = 0$ and $\omega t = \pi$ (Figure 3.12b). Due to the nonzero $L_s$, during commutation, all four diodes conduct and thus $V_d = 0$. For $\omega t < 0$, $D_3D_4$ conduct while after commutation ($\omega t > u$), only $D_1D_2$ are on.

As $V_d = 0$, the source voltage during commutation is dropped solely across inductance $L_s$

$$V_s\sqrt{2}\sin\omega t = \omega L_s\frac{di_s}{d(\omega t)} \tag{3.21}$$

Through integration for the commutation interval (0,u)

$$\int_0^u V_s\sqrt{2}\sin\omega t\cdot d(\omega t) = \omega L_s\cdot\int_{-I_d}^{I_d} di_s = 2\omega L_s\cdot I_d \tag{3.22}$$

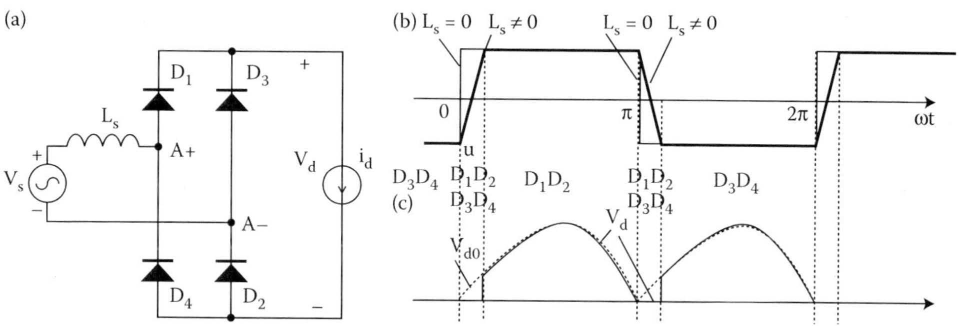


**FIGURE 3.12**
Current commutation in single-phase rectifier with $I_d$ = ct: (a) equivalent circuit; (b) source current; (c) rectified voltage.

We find

$$\cos u = 1 - \frac{2\omega L_s}{V_s\sqrt{2}} \cdot I_d \tag{3.23}$$

Now the average d.c. voltage $V_d$ is

$$V_d = V_{d0} - \frac{2\omega L_s}{\pi} \cdot I_d \tag{3.24}$$

where

$$V_{d0} = \frac{2}{2\pi} \cdot \int_0^\pi V_s\sqrt{2}\sin\omega t \cdot d(\omega t) = \frac{2\sqrt{2}}{\pi} V_s = 0.9 V_s \tag{3.25}$$

is the ideal ($L_s = 0$) average d.c. voltage (Figure 3.12c).

So the source inductance $L_s$ produces a reduction in the d.c. output voltage for constant d.c. output current. The current commutation is not instantaneous and during the overlapping period angle u, all four diodes are conducting.

## 3.5 Three-Phase Diode Rectifiers

In industrial applications, three-phase a.c. sources are available, so three-phase rectifiers seem the obvious choice (Figure 3.13).

The load resistance $R_L$ with a filtering capacitor $C_d$ may be replaced by a constant d.c. current source $I_d$. Using the same rationale as in the previous paragraph, we obtain

$$V_d = V_{d0} - \frac{3\omega L_s}{\pi} \cdot I_d \tag{3.26}$$

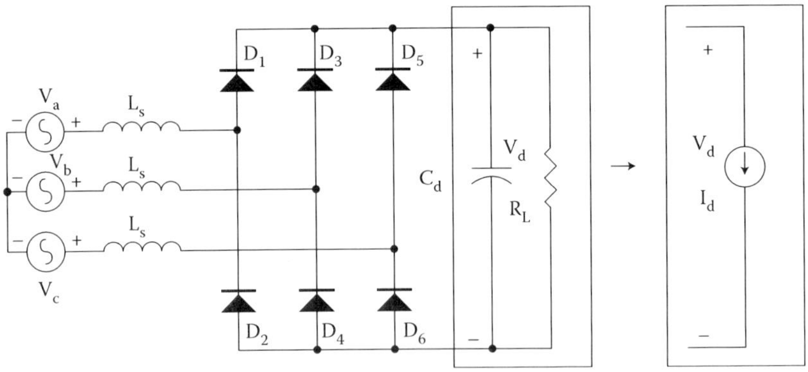


**FIGURE 3.13**
Three-phase diode rectifier.

$$\cos u = 1 - \frac{2\omega L_s}{V_{LL}\sqrt{2}} \cdot I_d \tag{3.27}$$

with $V_{d0} = (3\sqrt{2}/\pi)V_{LL}$, where $V_{LL}$ is the line voltage (rms).

The corresponding waveforms for $L_s = 0$ are shown in Figure 3.14 and for $L_s \neq 0$ in Figure 3.15.

For nonzero $L_s$, a reduction of output d.c. voltage (3.26) is accompanied by all three phases conducting during the commutation angle u (Figure 3.15).

On the other hand, for constant d.c. voltage (infinite capacitance $C_d$), as for the single-phase rectifier, the source current waveform is as in Figure 3.16.

We should note that in both cases the source currents are distorted and thus current harmonics are present in the power source.

Both current source and d.c. output voltage (current) harmonics in actual single-phase rectifiers are higher than in three-phase rectifiers.

**EXAMPLE 3.2: COMMUTATION OVERLAPPING ANGLE U**

For a single-phase or three-phase a.c. system (star connection) with phase voltage $V_s(t) = 120\sqrt{2} \cdot \sin 376t$, calculate the commutation angles u, ideal no-load voltage, and load voltage of a single- or three-phase diode rectifier delivering a constant d.c. current $I_d = 10$ A for the source inductance $L_s = 5$ mH.

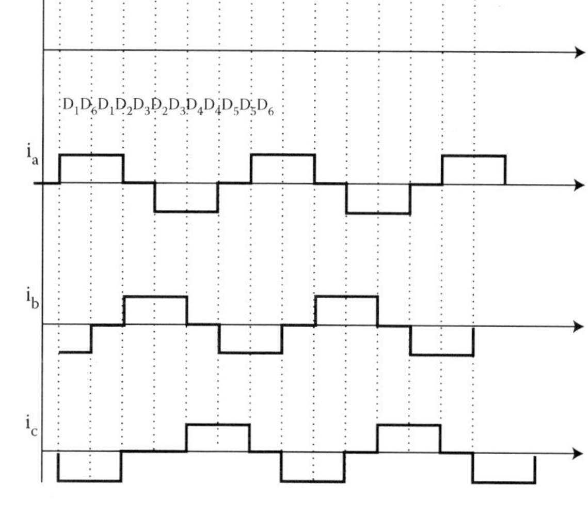


**FIGURE 3.14**
Three-phase ideal waveforms for $L_s = 0$.

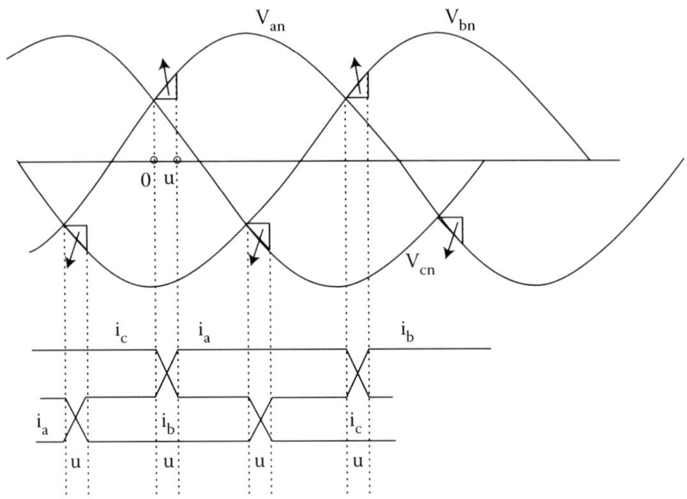


**FIGURE 3.15**
Three-phase current commutation with $L_s \neq 0$.

**Solution**

For the single-phase diode rectifier, using Equation 3.23

$$\cos u = 1 - \frac{2\omega L_s}{V_s\sqrt{2}} \cdot I_d = 1 - \frac{2 \cdot 367 \cdot 5 \cdot 10^{-3}}{120\sqrt{2}} \cdot 10 = 0.783 \tag{3.28}$$

$u = 22.727°$.

The ideal no-load voltage $V_{d0}$ (3.25) is

$$V_{d0} = \frac{2\sqrt{2}}{\pi} V_s = \frac{2\sqrt{2}}{\pi} 120 = 108\,V \tag{3.29}$$

$$V_d = V_{d0} - \frac{2\omega L_s}{\pi} \cdot I_d = 108 - \frac{2 \cdot 367 \cdot 5 \cdot 10^{-3}}{\pi} \cdot 10 = 96.312\,V \tag{3.30}$$

For the three-phase diode rectifier (3.27), u is

$$\cos u = 1 - \frac{2 \cdot 367 \cdot 5 \cdot 10^{-3}}{120\sqrt{2} \cdot \sqrt{3}} \cdot 10 = 0.8746 \tag{3.31}$$

$u = 12.22°$.

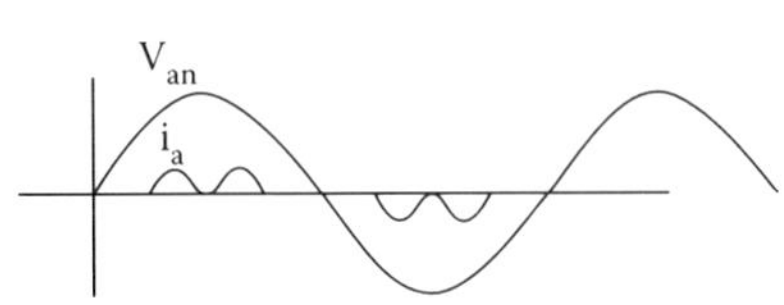


**FIGURE 3.16**
Three-phase rectifier with finite $L_s$ and infinite $C_d$ ($V_d = ct$)—the source current and voltage.

$V_{d0}$ and $V_d$ (from Equation 3.26) are

$$V_{d0} = \frac{3\sqrt{2}}{\pi} V_{LL} = \frac{3\sqrt{2}}{\pi} 120\sqrt{3} = 279.66\,V \tag{3.32}$$

$$V_d = V_{d0} - \frac{3\omega L_s}{\pi} \cdot I_d = 279.66 - \frac{3 \cdot 367 \cdot 5 \cdot 10^{-3}}{\pi} \cdot 10 = 262.128\,V \tag{3.33}$$

Thus the filtering capacitor $C_d$ is notably smaller in three-phase than in single-phase diode rectifiers.

NOTE: This rather detailed introduction to diode rectifiers has been given because they are used frequently as source-side PECs in most electrical drives while they show a strong indication of line current harmonics and commutation aspects, to be met with in all other PECs.

## 3.6 Phase-Controlled Rectifiers (A.C.–D.C. Converters)

Phase-controlled rectifiers—a.c.–d.c. converters—are used to provide controlled d.c. output to directly supply d.c. brush motors or as line-source (first-stage) PECs into two-stage a.c.–d.c.–a.c. converters for a.c. drives.

In principle, phase-controlled rectifiers might be fully controlled or semicontrolled. A rather complete survey of various phase-controlled rectifiers is given in Table 3.1.

It is important to note the power range and quadrant operation of various configurations. We also should note that besides thyristors, GTOs or IGBTs may be used. For high powers (MW and tens of MW), and as a.c. side converters in a.c.–d.c.–a.c. converters for a.c. motors, special configurations are used. They will be treated in their respective chapters. Also, unity input PF configurations are available.

Principal details, numerical examples, or digital simulation results on various rectifier configurations will occur in pertinent chapters on d.c. brush or large-power a.c. motors.

## 3.7 D.C.–D.C. Converters (Choppers)

Choppers are d.c.–d.c. switch-mode converters with unipolar or bipolar current output capability. They are widely used in d.c. brush motor drives in single-phase output unipolar current configurations (Table 3.2) and for SRMs in multiphase configurations (Figure 3.17).

Multiphase choppers for SRMs of various configurations have been proposed and they will be treated in more detail in Chapter 12.

If an a.c. source is available, a diode rectifier and filter are used in front of all choppers (Figure 3.17).

**TABLE 3.1**

Phase-Controlled Rectifier Circuits

| Circuit Type | | Power Range (kW) | Ripple Frequency | Quadrant Operation |
|---|---|---|---|---|
| 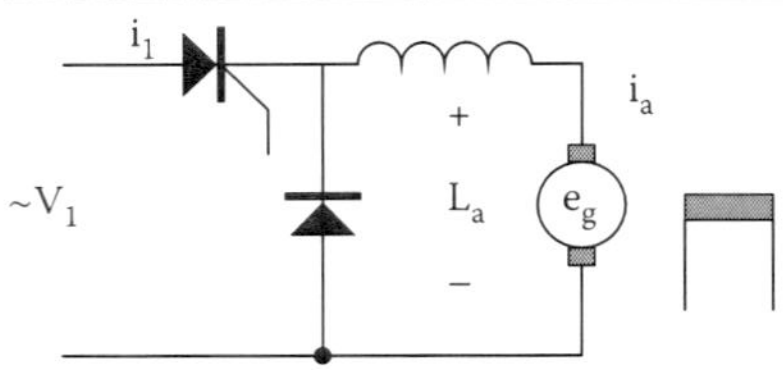  | Half wave single phase | Below 0.5 | $f_s$ | 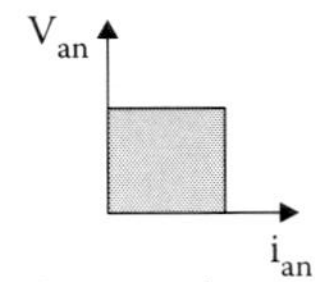  One quadrant |
| 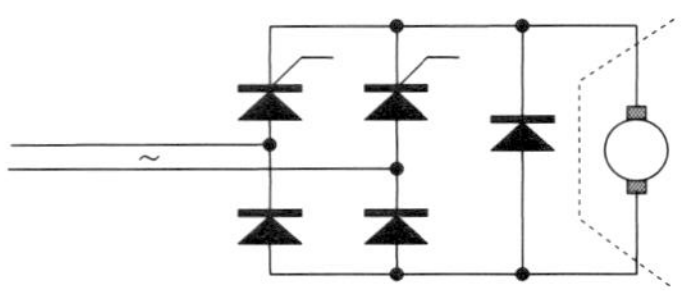  | Half wave three phase | Up to 50 | $3f_s$ | 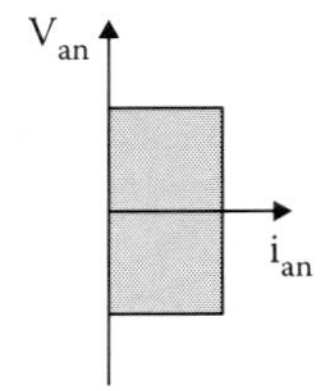  Two quadrant |
| 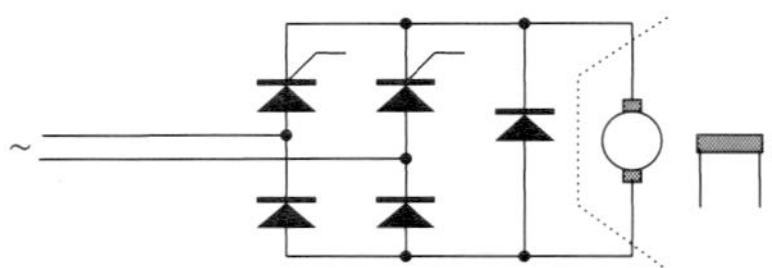  | Semiconverter single phase | Up to 75 | $2f_s$ | 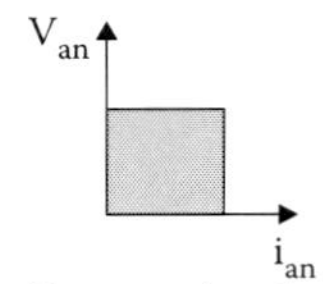  One quadrant |

(*Continued*)

**TABLE 3.1 (*Continued*)**

Phase-Controlled Rectifier Circuits

| Circuit Type | | Power Range (kW) | Ripple Frequency | Quadrant Operation |
|---|---|---|---|---|
| 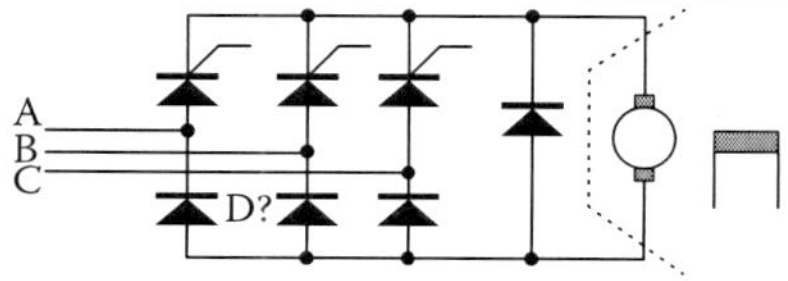  | Semiconverter three phase | Up to 100 | $3f_s$ | 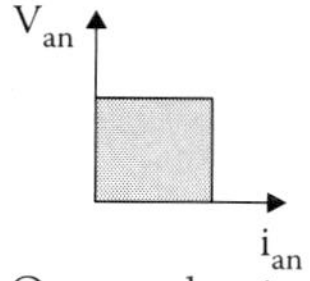  One quadrant |
| 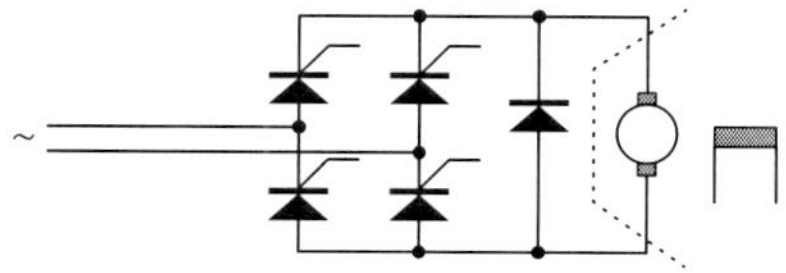  | Full converter single phase | Up to 75 | $2f_s$ | 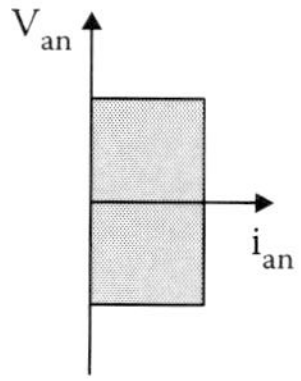  Two quadrant |
| 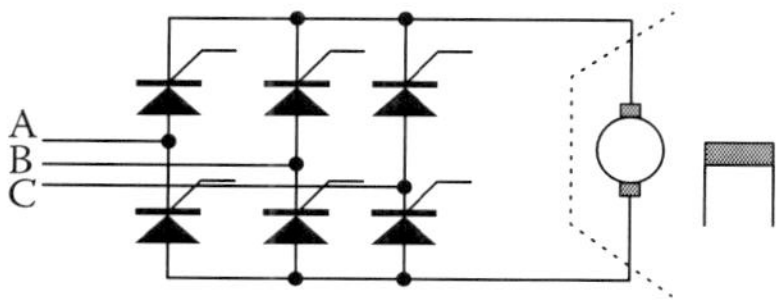  | Full converter three phase | Up to 150 | $6f_s$ | 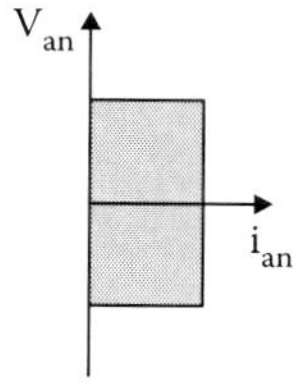  Two quadrant |

(*Continued*)

**TABLE 3.1 (*Continued*)**
Phase-Controlled Rectifier Circuits

| Circuit Type | | Power Range (kW) | Ripple Frequency | Quadrant Operation |
|---|---|---|---|---|
| 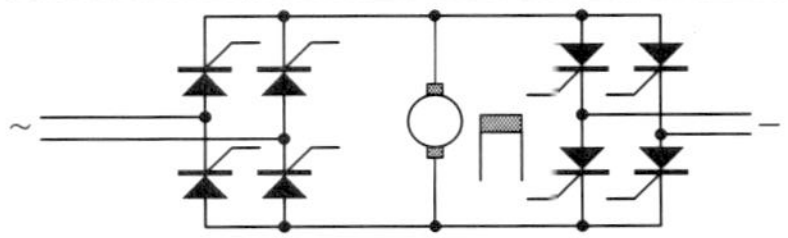 | Dual converter single phase | Up to 15 | $2f_s$ | 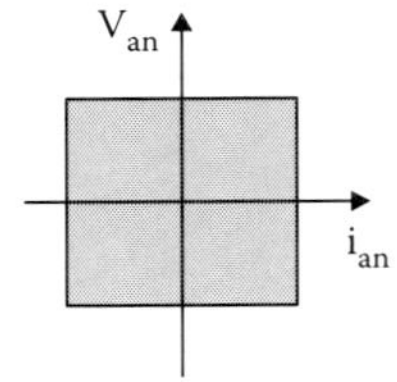 Four quadrant |
| 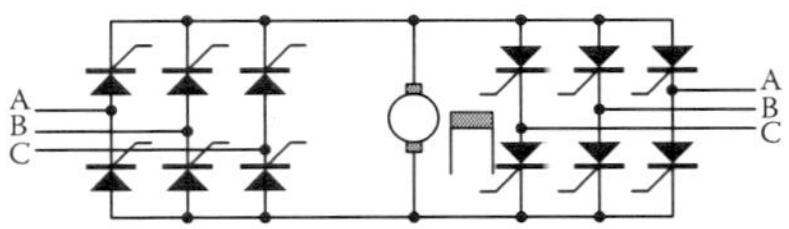 | Dual converter three phase | Up to 1500 | $6f_s$ | 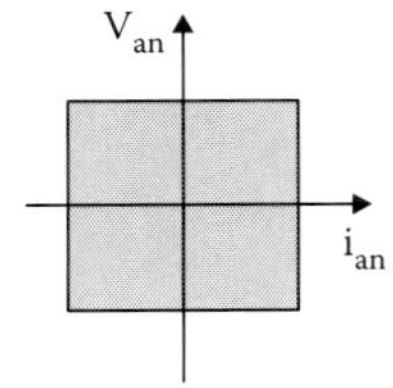 Four quadrant |

**TABLE 3.2**

Single-Phase Chopper Configurations for D.C. Brush Motors

| Type | Chopper Configuration | $e_a - I_a$ Characteristics | Function |
|---|---|---|---|
| First quadrant (step-down) choppers | 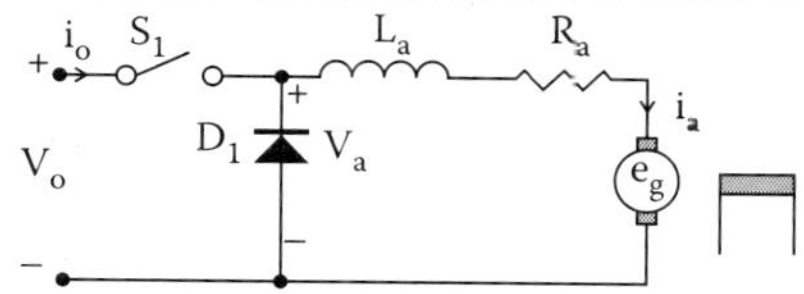 | 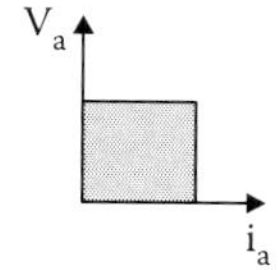 | $V_a = V_0$ for $S_1$ on<br>$V_a = 0$ for $S_1$ off and $D_1$ on |
| Second quadrant, regeneration (step-up) chopper | 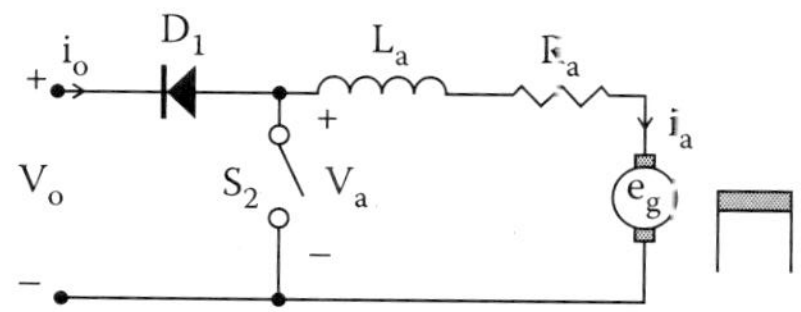 | 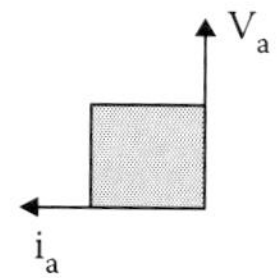 | $V_a = 0$ for $S_2$ on<br>$V_a = V_0$ for $S_2$ off and $D_2$ on |
| Two-quadrant chopper | 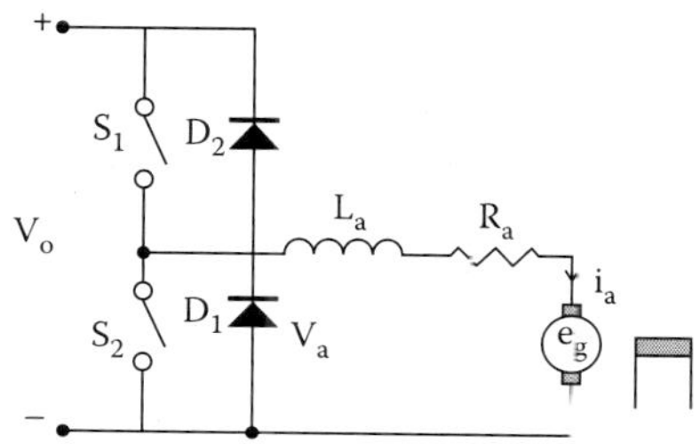 | 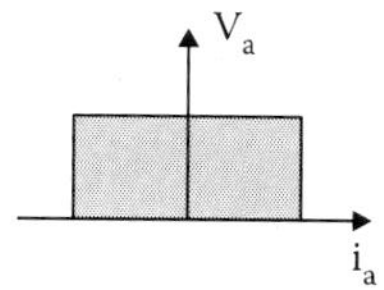 | $e_a = V_0$ for $S_1$ or $D_2$ on<br>$e_a = V_0$ for $S_2$ or $D_1$ on<br>$i_a > 0$ for $S_1$ or $D_1$ on<br>$i_a < 0$ for $S_2$ or $D_2$ on |
| Two-quadrant chopper | 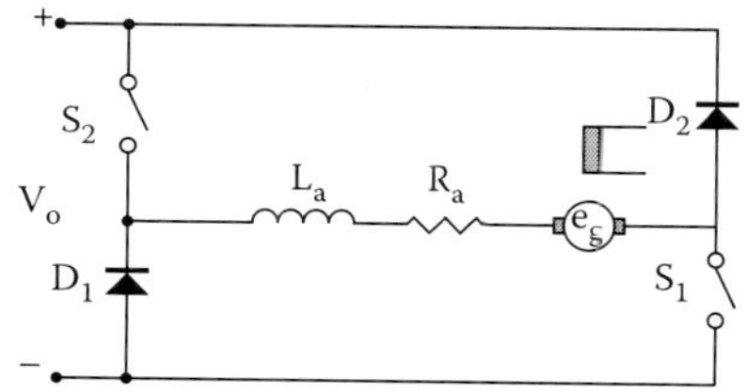 | 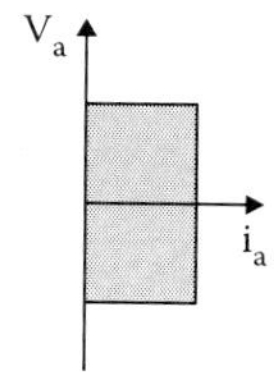 | $V_a = +V_0$ for $S_1$ and $S_2$ on<br>$V_a = -V_0$ for $S_1$ and $S_2$ off and $D_1$ and $D_2$ on |

*(Continued)*

**TABLE 3.2 (*Continued*)**

Single-Phase Chopper Configurations for D.C. Brush Motors

| Type | Chopper Configuration | $e_a - I_a$ Characteristics | Function |
|---|---|---|---|
| Four-quadrant chopper | 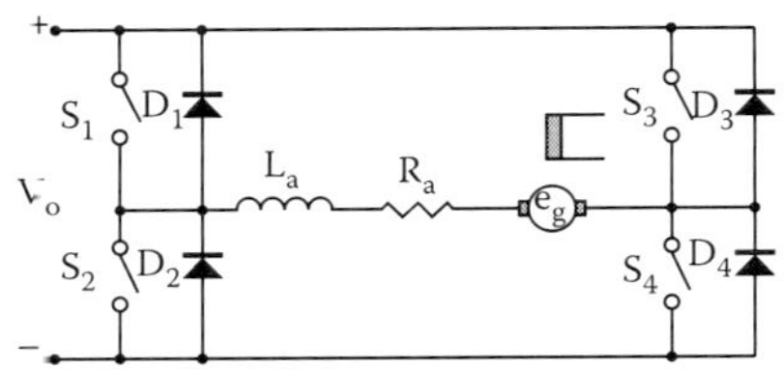 | 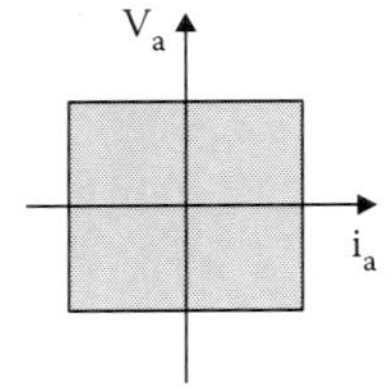 | $S_4$ on and $S_3$ off $S_1$ and $S_2$ operated<br>$V_a > 0$ $i_a$—reversible<br>$S_2$ on and $S_1$ off $S_3$ and $S_4$ operated<br>$V_a < 0$ $i_a$—reversible |

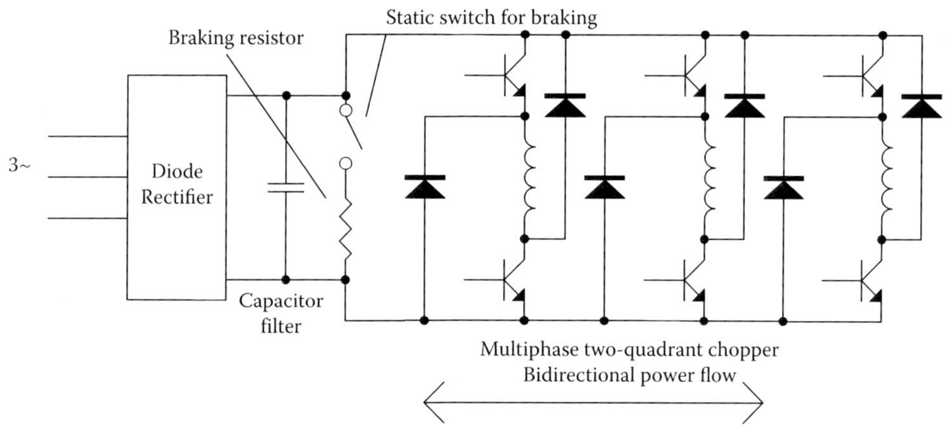


**FIGURE 3.17**
Multiphase d.c.–d.c. converters for switched reluctance motors.

## 3.8 D.C.–A.C. Converters (Inverters)

There are two main types of d.c.–a.c. converters (inverters): voltage-source inverters (Figure 3.18) and current-source inverters (Figure 3.19), in the sense that the input source is of constant voltage or constant current character.

The six switch count converters in Figures 3.18 and 3.19 produce two nonzero ($\pm V_{dc}$, $\pm I_{dc}$) levels in the line voltage and currents, therefore their characterization as two level inverters/converters. Multilevel PWM converters are to be used for high-power electric drives [2].

Voltage-source inverters are built in general with IGBTs or GTOs while current-source inverters are in general built with thyristors and GTOs and refer to high power levels (MW level and higher).

Inverters may be single phase or multiphase and they deliver bipolar current waveforms and allow for bidirectional power flow. They are used for a.c. motor drives.

When the a.c. motor exhibits a lagging PF (induction motor), capacitors and diodes are required for successful commutation of thyristors in the current-source inverter. In contrast, when the motor exhibits a leading PF (overexcited synchronous motors), induced voltage (load) commutation is accomplished and thus the capacitors and the diodes are eliminated.

Current-source inverters allow for bidirectional power flow (Figure 3.19). The energy retrieved from the motor to the voltage-source converter has to be either returned to the power source or dumped into a braking resistor (Figure 3.18)—when the source-side converter does not allow for bidirectional power flow.

The phase-controlled rectifier allows for bidirectional power flow that is so necessary during fast braking of high inertia loads. So, from this point of view, the configuration in Figure 3.19 is superior to that of Figure 3.18.

However, switching frequency is smaller with current-source inverters and thus the motor current waveform is distorted, leading to a larger derating of the motor to avoid overheating.

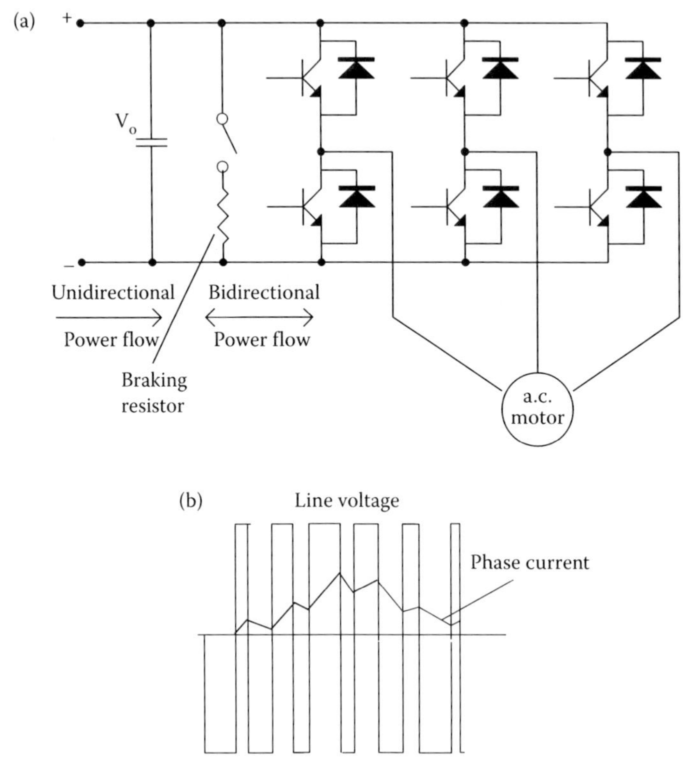


**FIGURE 3.18**
Voltage-source PWM inverter: (a) basic configuration; (b) output waveforms.

On the other hand, with both a.c.–d.c.–a.c. converters, unity input PF and full bidirectional power flow are possible only with configurations such as those in Figure 3.20.

The diode rectifier works during motoring while the IGBTs work during generating when the source-side converter will work as a PWM voltage-source inverter. Unity PF and quasi-sinusoidal source currents and voltages may be obtained.

Though the phase-controlled rectifier—current-source inverter configuration (Figure 3.19) provides for bidirectional power flow, the input PF decreases with rectifier d.c. voltage (motor speed) reduction. To provide for unity PF and sinusoidal input and output waveforms, the structure in Figure 3.21 with GTOs (or IGBTs) has been introduced.

The capacitors on the a.c. source-side converter are commutated properly by an additional leg of GTOs. The same is done on the current-source side. Thus the voltages and currents are ideally sinusoidal both on the source side and on the motor side through PWM. Also, the unity PF on the source side is performed with bidirectional power flow. Note, however, that rather large capacitors and filters are required while the number of PESs is 12 in Figure 3.20 and 16 in Figure 3.21.

All the PECs introduced so far are characterized by hard switching of PESs—at nonzero voltage or current. As in PECs, both conducting and commutation losses count, when switching frequency increases so does commutation loss.

A new breed of PECs, with a very rich literature, called soft-switching (or resonant) PECs, that provides commutation at zero voltage (for IBGTs) or zero current (for GTOs) has been proposed; it boosts the switching frequency 1 order of magnitude [1]. Another breed would be Z-converters/self-boosting [3].

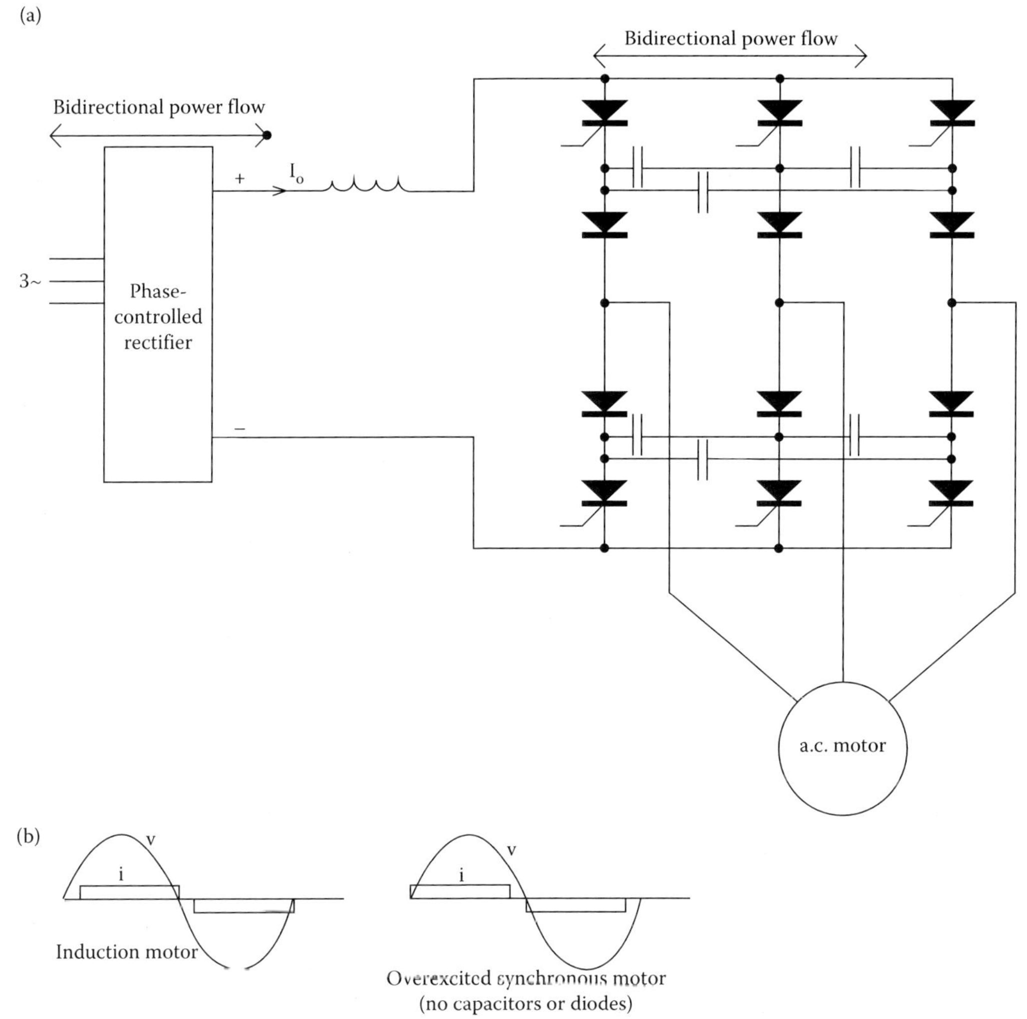


**FIGURE 3.19**
Current-source inverter (a) basic configuration; (b) ideal output waveforms (i—current, v—voltage).

As soft-switching PECs are not yet commercial for drives we do not pursue them further here. Besides a.c.–d.c.–a.c. converters, there are also direct a.c.–a.c. converters.

## 3.9 Direct A.C.–A.C. Converters

Direct a.c.–a.c. converters are used in industry for high-power synchronous and induction motor drives (both for frequencies lower than that of the power source [4]) as cycloconverters.

Cycloconverters (Figure 3.22) represent, in the six-pulse configuration, a group of two back-to-back fully controlled rectifiers for each phase.

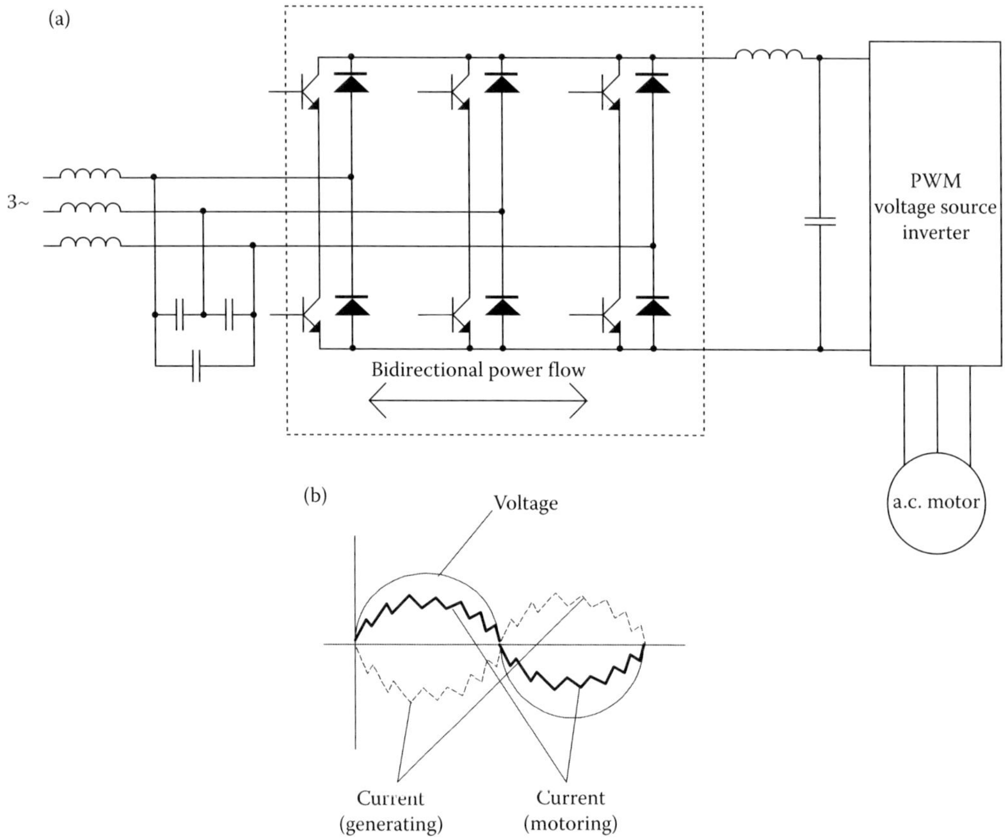


**FIGURE 3.20**
Bidirectional power flow (dual) a.c.–d.c. converter with unity power factor (a) and sinusoidal inputs (b).

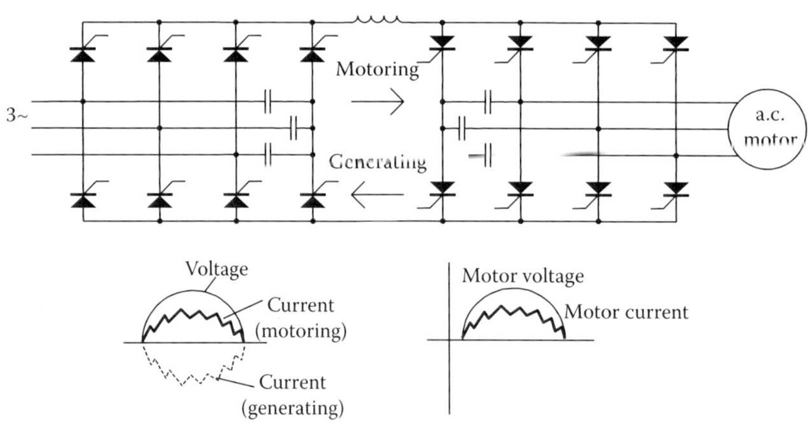


**FIGURE 3.21**
The a.c.–d.c.–a.c. converter with bidirectional power flow and unity input power factor.

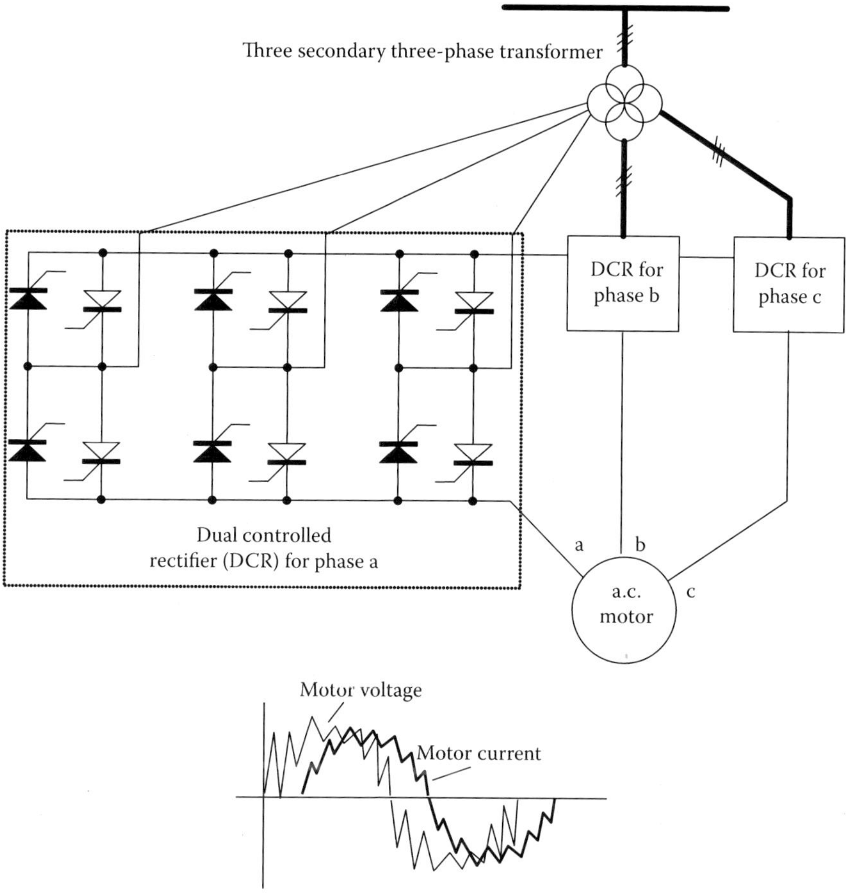


**FIGURE 3.22**
Six-pulse cycloconvertor for a.c. motor drives.

The source voltage and current waveforms are both ideally sinusoidal. The commutation of PESs is provided by the power source and the maximum output frequency is about one-third of the input frequency. More involved configurations allow higher output frequencies [4].

A general concept of direct a.c.–a.c. converters called a matrix converter has been proposed in 1980 [5–8]. Matrix converters have just reached the markets.

The matrix converter (Figure 3.23) is a matrix of switches that allows for the connection of any input to any output phase.

The PWM techniques for matrix converters do in essence provide for a fictitious diode bridge rectifier d.c. link (Figure 3.24b) and then recompose the output line voltage ($v_{ab}$, Figure 3.24c) by pulse width modulating (PWM) this fictitious d.c. link voltage. One switch in each of the three banks conducts at any instant, so there are $3^3 = 27$ switching states; only lateral line switches are not closed in order to avoid line shorting.

The nine bidirectional power switches may be built either with two IGBTs and two fast switching diodes or from one IGBT and four switching diodes.

The $L_FC_F$ filter is essential for the commutation of a.c. switches when the inductive load current has to be transferred from one line to another; it also serves to filter the line current harmonics.

The matrix converter is bidirectional in principle, with the line current almost sinusoidal.

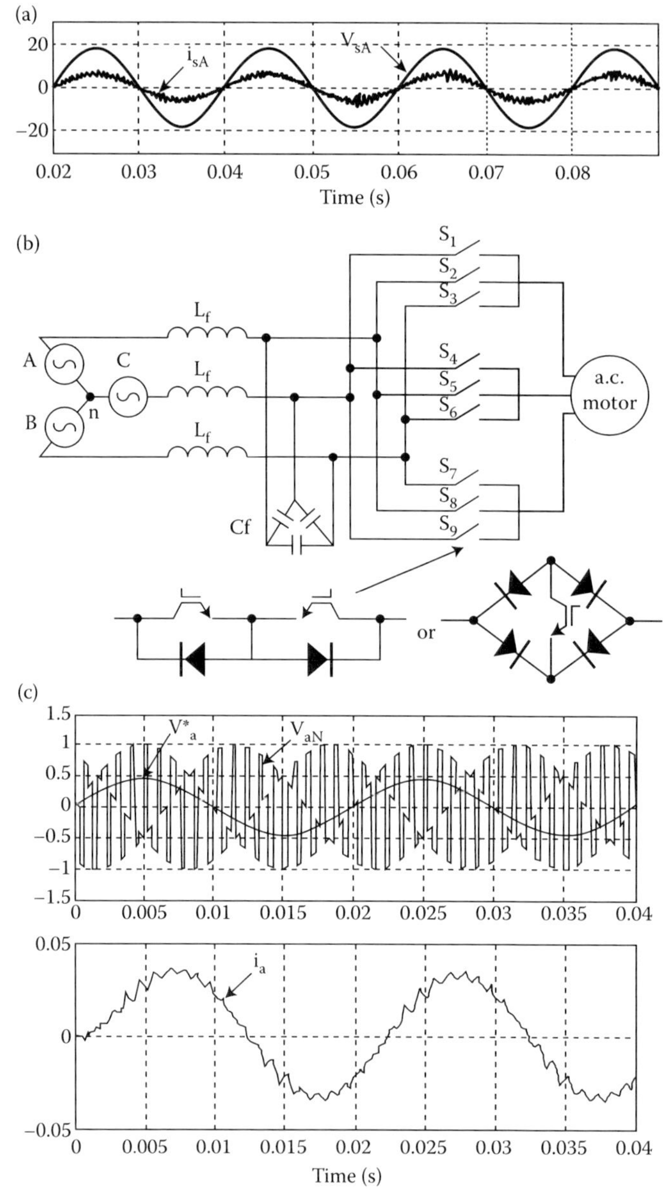


**FIGURE 3.23**
(a) Input waveforms; (b) three phase to three-phase matrix converter; (c) output waveforms.

There are no constraints on the output frequency but the maximum voltage gain is 0.86 and thus the motor has to be sized at lower voltage and higher current.

### 3.9.1 Low-Cost PWM Converters

While it has been proved that the six-leg PWM voltage-source converter (Figure 3.18) is the best choice for a.c. sinusoidal current output in three-phase loads [9], for single-phase loads,

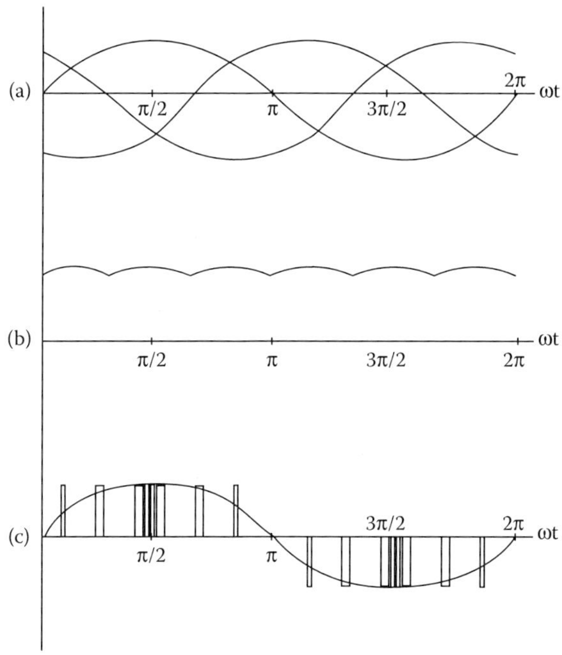


**FIGURE 3.24**
Matrix converter waveforms: (a) input line voltage; (b) fictitious diode rectifier bridge like d.c. voltage link; (c) output line voltage ($v_{ab}$).

lower power switch count converters have been analyzed [9]. As none of them is without notable demerits, we represent here just one of them (Figure 3.25), of low cost, eventually suitable for split-phase capacitor-run a.c. (induction and PM synchronous) small motors, for home appliances (HVAC: heating, ventilating, air conditioning).

Only half of the d.c. link voltage is applied to the load, so the voltage rating of IGBTs is reduced but, for the same load power, their current rating is increased. As the capacitor $C_L$ in the load is fixed, the lowest frequency of the converter is also limited to perhaps about 6–10 Hz. Fortunately, in HVAC applications, this is acceptable.

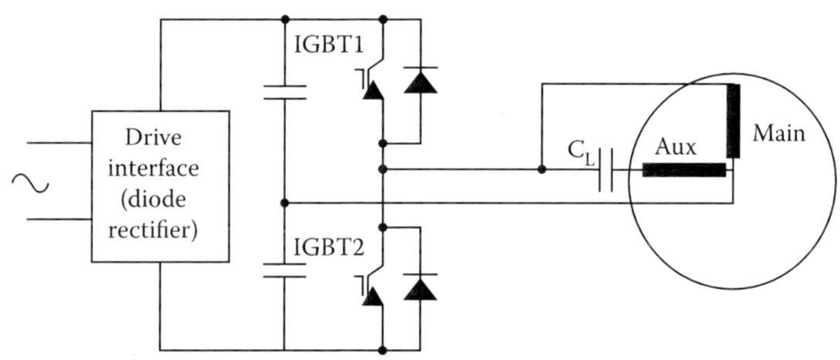


**FIGURE 3.25**
A B2 inverter supplying a single-phase load.

## 3.10 Summary

- A modern electric drive, capable of variable speeds, comprises, in general, a motor, a multistage PEC, and a controller.
- PECs process the power supplied to the motor and are classified in many ways.
- Basically, the input–output waveforms are most important. PECs may be single stage or two stage. Single-stage converters are a.c.–d.c. or d.c.–a.c. or direct a.c.–a.c.
- Two-stage PECs are a.c.–d.c.–a.c. with an intermediate d.c. voltage-type or current-type link. Most commercial PECs are hard switched, that is, the PESs are turned on (or off) when the voltage (current) is nonzero.
- Soft-switching (resonant) PECs make use of soft switching (under zero voltage or current) and thus are characterized by reduced commutation losses for given switching frequency. Conversely, they allow switching frequencies an order of magnitude higher than for hard switching.
- The various PESs are characterized by voltage, current, dv/dt and di/dt limitations, and a certain switching frequency limit.
- The line-frequency diode rectifiers presented in some detail in this chapter show the strong influence of source inductance on d.c. output voltage reduction with rectified current (and highly distorted source current) though the DPF is above 0.9 in all cases.
- Filters on the source side are required to attenuate the current harmonics in the a.c. power source. These aspects are common to all PECs, as will be seen in subsequent chapters.
- Large power drives prompted vigorous R&D recent efforts [10–12].

## 3.11 Problems

3.1 An ideal single-phase rectifier with zero source inductance $L_s$ (Figure 3.26) produces constant d.c. current output $I_d = 50$ A.

a. Show the source current and load current waveforms.

b. With a sinusoidal voltage $V_s = 120$ V (rms) at 60 Hz, calculate the d.c. output voltage $V_d$ and the d.c. power.

c. For case b, calculate the source current fundamental, the DPF, and the PF.

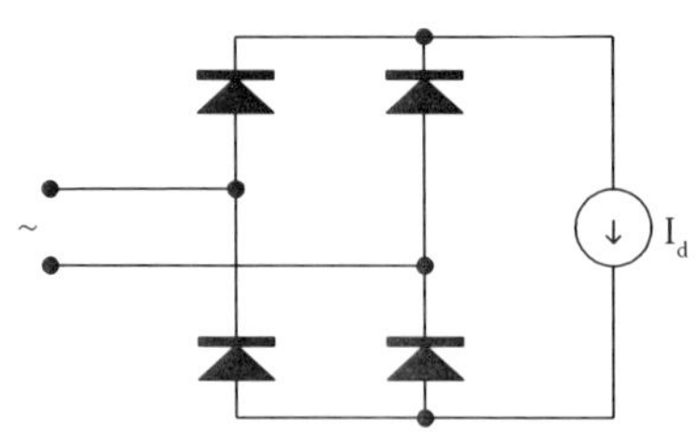


**FIGURE 3.26**
Ideal single-phase diode rectifier with zero source inductance ($L_s = 0$) and constant d.c. current $I_d$.

3.2 A single-phase diode rectifier with constant output voltage $V_d = 120$ V is fed from an a.c. supply whose voltage is $V_s(t) = 120\sqrt{2} \cdot \sin 376t$. For the source inductance $L_s = 1$ mH, calculate the time dependence of the d.c. output current.

3.3 A three-phase diode rectifier has $V_s(t) = 120\sqrt{2} \cdot \sin 376t$ and the source inductance $L_s = 5$ mH. The maximum commutation overlapping angle is $u = 30°$. For this situation, calculate the rectifier current $I_d$ and the ideal and actual d.c. output voltage $V_d$.

## References


1. N. Mohan, T.M. Underlund, W.P. Robbins, *Power Electronics*, second edition, John Wiley & Sons Inc., New York, 1995.
2. L. Gyugyi, B.R. Pelly, *Static Power Frequency Converters*, John Wiley & Sons Inc., New York, 1976.
3. A. Alessia, M.G.B. Venturini, Analysis and design of optimum-amplitude nine-switch direct ac-ac converter, *IEEE Trans.*, vol. PE-4, no. 1, 1989, pp. 101–112.
4. M. Milenovic, F. Mihalic, A switching matrix analysis of three phase boost rectifier, *Record of IEEE-IAS-1996 Annual Meeting*, vol. 2, 1996, pp. 1218–1224 (IEEExplore).
5. S. Bernet, K. Bernet, T.A. Lipo, The auxiliary resonant pole matrix converter—A new topology for high power applications, *Record of IEEE-IAS-1996 Annual Meeting*, vol. 2, 1996, pp. 1242–1249.
6. L. Huber, D. Borojevic, Space vector modulated three phase to three phase matrix converter with input power factor correction, *IEEE Trans.*, vol. IA-31, no. 6., 1995, pp. 1234–1246.
7. T. Sawa, T. Kuine, Motor drive technology—History and vision for the future, keynote address at *IEEE–PESC*, Aachen, Germany, 2004.
8. C. Klumpner, F. Blaabjerg, Fundamental of the matrix converter technology, *Control in Power Electronics*, Chapter 3, eds. M. Kazmierkowski, R. Khrishnan, F. Blaabjerg, Academic Press, USA, 2002.
9. H. Kragh, Modeling, analysis and optimization of power electronic circuits for low cost drives, PhD thesis, IET, Aalborg University, Denmark, 2000.
10. B. Wu, *High Power Converters and AC Drives*, IEEE Press/Wiley Interscience, New York, 2006.
11. D. Li., P. Ch. Loh, M. Zhu, F. Gao, F. Blaabjerg, Enhanced—Boost Z-source inverters with alternate—Cascaded switched and tapped—inductor cells, *IEEE Trans.*, vol. PE-60, no. 9, 2013, pp. 3567–3578.
12. J. Rodriguez, M. Rivera, J.W. Kolar, P.W. Wheeler, A review of control and modulation methods for matrix converters, *IEEE Trans.*, vol. IE-59, no. 1, 2012, pp. 58–70.

# 4

# D.C. Brush Motors for Drives

D.c. brush or d.c. commutator machines have traditionally been used in variable speed drives in the range of low (a few watts) to medium, 10 MW, power ratings at low speeds. The popularity of d.c. brush motors in variable speed drives is primarily due to the lower cost of a single-stage (rectifier) PEC required for two-quadrant operation (one direction of motion). Four-quadrant operation, however, implies rather involved PEC configurations and controls.

In this chapter, we will discuss d.c. brush motor basic topologies, state-space equations, steady-state curves, losses, and transfer functions, as they will be used in subsequent chapters on d.c. brush motor drives.

## 4.1 Basic Topologies

As in any electric motor, the d.c. brush motor has two main parts: the stator (fixed) part and the rotor (movable) part. The rotor may be cylindrical (Figure 4.1a) or disk shaped and contains a symmetric winding made of identical coils connected in series to the insulated copper sectors of the so-called mechanical commutator. The unipolar current injected through the brushes is converted into bipolar current in the rotor coils through the commutator copper sectors in pace with the rotor position. The mechanical commutator is, in fact, an inverter (d.c.–a.c. converter) which changes the frequency from zero to $f_n = p \cdot n$, where 2p is the number of stator and rotor poles (semiperiods) and n is the rotor speed.

Electromagnetic d.c. excitation (EE) may be replaced by PMs (Figures 4.1b and 4.2). High-energy PMs may be replaced by a constant excitation (field) current fictitious (or superconducting) coil. The PM stator is ideally lossless as the PM magnetization (or demagnetization) losses are fairly small.

The disk rotor does not have an iron core and thus the rotor winding inductance is fairly small. Moreover, the design current density is higher than for coils in iron core slots, due to direct air exposure. Consequently, the electrical time constant of the disk d.c. brush motor is the lowest known (approximately 1 ms for 1 kW, 3000 rpm motors).

Unfortunately, the power per unit mass of a disk d.c. brush motor is limited, by the mechanical fragility of the "aircore" rotor, to 2–3 kW and less than 6000 rpm. On the other hand, the cylindrical d.c. brush motor is power limited by the commutator to 10 MW at low speeds for rotors with coils in slots of laminated iron cores. "Air core" cylindrical rotor windings used for small powers are now proposed to improve commutation (due to inductance reduction) and thus eventually increase the power above 10 MW at very low speeds.

For details on d.c. brush motors, refer to Reference 1.

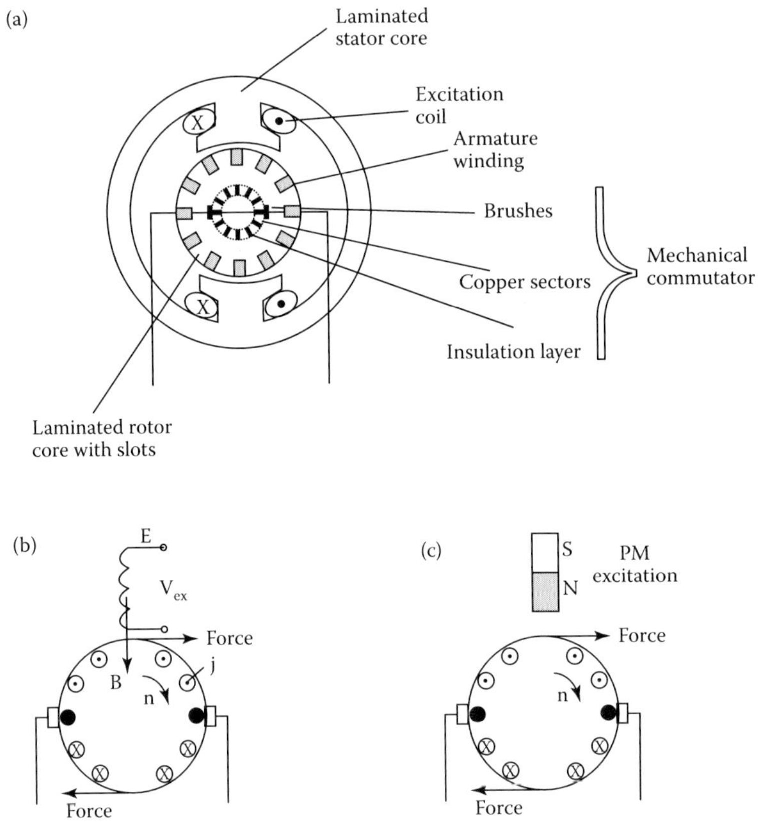


**FIGURE 4.1**
Cylindrical d.c. brush motor: (a) topology; (b) schematics with electromagnetic excitation; and (c) schematics with PMs.

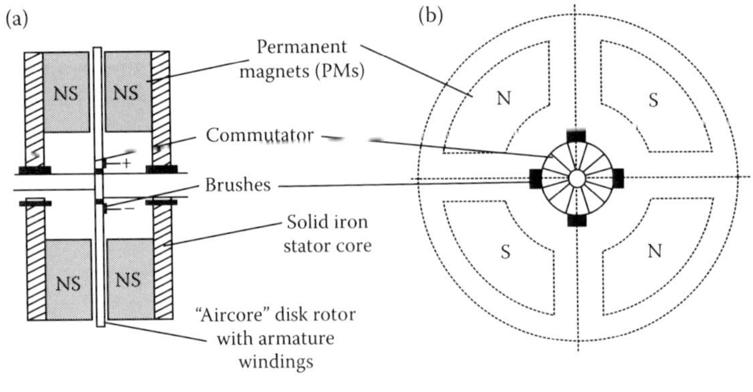


**FIGURE 4.2**
Disk d.c. brush motor (2p = 4 poles): (a) cross section; (b) axial view.

## 4.2 Motion-Induced Voltage (e.m.f.)

As the rotor conductors (Figure 4.1) move in the field produced by the field current (or the PM), an e.m.f. is induced in each rotor coil. The rotor coils are connected in series between neighboring brushes through the copper sectors of the mechanical commutator.

As seen from Figure 4.1, there are at least two current paths in parallel. In general, there are 2a paths. The number of current paths depends on the number of poles, 2p, and the type of the armature winding (lap or wave type). The total number of conductors per rotor periphery is N and the flux linkage per pole is (half-period of excitation field in the airgap) $\lambda_p$.

The motion-induced voltage in the rotor, measured at the brushes, E, is also proportional to the rotor speed n.

Finally

$$E = K_e \cdot n \cdot \lambda_p; \quad K_e = \frac{p}{a} \cdot N \tag{4.1}$$

The pole flux linkage $\lambda_p$ is proportional to the average flux density in the airgap per pole $B_{gav}$, to the pole pitch $\tau$, and to the core stack length L

$$\lambda_p = B_{gav} \cdot \tau \cdot L \tag{4.2}$$

As seen from Figure 4.1 the motion-induced voltage (e.m.f.) in each coil does not change sign under one pole as the fixed brushes collect the voltage of the coils temporarily located (though in continuous motion) under one pole (field) polarity. The armature current changes polarity when the respective coil is short circuited by the brushes.

This phenomenon is called mechanical commutation of currents. For more details on the d.c. brush motor topology, principle, performance, and design, see References 1–4.

## 4.3 Performance Equations: d-q Model

The d.c. brush motor schematics in Figure 4.1 has ideally an electrical 90° spatial phase difference between the stator (excitation, or PM) magnetic field and rotor (armature) magnetic field (brush [neutral]) axis. Consequently, in the absence of magnetic saturation, there is no interaction (transformer-induced voltage) between excitation and armature windings. The excitation (PM) circuit may be termed the field circuit while the rotor (armature) winding may be termed the torque circuit. *Thus the d.c. brush motor allows for separate (decoupled) control of field and torque currents (or torque)* which is an extraordinary built-in property of d.c. brush motor. Though the stator excitation winding may also be series-connected to the brushes, separate excitation is considered in what follows.

There is an interaction between the stator and rotor windings only through motion voltage, E, induced in the rotor by the stator excitation current. The state-space equations of a d.c. brush motor are

$$V_{ex} = R_e \cdot i_e + L_e \cdot \frac{di_e}{dt}$$
$$V_a = R_a \cdot i_a + E + L_a \cdot \frac{di_a}{dt} \tag{4.3}$$

The pole flux, $\lambda_p$, in the rotor windings is dependent on $i_e$, in the absence of magnetic saturation, and also on $i_a$ at high rotor currents if magnetic saturation occurs.

So, in general

$$\lambda_p = G \cdot i_e \tag{4.4}$$

where G is the "motion-produced" inductance between the stator and rotor orthogonal windings. The electromagnetic torque $T_e$ expression is obtained from the electromagnetic power $P_e$:

$$T_e = \frac{P_e}{2\pi n}; \quad P_e = E \cdot i_a \tag{4.5}$$

Thus

$$T_e = \frac{E \cdot i_a}{2\pi n} = \frac{K_e}{2\pi} \cdot \lambda_p \cdot i_a \tag{4.6}$$

The d.c. brush motor parameters are $R_e$, $R_a$, as resistances and $L_e$, G, $L_a$ as inductances. We should add the inertia, J, as defined through the motion equation

$$J2\pi \cdot \frac{dn}{dt} = T_e - T_{load} - B \cdot n; \quad \frac{d\theta_r}{dt} = 2\pi n \tag{4.7}$$

where $T_{load}$ is the load torque and B is the friction torque coefficient.

## 4.4 Steady-State Motor Characteristics

Steady state means constant speed (dn/dt = 0) and constant currents (i.e., $i_e$ = ct, $i_a$ = ct). The steady-state voltage equations, obtained from Equations 4.1 through 4.3 are

$$V_{ex} = R_e \cdot i_e \tag{4.8}$$

$$V_a = R_a \cdot i_a + K_e \cdot n \cdot \lambda_p \tag{4.9}$$

$$T_e = \frac{K_e}{2\pi} \cdot \lambda_p \cdot i_a = T_{load} + B \cdot n \quad (4.10)$$

The main characteristic of a motor is the torque ($T_e$) versus speed (n, or $\Omega_r$) curve called the mechanical characteristic.

From Equations 4.8 through 4.10

$$V_a = K_e \cdot n \cdot \lambda_p + R_a \cdot \frac{T_e \cdot 2\pi}{K_e \cdot \lambda_p} \quad (4.11)$$

Modifying the speed, for a given electromagnetic torque ($T_e$) may be done through

- Voltage, $V_a$, control
- Flux, $\lambda_p$, control

As apparent from Equation 4.11, it is also possible to add a resistance in series with $R_a$, to modify speed. This is, however, an energy-consuming method, which is to be avoided in modern electric drives.

The torque speed curves obtained through voltage and flux control, all straight lines, are shown in Figure 4.3 in relative units.

In Figure 4.3, $n_b$, $T_{eb}$, and $\lambda_{pb}$ are base-speed, base-torque, and base-flux values (continuous duty, rated [maximum] voltage).

Above the base-speed flux weakening at constant armature voltage $V_a$ (limited by the PEC capabilities) is performed, in general, at constant power $P_e$ up to $n_{max}/n_b = 2–3$. Obviously, in PM brush motors, flux weakening is not possible.

Also it is evident that both speed control methods illustrated in Figure 4.3 are of the high energy conversion ratio type. So far we have not discussed motor losses though they are crucial in defining the capability of an electric drive.

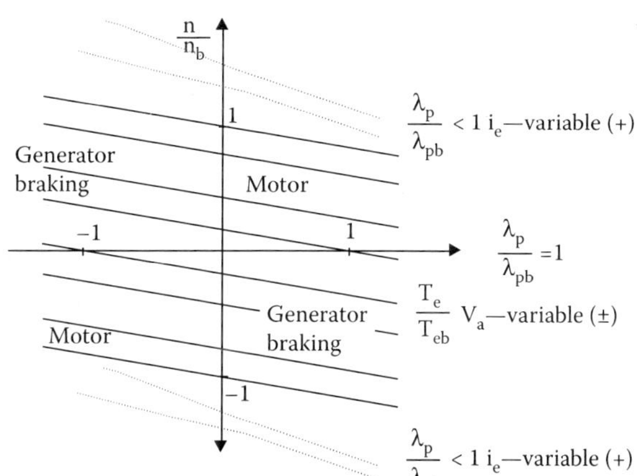


**FIGURE 4.3**
The torque speed curves for variable speed.

## 4.5 D.C. Brush Motor Losses

The performance equations already show two kinds of losses in the motor: the armature (copper) losses represented by $R_a$ and the friction (mechanical) losses $p_{mec}$

$$p_{mec} \approx B \cdot \omega_r \tag{4.12}$$

In reality, the friction (mechanical) losses are a complex function of speed, depending on application. For example, in an electric train, the wheel-track and head and lateral surface air-drag losses all constitute load at constant speed and have complex mathematical expressions.

The N–S–N–S pole sequence in the stator poles field produces hysteresis and eddy current losses in the rotor iron core. They are called core losses $p_{iron}$

$$p_{iron} \approx [C_h(p\,n) + C_e(p\,n)^2] \cdot B_{iron}^2 \cdot GC \tag{4.13}$$

The first term represents the hysteresis losses while the second takes care of the eddy current losses. $B_{iron}$ is the flux density in the rotor core and GC is the respective core weight.

The slotted rotor core has two main zones: the teeth and back core (or yoke). Further, the slot opening presence leads to stator-pole shoe core loss due to rotor (armature) winding m.m.f. (magnetomotive force). These are called additional (surface) core losses, $p_{add}$.

Now the armature resistance $R_a$ includes, in principle, the brush, brush–commutator contact, and the equivalent resistance of commutator sectors. As the brush–commutator contact surface electrical resistance depends on many factors such as speed n, current $i_a$, brush-spring tension, especially in low-voltage motors, the commutator losses, $p_{com}$, are to be calculated separately:

$$p_{com} = \Delta V_b(i_a, n) \cdot i_a \tag{4.14}$$

Finally, the excitation winding losses, $p_{ex}$, have to be considered

$$p_{ex} = R_e \cdot i_e^2 \tag{4.15}$$

Figure 4.4 summarizes the breakdown of motor losses as developed above.

The efficiency $\eta$ is

$$\eta = \frac{P_{output}}{P_{input}} \tag{4.16}$$

#### EXAMPLE 4.1: STEADY-STATE PERFORMANCE

A d.c. motor with separate excitation has the following data: rated power $P_n = 3$ kW, rated (base) speed $n = 1200$ rpm, $p_{com} = 0.5\%P_n$, $p_{add} = 0.5\%P_n$, $p_{iron} = p_{mec} = 1\%P_n$, and $p_{co} = 4\%P_n$, at rated (maximum) voltage $V_{an} = 110$ V. The excitation losses are neglected.

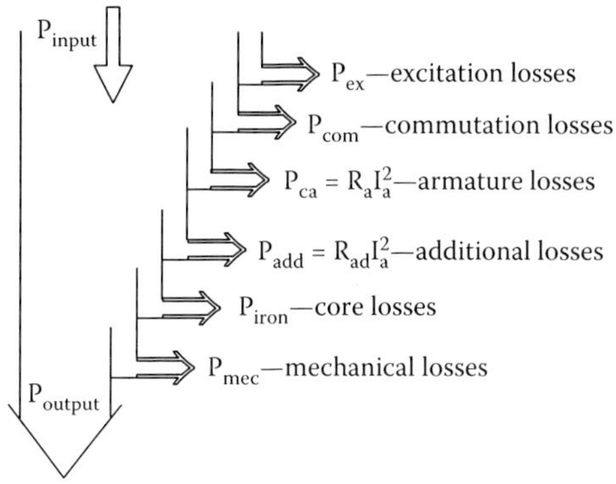


**FIGURE 4.4**
Loss breakdown in d.c. brush motors.

Calculate:

a. Total losses and rated efficiency, $\eta_n$
b. Rated current, $I_n$, and armature resistance, $R_a$
c. Brush voltage drop, $\Delta V_b$
d. Motion-induced voltage, $E_n$
e. Rated (base) electromagnetic torque, $T_{eb}$
f. Shaft torque, $T_{load}$
g. Ideal no-load speed, $n_0$
h. Armature voltage, $V_{ag}$, required to produce rated generator braking torque at rated speed

**Solution**

a. The total losses $\Sigma p_i$ are

$$\sum p_i = p_{co} + p_{com} + p_{add} + p_{mec} + p_{iron}$$

$$= (4.0 + 0.5 + 0.5 + 1 + 1) \cdot \frac{3000}{100} = 210\,\text{W} \tag{4.17}$$

The efficiency

$$\eta_n = \frac{P_n}{P_n + \sum p_i} = \frac{3000}{3000 + 210} = 0.9345 \tag{4.18}$$

b. The rated current $I_n$ is

$$I_n = \frac{P_n}{\eta_n \cdot V_n} = \frac{3000}{0.9345 \cdot 110} = 29.18\,\text{A} \tag{4.19}$$

The armature resistance $R_a$ is

$$R_a = \frac{p_{co}}{I_n^2} = \frac{0.04 \cdot 3000}{29.18^2} = 0.140\,\Omega \tag{4.20}$$

c, d. From the voltage equation (4.9), adding the brush voltage drop $\Delta V_b$, we obtain

$$E_n = V_n - R_a \cdot I_n - \Delta V_b \tag{4.21}$$

with

$$\Delta V_b = \frac{p_{com}}{I_n} = \frac{0.005 \cdot 3000}{29.18} = 0.514\,V \tag{4.22}$$

From Equation 4.9

$$E_n = 110 - 0.14 \cdot 29.18 - 0.514 = 105.4\,V \tag{4.23}$$

e. Rated electromagnetic torque, $T_{en}$, is calculated from Equation 4.6

$$T_{en} = \frac{E_n \cdot I_n}{2\pi n} = \frac{105.4 \cdot 29.18}{2\pi \cdot (1200/60)} = 24.48\,N\,m \tag{4.24}$$

f. The shaft torque, $T_{load}$, comes directly from rated power $P_n$ and speed $n_n$

$$T_{load} = \frac{P_n}{2\pi n_n} = \frac{3000}{2\pi \cdot (1200/60)} = 23.88\,N\,m \tag{4.25}$$

g. The ideal no-load speed, $n_0$, corresponds to zero armature current in the voltage equation (4.9)

$$V_{an} = K_e \cdot n_0 \cdot \lambda_p \tag{4.26}$$

but $K_e \cdot \lambda_p$ is constant

$$K_e \cdot \lambda_p = \frac{E_n}{n_n} = \frac{105.4}{(1200/60)} = 5.27\,Wb \tag{4.27}$$

Consequently

$$n_0 = \frac{V_{an}}{K_e \cdot \lambda_p} = \frac{110}{5.27} = 20.8728\,rps = 1252.37\,rpm \tag{4.28}$$

h. For regenerative braking, the armature current becomes negative $I_{gn} = -I_n$; also $\Delta V_b = -\Delta V_b$, and $E_n$ remains the same; thus, from Equation 4.9

$$V_{ag} = E_n - R_a \cdot i_a - \Delta V_b = 105.4 - 0.14 \cdot 29.18 - 0.514 = 100.8\,V \tag{4.29}$$

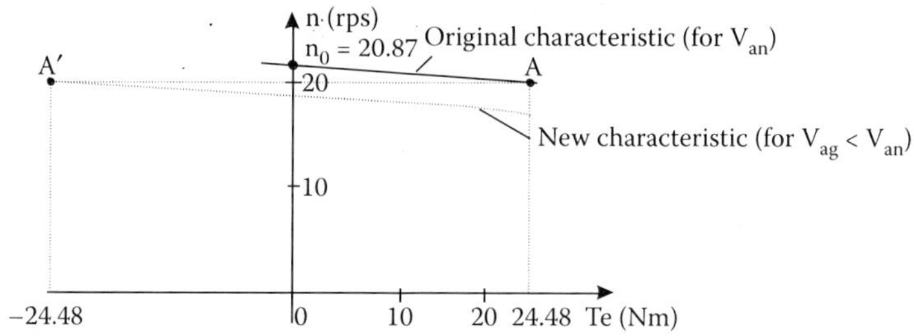


**FIGURE 4.5**
Torque speed curves when switching from motoring to generating.

So the voltage produced by the PEC should be simply reduced below the e.m.f. $E_n$ level to produce regenerative braking for a given speed. The operating point moves from A in the first quadrant to A′ in the second quadrant (Figure 4.5).

## 4.6 Varying the Speed

**EXAMPLE 4.2**

For the motor in Example 4.1 and assuming that the mechanical losses are proportional to speed squared, while core losses depend, in addition, on flux squared

a. Calculate the voltage $V_a$ required and efficiency at $n_n/2 = 600$ rpm for motoring at rated current.
b. For rated power and rated current at $2n_n = 2400$ rpm, calculate the flux-weakening ratio and efficiency.

**Solution**

a. From Equation 4.21

$$V_a = R_a \cdot i_a + E_n \cdot \frac{n}{n_n} + \Delta V_b$$

$$= 0.14 \cdot 29.18 + 105.4 \cdot \frac{60}{1200} + 0.514 = 57.2992\,\text{V} \tag{4.30}$$

The mechanical and core losses are reduced 4 times (as the speed is halved) while the commutator losses remain the same as the current remains constant, $I_n$. So the total loss $\sum p_i$ is

$$\sum p_i = 0.04P_n + (0.01 + 0.01) \cdot P_n \cdot \left(\frac{1}{2}\right)^2 + (0.005 + 0.005) \cdot P_n$$

$$= (0.04 + 0.005 + 0.01) \cdot 3000 = 165\,\text{W} \tag{4.31}$$

The input power, $P_{input}$, is

$$P_{input} = V_a \cdot I_n = 57.2992 \cdot 29.18 = 1672\,W \tag{4.32}$$

The efficiency η is

$$\eta = \frac{P_{input} - \sum p_i}{P_{input}} = \frac{1672 - 165}{1672} = 0.90 \tag{4.33}$$

It should be noted that even for half the rated speed, at full torque, the efficiency remains high. Consequently, varying speed through varying armature voltage is a high-efficiency method.

b. When raising the speed above the base speed, at rated current and voltage, the e.m.f. remains the same as in Equation 4.22: $E = E_n = 105.4$ V.
As the speed is doubled n = 2400 rpm the new flux level $\lambda'_p$ is

$$\frac{\lambda'_p}{\lambda_p} = \frac{n_n}{n} = \frac{1200}{2400} = \frac{1}{2} \tag{4.34}$$

Thus the flux is halved and so is the electromagnetic torque $T_e$:

$$\frac{T_e}{T_{en}} = \frac{\lambda'_p}{\lambda_p} = \frac{1}{2} \tag{4.35}$$

Now the losses, $\Sigma p'$, are

$$\sum p' = R_a \cdot I_n^2 + p_{add} + p_{iron} + p_{mec} + p_{com}$$

$$= 0.14 \cdot 29.18^2 + \left(0.005 + (0.01 + 0.01) \cdot \left(\frac{2400}{1200}\right)^2 + 0.005\right) \cdot 3000 = 300\,W \tag{4.36}$$

As the input power, $P_{input}$ is still

$$P_{input} = V_n \cdot I_n = 110 \cdot 29.18 = 3209.8\,W \tag{4.37}$$

The efficiency η is

$$\eta = \frac{P_{input} - \sum p'}{P_{input}} = \frac{3209.8 - 300}{3209.8} = 0.9065 \tag{4.38}$$

Again, the efficiency is high.

The above results are summarized in Figure 4.6.

Note that the electromagnetic power for points A and C is the same, while the electromagnetic torque for points A and B is the same. A 3 to 1 speed range ($n_{max}/n_b$) for constant power and current is quite feasible.

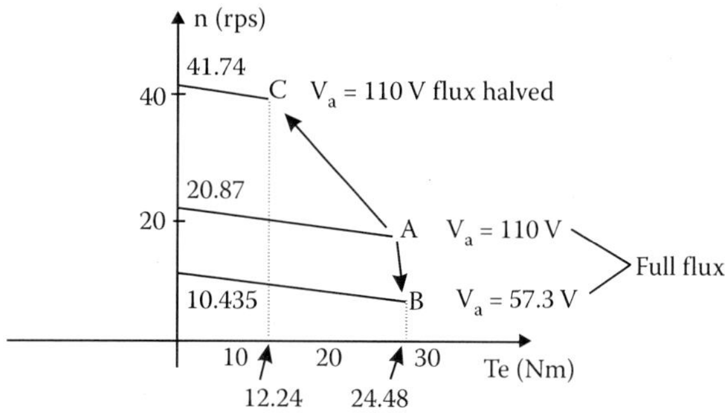


**FIGURE 4.6**
Torque speed curves for varying speed from 1200 rpm (A) to 600 rpm (B) and, respectively, to 2400 rpm (C) for constant current.

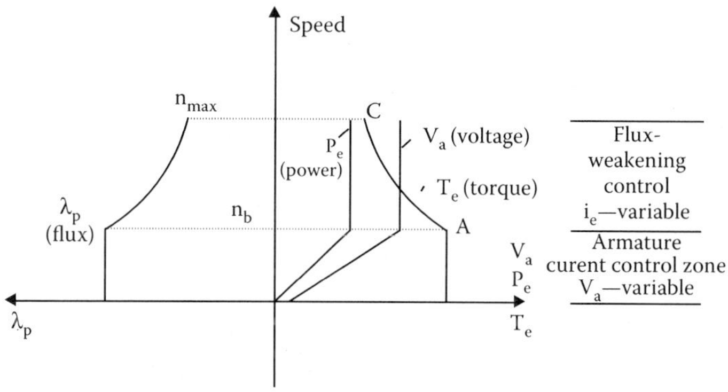


**FIGURE 4.7**
Speed/torque, power, voltage envelopes for constant current.

We may infer from here that below rated (base) speed, the torque may be maintained constant for variable armature voltage, while above rated (base) speed, the electromagnetic power may be maintained constant through flux weakening at constant armature voltage as shown schematically in Figure 4.7.

NOTE: PM d.c. brush motors do not allow flux weakening and thus $n_b = n_{max}$.

There are applications with a constant power-wide speed zone (from $n_b$ to $n_{max}$: 2–4$n_b$), such as in spindle drives, where only electromagnetic excitation is adequate.

## 4.7 Transient Operation for Constant Flux

Constant flux means constant excitation (field) current or PM excitation. Consequently, in the performance equations (4.1) through (4.7), $di_e/dt = 0$ ($\lambda_p$ = ct). Torque equation (4.6) becomes

$$T_e = \frac{K_e}{2\pi} \cdot \lambda_p \cdot I_a = K_T \cdot I_a \tag{4.39}$$

Similarly, the e.m.f. equation (4.3) is

$$E = K_E \cdot n; \quad K_T = \frac{K_E}{2\pi} \tag{4.40}$$

$K_T$ is called the torque/current constant and in reality is constant only when the magnetic saturation level is constant or in the absence of magnetic saturation.

The variables, in performance equations, are now the armature current, $i_a$, and the speed, n. Equations 4.1 through 4.7 reduce to

$$\begin{aligned} V_a &= R_a \cdot i_a + L_a \cdot \frac{di_a}{dt} + K_E n \\ 2\pi J \cdot \frac{dn}{dt} &= \frac{K_E}{2\pi} \cdot i_a - T_{load} - B \cdot n \\ \frac{d\theta_r}{dt} &= 2\pi n \end{aligned} \tag{4.41}$$

As Equation 4.41 constitutes a system of linear differential equations no product of variables $i_a$, n, $\theta_r$, Laplace transform may be used. For zero initial conditions, we obtain

$$\begin{aligned} \tilde{V}_a &= (R_a + sL_a) \cdot \tilde{i}_a + K_E \tilde{n} \\ s\tilde{n} &= \frac{1}{2\pi J} \cdot \left( \frac{K_E}{2\pi} \cdot \tilde{i}_a - \tilde{T}_{load} - B \cdot \tilde{n} \right) \\ s\tilde{\theta}_r &= 2\pi\tilde{n} \end{aligned} \tag{4.42}$$

Equations 4.42 suggests the block diagram of Figure 4.8, valid for small or large transients. So, under constant flux (field current), the d.c. brush motor has two input variables $\tilde{V}_a$ and $\tilde{T}_{load}$ and three state variables $\tilde{i}_a$, $\tilde{n}$, and $\tilde{\theta}_r$. Only with a single input—either voltage $\tilde{V}_a$ for speed control ($\tilde{T}_{load} = 0$) or load torque $\tilde{T}_{load}$ for torque control ($\tilde{V}_a = 0$)—cascaded transfer functions can be obtained

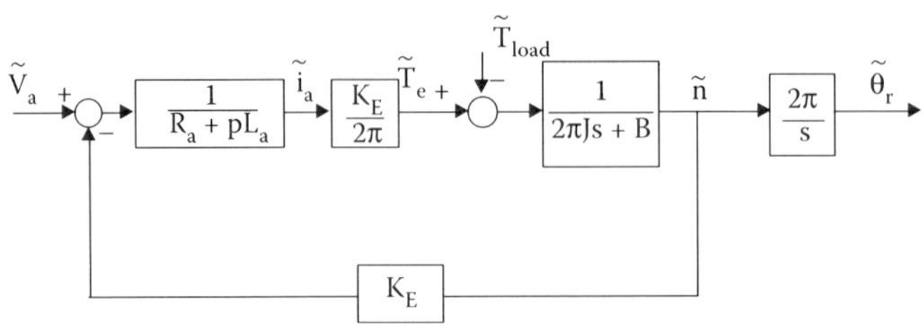


**FIGURE 4.8**
Constant flux block diagram of d.c. brush motor.

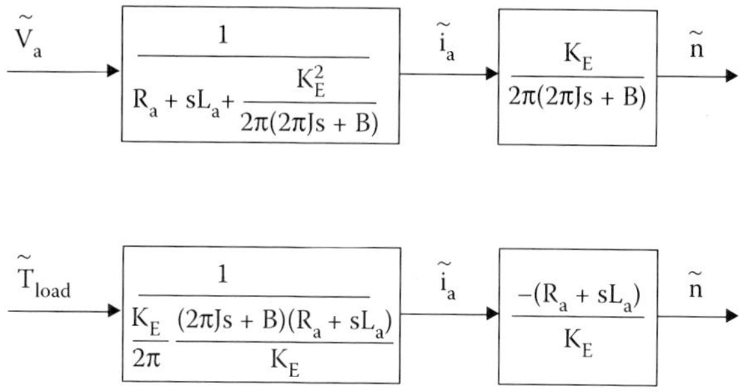


**FIGURE 4.9**
Cascaded transfer function for constant flux (a) for speed control; (b) for torque control.

$$\tilde{i}_a = \frac{\tilde{V}_a}{(R_a + sL_a) + K_E \cdot (K_E/2\pi) \cdot (1/(2\pi Js + B))}; \quad \tilde{n} = \frac{K_E \cdot \tilde{i}_a}{2\pi \cdot (2\pi Js + B)}; \quad \tilde{T}_{load} = 0 \tag{4.43}$$

or

$$\tilde{i}_a = \frac{\tilde{T}_{load}}{(K_E/2\pi) + ((2\pi Js + B) \cdot (R_a + sL_a)/K_E)}; \quad \tilde{n} = \frac{-(R_a + sL_a)\tilde{i}_a}{K_E}; \quad \tilde{V}_a = 0 \tag{4.44}$$

The cascaded block diagrams described by Equations 4.43 and 4.44 are shown in Figure 4.9.

The cascaded block diagrams are instrumental in designing the torque, speed, or position control closed loops as shown in Chapter 7. The transfer functions $\tilde{i}_a/\tilde{V}_a$ and $\tilde{i}_a/\tilde{T}_{load}$ have (Figure 4.9) the same poles (denominator's zeros) as a motor property:

$$2\pi \cdot (2\pi Js + B) \cdot (R_a + sL_a) + K_E^2 = 0 \tag{4.45}$$

Both poles $s_{1,2}$ have a negative real part, so the response is always stable.

For $B = 0$ $s_{1,2}$, we have the simplified form

$$s_{1,2} = \frac{-1 \pm \sqrt{1 - 4\tau_e/\tau_{em}}}{2\tau_e} \tag{4.46}$$

with

$$\tau_e = \frac{L_a}{R_a}; \quad \tau_{em} = \frac{4\pi^2 J R_a}{K_E^2} \tag{4.47}$$

where $\tau_e$ is the electrical time constant while $\tau_{em}$ is the electromechanical time constant. As seen from Equation 4.46 for $4\tau_e \le \tau_{em}$, the response is *aperiodically stable*, while for $4\tau_e > \tau_{em}$, the response of the motor is *periodically stable.*

Low inertia fast response drives qualify for $4\tau_e > \tau_{em}$ and thus open-loop motor response is periodically stable.

## 4.8 PM D.C. Brush Motor Transients

**EXAMPLE 4.3**

A PM d.c. brush motor has the following data: $V_n = 110$ V, $I_n = 10$ A, $R_a = 0.5\ \Omega$, $n_n = 1200$ rpm, $\tau_e = 2$ ms. For a 10 V step voltage increase, calculate the speed response for $J = 0.005$ kg m² and $J' = 0.05$ kg m², for constant load torque ($\tilde{T}_{load} = ct$) and $B = 0$ (no friction torque).

**Solution**

From the voltage equation under steady state (4.20), we may calculate the e.m.f. $E_n$

$$E_n = V_n - R_a \cdot I_n = 110 - 0.5 \cdot 10 = 105\,V \tag{4.48}$$

From Equation 4.40

$$K_E = \frac{E_n}{n_n} = \frac{105}{(1200/60)} = 5.25\,Wb \tag{4.49}$$

Also from Equation 4.40

$$K_T - \frac{K_E}{2\pi} \tag{4.50}$$

The electromagnetic torque $T_{em}$ (4.39) is

$$T_{em} = K_T \cdot I_n = 0.836 \cdot 10 = 8.36\,Nm \tag{4.51}$$

Under steady state, the load torque $T_{load}$ is equal to the motor torque $T_{em}$.

In our case, it means that the initial and final current values are the same. Eliminating the current $i_a$ in Equation 4.41, we obtain

$$\tau_e \tau_{em} \frac{d^2 n}{dt^2} + \tau_{em} \frac{dn}{dt} + n = \frac{V_a}{K_E} - \frac{T_{load} \cdot R_a}{K_T \cdot K_E} \tag{4.52}$$

The roots of the characteristic equation 4.52 are evidently equal to $s_{1,2}$ of Equation 4.46. With $\tau_e = 2$ ms

$$\tau_{em} = \frac{4\pi^2 J R_a}{K_E^2} = \begin{cases} \dfrac{4\pi^2 0.005 \cdot 0.5}{5.25^2} = 3.577 \cdot 10^{-3}\,s \\[2ex] \dfrac{4\pi^2 0.05 \cdot 0.5}{5.25^2} = 35.77 \cdot 10^{-3}\,s \end{cases} \tag{4.53}$$

Consequently (from Equation 4.46)

$$s_{1,2} = \frac{-1 \pm \sqrt{1 - 4 \cdot 2/3.577}}{4 \cdot 10^{-3}} = -250 \pm j280; \quad \text{for } \tau_{em} = 3.577\,\text{ms} \tag{4.54}$$

and

$$s_{1,2} = \frac{-1 \pm \sqrt{1 - 4 \cdot 2/35.77}}{4 \cdot 10^{-3}} = -250 + 218.8; \quad \text{for } \tau_{em} = 35.77\,\text{ms}$$

The final value of speed is obtained from Equation 4.52 with zero derivatives

$$(n)_{t=\infty} = \frac{(V_a)_{t=\infty}}{K_E} - \frac{T_{load} \cdot R_a}{K_T K_E} = \frac{(110 + 10)}{5.25} - \frac{8.36 \cdot 0.5}{0.836 \cdot 5.25}$$
$$= 21.904\ \text{rps} = 1314.27\ \text{rpm} \tag{4.55}$$

The solution of Equation 4.52 for the two cases is

$$n(t) = (n)_{t=\infty} + A \cdot e^{-250t} \cdot \cos(280t + \gamma); \quad \text{for } \tau_{em} = 3.577\,\text{ms}$$

and

$$n(t) = (n)_{t=\infty} + A_1 \cdot e^{-468.8t} + A_2 \cdot e^{-31.2t}; \quad \text{for } \tau_{em} = 35.77\,\text{ms} \tag{4.56}$$

The initial conditions refer to the fact that

$$(n)_{t=0} = 20\ \text{rps}; \quad \left(\frac{dn}{dt}\right)_{t=0} = 0 \tag{4.57}$$

Finally,

$$A = 2.55; \quad \gamma = -41.76^\circ$$
$$A_1 = 0.136; \quad A_2 = -2.040 \tag{4.58}$$

The two speed responses are drawn in Figure 4.10.

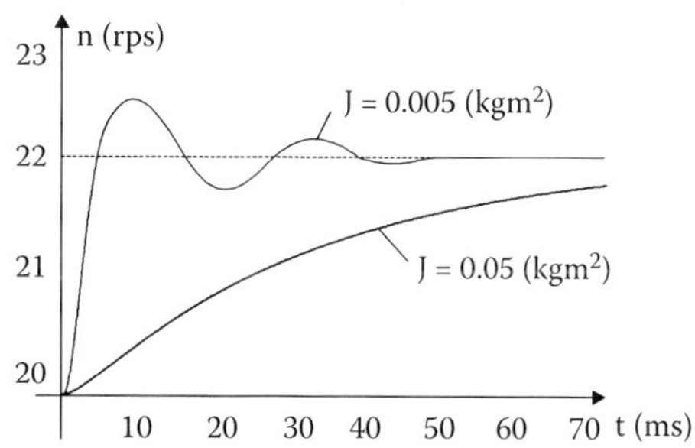


**FIGURE 4.10**
Speed responses to a step voltage increase from 110 to 120 V at constant load torque.

## 4.9 Transient Operation for Variable Flux

Variable flux means variable excitation current. This time the complete set of performance equations (4.1) through (4.7) has to be used. Even in the absence of magnetic saturation, when $\lambda_p = G \cdot i_e$, there are products of variables ($i_e \cdot i_a$ or $i_e \cdot n$) which render the system nonlinear

$$\begin{aligned}\frac{di_e}{dt} &= \frac{V_{ex} - R_e \cdot i_e}{L_e}\\ \frac{di_a}{dt} &= \frac{V_a - R_a \cdot i_a - n \cdot G \cdot i_e}{L_a}\\ \frac{dn}{dt} &= \frac{1}{2\pi J}\left(\frac{G \cdot i_e}{2\pi} i_a - T_{load} - B \cdot n\right)\end{aligned} \tag{4.59}$$

The position $\theta_r$ is left out as speed control is primarily performed with PM d.c. brush motor drives.

Through numerical methods, such as Runge–Kutta–Gill, etc., the system (4.59) may be solved for large signal variations $V_{ex}$, $V_a$, and $T_{load}$.

However, for the closed-loop control, design linearization around a steady-state point is standard:

$$\begin{aligned}&V_{ex} = V_{ex0} + \Delta V_{ex}; \quad V_a = V_{a0} + \Delta V_a; \quad i_a = i_{a0} + \Delta i_a;\\ &i_{ex} = i_{ex0} + \Delta i_{ex}; \quad T_{load} = T_{load0} + \Delta T_l; \quad n = n_0 + \Delta n\end{aligned} \tag{4.60}$$

For the initial steady-state point, d/dt = 0 in Equation 4.59

$$\begin{aligned}V_{ex0} &= R_e \cdot i_{e0}\\ V_{a0} &= R_a \cdot i_{a0} + n_0 \cdot G \cdot i_{e0}\\ \frac{G \cdot i_{e0}}{2\pi} i_{a0} &= T_{load0} + B \cdot n_0\end{aligned} \tag{4.61}$$

From Equation 4.59 with Equations 4.60 and 4.61, we get the matrix

$$\begin{vmatrix}\Delta V_{ex}\\ \Delta V_a\\ \Delta T_l\end{vmatrix} = \begin{vmatrix} R_e + sL_e & 0 & 0\\ n_0 \cdot G & R_a + sL_a & G \cdot i_{e0}\\ \dfrac{G \cdot i_{a0}}{2\pi} & \dfrac{G \cdot i_{e0}}{2\pi} & -(J2\pi s + B)\end{vmatrix} \cdot \begin{vmatrix}\Delta i_e\\ \Delta i_a\\ \Delta n\end{vmatrix} \tag{4.62}$$

It is now evident, again, that the excitation circuit is decoupled from the armature circuit. The eigenvalues of matrix (4.62) are obtained by solving its determinant equation

$$(R_e + sL_e) \cdot \left( (R_a + sL_a)(J2\pi s + B) + \frac{G^2 \cdot i_{e0}^2}{2\pi} \right) = 0 \tag{4.63}$$

The first root of (4.63) $s_0$ is related to excitation

$$s_0 = -\frac{L_e}{R_e} \tag{4.64}$$

The other two roots $s_{1,2}$ are identical to those obtained for constant flux ((4.43) and (4.44)). The field current $i_e$, when varied through $\Delta V_{ex}$, produces a notable delay in the current $\Delta i_a$ and speed $\Delta n$ response and thus should be avoided when fast transients are required. On the other hand, variable flux is useful for extending the speed/torque envelope for a given armature voltage $V_{an}$.

## 4.10 Speed/Excitation Voltage Transfer Function

**EXAMPLE 4.4**

For a d.c. brush motor with separate excitation, find the speed to excitation voltage transfer function based on the following numerical data: $i_{e0} = 5$ A, $R_e = 1\ \Omega$, $L_e = 1$ H, $R_a = 0.1\ \Omega$, $L_a = 5$ mH, $I_{a0} = 100$ A, $J = 1$ kgm², $n_0 = 1200$ rpm, $B = 0$, $V_{a0} = 210$ V.

**Solution**

The required transfer function may be obtained from Equation 4.62, with $\Delta V_a = 0$ and $\Delta T_l = 0$, by eliminating $\Delta i_e$ and $\Delta i_a$:

$$\frac{dn}{dv_{ex}} = \frac{G \cdot i_{a0} \cdot (R_a + sL_a) - n_0 \cdot G^2 \cdot i_{e0}}{(R_e + sL_e) \cdot [G^2 \cdot i_{e0}{}^2 + 4\pi^2 Js \cdot (R_a + sL_a)]} \tag{4.65}$$

In Equation 4.65, we have all the parameters with the exception of G:

$$E_0 = V_{a0} - R_a \cdot i_{a0} = 210 - 0.1 \cdot 100 = 200\,\text{V} \tag{4.66}$$

$$G = \frac{E_0}{n \cdot i_{e0}} = \frac{200}{(1200/60) \cdot 5} = 2\text{H} \tag{4.67}$$

Finally,

$$\frac{dn}{dV_{ex}} = \frac{2 \cdot 100 \cdot (0.1 + 0.005s) - 20 \cdot 2^2 \cdot 5}{(1 + 1 \cdot s)(2^2 \cdot 5^2 + 4\pi^2 \cdot 1 \cdot s \cdot (0.1 + 0.005s))}$$

$$= \frac{s - 380}{(1 + s)(100 + 3.4438s + 0.1972s^2)} \tag{4.68}$$

## 4.11 D.C. Brush Series Motor

The d.c. brush series motor schematics is shown in Figure 4.11. The excitation and armature currents are equal to each other unless an additional resistance $R_{ead}$ is connected in parallel to the excitation circuit to produce flux weakening.

With $R_{ead} = \infty$, the governing equations in terms of symbols shown in Figure 4.11 are

$$V_a = (R_a + R_{es}) \cdot i_a + (L_a + L_{es})\frac{di_a}{dt} + nGi_a$$
$$T_e = \frac{Gi_a^2}{2\pi} = J2\pi\frac{dn}{dt} + T_{load} + B \cdot n \quad (4.69)$$

As it contains products of variables, the system (4.69) is nonlinear and thus solving it directly may be performed only through numerical methods.

For small perturbations, we may, however, linearize the equations

$$V_a = V_{a0} + \Delta V_a; \quad T_{load} = T_{load0} + \Delta T_l;$$
$$n = n_0 + \Delta n; \quad i_a = i_{a0} + \Delta i_a \quad (4.70)$$

to find

$$\begin{vmatrix} \Delta V_a \\ \Delta T_l \end{vmatrix} = \begin{vmatrix} R_a + R_{es} + n_0 \cdot G + s(L_a + L_{es}) & Gi_{a0} \\ 2\frac{G}{2\pi} i_{a0} & -(2\pi Js + B) \end{vmatrix} \cdot \begin{vmatrix} \Delta i_a \\ \Delta n \end{vmatrix} \quad (4.71)$$

The determinant of Equation 4.71 leads to the eigen values of the system

$$(R_a + R_{es} + n_0 \cdot G + s \cdot (L_a + L_{es})) \cdot (2\pi Js + B) + 2\frac{G^2}{2\pi} \cdot i_{a0}^2 = 0 \quad (4.72)$$

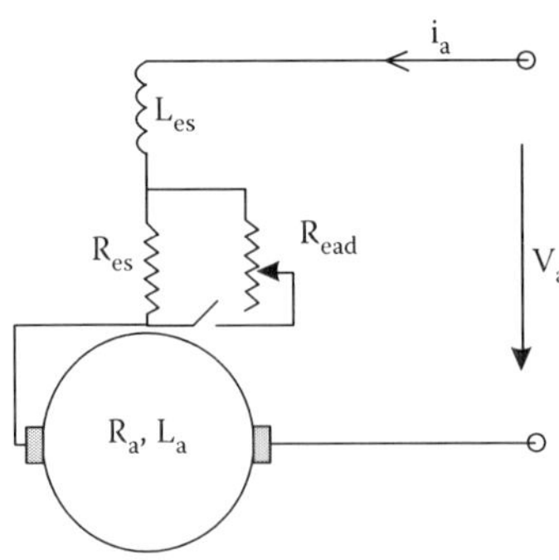


**FIGURE 4.11**
The d.c. brush series motor.

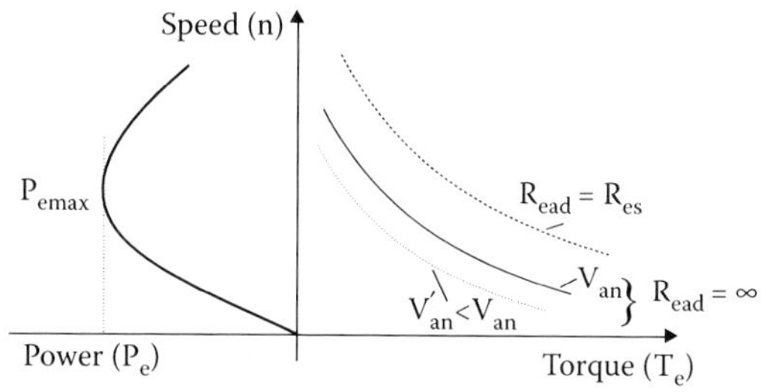


**FIGURE 4.12**
Torque speed curve of d.c. brush series motor.

The apparent electrical time constant in Equation 4.72, if compared to the constant flux case (4.45), depends on speed through the term $n_0G$. At zero speed as $L_{es} > L_a$, and $R_{es} < R_a$, the electrical time constant is again larger than for the constant flux case when $\tau_e = L_a/R_a$.

Fast torque response is thus not easily expected from the d.c. brush series motor. However, it does not require a separate source to supply the field current, and flux weakening is simply possible through the shunt resistor $R_{ead}$ (Figure 4.11).

For steady state, the torque speed curve may be obtained from Equation 4.67 with $d/dt = 0$.

$$i_a = \frac{V_a}{(R_a + R_{es}) + n \cdot G} = \sqrt{\frac{2\pi}{G} T_e} \tag{4.73}$$

At zero current (torque), the ideal speed is infinite and thus the torque speed curve is called mild (Figure 4.12).

Reducing the speed may be accomplished through $V_a$ reduction, as done with d.c. brush motors with separate excitation.

Flux weakening may be performed through $R_{ead}$ to obtain speeds above the rated (base) value for full armature voltage $V_{an}$. The electromagnetic power $P_e$ is

$$P_e = T_e \cdot 2\pi n = \frac{V_a^2 \cdot G \cdot n}{(R_a + R_{es} + nG)^2} \tag{4.74}$$

Over a certain speed range, electromagnetic power does not vary much and is, anyway, limited. This is a unique characteristic of the d.c. brush series motor that is so beneficial in transportation applications where the on-board installed power level is limited.

## 4.12 A.C. Brush Series Motor

Known also as the universal motor, the a.c. brush series motor is still used extensively in some home appliances such as washing machines, kitchen robots, and vacuum cleaners. It is also predominant in handheld (portable) tools and is fabricated up to 30,000 rpm at 1 kW, in general. Despite the brush burden, the universal motor survived due to its low cost/performance [5].

Topologically, the universal motor is very similar to the d.c. brush series motor (Figure 4.13) but the stator is made of stamped laminations because the stator excitation coils are a.c. fed. The commutation of the a.c. current at brushes is more difficult in comparison with d.c. brush motors, due to the a.c. transformer-type (speed independent)—additional—e.m.f. induced in the commutating coil in the rotor, by the stator excitation a.c. current.

Because the stator excitation coils are series-connected to the brushes, the field current and rotor current are equal to each other.

Also, as there is a.c. current at brushes and in the stator, phasors may be used to investigate the machine steady state, despite the fact that in the rotor, the current has two frequencies $\omega_r \pm \omega_1$ ($\omega_1$—stator frequency, $\omega_r = 2\pi np$—electric rotor angular speed).

The voltage equation, valid also for transients, is straightforward:

$$(R_a + R_{es}) \cdot i - v = E_{pulse} - E_{rot} \tag{4.75}$$

$E_{pulse}$ is the transformer (self-induced) voltage in the stator and rotor inductances $L_{es}$ and $L_a$:

$$E_{pulse} = -(L_{es} + L_a)\frac{di}{dt} \tag{4.76}$$

The rotational-induced voltage (the motion e.m.f.) $E_{rot}$ is, as for the d.c. brush machine, in phase with the stator current:

$$E_{rot} = k_e k_\Phi ni \tag{4.77}$$

It is evident that only the motion e.m.f. produces electromagnetic power $P_{elm}$:

$$P_{elm} = E_{rot} i = T_e 2\pi n \tag{4.78}$$

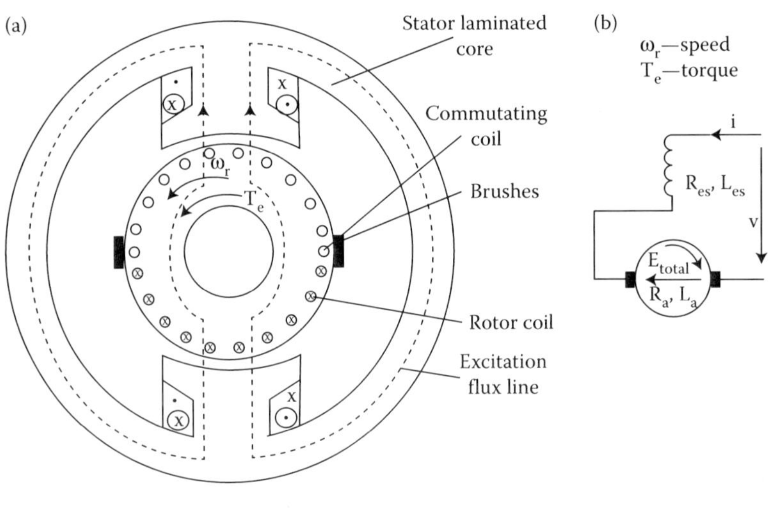


**FIGURE 4.13**
The two pole a.c. brush series motor: (a) cross section; (b) equivalent scheme.

So the instantaneous torque $T_e$ (from Equation 4.78 with Equation 4.77) is

$$T_e = k_e k_\Phi \frac{i^2}{2\pi} \tag{4.79}$$

Now for steady state, the stator voltage and current are sinusoidal:

$$v = V\sqrt{2}\cos\omega_1 t; \quad i = I\sqrt{2}\cos(\omega_1 t - \varphi_1) \tag{4.80}$$

Consequently with I from Equation 4.80, the instantaneous torque $T_e$ is

$$T_e = \frac{k_e k_\Phi I^2}{2\pi}\left[1 + \cos(2\omega_1 t - \varphi_1)\right] \tag{4.81}$$

So the instantaneous torque, for steady state, has a constant component that is the average torque and an a.c. (pulsating) component at $2\omega_1$ (Figure 4.14a). The $2\omega_1$ torque pulsations mean vibrations, noise, and also additional losses.

For steady state, from now on, we make use of the phasors concept:

$$\underline{V} = V\sqrt{2}\cdot e^{j\omega_1 t}; \quad \underline{I} = I\sqrt{2}\cdot e^{j(\omega_1 t - \varphi_1)} \tag{4.82}$$

With this denomination, the voltage equation (4.75) turns into

$$\begin{aligned} \underline{V} &= (R_{es} + R_a)\underline{I} + j\omega_1(L_{es} + L_a)\underline{I} + k_e k_\Phi n \underline{I} \\ R_{ae} &= R_a + R_{es}; \quad \omega_1(L_a + L_{es}) = X_{ae} \end{aligned} \tag{4.83}$$

We may also include the stator and rotor core losses by replacing the total reactance $X_{ae}$ to a series impedance $\underline{Z}_{ae}$:

$$\underline{Z}_{ae} \approx R_{core} + jX_{ae} \tag{4.84}$$

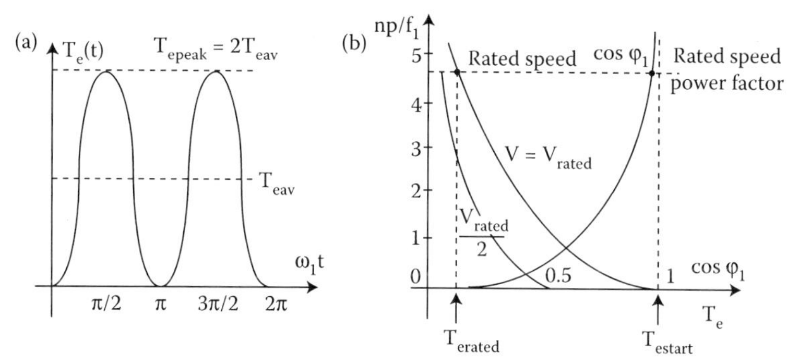


**FIGURE 4.14**
A.c. brush series motor steady-state characteristics: (a) instantaneous torque; (b) torque and power factor versus relative speed $np/f_1$.

Consequently, with core losses included, voltage equation (4.83) becomes

$$\underline{V} = (\underline{R}_{ae} + \underline{Z}_{ae} + k_e k_\Phi n)\underline{I} \tag{4.85}$$

The average torque is now evident:

$$T_{eav} = \frac{(k_e k_\Phi n \cdot I) \cdot I^*}{2\pi n} = \frac{k_e k_\Phi I^2}{2\pi} \tag{4.86}$$

The dependence of torque on speed n is very similar to the case of the d.c. brush series motor (Figure 4.14b) but with the torque reduced by the presence of total reactance $X_{ae}$.

Also, the power factor $\cos \varphi_1$ is

$$\cos \varphi_1 = \frac{1}{\sqrt{(X_{ae}/R_{ae} + k_e k_\Phi n + R_{core})^2 + 1}} \tag{4.87}$$

The larger the speed n (or the $np/f_1$ ratio), the larger the power factor (Figure 4.14b). A ratio $np/f_1 > 6/1$ is common to vacuum cleaners and for kitchen robots, with only 3/1 in washing machines.

In general, the universal motor is designed for a power factor above 0.9 for rated speed, which, in low power drives is a good asset in the absence of capacitors (required for the brushless capacitor—split-phase induction motors). For a comprehensive analysis, see Reference 5.

Speed control may be approached simply by an a.c.–a.c. voltage changer that makes use of a single bidirectional thyristor module (called triac). Torque equation (4.86) illustrates its quadratic dependence on the voltage amplitude.

The triac is a very-low-cost device with simple controllability but it produces notable line current harmonics, which have to be filtered. Also, when the voltage is decreased, the DPF is reduced.

**EXAMPLE 4.5: THE UNIVERSAL MOTOR**

Let us consider a universal motor for a home appliance supplied at 120 V a.c. (rms), 60 Hz, that produces $P_n = 600$ W at 18,000 rpm. It has two poles. The core loss is equal to half the copper loss and the mechanical loss is 2% of rated power. The efficiency is $\eta = 0.9$ and the power factor $\cos \varphi_n = 0.97$. Calculate:

a. The rated current
b. The total winding resistance $R_{ae}$ and the core loss resistance $R_{core}$
c. The motion-induced e.m.f. $E_{rot}$
d. The total inductance $L_{ae}$
e. The electromagnetic average torque
f. The shaft torque
g. The ratio of speed to frequency $n_n p/f_1$
h. The starting current and torque
i. The starting current when fed at 120 V d.c.

**Solution**

a. From the definition of efficiency as output to input power:

$$\eta_n = \frac{P_n}{V_n I_n \cos \varphi_n}$$

The rated current $I_n$ is

$$I_n = \frac{600}{120 \cdot 0.97 \cdot 0.9} = 5.7273\,\text{A}$$

b. The total losses in the machine $\Sigma p$ is

$$\Sigma p = \frac{P_n}{\eta_n} - P_n = 66.66\,\text{W}$$

With mechanical losses:

$$p_{mec} = 0.02 P_n = 12\,\text{W}$$

The winding plus iron losses are

$$p_{copper} + p_{core} = R_{ae} I_n^2 + R_{core} I_n^2 = \Sigma p - p_{mec} = 66.66 - 12 = 54.66\,\text{W}$$

With the core losses as half the winding losses

$$R_{ae} I_n^2 = \frac{2}{3} \cdot 54.66; \quad R_{ae} = \frac{54.66 \cdot 2}{3} \frac{1}{5.7273^2} = 1.111\,\Omega$$

$$R_{core} = \frac{1}{2} R_{ae} \approx 0.555\,\Omega$$

From Equation 4.87

$$I_n = \frac{U_n \cos\varphi_n}{R_{ae} + R_{core} + k_e k_\Phi n}$$

$$\tan\varphi_n = \frac{X_{ae}}{R_{ae} + R_{core} + k_e k_\Phi n}$$

So

$$R_{ae} + R_{core} + k_e k_\Phi n = \frac{120 \cdot 0.97}{5.7273} = 20.3237\,\Omega$$

c. The motion-induced e.m.f.

$$E_{rot} = k_e k_\Phi n I_n = (20.3237 - 1.111 - 0.555) \cdot 5.7273 = 106.858\,\text{V}$$

Finally, the machine reactance $X_{ae}$ from the impedance definition is

$$X_{ae} = \frac{V_n}{I_n} \sin\varphi_n = \frac{120}{5.7273} 0.243 = 5.0936\,\Omega$$

d. So the machine inductance

$$L_{ae} = \frac{X_{ae}}{\omega_1} = \frac{5.0936}{2\pi 60} = 0.0135\,\text{H}$$

e. The average torque $T_{eav}$ comes from Equation 4.79:

$$T_{eav} = \frac{k_e k_\Phi I_n^2}{2\pi} = \frac{E_{rot} I_n}{2\pi \cdot n_n} = 0.3248\,\text{N m}$$

f. The shaft torque $T_{shaft}$ is approximately

$$T_{shaft} \approx T_{eav} - \frac{p_{mec}}{2\pi n_n} = 0.3248 - \frac{12}{2\pi \cdot 300} = 0.31847\,\text{N m}$$

g. The ratio $n_n p/f_1 = (18{,}000/60)/60 = 5/1$ explains the good power factor (DPF) of the machine.

h. The starting current is obtained again from Equation 4.85 but with $n = 0$:

$$I_{start} = \frac{V_1}{\sqrt{(R_{ac} + R_{core})^2 + X_{ae}^2}} = 22.39\,\text{A}$$

The average starting torque $(T_{eav})_{start}$ is (4.79)

$$(T_{eav})_{start} = T_{eav}\left(\frac{I_{start}}{I_n}\right)^2 = 4.9646\,\text{N m}$$

**NOTE:** In reality, for this high current, the magnetic flux saturates heavily and thus the ideal starting torque above is reduced considerably.

i. For d.c. at start, there is no a.c. (pulse) voltage across the machine inductances ($X_{ae} = 0$) and $R_{core} = 0$ (no core losses in d.c.).

$$I_{start\,d.c.} = \frac{V_{dc}}{R_{ae}} = \frac{120}{1.111} = 108\,\text{A d.c.}$$

So, the a.c. brush series motor is to be fed from a much smaller d.c. voltage if the starting current is to be limited to reasonable values (3–5 times the rated value).

**NOTE:** In conjunction with PECs, the d.c. brush series motor is still widely used in standard electric propulsion systems for urban, interurban, or water transportation and some heavy-duty off-highway vehicles. Though the a.c. drives are taking over the electric propulsion technologies, we felt it was useful to devote two pages to the d.c. brush series motor, the workhorse of electric propulsion in the twentieth century, not in the twenty-first century, however. A similar fate faces the a.c brush series motor.

## 4.13 Summary

- The d.c. brush motor may be electromagnetically or PM excited in the stator. For variable speed drives, due to current pulse additional losses for PEC control, both the stator and rotor cores are laminated.
- For the axial airgap disk-rotor PM d.c brush machine, the rotor windings are in air and thus the core losses in the machine are very small and the electrical time constant is small (≤1 ms in the kW power range), allowing for fast current (torque) control.
- Separately excited or PM machines, for constant field current (or PM), are second-order linear systems and, due to the feedback of e.m.f., they provide stable responses in speed or current, either periodic or aperiodic.
- The variation of field current introduces an additional, large delay in the response, corresponding to the excitation circuit time constant. Still the responses in speed and current are stable.
- The inherent decoupling between excitation and armature windings allows for quick armature current (torque) control for constant field current.
- The d.c. brush series motor has a mild torque speed curve while the one with separate excitation has a rigid, linear, torque speed curve. The former also proved to be very useful in the electrical propulsion of various vehicles.
- The a.c. brush series (universal) motor still enjoys sizeable markets in home appliances and hand-held tools at high speeds, due to low cost with simple speed control.

## 4.14 Problems

4.1 For the d.c. brush motor in Example 4.1, add the fact that the rated field current is $I_n = 1$ A and $V_{excn} = 110$ V, and determine:

a. The voltage $V_a$ at standstill for rated current

b. The field current at 3600 rpm and rated current, and the corresponding torque, and total input power

4.2 For the PM d.c. brush motor in Example 4.3, determine the time variation of armature current for the 10 V step in the armature voltage for constant load torque.

4.3 A d.c. brush series motor for a light rail urban transportation system has the data: $V_{an} = 800$ V (d.c.), $P_n = 100$ kW, $n_n = 1200$ rpm, rated efficiency $\eta_n = 0.92$, $p_{mec} = p_{iron} = 0.015\ P_n$, $R_{es} = R_a$. The commutator and additional losses are neglected. Calculate:

a. The rated current, $I_n$

b. Armature and excitation resistances, $R_a$ and $R_{es}$

c. The rated e.m.f., $E_n$, and electromagnetic torque, $T_{en}$

## References

1. M.G. Say, D.c. machines. In S.A. Nasar (ed.), *Handbook of Electric Machines*, Chapter 5, New York: McGraw-Hill Book Company, 1987.
2. W. Leonhard, *Control of Electric Drives*, first, second, third editions, Berlin, Heidelberg: Springer-Verlag, 1985, 1996, 2001.
3. P.C. Sen, *Thyristor DC Drives*, New York: John Wiley and Sons, 1981.
4. S.A. Nasar, I. Boldea, L. Unnewehr, *Permanent Magnet, Reluctance and Self-Synchronous Motors*, Chapter 3, Boca Raton, Florida: CRC Press, 1993.
5. A. Di Gerlando, R. Perini, G. Rapi, Equivalent circuit for the performance analysis of universal motors, *IEEE Trans.* vol. EC-19, no. 1, 2004.

# 5

# Controlled Rectifier D.C. Brush Motor Drives

## 5.1 Introduction

Rectifiers are phase-controlled a.c. to d.c. static power converters. The rectifier provides variable d.c. voltage to the d.c. brush motor. Thyristors, bipolar transistors, insulated gate bipolar transistors (IGBTs) or MOSFETs may be used as PESs in the converter.

In general, the commutation process is natural, from the d.c. source, without any additional circuitry. However, for improved PF, forced commutation is used frequently. Phase-controlled power electronic converters are broadly classified as a.c.–d.c. single-phase or three-phase converters [1–2]. Their main configurations are shown again in Table 5.1 for convenience (after their classification in Chapter 3).

Half-wave single-phase and semiconverters have one polarity of output voltage $e_{av}$ and current $i_{av}$. That is, they work in one quadrant. Full converters work in two quadrants: the output voltage $e_{av}$ may be positive or negative while the output current remains positive. Only dual converters can operate in four quadrants.

When the PESs are blocked, the stored energy dissipates through the freewheeling diodes in half-wave and semiconverters. In a single-phase half-wave converter, the motor current is discontinuous unless a high additional inductance is added, while for the other converters, the output current may be either continuous or discontinuous. In three-phase converters, the motor current is mostly continuous.

## 5.2 Performance Indices

As the d.c. motor current, when fed from phase-controlled a.c.–d.c. converters, is not constant and a.c. supply current is not sinusoidal, adequate performance indexes for the motor–converter combination should be defined.

The main performance indices related to the motor are

- The torque–speed characteristic
- Nature of motor current—continuous or discontinuous
- Average motor current $I_a$:

$$I_a = \frac{1}{T} \cdot \int_{t_1}^{t_1+T} i_a dt \tag{5.1}$$

**TABLE 5.1**
Phase-Controlled Rectifier Circuits

| Circuit | Type | Power Range | Ripple Frequency | Quadrant Operation |
|---|---|---|---|---|
| 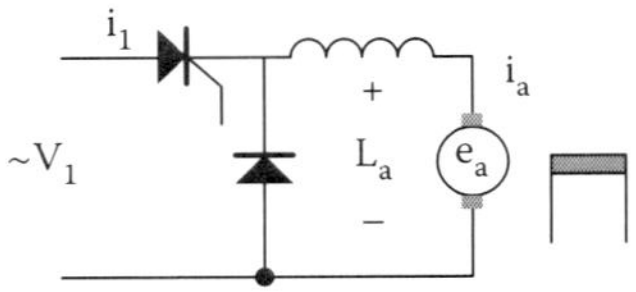 | Half-wave single phase | Below 0.5 kW | $f_s$ | 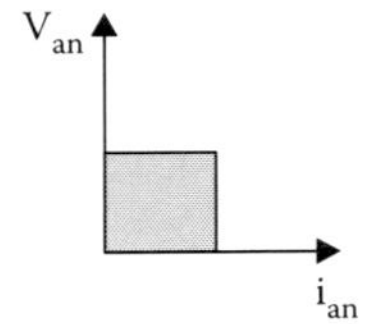 One quadrant |
| 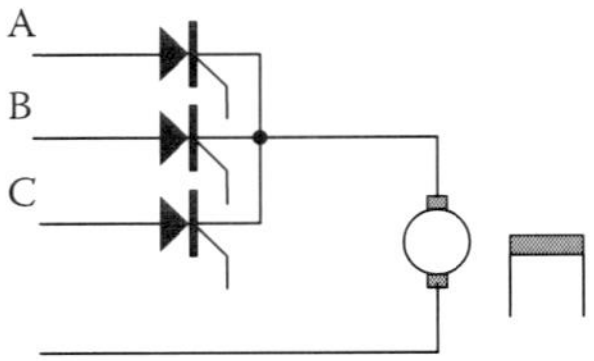 | Half-wave three phase | Up to 50 kW | $3f_s$ | 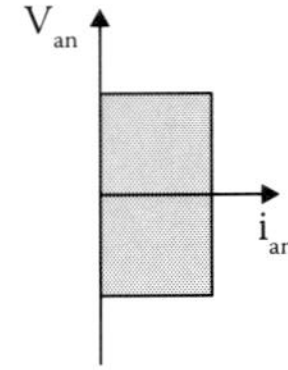 Two quadrant |
| 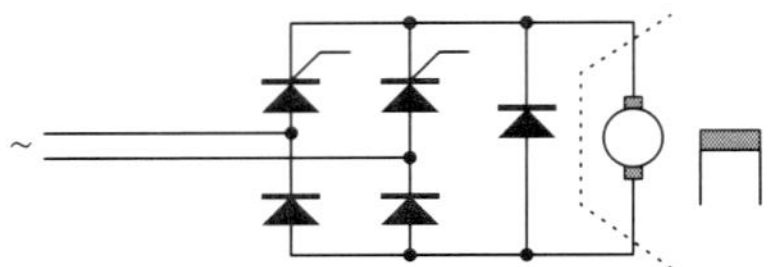 | Semiconverter single phase | Up to 15 kW | $2f_s$ | 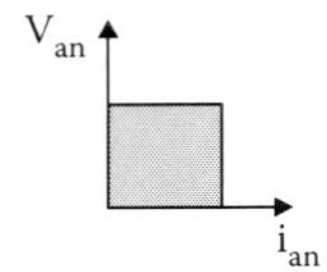 One quadrant |
| 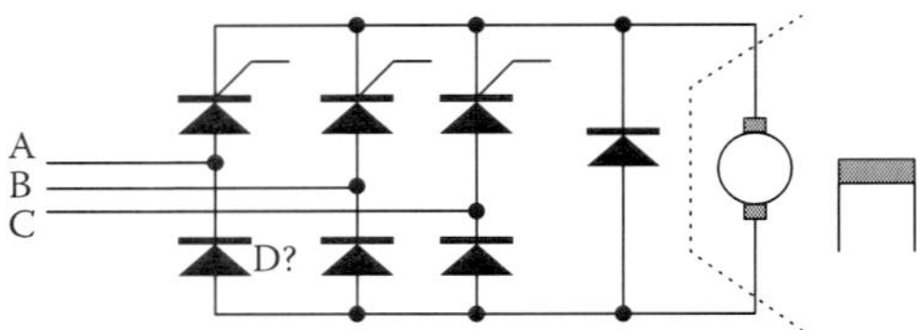 | Semiconverter three phase | Up to 100 kW | $3f_s$ | 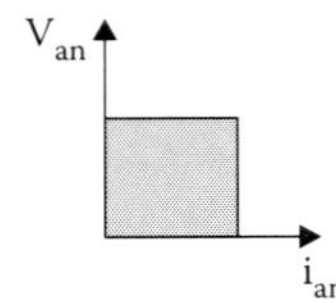 One quadrant |

*(Continued)*

**TABLE 5.1** (***Continued***)

Phase-Controlled Rectifier Circuits

| Circuit | Type | Power Range | Ripple Frequency | Quadrant Operation |
|---|---|---|---|---|
| 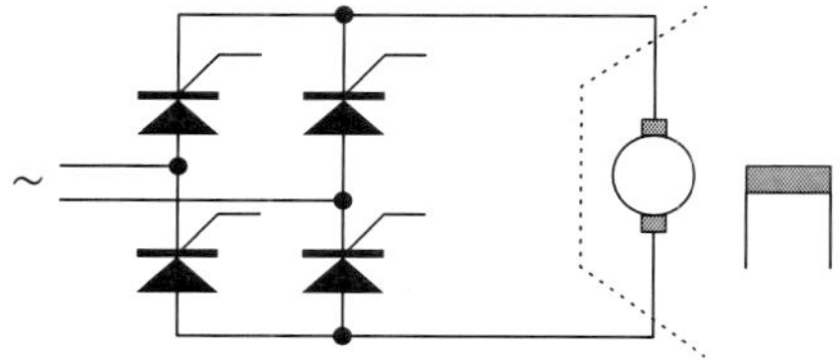 | Full converter single phase | Up to 15 kW | $2f_s$ | 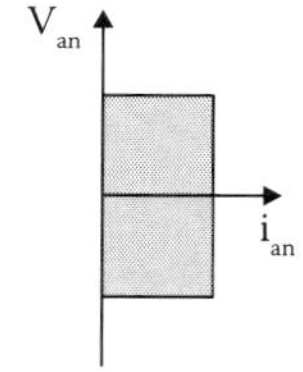 Two quadrant |
| 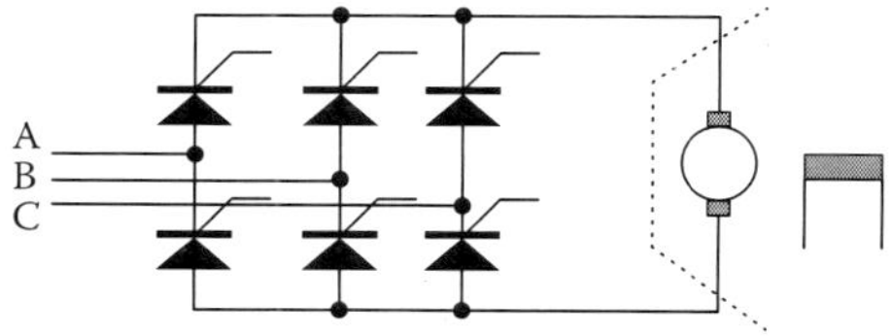 | Full converter three phase | Up to 1500 kW | $6f_s$ | 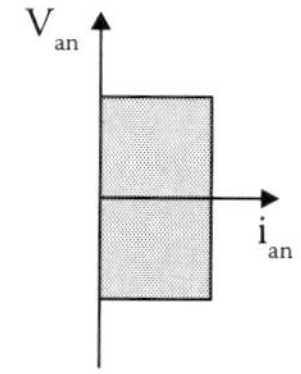 Two quadrant |
| 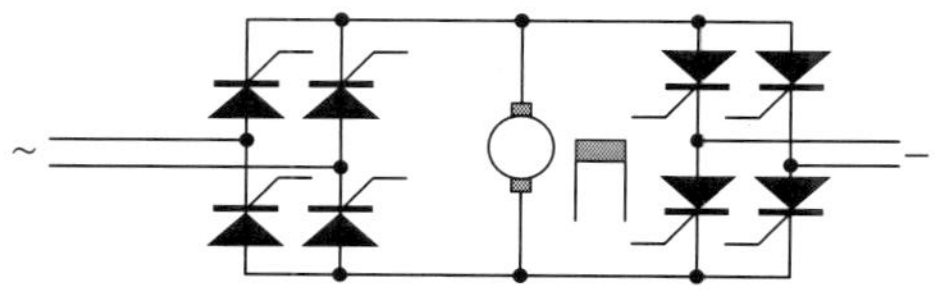 | Dual converter Single phase | Up to 15 kW | $2f_s$ | 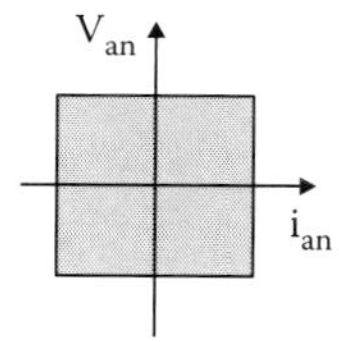 Four quadrant |
| 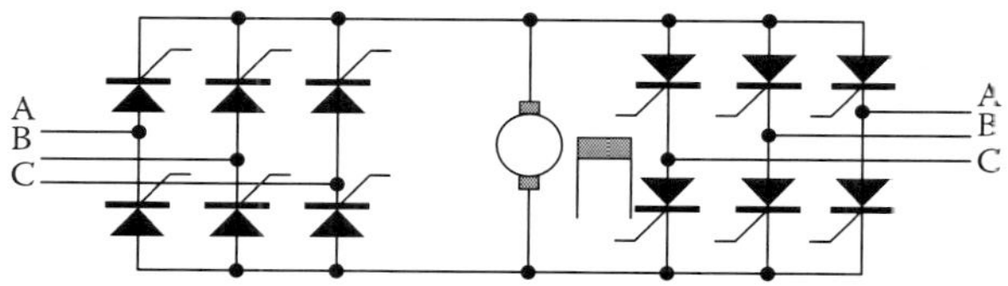 | Dual converter three phase | Up to 1500 kW | $6f_s$ | 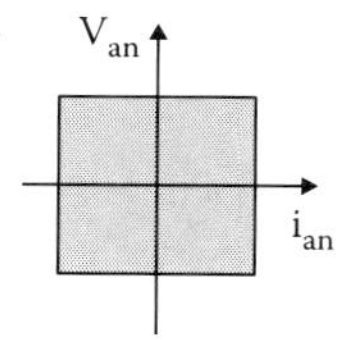 Four quadrant |

where $i_a$ is the instantaneous armature current and T is the time period of one cycle of $i_a$ variation.

- Rms motor current $I_{ar}$

$$I_{ar} = \sqrt{\frac{1}{T} \cdot \int_{t_1}^{t_1+T} i_a^2 dt} \tag{5.2}$$

As known, the rms current squared is proportional to the heat produced in the armature winding.

- Peak motor current $i_{ap}$; the mechanical commutator stress depends on the peak value of the current.

The main performance related to the input (a.c.) source are

- Input power factor

$$PF = \frac{\text{mean input power}}{\text{rms input volt} \times \text{amperes}} \tag{5.3}$$

If the supply voltage is a pure sinusoid, only the fundamental input current will produce the mean input power and thus

$$PF = \frac{V \cdot I_1 \cdot \cos\phi_1}{V \cdot I} \tag{5.4}$$

where V is the rms supply phase voltage, I is the rms supply phase current, $I_1$ is the rms fundamental component of a.c. supply current, and $\varphi_1$ is the angle between supply voltage and current fundamentals.

- Input displacement power factor (DPF) or the fundamental PF

$$DPF = \cos\phi_1 \tag{5.5}$$

- Harmonic factor (HF)

$$HF = \frac{\sqrt{(I^2 - I_1^2)}}{I_1} = \frac{\sqrt{\sum_{n=2}^{\infty} I_n^2}}{I_1} = \frac{I_h}{I_1} \tag{5.6}$$

where $I_h$ is the rms value of the net harmonic current.

The above performance indices are somewhat similar to those for diode rectifiers (Chapter 3). They are summarized here only for convenience.

The basic motor equations are

$$\begin{aligned} v_a &= R_a \cdot i_a + L_a \cdot \frac{di_a}{dt} + e_g; \quad e_g = K_e \cdot \lambda_p \cdot n \\ 2\pi J \frac{dn}{dt} &= T_e - T_{load}; \quad T_e = \frac{e_g \cdot i_a}{2\pi n} = \frac{K_e}{2\pi} \cdot \lambda_p \cdot i_a \end{aligned} \tag{5.7}$$

where $v_a$, $i_a$ are the motor input voltage and current, $e_g$ is the motion-induced voltage, $R_a$, $L_a$ are the armature resistance and inductance, J is the inertia, $T_e$ is the motor torque, and $T_{load}$ is the load torque. For the steady state, $dn/dt = 0$, while in general $di_a/dt \neq 0$ as the armature current is not constant in time with phase-controlled converter supplies. To facilitate a quick assimilation of so many controlled rectifier configuration performance, we will proceed directly through numerical examples.

## 5.3 Single-Phase PES-Controlled Rectifier

A d.c. brush motor with separate excitation with the data: $K_e\lambda_p = 2$ Wb, $R_a = 5\ \Omega$, $L_a = 0.1$ H is fed through a thyristor (Figure 5.1) from a single-phase a.c. source whose voltage is $V = V_1 \cdot \sin\omega_1 t = 120\sqrt{2}\sin 376.8t$. The motor speed is considered constant at $n = 750$ rpm.

a. Calculate the motor current $i_a$ time variation for a delay angle $\alpha = +30°$.
b. How the motor voltage $V_a$ does vary in time?
c. How does $i_a$ vary in time in the presence of a freewheeling diode in parallel with the motor armature (Figure 5.1)? The thyristor and the diode are considered as ideal switches.

### *Solution*

a. For $\omega t > \alpha$ (Figure 5.1) with discontinuous current:

$$V_a(t) = R_a i_a + L_a \frac{di_a}{dt} + K_e\lambda_p n \tag{5.8}$$

with the initial condition $i_a = 0$ for $\omega_1 t = \alpha = \pi/6$.

The steady-state solution is

$$i_{ap} = A + B \cdot \cos\omega_1 t + C \cdot \sin\omega_1 t \tag{5.9}$$

and finally

$$i_{ap} = -\frac{K_e \cdot \lambda_p \cdot n}{R_a} + \frac{V_1}{\sqrt{R_a^2 + \omega_1^2 \cdot L_a^2}} \cdot \sin(\omega_1 t - \gamma)$$
$$\gamma = \tan^{-1}(\omega_1 L_a / R_a) \tag{5.10}$$

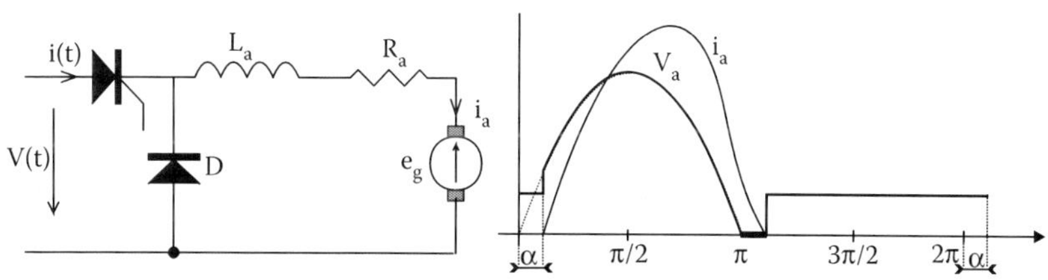


**FIGURE 5.1**
A d.c. brush motor supplied through a thyristor.

The complete solution:

$$i_a(t) = i_{ap} + A \cdot e^{-tR_a/L_a} \tag{5.11}$$

For $t = t_1 = \alpha/\omega_1 = \pi/(6.2\pi \cdot 60) = 1.3888 \cdot 10^{-3}$ s, $i_a(t) = 0$ and thus, from Equation 5.10:

$$\gamma = \tan^{-1}\frac{2 \cdot \pi \cdot 60 \cdot 0.1}{5} = 82.44^\circ$$
$$\frac{V_1}{\sqrt{R_a^{\,2} + \omega_1^{\,2} \cdot L_a^{\,2}}} = \frac{120}{\sqrt{5^2 + (2 \cdot \pi \cdot 60 \cdot 0.1)^2}} = 3.157 \text{ A} \tag{5.12}$$

and

$$\frac{K_e \cdot \lambda_p \cdot n}{R_a} = \frac{2 \cdot 12.5}{5} = 5 \text{ A} \tag{5.13}$$

$$A = \left[\frac{K_e \cdot \lambda_p \cdot n}{R_a} - \frac{V_1}{\sqrt{R_a^{\,2} + \omega_1^{\,2} \cdot L_a^{\,2}}} \cdot \sin\left(30^\circ - \gamma\right)\right] \cdot e^{1.3888 \cdot 10^{-3} \cdot 5/0.1}$$
$$= \left[5 - 3.157 \sin(-52.44^\circ)\right] \cdot 1.07186 = 8.04 \text{ A} \tag{5.14}$$

Finally,

$$i_a(t) = 8.04 \cdot e^{-50t} - 5 + 3.157 \cdot \sin(377t - 1.43812) \tag{5.15}$$

Equation 5.15 is valid for $\omega_1 t > \pi/6$ until the current becomes zero soon after $\omega_1 t = \pi$, as shown in Figure 5.1.

b. The motor voltage is equal to the source voltage as long as the thyristor is on and becomes equal to minus the e.m.f.: $-e_a = K_e \cdot \lambda_p \cdot n = 2 \cdot 12.5 = 25$ V, when the motor current is zero (Figure 5.1).

c. In presence of the freewheeling diode D, the latter starts conducting when V(t) becomes negative ($\omega_1 t = \pi$).

   From now on, the motor current $i'_a$ flows through the diode until it becomes zero

$$0 = R_a \cdot i'_a + L_a \cdot \frac{di'_a}{dt} + K_e \cdot \lambda_p \cdot n \tag{5.16}$$

With

$$i'_a(\pi) = i_a(\pi) \tag{5.17}$$

$$i'_a = -\frac{K_e \cdot \lambda_p \cdot n}{R_a} + A' \cdot e^{-t'R_a/L_a} \tag{5.18}$$

$$i'_a(\pi) = i_a(\pi) = \frac{-2 \cdot 12.5}{5} + A' \cdot e^{-50/120} = 3.42729\ \text{A} \tag{5.19}$$

$$A' = 12.78319\ \text{A}$$

Thus

$$i'_a = -5 + 12.783 \cdot e^{-50t'} \tag{5.20}$$

The current $i'_a(t) = 0$ for $\omega_1 t = 360° + 45.5°$; so the current is continuous.

The motor voltage $V_a$ becomes zero during the time interval when the freewheeling diode is conducting (180° to 360° + 30°).

## 5.4 Single-Phase Semiconverter

A d.c. motor supplied through a single-phase semiconverter has a constant speed and the motor current is constant in time, $I_a = 3$ A, $R_a = 5\ \Omega$, $K_e\Phi = 2$ Wb, $v = 120\sqrt{2}\sin 120\pi t$ with $\alpha = \pi/4$.

a. Determine the time waveforms of voltages at motor terminals and along the thyristors and diodes and the corresponding current waveforms.
b. Calculate the rms values of currents through diodes, thyristors, and from the a.c. supply.
c. Determine the motor voltage average value dependence and its maximum value.
d. Calculate the rms of a.c. supply current fundamental and its phase shift with respect to the a.c. supply voltage.

### *Solution*

a. As the filter inductance $L_a$ is large, the motor current $i_a$ is considered constant (Figure 5.2). The thyristors $T_1$ and $T_3$ conduct for 180° region until the other side is triggered while the diodes $D_1$, $D_3$ conduct for half-wave of the source voltage (Figure 5.2d). All currents are rectangular (Figure 5.2d,e) while the motor voltage is either positive or zero (Figure 5.2b).
b. The rms value of thyristor current is

$$I_{T1,3} = \sqrt{\frac{1}{2\pi}\int_0^{2\pi} i_{T_1}{}^2 d(\omega_1 t)} = \sqrt{\frac{1}{2\pi}\int_\alpha^{\pi} i_a{}^2 d(\omega_1 t)}$$

$$= I_a\sqrt{\frac{\pi}{2\pi}} = 3 \cdot 0.707 = 2.121\ \text{A} \tag{5.21}$$

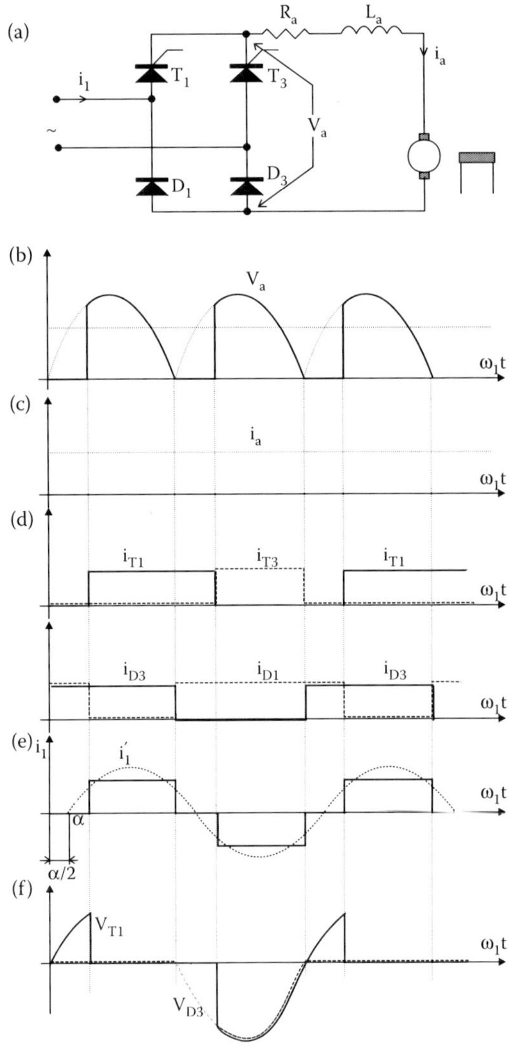


**FIGURE 5.2**
D.c. brush motor supplied through a single-phase semiconverter (with large inductance $L_a$). (a) 1-phase half-wave thyristor rectifier, (b) output voltage, (c) thristor currents, (d) diode currents, (e) ideal input current, (f) thyristor T1 voltage.

The diode current $I_{D_{1,3}}$

$$I_{D_{1,3}} = \sqrt{\frac{1}{2\pi}\int_0^{\pi+\alpha} i_a^2 d(\omega_1 t)} = I_a\sqrt{\frac{\pi}{2\pi}} = 3\cdot 0.707 = 2.121 \text{ A} \tag{5.22}$$

The rms primary (supply current)

$$I_1 = \sqrt{\frac{1}{\pi}\int_\alpha^{\pi} i_a^2 d(\omega_1 t)} = I_a\sqrt{\frac{\pi-\alpha}{\pi}} = 3\cdot 0.866 = 2.598 \text{ A} \tag{5.23}$$

c. The motor average voltage $V_{av}$ is

$$V_{av}=\frac{1}{\pi}\int_{\alpha}^{\pi}v_a d(\omega_1 t)=\frac{1}{\pi}\int_{\alpha}^{\pi}V\sqrt{2}\sin\omega_1 t\cdot d(\omega_1 t)=\frac{V_1\sqrt{2}}{\pi}(1+\cos\alpha) \tag{5.24}$$

For $\alpha=\pi/4$

$$(V_{av})_{\alpha=\pi/4}=\frac{120\sqrt{2}}{\pi}(1+0.707)=91.98\text{ V} \tag{5.25}$$

The maximum average voltage is obtained for $\alpha=0$

$$(V_{av})_{\alpha=0}=\frac{120\sqrt{2}}{\pi}(1+1)=107.77\text{ V} \tag{5.26}$$

and corresponds to the full bridge diode rectifier (Chapter 3).

d. The a.c. supply current harmonics $I_{1v}$ is

$$I_{1v}=\frac{1}{\pi}\int_{0}^{2\pi}i_1\sin v\omega_1 t\cdot d(\omega_1 t)=\frac{4}{\pi v}I_a\cos^2\frac{v\alpha}{2} \tag{5.27}$$

The fundamental is

$$I_{1(1)}=\frac{4}{\pi}3\cos\frac{2\pi}{4\cdot 2}=3.26\text{ A} \tag{5.28}$$

The input displacement power factor DPF = cos $\varphi_1$ refers to the cosine of the angle between the a.c. source voltage and current fundamentals. As seen from Figure 5.2b,e $\varphi_1=\alpha/2=\pi/8$. Thus

$$\text{DPF}=\cos\phi_1=\cos\pi/8=0.9238 \tag{5.29}$$

On the other hand, the PF (5.4) is

$$\text{PF}=\frac{I_{1(1)}\cdot\cos\phi_1}{\sqrt{2}\cdot I_1}=\frac{3.26\cdot 0.9238}{1.41\cdot 2.598}=0.82 \tag{5.30}$$

## 5.5 Single-Phase Full Converter

A d.c. motor of 7 kW and 1200 rpm rating with separate excitation is supplied through a single-phase full converter as shown in Figure 5.3a. The other data are $R_a=0.2\ \Omega$, rated current $I_{ar}=40$ A, $K_e\lambda_p=10$ Wb; the a.c. supply voltage (rms) is 260 V.

a. Draw the voltage and current waveforms for steady state and finite motor armature inductance and $\alpha=45°$ and $\alpha=135°$.

b. For a firing angle of $\alpha = 30°$ and rated motor current (rectifier regime), calculate motor speed, torque, and supply PF, neglecting the motor current ripple.
c. By reversing the field current, the motor back e.m.f. $E_g$ is reversed; for rated current, calculate converter firing angle $\alpha'$ and the power fed back to the supply.

### *Solution*

a. The waveforms of currents and voltages are shown in Figure 5.3. It should be noted that because the motor inductance is not very high, the armature current and speed pulsate ($E_g$) with time.

   Motor action is obtained for $\alpha = 45°$ when both the motor and converter average input powers are positive. For generator action ($\alpha = 135°$), the motor current remains positive but the motor average voltage $V_a$ is negative. Thus both the motor and converter input average powers are negative for generating.
b. As the motor current is considered ripple free, we use only the average values of voltage and current while the speed is also constant.

   The average motor voltage $V_{av}$ is

$$V_{av} = \frac{1}{\pi}\int_{\alpha}^{\pi+\alpha} V\sqrt{2}\sin\omega_1 t\cdot d(\omega_1 t) = \frac{2\sqrt{2}}{\pi}V\cos\alpha$$

$$= \frac{2\sqrt{2}}{\pi}260\cos 30^{\circ} = 202.44\text{ V} \tag{5.31}$$

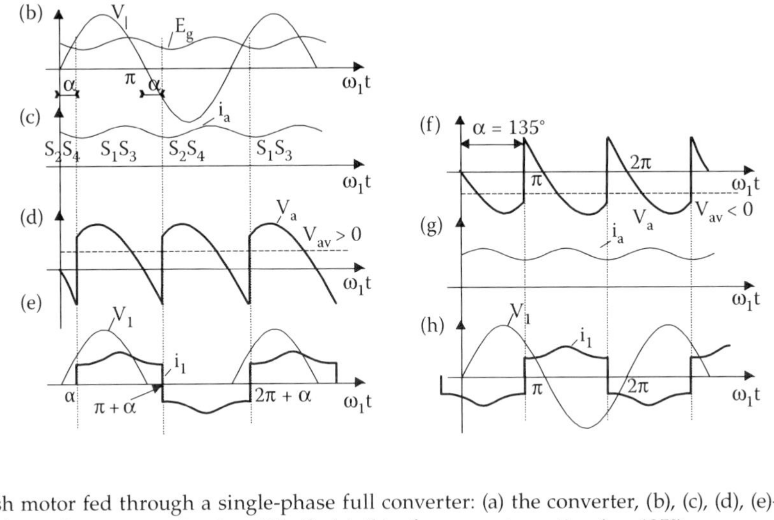


**FIGURE 5.3**
The d.c. brush motor fed through a single-phase full converter: (a) the converter, (b), (c), (d), (e)—voltage and current waveforms for motor action ($\alpha = 45°$), (f), (g), (h)—for generator action ($\alpha = 135°$).

The motor torque $T_e$ is

$$T_e = \frac{K_e \lambda_p}{2\pi} I_a = \frac{10}{2\pi} 40 = 63.6942 \text{ N m} \tag{5.32}$$

The motor speed n is

$$n = \frac{V_{av} - R_a I_a}{K_e \lambda_p} = \frac{202.44 - 0.2 \cdot 40}{10} = 19.444 \text{ rps} = 1166.64 \text{ rpm} \tag{5.33}$$

As the primary current is now rectangular (40 A), the rms current is $I_1 = 40$ A. The power from the supply, neglecting the losses, is $P_s = V_{av} \cdot I_a = 202.44 \cdot 40 = 8097.6$ W. Thus, the supply PF

$$PF = \frac{P_s}{V_1 \cdot I_1} = \frac{8097.6}{260 \cdot 40} = 0.7786 \tag{5.34}$$

c. For generator action, the polarity of induced voltage should be reversed

$$e_g = -K_e \lambda_p n = -10 \cdot 19.44 = -194.44 \text{ V} \tag{5.35}$$

Thus the motor voltage $V_a$ becomes

$$V_a = e_g + R_a I_a = -194.44 + 0.2 \cdot 40 = -186.44 \text{ V} \tag{5.36}$$

Finally,

$$\alpha = \cos^{-1}\left[\frac{V_a \pi}{2\sqrt{2}V}\right] = \cos^{-1}\left[\frac{-186.44 \cdot \pi}{2\sqrt{2} \cdot 260}\right] \approx 143^\circ \tag{5.37}$$

The regenerated power

$$P_{sg} = V_a I_a = 186.44 \cdot 40 = 7457.6 \text{ W} \tag{5.38}$$

As for this regime, either the flux $\lambda_p$ or the speed n changes sign, the motor torque opposes the motion providing generator braking.

In the process, the motor speed and e.m.f. decrease and, to keep the current constant, the firing angle $\alpha > 90°$ in the converter should be modified accordingly.

For the motor as above, having $L_a = 2$ mH, calculate the following:

a. For $\alpha = 60°$ and constant speed n = 1200 rpm, draw the voltage and current waveforms knowing that the motor current is discontinuous.
b. Calculate the motor current waveform for n = 600 rpm (Figure 5.4).

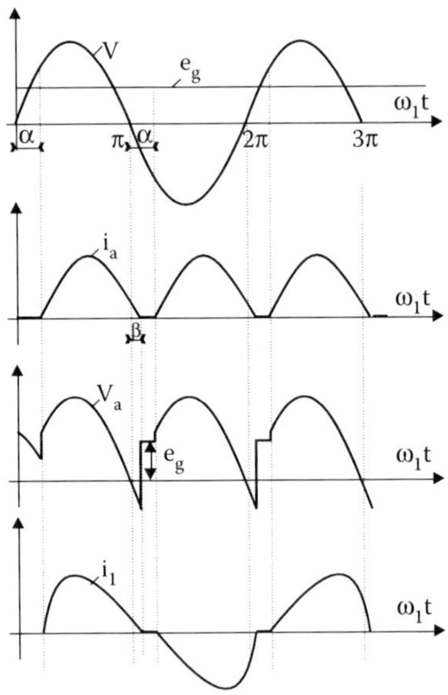


**FIGURE 5.4**
D.c. brush motor supplied from a full converter—discontinuous current mode.

### *Solution*

a. As the motor electrical time constant $\tau_e = L_a/R_a = 2.10^{-3}/0.2 = 10$ ms is small, when the voltage becomes zero (for $\omega_1 t = \pi$), the motor current decays quickly to zero along the angle β (from π to π + β) with β < α and thus the current is discontinuous. As α increases, for a given speed n (and e.m.f. $e_g$), the average voltage $V_{av}$ increases with respect to the case of continuous current (Figure 5.4).

   The discontinuity of current contributes to a sluggish dynamic response as during zero current, motor internal torque control is lost. The situation occurs especially at low torques (and currents). Special measures such as flux weakening (reducing $\lambda_p$) for low torques leads to reduced α at high speeds, avoiding the discontinuity in current.

b. With the speed considered constant, the voltage equation is

$$V\sqrt{2}\sin\omega_1 t = R_a i_a + L_a \cdot \frac{di_a}{dt} + K_e \lambda_p n \tag{5.39}$$

with $i_a = 0$ for $\omega_1 t_1 = \alpha$, the solution of above equation is

$$i_a = A \cdot e^{-tR_a/L_a} - \frac{K_e \lambda_p n}{R_a} + \frac{V\sqrt{2}\sin(\omega_1 t - \varphi_1)}{\sqrt{R_a^2 + \omega_1^2 L_a^2}} \tag{5.40}$$

$$\varphi_1 = \tan^{-1}(\omega_1 L_a / R_a)$$

for $t = t_1 = \alpha_0/\omega_1$; $(i_a)_{t_1} = 0$ and thus

$$\varphi_1 = \tan^{-1}\left(\frac{2\pi 60\cdot 2\cdot 10^{-3}}{0.2}\right) = 75.136^{\circ} \tag{5.41}$$

$$i_a = 1482\cdot e^{-100t} - 1000 + 470\sin(\omega_1 t - \phi_1) \tag{5.42}$$

The current becomes zero again for $\omega_1 t_2 = \pi + \beta$

$$0 = 1482\cdot e^{-100(\pi+\beta)/2\pi 60} - 1000 + 470\sin(\pi + \beta - 1.308) \tag{5.43}$$

The solution for $\beta$ is $\beta \approx 16.2°$.

## 5.6 Three-Phase Semiconverter

Consider a three-phase semiconverter (Figure 5.5) supplying a d.c. motor. The armature current is considered ripple free and the commutation is instantaneous.

a. Draw the waveform of output voltage for $\alpha = 30°$.
b. Calculate the average (rms) voltage as a function of $\alpha$.
c. For a line voltage (rms) $V_L = 220$ V, f = 60 Hz, $R_a = 0.2\ \Omega$, $K_e\lambda_p = 4$ Wb, and an armature current $I_a = 50$ A, determine the motor speed for $\alpha = 30°$.

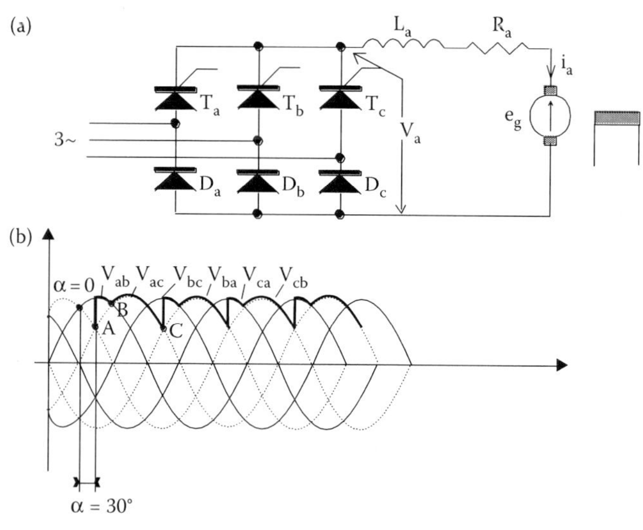


**FIGURE 5.5**
D.c. brush motor supplied through a three-phase semiconverter. (a) Electric circuit, (b) voltage wave forms.

*Solution*

a. As seen from Figure 5.5b if the thyristor $T_a$ is turned on with a delay angle $\alpha = 30°$ (point A) from the moment corresponding to $\alpha = 0$, it will conduct together with the diode $D_b$ for 30° as long as $V_{ab}$ is positive and higher than the other line voltages. At point B, $T_a$ will conduct together with the diode $D_c$ for 90° as $V_{ac} > V_{ab}$. At point C, after 120° of conduction, the thyristor $T_a$ is turned off and $T_b$ turned on and so on.

b. The expression for average voltage is

$$V_{av} = \frac{3}{2\pi}\left[\int_{\alpha}^{\pi/3} \sqrt{2}V_L \sin(\omega_1 t + \pi/3)d(\omega_1 t) + \int_{\pi/3}^{\alpha+2\pi/3} \sqrt{2}V_L \sin(\omega_1 t)d(\omega_1 t)\right]$$

$$= \frac{3\sqrt{2}V_L}{2\pi}(1+\cos\alpha) \tag{5.44}$$

For $\alpha > 60°$ the integration interval is from $\alpha$ to $\pi$ with the same final result. The output average voltage $V_{av}$ may not be negative and thus the converter may not be used as an inverter, confirming the single quadrant operation.

c. The motor voltage equation for ripple-free current is

$$\frac{3\sqrt{2}V_L}{2\pi}(1+\cos\alpha) = R_a I_a + K_e\lambda_p n \tag{5.45}$$

$$\frac{3\cdot\sqrt{2}\cdot 220}{2\pi}\left(1+\cos\frac{\pi}{6}\right) = 0.2\cdot 50 + 4\cdot n \tag{5.46}$$

$$n = 66.66 \text{ rps} = 4000 \text{ rpm} \tag{5.47}$$

## 5.7 Three-Phase Full Converter: Motor Side

A three-phase full converter (Figure 5.6) supplying a d.c. motor has the data: line voltage $V_L = 220$ V (rms), $K_e\lambda_p = 10$ Wb, $R_a = 0.2\ \Omega$, and neglect the ripples in the motor current.

a. For $L_s$ equal to zero, calculate the output average voltage as a function of the delay angle $\alpha$, and for $\alpha = 30°$, determine the motor speed for $I_a = 50$ A.
b. Considering the a.c. source inductance $L_s \neq 0$, determine the expression of the output voltage as a function of $\alpha$ and $I_a$ and calculate the motor speed for $I_a = 50$ A, $\alpha = 30°$, and $L_s = 1$ mH.
c. Calculate the overlapping angle at commutation for case b.

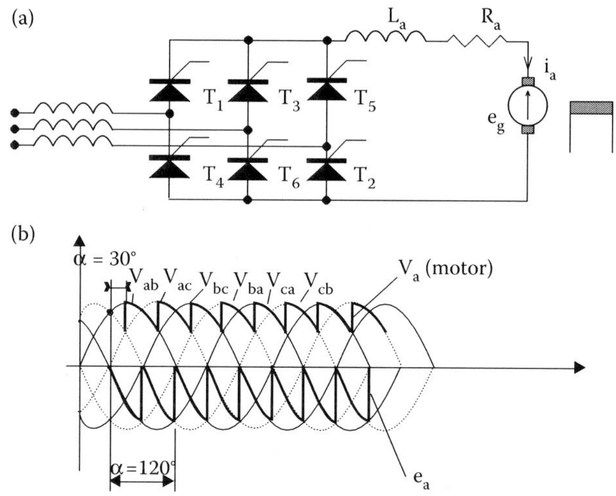


**FIGURE 5.6**
D.c. brush motor fed from a three-phase full converter. (a) Electric circuit, (b) voltage waveforms.

### *Solution*

a. The principle of operation is quite similar to that of the previous case. However, only a 60° conduction of each pair of thyristors occurs as shown in Figure 5.6.
   In the absence of a.c. source inductances $L_s$ ($L_s = 0$), the commutation is instantaneous. The average output voltage $V_{av}$ is

$$V_{av} = \frac{3}{\pi}\int_{\alpha+\pi/3}^{\alpha+2\pi/3} \sqrt{2}V_L \sin\omega_1 t \cdot d(\omega_1 t) = \frac{3\sqrt{2}V_L}{\pi}\cos\alpha \tag{5.48}$$

   Consequently, $V_{av} < 0$ ($V_{av} > 0$) depending on α. For $\alpha < 90°$ the average output voltage $V_{av}$ is positive, that is rectifier mode; for $\alpha > 90°$ $V_{av} < 0$ and thus the inverter mode is obtained. The motor current remains positive irrespective of α. A two-quadrant operation is obtained.
   The motor voltage equation is

$$\frac{3\sqrt{2}V_L}{\pi}\cos\alpha = R_a I_a + K_e \lambda_p n \tag{5.49}$$

$$n = \left(\frac{3\cdot 220\sqrt{2}\cos(\pi/6)}{\pi} - 0.2\cdot 50\right)/10 = 24.74 \text{ rps} = 1484.58 \text{ rpm} \tag{5.50}$$

b. The effect of a.c. source inductances $L_s$ on current commutation is shown in Figure 5.7 for the case of constant motor current (similar to the diode full converter [Chapter 3]). The commutation between $T_1$ and $T_5$ is not instantaneous anymore;

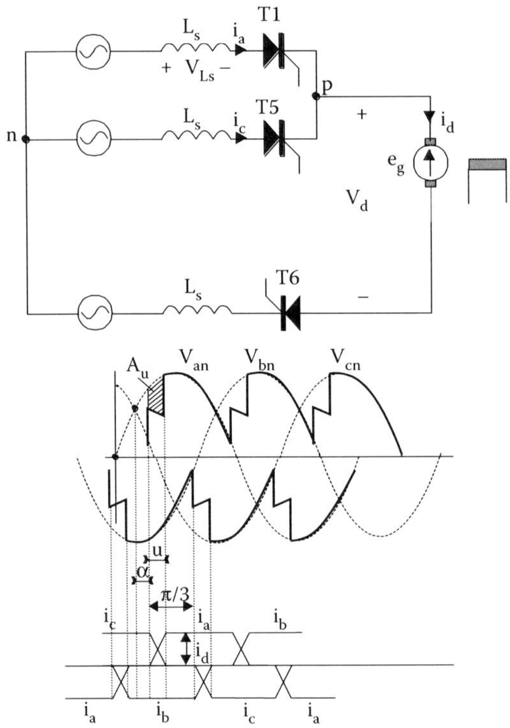


**FIGURE 5.7**
The current commutation in the presence of a.c. source-side inductances $L_s$.

an overlapping angle u occurs. The effect of actual commutation is, in fact, a reduction of output voltage $V_d$ determined by the area $A_u$ when a kind of short circuit occurs between phases a and c.

$$A_u = \int_{\alpha}^{\alpha+u} V_{Ls} d(\omega_1 t) = \int_{\alpha}^{\alpha+u} L_s \cdot \frac{di_a}{dt} d(\omega_1 t) = \omega L_s I_d \tag{5.51}$$

This is so since the current in phase a is zero for $\omega_1 t = \alpha$ and equal to $I_d$ for $\omega_1 t = \alpha + u$. The average voltage is reduced by $3/\pi A_u$.

Finally, the average voltage is equal to that already found for $L_s = 0$: $-3/\pi\ A_u$

$$V_d = \frac{3\sqrt{2}V_L}{\pi}\cos\alpha - \frac{3}{\pi}\omega_1 L_s I_d \tag{5.52}$$

As we can see, the angle u is not required to calculate the output voltage $V_d$. However, it is needed to ensure reliable operation in the inverter mode ($\alpha > 90°$). The second Kirchoff law for phases a and c during commutation provides the equation

$$V_{an} - L_s \cdot \frac{di_a}{dt} = V_{cn} - L_s \cdot \frac{di_c}{dt} \tag{5.53}$$

$$\text{with } i_c + i_a = i_d, \quad \text{that is } \frac{di_a}{dt} = -\frac{di_c}{dt} \tag{5.54}$$

Thus

$$V_{an} - V_{cn} = V_{ac} = 2L_s \cdot \frac{di_a}{dt} \tag{5.55}$$

$$\text{with } V_{ac} = \sqrt{2}V_L \sin\omega_1 t \tag{5.56}$$

$$\int_{\alpha}^{\alpha+u} V_{ac} d(\omega_1 t) = 2\omega_1 L_s I_d \tag{5.57}$$

and finally

$$\cos(\alpha + u) = \cos\alpha - \frac{2\omega_1 L_s I_d}{\sqrt{2}V_L} \tag{5.58}$$

Thus, knowing $\alpha$ and $I_d$, we may calculate the overlapping angle u.

In the inverter mode ($\alpha > 90°$), the commutation should be finished before $\alpha + u = \pi$ in order to allow the turnoff time $t_{off}$ required for the recombination of charged particles in the thyristors ($(\pi - (\alpha + u))/\omega_1 > t_{off}$) for negative voltage along the turning-off thyristors.

The rectifier voltage $V_d$ for $\alpha = \pi/6$ and $I_d = 50$ is

$$V_d = \frac{3\sqrt{2}\cdot 220}{\pi} \cos\frac{\pi}{6} \; \frac{3}{\pi} \; 2 \cdot \pi \; 60 \cdot 50 \; 10^{-3} = 239.43 \text{ V} \tag{5.59}$$

$$\begin{gathered} V_d = R_a I_a + K_e \lambda_p n \\ 239.43 = 0.2 \cdot 50 + 10 \cdot n; \quad n = 22.943 \text{ rps} = 1376.58 \text{ rpm} \end{gathered} \tag{5.60}$$

The reduction of output voltage for the same $\alpha$ due to commutation with 18 V has contributed to a notable reduction in speed, for 50 A, from 1484.58 to 1376.58 rpm.

c. To calculate the overlapping angle u, we use the expression derived above:

$$\begin{gathered} \cos(\alpha + u) = \cos\alpha - \frac{3\omega_1 L_s I_d}{\sqrt{2}V_L} = \cos\frac{\pi}{6} - \frac{3 \cdot 2\pi \cdot 60 \cdot 50 \cdot 10^{-3}}{\sqrt{2} \cdot 220} \\ 30^{\circ} + u = 46.78^{\circ}; \quad u = 16.78^{\circ} \end{gathered} \tag{5.61}$$

A considerable value for u has been obtained.

## 5.8 Three-Phase Full Converter: Source-Side Aspects

For the three-phase full converter and d.c. motor in the previous example with $L_s = 0$ for $\alpha = 0°$, $\alpha = 45°$ with $I_d = 50$ A, determine:

a. The waveforms of a.c. source current.
b. The rms of fundamental source current and of the total source current.
c. The DPF for $L_s = 0$ and $L_s = 1$ mH for $\alpha = 45°$.
d. Calculate the line voltage and voltage distortion due to $L_s = 1$ mH and $\alpha = 45°$.

### *Solution*

a. As the motor armature current is considered constant, the a.c. source current is rectangular as shown in Figure 5.8a and b.

The presence of a.c. source inductances $L_s \neq 0$ leads to the overlapping angle $u \neq 0$.

b. The rms value of the current fundamental, $I_{a1}$, is

$$I_{a1} = \frac{2\sqrt{3}}{\pi} \cdot \frac{I_d}{\sqrt{2}} = \frac{\sqrt{6}}{\pi} \cdot I_d = 0.78 \cdot 50 = 39 \text{ A} \quad (5.62)$$

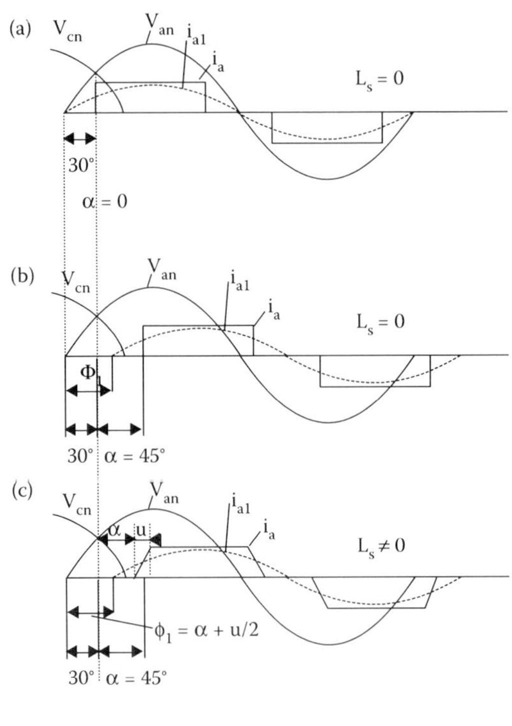


**FIGURE 5.8**
A.c. source currents in a three-phase full converter and constant motor current: (a) for $\alpha = 0$, $L_s = 0$; (b) for $\alpha = 45°$, $L_s = 0$; (c) for $\alpha = 45°$, $L_s = 1$ mH.

For the total a.c. source current, which is made of rectangular 120° wide "blocks"

$$I_a = \sqrt{\frac{2}{3}} \cdot I_d = 0.816 \cdot 50 = 40.8 \text{ A} \tag{5.63}$$

c. In the absence of $L_s$, the DPF angle is equal to $\alpha$ and thus the DPF is

$$DPF = \cos\varphi_1 = \cos\alpha = \begin{cases} 1 & \text{for } \alpha = 0 \\ 0.709 & \text{for } \alpha = 45^\circ \end{cases} \tag{5.64}$$

In the presence of $L_s$, the DPF is approximately

$$DPF = \cos\left(\alpha + \frac{u}{2}\right) \tag{5.65}$$

To calculate u for $\alpha = 45°$, we use Equation 5.58:

$$\begin{aligned} \cos(\alpha + u) &= \cos\alpha - \frac{3\omega_1 L_s I_d}{\sqrt{2}V_L} = 0.707 - \frac{3 \cdot 2\pi \cdot 60 \cdot 50 \cdot 10^{-3}}{\sqrt{2} \cdot 220} = 0.5248 \\ 45 + u &= 58.34^\circ; \quad u = 13.34^\circ \end{aligned} \tag{5.66}$$

Finally,

$$DPF = \cos\left(45 + \frac{13.34}{2}\right) = 0.62$$

The current commutation in the presence of $L_s$ produces a further reduction of the DPF.

In general, the DPF decreases with $\alpha$ increasing, which constitutes a notable disadvantage of phase delay a.c.–d.c. converters. Special measures are required to improve DPF with $\alpha$ increasing.

d. The a.c. source current overlapping during commutation produces notches in the line voltage. From Figure 5.7, we may obtain $V_{ab} = V_{an} – V_{bn}$ of the waveform shown in Figure 5.9.

Considering u as small, the deep notch depth is considered equal to $\sqrt{2}V_L \sin\alpha$ and thus the notch width u is approximately

$$u = \frac{A_u}{\text{notch depth}} = \frac{2\omega_1 L_s I_d}{\sqrt{2}V_L \sin\alpha} \tag{5.67}$$

The depth of shallow notches is considered half of that of deep notches. IEEE standard 519–1981 suggests the limitation of line notches to 250 μs (5.4. electrical degrees) and of deep notch depth to 70% of rated peak line voltage in order to perform satisfactorily.

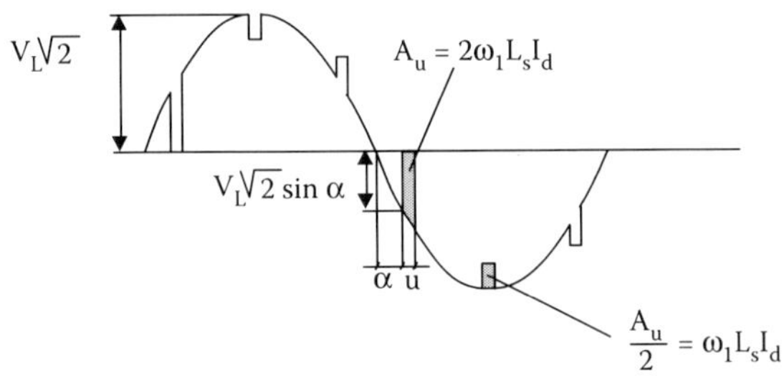


**FIGURE 5.9**
A.c. source-line voltage notches due to $L_s$ (commutation).

Special filtering is required to cope with more recent standards. The voltage distortion due to notches depends on the harmonics currents $I_v$ and the a.c. source inductance $L_s$

$$\text{voltage\%THD} = \frac{\sqrt{\left[\sum_{v>1} (I_v \cdot v\omega_1 L_s)^2\right]}}{V_{\text{phase (fundamental)}}} \cdot 100 \tag{5.68}$$

with $I_v \approx I_{a1}/v = \sqrt{6}/\pi \cdot I_d/v$ due to the almost rectangular form of a.c. source current.

## 5.9 Dual Converter: Four-Quadrant Operation

A high-power d.c. motor drive sometimes has to undergo four-quadrant operation. Two full converters are connected back to back for this purpose (Figure 5.10).

a. Assuming that the converters are ideal and produce pure d.c. output voltages with one converter as rectifier and the other as inverter, calculate the relationship between the delay angles $\alpha_1$ and $\alpha_2$ in the two converters.
b. With $\alpha_1 + \alpha_2$ as above, calculate the circulating current between the two converters and show the voltage and current actual output waveforms. The numerical data are $V_L = 220$ V, $\omega_1 = 377$ rad/s, $L = 10$ mH, $\alpha_1 = 60°$.

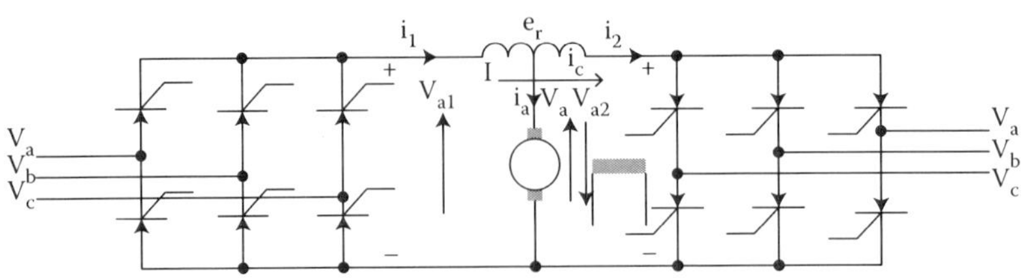


**FIGURE 5.10**
Dual converter with circulating current supplying a d.c. brush motor.

### *Solution*

a. In an ideal dual converter, the voltages produced by the two full converters should be equal and opposite.

By now we know that

$$V_{a1} = V_{max} \cdot \cos\alpha_1$$
$$V_{a2} = V_{max} \cdot \cos\alpha_2 \tag{5.69}$$

with $V_a = V_{a1} = -V_{a2}$, it follows that $\cos\alpha_1 + \cos\alpha_2 = 0$.

Hence, $\alpha_1 + \alpha_2 = 180°$ (Figure 5.11). In the ideal converter, the load voltage is equal to the converter output voltages and thus the current may flow equally through either converter.

b. In a real nonideal converter, each converter produces a voltage with ripples. The ripple voltages of the two converters are out of phase (Figure 5.12). The instantaneous voltage difference produces a circulating current which is limited through a reactor L.

With

$$V_{a,b,c} = \frac{V_L\sqrt{2}}{\sqrt{3}}\sin\left[\omega_1 t - (i-1)\frac{2\pi}{3}\right] \tag{5.70}$$

during the interval

$$\frac{\pi}{6} + \alpha_1 < \omega_1 t < \frac{\pi}{6} + \alpha_1 + \frac{\pi}{3} \tag{5.71}$$

$$V_{a1} = V_a - V_b$$
$$V_{a2} = -(V_c - V_b) \tag{5.72}$$
$$e_r = V_{a1} - V_{a2} = V_a + V_c - 2V_b = -3V_b$$

The circulating current $i_c$ is

$$i_c = \frac{1}{\omega_1 L}\int_{\alpha_1+(\pi/6)}^{\omega_1 t} e_r dt = \frac{\sqrt{6}V_L}{\omega_1 L}\left[\cos\left(\omega_1 t - \frac{2\pi}{3}\right) - \cos\left(\alpha_1 - \frac{\pi}{2}\right)\right] \tag{5.73}$$

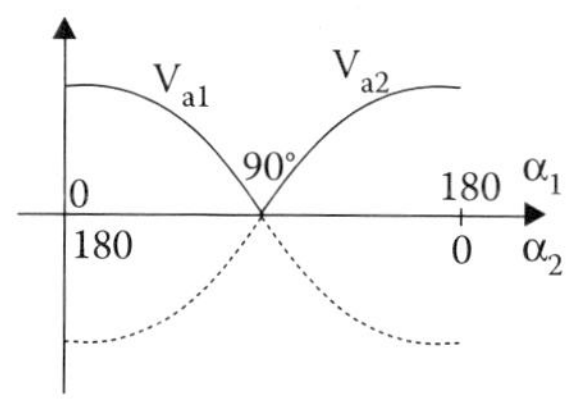


**FIGURE 5.11**
Ideal dual converter voltages.

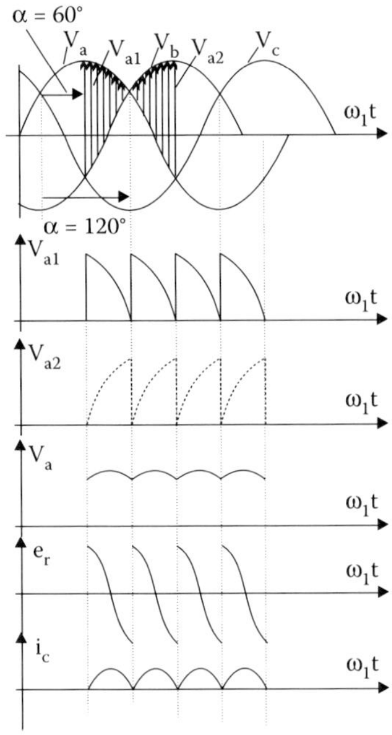


**FIGURE 5.12**
The real dual converter.

When the motor current is zero, the converter currents are equal to the circulating current $i_1 = i_2 = i_c$ ($i_a = 0$). Consequently, the converters have a continuous current though the load current is zero. However, for $\alpha_1 = 60°$ and $\alpha_2 = 120°$, the peak value of circulating current (Equation 5.73) occurs at $\omega_1 t = 2\pi/3$:

$$(i_c)_{peak} = \frac{V_L\sqrt{6}}{\omega_1 L}\left[1-\cos\frac{\pi}{6}\right] = \frac{220\sqrt{6}}{2\cdot\pi\cdot 60\cdot 10^{-2}}[1-0.867] = 19.02\text{ A} \tag{5.74}$$

If the load current $i_a$ is constant (no ripples), the first converter ($\alpha_1 = 60°$) carries $i_a + i_c$ while the second converter ($\alpha_2 = 120°$) has the circulating current $i_c$ only. Thus, the first converter is "overloaded" with the circulating current. However, for low load current, the discontinuous current mode in the converters is avoided as shown above. This could be an important advantage in terms of control performance.

## 5.10 A.C. Brush Series (Universal) Motor Control

The universal motor is a.c. voltage supplied but it is still a brush (mechanical comutator) series connected motor (Chapter 4), (Figure 5.13).

The motor voltage equation is (Chapter 4)

$$L_{ae}\frac{di}{dt} \approx v - (R_{ae} + R_{core} + k_e k_\Phi n)i; \quad T_e = \frac{k_e k_\Phi i^2}{2\pi} \tag{5.75}$$

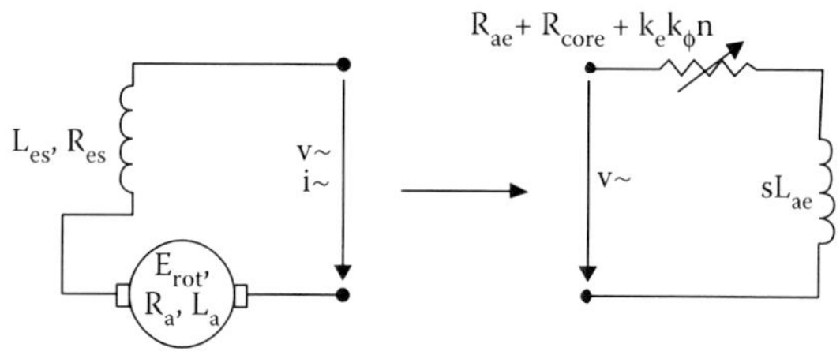


**FIGURE 5.13**
The equivalent scheme of universal motor.

The mechanical equation is straightforward

$$J2\pi\frac{dn}{dt}\approx\frac{k_ek_\Phi i^2}{2\pi}-Bn-T_{load} \tag{5.76}$$

All core losses are lumped into the stator while in the motion equation they are left out for simplicity. Magnetic saturation of the magnetic circuit is considered constant or absent ($L_{ae}$ = ct, $k_\Phi$ = const.).

The investigation of transients with any input voltage waveform and load torque perturbation is rather straightforward for such a second-order system.

The motion-induced voltage (e.m.f.), $E_{rot}$, may be considered as an additional, variable resistance voltage drop ($k_ek_\Phi n_i$) (5.75).

The torque is essentially proportional to current squared and, for a given voltage, it is proportional to voltage squared.

An a.c. voltage changer is required.

At start (n = 0), the machine is represented by a small resistance $R_{ae}+R_{core}$ plus the inductance $L_{ae}$. So the electric circuit of the machine at start is strongly inductive.

On the other end, at high speed, due to the large e.m.f. ($E_{rot}$)—Equation 5.75—the machine equivalent circuit is notably resistive.

The equivalent circuit is composed simply of an equivalent inductance $L_{ae}$—relatively constant—and a resistance $R_{en}=(R_{ae}+R_{core}+k_ek_\Phi n)$ which strongly increases with speed (Figure 5.13).

A typical a.c. voltage changer—which may be used also as a power switch—may be obtained by using two antiparalleled thyristors, which may be assembled into a single bidirectional power switch—the Triac (Figure 5.14).

The Triac is turned on by applying a short resistive current pulse on the thyristor gate. The thyristor will turn off when the current decays naturally to zero. Then the thyristor for the negative voltage polarity, the antiparalleled thyristor, is turned on. As the turn-on angle α—with respect to zero crossing of the voltage waveform—increases, the average voltage decreases. But the average voltage also depends on the machine equivalent resistance $R_e$, which increases with speed.

Let us consider separately the two voltage polarities (Figure 5.14).

The difference between supply voltage v(t) and the resistance voltage $R_e(n)i$ is equal to the voltage drop along the machine inductance $L_{ae}$:

$$v_{Lae}(t)=L_{ae}\frac{di}{dt}=v_s(t)-R_e(n)i \tag{5.77}$$

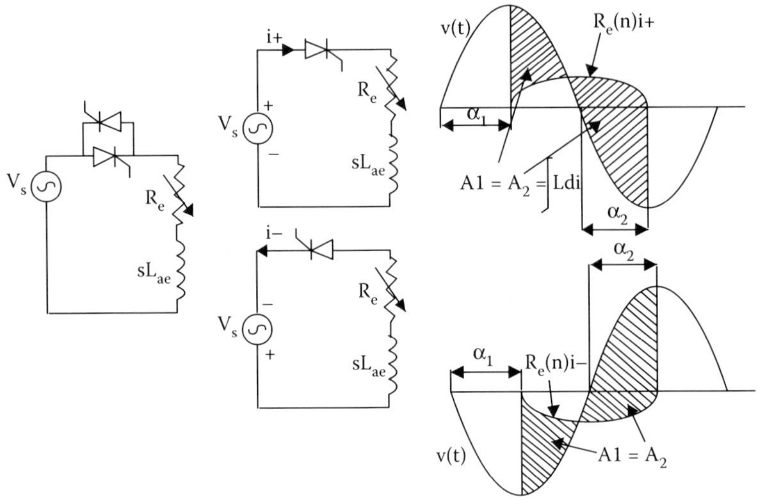


**FIGURE 5.14**
The Triac variac and the current.

$$\int_{\alpha 1}^{\pi+\alpha 2} v_{Lae}(t)dt = 0 \tag{5.78}$$

The current becomes zero when areas A1 and A2 on Figure 5.14 become equal to each other.

It is evident that for the universal motor the current becomes zero after the voltage changes polarity by the angle $\alpha_2$. However, this delay angle decreases with speed as the circuit becomes more and more resistive as the e.m.f. increases.

There may be situations with $\alpha_2 > \alpha_1$ and $\alpha_2 < \alpha_1$. In the first case, the current will be continuous while in the second case it will be discontinuous.

In general, $\alpha_1$ has to be larger than the DPF angle $\varphi_1$ of the equivalent circuit, to secure the fundamental output voltage control by the Triac.

So, at low speed, $\alpha_1$ should be large to reduce the voltage fundamental and will decrease with speed. But the output voltage is full of harmonics as only parts of the sinusoid are active. Also there are harmonics in the machine (input) current (Figure 5.15). An input power filter is required.

For constant speed and given delay angle $\alpha_1$, Equation 5.77 has an analytical solution:

$$i = Ae^{-R_e(n)/L_{ae}t} + B\sin(\omega_1 t - \phi_1);\quad B = \frac{V\sqrt{2}}{\sqrt{R_e^2(n) + \omega_1^2 L_{ae}^2}};\quad \phi_1 = \tan^{-1}\frac{\omega_1 L_{ae}}{R_e(n)} \tag{5.79}$$

with $i = 0$ at $\omega_1 t = \alpha_1$.

Consequently

$$A = -B\sin(\alpha_1 - \phi_1)e^{R_e(n)\alpha_1/L_{ae}\omega_1} < 0 \tag{5.80}$$

So, $\alpha_1 > \varphi_1$ to secure predominantly positive sinusoidal current for positive voltage polarity.

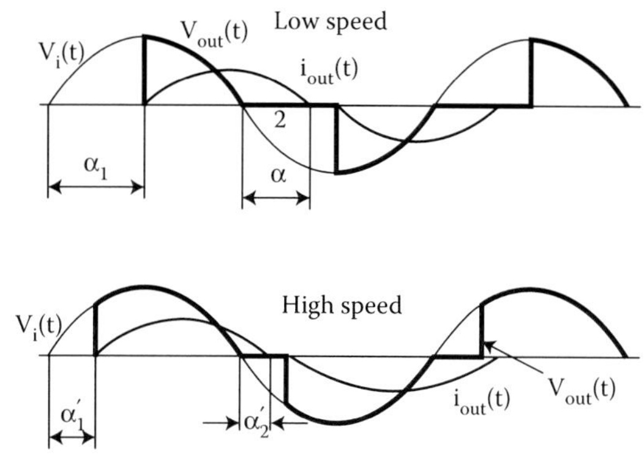


**FIGURE 5.15**
Output voltage and current waveforms at low and high speed.

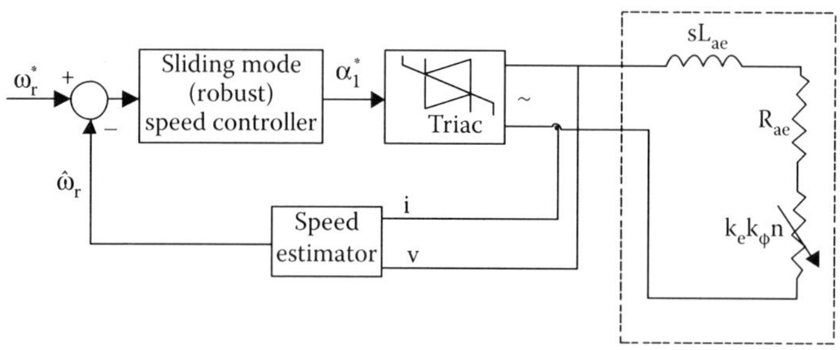


**FIGURE 5.16**
Generic control system for universal motors.

The angle $\alpha_2$ within negative polarity of voltage where the current decays to zero is obtained by applying Equation 5.78. The obtained nonlinear equation may be solved numerically for $\alpha_2$.

It is also evident that the output voltage fundamental $V_{1out}$ is nonlinearly dependent on $(\pi/2 - \alpha_1)$ or on $(1 - \sin \alpha_1)$.

A robust speed controller is required to overcome this difficulty.

A generic control system for the universal motor is presented in Figure 5.16.

A speed estimator is required for closed-loop control. In less demanding applications, a well-commissioned feedforward open-loop control $\alpha_1^*(\omega_r^*)$ with limited reference speed ramps may be used to avoid the speed estimator and the speed loop control.

*Note*: The low cost of universal motor drive at low power levels has secured its presence today in some home appliances and hand-held tools (with natural limited operation life) despite the occurrence of brushless drives. For more details, see Reference 8.

## 5.11 Summary

- Controlled rectifiers, also known as phase delay rectifiers, of various basic configurations with zero and nonzero source inductances have been presented in interaction with d.c. brush motors at constant speed.

- Single-phase and three-phase full converters provide for bidirectional power flow for positive output current and positive and negative average output voltages. Consequently, for positive speeds, regenerative braking is possible only if the field current direction is changed as negative e.m.f. is required.
- Dual converters with circulating currents, requiring 12 PESs in three-phase configurations provide for four-quadrant operation while avoiding the discontinuous current mode at the price of overloading one of the two converters.
- The source inductance produces an overlapping during phase commutation, which results in a kind of resistive-like output voltage drop in the rectifier.
- The PF decreases with a decrease in d.c. output voltage in phase delay rectifiers. Forced commutation or special complex configurations (for high powers) may solve this problem [3–6].
- All controlled rectifiers produce current harmonics and voltage notches on the a.c. source-side and special input filters are required to reduce them to acceptable standardized levels.

## 5.12 Problems

5.1 A three-phase rectifier with controlled null (Figure 5.17) supplies a load made of a resistance $R_s$ and an inductance $L_s$. The a.c. supply phase voltage is 120 V (rms), $R_s = 10\ \Omega$, and the transformer ratio $K = w_1/w_2 = 2$. For $L_s = 0$ and $L_s = \infty$, determine:

a. The output and transformer secondary voltages and current waveforms for the delay angle $\alpha_1 = \pi/3$

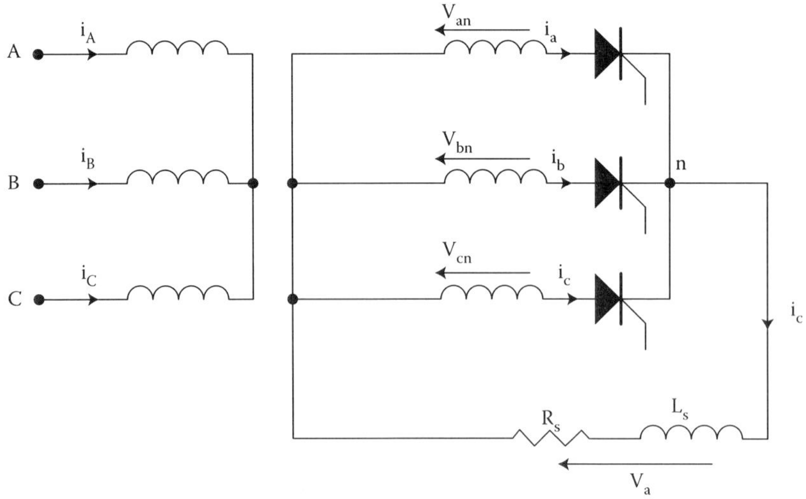


FIGURE 5.17
Three-phase rectifier with controlled null.

b. The average values of output voltage and current
c. The waveform of transformer primary current

5.2 A d.c. motor is supplied through a single-phase semiconverter with a freewheeling diode (Figure 5.18). Depict the waveforms of the induced voltage $e_g$ and motor voltage $V_a$ for discontinuous and continuous current.

5.3 For problem 5.2, find analytical expressions of armature current for discontinuous and continuous modes for constant speed, with saturation neglected. Find the condition to determine the minimum value of $L_a$ for which the current is still continuous. Determine an expression of average output voltage for the continuous current.

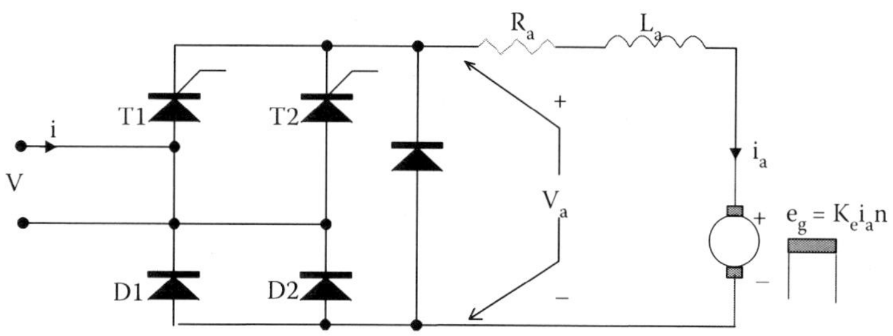


**FIGURE 5.18**
D.c. motor supplied through a single-phase semiconverter.

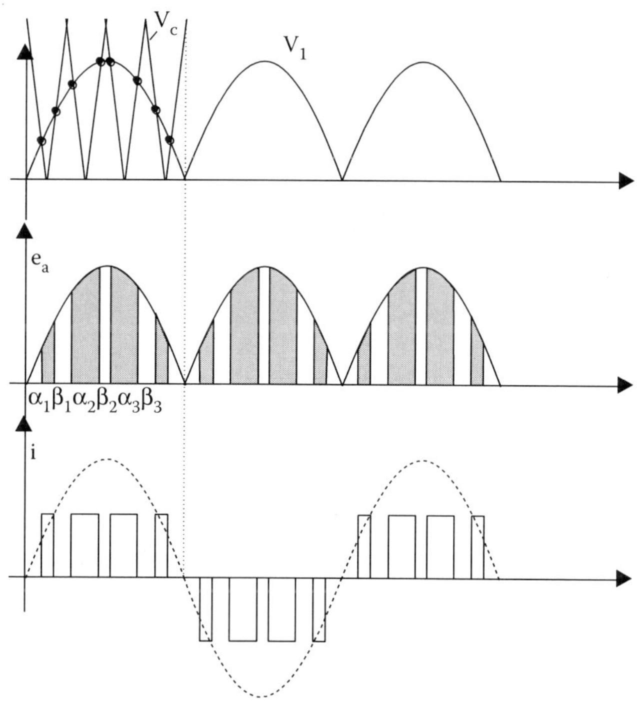


**FIGURE 5.19**
Single-phase full converter with PWM control.

5.4 A single-phase full converter [7] uses power transistors and is controlled through PWM with a triangular carrier with n pulses per semiperiod (Figure 5.19). Obtain the expressions of average output voltage, current harmonics, PF, DPF and current harmonic factor (HF).

## References


1. N. Mohan, T.M. Underland, W.P. Robbins, *Power Electronics*, second edition, Chapter 6, John Wiley & Sons, New York, 1995.
2. G.K. Dubey, *Power Semiconductor Controlled Drives*, Chapter 3, Prentice-Hall, Englewood Cliffs, New Jersey, 1989.
3. T. Ohnuki, O. Miyashita, T. Haneyoski, E. Ohtsuji, High power factor PWM rectifiers with an analog PWM prediction controller, *IEEE Trans.*, vol. PE-11, no. 3, 1996, pp. 460–465.
4. M.O. Eissa, S. Leeb, G.C. Verghese, A.M. Stankovic, Fast controller for unity power factor PWM rectifier, *IEEE Trans.*, vol. PE-11, no. 1, 1996, pp. 1–6.
5. N. Akagi, New trends in active filters for power conditioning, *IEEE Trans*, vol. IA-32, no. 6, 1996, pp. 1312–1327.
6. I.O. Krah, I. Holtz, Total compensation of line-side switching harmonics in converter-fed A.C. locomotives, *IEEE Trans.*, vol. 31, no. 6, 1995, pp. 1264–1273.
7. M. Malinowski, M. Kazmierkowski, Control of three phase PWM rectifiers, *Control in Power Electronics*, Chapter 11, eds. M. Kazmierkowski, R. Khrishnan, F. Blaabjerg, Academic Press, 2002.
8. A. di Gerlando, R. Perini, A model of the operation analysis of high speed universal motor with triac regulated mains voltage supply, *Proceedings of the Symposium on Power Electronic and Electric Drives, Automation Motion*, Ravello, Italy, 2002, pp. C407–C412.

# 6

# Chopper-Controlled D.C. Brush Motor Drives

## 6.1 Introduction

The d.c. chopper is a d.c. to d.c. PEC with forced commutation. It is used for armature voltage control in d.c. brush motor drives. D.c. sources to supply d.c. choppers are batteries or diode rectifiers with output filters so typical for urban electric transportation systems or to low-power d.c. brush motor drives. Thyristors, bipolar power transistors, MOSFETs, or IGBTs are used in d.c. choppers.

The basic configurations are shown in Table 6.1 and they correspond to single-, two-, or four-quadrant operation.

The first-quadrant chopper (Figure 6.1) is operated by turning on the PES for the interval $t_{on}$, when the supply voltage is connected to the load. During the interval $t_{off}$, when the main switch is off, the load current flows through the freewheeling diode $D_1$. The output voltage $e_a$ is shown in Figure 6.1.

The average voltage $V_{av}$ is

$$V_{av} = e_a \cdot \frac{t_{on}}{T} \leq V_0 \tag{6.1}$$

That is, a step-down chopper.

Constant frequency (constant T) control is preferred in order to improve the input filter operation and reduce the possibility of discontinuous current mode (Figure 6.1b) operation.

The voltage equation for constant speed is

$$V_0 = R_a \cdot i_a + L_a \cdot \frac{di_a}{dt} + e_g; \quad e_g = K_e \cdot \lambda_p \cdot n \quad \text{for } 0 \leq t \leq t_{on} \tag{6.2}$$

$$\begin{gathered} 0 = R_a \cdot i_a + L_a \cdot \frac{di_a}{dt} + e_g; \quad t_{on} \leq t \leq t_1; \quad i_a(t_1) = 0 \\ t_1 < T \text{ for discontinuous mode} \end{gathered} \tag{6.3}$$

For continuous current mode $t_1 = T$ and $i_a(T) = i_a(0) \neq 0$ for steady state.

For the d.c. brush series motor

$$e_g = K_{ei} \cdot i_a \cdot n + K_{rem} \cdot n \tag{6.4}$$

**TABLE 6.1**

Single-Phase Chopper Configurations for the d.c. Brush Motors

| Type | Chopper Configuration | $e_a$–$i_a$ Characteristics | Function |
|---|---|---|---|
| First-quadrant (step-down) choppers | 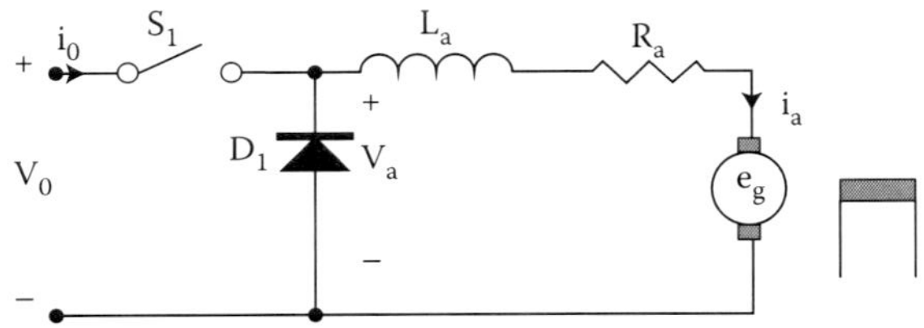  | 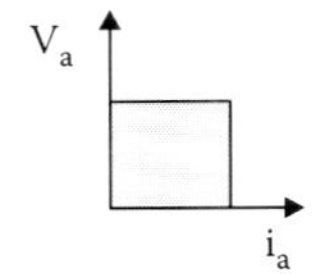  | $V_a = V_0$ for $S_1$ on<br>$V_a = 0$ for $S_1$ off and $D_1$ on |
| Second-quadrant, regeneration (step-up) chopper | 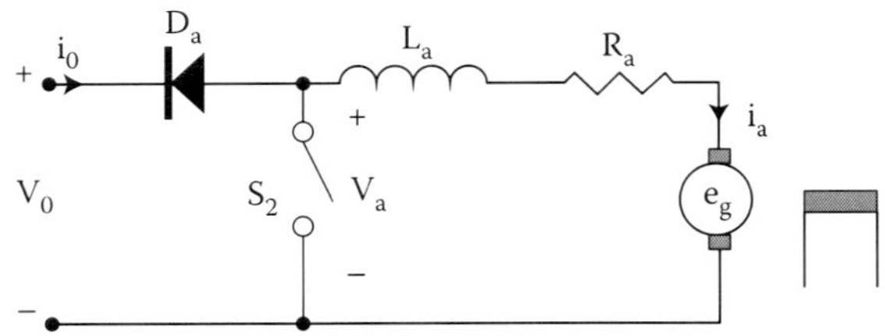  | 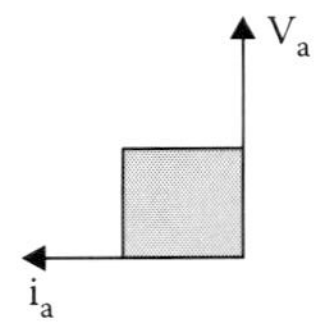  | $V_a = 0$ for $S_2$ on<br>$V_a = V_0$ for $S_2$ off and $D_2$ on |
| Two-quadrant chopper | 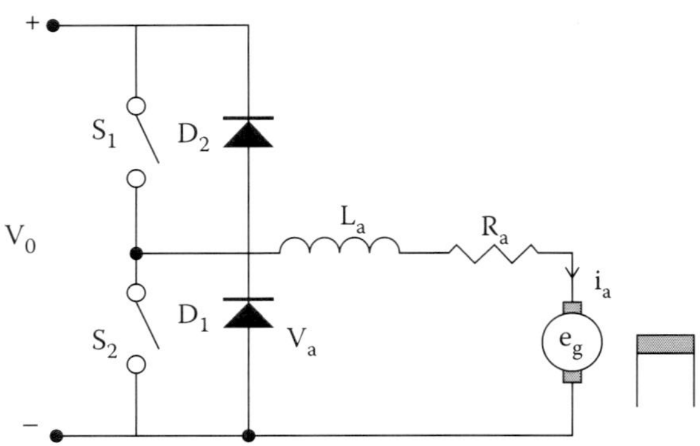  | 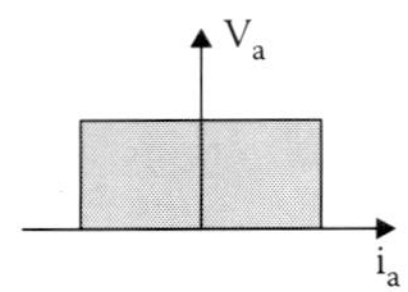  | $e_a = V_0$ for $S_1$ or $D_2$ on<br>$e_a = V_0$ for $S_2$ or $D_1$ on<br>$i_a > 0$ for $S_1$ or $D_1$ on<br>$i_a < 0$ for $S_2$ or $D_2$ on |

(*Continued*)

**TABLE 6.1 (*Continued*)**

Single-Phase Chopper Configurations for the d.c. Brush Motors

| Type | Chopper Configuration | $e_a$–$i_a$ Characteristics | Function |
|---|---|---|---|
| Two-quadrant chopper | 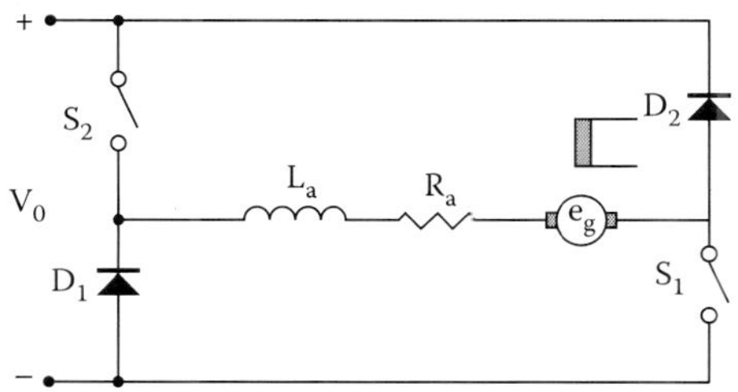 | 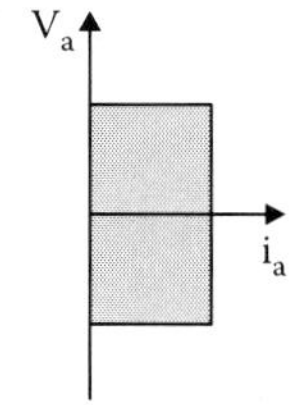 | $V_a = +V_0$ for $S_1$ and $S_2$ on<br>$V_a = -V_0$ for $S_1$ and $S_2$ off and $D_1$ and $D_2$ on |
| Four-quadrant chopper | 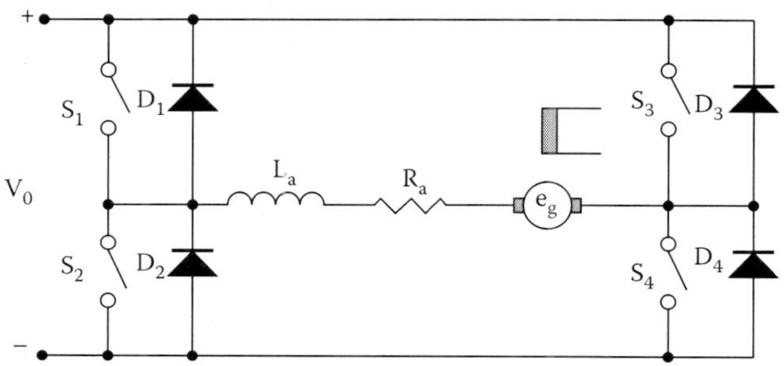 | 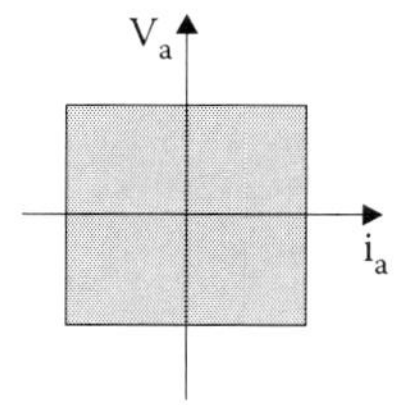 | $S_4$ on and $S_3$ off $S_1$ and $S_2$ operated<br>$V_a > 0$ $i_a$—reversible<br>$S_2$ on and $S_1$ off $S_3$ and $S_4$ operated<br>$V_a < 0$ $i_a$—reversible |

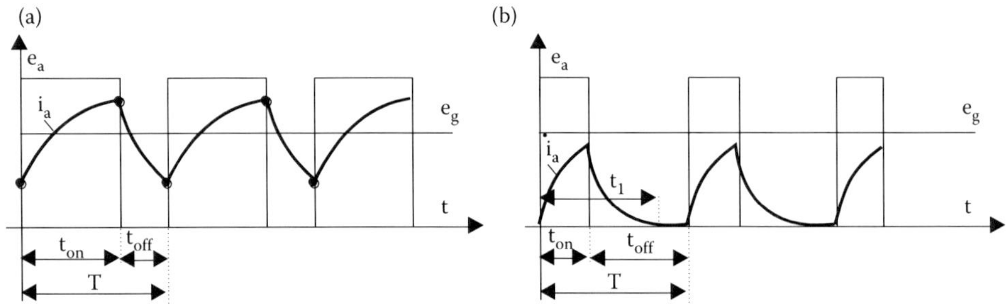


**FIGURE 6.1**
First-quadrant chopper operation: (a) continuous mode, (b) discontinuous mode.

In Equation 6.4, $K_{rem}$ refers to the remnant flux while the magnetization curve of the machine is considered linear.

The average output voltage for the discontinuous mode may be determined noting that the motor voltage is then equal to its e.m.f. $e_g$.

$$V_{av} = V_0 \frac{t_{on}}{T} + e_g \cdot \frac{T - t_1}{T}; \quad t_1 \le T \tag{6.5}$$

The output current expressions are obtained from Equations 6.2 and 6.3

$$i_a = A \cdot e^{-t \cdot (R_a/L_a)} + \frac{V_0 - e_g}{R_a}; \quad \text{for } 0 \le t \le t_{on} \tag{6.6}$$

$$i_a' = A' \cdot e^{-(t - t_{on}) \cdot (R_a/L_a)} - \frac{e_g}{R_a}; \quad \text{for } t_{on} \le t \le t_1 \tag{6.7}$$

The continuity condition is

$$i_a(t_{on}) = i_a'(t_{on}) \tag{6.8}$$

The average output current $i_{av}$ is

$$i_{av} = \frac{\int_0^{t_{on}} i_a dt + \int_{t_{on}}^{t_1} i_a' dt}{T} \tag{6.9}$$

For the second-quadrant chopper (Table 6.1) the d.c. motor e.m.f. $e_g$ with $S_2$ on produces a current rise in inductance $L_a$

$$R_a \cdot i_a + L_a \cdot \frac{di_a}{dt} = -e_g; \quad \text{for } 0 \le t \le t_{on}; \quad i_a(0) = 0 \tag{6.10}$$

When $S_2$ is turned off, the energy stored in the inductor is sent back to the source as long as $V_0 < V_a$

$$e_g - V_0 = R_a \cdot (-i_a') + L_a \cdot \frac{d(-i_a')}{dt}; \quad t_{on} \le t \le T \tag{6.11}$$

with the solution

$$i_a = -\frac{e_g}{R_a} + B \cdot e^{-t \cdot (R_a/L_a)} + i_{a0} \tag{6.12}$$

$$i_a' = +\frac{V_0 - e_g}{R_a} + B' \cdot e^{-(t - t_{on}) \cdot (R_a/L_a)} \tag{6.13}$$

The boundary conditions are

$$i_a(t_{on}) = i_a'(t_{on}),\ i_a(0) = i_{a0} \quad \text{and} \quad i_a'(T) = i_{a0} \tag{6.14}$$

It is thus possible with $e_g < V_0$ to retrieve the energy back from the d.c. brush motor by using the inductor $L_a$ as an energy sink (Figure 6.2).

The two-quadrant chopper (Table 6.1) is a combination of one first and one second-quadrant chopper. Finally, two-quadrant choppers are combined to obtain a four-quadrant chopper.

As the chopper is an on–off switch, the source current is chopped (Figure 6.3). This makes the peak input power demand high. Also, the supply current (Figure 6.3) has harmonics that produce voltage fluctuations, signal interference, etc.

An LC input filter (Figure 6.4) will provide a path for the ripple current such that only (approximately) the average current is drawn from the supply.

The nth harmonic current $i_n$ in the supply (Figure 6.4b) is

$$i_n = \frac{X_c/n}{(nX_L - X_C/n)} \quad I_{sn} = \frac{I_{sn}}{(nf_{ch}/f_r)^2 - 1} \tag{6.15}$$

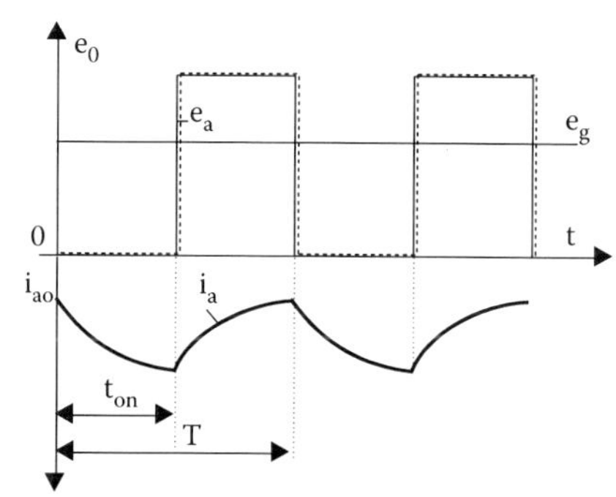


**FIGURE 6.2**
Second-quadrant chopper operation.

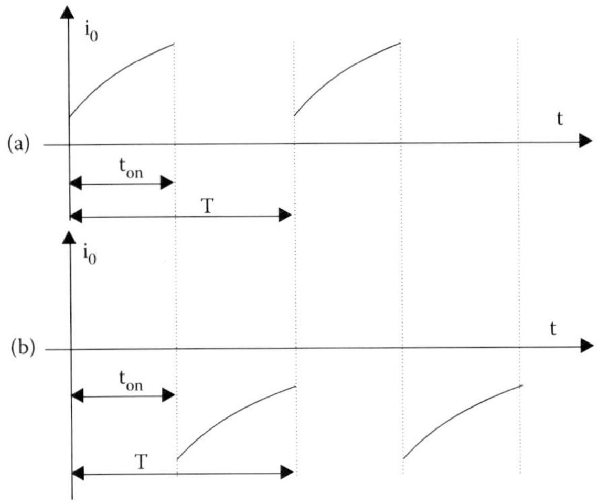


**FIGURE 6.3**
Source-current waveforms: (a) first-quadrant operation, (b) second-quadrant operation.

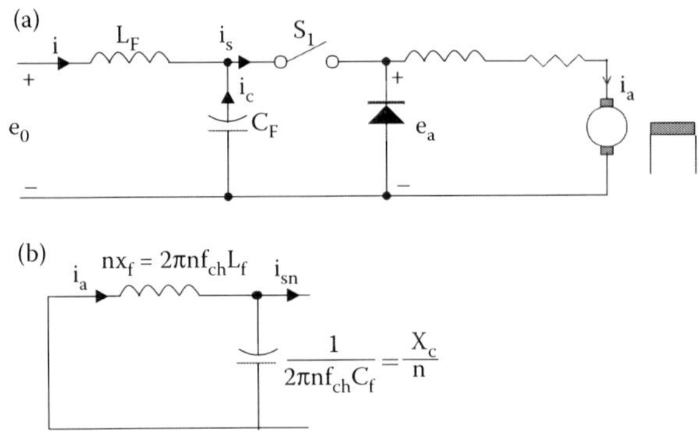


**FIGURE 6.4**
First-quadrant chopper with LC input filter: (a) basic circuit, (b) equivalent circuit for nth harmonic.

where $f_{ch}$ is the chopping frequency ($f_{ch} = 1/T$) and $f_r$ is the resonance frequency of the filter $f_r = 1/2\pi\sqrt{LC}$. To avoid resonance, $f_{ch} \geq (2-3)f_r$. Given the variety of configurations in Table 6.1, we will proceed directly with numerical examples in parallel with theoretical development to facilitate quick assimilation of knowledge.

## 6.2 First-Quadrant (Step-Down) Chopper

A d.c. brush motor with permanent magnet excitation with the data $R_a = 1\ \Omega$, $K_e\lambda_p = 0.055$ Wb/rpm, is fed through a first-quadrant chopper (Table 6.1) from a 120 V d.c. supply at a constant (ideal) armature current of 10 A.

Determine:

a. The range of duty cycle $\alpha$ from zero to maximum speed
b. The range of speed

***Solution***

a. The average output voltage $V_a$ is

$$V_a = V_0 \cdot \alpha = 120\alpha \tag{6.16}$$

At standstill, n = 0 and thus

$$V_a = R_a i_a = 1 \cdot 10 = 10\ \text{V}$$

$$\text{Thus } \alpha_{min} = \frac{(V_a)_{n=0}}{V_0} = \frac{10}{120} = \frac{1}{12} \tag{6.17}$$

For maximum speed, the voltage is 120 V ($\alpha = 1$). Consequently, $\alpha$ varies from 1/12 to 1.

b. The voltage equation for maximum speed is

$$V_{a_{max}} = R_a I_a + K_e \lambda_p n_{max} \tag{6.18}$$

$$n_{max} = \frac{120 - 1 \cdot 10}{0.055} = 2000\ \text{rpm} \tag{6.19}$$

The speed range is thus from 0 to 2000 rpm.

c. For the d.c. brush motor and chopper as above and $\alpha = 0.3$, calculate the actual armature current waveform, its average value, and voltage average value at n = 1600 rpm, for the chopping frequency $f_{ch} = 50$ Hz. Determine the chopping frequency for which the limit between discontinuous and continuous current is reached at the same $t_{on}$ as above.

***Solution***

We now apply the armature current expressions (6.6) and (6.7) first for $f_{ch} = 50$ Hz. The turn-on time interval

$$t_{on} = \frac{1}{f_{ch}} \cdot \alpha = \frac{0.3}{50} = 6 \cdot 10^{-3}\,\text{s} = 6\,\text{ms} \tag{6.20}$$

with

$$T = \frac{1}{f_{ch}} = \frac{1}{50} = 0.02\ \text{s} = 20\ \text{ms}$$

$$i_a = A \cdot e^{-t(1/5 \cdot 10^{-3})} + \frac{120 - 0.055 \cdot 1600}{1} = A \cdot e^{-200t} + 32$$
$$i'_a = A' \cdot e^{-200(t - 6 \cdot 10^{-3})} - \frac{88}{1} \tag{6.21}$$

Assuming discontinuous current mode

$$(i_a)_{t=0} = 0; \quad A = -32 \tag{6.22}$$

Also

$$(i'_a)_{t_{on}} = (i_a)_{t_{on}}; \quad A' - 88 = 32(1 - e^{-200 \cdot 6 \cdot 10^{-3}}) = 22.36$$
$$A' = 110.36 \tag{6.23}$$

The current $i'_a$ becomes zero at $t = t_1$

$$A' e^{-200(t_1 - 6 \cdot 10^{-3})} - 88 = 0; \quad 110.36 \cdot e^{-200(t_1 - 6 \cdot 10^{-3})} - 88 = 0$$
$$t_1 = 7.132 \cdot 10^{-3}\,\mathrm{s} < T = 20\,\mathrm{ms} \tag{6.24}$$

Thus, indeed the current is discontinuous (Figure 6.5).
The average current $i_{av}$ is

$$i_{av} = \frac{1}{T}\left[\int_0^{t_{on}} i_a dt + \int_{t_{on}}^{t_1} i'_a dt\right]$$
$$= \frac{1}{2 \cdot 10^{-2}}\left[32 \int_0^{6 \cdot 10^{-3}} (1 - e^{-200t}) dt + \int_{6 \cdot 10^{-3}}^{7.132 \cdot 10^{-3}} (110.36 \cdot e^{-200(t - 6 \cdot 10^{-3})} - 88) dt\right] \tag{6.25}$$
$$i_{av} = 7.8452\ \mathrm{A}$$

The chopping frequency for the limit between the discontinuous and continuous current is obtained for

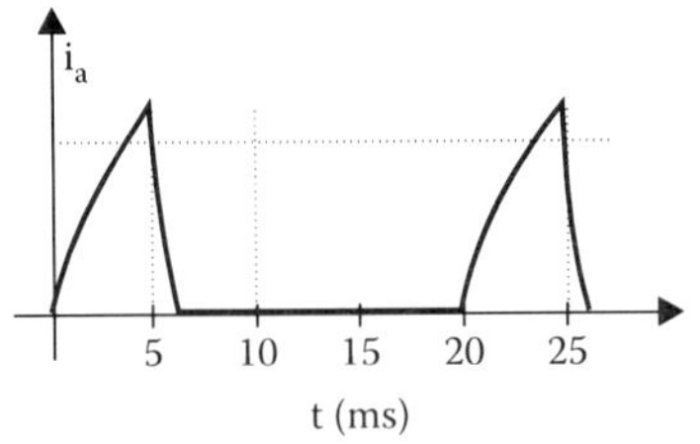


**FIGURE 6.5**
Discontinuous current.

$$t_1 = T_c = 7.132 \cdot 10^{-3}; \quad f'_{ch} = \frac{1}{T_c} = \frac{1}{7.132 \cdot 10^{-3}} = 140.2 \text{ Hz} \tag{6.26}$$

In this case, $\alpha$ becomes

$$\alpha_c = \frac{t_{on}}{T_c} = \frac{6 \cdot 10^{-3}}{7.132 \cdot 10^{-3}} = 0.8412 \tag{6.27}$$

As the current is discontinuous, the average voltage is from Equation 6.5

$$V_{av} = V_0 \frac{t_{on}}{T} + e_g \frac{(T - t_1)}{T} = 120 \cdot 0.3 + 88 \cdot \frac{20 - 7.132}{20}$$
$$= 40 + 56.619 = 96.619 \text{ V} \tag{6.28}$$

## 6.3 Second-Quadrant (Step-Up) Chopper for Generator Braking

A d.c. brush motor with PM excitation is fed through a second-quadrant chopper for regenerative braking (Table 6.1) (Figure 6.6).

The motor data are $R_a = 1\ \Omega$; $L_a = 20$ mH; $e_g = 80$ V (given speed). The supply voltage $V_0$ is 120 V d.c. and $t_{on} = 5.10^{-3}$ s.

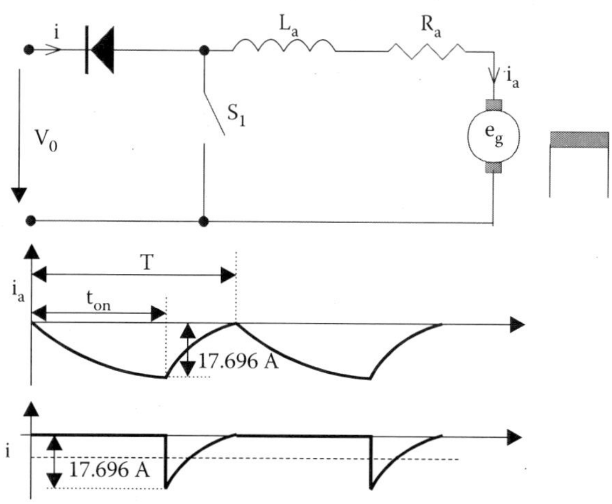


**FIGURE 6.6**
The step-up second-quadrant chopper.

Determine:

a. The waveform of motor current for zero initial current
b. The waveform of source current
c. The maximum average power generated

***Solution***

a. The current waveforms are as shown in Figures 6.2 and 6.3 with $i_{a0} = 0$.

$$i_a = -\frac{e_g}{R_a} + B \cdot e^{-t(R_a/L_a)}; \quad 0 < t < t_{on} \tag{6.29}$$

b.

$$i'_a = \frac{V_0 - e_g}{R_a} + B' \cdot e^{-(t-t_{on})(R_a/L_a)}; \quad t_{on} < t < T \tag{6.30}$$

The boundary conditions are

$$(i_a)_{t=0} = 0; \quad (i_a)_{t=t_{on}} = (i'_a)_{t=t_{on}}; \quad (i'_a)_{t=T} = 0 \tag{6.31}$$

The unknowns are B, B′, and T. Consequently,

$$B = \frac{e_g}{R_a} = \frac{80}{1} = 80 \tag{6.32}$$

$$80 \cdot e^{-5 \cdot 10^{-3} \cdot 50} - 80 = \frac{(120 - 80)}{1} + B'; \quad B' = -57.696 \tag{6.33}$$

$$\begin{aligned} &40 - 57.696 \cdot e^{-(T-t_{on}) \cdot 50} = 0; \quad T - t_{on} = 7.326 \cdot 10^{-3}\ \text{s} \\ &\text{thus, } T = (7.326 + 5) \cdot 10^{-3}\ \text{s} = 12.326 \cdot 10^{-3}\ \text{s} \end{aligned} \tag{6.34}$$

Note that the source current occurs during the $S_1$ turn-off and is negative, proving the regenerative operation.

The average source current $i_{av}$ is

$$\begin{aligned} i'_{av} &= \frac{1}{T} \int_{t_{on}}^{T} i'_a dt = \frac{1}{T} \left[ \frac{V_0 - e_g}{R_a} (T - t_{on}) - \frac{L_a}{R_a} B' \left( e^{-(T-t_{on})(R_a/L_a)} - 1 \right) \right] \\ &= \frac{1}{12.326 \cdot 10^{-3}} \left[ \frac{40}{1} \cdot 7.326 \cdot 10^{-3} + 20 \cdot 10^{-3} \cdot 57.696 \cdot \left( e^{-7.326 \cdot 10^{-3} \cdot 50} - 1 \right) \right] \end{aligned} \tag{6.35}$$

$$i'_{av} = -4.938\ \text{A}$$

c. The average power regenerated $P_{av}$ is

$$P_{av} = -i'_{av} \cdot V_a = 4.938 \cdot 120 = 592.56 \text{ W} \tag{6.36}$$

## 6.4 Two-Quadrant Chopper

Consider a two-quadrant chopper (Figure 6.7) supplying a d.c. brush motor with separate excitation. The load current varies between $I_{max} > 0$ and $I_{min} < 0$.

Determine:

a. The voltage and current waveforms for the load current varying between an $I_{max} > 0$ and $I_{min} < 0$ with $I_{max} > |I_{min}|$.
b. Derive the expression of the conducting times $t_{d2}$ and $t_{d1}$ of the diodes $D_1$ and $D_2$ for case a.

### *Solution*

a. Let us first draw the load current which varies from a positive maximum to a negative minimum (Figure 6.8). The conduction interval for each of the four switches $T_1$, $D_2$, $D_1$, and $T_2$ are

$$\begin{aligned} &T_1 \quad \text{for } t_{d2} < t < t_c \\ &T_2 \quad \text{for } t_{d1} < t < T \\ &D_1 \quad \text{for } t_c < t < t_{d1} \\ &D_2 \quad \text{for } 0 < t < t_{d2} \end{aligned} \tag{6.37}$$

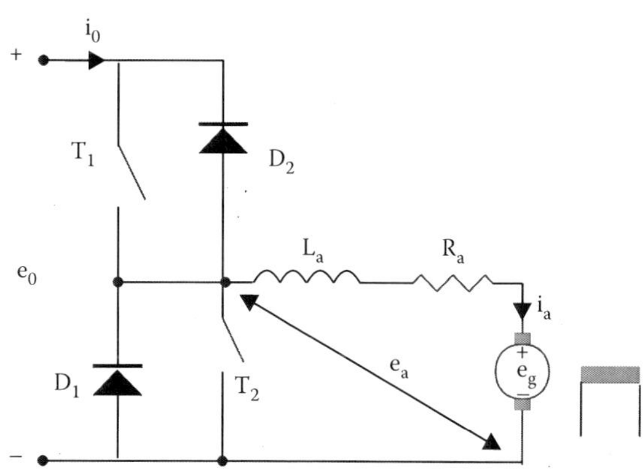


**FIGURE 6.7**
Two-quadrant chopper supplying a separately excited d.c. brush motor.

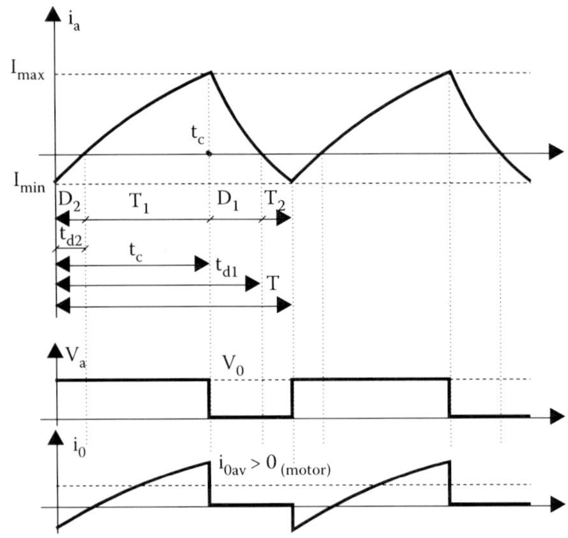


**FIGURE 6.8**
Voltage and current waveforms of two-quadrant chopper-fed d.c. motor.

b. The equations for the current are

$$V_0 - e_g = R_a i_a + L_a \frac{di_a}{dt}; \quad \text{for } 0 < t < t_c \tag{6.38}$$

$$-e_g = R_a i'_a + L_a \frac{di'_a}{dt}; \quad \text{for } t_c < t < T \tag{6.39}$$

with the solutions

$$i_a = \frac{V_0 - e_g}{R_a} + A \cdot e^{-t(R_a/L_a)}; \quad 0 < t \le t_c \tag{6.40}$$

$$i'_a = -\frac{e_g}{R_a} + A' \cdot e^{-(t-t_c)(R_a/L_a)}; \quad t_c < t \le T \tag{6.41}$$

with the boundary conditions $i_a(0) = i_{min}$, $i'_a(T) = i_{min}$, and $i_a(t_c) = i'_a(t_c)$.
The unknowns A, A′, and $t_c$ are obtained from

$$A = I_{min} - \frac{e_0 - e_g}{R_a} \tag{6.42}$$

$$A' = I_{max} + \frac{e_g}{R_a} \tag{6.43}$$

$$T - t_c = -\frac{L_a}{R_a} \cdot \ln\left[\frac{(A' - (e_0/R_a))}{A}\right] \tag{6.44}$$

Consider a d.c. brush motor whose data are $e_0 = 120$ V, $R_a = 1\ \Omega$, $L_a = 5$ mH, $e_g = 80$ V, $I_{max} = 5$ A, $I_{min} = -2$ A, and chopping frequency $f_{ch} = 0.5$ kHz. For the two-quadrant chopper as above, calculate:

a. The $t_c/T = \alpha_{on}$ ratio.
b. The conducting intervals of the four switches.

### *Solution*

a. The constants A, A′, $t_c$ expressions developed above ((6.42)–(6.44)) yield

$$A = I_{min} - \frac{e_0 - e_g}{R_a} = -2 - \frac{120 - 80}{1} = -42\ \text{A} \tag{6.45}$$

$$A' = I_{max} + \frac{e_g}{R_a} = 5 + \frac{80}{1} = 85\ \text{A} \tag{6.46}$$

$$T = \frac{1}{f_{ch}} = \frac{1}{500}\,\text{s} = 0.002\text{s} = 2\ \text{ms} \tag{6.47}$$

$$\begin{aligned} T - t_c &= -\frac{L_a}{R_a} \cdot \ln\left[\frac{(A' - (e_0/R_a))}{A}\right] \\ &= -\frac{5 \cdot 10^{-3}}{1} \cdot \ln\left[\frac{(85 - (120/1))}{(-42)}\right] = 0.9116 \cdot 10^{-3}\ \text{s} \\ t_c &= 2 \cdot 10^{-3} - 0.9116 \cdot 10^{-3} = 1.0884 \cdot 10^{-3}\ \text{s} \end{aligned} \tag{6.48}$$

b. The conducting interval of $D_2$, $t_{d2}$, corresponds to $i_a = 0$

$$\begin{aligned} &\frac{V_0 - e_g}{R_a} + A \cdot e^{-t_{d2}(R_a/L_a)} = 0 \\ &t_{d2} = -\frac{L_a}{R_a} \cdot \ln\left[\frac{V_0 - e_g}{-A \cdot R_a}\right] = \frac{-5 \cdot 10^{-3}}{1} \ln\left(\frac{120 - 80}{-(-42) \cdot 1}\right) = 0.2439 \cdot 10^{-3}\ \text{s} \end{aligned} \tag{6.49}$$

Thus, the main switch $T_1$ conducts for a time interval

$$t_c - t_{d2} = (1.0884 - 0.2439) \cdot 10^{-3} = 0.84445 \cdot 10^{-3}\ \text{s} \tag{6.50}$$

To calculate the conducting time of the diode $D_1$, we apply the condition $i'_a(t_{d1}) = 0$:

$$\frac{-e_g}{R_a} + A' \cdot e^{-(t_{d1}-t_c)(R_a/L_a)} = 0 \tag{6.51}$$

$$t_{d1} - t_c = -\frac{L_a}{R_a} \cdot \ln\left[\frac{e_g}{A' \cdot R_a}\right] = \frac{-5 \cdot 10^{-3}}{1} \ln\left(\frac{80}{85 \cdot 1}\right) = 0.303 \cdot 10^{-3}\,\text{s} \tag{6.52}$$

Consequently, the diode $D_1$ conducts for 0.303 ms. Finally, the static switch $T_2$ conducts for the time interval

$$T - t_{d1} = (2 - 1.0884 - 0.303) \cdot 10^{-3} = 0.6084 \cdot 10^{-3}\,\text{s} \tag{6.53}$$

NOTE: As seen above, the two-quadrant operation of the chopper resides in the variation of $t_c/T$ as the main switches command signals last $t_c$ and $T-t_c$ intervals, respectively, though they conduct less time than that, allowing the diodes $D_2$ and $D_1$ to conduct. The two-quadrant chopper has the advantage of natural (continuous) transition from motor to generator action.

## 6.5 Four-Quadrant Chopper

A d.c. brush motor with separate excitation is fed through a four-quadrant chopper (Table 6.1). Show the waveforms of voltage and current in the third and fourth quadrants.

*Solution*

The basic circuit of a four-quadrant chopper is shown in Figure 6.9.

If $T_4$ is on all the time, $T_1$–$D_1$ and $T_2$–$D_2$ provide first- and second-quadrant operations, respectively, as shown in previous paragraphs. With $T_2$ on all the time and $T_3$–$D_3$ and $T_4$–$D_4$, respectively, the third- and fourth-quadrant operations is obtained (Figure 6.10). So, in fact, we have two two-quadrant choppers acting in turns.

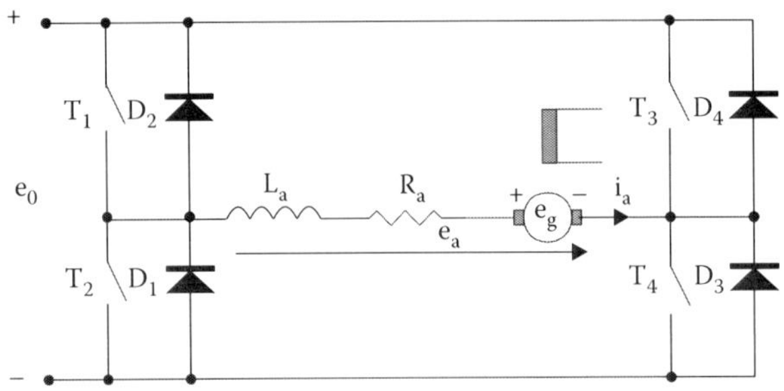


**FIGURE 6.9**
D.c. brush motor fed through a four-quadrant chopper.

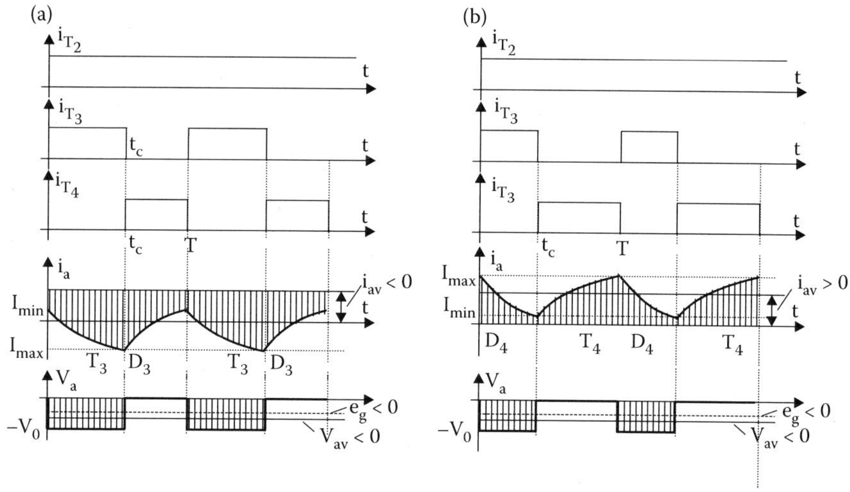

**FIGURE 6.10**
Four-quadrant chopper supplying a d.c. brush motor: (a) third quadrant: $i_{av} < 0$, $V_{av} < 0$; (b) fourth quadrant: $i_{av} > 0$, $V_{av} < 0$.

However, only two out of four main switches are turned on and off with the frequency $f_{ch}$ while the third main switch is kept on all the time and the fourth one is off all the time.

Four-quadrant operation is required for fast response reversible variable speed drives.

As expected, discontinuous current mode is also possible but it should be avoided by increasing the switching frequency $f_{ch}$ or adding an inductance in series with the motor.

Let us assume that a d.c. brush motor, fed through a four-quadrant chopper, works as a motor in the third quadrant (reverse motion). The main data are $V_0 = 120$ V, $R_a = 0.5\ \Omega$, $L_a = 2.5$ mH, rated current $I_{an} = 20$ A; rated speed $n_n = 3000$ rpm; separate excitation.

a. Calculate the rated e.m.f., $e_g$, and rated electromagnetic torque, $T_e$.
b. For $n = -1200$ rpm and rated average current ($i_{av} = -I_{an}$), determine the average voltage $V_{av}$, $t_c/T = \alpha_{on}$, and maximum and minimum values of motor current $I_{max}$ and $I_{min}$ for 1 kHz switching frequency.

### *Solution*

a. The motor voltage equation for steady state is

$$V_{av} = R_a i_a + e_g \tag{6.54}$$

for rated values $V_{av} = V_0 = 120$ V, $i_a = i_{an} = 20$ A, thus

$$e_{gn} = K_e \lambda_p n_n = V_{av} - R_a i_a = 120 - 20 \cdot 0.5 = 110 \text{ V} \tag{6.55}$$

$$K_e \lambda_p = \frac{e_{gn}}{n_n} = \frac{110}{50} = 2.2 \text{ Wb} \tag{6.56}$$

$$T_e = \frac{K_e\lambda_p}{2\pi} I_{an} = \frac{2.2}{2\pi} \cdot 20 = 7\,\text{N m} \tag{6.57}$$

b. The motor equation in the third quadrant is

$$V_{av} = R_a i_a + e_g = 0.5 \cdot (-20) + 2.2 \cdot (-20) = -54\text{ V}, \tag{6.58}$$

the conducting time $t_c$ for $T_3$ (Figure 6.10a) is

$$\frac{t_c}{T} = \frac{V_{av}}{-V_0} = \frac{-54}{-120} = 0.45 \tag{6.59}$$

$$t_c = T \cdot 0.45 = \frac{1}{f_{ch}} \cdot 0.45 = \frac{1}{10^3} \cdot 0.45 = 0.45 \cdot 10^{-3}\text{s} \tag{6.60}$$

From Equations 6.40 and 6.41, the motor current variation (Figure 6.10a) is described by

$$i_a = \frac{V_0' - e_g}{R_a} + A \cdot e^{-t(R_a/L_a)}; \quad 0 < t \le t_c \tag{6.61}$$

$$i_a' = -\frac{e_g}{R_a} + A' \cdot e^{-(t-t_c)(R_a/L_a)}; \quad t_c < t \le T \tag{6.62}$$

The current continuity condition ($i_a(t_c) = i_a'(t_c)$) provides

$$t_c = -\frac{L_a}{R_a} \cdot \ln\left[\frac{(A' - (V_0'/R_a))}{A}\right] \tag{6.63}$$

The second condition is obtained from the average current expression

$$i_{av} = \frac{1}{T}\left[\int_0^{t_c} i_a dt + \int_{t_c}^{T} i_a' dt\right]$$

$$= \frac{1}{T}\left\{\frac{V_0' - e_g}{R_a} t_c - \frac{e_g}{R_a}(T - t_c) + \frac{L_a}{R_a}\left[\left(1 - e^{-t_c(R_a/L_a)}\right)A + A'\left(1 - e^{-(T-t_c)(R_a/L_a)}\right)\right.\right. \tag{6.64}$$

From Equations 6.63 and 6.64, we obtain

$$\frac{(A' - (V_0'/R_a))}{A} = e^{-t_c(R_a/L_a)}$$

$$V_0' = -V_0; \quad e_g = K_e\lambda_p n = 2.2 \cdot (-20) = -44\text{ V} \tag{6.65}$$

$$\frac{(A' + ((-120)/0.5))}{A} = e^{-0.45 \cdot 10^{-3}(0.5/2.5 \cdot 10^{-3})} = 0.914$$

$$-20 = 10^3\left\{\frac{-120-(-44)}{0.5}0.45\cdot 10^{-3} - \frac{(-44)}{0.5}0.55\cdot 10^{-3}\right.$$

$$\left.+\frac{2.5\cdot 10^{-3}}{0.5}\left[\left(1-e^{-0.45\cdot 10^{-3}(0.5/2.5\cdot 10^{-3})}\right)A + A'\left(1-e^{-0.55\cdot 10^{-3}(0.5/2.5\cdot 10^{-3})}\right)\right]\right\} \tag{6.66}$$

$$-20 = -20 + 0.43A + 0.5205\ A' \tag{6.67}$$

$$0.43A + 0.5205A' = 0 \tag{6.68}$$

$$A' + 240 = 0.914\ A \tag{6.69}$$

$$A = 137.62;\ A' = -113.92 \tag{6.70}$$

Now we may calculate $I_{min} = i_a(0)$

$$I_{min} = A + \frac{V_0' - e_g}{R_a} = 137.92 + \frac{-120-(-44)}{0.5} = -15.08\ A \tag{6.71}$$

Also $I_{max} = i_a'(t_c)$

$$I_{max} = A' - \frac{e_g}{R_a} = -113.92 + \frac{-(-44)}{0.5} = -25.92\ A \tag{6.72}$$

## 6.6 Input Filter

A first-quadrant chopper with an L–C input filter supplies a d.c. brush motor with PM excitation under constant current start-up.

a. Demonstrate that maximum rms ripple current in the chopper current $i_{ch}$ occurs at a duty cycle $\alpha = 0.5$.
b. For $f_{ch} = 400$ Hz, $I_a = 100$ A, the rms fundamental (a.c.) current allowed in the supply is 10% of d.c. supply current. Capacitors of 1 mF which can take 5 A rms ripple current are available. Determine $L_f$ and $C_f$ of the filter for $f_{ch} > 2f_r$.
c. For case b, calculate the d.c., first and third harmonics of the supply current.

*Solution*

a. The chopper configuration (Figure 6.11) provides square current pulses for $i_s$ of width $\alpha$.

   Thus, the d.c., $I_{sdc}$, rms $I_{srms}$, and ripple $I_{sripple}$ components of chopper currents are

$$I_{sdc} = I_a \cdot \alpha \tag{6.73}$$

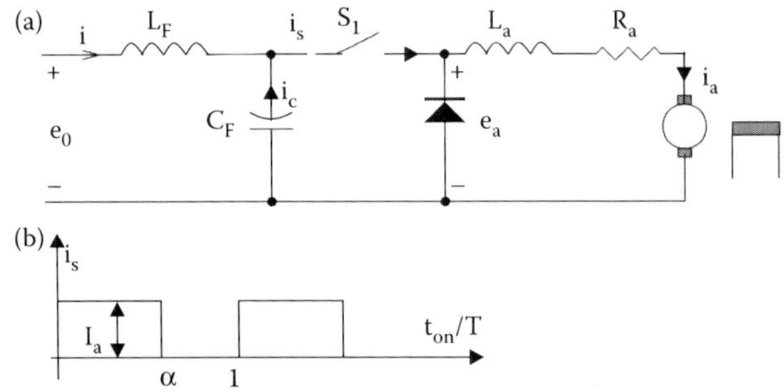


**FIGURE 6.11**
First-quadrant chopper with $L_fC_f$ filter: (a) basic circuit, (b) chopper input current.

$$I_{srms} = \sqrt{\int_0^{\alpha} I_a^2 d\alpha} = I_a\sqrt{\alpha} \tag{6.74}$$

$$I_{sripple} = \sqrt{(I_a\sqrt{\alpha})^2 - (I_a\alpha)^2} = I_a\sqrt{(\alpha - \alpha^2)} \tag{6.75}$$

The maximum ripple current is obtained for $dI_{sripple}/d\alpha = 0$, which leads to $\alpha = 1/2$. The filter design will be performed for this worst case.

b. To design the filter we first need the chopper current harmonics content, which, for $\alpha = 0.5$, is

$$i_s = I_{sdc} + \frac{4}{\pi}\cdot\frac{I_a}{2}\left(\frac{\sin\omega t}{1} + \frac{\sin 3\omega t}{3} + \frac{\sin 5\omega t}{5}\right) \tag{6.76}$$

with

$$I_{sdc} = I_a \cdot \alpha = 100\cdot\frac{1}{2} = 50 \text{ A} \tag{6.77}$$

$$(i_{s_1})_{rms} = \frac{4\cdot 100}{\pi\cdot 2\cdot 1}\cdot\frac{1}{\sqrt{2}} = 45.1733 \text{ A} \tag{6.78}$$

$$(i_{s_3})_{rms} = \frac{4\cdot 100}{\pi\cdot 2\cdot 3}\cdot\frac{1}{\sqrt{2}} = 15.05 \text{ A} \tag{6.79}$$

$$(i_{s_5})_{rms} = \frac{4\cdot 100}{\pi\cdot 2\cdot 5}\cdot\frac{1}{\sqrt{2}} = 9.03 \text{ A} \tag{6.80}$$

The input (source) a.c. current is not to surpass 10% of d.c. input current, that is, $I_1 = I_{sdc}\cdot(10/100) = 50\cdot(10/100) = 5$ A.

According to Equation 6.15, for n = 1, we obtain

$$I_1 = \frac{x_C}{x_L - x_C} \cdot I_{s1}; \quad 5 = \frac{x_C}{x_L - x_C} \cdot 45 \quad \text{or} \quad x_L = 10x_C \tag{6.81}$$

Also, the fundamental capacitor current $I_{C1}$ is

$$I_{C1} = \frac{x_L}{x_L - x_C} \cdot I_{s1} = \frac{10x_C}{10x_C - x_C} \cdot 45 = 50\ \text{A} \tag{6.82}$$

As each 1 mF capacitor can take 5 A, 10 such capacitors in parallel are needed and thus $C_f$ = 10 mF. On the other hand, the reactance $x_L$ is

$$x_L = 10x_C = 10 \cdot \frac{1}{2 \cdot \pi \cdot 400 \cdot 10 \cdot 10^{-3}} = 0.398\ \Omega \tag{6.83}$$

$$L_f = \frac{x_L}{2 \cdot \pi \cdot 400} = \frac{0.398}{2 \cdot \pi \cdot 400} = 0.15847 \cdot 10^{-3}\ \text{H} \tag{6.84}$$

The filter resonance frequency $f_r$ is

$$f_r = \frac{1}{2 \cdot \pi \sqrt{L_f C_f}} = \frac{1}{2 \cdot \pi \sqrt{0.15847 \cdot 10^{-3} \cdot 10^{-2}}} = 127\ \text{Hz} \tag{6.85}$$

The ratio between the chopper switching frequency $f_{ch}$ and the filter resonance frequency $f_r$ is

$$\frac{f_{ch}}{f_r} = \frac{400}{127} = 3.15$$

c. The same a.c. current components (6.15) (after filtering) are

$$I_1 = \frac{I_{s1}}{(f_{ch}/f_r)^2 - 1} \approx \frac{45}{3.15^2 - 1} = 5\ \text{A} \tag{6.86}$$

$$I_3 = \frac{I_{s1}}{(3f_{ch}/f_r)^2 - 1} \approx \frac{15}{(3 \cdot 3.15)^2 - 1} = 0.17\ \text{A} \tag{6.87}$$

$$I_5 = \frac{I_{s1}}{(5f_{ch}/f_r)^2 - 1} \approx \frac{9}{(5 \cdot 3.15)^2 - 1} = 0.036\ \text{A} \tag{6.88}$$

As noted the $L_fC_f$ filter produces a drastic reduction of source-current harmonics.

## 6.7 Basic PM D.C. Motor Closed-Loop Drive/ MATLAB®–Simulink® (Available Online)

Simulation results of a d.c. motor drive with a four-quadrant chopper are presented. The motor model was integrated in a block (d.c. brush motor in Figure 6.12). Changing of motor parameters is done by clicking on their block. A dialogue box appears and you can change them by modifying their default values. This motor model block includes the possibility of adding an extra inductance in the motor circuit ($L_{add}$).

### 6.7.1 Description of the Block Diagrams

The drive system consists of PI speed controller (Ki = 1, Ti = 0.1), PI torque controller (Ki = 50, Ti = 0.0005), and motor blocks. The study examines the system behavior during starting, load perturbation (at 0.2 s) and speed reversal with no load (at 0.5 s) and another load perturbation (at 0.6 s).

The integration step (10 μs) can be modified from Simulink's *Simulation/Parameters*. The chopper frequency is 20 kHz and this block input is the $t_c/T_e$ ratio.

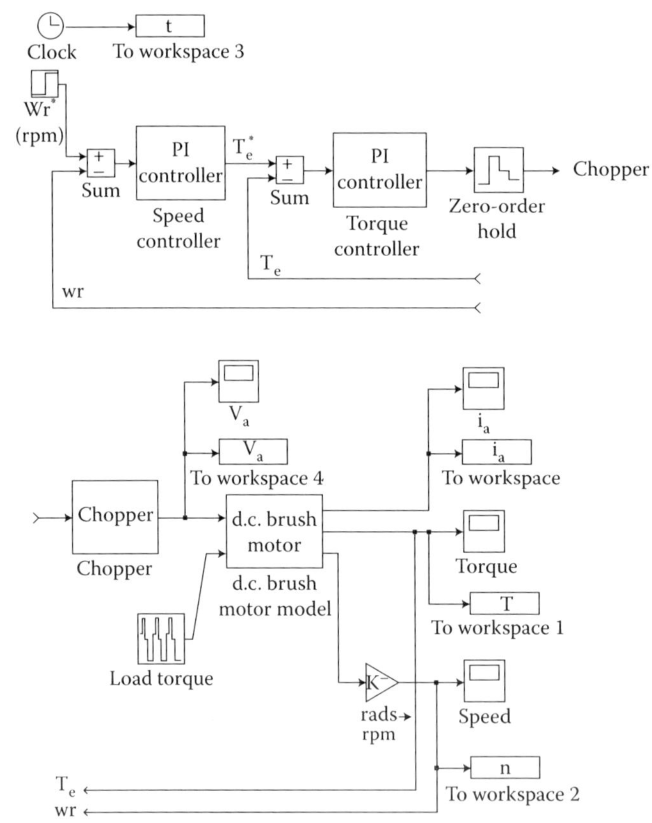


**FIGURE 6.12**
D.c. brush motor drive block diagram.

To find out the structure of each block presented above, unmask it (*Options/Unmask*). Each masked block contains a brief description of that block (inputs/outputs/parameters).

The block diagram of the electric drive system is presented in Figure 6.12.

The motor used for this simulation has the following parameters: $V_{dc} = 120$ V, $I_n = 20$ A, $n_n = 3000$ rpm, $R_s = 0.5\ \Omega$, $L_a = 0.0025$ H, $J = 0.001$ kg m$^2$, $K_e\lambda_p = 2.2$ Wb.

Figures 6.13 through 6.16 represent the speed (Figure 6.13), torque (Figure 6.14) responses and current (Figure 6.15) and voltage (Figure 6.16) waveforms, for the *starting process and load torque (8 Nm) applied at 0.2 s, and speed reversal with no load at 0.5 s and load torque applied at 0.6 s.*

Fast response with rather low current ripple is obtained due to the rather high (20 kHz) switching frequency for a 5 ms electrical time constant d.c. brush PM motor four-quadrant drive.

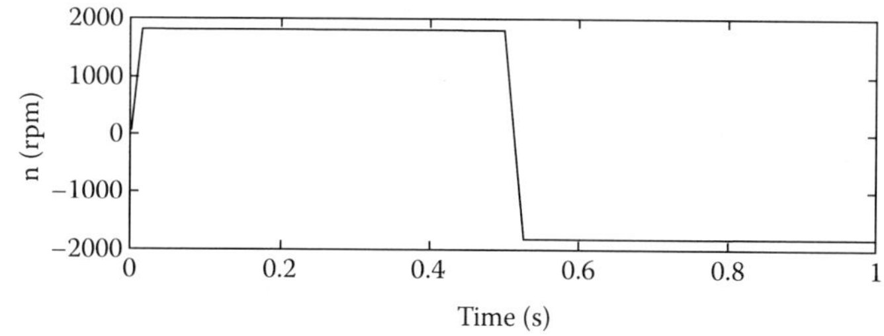


**FIGURE 6.13**
Speed transient response.

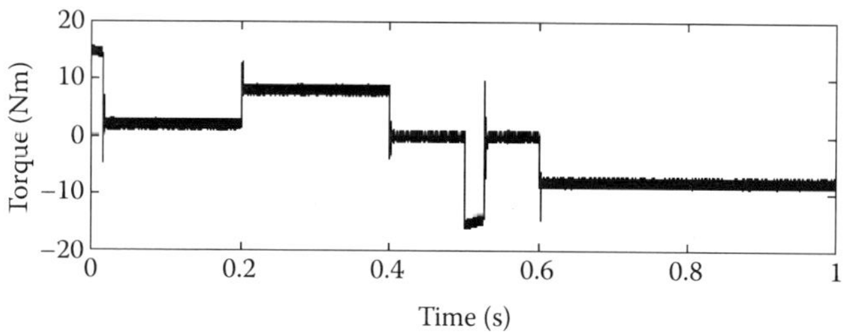


**FIGURE 6.14**
Torque response.

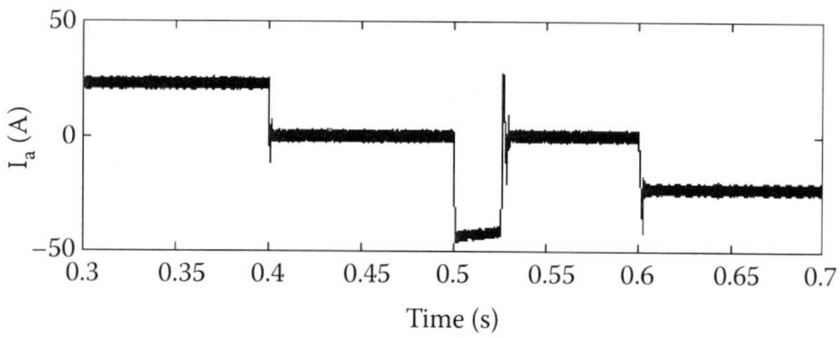


**FIGURE 6.15**
Current waveform ($i_a$).

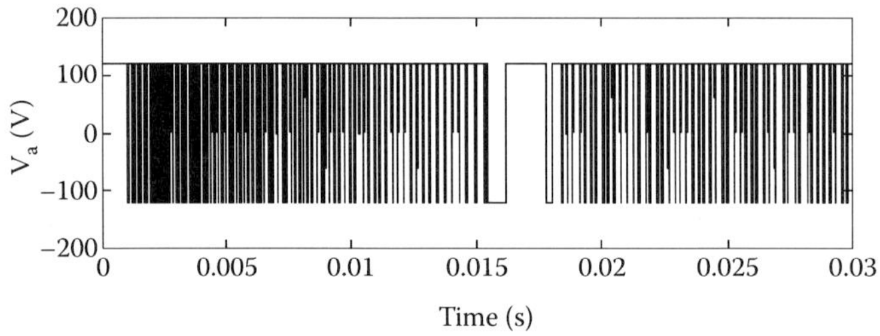


**FIGURE 6.16**
Voltage waveform ($V_a$).

## 6.8 Summary

In this chapter, single-, two-, or four-quadrant chopper basic configurations in interaction with the d.c. brush motor (for constant speed) have been introduced.

- In all configurations the PESs have been considered ideal—with instantaneous hard commutation (switching).
- Resonant d.c.–d.c. converters have not been treated. The interested reader is invited to study the literature [1].
- Choppers, as rectifiers, face the same problem of continuous–discontinuous current mode switching.
- Under discontinuous current mode, for low-voltage low-inductance motors and low switching frequency (thyristors), the dependence of output voltage on the on/off time ratios of PESs renders the control sluggish. Increasing the switching frequency or including a series inductance solves the problem.
- The input current has harmonics and thus an input filter is required to reduce them to acceptable limits.
- As the current in the motor pulsates, care must also be exercised in assessing the armature copper losses and the eventual derating of the motor for a certain application, when chopper-fed.
- The interaction between the chopper and the motor at constant speed discussed in this chapter should be continued with closed-loop control to produce a variable speed drive. This will be done in Chapter 7, though Section 6.7 anticipates it.

## 6.9 Problems

6.1 A step-down (buck) first-quadrant chopper is shown in Figure 6.17.

a. Explain the principle and draw the inductance voltage and current waveforms for continuous current mode.
b. Calculate the output voltage and current for a resistive load as a function of duty ratio α, considering ideal switching and zero losses.

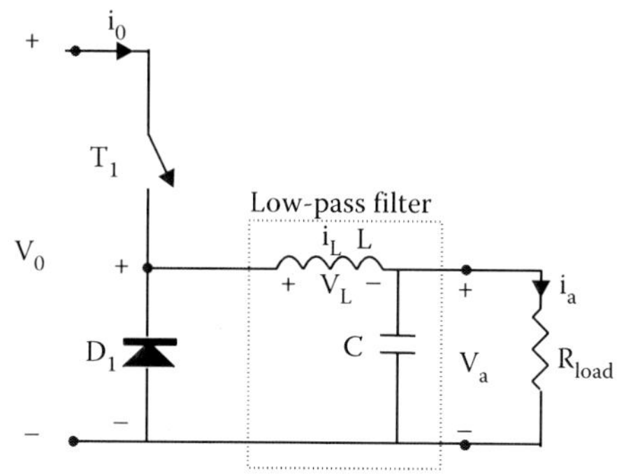


**FIGURE 6.17**
Step-down (buck) chopper.

6.2 For Problem 6.1 find α for the limit between continuous and discontinuous current mode and the average output voltage for discontinuous current mode. Plot the average output voltage versus average output current both for continuous and discontinuous modes.

6.3 A step-up (boost) first-quadrant chopper is shown in Figure 6.18. Considering ideal elements determine:

a. The voltage and current waveforms for continuous current mode.
b. The boundary between continuous and discontinuous current mode.
c. The output voltage ripple for continuous current operation.

6.4 A four-quadrant chopper supplying a d.c. brush motor with PM excitation is controlled through PWM with bipolar voltage switching, Figure 6.19, where ($T_{A+}$, $T_{B-}$) and ($T_{A-}$, $T_{B+}$) are controlled simultaneously. Determine:

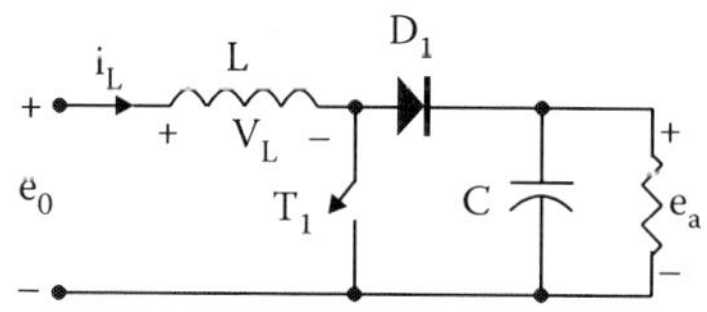


**FIGURE 6.18**
Step-up (boost) chopper.

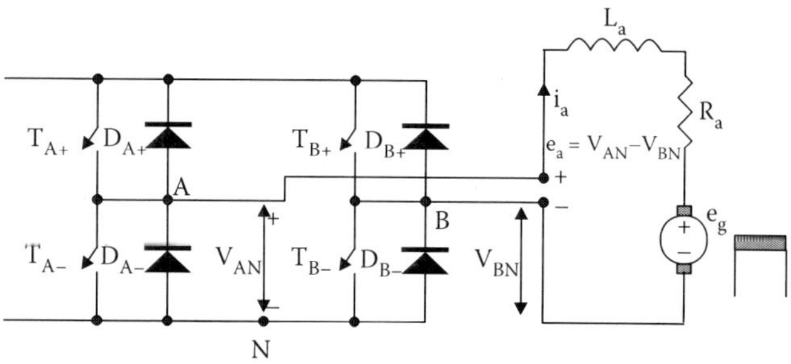


**FIGURE 6.19**
Four-quadrant chopper with d.c. brush motor.

a. The voltage and current waveforms for positive control voltage $V_{control}$.

b. The voltage and current waveforms for negative control $V_{con}$.

6.5 A d.c. series brush motor has the data: $L_{at} = 20$ mH, $R_{at} = 1\ \Omega$, rated voltage $V_n = 120$ V, rated current $I_{an} = 10$ A, rated speed $n_n = 1800$ rpm. The remnant flux-induced voltage at rated speed is $e_{grem} = 5$ V. When fed from a first-quadrant step-down chopper, at 120 $V_{dc}$ with $\alpha = 0.3$, $f_{ch} = 50$ Hz at rated speed, calculate the armature current waveform, average current, and average voltage.

6.6 Solve Problem 6.5 with MATLAB–Simulink or PSpice simulation programs.

## Reference

1. N. Mohan, T.M. Undeland, W.P. Robbins, *Power Electronics*, second edition, Chapter 7, John Wiley & Sons, New York, 1995.

# 7

# Closed-Loop Motion Control in Electric Drives

## 7.1 Introduction

By motion control we mean torque, speed, or position control. Motion control systems are characterized by precision, response quickness, and immunity to parameter detuning, torque and inertia perturbations, and energy conversion rates. Motion control through electric motors and PECs may be approached by the theory and practice of linear and nonlinear, continuous or discrete control systems. Control systems is a field of science in itself and in what follows only a few solutions of practical interest for motion control will be introduced with some application examples, from simple to complex.

To simplify the treatment and to include the permanent magnet (PM) d.c. brush motor control, the latter will be considered for various motion control systems. PM d.c. brush motors are characterized by a low electrical time constant, $\tau_e = L/R$, of a few milliseconds or less. The armature (torque) current is fully decoupled from the PM field because of the orthogonality of the armature and PM fields, both at standstill and for any rotor speed.

As shown in later chapters, vector control of a.c. motors also decouples flux and torque control if the orientation to the flux linkage is secured and flux amplitude is kept constant. Consequently, vector-controlled a.c. motors are similar to d.c. brush motors and thus the application of various motion control systems to the d.c. motor holds notable generality, while also eliminating the necessity of a separate chapter on closed-loop control of brush d.c. motors.

To start with, the cascaded motion linear control is presented through numerical examples.

## 7.2 Cascaded Motion Control

A typical cascaded motion control system comprises three loops: one for torque, one for speed, and one for position (Figure 7.1). Ideally the system has independent position and speed sensors though, in general, only one is used—the speed sensor for speed control and the position one for position control—with the speed in the latter case estimated from the position information.

The PM d.c. brush motor equations are

$$V = Ri + L\frac{di}{dt} + \lambda_{PM}\omega_r \tag{7.1}$$

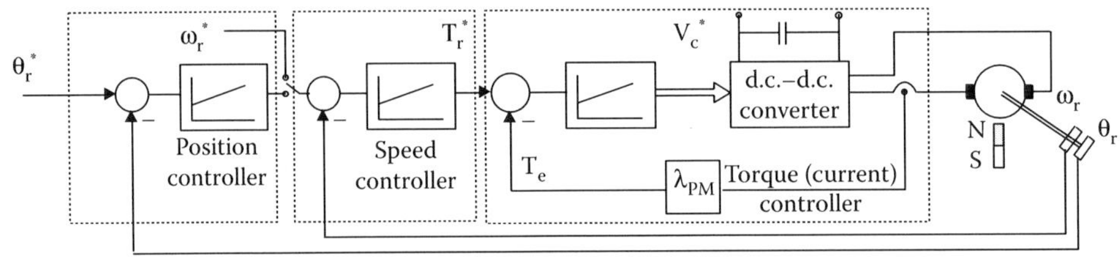


**FIGURE 7.1**
Typical cascaded motion control.

$$J\frac{d\omega_r}{dt} = T_e - T_L \tag{7.2}$$

$$\frac{d\theta_r}{dt} = \omega_r \tag{7.3}$$

$$T_e = \lambda_{PM} I \tag{7.4}$$

### 7.2.1 Torque Loop

The torque control for constant flux linkage ($\lambda_{PM}$) means (with core loss neglected) armature current control for a PM d.c. brush motor. Fast current control also provides for fast current protection. The design of the torque loop implies knowing the load torque $T_L$. Alternatively, $T_L$ = constant and, after the design is completed, the influence of load torque perturbations on loop stability is investigated and adequate corrections added, if required.

For constant (or zero) load torque, the PM d.c. brush motor current/voltage transfer function, from Equations 7.1 through 7.4, becomes (see Equation 4.43)

$$H_V(s) = \frac{i(s)}{V(s)} = \frac{s\tau_{em}}{(s^2\tau_{em}\tau_e + s\tau_{em} + 1)R} \tag{7.5}$$

where

$$\tau_{em} = \frac{JR}{{\lambda_{PM}}^2} \tag{7.6}$$

is the electromechanical time constant of the motor.

The superaudible frequency chopper may be modeled through its gain $K_c$ as its delay may be neglected. The torque constant $K_T = T_e/I = \lambda_{PM}$ and the current sensor amplification is $K_i$.

The typical torque controller is proportional–integral (PI) type with the gain $K_{si}$ and the time constant $\tau_{si}$. The block diagram is shown in Figure 7.2. The theory of linear systems offers numerous design approaches to PI controllers [1] both for continuous and discrete implementation.

In what follows, we are using the critical frequency $\omega_c$ and phase margin $\varphi_c$ contraints for the open-loop transfer function A(s) of the system in Figure 7.2.

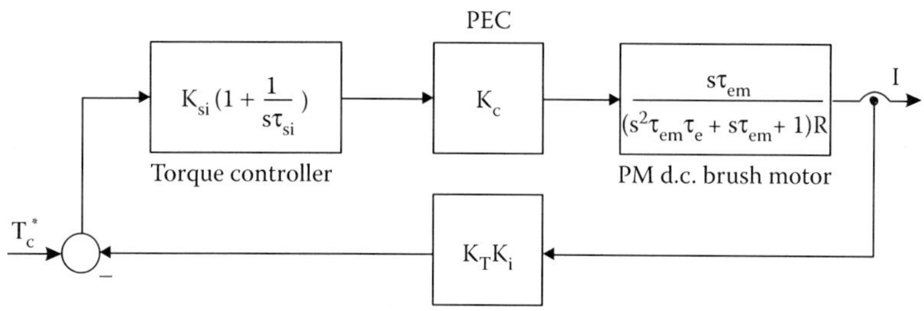


**FIGURE 7.2**
PI torque loop for a PM d.c. brush motor.

$$A(s) = \frac{K_{si}(1+s\tau_{si})}{s\tau_{si}R}\frac{K_c K_T K_i s\tau_{em}}{s^2\tau_{em}\tau_e + s\tau_{em} + 1} \tag{7.7}$$

The critical frequency $\omega_c$ should be high—up to 1–2 kHz—to provide fast torque (current) control.

Let us take as a numerical example a PM d.c. brush motor with the data: $V_n = 110$ V, $P_n = 2$ kW, $n_n = 1800$ rpm, $R = 1\ \Omega$, $L = 20$ mH, $K_T = 1.1$ Nm/A, $\tau_{em} = 0.1$ s, $K_c = 25$ V/V, $K_i = 0.5$ V/A, critical frequency $f_c = 500$ Hz, and the phase margin $\varphi_c = 47°$.

The phase margin $\varphi_c$ of A(s) from Equation 7.7, for the critical frequency $\omega_c = 2\pi f_c$, is

$$\begin{aligned}\phi_c &= +180° + \mathrm{Arg}[A(j\omega_c)] \\ &= +180° + \tan^{-1}(\omega_c\tau_{si}) - \tan^{-1}\left(\frac{\omega_c\tau_{em}}{1-\omega_c^2\tau_{em}\tau_e}\right)\end{aligned} \tag{7.8}$$

Consequently

$$\tan^{-1}(\omega_c\tau_{si}) = -180° + 47° + \tan^{-1}\frac{2\pi 500\cdot 0.1}{1-(2\pi 500)^2\cdot 0.1\cdot 0.02} = 46° \tag{7.9}$$

And thus,

$$\tau_{si} = \frac{\tan 46°}{2\pi 500} = 0.3075 \text{ ms} \tag{7.10}$$

The gain of the torque controller $K_{si}$ may be calculated from the known condition:

$$|A(j\omega_c)| = 1 \tag{7.11}$$

Finally,

$$K_{si} = \frac{0.3075\cdot 10^{-3}\cdot 1}{25\cdot 1.1\cdot 0.5\cdot 0.1}\cdot\sqrt{\frac{(10^3\pi 0.1)^2 + (1-10^6\pi^2 0.1\cdot 0.01)^2}{1+0.965^2}} = 2.205 \tag{7.12}$$

### 7.2.2 Speed Loop

In numerous applications, speed control is required. The torque loop is still there to limit the current, quicken the response, and reduce the gain in the current loop. The block diagram of a system for speed control, with the torque (current) loop included, is shown in Figure 7.3.

Assuming that the torque loop has been already designed as above, only the speed-loop design will be discussed here. The block diagram in Figure 7.3 may be restructured as in Figure 7.4, where $K_\omega$ is the gain of the speed sensor.

For the speed-loop design, we may proceed as above for the torque loop, but with a critical frequency $f_{c\omega} = 100$ Hz and a phase margin $\varphi_{c\omega} = 60°$ with $K_\omega = 0.057$ V s/rad. The final results are $\tau_{s\omega} = 2.87$ ms and $K_{s\omega} = 2.897 \cdot 10^3$. The amplification $K_{s\omega}$ is rather high but the phase margin is ample and produces a reasonably low overshoot and well-damped oscillatory response.

In a similar manner, we may proceed to design and add the position loop for completion of the cascaded controller. It is known that such cascaded linear controllers do not excel in terms of robustness to parameter, inertia, or load changes. Additional measures are required and they will be treated later on in this chapter.

For now, however, we are treating the digital position control through a single digital filter rather than a cascaded controller, to broaden the spectrum of exposed motion control practical approaches.

### 7.2.3 Digital Position Control

The basic configuration of a digital position control system is shown in Figure 7.5.

Figure 7.5 features the motor-converter transfer function, a digital–analog converter (DAC), a sample-and-hold (ZOH), and digital filter D(z) which represents, in fact, the

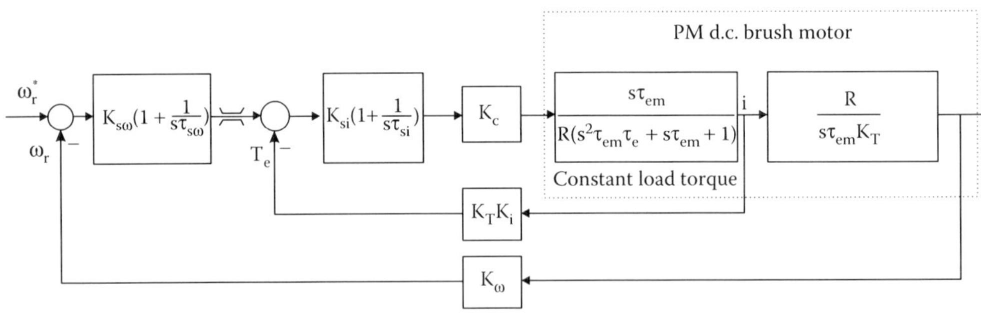


**FIGURE 7.3**
Speed control with torque (current) inner loop.

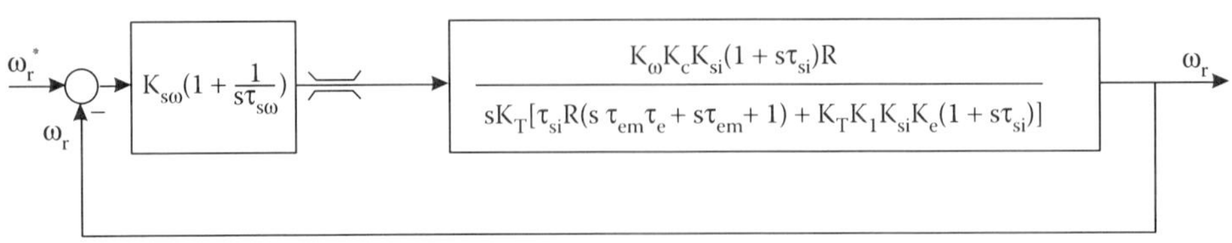


**FIGURE 7.4**
Simplified speed-loop block diagram.

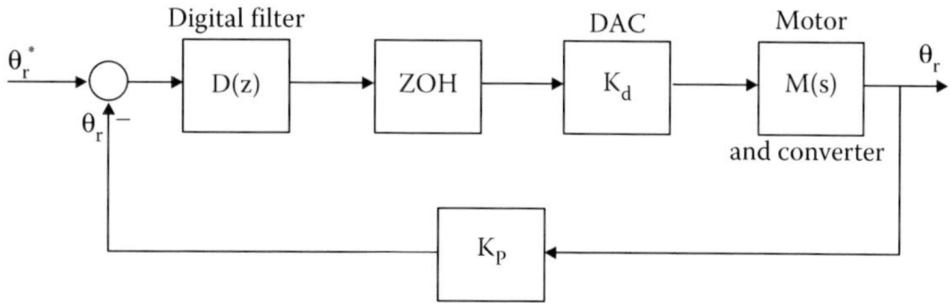


**FIGURE 7.5**
Basic digital position control system.

position controller. Only position feedback is included as obtained from an encoder with N pulses per revolution which, through the division of the two 90° phase-shifted pulse trains, produces 4N pulses per revolution and a bit for motion direction.

Suppose only an 8-bit DAC with $V_d = \pm 10$ V is used. Consequently its amplification $K_d$ is

$$K_d = \frac{2V_d}{2^n} = \frac{20}{2^8} = \frac{20}{256}\,\text{V/pulse} \tag{7.13}$$

The amplification of the encoder $K_p$ is

$$K_p = \frac{4N}{2\pi}\,\text{pulse/rad} \tag{7.14}$$

The position control is performed in a digital signal processor (DSP) with the discretization period of T. The position error $\varepsilon(KT)$ is

$$\varepsilon(KT) = \theta_r^*(KT) - \theta_r(KT) \tag{7.15}$$

The position controller filters this error to produce the command Y(KT) which is periodically applied through the DAC

$$V_C = K_d \cdot Y(KT) \tag{7.16}$$

The command voltage $V_c$ is kept constant during the discretization period T, the effect being known as ZOH. The digital filter may be expressed by finite difference equations of the type [2–3]

$$Y(KT) = 2\varepsilon(KT) - \varepsilon((K-1)T) \tag{7.17}$$

z-transform may be used to model Equation 7.17. The z-transform shifting property is

$$Y(K-m) \rightarrow z^{-m}f(z) \quad \text{with } Y(K) \rightarrow f(z) \tag{7.18}$$

Consequently, Equation 7.17 becomes

$$Y(z) = 2\varepsilon(z) - z^{-1}\varepsilon(z) \tag{7.19}$$

The D(z) is thus

$$D(z) = \frac{Y(z)}{\varepsilon(z)} = \frac{2z-1}{z} \tag{7.20}$$

Let us now prepare for a numerical example by specifying the PM d.c. brush motor-converter transfer function with the electrical time constant $\tau_e$ neglected as it is very low for an axial airgap configuration motor, for example,

$$M(s) = \frac{\theta_s(s)}{V_C(s)} = \frac{K_c}{s(1+s\tau_{em})K_T} \tag{7.21}$$

The data for this new numerical example are $K_T = 0.1$ Nm/A, $R = 1\ \Omega$, $L = 0$ H, $J = 10^{-3}$ kg m$^2$, $K_c = 5$ V/V with the critical frequency $\omega_c = 125$ rad/s and the phase margin $\varphi_{cp} = 45°$.

The problem may be solved adopting a continual filter (controller) and finally "translating" it into a discrete form.

The motor-converter transfer function M(s) is

$$M(s) = \frac{50}{s(1+s\cdot 0.1)}; \quad \tau_{em} = \frac{RJ}{K_T^2} = \frac{1\cdot 10^{-3}}{0.1^2} = 0.1\ \text{s} \tag{7.22}$$

The DAC amplification is $K_d = 10/128$ (8 bit). The encoder produces N = 500 pulses/revolution which provides in fact 4N = 2000 pulses/revolution, and has the amplification $K_p$:

$$K_p = \frac{4N}{2\pi} = \frac{2000}{2\pi} = 318\ \text{pulses/rad} \tag{7.23}$$

The discretization (sampling) time is $T = 10^{-3}$ s. Thus the ZOH transfer function is

$$ZOH(s) \approx e^{-\frac{sT}{2}} = e^{-s\cdot 5\cdot 10^{-4}} \tag{7.24}$$

The various transfer functions in Figure 7.5 are united in a unique transfer function H(s)

$$H(s) = K_d \cdot K_p \cdot ZOH(s) \cdot M(s) \tag{7.25}$$

or

$$H(s) = \frac{1242.8\cdot e^{-s\cdot 5\cdot 10^{-4}}}{s(1+s\cdot 0.1)} \tag{7.26}$$

The phase angle of H(s) for $\omega_c$ is

$$\varphi_H = -\frac{\omega_c T}{2} \cdot \frac{180°}{\pi} - 90° - \tan^{-1}(0.1 \cdot 125) = -179° \tag{7.27}$$

To produce a phase margin $\varphi_c$ of 45°, the digital filter G(z) has to produce a phase anticipation $\varphi_D$ of 44°. A tentative G(s) lead-lag filter is

$$G(s) = K\frac{s+\omega_1}{s+\omega_2} \tag{7.28}$$

The maximum anticipation is obtained for $\omega_c = (\omega_1\omega_2)^{0.5}$. The filter phase angle $\varphi_D$ is

$$\varphi_D = \tan^{-1}\left(\frac{\omega_c}{\omega_1}\right) - \tan^{-1}\left(\frac{\omega_c}{\omega_2}\right) \tag{7.29}$$

with $\omega_c = 125$ rad/s, $\varphi_D = 44°$, $\omega_1 = 53.3$ rad/s, and $\omega_2 = 293.15$ rad/s.

Finally, the amplification K of G(s) is obtained from the condition that the amplitude of the open-loop transfer function of the system |H(s) G(s)| be equal to unity for critical frequency. With the above data K = 2.96.

To discretize the G(s) transfer function, the correspondence between z and s is considered

$$s = \frac{2}{T}\frac{z-1}{z+1} = 2000\frac{z-1}{z+1} \tag{7.30}$$

Consequently G(z) is

$$G(z) = 2.65\frac{z-0.947}{z-0.744} = \frac{Y(z)}{\varepsilon(z)} \tag{7.31}$$

or

$$Y(z) - 0.744 \cdot z^{-1}Y(z) = 2.65\varepsilon(z) - 2.51 \cdot z^{-1}\varepsilon(z) \tag{7.32}$$

Finally the output of the digital filter Y(k) is

$$Y(k) = 0.744 \cdot Y(k-1) + 2.65\varepsilon(k) - 2.51 \cdot \varepsilon(k-1) \tag{7.33}$$

### 7.2.4 Positioning Precision

It is known that the positioning precision is influenced by the friction torque. Let us consider a friction torque $T_f = 0.05$ Nm.

The speed is small around the target position and thus the motion-induced voltage of the motor may be neglected

$$V = R \cdot I = R\frac{T_f}{K_T} = \frac{1 \cdot 0.05}{0.1} = 0.5\,V \tag{7.34}$$

The command voltage of the static power converter $V_c$ is

$$V_c = \frac{V}{K_c} = \frac{0.5}{5} = 0.05 \text{ V} \tag{7.35}$$

The DAC input $N_c$ is

$$N_c = \frac{V_c}{K_d} = \frac{0.05}{(10/128)} = 0.64 \text{pulses} \tag{7.36}$$

The gain $K_0$ of the digital filter G(z) is obtained for $Z = 1$

$$K_0 = 2.65\left(\frac{1-0.947}{1-0.7441}\right) = 0.548 \tag{7.37}$$

Consequently, the input of the digital filter $N_e$ is

$$N_e = \frac{N_c}{K_0} = \frac{0.64}{0.548} = 1.33 \text{pulses} \tag{7.38}$$

The closest integer is $N_{ei} = 2$ pulses and thus the position error $\Delta\theta_r$ caused by the above friction torque is

$$\Delta\theta_r = \frac{N_e \cdot 360°}{4N} = \frac{2 \cdot 360}{2000} = 0.36° \tag{7.39}$$

The positioning precision may be improved through diverse methods such as increasing the encoder number of pulses per revolution N or by a feedforward signal proportional to the torque perturbation. Optical encoders with up to 20,000 pulses/revolution are available today.

Industrial position controllers in digital implementation are, in general, of P type but contain a speed feedforward signal and an inner PI speed loop besides the torque loop (Figure 7.6) [4].

It should be noted that both $\hat{\omega}_r^*$ and $\hat{\omega}_r$—estimated reference speed and actual speed—are calculated through the position time derivative and thus contain noise and errors.

In general, such a controller responds well to slow dynamics commands, well below the frequency of the position controller (0.5–10 Hz). For this reason, in order to follow correctly the targeted position during transients, in the presence of torque and inertia changes, state-space control is used where the speed and acceleration are estimated adequately from the measured position.

## 7.3 State-Space Motion Control

State-space motion control which makes use of a few variables has the advantage that the scales are directly related to the capacity to reject torque perturbations (to mechanical rigidity). In such systems there is a separation of position tracking commands from those related to perturbation rejection. To yield zero position tracking errors, a positive torque reaction is used. That does not affect the stability to perturbations but poses problems to torque dynamics and estimation precision (Figure 7.7).

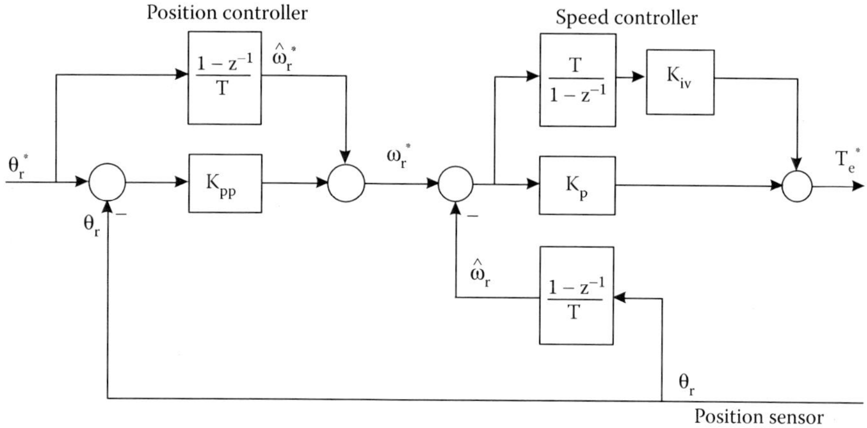


**FIGURE 7.6**
Standard digital position control system with inner speed loop.

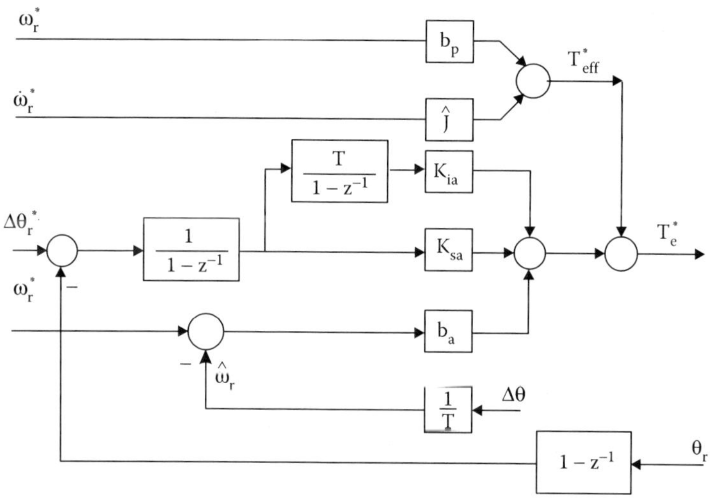


**FIGURE 7.7**
State-space position controller with zero tracking error.

It should be noted that the friction torque coefficient $b_p$ (Figure 7.7) as well as the inertia $\hat{J}$ have to be known with good precision. The estimated speed ($\hat{\omega}_r$), the reference speed ($\omega_r^*$), and the reference acceleration ($\dot{\omega}_r^*$) are all given as time functions

$$\hat{\omega}_r = \frac{\theta_r(k) - \theta_r(k-1)}{T} \tag{7.40}$$

The speed estimation with zero delay may be done through an observer (Figure 7.8) more complicated than the simple reference equation in Equation 7.40.

However, even such an observer is sensitive to torque dynamics and to torque estimation precision. The immunity to torque perturbations in position control may be further

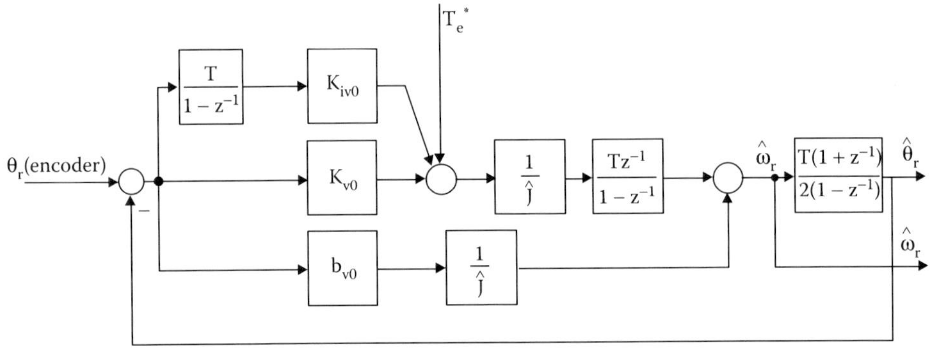


FIGURE 7.8
Zero delay speed observer.

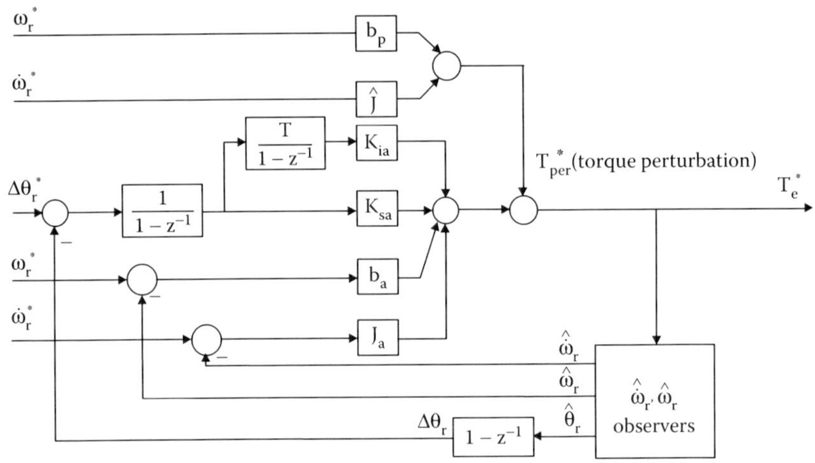


FIGURE 7.9
Position controller with observers and loops for speed and acceleration and reference torque feedforward signal.

increased, especially at low speeds, through an acceleration feedback. The required acceleration observer is similar to the speed observer in Figure 7.8, but is also provided with a speed loop whose reference is the output of the speed observer itself. Consequently, the position control with observers and loops for speed and acceleration thus obtained (Figure 7.9) shows an increased dynamics rigidity. Even in this case, problems remain related to the knowledge of instantaneous torque and inertia $\hat{J}$. In what follows the estimation of torque perturbations is approached.

## 7.4 Torque Perturbation Observers

The response of the cascaded motion control systems, designed for given (constant) torque and rated motor parameters, proved to be strongly influenced by torque, inertia, and parameter variations. Increasing the gain would make the system more rigid but it could excite low-frequency mechanical resonant modes which may lead to instabilities. The reference torque feedforward signal produces similar effects. Motion control (in robotics, etc.) requires, on the other hand, even more increased dexterity and thus a variable dynamic rigidity. A robot that rotates a camshaft requires a force loop while another that plants electronic components implies refined motion control. In these latter cases, there is no urgent need for control rigidity but the robustness of response is also lost if the gains would be reduced in the standard controller. This is how a variable rigidity controller becomes necessary. To reach this goal use is made of total perturbation observers (of torque and parameters) or partial ones (of torque) [5].

To simplify the presentation, a single degree of freedom system is considered. Its motion equation is

$$J\frac{d\omega_r}{dt} = T_e - T_L; \quad T_L = T_{in} + T_{Lo} + T_{fr} \tag{7.41}$$

where $T_{in}$ is the inertia torque, $T_{Lo}$ is the load torque, and $T_{fr}$ is the friction torque.

The electromagnetic torque of the motor $T_e$ (of the PM d.c. brush or of vector-controlled a.c. motors) has the simple expression

$$T_e = K_T \cdot I^* \tag{7.42}$$

where I* is the reference torque current and, consequently, the torque (current) controller is considered to be hyper-rapid in response. The parameters in Equations 7.41 and 7.42 are J and $K_T$, and they deviate from their rated values $J_n$ and $K_{Tn}$:

$$J = J_n + \Delta J \tag{7.43}$$

$$K_T = K_{Tn} + \Delta K_T \tag{7.44}$$

In terms of torque, the variations of ΔJ and $\Delta K_T$ are $\Delta Js\omega_r$ and $\Delta K_T I^*$, respectively. Consequently, the torque perturbation $T_{per}$ is

$$T_{per} = T_L + \Delta Js\omega_r - \Delta K_T \cdot I^* \tag{7.45}$$

Making use of Equations 7.43 through 7.45 in Equation 7.41 leads to

$$T_{per} = K_{Tn} \cdot I^* - J_n s\omega_r \tag{7.46}$$

Based on Equation 7.46, the torque perturbation may be rather simply calculated. To avoid the time derivative in Equation 7.46, a change is operated and a low pass filter added (Figure 7.10). If the critical frequency, a, of the low pass filter is sufficiently high, the estimated torque perturbation $T_{per}$ is very close to the actual one. The torque perturbation $T_{per}$ may also be used as a positive reaction (Figure 7.11) to replace the same quantity, but recalculated with the reference values.

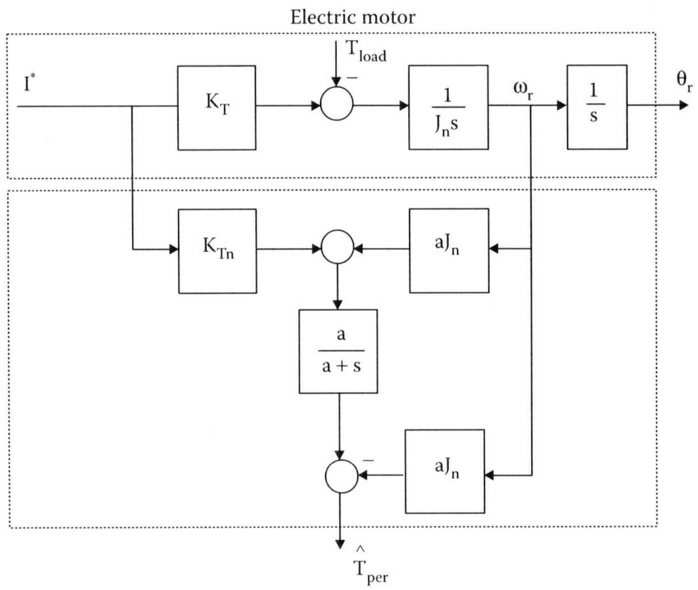


**FIGURE 7.10**
Torque perturbation estimator based on speed information.

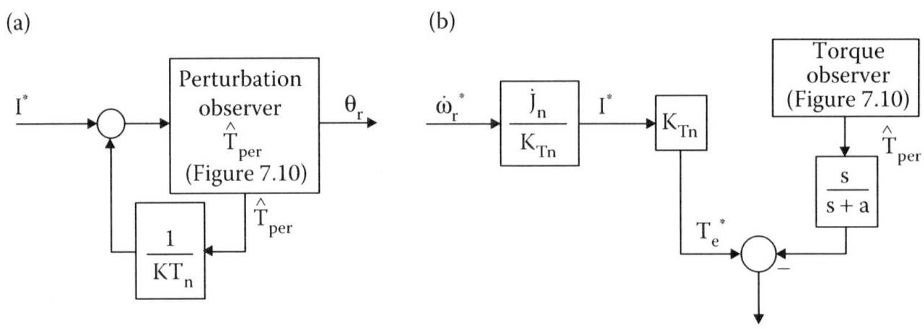


**FIGURE 7.11**
Torque perturbation reaction: (a) positive reaction; (b) negative reaction at the reference $T_e^*$.

## 7.5 Path Tracking

Making use of torque perturbation ($T_{per}$) leads to increased control rigidity and thus to the possibility of close trajectory (path) tracking. For path tracking the reference position in two (three) previous time steps is known and thus reference acceleration and speed signals may be calculated and used in the control system.

Stabilizing the position response is accomplished by at least two poles with negative real parts which could be made dependent on the gains of the position and speed loops $K_1$ and $K_2$. This solution plus the inverse system may lead to a robust path tracking control system (Figure 7.12—for multivariable systems).

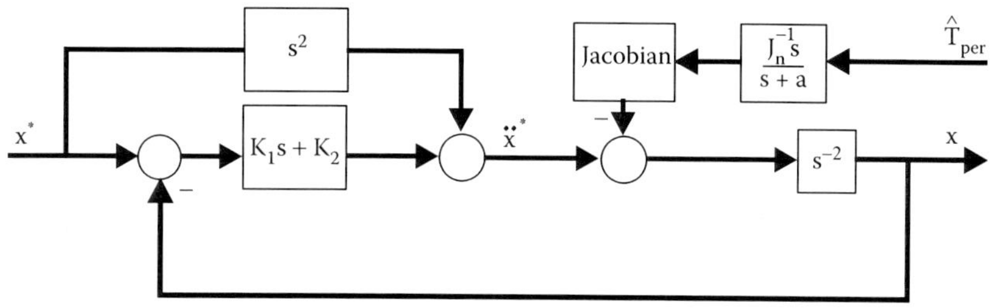


**FIGURE 7.12**
Position tracking multivariable system (J—Jacobian matrix).

The relationship between the actual and reference position (Figure 7.12) is

$$x = x^* - \text{Jacobian}\frac{s\hat{T}_{per}}{(s+a)J_n(s^2+K_1s+K_2)} \tag{7.47}$$

## 7.6 Force Control

The above procedure may be used to control the force without its direct measurement (Figure 7.13).

The rigidity $K_e$ and the viscosity coefficient $D_e$ model the medium where the force acts. The speed feedback amplification $K_1$ provides for stability. It may well be said that the external loop in Figure 7.13 adjusts the motion rigidity by generating the reference acceleration $\ddot{x}^*$ while the interior loop provides for force tracking, yielding a robust response.

Applying such robust motion control systems also implies, as shown above, speed calculation from the measured position [5]. Results obtained at low speed show remarkable performance (Figure 7.14) [5].

We should also mention that so far inertia was considered known. In reality even this parameter has to be estimated and this is why the block diagrams so far contain the

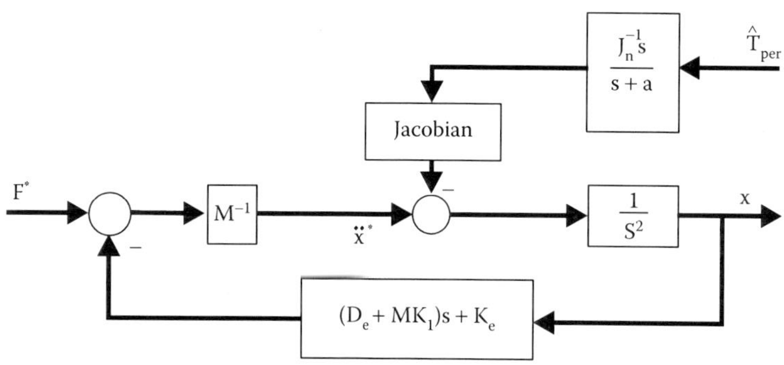


**FIGURE 7.13**
Multivariable force control.

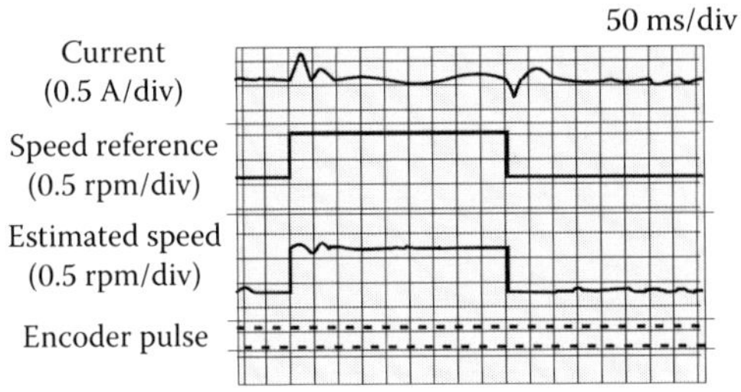


**FIGURE 7.14**
Motion control at low speeds with a speed observer with average speed input (4N = 5000 pulses/rev, with a pole z = 0.3). (From Ohnishi, K., Matsui, N., Hori, Y., *IEEE Proc.*, 82(8), 1253–1265, 1994.)

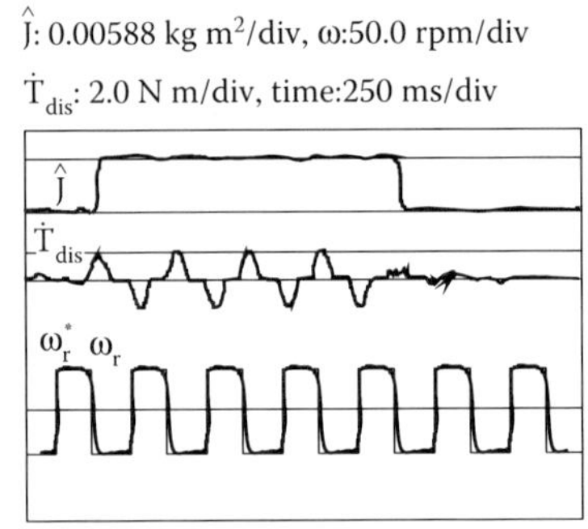


**FIGURE 7.15**
Inertia identification (based on the time interval between encoder pulses $T_2 = 100\ \mu s$, with the speed estimation time $T_1 = 5$ ms; observer poles: z = 0.6, M = 50). (From Ohnishi, K., Matsui, N., Hori, Y., *IEEE Proc.*, 82(8), 1253–1265, 1994.)

estimated inertia $\hat{J}$ [4–6]. To estimate $\hat{J}$ we may use the motion equation in a few moments in time

$$J\dot{\omega}_r(k) = K_{Tn} \cdot I(k) - T_{per}(k) \tag{7.48}$$

$$J\dot{\omega}_r(k-1) = K_{Tn} \cdot I(k-1) - T_{per}(k-1) \tag{7.49}$$

Making use of the acceleration and perturbation torque estimators, we may calculate—through averaging methods (such as the least-squares recursive method)—in real time, the inertia. Consequently, a fully robust motion controller performance is obtained (Figure 7.15) [5].

Though the series of procedures for robust control presented under the umbrella of state-space control seems to exhaust the subject, other less complex but high-performing nonlinear control methods have been proposed. Among them we will treat in some detail the variable structure (sliding-mode) systems, fuzzy control systems, and neural-network control systems.

## 7.7 Sliding-Mode Motion Control

Sliding-mode motion controllers force the system to move on a given surface in the state-space (plane) [7]. The simplest "surface" in the state plane is a straight line. The equation of the sliding mode functional $s_s$ is

$$s_s = c\varepsilon + \frac{d\varepsilon}{dt}; \quad \varepsilon = x^* - x \tag{7.50}$$

where $\epsilon$ is the error of the controlled variable x. For $s_s = 0$, a straight line is obtained.

According to Figure 7.16, in the initial moment, the system starts from point A with an error $\varepsilon_i = OA$ and its derivative is zero. At A, $s_s > 0$. By forcing the command variable (voltage or current) in the positive direction, the motor has to be capable of jumping up to the straight line $s_s = 0$.

Once this is done, the positive and negative commands (±V) alternate depending on the sign of $s_s$ (at a constant switching frequency or with a hysteresis band $h_s$) to run the system to origin where a new steady-state point is reached

$$\begin{aligned} V_c &= +V_0 \quad \text{for } s_s > h_s \\ V_c &= 0 \quad \text{for } |s_s| < h_s \\ V_c &= -V_0 \quad \text{for } s_s < -h_s \end{aligned} \tag{7.51}$$

Such a simple forcing process done through the switching of the command variable between three discrete values leads, however, to oscillations around the target position (chattering) unless the switching frequency is high enough. The system's behavior seems to depend only on the constant c and thus only the limits within which the motor parameters and load torque vary are to be known in order to allow for sliding along the straight line $s_s = 0$. However, the sliding-modes existence conditions also have to be fulfilled [7].

$$s_s \cdot \dot{s}_s < 0 \tag{7.52}$$

The simplifications introduced by the sliding-mode robust control seem notable but operation without chattering requires some additions such as a perturbation observer [8].

The block diagram of a motion (position) control system with sliding modes for a robot is shown in Figure 7.17.

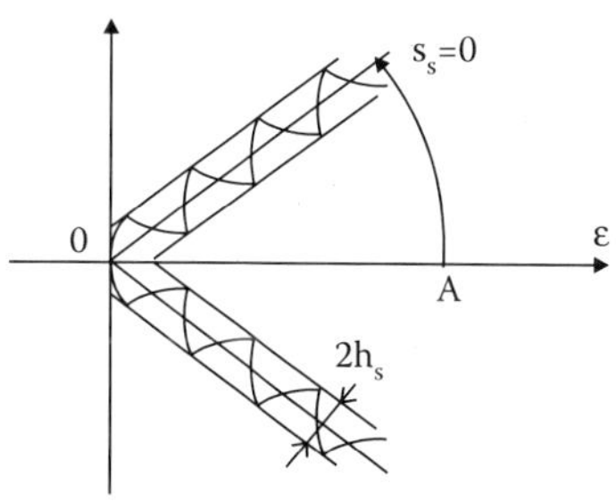


**FIGURE 7.16**
Sliding-mode functional.

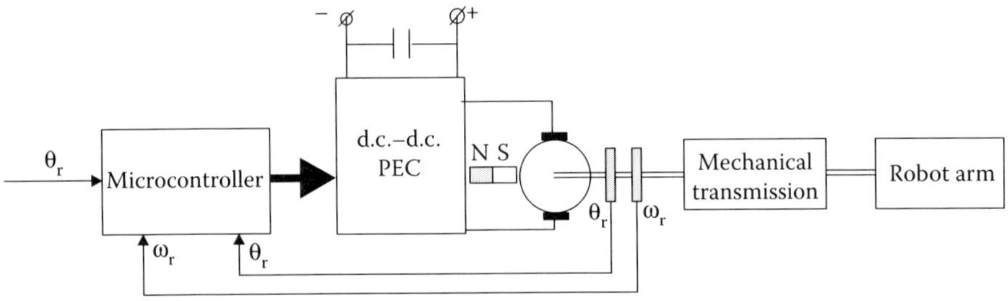


**FIGURE 7.17**
Sliding-mode motion control system for a robot arm.

The mathematical model of the PM d.c. brush motor for position control, accounting for the mechanical transmission ratio $K_{mT}$, the friction torque coefficient D, and the total load torque $T_{per}$, has a second order only if the motor inductance is considered zero

$$\ddot{\theta}_r = \frac{-(K_E K_T + DR)\dot{\theta}_r}{RJ} + \frac{K_E K_{mT} V_a}{RJ} - \frac{K_{mT}}{J} T_{per} \tag{7.53}$$

where D is the friction coefficient, J is the total inertia referred to the motor shaft, $K_E$ is the PM motion-induced voltage coefficient ($E = K_E \dot{\theta}_r$), $K_T$ is the motor torque coefficient ($T_e = K_T I$; $K_T = \lambda_{PM}$), and $V_a$ is the motor armature voltage.

Two new variables $x_1$ and $x_2$ are introduced

$$x_1 = \overset{*}{\theta}_r - \theta_r \quad \text{and} \quad x_2 = \frac{dx_1}{dt} \tag{7.54}$$

with

$$V = aV_a; \quad a = \frac{K_E K_{mT}}{JR}$$
$$b = \frac{DR + K_E K_T}{JR}; \quad f = \frac{K_{mT} T_{per}}{J} \tag{7.55}$$

Equation 7.53 becomes

$$\dot{x}_1 = x_2$$
$$\dot{x}_2 = -bx_2 - V + f \tag{7.56}$$

The sliding-mode command voltage is chosen in the form [9]

$$V_c = \gamma x_1 + K_f \operatorname{sgn}(s_s) \tag{7.57}$$

with

$$\gamma = \begin{cases} \alpha & \text{if } s_s x_1 > 0 \\ \beta & \text{if } s_s x_1 < 0 \end{cases} \tag{7.58}$$

The state functional is given by Equation 7.50 while the sliding-mode existence condition is given by Equation 7.52. The existence condition Equation 7.52 leads to the inequalities

$$\alpha > c(b - c) \tag{7.59}$$

$$\beta < c(b - c) \tag{7.60}$$

$$K_f > |f|_{max} \tag{7.61}$$

These conditions secure safe jumping of the system to the straight line of the sliding-mode functional. We call $K_f$ a dither signal and it should be greater than a quantity proportional to the largest torque perturbation. Unfortunately once the straight line is reached, the term in $K_f$ in the command voltage (7.57) causes undesirable vibrations.

Figures 7.18 and 7.19 show results obtained with such a control system implemented on a DSP (NEC: μPD77230, 32 bit) with a sampling time of 200 μs, 20 kHz switching frequency, for a SCARA robot with two axes [8].

The perturbation, which in the first case ($K_f$ small, Figure 7.18) produces a nonzero steady-state position error, is caused by pulling down the robot arm with 300 g hanged through a spring (about 20% loading). The undesirable mechanical vibrations are notable.

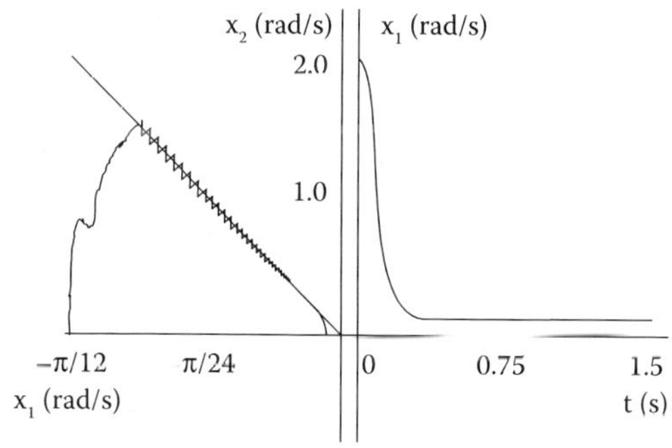


**FIGURE 7.18**
Sliding-mode control (7.57) with low $K_f$.

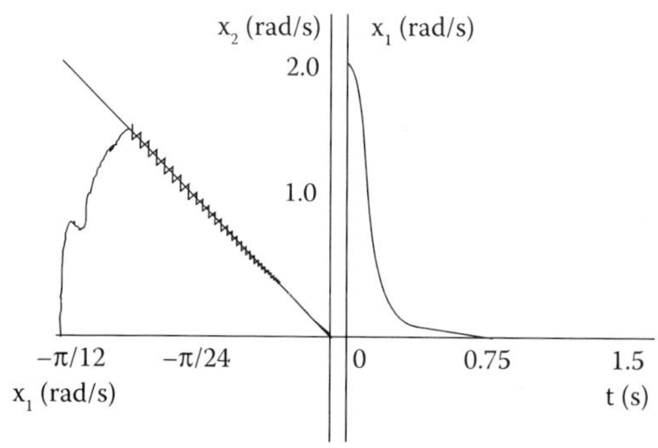


**FIGURE 7.19**
Sliding-mode control with large $K_f$.

Elimination of such vibrations is accomplished through the contribution of a perturbation observer with a reduction of $K_f$. If stepwise perturbations are assumed, the equations are

$$\dot{T}_{per} = 0$$
$$\dot{x}_2 = \frac{K_{mT}T_{per}}{J} - b\hat{x}_2 - V \tag{7.62}$$

It may be proved that the system of Equation 7.62 is observable and, to calculate (observe) $T_{per}$ and $x_2$, a complete observer may be used

$$\dot{\hat{T}}_{per} = K_1(x_2 - \hat{x}_2) \tag{7.63}$$

$$\dot{\hat{x}}_2 = \frac{K_{mT}\hat{T}_{per}}{J} - b\hat{x}_2 - V + K_2(x_2 - \hat{x}_2) \tag{7.64}$$

The two amplifications $K_1$ and $K_2$ may be calculated by the method of pole allocation. Complex poles with negative real part are assumed. The positive perturbation reaction $V_0$ becomes

$$V_0 = \frac{K_{mT}T_{per}}{J} \tag{7.65}$$

Finally, the complete command voltage $V_{cc}$ is (Figure 7.20)

$$V_{cc} = V_c + V_0 = \gamma x_1 + K_f \operatorname{sgn}(s_s) + V_0 \tag{7.66}$$

The stability of such a nonlinear system may be studied through Lyapunov's theorem. A Lyapunov function $V(s_s, \Delta x_2, \Delta T_{per})$ is first defined as

$$V(s_s, \Delta x_2, \Delta T_{per}) = \frac{1}{2}s_s^2 + \frac{1}{2}\left[\frac{1}{K_1}(\Delta T_{per})^2 + \frac{J}{K_{mT}}(\Delta x_2)^2\right] \tag{7.67}$$

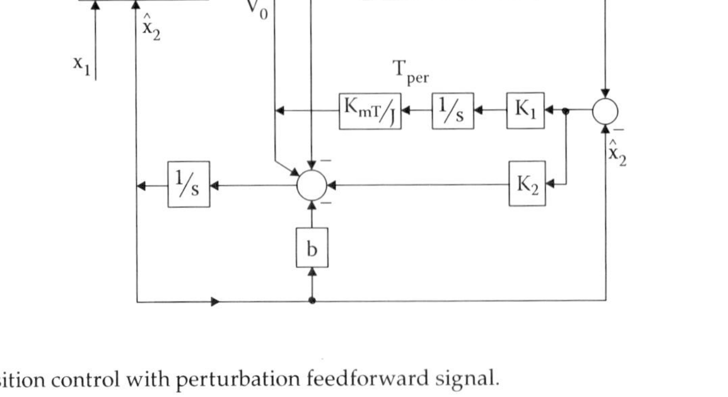


**FIGURE 7.20**
Sliding-mode position control with perturbation feedforward signal.

The sufficient condition for stability is satisfied if $\dot{V} < 0$

$$\dot{V} = (c-b)|\hat{s}_s|^2 - [c(c-b)+\gamma]\hat{s}_s\Delta\hat{x}_1$$
$$- \left[K_f - K_2\Delta\hat{x}_2 \ \mathrm{sgn}(\hat{s}_s)\right]|\hat{s}_s| - (b+K_2)\frac{J}{K_{mT}}(\Delta\hat{x}_2)^2 < 0 \quad (7.68)$$

If each term in Equation 7.68 is negative, then $\dot{V} < 0$

$$\begin{array}{ll} c \le b; & \alpha > c(b-c) \\ \beta < c(b-c); & K_f > K_2|\Delta\hat{x}_2|_{max} \\ K_1 > 0; & K_2 > -b \end{array} \quad (7.69)$$

In case of a high step-torque perturbation ($K_{mT}T_{per}/J = 0.15$, $K_f = 0.01$), $K_f$ becomes too small to fulfill Equation 7.69 and thus the controller regains stability only after $\Delta x_2$ increases notably (Figure 7.21) [8].

Because the perturbation observer is of the step form, a ramp perturbation causes errors in the estimated speed $x_2$ and thus the last two stability conditions in Equation 7.69 have to be reformulated as [8]

$$K_f > \frac{K_2}{K_1}T_{per}^{ramp}; \quad T_{per}^{ramp}(s) = \frac{T_{per}^{ramp}}{s} \quad (7.70)$$

The stability analysis for this case leads to new conditions related to $K_f$

$$cT < 1 \quad (7.71)$$

$$TK_f < |c\hat{x}_1(k+1) + \hat{x}_2(k+1)| \quad (7.72)$$

The same control system may be extended to position tracking when the reference position, speed, and acceleration, $\theta_r^*$, $\dot{\theta}_r^*$, $\ddot{\theta}_r^*$ are known time functions. In this case we invert the system (7.54) with Equations 7.55 and 7.56

$$V_a = \frac{1}{a}(b\dot{\theta}_r + \ddot{\theta}_r^* + hx_2 + V_{cc}) \quad (7.73)$$

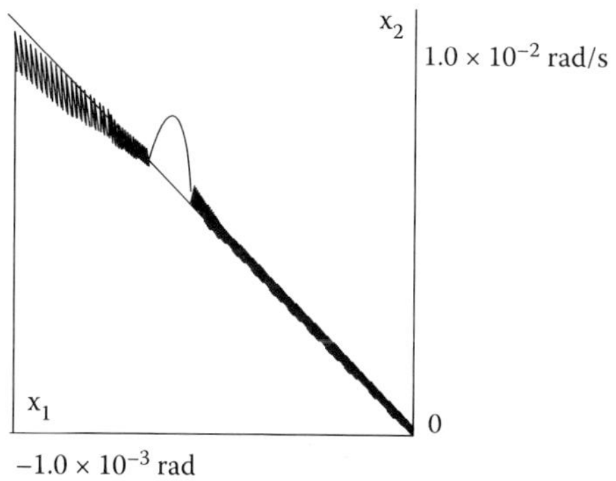


**FIGURE 7.21**
System response to stepwise perturbation ($K_{mT}T_{per}/J = 0.15$, $K_f = 0.01$).

Comparing Equation 7.53 with Equation 7.73 yields

$$\dot{x}_2 = -hx_2 - V_{cc} + \frac{K_{mT}T_{per}}{J} \tag{7.74}$$

Equation 7.73 represents the new control law which embeds Equation 7.66 and adds three new terms for good path tracking. Figure 7.22 presents a triangle drawn by the robot SCARA (as a plotter) in the presence of a 20% torque perturbation (obtained through 300 g hanged on the robot arm through a mechanical spring).

The maximum speed is 0.5 m/s [8], $\alpha = 2550$, $\beta = -2550$, $c = 40$, $h = 100$, the observer poles are $p_{1,2} = Lb(-\cos\gamma + j\sin\gamma)$ with $L = 30$, $\gamma = 30°$.

Practical experience has shown that if in the trajectory planning high accelerations ($\theta_r{}^*$) are allowed, vibrations occur in the system's response. A quadratic function E, to minimize along the trajectory, may be chosen [9]

$$E = \int_0^t \left\{[f(x,y)]^2 + A_1(\ddot{\theta}^*_{r1})^2 + A_2(\ddot{\theta}^*_{r2})^2\right\} \cdot dt \tag{7.75}$$

$$A_1 > 0; \quad A_2 > 0$$

with

$$x = L_1 \cos\theta^*_{r1} + L_2 \cos(\theta^*_{r1} + \theta^*_{r2}) \tag{7.76}$$

$$y = L_1 \sin\theta^*_{r1} + L_2 \sin(\theta^*_{r1} + \theta^*_{r2}) \tag{7.77}$$

where f(x,y) is the required trajectory in plane.

The scope of the optimization process is to determine $\theta^*_{r1}$, $\dot{\theta}^*_{r1}$, $\ddot{\theta}^*_{r1}$, and $\theta^*_{r2}$, $\dot{\theta}^*_{r2}$, $\ddot{\theta}^*_{r2}$ along the two axes as time functions such that the penalty function is minimum. Through variational methods combined with spline trajectory approximations, the above problem may

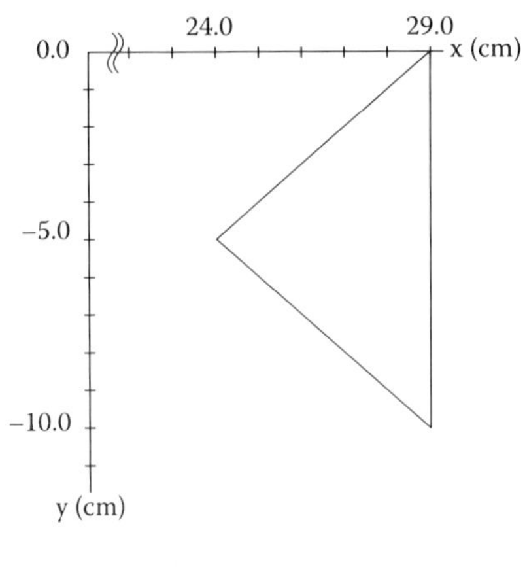


**FIGURE 7.22**
Triangular trajectory in the presence of a 20% torque perturbation.

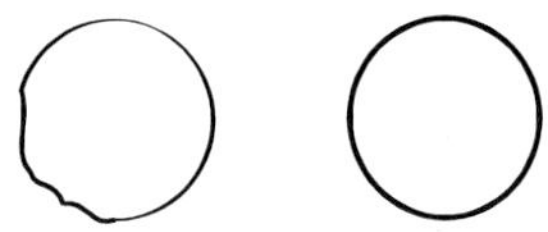

**FIGURE 7.23**
The circle drawn by SCARA robot without and with the reduction of reference acceleration (total time 1.6 s.) (From Kawamura, A., Itoh, H., Sakamoto, K., *IEEE Trans.*, 30(2), 456–461, 1994.)

be solved iteratively. Figure 7.23 shows the results obtained with SCARA robot in drawing a circle without and with acceleration reduction.

Far from exhausting the subject, the sliding-mode motion control with torque perturbation compensation, as applied to robotics, merely suggests the complexity of motion robust control even with the motor electrical time constant (constants) considered zero.

For systems with vague mathematical models, new motion control systems such as fuzzy systems and neural network (NN) systems have been recently proposed. Their first implementation has been done on PM d.c. brush motors [10] and then on a.c. motors [11,12].

## 7.8 Motion Control by Fuzzy Systems

Fuzzy systems stem from a logic which treats vaguely known plants by membership functions (MFs) with values between 0 and 1.

In fuzzy sets, based on fuzzy logic, an object (variable) has a membership degree to a given set with values between 0 and 1 [13].

A fuzzy variable has linguistic values, for example, LOW, MEDIUM, HIGH, which may be defined through bell shape (Gauss) MFs with gradual variation (Figure 7.24).

The MFs are, in general, either symmetric or nonsymmetric, triangular or trapezoidal.

At 750 rpm (Figure 7.24), for example, the variable belongs with 50% (MF = 0.5) to LOW and with 50% (MF = 0.5) to MEDIUM. All possible values of a variable constitute the universe of discourse.

The properties of Boolean theory remain valid for fuzzy sets. Reunion (corresponding to OR) is

$$\mu_{A\cup B}(x) = \max[\mu_A(x), \mu_B(x)] \tag{7.78}$$

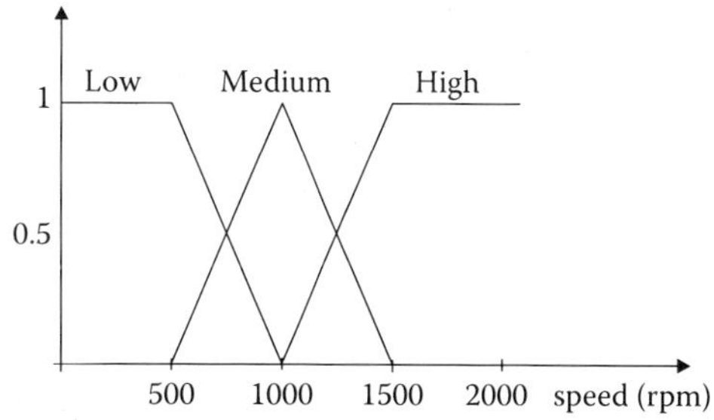


**FIGURE 7.24**
Fuzzy sets defined through MFs.

Intersection (corresponding to AND) writes

$$\mu_{A \cap B}(x) = \min[\mu_A(x), \mu_B(x)] \tag{7.79}$$

The negation (corresponding to NOT) is

$$\mu_A(x) + \mu_{\Gamma A}(x) = 1 \tag{7.80}$$

Triangular MFs OR, AND, and NOT are illustrated in Figure 7.25.

While conventional control systems are based on the mathematical model of the plant, fuzzy control is based on the intuition and experience of the human operator. And thus, besides self-tuning adaptive systems, reference model adaptive systems, variable structure systems, for plants with a vaguely known model, fuzzy control is clearly opportune and adequate.

In essence, implicitly, fuzzy motion control is self-adaptive and thus its robustness becomes apparent.

The command law in fuzzy systems is of the form

$$\text{IF } x = A \quad \text{AND} \quad y = B \quad \text{THEN} \quad z = C \tag{7.81}$$

where x, y, z are fuzzy variables where the universe of discourse is A, B, C. For example, for the speed control of a PM d.c. brush motor, the fuzzy variables are the speed error $E = \omega_r^* - \omega_r$ and its derivative $CE = \dot{\omega}_r^* - \dot{\omega}_r$. Let us remember that the same variables have been used to express the sliding-mode functional. The rule number one may be of the form: if E is zero (Z) and the error derivative CE is negative small (NS), then the increment $\Delta V_1$ of the command voltage is NS.

The linguistic variables Z, NS, and ΔV are defined through membership symmetric functions (Figure 7.26).

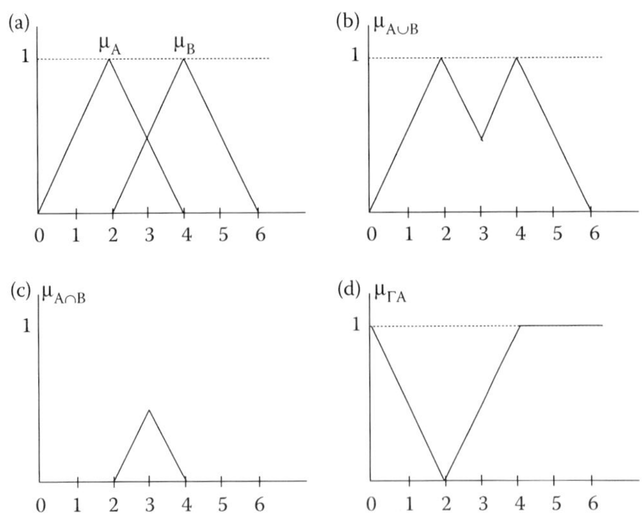


**FIGURE 7.25**
(a) Fuzzy sets A and B; (b) reunion; (c) intersection; and (d) negation.

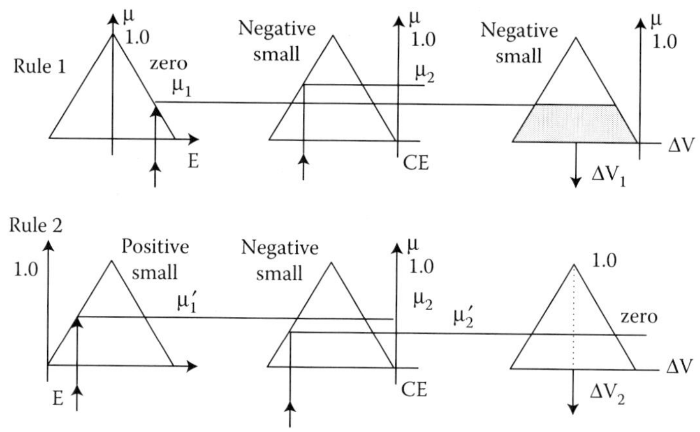


**FIGURE 7.26**
The MAX–MIN (SUP-MIN) fuzzy rule composition.

The command law $\Delta V_1$ may be obtained graphically but two rules may emerge, the second with the output $\Delta V_2$. The effective control will be an average between two outputs $\Delta V_1$ and $\Delta V_2$. First, a basis of fuzzy rules is usually constituted (Table 7.1) and the corresponding MFs are described (Figure 7.27).

The composition operation refers to the calculus of the command law. Among the composition methods we mention the MAX-MIN (SUP MIN) and MAX-DOT. The MAX–MIN method already illustrated in Figure 7.26 is written in the form

$$\mu_u(u) = SUP_x[\min \mu_x(x) \cdot \mu_R(x,u)] \tag{7.82}$$

Consequently, the membership function (MF) for each rule is given by the minimum value (MIN) and the combined output is given by the supreme maximum of all rules. The general structure of a single input fuzzy motion controller is given in Figure 7.28.

The command variable (voltage or current in our example) is determined based on the two variables e and ce (in relative values) calculated from the absolute values E and CE by division through the corresponding amplifications.

**TABLE 7.1**
Fuzzy Logic Variables

| CE/E | NB | NM | NS | Z | PS | PM | PB |
|---|---|---|---|---|---|---|---|
| NB | NVB | NVB | NVB | NB | NM | NS | Z |
| NM | NVB | NVB | NB | NM | NS | Z | PS |
| NS | NVB | NB | NM | NS | Z | PS | PM |
| Z | NB | NM | NS | Z | PS | PM | PB |
| PS | NM | NS | Z | PS | PM | PB | PVB |
| PM | NS | Z | PS | PM | PB | PVB | PVB |
| PB | Z | PS | PM | PB | PVB | PVB | PVB |

*Note:* Z: zero, PS: positive small, PM: positive medium, PB: positive big, PVB: positive very big, NS: negative small, NM: negative medium, NB: negative big, NVB: negative very big.

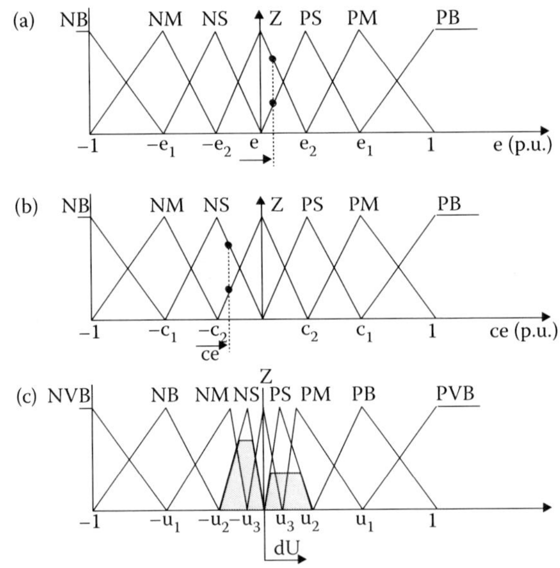


**FIGURE 7.27**
The fuzzy MFs for speed control. (a) Speed error; (b) acceleration error; and (c) command increment.

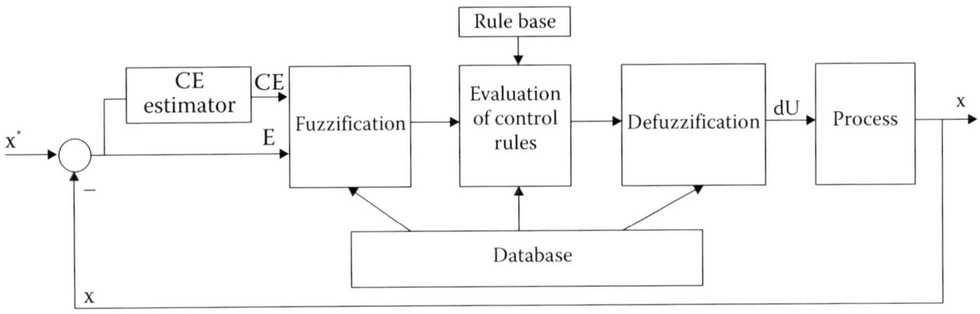


**FIGURE 7.28**
Typical single input fuzzy control system.

Fuzzification may be done by taking definite values of the variables to calculate the MF in these cases (Figures 7.26 and 7.27). Defuzzification may be done in quite a few ways; one of which is the centroid method by which the command variable $U_0$ is obtained through the gravity center of the fuzzy function U (Figure 7.26).

$$U_0 = \frac{\int U\mu_u(u)\cdot du}{\int \mu_u(u)\cdot du} \tag{7.83}$$

and the height method, where the centroid of each function first and, then, a weighted medium value of the heights is calculated

$$U_0 = \frac{\sum_{i=1}^{n} U_i\mu_u(u_i)}{\sum_{i=1}^{n} \mu_u(u_i)} \tag{7.84}$$

Though the fuzzy control looks simple, mathematical implications increases the implementation time. Recently, neural-fuzzy sets that allow the rapid selection of the MFs, rules and command signals have been introduced.

Besides the rule-based method (for fuzzy systems), a relational method has been proposed in Reference 12. According to the relational method, regions look for linear combinations of input functions with W and H values.

The rule number one may be defined: IF W is MEDIUM and H is MEDIUM, then

$$I_s = A_{01} + A_{11}W + A_{21}H \tag{7.85}$$

The coefficients $A_{ij}$ may be determined through linear regression and then adjusted through observation and simulation. Finally, the linear equation of the outputs (7.85) is defuzzified, that is a weighted average of the components $I_{s1}$ and $I_{s2}$ by the MFs $\mu_1$ and $\mu_2$ is calculated

$$I_s = \frac{I_{s1}\mu_1 + I_{s2}\mu_2}{\mu_1 + \mu_2} \tag{7.86}$$

To control the speed of a PM d.c. brush motor (with the rule-based method, Table 7.1), the command signal is

$$U(k) = U(k-1) + GU \cdot du \tag{7.87}$$

where GU is the voltage amplification and du the command voltage increment expressed in relative values and calculated during defuzzification through the height method. For a PM d.c. brush motor of 1.84 kW, 1800 rpm, $J = 0.0465$ kg m$^2$, $R = 0.6\ \Omega$, $L = 8$ mH, $K_E = 0.55$ V s/rad, $T_s = k_L\omega_r^2$, $K_T = 2.78 \times 10^{-4}$ Nm s$^2$/rad, the system's response to a step-torque perturbation (Figure 7.29), and then to a step increase in inertia (Figure 7.30) demonstrates a remarkable robustness.

If for high speeds the performance is good, at low speed problems could occur and may be solved through a new set of fuzzy rules and commands.

## 7.9 Motion Control through NNs

Neural networks (NNs) represent systems of interconnections between artificial neurons which emulate the neuron system of the human brain. NN constitutes a more general form of artificial intelligence than the expert systems or fuzzy sets [11]. The model of an artificial neuron is as in Figure 7.31.

Each input continuous signal is passed through weighting synapses—positive (exciters) or negative (inhibitors)—toward the summation mode, then, through a nonlinear transfer function, to output. The transfer function may be stepwise, limit form, sign form (sgn), linear with a limit or nonlinear, of the sigmoid form, for example

$$y = \frac{1}{1 + e^{-\alpha_s x}} \tag{7.88}$$

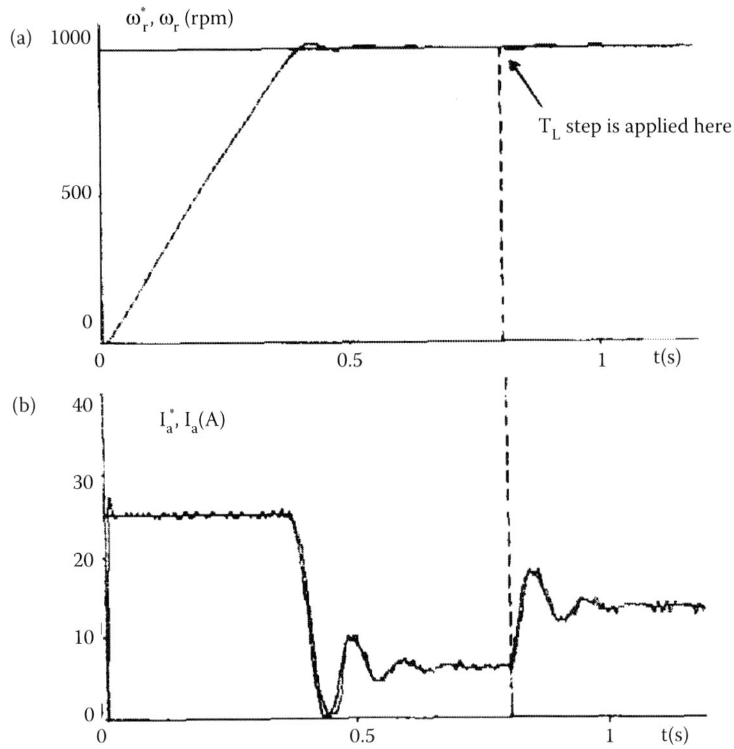


**FIGURE 7.29**
A fuzzy controller; the response to torque perturbation after step speed response: (a) speed, (b) current. (From Sousa, G.D., Bose, B.K., *IEEE Trans.*, 30(1), 34–44, 1994.)

For large $\alpha_s$, y gets close to a step function and varies between 0 (for $x = -\infty$) to 1 (for $x = +\infty$). Though it is still a mystery how neurons are interconnected in the human brain, more than 60 models of artificial NNs have been proposed to be applied in science and technology. They may be classified in NNs with positive and negative reaction. The first ones are dominant so far. Such a positive reaction model with three layers—input, hidden (screened), and output—is shown in Figure 7.32.

The relationships between the input and output variables are of the form

$$[V]_b = [W]_{ba}[X]_a \tag{7.89}$$

$$[Y]_c = [W]_{cb}[V]_b \tag{7.90}$$

The 25 weights $W_{i,j}$ (on Figure 7.32) have to be calculated through an iterative educational process by comparing a large number of desired types of output signals with the actual ones.

The back propagation educational algorithm, which has received rather wide acceptance lately, starts with a few initial values and, based on a penalty function (such as the minimum of the sum of squared errors), continues through the descent gradient method and changes the weights $W_{i,j}$.

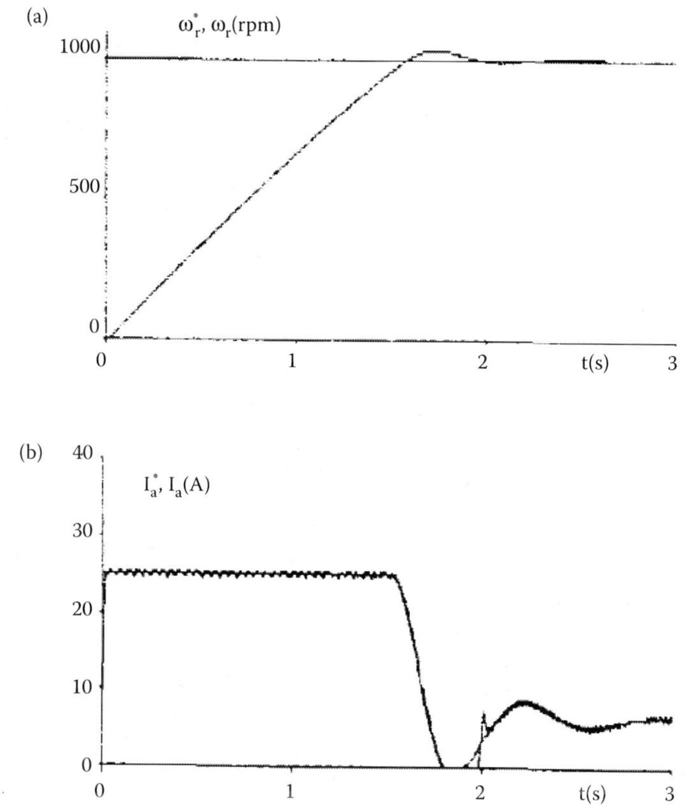


**FIGURE 7.30**
Fuzzy speed controller; the response to inertia variation: (a) speed, (b) current. (From Sousa, G.D., Bose, B.K., *IEEE Trans.*, 30(1), 34–44, 1994.)

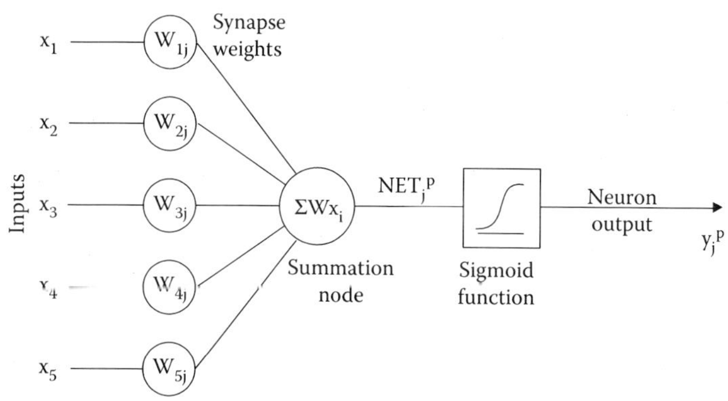


**FIGURE 7.31**
The structure of an artificial neuron.

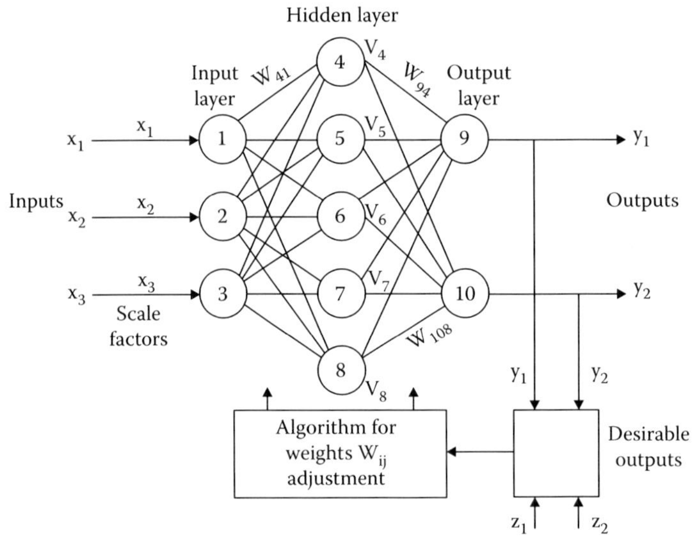


**FIGURE 7.32**
Positive reaction neural model.

For the case of a single layer (Figure 7.31) the network equations, based on Equations 7.89 and 7.90, become

$$NET_j^p = \sum_1^N W_{ij} X_i \tag{7.91}$$

$$Y_i^p = f_i(NET_j^p) \tag{7.92}$$

For the pth input signal, the squared error $E_p$ for all the neurons of input layer is

$$E_p = \frac{1}{2}(d^p - y^p)^2 = \frac{1}{2}\sum_{j=1}^{s}(d_j^p - y_j^p)^2 \tag{7.93}$$

For all the input signals the resultant error E is

$$E = \sum_{p=1}^{P} E_p = \frac{1}{2}\sum_{p=1}^{P}\sum_{j=1}^{s}(d_j^p - y_j^p)^2 \tag{7.94}$$

The weights $W_{i,j}$ will be adjusted until E reaches a minimum

$$W_{ij}(t+1) = W_{ij}(t) + \eta\left(\frac{\partial E_p}{\partial W_{ij}(t)}\right) + \alpha[W_{ij}(t) - W_{ij}(t-1)] \tag{7.95}$$

The iterative procedure propagates the error backwards and provides for a global minimum by adequately choosing the educational rate η and the coefficient α

$$\eta(t+1) = U \cdot \eta(t); \quad U > 0 \tag{7.96}$$

Special education algorithms with backpropagation are now available.

## 7.10 Neuro-Fuzzy Networks

The neuro-fuzzy networks do the calculations, required in the fuzzy systems by the rule-based or relational methods, through which the rules and MFs are identified.

For the fuzzy speed control (Figure 7.27) a neuro-fuzzy network is given in Figure 7.33.

In essence, the fuzzy control, with the relational method used for command signal calculation, is done through the neuro-fuzzy controller. The quantities $W_c$ and $W_g$ create the space between the MFs and, respectively, their ramp. Nine rules occur. The linear functions (for the relational method) are given on the lower part of Figure 7.33 while defuzzification is shown on the right side of the same figure.

Besides positive reaction, negative reaction NNs [14] have been proposed.

### 7.10.1 Applications of Neural Networks

It is also possible to use NN to selectively eliminate harmonics for pulse width modulation in static power converters in the sense that the NN may be educated through the voltage

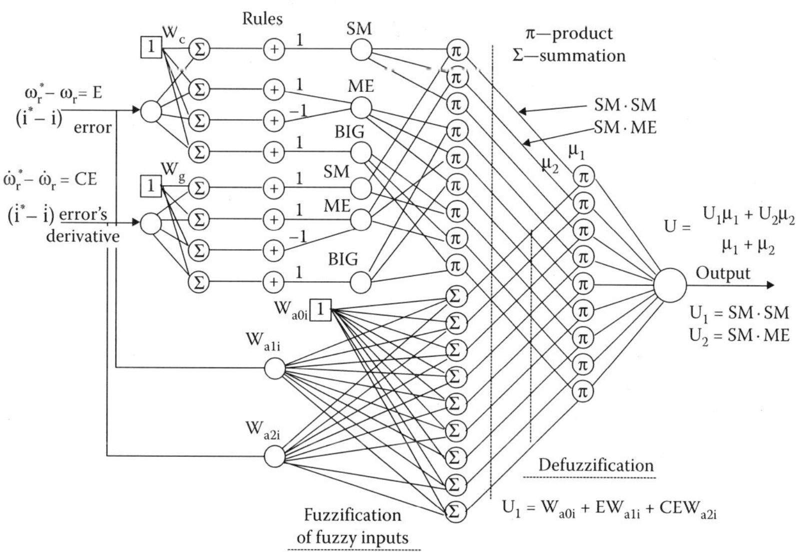


**FIGURE 7.33**
A neuro-fuzzy speed controller.

chopping angles. Education of NN may be done "on line" or "off line" with applications in diagnosis and monitoring and in intelligent motion control. Also flux linkage, torque, or active power may be estimated through NN, based on measured voltages and currents [15]. Other applications include a.c. current controllers for a.c. motors [16] or d.c. brush motor control [17].

The NN may be implemented through software in microcontrollers or on special hardware. A few dedicated IC chips for the scope are already on the market such as INTEL 8017 NxETANN, Microdevice MD 1220NBS, and Neural Semiconductor NUSU32.

## 7.11 Summary

This chapter presented a synthesis of classical—cascaded—and nonlinear motion closed-loop control systems for electrical drives.

- The cascaded—torque, speed—controllers are presented in detail with the PECs simulated by simple gains.
- The digital position controller performed through a single digital filter is illustrated by an example where the PEC is simulated by a gain and a zero-order hold.
- The nonlinear motion control systems—a science in itself—is introduced here only through a few methods such as state-space control with torque and speed observers, sliding mode control, fuzzy control, and NNs control with recent results from literature, all related to the PM d.c. brush motor drives. Our introduction to this subject is only a starting point for the interested reader who should pursue the selected literature at the end of this chapter.
- Though applied only to d.c. brush motors the issues in this chapter are far more general and may be applied to vector-controlled a.c. motors.
- New control options such as super-twisting sliding mode control [18] or predictive control [19] have been proposed recently, with very promising results.

## 7.12 Problems

7.1 Calculate the response of the current controller in Section 7.2.1 for a step increase of reference current of 10 A.

7.2 A d.c. brush motor with the data $R_a = 0.25\ \Omega$, rated current 20 A, rated (source) voltage $e_0 = 110$ V, rated speed $n_n = 1800$ rpm is fed through a power transistor chopper in the continuous (constant) current mode. The inertia $J = 0.1$ kg m$^2$. The mechanical time constant $\tau_m = 2\pi J/B = 10$ s. The electrical time constant $\tau_e = 40$ ms and thus the electrical phenomena are much faster than the mechanical ones. The drive has a proportional (P) current controller with a current sensor having $K_i = 0.5$ V/A ratio and a PI speed controller. Design the speed controller—find $K_{sn}$ and $\tau_{sn}$—for a damping ratio of $1/\sqrt{2}$ and a natural frequency $\omega_n = 10$ rad/s. The speed sensor gain is $K_n = 0.1$ V s/rot.

7.3 For the speed controller d.c. brush motor system in Problem 7.2 find the speed–load torque transfer function and show that the steady-state error in speed to a torque perturbation is zero.

7.4 A d.c. brush series motor has the data: total (armature + field) resistance $R_{at} = 1\ \Omega$, total inductance $L_{at} = 40$ mH, rated voltage $V_n = 120$ V, rated speed $n_n = 1800$ rpm, rated current 10 A. The remnant flux-induced voltage at rated speed is $e_{rem} = 5$ V and the magnetizing curve is otherwise linear and $J = 10^{-2}$ kg m$^2$.

a. Determine the open-loop transfer functions $i_a(p)/e_a(p)$, $n(p)/i_a(p)$ for zero load torque variation, after linearization around the rated point ($e_a$—terminal voltage).

b. Determine the open-loop transfer functions $i_a(p)/T_L(p)$, $n(p)/i_a(p)$ for zero input voltage variation, after linearization around the rated point.

## References


1. B. Kisacanin, G.C. Agarwal, *Linear Control Systems*, Kluwer Academic Press, New York, 2001.
2. P. Katz, *Digital Control Systems Using Microprocessors*, Prentice-Hall, Englewood Cliffs, New Jersey, 1982.
3. H.F. Vanlandingham, *Introduction to Digital Control Systems*, Macmillan, New York, 1985.
4. R. Lorenz, T.A. Lipo, D.W. Novotny, Motion control with induction motors, *IEEE Proc.*, vol. 82, no. 8, 1994, pp. 1215–1240.
5. K. Ohnishi, N. Matsui, Y. Hori, Estimation, identification and sensorless control in motion control systems, *IEEE Proc.*, vol. 82, no. 8, 1994, pp. 1253–1265.
6. I. Awaya et al., New motion control with inertia identification using disturbance observer, *Record of IEEE – IECON*, vol. 1, 1992, pp. 77–80 (IEEExplore).
7. V.I. Utkin, Variable structure system with sliding mode, *IEEE Trans.*, vol. AC-27, no. 2, 1977, pp. 212–222.
8. A. Kawamura, H. Itoh, K. Sakamoto, Chattering reduction of disturbance observer based sliding mode control, *IEEE Trans.*, vol. IA-30, no. 2, 1994, pp. 456–461.
9. K. Sakamoto, A. Kawamura, Trajectory planning using optimum solution of variational problem, *Power Conversion Conference*, Yokohama, 1993, pp. 666–671.
10. G.D. Sousa, B.K. Bose, A fuzzy set theory based control of a phase—Controlled converter DC machine drive, *IEEE Trans.*, vol. IA-30, no. 1, 1994, pp. 34–44.
11. B.K. Bose, Expert system, fuzzy logic and neural network applications in power electronics and motion control, *IEEE Proc.*, vol. 82, no. 8, 1994, pp. 1303–1323.
12. T. Takagi, M. Sugero, Fuzzy identification systems and its applications to modeling and control, *IEEE Trans.*, vol. SMC-15, no. 1, 1985, pp. 116–132.
13. L.A. Zadeh, Fuzzy sets, *Inf. Control*, vol. 8, 1965, pp. 338–343.
14. California Scientific Software, *Introduction to Neural Networks*, Grass Valley, California, 1991.
15. M.G. Simoes, B.K. Bose, Feedback signal estimation by neural network, *Record of IEEE-IAS-1994 Annual Meeting*, 1994 (IEEExplore).
16. M.R. Buhland, R.D. Lorentz, Design and implementation of neural networks for digital current regulation of inverter drive, *Record of IEEE-IAS-1991 Annual Meeting, Part 1*, IEEExplore, 1991, pp. 415–423 (IEEExplore).

17. S. Weerasooriya, M.A. El-Sharkawi, Identification and control of a DC motor using back—Propagation neural networks, *IEEE Trans.*, vol. EC-6, no. 6, 1991, pp. 663–669.
18. C. Lascu, I. Boldea, F. Blaabjerg, Super-twisting sliding mode control of torque and flux in PMSM drives, *Industrial Electronics Society, IECON 2013-39th Annual Conference of the IEEE*, IEEExplore, 2013, pp. 3171–3176 (IEEExplore).
19. J. Rodriguez, M.A. Perez, H. Young, H. Abu-Rub, Model predictive speed control of electric machines *Power Electronics*, Chapter 19, eds. H. Abu-Rub, M. Malinowski, K. Al-Haddad, IEEE Press and Wiley, 2014.

# 8

# *Induction Motors for Drives*

The induction motor is considered to be the workhorse of industry. It is an a.c. motor, either three-phase or (for low powers) single-phase. Industrial (conventional) induction motors are supplied from constant voltage and frequency industrial power grids for rather constant speed operation. For variable speed drives, induction motors are fed from PECs at variable voltage amplitude and frequency.

Like the d.c. motor, an induction motor consists of a stator (the fixed part) and a rotor (the moving part) mounted on mechanical bearings and separated from the stator by an airgap.

## 8.1 Stator and Its Traveling Field

The stator consists essentially of a magnetic core made up of punchings (laminations)—0.1–0.5 mm thick—carrying slot-embedded coils. These coils are interconnected in a certain fashion to constitute the so-called a.c. armature (primary) winding (Figure 8.1).

The three-phase windings may be placed in slots in a single layer (Figure 8.1) or in two layers. All the coils are, in general, identical and the span is close or equal to what is called the pole pitch $\tau$, or the half-period of m.m.f. of that coil or phase. The number of poles per periphery is denoted as 2p.

Under each pole there are three zones, one for each phase. Each phase zone per pole contains q slots (q = 2–8, integer in general). The consecutive phase zones of the same polarity, $a_+a_+$, $b_+b_+$ in Figure 8.1, are spatially shifted by a geometrical angle of $2\pi/3p$ which corresponds to 2/3 of pole pitch of the winding.

As one pole pitch corresponds to a semiperiod (180° electrical degrees), the electrical angle $\alpha_e$ is related to the mechanical angle $\alpha_g$ by

$$\alpha_e = p \cdot \alpha_g \tag{8.1}$$

The m.m.f. of each phase coil has a stepwise waveform (Figure 8.2) which is assimilated by a sinusoidal distribution with the semiperiod $\tau$ (pole pitch). There are also harmonics which, in general, produce parasitic torques, augmented by the rotor and stator slot openings.

As the phase current is sinusoidal, the phase m.m.f. fundamental $F_{a1}(x, t)$ may be written as

$$F_{a1}(x,t) = F_{a1m} \cdot \sin\frac{\pi}{\tau}x \cdot \sin\omega_1 t \tag{8.2}$$

$$i_a(t) = I\sqrt{2}\sin\omega_1 t \tag{8.3}$$

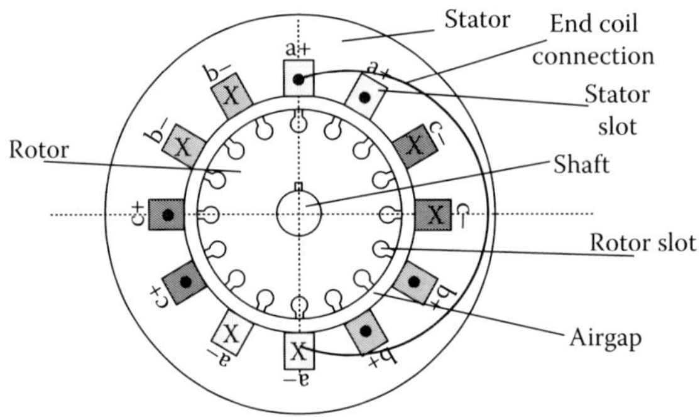


**FIGURE 8.1**
Cross section of an induction motor with two poles.

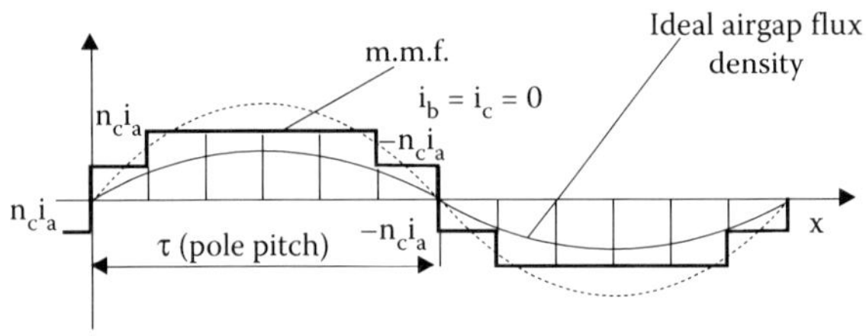


**FIGURE 8.2**
The m.m.f. and airgap flux density of phase a.

As the other two phases b, c are space shifted by an electrical angle of $2\pi/3$ and their currents $i_b$, $i_c$ (in time) by $2\pi/3$, the m.m.f. of phases b and c are

$$F_{b1}(x,t) = F_{a1m} \cdot \sin\left(\frac{\pi}{\tau}x - \frac{2\pi}{3}\right) \cdot \sin\left(\omega_1 t - \frac{2\pi}{3}\right) \quad (8.4)$$

$$F_{c1}(x,t) = F_{a1m} \cdot \sin\left(\frac{\pi}{\tau}x + \frac{2\pi}{3}\right) \cdot \sin\left(\omega_1 t + \frac{2\pi}{3}\right) \quad (8.5)$$

The resultant stator m.m.f. fundamental $F_{s1}(x, t)$ is the sum of $F_{a1}$, $F_{b1}$, and $F_{c1}$

$$F_{s1}(x,t) = \frac{3}{2} F_{a1m} \cdot \cos\left(\frac{\pi}{\tau}x - \omega_1 t\right) \quad (8.6)$$

This is evidently a wave traveling along the rotor periphery with the linear speed $U_s$ obtained from

$$\frac{\pi}{\tau}x - \omega_1 t = \text{const.} \quad (8.7)$$

Its increment should be zero

$$\frac{\pi}{\tau}dx - \omega_1 dt = 0$$

Finally

$$U_s = \frac{dx}{dt} = \tau \cdot \frac{\omega_1}{\pi} = 2\tau f_1 \quad (8.8)$$

where $f_1$ is the primary frequency.

As the airgap is uniform, neglecting the slot openings, the airgap flux density (for zero rotor currents) $B_{g10}(x, t)$ is

$$B_{g10}(x,t) \approx \mu_0 \cdot \frac{F_{s1}(x,t)}{g_e} \quad (8.9)$$

where $g_e$ is an equivalent airgap accounting globally for the slot openings and stator and rotor core magnetic saturation.

Consequently, the three-phase stator winding produces—for zero rotor current—a traveling field in the airgap, with the linear speed $U_s = 2\tau f_1$. As the peripheral speed is related to the angular speed $n_1$ and the stator bore $D_i$ by

$$U_s = \pi \cdot D_i \cdot n_1 = 2p \cdot \tau \cdot n_1 \quad (8.10)$$

the angular speed of the traveling field $n_1$ is

$$n_1 = \frac{f_1}{p} \quad (8.11)$$

$n_1$ is also called the synchronous speed since for this speed of the rotor no voltages are induced in the rotor windings.

## 8.2 Cage and Wound Rotors Are Equivalent

The rotor consists of a laminated core with uniform slotting, accommodating either aluminum (copper) bars short-circuited by end-rings (the squirrel cage) (Figure 8.3a) or a three-phase winding (as in the stator) connected to some copper rings and stator-fixed brushes—the wound rotor (Figure 8.3b).

It has been demonstrated that a symmetric cage (with round bars) may be modeled by an equivalent three-phase winding, that is, a wound rotor.

A PEC or variable resistor may be connected to the wound rotor brushes.

## 8.3 Slot Shaping Depends on Application and Power Level

Stator slots are either semiclosed (Figure 8.4a) for low and medium power (hundreds of kW) or open (Figure 8.4b) above hundreds of kW when preformed coils are introduced in

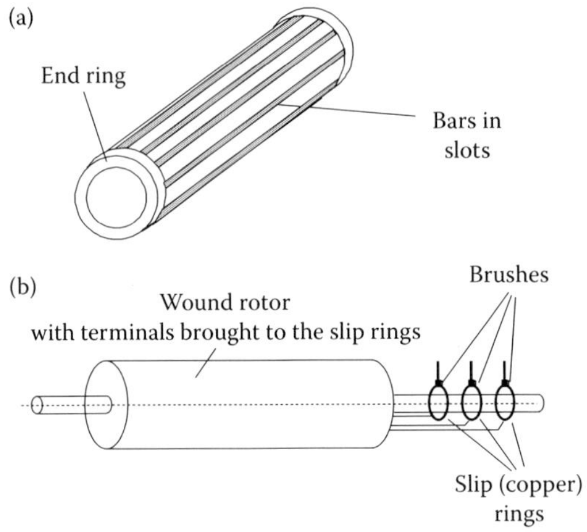


**FIGURE 8.3**
Induction motor rotors: (a) cage-type rotor; (b) wound rotor.

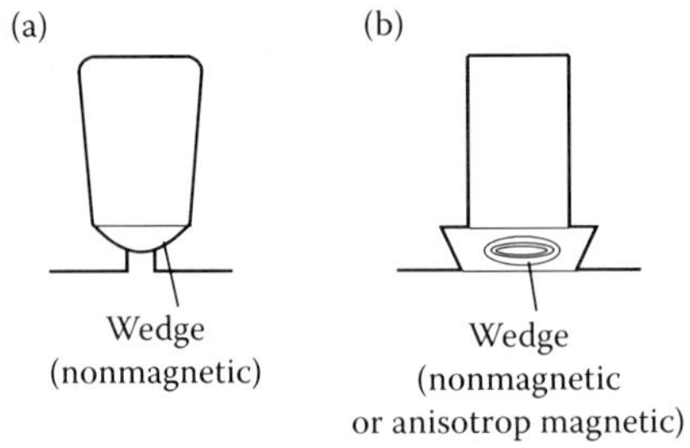


**FIGURE 8.4**
Stator or wound rotor slots: (a) for low power (semiclosed slots); (b) for high power (open slots).

slots. Rotor slots for wound rotors are in general semiclosed if the stator ones are open to allow for a rather small airgap (below 2 mm) even for high powers (MW and above). Cage rotor slot shapes depend on the power and speed levels and on starting torque requirements in constant frequency-fed (industrial) applications.

Round semiclosed slots (Figure 8.5a) do not exhibit notable skin effect at start and may be used for constant frequency-fed low-power low starting torque motors or for variable speed when skin effect is to be avoided.

Skin effect is the concentration of current in the rotor bar toward the upper part of the rotor bar at high rotor frequency (beginning with constant frequency $f_1$, at standstill when the rotor frequency $f_2 = f_1$). The consequence is an apparent increase in rotor resistance and a less important slot leakage inductance reduction.

Double cages are used in the medium power range to reduce the starting current and increase the starting torque (Figure 8.5e and f). Skin effect (deep bar) or double cages imply higher rotor resistance and losses at rated speed and thus are to be avoided in variable speed drives. In variable speed (frequency) drives, slots as in Figure 8.5g are proposed to reduce the rotor surface losses.

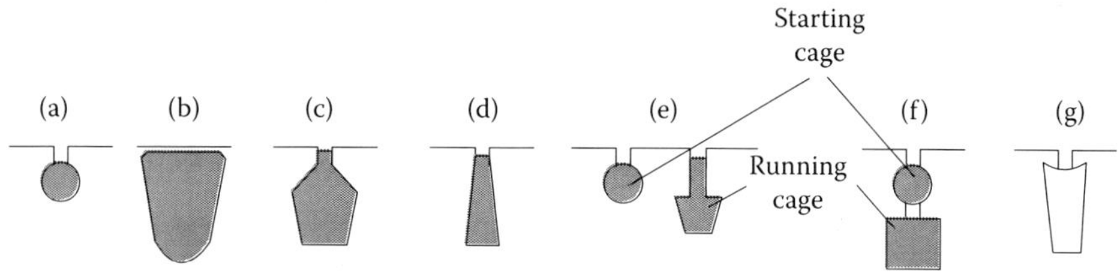


**FIGURE 8.5**
Rotor slots: (a) semiclosed—round for low-power and variable frequency motors, (b) closed—for low-noise or high-speed motors, (c) semiclosed with moderate skin effect and starting torque—constant frequency, (d) semiclosed with high skin effect for high starting torque—constant frequency, (e, f) double cage—for superhigh starting torque—constant frequency, (g) for high-speed inverter-fed motors.

We entered this discussion as many existing motors are now provided with PECs and thus care must be exercised about performance. Moreover, for constant frequency (speed) operation, the efficient motor category has been introduced—though at lower starting torque and higher starting current. These efficient motors have a short payback period and are also more adequate in PEC-fed variable speed drives.

## 8.4 Inductance Matrix

An electric machine is a system of electric and magnetic circuits that are coupled magnetically and electrically. It may be viewed as an assembly of resistances, self-inductances, and mutual inductances. We now briefly discuss these inductances.

As a symmetrical rotor cage is equivalent to a three-phase winding, we will consider the wound rotor induction motor case (Figure 8.6).

There are six circuits (phases) and each has a self-inductance and mutual inductances. Let us note that the self-inductance of phases a, b, c, $a_r$, $b_r$, $c_r$ do not depend on rotor position (slot openings are neglected). They have a main part $L_{ms}$, $L_{mr}$ which corresponds to the flux paths that cross the airgap and embrace all windings on the stator and rotor, and

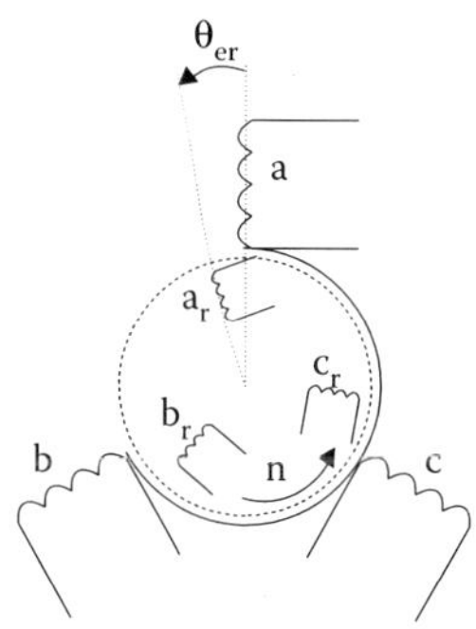


**FIGURE 8.6**
Three-phase induction motor with equivalent wound rotor.

leakage parts $L_{ls}$ and $L_{lr}$ which correspond to the flux paths in the slots, coil end connections, mostly in air, which do not embrace rotor and stator windings.

Let us note that the airgap flux density of each phase m.m.f. is basically sinusoidal along the rotor periphery, and any coupling inductance between phases varies cosinusoidally with the electrical angle of the two windings. Also, mutual inductances on the stator $L_{ab}$, $L_{bc}$, $L_{ac}$ and, respectively, on the rotor $L_{a_rb_r}$, $L_{a_rc_r}$, $L_{b_rc_r}$, though referring mainly to flux paths through the airgap, do not depend on rotor position

$$L_{ab} = L_{bc} = L_{ca} = L_{ms}\cos\frac{2\pi}{3} = -\frac{L_{ms}}{2} \tag{8.12}$$

$$L_{a_rb_r} = L_{b_rc_r} = L_{c_ra_r} = L_{mr}\cos\frac{2\pi}{3} = -\frac{L_{mr}}{2} \tag{8.13}$$

Finally, the stator–rotor coupling inductance $L_{sr}$ depends on rotor position through $\cos\theta_{er}$ as the airgap flux density in the airgap produced by any stator phase is rather sinusoidal

$$L_{sr}(\theta_{er}) = L_{ms}\cdot\frac{W_{re}}{W_{se}}\cos\theta_{er} \tag{8.14}$$

$W_{re}$, $W_{se}$ are the equivalent turns ratio in the rotor and stator phases

$$\frac{W_{re}}{W_{se}} = \sqrt{\frac{L_{mr}^{r}}{L_{ms}}} = K_{rs} \tag{8.15}$$

as self-inductances depend on the number of turns squared.

Consequently, $L_{sr}(\theta_{er})$ is

$$L_{sr}(\theta_{er}) = L_{srm}\cos\theta_{er} \tag{8.16}$$

$$L_{srm} = \sqrt{L_{ms}\cdot L_{mr}^{r}} \tag{8.17}$$

The inductances may be now assembled into the so-called matrix inductance $\left[L_{a,b,c,a_r^r,b_r^r,c_r^r}(\theta_{er})\right]$

$$\left[L_{a,b,c,a_r^r,b_r^r,c_r^r}(\theta_{er})\right] = \begin{bmatrix} & a & b & c & a_r & b_r & c_r \\ a & L_{aa} & L_{ab} & L_{ac} & L_{a_ra} & L_{b_ra} & L_{c_ra} \\ b & L_{ab} & L_{bb} & L_{bc} & L_{a_rb} & L_{b_rb} & L_{c_rb} \\ c & L_{ac} & L_{bc} & L_{cc} & L_{a_rc} & L_{b_rc} & L_{c_rc} \\ a_r & L_{a_ra} & L_{a_rb} & L_{a_rc} & L_{a_ra_r} & L_{a_rb_r} & L_{a_rc_r} \\ b_r & L_{b_ra} & L_{b_rb} & L_{b_rc} & L_{a_rb_r} & L_{b_rb_r} & L_{b_rc_r} \\ c_r & L_{c_ra} & L_{c_rb} & L_{c_rc} & L_{a_rc_r} & L_{b_rc_r} & L_{c_rc_r} \end{bmatrix} \tag{8.18}$$

where

$$L_{aa} = L_{bb} = L_{cc} = L_{ls} + L_{ms}; \quad L_{ab} = -\frac{L_{ms}}{2}; \quad L_{ac} = -\frac{L_{ms}}{2}$$

$$L_{a_r a} = L_{b_r b} = L_{c_r c} = L_{srm} \cos\theta_{er}; \quad L_{a_r a_r} = L_{b_r b_r} = L_{c_r c_r} = L^r_{lr} + L^r_{mr}$$

$$L_{c_r a} = L_{a_r b} = L_{b_r c} = L_{srm} \cos\left(\theta_{er} - \frac{2\pi}{3}\right)$$

$$L_{c_r b} = L_{a_r c} = L_{b_r a} = L_{srm} \cos\left(\theta_{er} + \frac{2\pi}{3}\right)$$

$$L_{a_r b_r} = -\frac{L^r_{mr}}{2}; \quad L_{b_r c_r} = -\frac{L^r_{mr}}{2}$$

## 8.5 Reducing the Rotor to Stator

It is useful to replace the actual rotor winding by an equivalent one with the same performance (and losses) but having the same number of turns per phase as in the stator.

Evidently, for this case, the maximum mutual inductance $L_{srm}$ becomes equal to $L_{ms} = L_{mr}$

$$L_{srm} = L_{ms} \tag{8.19}$$

Denoting the currents and voltages in the rotor with $i^r_{ar}$, $i^r_{br}$, $i^r_{cr}$, $V^r_{ar}$, $V^r_{br}$, $V^r_{cr}$ and the ones reduced to the stator as $i_{ar}$, $i_{br}$, $i_{cr}$, $V_{ar}$, $V_{br}$, $V_{cr}$

$$\frac{i_{ar}}{i^r_{ar}} = \frac{i_{br}}{i^r_{br}} - \frac{i_{cr}}{i^r_{cr}} - K_{rs} \tag{8.20}$$

To conserve the rotor input power

$$\frac{V_{ar}}{V^r_{ar}} = \frac{V_{br}}{V^r_{br}} = \frac{V_{cr}}{V^r_{cr}} = \frac{1}{K_{rs}} \tag{8.21}$$

For equal winding losses and leakage magnetic energy

$$\frac{r_r}{r^r_r} = \frac{L_{lr}}{L^r_{lr}} = \frac{1}{K^2_{rs}} \tag{8.22}$$

So the new inductance matrix $\left[L_{a,b,c,a_r,b_r,c_r}(\theta_{er})\right]$ is similar to that in Equation 8.18, but with $L^r_{mr} \to L_{ms}$, $L_{srm} \to L_{ms}$.

**NOTE:** For a cage rotor, the expression of the rotor to stator reduction factor $K_{rs}$ of Equation 8.15 is more complicated but, as we do not have access to the cage current, dealing directly with the reduced parameters is common practice. For cage motors, $K_{rs}$ is thus required only for motor design purposes.

We may now pursue the mathematical model in phase coordinates (variables).

## 8.6 Phase Coordinate Model Goes to Eighth Order

In matrix form, with rotor reduced to the stator, the voltage–current equations—in stator coordinates for the stator and in rotor coordinates for the rotor—are

$$[V] = [r]\cdot[i] + \frac{d}{dt}[\lambda] \tag{8.23}$$

$$[\lambda] = \left[L_{a,b,c,a_r,b_r,c_r}(\theta_{er})\right]\cdot[i] \tag{8.24}$$

$$[r] = \text{Diag}[r_s, r_s, r_s, r_r, r_r, r_r] \tag{8.25}$$

$$[V] = [V_a, V_b, V_c, V_{ar}, V_{br}, V_{cr}]^T \tag{8.26}$$

$$[i] = \left[i_a, i_b, i_c, i_{ar}, i_{br}, i_{cr}\right]^T \tag{8.27}$$

Using Equation 8.24 in Equation 8.23—with $\theta_{er}$ variable in time in any case—we obtain

$$[V] = [r]\cdot[i] + [L]\frac{d[i]}{dt} + \frac{d[L]}{d\theta_{er}}\cdot[i]\cdot\frac{d\theta_{er}}{dt} \tag{8.28}$$

where

$$\frac{d\theta_{er}}{dt} = \omega_r = p\Omega_r \tag{8.29}$$

$\Omega_r$ is the mechanical angular speed ($\Omega_r = 2\pi n$).

In the absence of magnetic saturation, multiplying Equation 8.28 by $[i]^T$ yields

$$[i]^T\cdot[V] = [i]^T\cdot[r]\cdot[i] + \frac{d}{dt}\frac{1}{2}[[L]\cdot[i]\cdot[i]^T] + \frac{1}{2}\cdot[i]^T\cdot\frac{d[L]}{d\theta_{er}}\cdot[i]\cdot\omega_r \tag{8.30}$$

The first right term represents the winding losses, the second, the stored energy variation, and the third, the electromagnetic power $P_e$:

$$P_e = T_e\cdot\Omega_r = \frac{1}{2}[i]^T\cdot\frac{d[L]}{d\theta_{er}}[i]\cdot\omega_r \tag{8.31}$$

Finally, the electromagnetic torque $T_e$ is

$$T_e = \frac{1}{2}p\cdot[i]^T\cdot\frac{d[L]}{d\theta_{er}}[i] \tag{8.32}$$

The motion equations are

$$\frac{J}{p}\frac{d\omega_r}{dt} = T_e - T_{load}; \quad \frac{d\theta_{er}}{dt} = \omega_r \tag{8.33}$$

An eighth-order nonlinear model with time variable coefficients (inductances) has been obtained, while still neglecting the core losses. Only numerical methods provide a solution to it. This complex model is to be used directly only in special cases, when the computation effort is justified. Not so for electric drives, in general.

*Complex (space-phasor) variables are introduced to obtain a model with position (time)-independent coefficients.*

## 8.7 Space-Phasor Model

Let us introduce first the following notations [1–4]:

$$\begin{aligned} &a = e^{j(2\pi/3)}; \quad \cos\frac{2\pi}{3} = \mathrm{Re}[a]; \quad \cos\frac{4\pi}{3} = \mathrm{Re}[a^2] \\ &\cos\left(\theta_{er} + \frac{2\pi}{3}\right) = \mathrm{Re}\left[a \cdot e^{j\theta_{er}}\right]; \quad \cos\left(\theta_{er} + \frac{4\pi}{3}\right) = \mathrm{Re}\left[a^2 \cdot e^{j\theta_{er}}\right] \end{aligned} \tag{8.34}$$

Using Equation 8.34 in the flux expression (8.24), the phase a and $a_r$ flux linkages $\lambda_a$ and $\lambda_{ar}$ can be written as

$$\lambda_a = L_{ls} \cdot i_a + L_{ms} \cdot \mathrm{Re}[i_a + a \cdot i_b + a^2 \cdot i_c] + L_{ms} \cdot \mathrm{Re}\left[(i_{ar} + a \cdot i_{br} + a^2 \cdot i_{cr})e^{j\theta_{er}}\right] \tag{8.35}$$

$$\lambda_{ar} = L_{lr} \cdot i_{ar} + L_{ms} \cdot \mathrm{Re}[i_{ar} + a \cdot i_{br} + a^2 \cdot i_{cr}] + L_{ms} \cdot \mathrm{Re}\left[(i_a + a \cdot i_b + a^2 \cdot i_c)e^{-j\theta_{er}}\right] \tag{8.36}$$

We now introduce the following complex variables as space-phasors:

$$\bar{i}_s^s = \frac{2}{3} \cdot (i_a + a \cdot i_b + a^2 \cdot i_c) \tag{8.37}$$

$$\bar{i}_r^r = \frac{2}{3} \cdot (i_{ar} + a \cdot i_{br} + a^2 \cdot i_{cr}) \tag{8.38}$$

where

$$\mathrm{Re}\left[\bar{i}_s^s\right] = i_a - \frac{1}{3} \cdot (i_a + i_b + i_c) = i_a - i_0 \tag{8.39}$$

$$\mathrm{Re}\left[\bar{i}_r^r\right] = i_{ar} - \frac{1}{3} \cdot (i_{ar} + i_{br} + i_{cr}) = i_{ar} - i_0 \tag{8.40}$$

In symmetric transient and steady-state regimes and symmetric windings

$$i_a + i_b + i_c = 0; \quad i_{ar} + i_{br} + i_{cr} = 0 \tag{8.41}$$

With definitions (8.37) and (8.38), Equations 8.35 and 8.36 become

$$\lambda_a = L_{ls} \cdot \mathrm{Re}\left(\bar{i}_s^{\,s}\right) + L_m \cdot \mathrm{Re}\left(\bar{i}_s^{\,s} + \bar{i}_r^{\,s} \cdot e^{j\theta_{er}}\right); \quad L_m = \frac{3}{2} L_{ms} \tag{8.42}$$

$$\lambda_{ar}^{r} = L_{lr} \cdot \mathrm{Re}\left(\bar{i}_r^{\,s}\right) + L_m \cdot \mathrm{Re}\left(\bar{i}_r^{\,r} + \bar{i}_s^{\,s} \cdot e^{-j\theta_{er}}\right) \tag{8.43}$$

If to Equations 8.42 and 8.43 we add similar equations for phases b, $b_r$, c, and $c_r$, Equation 8.23 becomes

$$\bar{V}_s^{s} = r_s \cdot \bar{i}_s^{s} + \frac{d\bar{\lambda}_s^{s}}{dt} = r_s \cdot \bar{i}_s^{s} + L_s \cdot \frac{d\bar{i}_s^{s}}{dt} + L_m \frac{d\left(\bar{i}_r^{r} e^{j\theta_{er}}\right)}{dt} \tag{8.44}$$

$$\bar{V}_r^{r} = r_r \cdot \bar{i}_r^{r} + \frac{d\bar{\lambda}_r^{r}}{dt} = r_r \cdot \bar{i}_r^{r} + L_r \cdot \frac{d\bar{i}_r^{r}}{dt} + L_m \frac{d\left(\bar{i}_s^{s} e^{-j\theta_{er}}\right)}{dt} \tag{8.45}$$

with

$$L_s = L_{ls} + L_m; \quad L_r = L_{lr} + L_m \tag{8.46}$$

and

$$\bar{V}_s^{s} = \frac{2}{3} \cdot (V_a + a \cdot V_b + a^2 \cdot V_c); \quad \bar{V}_r^{r} = \frac{2}{3} \cdot (V_{ar} + a \cdot V_{br} + a^2 \cdot V_{cr}) \tag{8.47}$$

The complex variables $\bar{V}_s^s$, $\bar{V}_r^r$, $\bar{i}_s^{\,s}$, $\bar{i}_r^{\,r}$, $\bar{\lambda}_s^s$, $\bar{\lambda}_r^r$, are still represented in their respective coordinates (stator for stator, rotor for rotor). We may now use a rotation of the complex variables by the angle $\theta_b$ in the stator and by $\theta_b - \theta_{er}$ in the rotor to obtain a unique coordinate system at some speed $\omega_b$

$$\omega_b = \frac{d\theta_b}{dt} \tag{8.48}$$

$$\bar{\lambda}_s^{s} = \bar{\lambda}_s^{b} \cdot e^{j\theta_b}; \quad \bar{i}_s^{\,s} = \bar{i}_s^{\,b} \cdot e^{j\theta_b}; \quad \bar{V}_s^{s} = \bar{V}_s^{b} \cdot e^{j\theta_b} \tag{8.49}$$

$$\bar{\lambda}_r^{r} = \bar{\lambda}_r^{b} \cdot e^{j(\theta_b - \theta_{er})}; \quad \bar{i}_r^{\,r} = \bar{i}_r^{\,b} \cdot e^{j(\theta_b - \theta_{er})}; \quad \bar{V}_r^{r} = \bar{V}_r^{b} \cdot e^{j(\theta_b - \theta_{er})} \tag{8.50}$$

With the new variables, Equations 8.44 and 8.45 become

$$\begin{aligned} \bar{V}_s &= r_s \cdot \bar{i}_s + \frac{d\bar{\lambda}_s}{dt} + j \cdot \omega_b \cdot \bar{\lambda}_s \\ \bar{V}_r &= r_r \cdot \bar{i}_r + \frac{d\bar{\lambda}_r}{dt} + j \cdot (\omega_b - \omega_r) \cdot \bar{\lambda}_r \end{aligned} \tag{8.51}$$

with

$$\bar{\lambda}_s = L_s \cdot \bar{i}_s + L_m \cdot \bar{i}_r \tag{8.52}$$

$$\bar{\lambda}_r = L_r \cdot \bar{i}_r + L_m \cdot \bar{i}_s \tag{8.53}$$

Note that for clarity we dropped the superscript b.

The torque should be calculated from Equation 8.42 with the above notations:

$$T_e = \frac{3}{2} \cdot p \cdot \mathrm{Re}\left(j \cdot \bar{\lambda}_s \cdot \bar{i}_s^*\right) = -\frac{3}{2} \cdot p \cdot \mathrm{Re}\left(j \cdot \bar{\lambda}_r \cdot \bar{i}_r^*\right) \tag{8.54}$$

Equations 8.51 through 8.54 together with the equations of motion (8.33) constitute the complex variable or space-phasor model of the induction machine with single cage (or wound) rotor and with core loss neglected.

We may now decompose in plane the space-phasors along two orthogonal axes d and q moving at speed $\omega_b$ [5]:

$$\begin{aligned} &\bar{V}_s = V_d + j \cdot V_q; \quad \bar{i}_s = i_d + j \cdot i_q; \quad \bar{\lambda}_s = \lambda_d + j \cdot \lambda_q \\ &\bar{V}_r = V_{dr} + j \cdot V_{qr}; \quad \bar{i}_r = i_{dr} + j \cdot i_{qr}; \quad \bar{\lambda}_r = \lambda_{dr} + j \cdot \lambda_{qr} \end{aligned} \tag{8.55}$$

With Equation 8.55, the two voltage equations 8.51 become

$$\begin{aligned} &V_d = r_s \cdot i_d + \frac{d\lambda_d}{dt} - \omega_b \cdot \lambda_q \\ &V_q = r_s \cdot i_q + \frac{d\lambda_q}{dt} + \omega_b \cdot \lambda_d \\ &V_{dr} = r_r \cdot i_{dr} + \frac{d\lambda_{dr}}{dt} - (\omega_b - \omega_r) \cdot \lambda_{qr} \\ &V_{qr} = r_r \cdot i_{qr} + \frac{d\lambda_{qr}}{dt} + (\omega_b - \omega_r) \cdot \lambda_{dr} \\ &T_e = \frac{3}{2} p(\lambda_d i_q - \lambda_q i_d) = \frac{3}{2} p L_m (i_q i_{dr} - i_d i_{qr}) \end{aligned} \tag{8.56}$$

Also from Equations 8.49, 8.50, and 8.47

$$\begin{bmatrix} V_d \\ V_q \\ V_0 \end{bmatrix} = \left[P(\theta_b)\right] \cdot \begin{bmatrix} V_a \\ V_b \\ V_c \end{bmatrix} \tag{8.57}$$

$[P(\theta_b)]$ is the Park transformation

$$[P(\theta_b)] = \frac{2}{3} \cdot \begin{bmatrix} \cos(-\theta_b) & \cos\left(-\theta_b + \frac{2\pi}{3}\right) & \cos\left(-\theta_b - \frac{2\pi}{3}\right) \\ \sin(-\theta_b) & \sin\left(-\theta_b + \frac{2\pi}{3}\right) & \sin\left(-\theta_b - \frac{2\pi}{3}\right) \\ \frac{1}{2} & \frac{1}{2} & \frac{1}{2} \end{bmatrix} \tag{8.58}$$

The inverse of Park transformation is

$$[P(\theta_b)]^{-1} = \frac{3}{2}[P(\theta_b)]^T \tag{8.59}$$

A similar transformation is valid for rotor quantities with $\theta_b - \theta_{er}$ replacing $\theta_b$ in Equation 8.58.

It may be proved that the homopolar components $V_0$, $i_0$, $V_{0r}$, $i_{0r}$ have separate equations and do not interfere in the energy conversion process in the motor.

$$\begin{aligned} \bar{V}_0 &= r_s \cdot \bar{i}_0 + \frac{d\bar{\lambda}_{0s}}{dt}; \quad \bar{\lambda}_0 \approx L_{ls} \cdot \bar{i}_0 \\ \bar{V}_{0r} &= r_r \cdot \bar{i}_{0r} + \frac{d\bar{\lambda}_{0r}}{dt}; \quad \bar{\lambda}_{0r} \approx L_{lr} \cdot \bar{i}_{0r} \end{aligned} \tag{8.60}$$

Equations 8.56 through 8.60 represent the d-q0 model of the induction machine that operates with real (not complex) variables. The complex variable (space-phasor) and d-q model are equivalent as they are based on identical assumptions (symmetric sinusoidally distributed windings and constant airgap). The speed of the reference system $\omega_b$ is arbitrary as the airgap is uniform.

Up to now we rushed through equations to quickly obtain the complex variable (d-q) model of the induction machine; in what follows we insist on some graphical representations to facilitate the assimilation of this new knowledge.

**EXAMPLE 8.1: THE SPACE-PHASOR OF SINUSOIDAL SYMMETRIC CURRENTS**

Consider three symmetrical sinusoidal currents and show how their complex space-phasor $\bar{i}_s^s$ varies in time through six instants. Give a graphical description of this process in time.

**Solution**

The three-phase currents may be written as

$$i_{a,b,c} = I\sqrt{2} \cdot \cos\left(\omega_1 t - (i-1) \cdot \frac{2\pi}{3}\right); \quad i = 1,2,3 \tag{8.61}$$

The space-phasor in stator coordinates $\bar{i}_s^s$ is (8.37)

$$\bar{i}_s^s = \frac{2}{3} I\sqrt{2}\left[\cos\omega_1 t + e^{j(2\pi/3)}\cos\left(\omega_1 t - \frac{2\pi}{3}\right) + e^{j(4\pi/3)}\cos\left(\omega_1 t - \frac{4\pi}{3}\right)\right] \tag{8.62}$$

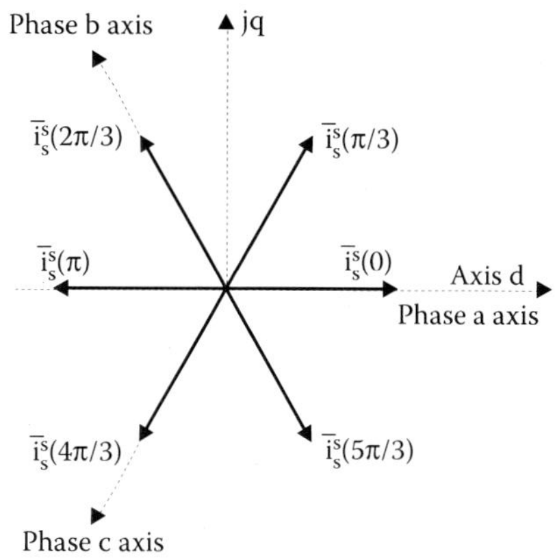


**FIGURE 8.7**
The space-phasor of sinusoidal three-phase currents.

with

$$e^{j(2\pi/3)} = \cos\frac{2\pi}{3} + j\cdot\sin\frac{2\pi}{3};\quad e^{j(4\pi/3)} = \cos\frac{4\pi}{3} + j\cdot\sin\frac{4\pi}{3} \tag{8.63}$$

Equation 8.62 becomes

$$\bar{i}_s^s = I\sqrt{2}[\cos\omega_1 t + j\cdot\sin\omega_1 t] = i_d + j\cdot i_q \tag{8.64}$$

The position of the space-phasor for $\omega_1 t = 0, \pi/3, 2\pi/3, \pi, 4\pi/3, 5\pi/3$ is shown in Figure 8.7.

It should be noted that the time "produces" the instantaneous values of currents while the definition of the space-phasor shows that each phase current instantaneous value is placed along the axis of the corresponding phase. So, in fact, the space-phasor travels in the d-q stator plane with the electrical angular speed equal to $\omega_1$.

The same concept is applied to flux linkages and thus their space-phasor is related to the traveling field in a.c. machines.

Also note that

$$\mathrm{Re}\left[\bar{i}_s^s\right] = i_a;\quad \text{if } i_a + i_b + i_c = 0 \tag{8.65}$$

## 8.8 Space-Phasor Diagram for Electrical Transients

The space-phasor model equations 8.51 through 8.53 may be represented on a space-phasor diagram in the d-q plane, with axes d and q rotating at speed $\omega_b = \omega_1$ (Figure 8.8): synchronous coordinates. Let us consider $\bar{V}_r^b = 0$.

As $\omega_1$ is the frequency of the actual stator voltages at steady state, for this latter case, $d/dt = 0$. In general, during steady state

$$\frac{d}{dt} = j(\omega_1 - \omega_b) \tag{8.66}$$

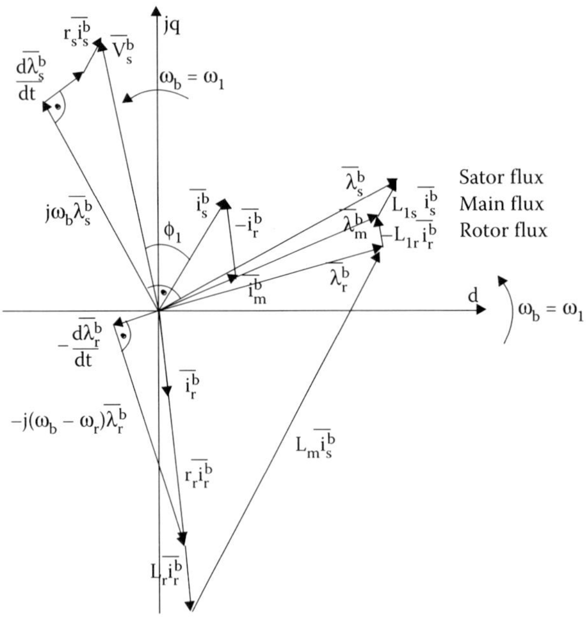


**FIGURE 8.8**
The space-phasor diagram of induction motor valid for transients (for steady d/dt = 0) in synchronous coordinates (cage rotor: $\bar{V}_r^b = 0$).

Also for steady state the rotor flux equation 8.51 becomes

$$\begin{aligned}\bar{V}_r^b &= r_r \cdot \bar{i}_r^{\,b} + j(\omega_1 - \omega_b) \cdot \bar{\lambda}_r^b + j(\omega_b - \omega_r) \cdot \bar{\lambda}_r^b \\ &= r_r \cdot \bar{i}_r^{\,b} + j(\omega_1 - \omega_r) \cdot \bar{\lambda}_r^b\end{aligned} \tag{8.67}$$

Clearly, when $\bar{V}_r^b = 0$ (short-circuited, cage, rotor) the rotor current and flux space-phasors are *orthogonal to each other*. The same condition may occur with additional adequate voltage $\bar{V}_r^s$ applied to wound rotors.

For this case, the torque expression (8.54) becomes

$$T_e = \frac{3}{2} p \cdot \lambda_r^b \cdot i_r^b; \quad \left|\bar{\lambda}_r^b\right| = ct \tag{8.68}$$

## 8.9 Electrical Transients with Flux Linkages as Variables

Equations 8.51 through 8.53 at constant speed allow for the elimination of rotor and stator currents $\bar{i}_s^{\,b}$ and $\bar{i}_r^{\,b}$.

With

$$\bar{i}_s^{\,b} = \sigma^{-1}\left(\frac{\bar{\lambda}_s^b}{L_s} - \frac{\bar{\lambda}_r^b \cdot L_m}{L_s L_r}\right) \tag{8.69}$$

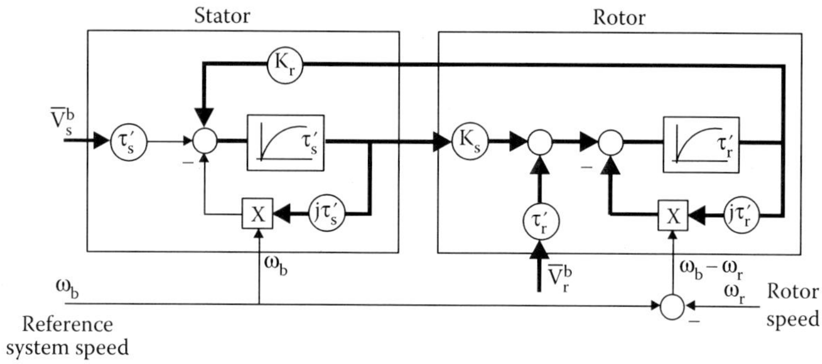


**FIGURE 8.9**
Structural diagram of induction motor with stator and rotor flux $\overline{\lambda}_s^b$ and $\overline{\lambda}_r^b$ as variables; random speed ($\omega_b$) coordinates.

$$\overline{i}_r^{\,b} = \sigma^{-1}\left(\frac{\overline{\lambda}_r^b}{L_r} - \frac{\overline{\lambda}_s^b \cdot L_m}{L_s L_r}\right) \tag{8.70}$$

$$\sigma = 1 - \frac{L_m^2}{L_s L_r} \tag{8.71}$$

Equations 8.51 through 8.53 become [6]

$$\tau_s' \frac{d\overline{\lambda}_s^b}{dt} + (1 + j \cdot \omega_b \cdot \tau_s') \cdot \overline{\lambda}_s^b = \tau_s' \overline{V}_s^b + K_r \overline{\lambda}_r^b \tag{8.72}$$

$$\tau_r' \frac{d\overline{\lambda}_r^b}{dt} + (1 + j \cdot (\omega_b - \omega_r) \cdot \tau_r') \cdot \overline{\lambda}_r^b = \tau_r' \overline{V}_r^b + K_s \overline{\lambda}_s^b \tag{8.73}$$

with

$$\begin{aligned} K_s &= \frac{L_m}{L_s}; \quad K_r = \frac{L_m}{L_r} \\ \tau_s' &= \tau_s \cdot \sigma; \quad \tau_r' = \tau_r \cdot \sigma \\ \tau_s &= \frac{L_s}{r_s}; \quad \tau_r = \frac{L_r}{r_r} \end{aligned} \tag{8.74}$$

$\tau_s$ and $\tau_r$ are the stator and rotor time constants, respectively, and $\tau_s'$ and $\tau_r'$ are transient time constants of the stator and rotor.

The structural diagram corresponding to Equations 8.72 and 8.73 is shown in Figure 8.9.

## 8.10 Complex Eigenvalues for Electrical Transients

Equations 8.72 and 8.73 imply a second-order complex variable system with only two complex eigenvalues corresponding to the determinant

$$\begin{vmatrix} \tau_s'\underline{s}+1+j\omega_b\tau_s' & -K_r \\ -K_s & \tau_r'\underline{s}+1+j(\omega_b-\omega_r)\tau_r' \end{vmatrix} = 0 \tag{8.75}$$

The complex eigen values $\underline{s}_{1,2}$ from Equation 8.75 depend on the reference system speed $\omega_b$ and on rotor speed $\omega_r$, but their real part is, in general, negative suggesting attenuated periodic response [4,6].

**EXAMPLE 8.2**

For an induction motor with the data $r_s = 0.5\ \Omega$, $r_r = 0.60\ \Omega$, $L_s = L_r = 0.08$ H, $L_m = 0.075$ H, $p = 2$ pole pairs, calculate the complex eigen values for constant speed, in synchronous coordinates ($\omega_1 = 2\pi 60$) for $n = 0$ and $n = 1800$ rpm.

**Solution**

The value of $\omega_b = \omega_1 = 2\pi 60$ rad/s. $\sigma$, $\tau_s'$, $\tau_r'$, $K_s$, $K_r$ are now calculated from Equations 8.71 through 8.74

$$\sigma = 1 - \frac{L_m^2}{L_s L_r} = 1 - \frac{0.075^2}{0.08^2} = 0.1211 \tag{8.76}$$

$$\tau_s' = \sigma\tau_s = \sigma\frac{L_s}{r_s} = 0.1211 \cdot \frac{0.08}{0.5} = 0.01937\,\text{s} \tag{8.77}$$

$$\tau_r' = \sigma\tau_r = \sigma\frac{L_r}{r_r} = 0.1211 \cdot \frac{0.08}{0.6} = 0.01614\,\text{s} \tag{8.78}$$

$$K_s = \frac{L_m}{L_s} = \frac{0.075}{0.08} = 0.9375 \tag{8.79}$$

$$K_r = \frac{L_m}{L_r} = \frac{0.075}{0.08} = 0.9375 \tag{8.80}$$

Also

$$\omega_r = 2\pi p n = 4\pi n \tag{8.81}$$

Now we rewrite Equation 8.75 in a canonical form

$$\begin{aligned} &\underline{s}^2\tau_s'\tau_r' + \underline{s}[(\tau_s'+\tau_r') + j\tau_s'\tau_r'(2\omega_b-\omega_r)] + (1+j\cdot\omega_b\cdot\tau_s')(1+j\cdot(\omega_b-\omega_r)\cdot\tau_r') - K_sK_r \\ &\quad = \underline{s}^2\cdot 0.01937\cdot 0.01614 + \underline{s}[(0.01937+0.01614) + j\cdot 0.01937\cdot 0.01614\cdot(2\cdot 376.8 - 4\pi n)] \\ &\qquad + (1+j\cdot 376.8\cdot 0.01937)\cdot(1+j\cdot(376.8-4\pi n)\cdot 0.01614) - 0.93752 = 0 \\ &\underline{s}_1 = -51.377 - j\cdot 369.2 \\ &\underline{s}_2 = -62.1754 - j\cdot 7.6031 \end{aligned} \tag{8.82}$$

It is evident that the roots s depend on speed n. The solving of the second-order equation is straightforward.

## 8.11 Electrical Transients for Constant Rotor Flux

By constant rotor flux we mean, in general, constant amplitude at $(\omega_1 - \omega_b)$ frequency. For synchronous coordinates it would mean zero frequency.

So constant rotor flux means that in Equations 8.72 and 8.73

$$\frac{d\bar{\lambda}_r^b}{dt} = j\cdot(\omega_b - \omega_r)\cdot\bar{\lambda}_r^b \tag{8.83}$$

Using Equation 8.83 in Equation 8.73

$$\bar{\lambda}_r^b = \frac{\tau_r'\bar{V}_r^b + K_s\bar{\lambda}_s^b}{1 + jS\omega_1\tau_r'} \tag{8.84}$$

It is more convenient to use synchronous coordinates for Equation 8.72:

$$\tau_s'\frac{d\bar{\lambda}_s^b}{dt} + (1 + j\cdot\omega_1\cdot\tau_s')\cdot\bar{\lambda}_s^b = \tau_s'\bar{V}_s^b + K_r\bar{\lambda}_r^b \tag{8.85}$$

Equations 8.84 and 8.85 lead to a simplification in the structural diagram of Figure 8.9 as the derivative term in the rotor disappears (Figure 8.10).

The system's order is reduced and thus the stator flux presents only one complex eigenvalue as may be inferred from Equations 8.84 and 8.85.

## 8.12 Steady State: It Is D.C. in Synchronous Coordinates

Steady state means, in general, that the three-phase voltages are symmetric and sinusoidal

$$V_{a,b,c} = V\sqrt{2}\cdot\cos\left(\omega_1 t - (i-1)\cdot\frac{2\pi}{3}\right); \quad i = 1,2,3 \tag{8.86}$$

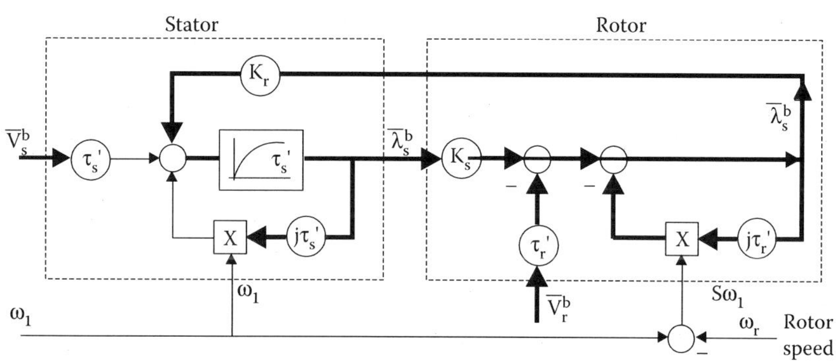


**FIGURE 8.10**
Structural diagram of induction motor with constant rotor flux and speed in synchronous coordinates $(\omega_b = \omega_1)$.

The voltage space-phasor in random coordinates is

$$\bar{V}_s^b = \frac{2}{3}[V_a(t) + V_b(t)\cdot e^{j(2\pi/3)} + V_c(t)\cdot e^{-j(2\pi/3)}]\cdot e^{-j\theta_b} \tag{8.87}$$

with Equation 8.86

$$\bar{V}_s^b = V\sqrt{2}[\cos(\omega_1 t - \theta_b) + j\cdot\sin(\omega_1 t - \theta_b)] \tag{8.88}$$

For steady state

$$\theta_b = \omega_b t + \theta_0 \tag{8.89}$$

Consequently

$$\begin{aligned}\bar{V}_s^b &= V\sqrt{2}[\cos[(\omega_1 - \omega_b)t - \theta_0] + j\cdot\sin[(\omega_1 - \omega_b)t - \theta_0]]\\ &= V\sqrt{2}\cdot e^{j[(\omega_1 - \omega_b)t - \theta_0]}\end{aligned} \tag{8.90}$$

It is obvious that, for steady state, the currents in the model must have the voltage frequency, which is $(\omega_1 - \omega_b)$. Once again for steady state we use Equation 8.51, with $d/dt = .j\cdot(\omega_1 - \omega_b)$, as inferred in Equation 8.66.

Consequently from Equation 8.51

$$\begin{aligned}\bar{V}_{s0}^b &= r_s\cdot\bar{i}_{s0}^b + j\cdot\omega_1\cdot\bar{\lambda}_{s0}^b\\ \bar{V}_{r0}^b &= r_r\cdot\bar{i}_{r0}^b + j\cdot S\omega_1\cdot\bar{\lambda}_{r0}^b;\quad S = 1 - \frac{\omega_r}{\omega_1}\end{aligned} \tag{8.91}$$

S is known as slip.

So the form of steady-state equations is independent of the reference speed $\omega_b$. What counts is the primary (stator) frequency $\omega_1$ and the actual rotor currents frequency $S\omega_1$.

Note that, for synchronous coordinates, steady state means $d/dt = j(\omega_b - \omega_1) = 0$, that is, *d.c. quantities*.

In the flux equations 8.52 and 8.53, we may separate the main (airgap) flux linkage $\bar{\lambda}_m$

$$\bar{\lambda}_s^b = L_{ls}\cdot\bar{i}_s^b + \bar{\lambda}_m;\quad \bar{\lambda}_m = L_m\cdot\left(\bar{i}_s^b + \bar{i}_r^b\right) = L_m\cdot\bar{i}_m^b \tag{8.92}$$

$$\bar{\lambda}_r^b = L_{lr}\cdot\bar{i}_r^b + \bar{\lambda}_m \tag{8.93}$$

Equations 8.91 through 8.93 lead to the standard equivalent circuit of Figure 8.11.

Magnetic saturation may be considered through $\lambda_m(i_m)$ functions approximating measurements or field distribution calculations [7].

NOTE (ON CORE LOSS): The core loss occurs mainly in the stator as $\omega_1 \gg S\omega_1$; that is, the frequency of magnetic fields is higher in the stator since the rated slip is $S_n = 0.08–0.005$ (decreasing with power). An equivalent resistance $r_m$ determined from experiments may be included in the equivalent circuit, Figure 8.11, to account for the core losses during steady state. For investigating the role of core loss during transients, additional stator windings in the d-q model have to be included (see Reference 7).

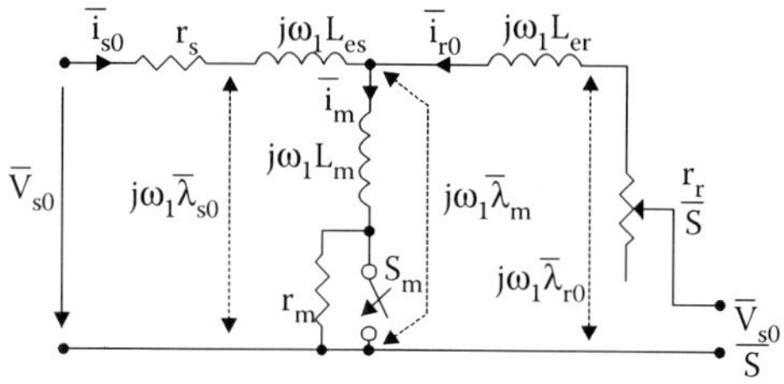


**FIGURE 8.11**
Space-phasor steady-state equivalent circuit of induction machine.

The torque expression (8.54) with Equation 8.91 and $V_{r0}^b = 0$ yields

$$T_e = -\frac{3}{2}p \cdot \mathrm{Re}\left(j\bar{\lambda}_r^b \cdot \bar{i}_r^{b*}\right) = \frac{3}{2}p \cdot \lambda_{r0} \cdot i_{r0} \tag{8.94}$$

From Equation 8.92

$$\bar{i}_{r0} = -j \cdot S\omega_1 \cdot \frac{\bar{\lambda}_{r0}}{r_r} \tag{8.95}$$

Consequently, the electromagnetic torque $T_e$ is

$$T_e = \frac{3}{2}p \cdot \frac{\lambda_{r0}^2}{r_r} \cdot S\omega_1 \tag{8.96}$$

Using the equivalent circuit of Figure 8.11 and Equation 8.95, we may obtain the conventional torque expression

$$T_e = \frac{3p}{\omega_1} \frac{V^2 \cdot r_r/S}{(r_s + c_1 r_r/S)^2 + \omega_1^2 (L_{ls} + c_1 L_{lr})^2} \tag{8.97}$$

with

$$c_1 \approx 1 + \frac{L_{ls}}{L_m} \tag{8.98}$$

## 8.13 No-Load Ideal Speed May Go under or over Conventional Value $\omega_1$

The no-load ideal speed (slip $S_0$) corresponds to zero torque, that is, Equation 8.91, zero rotor current

$$\bar{V}_{r0}^b = j \cdot S_0\omega_1 \cdot \bar{\lambda}_{r0}^b \tag{8.99}$$

Only for short-circuited rotor windings (or passive impedance at rotor terminals) $\left(\bar{V}_r^b = 0\right)$, the ideal no-load slip $S_0 = 0$ and $\omega_{ro} = \omega_1$.

When the induction machine is doubly fed $\left(\bar{V}_r^b \neq 0\right)$, the ideal no-load slip is different from zero and the no-load ideal speed is, in general

$$\omega_{r0} = \omega_1(1 - S_0) \tag{8.100}$$

The value and phase shift between $\bar{V}_{r0}^b$ and $\bar{\lambda}_{r0}^b$ could be arranged through a PEC supplying the wound rotor. So $S_0$ could be either positive or negative. Thus, only for $V_r^b = 0$ from Equation 8.96 the torque is positive (motoring) for $S > 0$ and negative (generating) for $S < 0$.

The doubly fed induction motor could operate either as a motor or as a generator below and above $\omega_{r0} = \omega_1$ $(S_0 = 0)$, provided that the PEC can produce bidirectional power flow between the wound rotor and the power grid. These doubly fed induction motor drives will be dealt with separately in Chapter 14 dedicated to high-power industrial drives.

### EXAMPLE 8.3

For steady state, calculate the stator voltage, stator flux, current, power factor, and torque of an induction motor at ideal no-load (synchronous) speed $\omega_1$ and $S = 0.02$.

The motor data are $r_s = 0.5\ \Omega$, $r_r = 0.6\ \Omega$, $L_s = L_r = 0.08$H, $L_m = 0.075$ H, $\lambda_{r0} = 0.8$ Wb, $\omega_1 = 2\pi 60$ rad/s, $p = 2$ pole pairs, $V_r^b = 0$ (cage rotor).

#### Solution

First, we have to calculate the initial conditions, which are implicitly steady state. Let us calculate speed $\omega_r$:

$$\omega_r = \omega_1(1 - S) \tag{8.101}$$

So

$$\omega_r = 2 \cdot \pi \cdot 60 \cdot (1 - 0.02) = 369.45\,\text{rad/s} \tag{8.102}$$

The torque ($T_e$) is (8.96)

$$(T_e)_{t=0} = \frac{3}{2} p \cdot \left(\lambda_r^b\right)^2 \cdot \frac{S\omega_1}{r_r} = \frac{3}{2} \cdot 2 \cdot (0.8)^2 \cdot 0.02 \cdot \frac{2\pi 60}{0.6} = 24.115\,\text{N m} \tag{8.103}$$

From Equations 8.72 and 8.73 with $d/dt = 0$ and $V_r^b = 0$, we may now calculate the stator flux $\bar{\lambda}_s^b$:

$$\bar{\lambda}_s^b = \bar{\lambda}_r^b \cdot \frac{1 + j \cdot S \cdot \omega_1 \cdot \tau_r'}{K_s} \tag{8.104}$$

Note that the motor parameters are as in Example 8.2 and thus $\tau_r' = 0.01614$ s, $\tau_s' = 0.01937$ s, $K_s = K_r = 0.9375$, $\sigma = 0.1211$.

$$\bar{\lambda}_s^b = \bar{\lambda}_r^b \cdot \frac{1 + j \cdot 0.02 \cdot 2\pi 60 \cdot 0.01614}{0.9375} = 0.8533 + j0.1038 \tag{8.105}$$

Let us consider the d axis along the rotor flux and thus $\bar{\lambda}_r^b = \lambda_r^b$.

Now from Equation 8.72 with $d\bar{\lambda}_s^b/dt = 0$ we may find the stator voltage $\bar{V}_s^b$:

$$\bar{V}_s^b = \frac{(1+j\omega_1\tau_s')\bar{\lambda}_s^b - K_r \cdot \bar{\lambda}_r^b}{\tau_s'}$$

$$= \frac{(1+j\cdot 2\pi 60\cdot 0.01937)\cdot 0.8\cdot(1.066+j0.1297) - 0.9375\cdot 0.8}{0.01937}$$

$$= -33.74 + j326.69; \quad V_s^b = 328.4\,\text{V} \tag{8.106}$$

The motor phase voltage (rms value)

$$V = \frac{V_s^b}{\sqrt{2}} = \frac{328.4}{1.41} = 232.91\,\text{V} \tag{8.107}$$

The stator current is obtained from Equation 8.69

$$\bar{i}_s^{\,b} = \frac{\left[\bar{\lambda}_s^b - \bar{\lambda}_r^b \cdot K_r\right]}{\sigma \cdot L_s}$$

$$= \frac{[0.8\cdot(1.066+j0.1297) - 0.8\cdot 0.9357]}{0.1211\cdot 0.08}$$

$$= 10.76 + j10.71 \tag{8.108}$$

The rotor space-phasor $\bar{i}_r^{\,b}$ is Equation 8.95

$$\bar{i}_r^{\,b} = -jS\omega_1\frac{\lambda_r^b}{r_r} = -j\cdot 0.02\cdot 2\pi 60\cdot\frac{0.8}{0.6} = -j10.05\,\text{A} \tag{8.109}$$

The amplitude of the stator current $i_s^b = 15.181\,\text{A}$. The results are illustrated by the space-phasor diagram in Figure 8.12.

The power factor angle $\varphi_1$ is

$$\varphi_1 = \arg\bar{V}_s^b - \arg\bar{i}_s^{\,b} = \frac{\pi}{2} + \tan^{-1}\frac{33.74}{326.69} - \tan^{-1}\frac{10.71}{10.76}$$

$$= 90° + 5.9° - 45.156° = 50.74° \tag{8.110}$$

Finally, $\cos\varphi_1 = 0.633$.

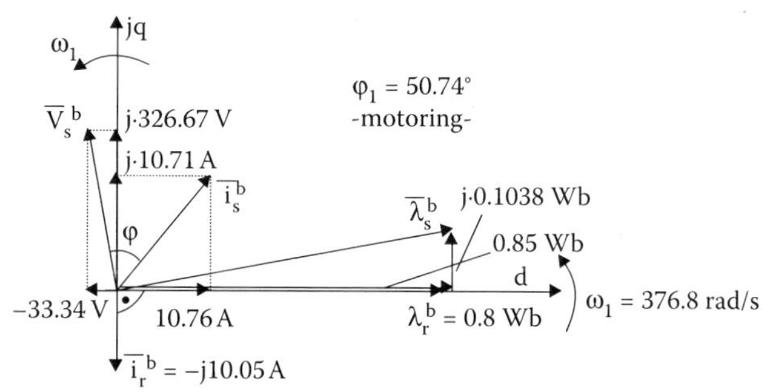


**FIGURE 8.12**
Steady-state space-phasor diagram in synchronous coordinates ($\omega_b = \omega_1$): d.c. quantities.

**EXAMPLE 8.4: LOSS BREAKDOWN**

A high-efficiency induction motor with cage rotor has the data: rated power $P_n = 5$ kW, rated line voltage (rms) $V_L = 220$ V (star connection), rated frequency $f_1 = 60$ Hz, number of pole pairs p = 2, core loss ($p_{iron}$) = mechanical loss $p_{mec}$ = 1.5% of $P_n$, additional losses $p_{add} = 1\%P_n$, stator per rotor winding losses $p_{cor}/p_{cos} = 2/3$, rated efficiency $\eta_n = 0.9$, and power factor $\cos\varphi_n = 0.88$.

Calculate all loss components, phase current (rms), then rated slip, speed, electromagnetic torque, shaft torque, and stator current (as space-phasors).

**Solution**

The loss breakdown diagram of the induction motor is shown in Figure 8.13.

The input power $P_{in}$ is

$$P_{in} = \frac{P_{out}}{\eta_n} = \frac{5000}{0.9} = 5555.55\,\text{W} \quad (8.111)$$

The phase current $I_n$ (rms) is

$$I_n = \frac{P_{in}}{\sqrt{3}\cdot V_L \cdot \cos\phi_n} = \frac{5555.55}{\sqrt{3}\cdot 220\cdot 0.88} = 16.58\,\text{A} \quad (8.112)$$

The total losses $\Sigma p_i$ are

$$\sum p_i = P_{in} - P_{out} = 5555.55 - 5000 = 555.55\,\text{W} \quad (8.113)$$

Consequently

$$p_{iron} = p_{mec} = 0.015 \cdot 5000 = 75\,\text{W} \quad (8.114)$$

$$p_{add} = 0.01 \cdot 5000 = 50\,\text{W} \quad (8.115)$$

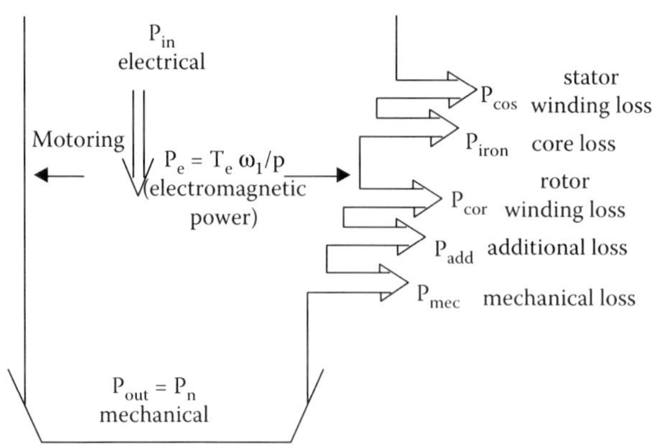


**FIGURE 8.13**
Induction motor energy conversion.

So

$$p_{cos} + p_{cor} = \sum p_i - p_{iron} - p_{mec} - p_{add} = 555.55 - 75 - 75 - 50 \approx 355\,\text{W} \quad (8.116)$$

$$p_{cos} + \frac{2}{3} p_{cos} = 335\,\text{W} \quad (8.117)$$

$$p_{cos} = 213.3\,\text{W}; \quad p_{cor} = 142\,\text{W} \quad (8.118)$$

The electromagnetic power $P_e$ is the active power that crosses the airgap

$$P_e = P_{in} - p_{cos} - p_{iron} = T_e \cdot \frac{\omega_1}{p} \quad (8.119)$$

$$P_e = 5555.55 - 213 - 75 = 5267\,\text{W} \quad (8.120)$$

$$T_e = \frac{5267}{2\pi 60} \cdot 2 = 27.956\,\text{N m} \quad (8.121)$$

The rotor winding loss $p_{cor}$ is

$$p_{cor} = S_n \cdot P_e \quad (8.122)$$

and thus the rated slip is

$$S_n = \frac{142}{5267} = 0.02696 \quad (8.123)$$

The rated speed $n_n$ is

$$n_n = \frac{f_1}{p}(1 - S_n) = \frac{60}{2}(1 - 0.02696) = 29.1912\,\text{rps} = 1751.472\,\text{rpm} \quad (8.124)$$

The shaft torque $T_n$ is calculated directly from mechanical power $P_n$

$$T_n = \frac{P_n}{2\pi n_n} = \frac{5000}{2\pi \cdot 29.1912} = 27.274\,\text{N m} < T_e \quad (8.125)$$

Finally, the amplitude of the stator current, $\bar{i}_s$ in synchronous coordinates (d.c. quantities) is

$$i_s = i_n \sqrt{2} = 16.58 \cdot 1.41 = 23.3778 \quad (8.126)$$

**NOTE:** To calculate all motor parameters—resistances and inductances—more data are required. This is beyond our scope in this example, however.

## 8.14 Motoring, Generating, A.C. Braking

The equivalent circuit for steady state (Figure 8.11) shows that the active power in the rotor (the electromagnetic power $P_e$) is

$$P_e = \frac{3}{2} \cdot I_{r0}^2 \cdot \frac{r_r}{S} = T_e \cdot \frac{\omega_1}{p}; \quad S = 1 - \frac{\omega_r}{\omega_1} \tag{8.127}$$

This expression is valid for the cage rotor $\left(V_r^b = 0\right)$.

*Motoring* mode is defined as the situation when the torque has the same sign as the speed

$$P_e > 0, T_e > 0; \quad \omega_r > 0 \Rightarrow 0 < S < 1 \tag{8.128}$$

For *generating* the torque is negative ($S < 0$) in Equation 8.127 but the speed is positive

$$P_e < 0, \ T_e < 0; \quad \omega_r > 0 \Rightarrow S < 0 \tag{8.129}$$

The generator produces braking ($T_e < 0$, $\omega_r > 0$) but the energy transfer direction in the motor is reversed. The energy is pumped back into the power source through the stator.

*Braking* is obtained when again ($T_e > 0$, $\omega_r < 0$) or ($T_e < 0$, $\omega_r > 0$) but the electromagnetic power is still positive

$$\begin{aligned} &P_e > 0; \quad T_e > 0; \quad \omega_r < 0 \Rightarrow S > 1 \\ &P_e > 0; \quad T_e < 0; \quad \omega_r > 0 \Rightarrow S > 1 \end{aligned} \tag{8.130}$$

We may synthesize the results on operation modes as in Table 8.1.

Using the torque expression (8.97) we may find the maximum torque for the critical slip

$$S_k = \frac{\pm c_1 r_r}{\sqrt{r_s^2 + \omega_1^2 (L_{ls} + c_1 L_{lr})^2}} \tag{8.131}$$

$$T_{ek} = \frac{3p}{\omega_1} \cdot \frac{V^2}{2c_1 \left(r_s \pm \sqrt{r_s^2 + \omega_1^2 (L_{ls} + c_1 L_{lr})^2}\right)} \tag{8.132}$$

**TABLE 8.1**

Operation Modes (Cage Rotor)

| | |
|---|---|
| $S$ | $-\infty - - - - - - 0 + + + + + 1 + + + + + + + \infty$ |
| $\omega_r$ | $+\infty + + + + + \omega_1 + + + + + 0 - - - - - - - - - \infty$ |
| $T_e$ | $0 - - - - - - - 0 + + + + + T_e$ (start) $+ + + + 0$ |
| $P_e$ | $- - - - - - - - 0 + + + + + + + + + + + + +$ |
| Mode | Generating Motoring A.C. braking |

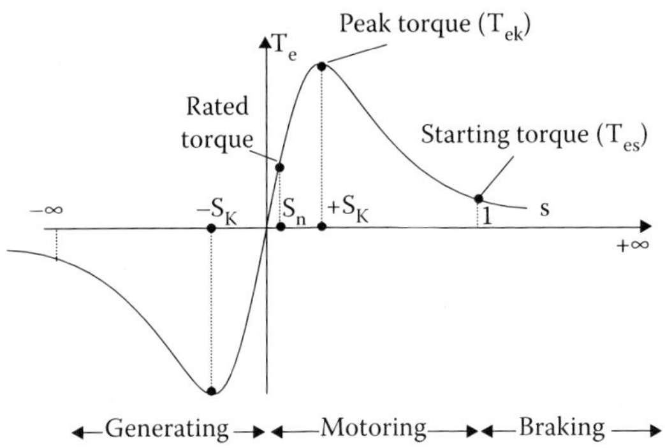


**FIGURE 8.14**
Torque/slip curve of induction motors for constant voltage and frequency.

The torque/slip curve is shown in Figure 8.14.

Using the definition of critical slip $S_k$ and torque $T_{ek}$ in the torque expression (8.97) we obtain, after some approximations, the Kloss formula

$$\frac{T_e}{T_{ek}} \approx \frac{2}{(S/S_k)+(S_k/S)} \tag{8.133}$$

## 8.15 D.C. Braking: Zero Braking Torque at Zero Speed

For moderate braking requirements, d.c. braking is commonly used in modern electrical drives. To calculate the d.c. braking torque, we redraw the space-phasor equivalent circuit in Figure 8.11, but with a d.c. stator current source space-phasor in Figure 8.15.

The electromagnetic torque is still computed from Equation 8.94:

$$T_e = \frac{3}{2}p\lambda_{rdc}i_{rdc} = -\frac{3}{2}p\frac{i_{rdc}^2 \cdot r_r}{\omega_r} \tag{8.134}$$

with

$$\bar{i}_{rdc}^s = \bar{i}_{sdc}^s \cdot \frac{L_m}{L_m + L_{lr} + j(r_r/\omega_r)} \tag{8.135}$$

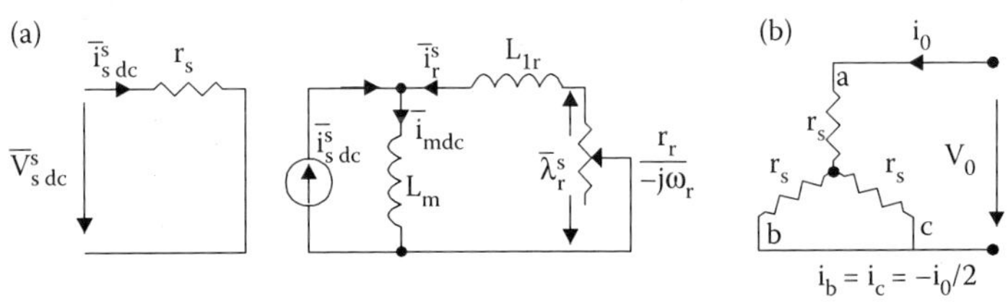


**FIGURE 8.15**
Equivalent circuit for d.c. braking—stator coordinates—in space-phasors; steady state.

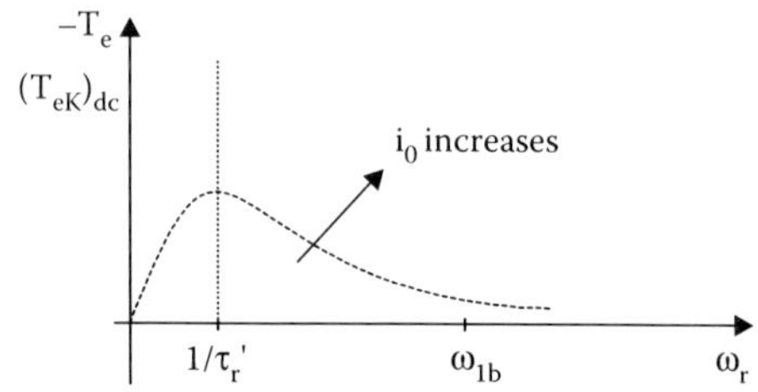


**FIGURE 8.16**
The d.c. braking torque of induction motors.

Finally

$$T_e = -\frac{3}{2}p\frac{L_m^2 \cdot i_{sdc}^2 \cdot \omega_r \cdot r_r}{(L_m + L_{lr})^2\omega_r^2 + r_r^2} \tag{8.136}$$

The peak torque is obtained for $\omega_{rk}$

$$\omega_{rk} = \frac{r_r}{L_m + L_{lr}} \approx \frac{1}{\tau_r} \tag{8.137}$$

and its value is

$$T_{ek} = -\frac{3}{2}p\frac{L_m^2 \cdot i_{sdc}^2}{2(L_m + L_{lr})} \tag{8.138}$$

The braking torque may be modified through the d.c. current level in the stator. The PEC can produce the phase connection in Figure 8.15b where the current complex variable in stator coordinates is

$$\bar{i}_{sdc}^s = \frac{2}{3}\left(i_a + i_b e^{j(2\pi/3)} + i_c e^{-j(2\pi/3)}\right) = \frac{2}{3}\left(i_0 - \frac{i_0}{2}2\cos\frac{2\pi}{3}\right) = i_0 \tag{8.139}$$

The torque speed curve for d.c. braking is shown in Figure 8.16. Note also that the rotor kinetic energy is dumped into the rotor resistor and that for zero speed the braking torque is zero. Also, above $\omega_{rk}$ (which is fairly small in high-efficiency motors), the torque is again rather small.

## 8.16 Speed Control Methods

Variable speed is required in many applications. It has to be performed at high energy conversion rates.

The no-load ideal speed $\omega_{r0}$ is (8.100 with 8.99)

$$\omega_{r0} = \omega_1\left[1 + \text{Imag}\left(\frac{\bar{V}_{r0}^b}{\omega_1\bar{\lambda}_{r0}^b}\right)\right]; \quad \omega_1 = \omega_r + \omega_2 \tag{8.140}$$

Evidently, for the cage rotor $\bar{V}_r^b = 0$ and thus $\omega_{r0} = \omega_1 = 2\pi f_1$ or

$$n_0 = \frac{\omega_{r0}}{2\pi p} = \frac{f_1}{p} \tag{8.141}$$

$\omega_2$ is the actual frequency of the rotor current (or of $\bar{V}_{r0}$ through rotor-side PEC). There are three essential methods to vary speed by changing the no-load ideal speed (as the rated slip is small) as suggested by (8.140)

- Stator frequency $f_1$ variation
- Pole number (2p) changing
- *Wound rotor supply* (or rotor frequency $f_2$ variation)

While pole number 2p changing involves either a separate stator winding or a special winding with a switch to change from $2p_1$ to $2p_2$ poles (Dahlander winding), the other two methods require variable frequency either in the stator or in the rotor, obtainable with PECs in the stator and, respectively, in the rotor.

For limited speed variation, the power rating of rotor-side PEC is smaller than that of the PEC in the stator. In high-power applications, the rotor PEC solution with a wound rotor induction motor is the one preferred for limited speed range (±30%) control.

Stator frequency control is far more frequently used, especially for wide speed control range. However, the level of flux depends on the current in the machine, especially on the magnetization current

$$\bar{I}_m = \bar{I}_s^b + \bar{I}_r^b \tag{8.142}$$

Consequently, it is crucial to control $I_m$ properly to avoid excessive magnetic saturation, while varying frequency $f_1$.

So we have to adopt either voltage $V_s$ and frequency $f_1$ coordinated control or current $\bar{I}_s$ (or $\bar{I}_m$) and frequency $f_1$ control. In general, $V_1/f_1$ coordinated change is applied. There is an infinity of possibilities to relate $\bar{V}_s(V_1)$ or $\bar{I}_s(\bar{I}_m)$ to frequency $f_1$ to obtain the desired performance. However, only three main methods reached the markets:

- $V_1/f_1$—scalar control
- Constant (controlled) rotor flux ($\lambda_r$)—vector control
- Constant (controlled) stator flux ($\lambda_s$)—vector control

Here, only the torque speed curves obtainable with the above methods are given. Note that in all these cases, the PEC voltage supplying the induction motor is voltage limited with the maximum voltage reached at base speed $\omega_b$.

## 8.17 $V_1/f_1$ Torque Speed Curves

$V_1/f_1$ control means that

$$V_1 = K_f \cdot f_1 \tag{8.143}$$

We may judge the torque speed curves obtained in this case through the critical slip $S_k$ and torque $T_{ek}$ of Equations 8.131 and 8.132. The critical slip increases notably with $f_1(\omega_1)$ reduction while the critical torque is only slightly decreased with $f_1$ decrease at frequencies above 5 Hz. Below this value the peak torque decreases dramatically if Equation 8.143 is applied (Figure 8.17). For a safe start $V = V_0 + K{\cdot}f_1$ is applied to compensate for the stator resistance drop $r_s i_s$

$$V_0 = c_0 r_s i_{sn} \tag{8.144}$$

$V_0$ is called the voltage boost and amounts to a few percent of rated voltage $V_n$, higher-for low-power motors.

Above the rated (base) speed, the voltage remains constant and invariably the critical torque decreases and, eventually, constant power is preserved up to a maximum frequency $f_{1max}$. Above base speed, if we neglect the voltage drop $r_s \bar{i}_s$ in Equation 8.91, the stator equation for steady state is

$$V_{s0} \approx j\omega_1 \bar{\lambda}_{s0}^b; \quad \omega_1 > \omega_b \tag{8.145}$$

So, for constant voltage $V_{s0}$, the stator flux $\lambda_{s0}$ and, consequently, the main flux $\lambda_m$ decrease with speed (frequency) increasing above $\omega_b$. This zone—$\omega_b$ to $\omega_{1max}$—is called the

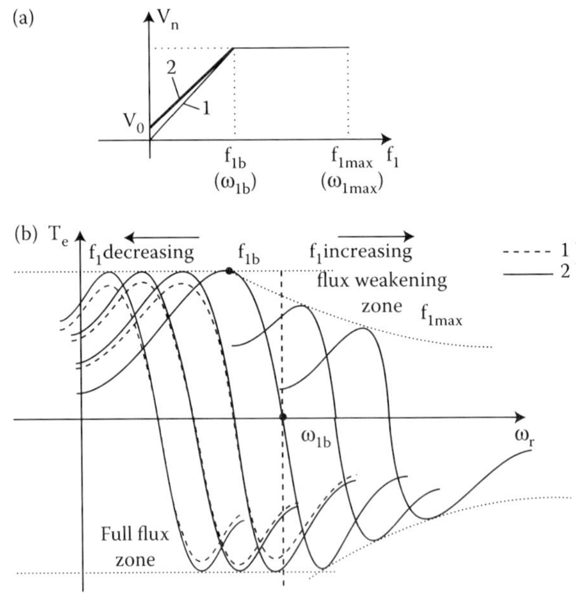


**FIGURE 8.17**
Torque speed curves for $V_1 = V_0 + K_f \cdot f_1$ ($V_1/f_1$ control): (a) $V_1/f_1$ dependencies, (b) $T_e/\omega_r$ curves.

flux-weakening zone. In many applications, constant power is required for a ratio $\omega_{1max}/\omega_b$ of 2–4. The entire motor design (sizing) depends on this requirement in terms of both electromagnetic and thermal loading.

$V_1/f_1$ drives are standard with low dynamics applications and moderate speed control range ($\omega_{1b}/\omega_{1min}$ = 10–15 or so) such as pumps or fans where the load torque is solely dependent on speed and thus an optimal $V_1/f_1$ relationship may be calculated off-line—for maximum efficiency or power factor—and implemented in the drive hardware.

## 8.18 Only for Constant Rotor Flux Torque Speed Curves Are Linear

The torque expression for constant rotor flux $\lambda_r^b$ is Equation 8.96

$$T_e = \frac{3}{2}p\frac{\lambda_{r0}^2(\omega_1 - \omega_r)}{r_r} \tag{8.146}$$

and it represents a straight line (Figure 8.18). This is ideal for speed (or torque) control purposes and led to (now widely accepted in industry) vector control. It is, however, to be noted that above base frequency $\omega_{1b}$, when the full voltage capability of PEC is reached, the value of rotor flux magnitude may not be maintained any more as the difference between the stator flux (8.145) and rotor flux amplitudes (8.104) is less than 15% in general (lower values for higher powers). So, above $\omega_{1b}$, the torque speed curves degenerate into the shapes obtained for $V_1/f_1$ control.

Four-quadrant operation presented in Figure 8.18 is obtainable as with $V_1/f_1$ control.

Negative frequency $\omega_1$ means negative (reverse) m.m.f. wave speed, to be obtained by changing the stator phase voltage sequence (from a b c to a c b).

## 8.19 Constant Stator Flux Torque Speed Curves Have Two Breakdown Points

Let us note from Equation 8.84 that the stator–rotor flux relationship for steady state (cage rotor) is $V_r^b = 0$:

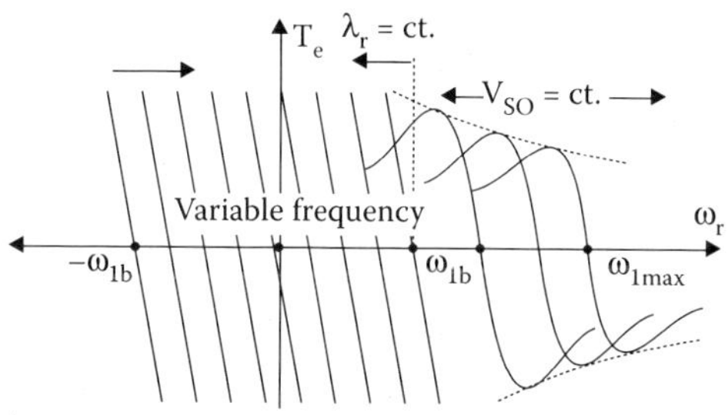


**FIGURE 8.18**
Torque speed curves for constant rotor flux $\lambda_r^b$ = ct up to base frequency $\omega_{1b}$; constant voltage and variable frequency above $\omega_{1b}$.

$$\bar{\lambda}_{s0} = \bar{\lambda}_{r0}\frac{(1+jS\omega_1\tau_r')}{K_s} \tag{8.147}$$

Consequently, the torque $T_e$ from Equation 8.146 becomes

$$T_e = \frac{3}{2}p\frac{K_s^2}{r_r}\frac{S\omega_1\lambda_{s0}^2}{1+(S\omega_1\tau_r')^2} \tag{8.148}$$

This expression has extreme (critical) values for

$$(S\omega_1)_k = \pm\frac{1}{\tau_r'} \tag{8.149}$$

$$T_{ek} = \frac{3}{2}p\frac{K_s^2\lambda_{s0}^2}{2r_r\tau_r'} \tag{8.150}$$

So the peak torque is independent of frequency as long as $\lambda_s$ amplitude may be realized, that is, below base (rated) frequency.

Above base frequency, according to Equation 8.145, the approximate peak torque would be

$$(T_{ek})_{\omega_1>\omega_{1b}} = \frac{3}{2}p\frac{K_s^2}{2r_r\tau_r'}\cdot\frac{V_{s0}^2}{\omega_1^2} \tag{8.151}$$

The torque speed curves are shown in Figure 8.19 for four-quadrant operation.

The peak torque is safely provided even for zero speed. Still the departure from linearity in the torque speed curve has to be dealt with when investigating electric drive transients and stability.

NOTE: In principle, PECs can also provide constant main (airgap) flux variable frequency control. So far this operation mode has not reached wide markets in variable speed drives.

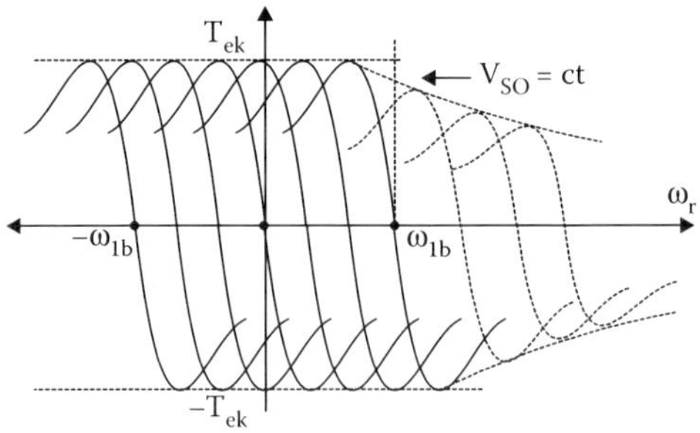


**FIGURE 8.19**
Torque speed curves for constant stator flux amplitude $\lambda_s$ up to $\omega_{1b}$ and constant voltage above $\omega_{1b}$.

## 8.20 Split-Phase Induction Motor

The split-phase induction motor is provided with two orthogonal stator windings: the main winding and the auxiliary winding.

This induction motor is, in general, fed from a single-phase a.c. supply of constant or variable voltage and frequency.

So, even for variable speed, it requires, essentially, a single-phase PWM inverter.

Connecting a capacitor in series with the auxiliary winding causes the current $I_a$ to lead the current in the main winding $I_m$ and thus the two orthogonal placed windings produce an elliptical magnetic field in the airgap, which has a traveling component. This explains the safe self-starting from the capacitor auxiliary to the main winding direction of rotation.

Switching the capacitor from auxiliary to main winding leads to the reversal of traveling field component in the airgap and thus the direction of motion may be reversed. For this latter case, the two windings should be identical (same number of turns and slots). For unidirectional motion, a higher than unity turns ratio $a = W_a/W_m$ is typical with $I_a < I_m$, as in many cases the auxiliary winding is turned off after starting the motor.

It is also common to use two capacitors: one (larger) for starting, and a smaller one for running.

It goes without saying that the power factor of the capacitor split-phase IM is very good, due to the presence of the capacitor.

Typical connections for dual-capacitor IM are shown in Figure 8.20.

The capacitor voltage vector is rectangular to auxiliary winding current vector. For symmetry conditions—90° electrical degrees between the $I_a$ and $I_m$ and equal ampere turns $W_m I_m = W_a I_a$—a pure traveling field is produced; once at zero speed with the capacitor $C_s$ and once at rated speed (load) with capacitor $C_n < C_s$ ($C_s/C_n \cong 4$–6).

At any other speed (slip S), the magnetic field in the airgap will have an additional (undesirable) inverse component, which produces a braking torque and additional losses.

As done for the three-phase motor, the m.m.f. of stator windings fundamental components are

$$
\begin{aligned}
F_m(\theta_{es}, t) &= F_{1m}\cos(\omega_1 t)\cos\theta_{es} \\
F_a(\theta_{es}, t) &= -F_{1a}\cos(\omega_1 t + \gamma_i)\sin\theta_{es}
\end{aligned}
\tag{8.152}
$$

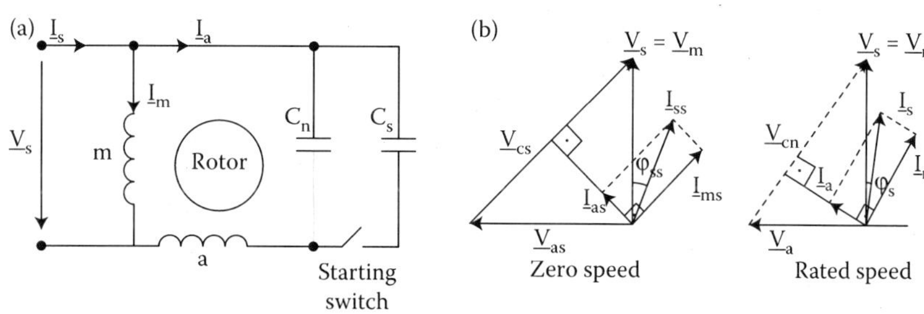


**FIGURE 8.20**
The dual-capacitor IM: (a) equivalent scheme, (b) phasor diagrams for zero and rated speed.

The ratio of the two m.m.f. amplitudes is

$$\frac{F_{1a}}{F_{1m}} = \frac{W_m k_{Wm1}}{W_a k_{Wa1}} \frac{I_m}{I_a} = \frac{1}{a} \frac{I_m}{I_a} \tag{8.153}$$

The total m.m.f. $F(\theta_{es}, t)$ may be decomposed in forward and backward waves

$$\begin{aligned} F(\theta_{es}, t) &= F_m(\theta_{es}, t) + F_a(\theta_{es}, t) \\ &= F_{1m} \cos(\omega_1 t) \cos\theta_{es} + F_{1a} \cos(\omega_1 t + \gamma_i) \sin\theta_{es} \\ &= \frac{1}{2} F_m C_f \sin(\theta_{es} - \omega_1 t - \beta_f) + \frac{1}{2} F_m C_b \sin(\theta_{es} + \omega_1 t - \beta_b) \end{aligned} \tag{8.154}$$

With

$$C_f = \sqrt{\left(1 + \frac{F_{1a}}{F_{1m}} \sin\gamma_i\right)^2 + \left(\frac{F_{1a}}{F_{1m}}\right)^2 \cos^2\gamma_i} \tag{8.155}$$

$$C_b = \sqrt{\left(1 - \frac{F_{1a}}{F_{1m}} \sin\gamma_i\right)^2 + \left(\frac{F_{1a}}{F_{1m}}\right)^2 \cos^2\gamma_i} \tag{8.156}$$

$$\sin\beta_f = \frac{1 + (F_{1a}/F_{1m})\sin\gamma_i}{C_f}; \quad \sin\beta_b = \frac{1 - (F_{1a}/F_{1m})\sin\gamma_i}{C_b} \tag{8.157}$$

It is evident that for $F_{1a}/F_{1m} \cong 1$ and the phase shift $\gamma_i = 90°$, $C_b = 0$ and thus the backward field (and torque) is zero. The slip is $S_+ = (\omega_1 - \omega_r)/\omega_1$ for the direct m.m.f. component and $S_- = 2 - S = (-\omega_1 - \omega_r)/(-\omega_1)$ for the backward (inverse) m.m.f. component.

For the steady state, the symmetrical (±) component model is straightforward, while for transients the d-q model in stator coordinates is very practical (Figure 8.21).

*The ± model*

The superposition principle is used

$$\underline{A}_m = \underline{A}_{m+} + \underline{A}_{m-}; \quad \underline{A}_a = \underline{A}_{a+} + \underline{A}_{a-} \tag{8.158}$$

and

$$\underline{A}_{m+} = \frac{1}{2}(\underline{A}_m - j\underline{A}_a); \quad \underline{A}_{m-} = \underline{A}_{m+}^* \tag{8.159}$$

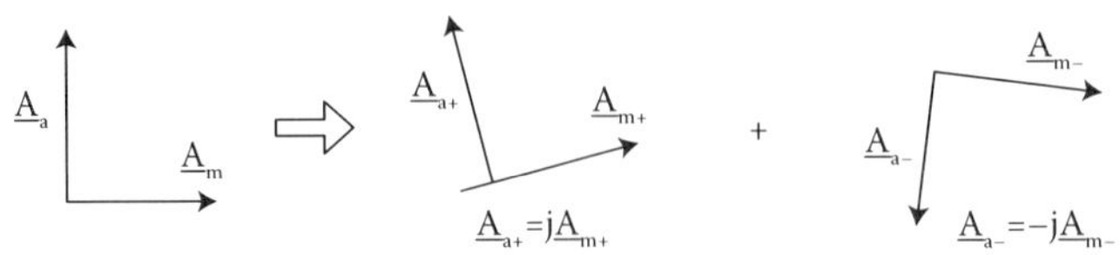


**FIGURE 8.21**
The ± model decomposition.

Now the machine behaves like two separate fictitious machines for the two components:

$$\underline{V}_{m+} = \underline{Z}_{m+}\underline{I}_{m+}; \quad \underline{V}_{m-} = \underline{Z}_{m-}\underline{I}_{m-}$$
$$\underline{V}_{a+} = \underline{Z}_{a+}\underline{I}_{a+}; \quad \underline{V}_{a-} = \underline{Z}_{a-}\underline{I}_{a-} \tag{8.160}$$

$$\underline{V}_{m} = \underline{V}_{m+} + \underline{V}_{m-}; \quad \underline{V}_{a} = \underline{V}_{a+} + \underline{V}_{a-} \tag{8.161}$$

The ± impedances $\underline{Z}_{m\pm}$ represent the total forward/backward impedances of the machine on a per phase basis, when the rotor cage is reduced to the main (m) and, respectively, auxiliary (a) winding (Figure 8.22).

The relationship of $\underline{V}_m$ and $\underline{V}_a$ voltages to the source voltage $\underline{V}_s$ are

$$\underline{V}_s = \underline{V}_m; \quad \underline{V}_a + (\underline{I}_{a+} + \underline{I}_{a-})\underline{Z}_a = \underline{V}_s \tag{8.162}$$

$\underline{Z}_a$ is the auxiliary impedance (a capacitor in general) added in series to the auxiliary phase for better starting and (or) running.

The torque $T_e$ expression has two terms: the forward and backward components.

$$T_e = T_{e+} + T_{e-} = \frac{2p_1}{\omega_1}\left[I_{rm+}^2 \frac{R_{rm}}{S} - I_{rm-}^2 \frac{R_{rm}}{2-S}\right] \tag{8.163}$$

When the auxiliary winding is open $Z_a = \infty$, $I_{rm+} = I_{rm-}$ and thus, at zero speed (S = 1) the torque is zero, as expected, and the machine does not start.

A typical mechanical characteristic $T_e(\omega_r)$ for a split-phase capacitor-run motor comprises the positive torque (with synchronism at $+\omega_1$) and the negative torque (with synchronism at $-\omega_1$) (Figure 8.23).

The backward (negative) torque is rather small but present at all speeds, except for the rated speed, where, by design, symmetry conditions are met. There are two symmetrization conditions that stem from $I_{m-} = 0$ which lead to a value of turns ratio a and of the capacitance C for given slip S [8, p. 851].

For the case of motor with same copper quantity in both stator windings, the symmetrization conditions are very simple [8]:

$$a = \frac{X_{m+}}{R_{m+}} = \tan\phi_+$$
$$X_c = \frac{1}{\omega C} = Z_+ \cdot a\sqrt{a+1} \tag{8.164}$$

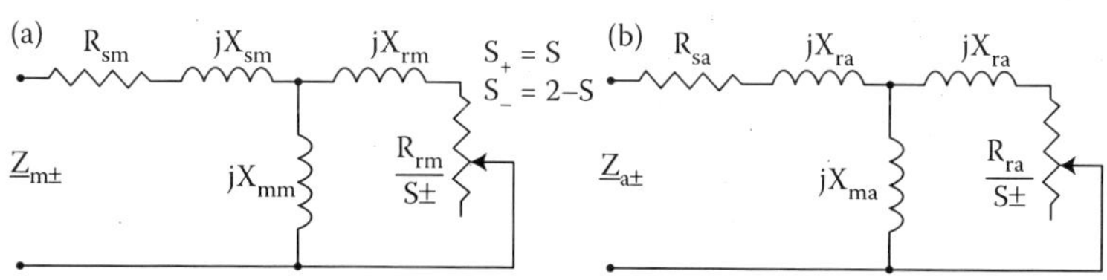


**FIGURE 8.22**
Equivalent symmetrical (±) impedance: (a) reduced to main winding; (b) reduced to auxiliary winding.

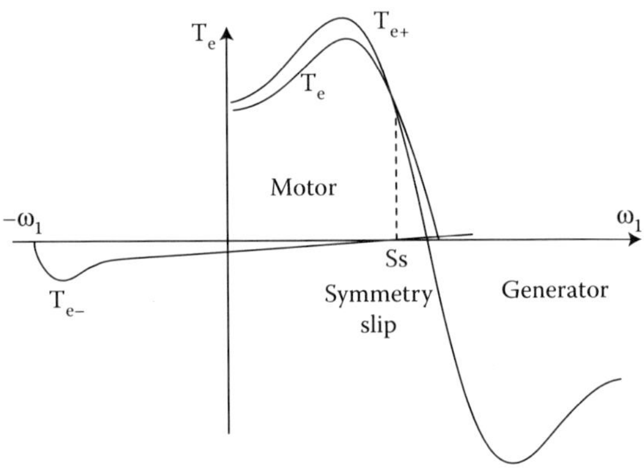


**FIGURE 8.23**
Typical mechanical characteristic of a split-phase capacitor-run induction motor.

So, for an existing motor (with fixed turns ratio a), there may not be any capacitor to symmetrize the motor at any slip (speed). Fortunately, perfect symmetrization is not necessary because the backward torque tends to be small for reasonably low rotor resistance (good efficiency) designed motors. The split-phase capacitor motor is widely used both at constant and variable speed for applications below 500 W, in general. More details on the control of this motor are presented in Chapter 9.

**EXAMPLE 8.5**

A capacitor-run split-phase IM with $C_a = 8\ \mu F$, $2p = 6$ pole, 230 V, 50 Hz, $n_n = 960$ rpm has the following main and auxiliary winding parameters: $R_{sm} = 34\ \Omega$, $R_{sa} = 150\ \Omega$, $a = 1.73$, $X_{sm} = 35.9\ \Omega$, $X_{sa} = a^2X_{sm}$, $X_{rm} = 29.32\ \Omega$, $R_{rm} = 23.25\ \Omega$, $X_m = 249\ \Omega$.

For the case of open auxiliary phase, calculate:

a. The ± impedances $\underline{Z}_{m+}$, $\underline{Z}_{m-}$ at rated slip
b. The equivalent circuit of the machine for this particular case
c. The main winding current components $I_{m+}$, $I_{m-}$ and the total current $I_m$
d. The ± torque components, the resultant torque and the mechanical power, efficiency and power factor

**Solution**

a. The rated slip

$$S_n = \frac{\omega_1 - \omega_r}{\omega_1} = \frac{2\pi 50 - 2\pi 3 \cdot 960/60}{2\pi 50} = 0.06; \quad 2 - S = 1.94 \tag{8.165}$$

From Figure 8.22, the ± impedances are

$$\underline{Z}_{m+} = R_{sm} + jX_{sm} + Z_{r+} = 34 + j35.9 + \frac{j249(23.25/0.06 + j29.32)}{23.25/0.06 + j(29.32 + 249)}$$

$$= 34 + j35.9 + 105.4 + j174 \tag{8.166}$$

$$\underline{Z}_{m-} = R_{sm} + jX_{sm} + Z_{r-} = 34 + j35.9 + \frac{j249(23.25/1.94 + j29.32)}{23.25/1.94 + j(29.32 + 249)}$$

$$= 34 + j35.9 + 9.6 + j27 \tag{8.167}$$

b. Let us note that because the auxiliary current $\underline{I}_a = 0$, from Equations 8.158 and 8.159

$$\underline{I}_{m+} = \underline{I}_{m-} = \frac{\underline{I}_m}{2}$$

Combining this with Equation 8.160

$$\underline{V}_m = \underline{V}_s = \underline{V}_{m+} + \underline{V}_{m-} = \underline{Z}_{m+}\underline{I}_{m+} + \underline{Z}_{m-}\underline{I}_{m-} = \underline{I}_m\left(\frac{\underline{Z}_{m+}}{2} + \frac{\underline{Z}_{m-}}{2}\right)$$

which means that we obtain a series equivalent circuit (Figure 8.24):

c. The main winding current components $I_{m+} = I_{m-}$ are then simply

$$I_{m+} = I_{m-} = \frac{V_s}{|\underline{Z}_{m+} + \underline{Z}_{m-}|} = \frac{230}{2 \cdot 165.2} = 0.696\,\text{A}$$

The main winding current $I_m = 2I_{m+} = 1.392$ A.
The power factor for $S = 0.06$, $\cos\varphi_1$ is

$$\cos\varphi_1 = \frac{\text{Real}(\underline{Z}_{m+} + \underline{Z}_{m-})}{|\underline{Z}_{m+} + \underline{Z}_{m-}|} = \frac{2 \cdot 92.8}{2 \cdot 165.2} = 0.5617!$$

d. The torque components in Equation 8.163 are alternatively

$$T_e = T_{e+} + T_{e-} = \frac{2p}{\omega_1}\left[I_{m+}^2\frac{R_{rm}}{S} - I_{m-}^2\frac{R_{rm}}{2-S}\right]$$

With $R_{rm+} = 105.4\ \Omega$ and $R_{rm-} = 9.6\ \Omega$ from $\underline{Z}_{m+}$ and $\underline{Z}_{m-}$ above

$$T_{e+} = \frac{2 \cdot 3}{2\pi 50}0.696^2 \cdot 105.4 = 0.9756\,\text{N m}$$

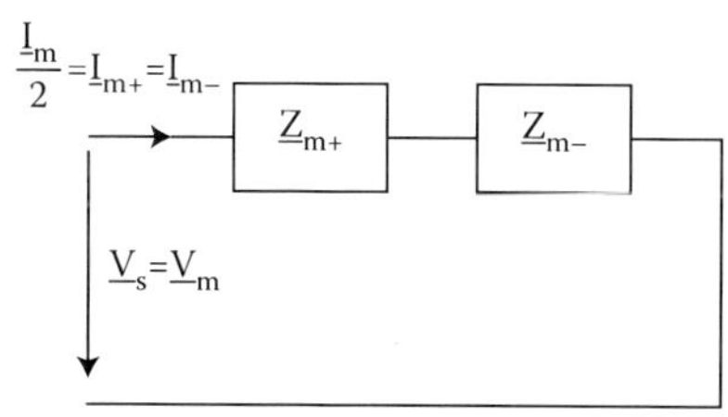


**FIGURE 8.24**
Single-phase ($\underline{I}_a = 0$) IM equivalent circuit.

$$T_{e-} = -\frac{2 \cdot 3}{2\pi 50} 0.696^2 \cdot 9.6 = -0.08886\,\text{N m}$$

So the total torque

$$T_e = T_{e+} + T_{e-} = 0.8687\,\text{N m}$$

The mechanical power $P_m$ is

$$P_m = T_e 2\pi 50 = 0.8687 \cdot 2\pi \cdot \frac{940}{60} = 85.47\,\text{W}$$

Neglecting the core and mechanical loss, the efficiency is

$$\eta = \frac{P_m}{P_m + p_{Co}} = \frac{85.47}{85.47 + 1.392^2 \cdot 34} = 0.5647$$

The power factor cos $\varphi_1$ is low, as calculated above because a capacitor is missing and the efficiency is low, mainly because the power is small.

The need to supply the auxiliary winding in series with a capacitor is evident.

NOTE: Even when the auxiliary phase winding is present, the performance computation process is as above, but a bit more tedious [8, p. 847].

## 8.21 Split-Phase Capacitor IM Transients

The split-phase IM is still used at a remarkable performance (efficiency) in driving pumps or compressors for household appliances, in the range of 100 W or more. Many other domestic tools, such as clothes washers, drills, and saws, make use of split-phase capacitor IMs in the range of hundreds of watts to 1 kW. They use a starting self-variable resistor, $R_{start}$, or a capacitor, $C_{start}$, in the auxiliary phase. The latter may be kept on during the on-load operation, to increase efficiency, but with a smaller running capacitor, $C_{run}$ ($C_{run} < C_{start}$) (Figure 8.25a). Yet another application may use a three-phase winding for unidirectional or reversible motion (Figure 8.25b and c).

For some cloth-washing machines, there is a three-phase 12-pole winding used for washing (at a low reversible speed) and a separate 2-pole orthogonal winding for spinning (at a high speed). Switching back and forth for the washing mode and for the drying mode implies important transients.

The three-phase connections (Figure 8.25b and c) may be reduced to orthogonal windings (Figure 8.25a) [8, Chapter 23]. So, here, only orthogonal windings are treated.

For better starting performance, 120° space-angle-shifted main and auxiliary windings may be used, but they also may be reduced to two orthogonal ones [8, Chapter 23].

Finally, the general case, when the main and auxiliary windings are orthogonal, but they use different number of slots and copper weights, should be treated in phase variables.

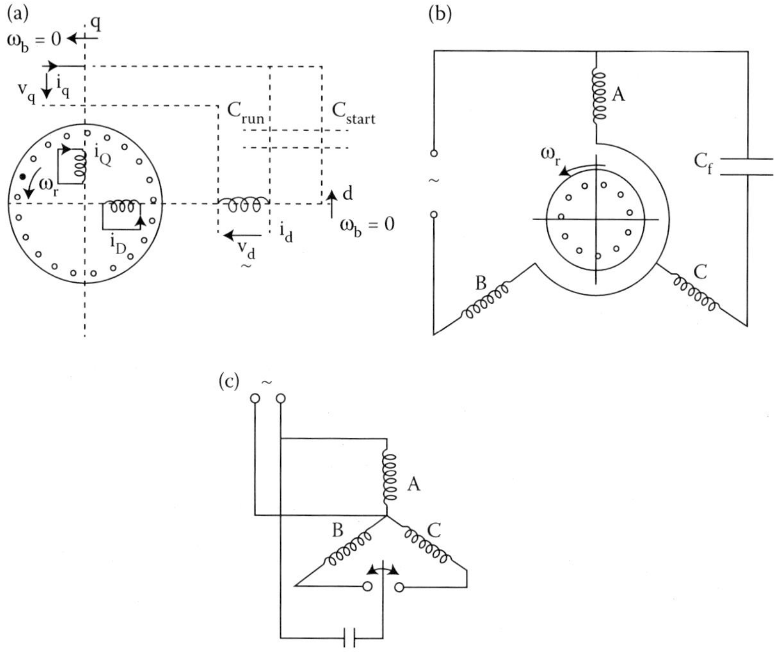


**FIGURE 8.25**
Split-phase capacitor IMs with (a) auxiliary phase, (b) three-phase, unidirectional, and (c) three-phase, bidirectional.

### 8.21.1 Phase Variable Model

In this case, the rotor is modeled by orthogonal windings, $d_r$ and $q_r$:

$$\left|I_{m,a,d_r,q_r}\right|\left|R_{m,a,d_r,q_r}\right|-\left|V_{m,a,d_r,q_r}\right|=-\frac{d}{dt}\left|\Psi_{m,a,d_r,q_r}\right| \tag{8.168}$$

$$\left|I_{m,a,d_r,q_r}\right|=\left|I_m, I_a, I_{d_r}, I_{q_r}\right|^T$$

$$\left|I_{m,a,d_r,q_r}\right|=\mathrm{Diag}(R_m, R_a, R_r, R_r) \tag{8.169}$$

$$\left|V_{m,a,d_r,q_r}\right|=\left|V_m, V_a, 0, 0\right|$$

$$\begin{vmatrix}\Psi_m\\ \Psi_a\\ \Psi_{dr}\\ \Psi_{qr}\end{vmatrix}=\begin{vmatrix}L_{ml}+L_m^m & 0 & L_{mr}\cos\theta_{er} & -L_{mr}\sin\theta_{er}\\ 0 & L_{al}+L_m^a & L_{ar}\sin\theta_{er} & L_{ar}\cos\theta_{er}\\ L_{mr}\cos\theta_{er} & L_{ar}\sin\theta_{er} & L'_{rl}+L_m^m & 0\\ -L_{mr}\sin\theta_{er} & L_{ar}\cos\theta_{er} & 0 & L'_{rl}+L_m^m\end{vmatrix}\cdot\begin{vmatrix}I_m\\ I_a\\ I_{dr}\\ I_{qr}\end{vmatrix} \tag{8.170}$$

$$T_e=p[I]\frac{\partial\left|L(\theta_{er})\right|}{\partial\theta_{er}}[I]^T \tag{8.171}$$

$$\frac{J}{p}\frac{d\omega_r}{dt}=T_e-T_{load};\quad \frac{d\theta_{er}}{dt}=\omega_r;\quad \omega_r=2\pi pn \tag{8.172}$$

We also should use the constraints

$$
\begin{aligned}
V_m(t) &= V_m\sqrt{2}\cos(\omega_1 t + \gamma_0) \\
V_m(t) &= V_a(t) + V_C(t)
\end{aligned}
\tag{8.173}
$$

and

$$
\frac{dV_C}{dt} = \frac{1}{C} I_a \tag{8.174}
$$

When a resistor is added to the auxiliary winding, $V_C$ is replaced by $R_{start}I_a$ and Equation 8.174 is eliminated. The system has a seventh order, with many variable coefficients, and may be solved by numerical methods, albeit within a large CPU time.

This general model gets simplified if the auxiliary winding and the main winding make use of the same copper weight when

$$
\begin{aligned}
&R_a = R_m a^2; \quad a = \frac{w_a k_{wa}}{(w_m k_{wm})} \\
&L_{al} = L_{ml} a^2 \\
&L_{ar} = L_{mr} a \\
&L_m^a = L_m^m a^2
\end{aligned}
\tag{8.175}
$$

In such a case, the d-q model in any coordinates may be used because the stator windings become almost symmetric under conditions (8.175), despite different number of turns.

### 8.21.2 d-q Model

Let us use stator coordinates and consider the d axis along the m (main) winding axis and the q axis along the auxiliary winding axis. There is no need to reduce the auxiliary winding to the main winding as long as Equation 8.175 is satisfied.

The d-q model equations are straightforward as the stator windings are already orthogonal, and, thus, no Park transformation is needed (for the phase machine connection in Figure 8.25b, the latter is instrumental [8, Chapter 23]):

$$
\begin{aligned}
&I_d R_m - V_d = -\frac{d\Psi_d}{dt}; \quad I_q R_a - V_q = -\frac{d\Psi_q}{dt} \\
&I_{dr} R_r = -\frac{d\Psi_{dr}}{dt} - \omega_r \Psi_{qr}; \quad I_{qr} R_r = -\frac{d\Psi_{qr}}{dt} + \omega_r \Psi_{dr} \\
&\Psi_d = L_{ml} I_d + L_{dm}(I_{dr} + I_d) \\
&\Psi_{dr} = L'_{rl} I_{dr} + L_{dm}(I_{dr} + I_d) \\
&\Psi_q = L_{al} I_q + L_{qm}\left(I_q + \frac{1}{a} I_{qr}\right) \\
&\Psi_{qr} = L'_{rl} I_{qr} + L_{qm}(a I_q + I_{qr})
\end{aligned}
\tag{8.176}
$$

Again

$$\frac{L_{qm}}{L_{dm}} = a^2 \text{ and } L_{dm} = L_m^m; \quad I_m = I_d; \quad I_a = I_q \tag{8.177}$$

$$V_d = V_m; \quad V_q = V_a = V_m - V_C(t); \quad \frac{dV_C}{dt} = \frac{I_q}{C} \tag{8.178}$$

$$T_e = -p(\Psi_{dr} I_{qr} - \Psi_{qr} I_{dr}) = pL_{dm}(aI_q I_{dr} - I_d I_{qr}) \tag{8.179}$$

Adding the motion equations 8.172, the complete model is obtained. The system's order is now 6 ($\theta_{er}$ [the rotor position] is irrelevant here).

NOTE: Again, the d-q model is valid only if conditions (8.175) are satisfied.

For more on the subject, see Reference 8, Chapters 24 through 28.

Magnetic saturation can be handled simply in the d-q model by the $L_m^m(I_m)$ function, with fluxes as variables. Magnetic saturation causes, under steady state, nonsinusoidal stator currents [8, Chapter 25].

## 8.22 Summary

- The three-phase induction motor is the workhorse of industry for powers from below 1 kW to 10 MW and up to over 100 MW in wound rotor configuration for pumped storage power plants.
- The a.c. stator and rotor windings are placed in slots. Skin effect in the rotor windings is put to work in constant speed high starting torque applications and is to be avoided in variable speed drives where variable voltage and frequency are provided by PECs.
- The inductances, between stator and rotor windings, vary sinusoidally with the rotor position and thus the voltage–current equations in phase coordinates are difficult to handle for transients even through numerical methods, in an eighth-order nonlinear system.
- The space-phasor model, equivalent to the d-q (orthogonal axis, or Park's) model, is characterized by constant coefficients.
- For steady state, the rotor current and flux space-phasors are spatially orthogonal to each other for the cage rotor.
- At constant speed the space-phasor model is a second-order system with two complex eigenvalues, one for the stator and the other for the rotor. They depend on rotor speed and on the reference system speed $\omega_b$, but their real part is negative in general.
- Under steady state, the space-phasor model has the same equations irrespective of the reference system speed.

- Losses in an induction motor are distributed both in the stator and in the rotor winding, core loss in the stator winding, and additional and mechanical losses in the rotor. The equivalent circuit allows for calculating the steady-state performance conveniently.
- The induction motor operation modes are motoring, generating, and a.c. braking (plugging). The torque speed curve shows two peak values (one for motoring and the other for generating), when constant voltage and frequency fed.
- When the motor stator is d.c. current fed and the rotor speed is nonzero, motion-induced voltages in the rotor produce rotor currents and as a consequence a braking torque. Its maximum is obtained at a small speed for cage rotor motors, as the rotor current (slip) frequency $\omega_2$ is equal to rotor speed $\omega_r$.
- Speed control methods include variable stator or rotor frequency and voltage through PECs. For limited speed control (±30% around ideal no-load speed) the variable rotor frequency supply method (through limited power PECs) is particularly suited for the scope, especially in high-power drives.
- Stator frequency variation is followed along three methods of practical interest: $V_1/f_1$ method, constant rotor flux, and constant stator flux methods.
- Only for constant rotor flux is the torque speed curve linear (as for a PM d.c. brush motor), that is, ideal for electric drives control. This method is known as vector control.
- Split-phase capacitor induction motors are used in general below 500 W in single-phase supply (home) applications.

## 8.23 Problems

8.1 Draw the complex root (eigenvalues) locus plot of a cage rotor induction motor with data as in Example 8.2 for eight different rotor speed values from 0 to 2400 rpm, in stator coordinates ($\omega_b = 0$).

8.2 For the induction motor with resistances and inductances as in Example 8.3, $p = 1$, $V_r^b = 0$, $n = 900$ rpm, rotor flux $\lambda_r^b = 1\,\text{Wb}$, torque level $T_e = 40$ N m, calculate primary frequency $\omega_1$, stator flux space-phasor amplitude in synchronous coordinates, stator voltage, stator current, and power factor. Align the d axis along the rotor flux and draw the space-phasor diagram with the corresponding numbers on it (as in Figure 8.12).

8.3 An induction motor has the data ideal no-load ($S = 0$) phase current (rms) $I_{on} = 5$ A, $p = 2$, no-load losses $p_0 = 200$ W, stator phase voltage $V = 120$ V (rms); $f_1 = 60$ Hz; starting current per phase ($\omega_r = 0, S = 1$) $I_{start} = 15I_{on}$ and starting power factor $\cos\varphi_{1s} = 0.3$; $L_{ls} = L_{lr}$; $r_{rstart} = 3r_r$. Using the equivalent circuit of Figure 8.11, calculate the stator and rotor resistances $r_s$, $r_{rstart}$ and the leakage inductances $L_{ls}$ and $L_{lr}$, the starting torque, core loss resistance $r_m$, and core loss $p_{iron}$, and main inductance $L_m$. Observing that we have a deep bar (skin effect) rotor and the rotor resistance during normal operation is $r_r = 1.2r_s$, calculate all the electrical losses in the machine for $S = 0.02$.

8.4 For an induction motor with the data: $r_s = 0.2\ \Omega$, $p = 2$, $r_r = 0.2\ \Omega$, $L_{lr} = L_{ls} = 5 \times 10^{-3}$ H, ($L_m = 0.1$ H), $f_1 = 60$ Hz, $V_L = 220$ V (rms) operating in regenerative braking at $S = -0.02$, calculate electromagnetic torque $T_e$ and the electric power retrieved if the core and mechanical losses are neglected.

8.5 For the induction motor with the data (resistances and inductances) of Problem 8.4, calculate the d.c. braking peak torque (for the d.c. connection of Figure 8.15b at $I_o = 4$ A), for $\omega_1 = 2\pi 10$ rad/s.

8.6 An induction motor works with a rotor flux level $\lambda_{r0} = 1$ Wb and has four poles ($2p = 4$) and the rotor resistance $r_r = 0.2\ \Omega$. For $\omega_1 = 2\pi 20$ rad/s, calculate and draw the torque speed curve. For $\tau_r' = 0.01$ s, $S = 0.1$, $K_s = 0.95$, determine the stator flux level $\lambda_{so}$ and, neglecting the stator resistance, the stator voltage $V_{s0}$ required.

8.7 An induction motor works at constant stator flux and variable frequency and has the data as in Problem 8.6. Determine the critical slip frequency and $(S\omega_1)_k$ and the critical (breakdown) torque for $\lambda_{s0} = 1.06$ Wb and draw the torque speed curve for $f_1 = 20$ Hz. For $f_1' = 60$ Hz, calculate the maximum flux level available for the same voltage as in the case of 20 Hz, and determine again the breakdown torque available, and the corresponding electromagnetic power.

## References


1. K.P. Kovacs, I. Racz, *Transient Regimes of AC Machines*, Springer-Verlag, 1995. [The original edition in German, in 1959.]
2. P. Vas, *Vector Control of AC Machines*, Oxford University Press, Oxford, UK, 1990.
3. I. Boldea, S.A. Nasar, *Vector Control of AC Drives*, Chapter 2, CRC Press, Boca Raton, Florida, 1992.
4. D.W. Novotny, T.A. Lipo, *Vector Control and Dynamics of AC Drives*, Oxford University Press, New York, 1996.
5. R.H. Park, Two reaction theory of synchronous machines: Generalized method of analysis, *AIEE Trans.*, vol. 48, 1929, pp. 716–730.
6. J. Holtz, On the spatial propagation of transient magnetic fields in ac machines, *IEEE Trans.*, vol. IA-32, no. 4, 1996, pp. 927–937.
7. I. Boldea, S.A. Nasar, Unified treatment of core losses and saturation in orthogonal axis model of electric machines, *Proc. IEE*, vol. 134, no. Part B, 1987, pp. 355–363.
8. I. Boldea, S.A. Nasar, *Induction Machine Design Handbook*, second edition, CRC Press, Taylor & Francis Group, New York, 2010.

# 9

# *PWM Inverter-Fed Induction Motor Drives*

## 9.1 Introduction

In this chapter, we will deal only with cage-rotor IM drives of small- and medium-power ratings provided with voltage-source PWM inverters for variable frequency and voltage amplitude.

*Open-loop coordinated voltage–frequency* $V_1/f_1$ *control or current–slip frequency control—called scalar control*—do not include a speed (or position) closed loop and have so far traditionally provided satisfactory variable speed with low dynamics and low-speed control precision.

Fast torque dynamics and speed control require closed-loop speed control and even torque closed-loop control. Linearizing the torque/speed characteristic, or torque/current characteristic, as in d.c. brush PM motors, would be ideal for the scope.

There are three main schemes to produce this linearization by intelligent manipulation of IM equations in space-phasors:

- Vector current and voltage control (VC) [1]
- Direct torque and flux control (DTFC) [2–4]
- Feedback linearization control (FLC) [5–7]

For closed-loop motion control, speed and (or) position sensors are to be used for wide speed control range (above 100 to 1), while below this value, no motion sensors are provided, with speed (position) calculated through observers based on measured motor voltages and (or) currents.

So we distinguish

- Control with motion sensors
- Control without motion sensors (sensorless)

The rapid development of powerful digital signal processors (DSPs) and ASICs led to a kind of universal IM drive that, on a menu basis, may work in various modes:

- $V_1/f_1$ with slip compensation with motion sensors or sensorless
- VC or DTFC or FLC with motion sensors or sensorless

As standard $V_1/f_1$ control is apparently losing momentum in industry, we will present it at the end of the chapter though historically it was the first sensorless (low dynamics)

IM drive. Consequently, we will deal first with the principle of VC, state observers implementation examples with motion sensors and sensorless. Then the same treatment will be applied to DTFC and FLC, and $V_1/f_1$ schemes.

## 9.2 VC: General Flux Orientation

VC implies independent (decoupled) control of flux-current and torque/current components of stator current through a coordinated change in the supply voltage amplitude, phase, and frequency. As the flux variation tends to be slow, especially with current control, constancy of flux should produce a fast torque/current (and torque) response, and finally fast speed (position) response. On the other hand, flux level control is essential to avoid magnetic saturation (and heavy core losses) and reduce the core losses at light loads through flux weakening; also, flux weakening is required to extend the speed/torque envelope.

There are three distinct flux space-phasors in the induction machine (Chapter 8, Equations 8.92 and 8.93): $\lambda_m^b$—airgap flux; $\lambda_s^b$—stator flux, and $\lambda_r^b$—rotor flux. Their relationships with currents are

$$\bar{\lambda}_m = L_m \cdot (\bar{i}_s + \bar{i}_r); \quad \bar{\lambda}_s = L_s \bar{i}_s + L_m \bar{i}_r; \quad \bar{\lambda}_r = L_r \bar{i}_r + L_m \bar{i}_s \tag{9.1}$$

VC could be performed with respect to any of these flux space-phasors by attaching the reference system d axis to the respective flux linkage space-phasor direction and by keeping its amplitude under surveillance. A general flux may be defined, with the above cases as a particular situation [8,9].

In order to facilitate comparisons between the three strategies, we introduce new, general, rotor variables $\bar{i}_{ra}$, $\bar{\lambda}_{ra}$, $r_{ra}$

$$\bar{i}_{ra} = \bar{i}_r / a; \quad \bar{\lambda}_{ra} = \bar{\lambda}_r \cdot a; \quad r_{ra} = r_r \cdot a^2 \tag{9.2}$$

and define a general flux

$$\bar{\lambda}_{ma} = a L_m (\bar{i}_s + \bar{i}_{ra}) \tag{9.3}$$

From the complex variable equations 8.50 through 8.53 (with d/dt = s)

$$\begin{aligned}
\bar{V}_s &= r_s \bar{i}_s + s\bar{\lambda}_s + j\omega_1 \bar{\lambda}_s \\
\bar{V}_r &= r_r \bar{i}_r + s\bar{\lambda}_r + j(\omega_1 - \omega_r)\bar{\lambda}_r \\
T_e &= \frac{3}{2} p \,\mathrm{Re}(j\bar{\lambda}_s \bar{i}_s^*)
\end{aligned} \tag{9.4}$$

with $\omega_1$ the general flux space-phasor electrical speed, $\bar{V}_r = 0$. Making use of Equations 9.1 through 9.3, we obtain

$$\bar{V}_s = [r_s + (s + j\omega_1)(L_s - aL_m)] \cdot \bar{i}_s + (s + j\omega_1) \cdot \bar{\lambda}_{ma} \tag{9.5}$$

$$0 = -[r_r + (s + jS_a\omega_1)(L_r - L_m / a)] \cdot \bar{i}_s + [r_r + (s + jS_a\omega_1)L_r] \cdot \frac{\bar{\lambda}_{ma}}{aL_m} \tag{9.6}$$

$$T_e = \frac{3}{2} p \,\mathrm{Re}\left(j\bar{\lambda}_{ma}\bar{i}_s^*\right) \tag{9.7}$$

with the slip $S_a$ defined for the general flux $\bar{\lambda}_{ma}$

$$S_a = \frac{\omega_1 - \omega_r}{\omega_1} \tag{9.8}$$

The complex variables are the stator current $\bar{i}_s$ and the general flux $\bar{\lambda}_{ma}$.
The reference system is attached to axis d along the general flux (Figure 9.1)

$$\bar{\lambda}_{ma} = \lambda_{ma}; \quad \bar{i}_s = i_{da} + ji_{qa} \tag{9.9}$$

According to Equation 9.9, the torque expression (9.7) becomes

$$T_e = \frac{3}{2} p\lambda_{ma} i_{qa}; \quad i_{qa} = i_s \sin\delta_{ia} \tag{9.10}$$

For motoring $T_e > 0$ (for $\omega_r > 0$) for $\delta_{ia} > 0$ and thus generating is obtained with $\delta_{ia} < 0$. *In other words, for direct (trigonometric) motion, the stator current leads the general flux for motoring and lags it for generating.*

Also from Equation 9.8 and Figure 9.1

$$\frac{d\gamma_a}{dt} = S_a\omega_a \tag{9.11}$$

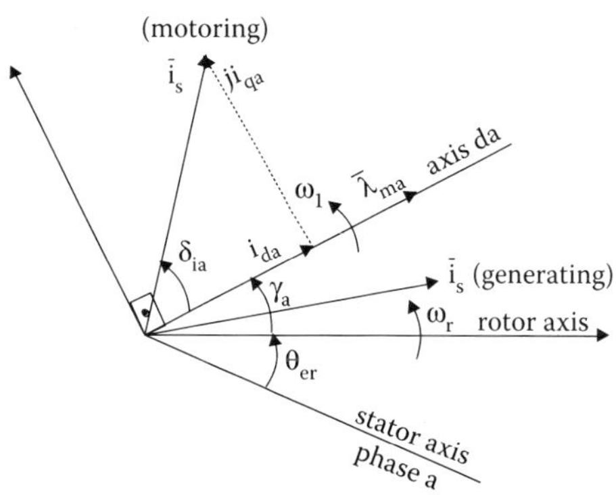


**FIGURE 9.1**
General flux orientation axes.

We may consider that $i_{qa}$ is the torque current component (9.10) while $i_{da}$ is the flux current component of stator current (Figure 9.1). *Their decoupled control is the essence of VC.* This is because, for constant $\lambda_{ma}$, torque control means $i_{qa}$ control and thus flux control means $i_{da}$ control.

## 9.3 General Current Decoupling

The input (command) variables in VC are the reference flux $\lambda^*_{ma}$ and the reference torque $T^*_e$. Consequently, the IM is a two-input system. The value of $\lambda^*_{ma}$ is either constant or it depends on speed or on torque, according to an *a priori* criterion of optimization, heavily related to application.

General current decoupling means to determine the reference current space-phasor $\bar{i}^*_s(i^*_{da}, i^*_{qa})$ based on reference flux $\lambda^*_{ma}$ and torque $T^*_e$. The reference torque may be the output of a position or speed closed-loop controller in speed (position) control drives or the required torque (speed or time dependent) in torque-control-only drives.

For general current decoupling, we make use of Equations 9.6 through 9.11, skipping the stator voltage equation, to obtain

$$S_a\omega_1 = \frac{i_{qa}(1+s\sigma_a\tau_r)}{((\lambda_{ma}/aL_m)-\sigma_a i_{da})\tau_r}; \quad \sigma_a = 1-\frac{L_m}{aL_r}; \quad \tau_r = \frac{L_r}{r_r} \tag{9.12}$$

$$(1+s\sigma_a\tau_r)i_{da} = (1+s\tau_r)\frac{\lambda_{ma}}{aL_m} + S_a\omega_1\sigma_a\tau_r i_{qa} \tag{9.13}$$

Equations 9.12 and 9.13 are illustrated in the general current decoupling network shown in Figure 9.2.

It becomes now possible, through PWM inverter control, to force the motor currents to follow $i^*_{da}$ and $i^*_{qa}$, in any flux orientation.

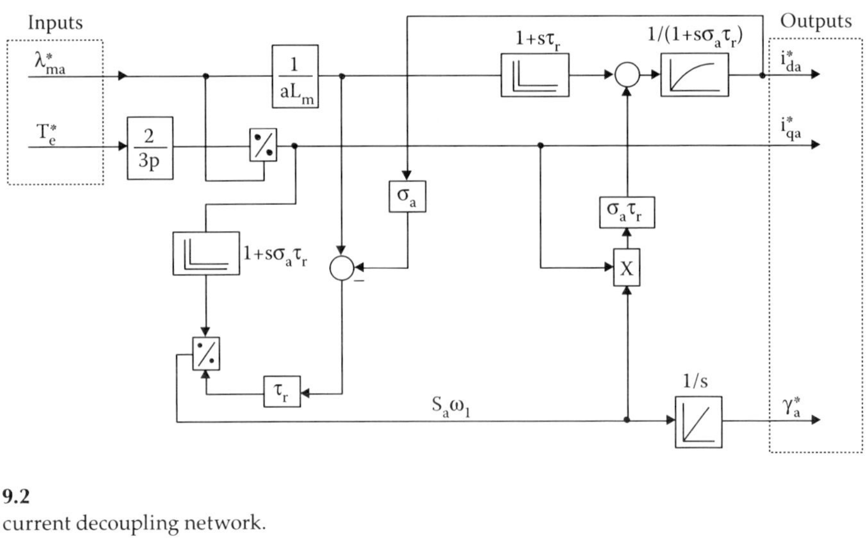


**FIGURE 9.2**
General current decoupling network.

It is interesting to see that for $\sigma_a = 0$

$$a = \frac{L_m}{L_r} \tag{9.14}$$

For $a = L_m/L_r$, from Equations 9.1 through 9.3, 9.12, and 9.13, we obtain

$$\begin{aligned}
&(\lambda_{ma})_{a=(L_m/L_r)} = \frac{L_m}{L_r}\lambda_r; \quad S_a\omega_1 = \frac{i_q}{i_d\tau_r} = \frac{i_{qa}}{i_{da}\tau_r/(1+s\tau_r)} \\
&i_{da} = (1+s\tau_r)\frac{\lambda_r}{L_m}; \quad T_e = \frac{3}{2}p\frac{L_m}{L_r}\lambda_r i_{qa} = \frac{3}{2}\lambda_r^{\,2}\frac{S_a\omega_1}{r_r}p
\end{aligned} \tag{9.15}$$

So only for rotor flux ($\lambda_r$) orientation the current decoupling network is simplified to the form in Figure 9.3.

For $a = L_s/L_m$, from Equations 9.1 through 9.3

$$(\bar{\lambda}_{ma})_{a=L_s/L_m} = \bar{\lambda}_s \tag{9.16}$$

So we end up with stator flux orientation but

$$(\sigma_a)_{a=L_s/L_m} = 1 - \frac{L_m^{\,2}}{L_s L_r} = \sigma \neq 0 \tag{9.17}$$

Consequently, in stator flux orientation, the current decoupling network retains the complicated form of Figure 9.2. A similar situation occurs with $a = 1$.

$$(\bar{\lambda}_{ma})_{a=1} = \bar{\lambda}_m; \quad (\sigma_a)_{a=1} = 1 - \frac{L_m}{L_r} \tag{9.18}$$

that is airgap flux orientation.

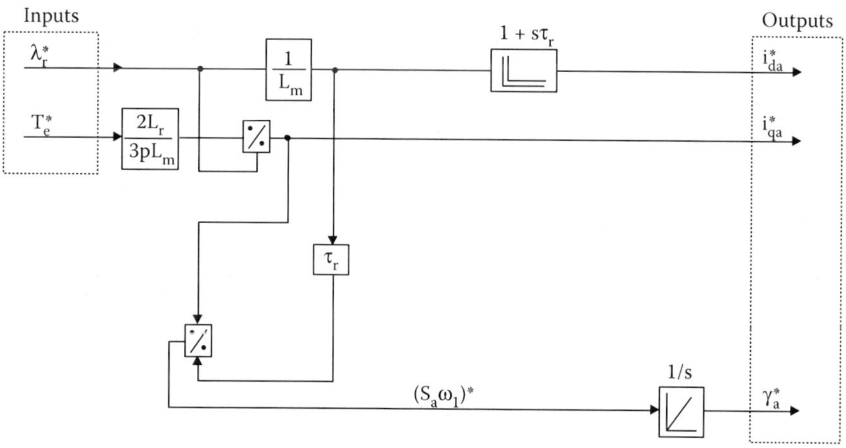


**FIGURE 9.3**
Current decoupling network in rotor flux orientation.

The current network is a feedforward (indirect) method to produce flux orientation. It presupposes a knowledge of machine parameters and an online computation effort (in DSP implementation) commensurable to the complexity of current decoupling network. The simpler it is, the better.

*This is why indirect VC is most adequate with rotor flux orientation.* However simple it is (Figure 9.3), the current decoupling network in rotor flux orientation depends heavily on motor parameters $L_m$, $L_r/L_m$, and $\tau_r$, which, in the end, depend on magnetic saturation level ($L_m$, $\tau_r$) and on rotor temperature ($\tau_r$) [10].

## 9.4 Parameter Detuning Effects in Rotor Flux Orientation Current Decoupling

Let us consider constant slip frequency

$$(S\omega_1) = (S\omega_1)^* \tag{9.19}$$

and constant stator current

$$i_s^* = i_s \tag{9.20}$$

But α and β, defined in Equations 9.21 and 9.22, vary with magnetic saturation and rotor temperature:

$$\frac{\tau_r}{\tau_r^*} = \alpha \tag{9.21}$$

$$\frac{L_m}{L_m^*} = \frac{L_r}{L_r^*} = \beta \tag{9.22}$$

For steady state, from Equation 9.15, we obtain

$$\begin{aligned}
&i_{da} = \frac{\lambda_r}{L_m}; \quad i_{da}^* = \frac{\lambda_r^*}{L_m} \\
&(S\omega_1) = (S\omega_1)^* = \frac{i_{qa}}{i_{da}\tau_r} = \frac{i_{qa}^*}{i_{da}^*\tau_r^*} \\
&T_e = \frac{3}{2}p\frac{L_m}{L_r}\lambda_r i_{qa} = \frac{3}{2}p\frac{L_m^2}{L_r}i_{da}i_{qa} = \frac{3}{2}p\lambda_r^2\frac{S_a\omega_1}{r_r} \\
&T_e^* = \frac{3}{2}p\frac{(L_m^*)^2}{L_r^*}i_{da}^* i_{qa}^*; \quad i_s^2 = (i_s^*)^2 = i_{da}^2 + i_{qa}^2 = (i_{da}^*)^2 + (i_{qa}^*)^2
\end{aligned} \tag{9.23}$$

From Equations 9.21 through 9.23 [10]

$$\frac{T_e}{T_e^*} = \alpha\beta\left[\frac{1+[(S_a\omega_1)^*\tau_r^*]^2}{1+[(S_a\omega_1)^*\alpha\tau_r^*]^2}\right] \tag{9.24}$$

$$\frac{\lambda_r}{\lambda_r^*} = \frac{L_m}{L_m^*}\frac{i_{da}}{i_{da}^*} = \sqrt{\frac{\beta}{\alpha}\frac{T_e}{T_e^*}} \tag{9.25}$$

where $T_e$ and $\lambda_r$ are the actual torque and rotor flux.

We may now illustrate the influence of $\alpha$ and $\beta$ on $T_e/T_e^*$ and $\lambda_r/\lambda_r^*$, for constant stator current $i_s^*$ and slip frequency $(S\omega_1)^*$, by considering

$$(S\omega_1)^*\tau_r^* = K \tag{9.26}$$

In general, $K \le 1$; introducing K in Equation 9.24 yields

$$\frac{T_e}{T_e^*} = \alpha\beta\left[\frac{1+K^2}{1+K^2\alpha^2}\right] \tag{9.27}$$

$$\frac{\lambda_r}{\lambda_r^*} = \beta\sqrt{\frac{1+K^2}{1+K^2\alpha^2}} \tag{9.28}$$

It is now clear that both saturation and temperature ($\alpha$, $\beta$) have a monotonous influence on the flux ratio while the influence of $\alpha$ on the torque ratio shows a maximum for $\alpha = 1/K$

$$\left(\frac{T_e}{T_e^*}\right)_{max} = \beta\frac{1+K^2}{2K} \ge \beta \tag{9.29}$$

Results for K = 0.5 and 1.0, $\beta = 1$ and $\alpha$ variable from 0.5 to 1.5 are shown in Figure 9.4.

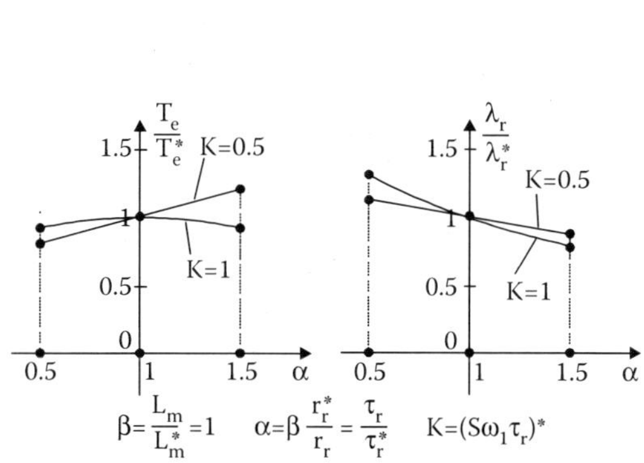


**FIGURE 9.4**
Actual/command torque ($T_e/T_e^*$) and rotor flux ($\lambda_r/\lambda_r^*$) versus rotor time constant detuning ratio $\alpha$, for current decoupling in rotor flux orientation.

*Being monotonous, the rotor flux detuning (error) may be used to correct the rotor time constant in rotor flux orientation indirect VC.*

The torque detuning is higher when the slip frequency $S\omega_1$ is smaller while the converse is true for rotor flux detuning.

Correcting the rotor time constant $\tau_r$, or for the slip frequency $(S\omega_1)^*$, to compensate for temperature and magnetic saturation detuning effects seems thus imperative in indirect vector current control, especially if reliable precise and quick torque control is required.

## 9.5 Direct versus Indirect Vector Current Decoupling

As seen above, the indirect current decoupling is either complicated (for stator flux orientation) or (and) strongly parameter dependent for rotor flux orientation.

To circumvent this difficulty, the current decoupling may be performed through direct closed-loop control of flux and torque based on their calculation through observers to produce the corresponding feedback (Figure 9.5).

In fact, parameter detuning problems are surfacing into observers but here they proved to be easier to handle.

## 9.6 A.C. versus D.C. Current Controllers

Once the reference d-q currents $i^*_{da}$, $i^*_{qa}$, and flux orientation angle $\theta_{er} + \gamma^*_a$ are known, we have to "translate" these commands into stator currents and to use current controllers to impose these currents through the PEC.

There are two ways toward this scope:

- Through a.c. current controllers
- Through d.c. (synchronous) current controllers [11]

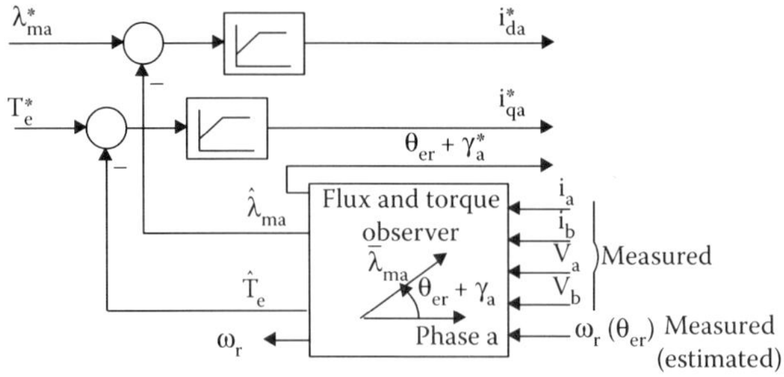


**FIGURE 9.5**
Direct (feedback) current decoupling with general flux orientation.

In any case, Park transformation is required (Chapter 8, Equation 8.57):

$$[P(\theta_{er}+\gamma_a)] = \frac{2}{3}\begin{bmatrix} \cos(-\theta_{er}-\gamma_a) & \cos\left(-\theta_{er}-\gamma_a+\frac{2\pi}{3}\right) & \cos\left(-\theta_{er}-\gamma_a-\frac{2\pi}{3}\right) \\ \sin(-\theta_{er}-\gamma_a) & \sin\left(-\theta_{er}-\gamma_a+\frac{2\pi}{3}\right) & \sin\left(-\theta_{er}-\gamma_a-\frac{2\pi}{3}\right) \end{bmatrix} \quad (9.30)$$

The current decoupling network in Figure 9.6 may be replaced by the direct (feedback) current decoupling (Figure 9.5) to obtain direct vector current control with a.c. current controllers. A.c. current controllers presuppose, as shown later in this chapter, high switching frequency to produce adequate phase current waveforms and lead to a kind of closed-loop control PWM in the inverter. On the other hand, open-loop PWM to produce close to sinusoidal phase voltages may work well even at lower switching frequencies to produce acceptable phase current waveforms due to the low-pass filter action of the induction machine.

The d.c. current controllers serve such a solution (Figure 9.7) and are load and frequency independent [11].

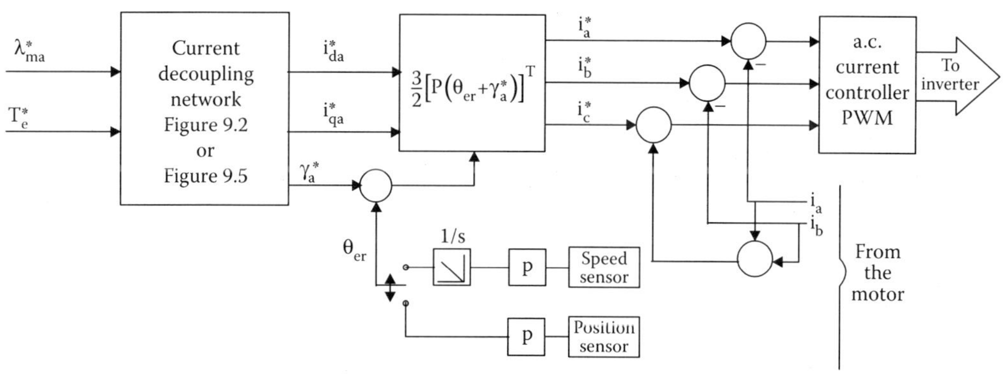


**FIGURE 9.6**
Indirect (or direct) vector current control with a.c. current controllers.

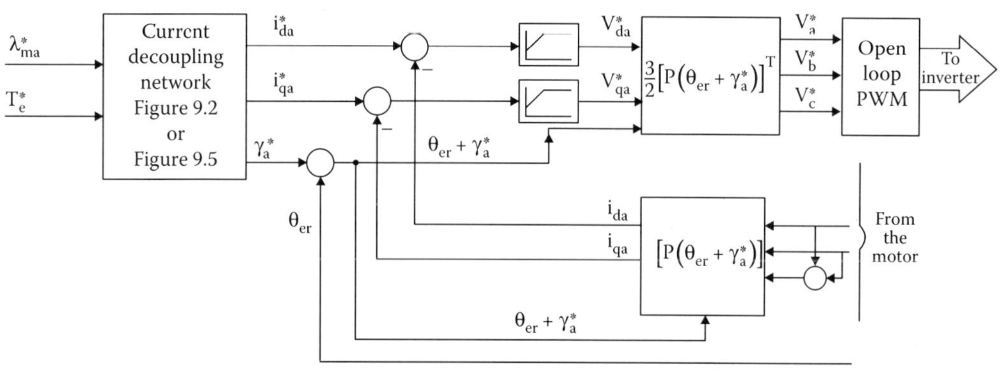


**FIGURE 9.7**
Indirect (or direct) vector current control with d.c. (synchronous) current controllers.

In presenting the vector current control, we assumed that the IM response is rather instantaneous and speed independent. However, in general, with a voltage-source PWM inverter, constant voltages are switched on and off and the current (torque) response depends on speed as the motion-induced voltage opposes current variation.

Also, above base speed, the voltage is constant and thus besides current decoupling, voltage decoupling is required to avoid the saturation of current (torque) controllers and thus avoid sluggish response of the drive.

## 9.7 Voltage Decoupling

Let us remember that so far we have not made use of the stator equation 9.5 summarized here for convenience:

$$\overline{V}_{sa} = [r_s + (s + j\omega_1)(L_s - aL_m)] \cdot \overline{i}_{sa} + (s + j\omega_1) \cdot \overline{\lambda}_{ma} \tag{9.31}$$

In Equation 9.31, $\omega_1$ is the instantaneous speed of the general flux space-phasor $\overline{\lambda}_{ma}$ and $\overline{i}_{sa}$ and $\overline{V}_{sa}$ are also written in general flux coordinates. Now, voltage decoupling means to calculate the voltage space-phasor $\overline{V}_{sa}$ required to "produce" the $\overline{i}^*_{sa}$ just calculated through the current decoupling network (Figure 9.8). It should be noted that the voltage decoupling network gets simplified only in stator flux orientation, when $a = L_s/L_m$ and $\overline{\lambda}_{ma} = \overline{\lambda}_s$. This simplified form is shown in Figure 9.9. Equation 9.31 becomes

$$\overline{V}_s = r_s \overline{i}_s + (s + j\omega_1)\overline{\lambda}_s \tag{9.32}$$

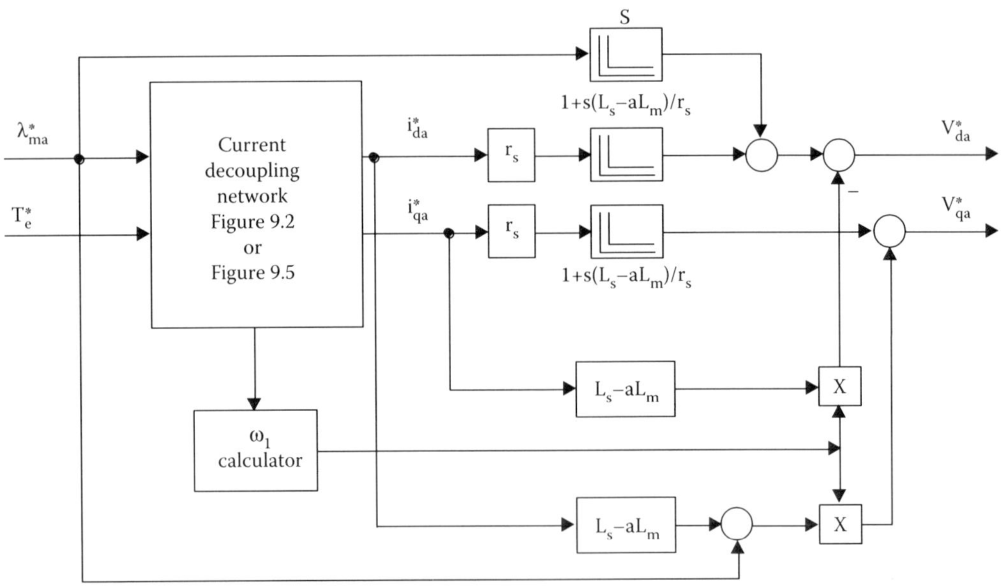


**FIGURE 9.8**
Voltage decoupling network for general flux orientation.

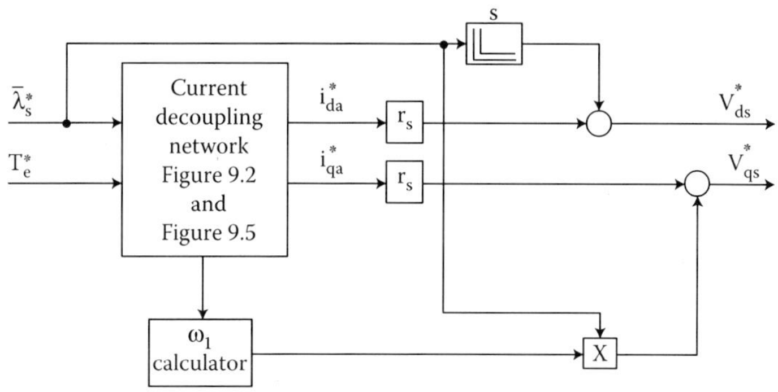


**FIGURE 9.9**
Voltage decoupling network for stator flux orientation.

It may thus be considered that, in vector current control, rotor flux orientation is less complicated, while, for voltage VC, stator flux orientation is to be preferred.

Note that even though in vector current control with d.c. current controllers (Figure 9.7) voltage (open loop) control (PWM) is performed, there is no e.m.f. (voltage) decoupling (compensation). Consequently, above base speed, at constant voltage, there is no guarantee that the drive will not lose control by requiring higher than available voltage levels.

To facilitate good control above base speed voltage decoupling is required. For implementation, in general, only the motion-induced voltage, E, is considered (s = 0 in Figure 9.8).

$$\bar{E} = j\omega_1[\bar{\lambda}_{ma} + (L_s - aL_m)\bar{i}_{sa}] = j\omega_1\bar{\lambda}_s \tag{9.33}$$

This way combined voltage–current VC is obtained. However, only d.c. current controllers (Figure 9.7) allow for a practical solution (Figure 9.10).

Vector voltage and d.c. current control performed in rotor or stator flux orientation (preferred for wide speed range control), implies, in fact, d.c. current closed-loop control at low

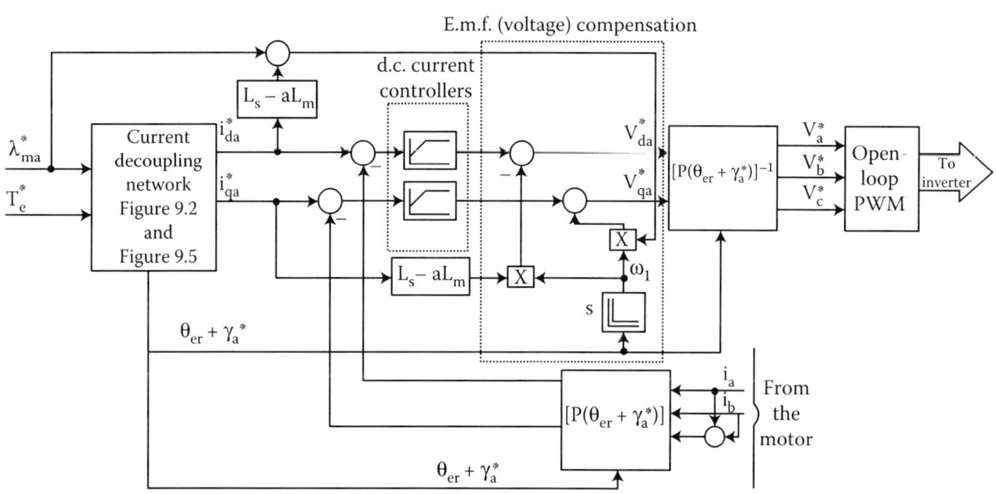


**FIGURE 9.10**
Indirect (or direct) combined vector voltage and d.c. current control in general flux orientation.

**TABLE 9.1**

Summary of Most Appropriate Strategies of Vector Control for Induction Motors

| Flux Orientation | Current Control | | Voltage and d.c. Current Control | Indirect Methods | Direct Methods | Constant Power Speed Range | |
|---|---|---|---|---|---|---|---|
| | a.c. | d.c. | | | | Small | Large |
| Rotor flux | X | XX | XX | XX | XXX | XXX | XX |
| Stator flux | | X | XXX | | XX | | XXX |
| Airgap flux | X | | | | X | X | |

*Note:* X—satisfactory; XX—good; XXX—very good.

speeds and then open-loop voltage control takes over at high speeds (above base speed), in the flux-weakening zone.

The voltage is limited to $V_{smax}$ by the inverter

$$\sqrt{\left(V_{da}^{*}\right)^{2}+\left(V_{qa}^{*}\right)^{2}} \leq V_{smax} \tag{9.34}$$

By now we have discriminated between the following main methods to perform VC:

- Indirect (feedforward)—with flux orientation calculated from reference flux and torque
- Direct (feedback)—with flux orientation calculated through observers from the motor voltage, current, and (sometimes) rotor speed

Any of these two methods may be implemented through:

- Vector current control—a.c. or d.c. (synchronous) current controllers
- Vector voltage-d.c. current control—with d.c. current controllers and e.m.f. (voltage) decoupling and open-loop (voltage) PWM (Table 9.1)

Vector voltage d.c. current control is, due to the feedforward e.m.f. compensation, less dependent on rotor time constant $\tau_r$ or on magnetic saturation, above 10% of base speed. Below that speed, if indirect current decoupling is used, the dependence of response on $\tau_r$ and saturation stays, as the e.m.f. is rather low. *At least for torque control, $\tau_r$ (rotor time constant) has to be adapted (corrected) if indirect VC methods are used.*

## 9.8 Voltage and Current Limitations for the Torque and Speed Control Range

Both motors and PECs are voltage and current limited or kVA limited. However, in terms of speed–torque envelope, it depends on how VC is performed to extract the most from the drive. This is especially so in constant power operation for speeds from $\omega_b$ to $\omega_{max}$ ($\omega_{max}/\omega_b = 2–4$) [12,13]. The voltage–current limits are easy trackable in rotor flux orientation.

For steady state (s = 0 in rotor flux orientation) from Equation 9.15

$$\lambda_r = L_m i_d; \quad i_q = S\omega_1\tau_r i_d \tag{9.35}$$

$$T_e = \frac{3}{2}p\frac{L_m^2}{L_r}i_d i_q \tag{9.36}$$

Equation 9.35 stresses the conjecture that $i_d$ is the flux current. Using Equation 9.33, we may obtain

$$\bar{\lambda}_s = \bar{\lambda}_{ma} + (L_s - aL_m)\bar{i}_{sa} \tag{9.37}$$

For rotor flux orientation $a = L_m/L_r$ and thus

$$L_s - \frac{L_m^2}{L_r} = (L_m + L_{se}) - \frac{L_m^2}{(L_m + L_{lr})} \approx L_{ls} + L_{lr} = L_{sc} \tag{9.38}$$

With Equations 9.15 and 9.38, Equation 9.37 becomes

$$\bar{\lambda}_s = \frac{L_m}{L_r}\lambda_r + L_{sc}\bar{i}_s = \frac{L_m^2}{L_r}i_d + \left(L_s - \frac{L_m^2}{L_r}\right)(i_d + ji_q)$$
$$= L_s i_d + jL_{sc}i_q = \lambda_d + j\lambda_q \tag{9.39}$$

The above mathematical manipulations under steady state are illustrated in Figure 9.11. Using Equation 9.39 in Equation 9.36, we obtain a new torque expression:

$$T_e = \frac{3}{2}p(L_s - L_{sc})i_d i_q \tag{9.40}$$

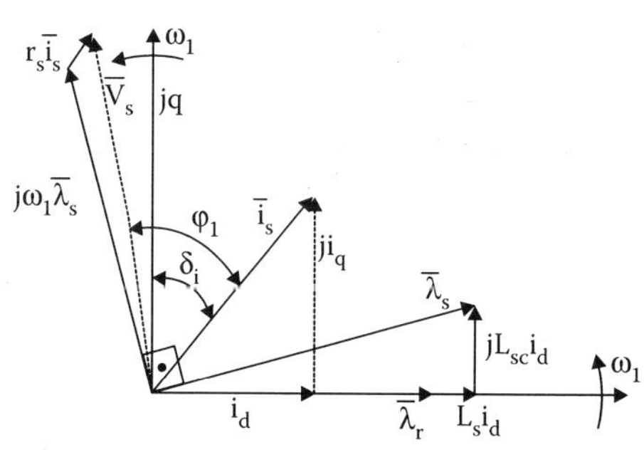

FIGURE 9.11
Stator flux and voltage in rotor flux coordinates.

*We may read Equation 9.40 as if the IM, under constant rotor flux is a reluctance synchronous motor with $L_s$ as the d axis inductance ($L_d$) and $L_{sc}$ as the q axis inductance ($L_q$): $L_s \gg L_{sc}$.*

Constant rotor flux is implicitly maintained during steady state. From Equations 9.31 and 9.39, the stator voltage becomes

$$\overline{V}_s = r_s(i_d + ji_q) + j\omega_1(L_s i_d + jL_{sc} i_q) \tag{9.41}$$

or

$$\begin{aligned} V_d &= r_s i_d - \omega_1 L_{sc} i_q \\ V_q &= r_s i_q + \omega_1 L_s i_d \end{aligned} \tag{9.42}$$

Note that $L_s = L_{ls} + L_m$ (the no-load inductance) is dependent on magnetic saturation through $L_m$ for the main flux path and through $L_{ls}$ for the leakage path.

The base speed $\omega_b$ is a matter of design options in the sense that the maximum flux level $\lambda_{smax}$ is related to motor size while the same motor has to provide continuous duty power (torque and power) at rated motor temperature and maximum available voltage $V_{smax}$:

$$\omega_b \approx \frac{V_{smax}}{\lambda_{smax}} \tag{9.43}$$

with

$$i_{dn}^2 + i_{qn}^2 = i_{sn}^2 \tag{9.44}$$

where $i_{sn}$ is the rated current.

For a short time, $1.5i_{sn}$ or even $2.0i_{sn}$ is available in most commercial drives. This, of course, means adequate motor and PEC rating. Well below the base speed, the drive is current limited in the sense that there is enough voltage reserve in the PEC, while around and above base speed, strict voltage limitations have to be observed for a reliable control.

One problem is how to produce, with $V_{smax}$ at base speed, not only $i_{sn}$ but $(1.5–2.0)i_{sn}$ and higher torque for short time intervals. To solve this problem, we have to calculate the maximum torque available at $V_{smax}$ and $\omega_1 = \omega_b$.

Neglecting $r_s$ ($r_s = 0$) in Equation 9.42, we obtain

$$\frac{V_{smax}^2}{\omega_b^2} = \lambda_{smax}^2 = ((L_s i_d)^2 + (L_{sc} i_q)^2) \tag{9.45}$$

The maximum torque $T_{ek}$ under these conditions is obtained from

$$\left(\frac{\partial T_e}{\partial i_d}\right) = 0 \tag{9.46}$$

with

$$i_q = \frac{1}{L_{sc}}\sqrt{V_{smax}^2 - \omega_b^2(L_s i_d)^2} \tag{9.47}$$

Finally, we obtain

$$i_{dk} = \frac{\lambda_{smax}}{\sqrt{2}L_s}; \quad i_{qk} = \frac{\lambda_{smax}}{\sqrt{2}L_{sc}} \tag{9.48}$$

$$T_{ek} = \frac{3}{2}p\left(1-\frac{L_{sc}}{L_s}\right)\cdot\frac{\lambda_{smax}^2}{2L_{sc}}; \quad i_{sk} = \frac{\lambda_{smax}}{\sqrt{2\left(L_s^2+L_{sc}^2\right)}} \tag{9.49}$$

*For high values of peak torques the short-circuit (transient) inductance should be small by design.* Note also that Equation 9.48 means

$$\lambda_{dk} = \lambda_{qk} = \frac{\lambda_{smax}}{\sqrt{2}}; \quad \frac{i_{dk}}{i_{qk}} = \frac{L_{sc}}{L_s} \tag{9.50}$$

$$(S\omega_1)_k = \frac{i_q}{i_d\tau_r} = \frac{L_s}{L_{sc}}\frac{r_r}{L_r} \approx \frac{r_r}{L_{sc}} \tag{9.51}$$

The power factor angle $\varphi_1$ (Figure 9.11) is

$$\phi_1 \approx \tan^{-1}\frac{i_d}{i_q} + \tan^{-1}\frac{L_{sc}i_q}{L_s i_d} \tag{9.52}$$

The peak current $i_{sk}$ determines the motor rating and sizing. The $i_d - i_q$ current components are limited by flux limitation as in Equation 9.45, which represents an ellipse whose span diminishes with increasing speed.

On the other hand, $i_d - i_q$ are limited by the current limitations $i_{sn}$ and $i_{sk}$ as in Equation 9.44, which depicts a circle (Figure 9.12).

Point $A_k$ corresponds to the peak torque $T_{ek}$ and current $i_{sk}$ at base speed $\omega_b$ and full voltage $V_{smax}$ and

$$\tan\delta_{ik} = \frac{i_{dk}}{i_{qk}} = \frac{L_{sc}}{L_s} \tag{9.53}$$

The corresponding power factor angle $\varphi_{1k}$ is

$$\varphi_{1k} = \tan^{-1}\left(\frac{i_{dk}}{i_{qk}}\right) + \tan^{-1}\left(\frac{L_{sc}}{L_s}\frac{i_{qk}}{i_{dk}}\right) = \tan^{-1}\left(\frac{L_{sc}}{L_s}\right) + 45° \tag{9.54}$$

*So, for maximum torque per given flux (given voltage and frequency), the power factor is slightly below 0.707 (cos $\varphi_{1k}$ < 0.707).*

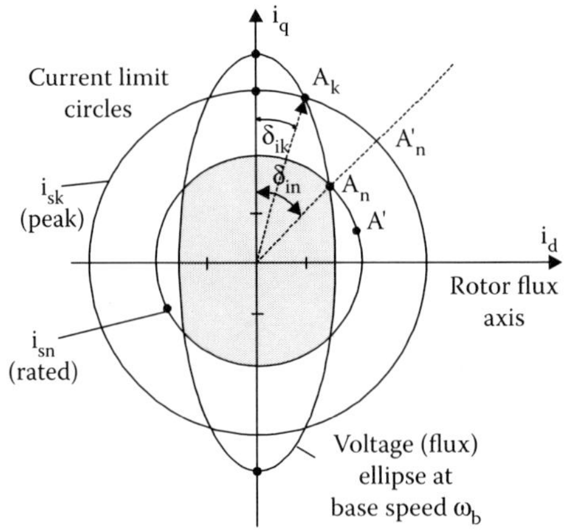


**FIGURE 9.12**
Current limit boundaries.

The current angle $\delta_{in}$ for rated (base) torque and current is to be obtained by observing Equation 9.40 at rated (base) torque $T_{eb}$ and Equation 9.44 at rated current $i_{sn}$.

In an existing motor, $L_s$ and $L_{sc}$ are given for a reasonable degree of saturation. If the values of $i_{dn}$, $i_{qn}$ correspond to point A′ in Figure 9.12 situated outside the hatched area, it means that they may not be met.

For a new motor, it seems reasonable to choose the rated conditions for the maximum power factor

$$\frac{\partial \varphi_1}{\partial (i_d/i_q)} = 0 \tag{9.55}$$

From Equations 9.54 and 9.55, it follows that

$$\tan \delta_{in} = \frac{i_{dn}}{i_{qn}} = \sqrt{\frac{L_{sc}}{L_s}} \tag{9.56}$$

$$(\cos \varphi_1)_{max} = \frac{1 - L_{sc}/L_s}{1 + L_{sc}/L_s} \tag{9.57}$$

The above rationale only introduces us to the IM design for variable speed drives. For constant power $P_e$, above base speed $\omega_b$ and for given stator current $i_s$, we should meet the conditions

$$P_{eb} = T_{eb} \cdot \frac{\omega_b}{p} = T_e \cdot \frac{\omega_1}{p} = \frac{3}{2}(L_s - L_{sc}) i_d i_q; \quad \omega_1 > \omega_b \tag{9.58}$$

and

$$i_s^2 \le i_d^2 + i_q^2; \quad S\omega_1 \le \frac{i_q}{i_d \tau_r} \tag{9.59}$$

Starting from the initial point ($A_n$ on Figure 9.12) for base speed, according to Equations 9.58 and 9.59, we have to calculate $i_d$, $i_q$, $S\omega_1$, for increasing frequency $\omega_1$ and with voltage limitation

$$(L_s i_d)^2 + (L_{sc} i_q)^2 \le (V_{s\max}/\omega_1)^2; \quad \omega_1 > \omega_b \tag{9.60}$$

When $\omega_1$ increases, so does $S\omega_1$, the limit being the critical value corresponding to maximum torque per flux condition (9.51). Beyond that point, at higher speeds, $S\omega_1$ remains constant, at its critical value: $(S\omega_1)_k \approx r_r/L_{sc}$. It is not advisable to start from $A_k$ ($\delta_{ik}$) because in that case $\delta_i$ remains equal to $\delta_{ik}$ and practically no constant power zone speed range is available. Consequently, placing ourselves at $A_n'$ (Figure 9.12), for example, for peak (short duration) torque and current, leaves us a zone for constant power. Thus the flux-weakening zone ($\omega_b > \omega_1$) may be divided in two zones (Figure 9.13). The first zone of constant electromagnetic power $P_{eb}$ ($\omega_b < \omega_1 < \omega_{max1}$) is characterized by variable current angle $\delta_i$ ($\delta_i$ decreasing from $\delta_{in}$ to $\delta_{ik}$) or $S\omega_1$ increasing from $(S\omega_1)_b$ to $(S\omega_1)_k$.

In the second zone, $\omega_{max1} < \omega_1 < \omega_{max2}$, the motor works at constant current angle $\delta_{ik}$ ($(S\omega_1)_k = r_r/L_{sc}$) and is not able (anymore) to produce constant power (Figure 9.13).

The lower the critical current angle $\delta_{ik}$, that is, the higher the critical slip frequency $(S\omega_1)_k$ or the smaller the short-circuit inductance $L_{sc}$ (for given rotor resistance), *the wider the constant power zone.*

Temperature and magnetic saturation variation introduce some departure from the above approximate analysis but this is beyond the scope of this chapter. In terms of torque transients, rotor flux and stator flux VC schemes produce similar performance in the flux-weakening zone, but the latter explains simpler the d.c. input voltage of the inverter.

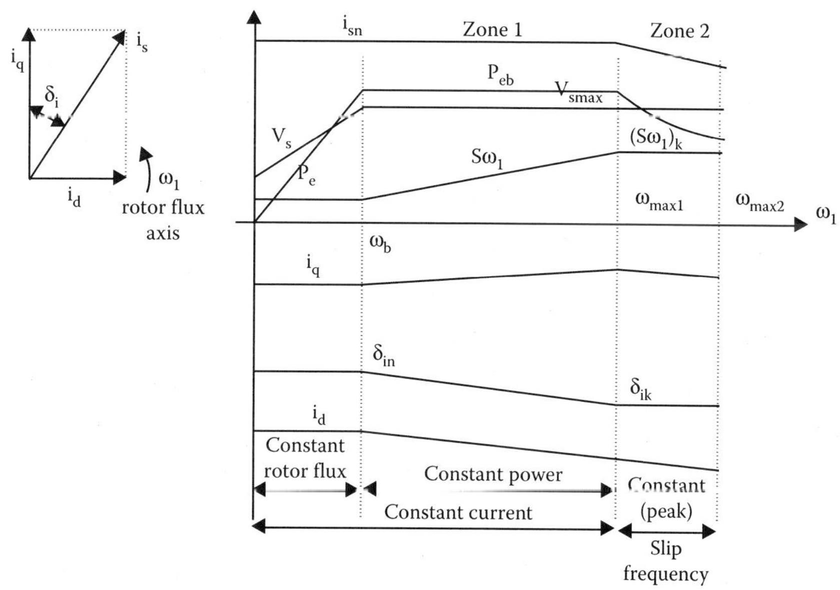


**FIGURE 9.13**
Flux-weakening zone and the constant power subzone (region).

**EXAMPLE 9.1**

Let us consider a standard IM with the following parameters: $r_s = r_r = 0.2\ \Omega$, $L_{ls} = L_{lr} = 0.005$ H, $L_m = 0.075$ H, rated line voltage 220 V (rms) (star), rated frequency $\omega_b = 2\pi 60$ rad/s, rated slip $S_n = 0.02$, $p = 2$ pole pairs. Determine the ideal no-load current $i_{on}$, for $i_{on} = i_{dn}/\sqrt{2}$, calculate $i_{qn}$ for rated slip, the rotor flux, stator flux components, and the rated electromagnetic torque $T_{eb}$, and power $P_{eb}$; calculate the maximum frequency $\omega_{max1}$ for which constant electromagnetic power $P_{eb}$ may be produced.

**Solution**

From the equivalent circuit (Chapter 8, Figure 8.11), for S = 0,

$$\overline{V}_{sn} = \bar{i}_{s0}(r_s + j\omega_b L_s) \tag{9.61}$$

Neglecting $r_s$

$$i_{s0} = i_{dn} = \frac{V_{sn}}{\omega_b L_s} = \frac{(220/\sqrt{3})\sqrt{2}}{2\pi 60 \cdot 0.08} = 5.948\ \text{A} \tag{9.62}$$

Consequently, the ideal no-load phase current $i_{on}$ (rms) is

$$i_{on} = \frac{i_{dn}}{\sqrt{2}} = \frac{5.948}{1.41} = 4.218\ \text{A} \tag{9.63}$$

The torque current $i_{qn}$—in rotor flux orientation—is (9.15)

$$i_{qn} = S_n \cdot \omega_1 \cdot \tau_r \cdot i_{dn} = 0.02 \cdot 2\pi 60 \cdot \frac{0.08}{0.2} \cdot 5.948 = 17.929\ \text{A} \tag{9.64}$$

The rotor flux, $\lambda_{rn}$ (9.15), is

$$\lambda_{rn} = L_m i_{dn} = 0.075 \cdot 5.948 = 0.4461\ \text{Wb} \tag{9.65}$$

The stator flux $\overline{\lambda}_s$ (9.39) is

$$\overline{\lambda}_{sn} = L_s i_{dn} + jL_{sc} i_{qn} = 0.080 \cdot 5.948 + j0.01 \cdot 17.929 = 0.4758 + j0.1793 \tag{9.66}$$

The rated electromagnetic torque $T_{eb}$ (9.40) is

$$\begin{aligned} T_{eb} &= \frac{3}{2} p(L_s - L_{sc}) i_{dn} i_{qn} \\ &= \frac{3}{2} \cdot 2 \cdot (0.08 - 0.01) \cdot 5.948 \cdot 17.929 = 22.3947\ \text{Nm} \end{aligned} \tag{9.67}$$

and the electromagnetic power $P_{eb}$ (9.58) is

$$P_{eb} = T_{eb} \cdot \frac{\omega_b}{p} = 22.3947 \cdot \frac{2\pi 60}{2} \approx 4220\ \text{W} \tag{9.68}$$

It is known that for $\omega_{max1}$ (Figure 9.13) the critical value of slip frequency is reached (9.51):

$$(S\omega_1)_k = \frac{r_r}{L_{sc}} = \frac{0.2}{0.01} = 20\,\text{rad/s} \tag{9.69}$$

The corresponding electromagnetic power $P_{eb}$ at $\omega_{max1}$ is (9.58)

$$P_{eb} = T_{eb} \cdot \frac{\omega_b}{p} = \frac{3}{2} p\left(1 - \frac{L_{sc}}{L_m}\right) \cdot \frac{V_{sn}^{\;2}}{\omega_{max1}^{\;2}} \cdot \frac{1}{2L_{sc}} \cdot \frac{\omega_{max1}}{p} \tag{9.70}$$

$$\omega_{max1} = V_{sn}^2 \cdot \frac{3(1-(L_{sc}/L_s))}{4L_{sc}P_{eb}} = \left(\frac{220\sqrt{2}}{\sqrt{3}}\right)^2 \frac{3(1-(0.01/0.08))}{4 \cdot 0.01 \cdot 4220} = 501.77\,\text{rad/s} \tag{9.71}$$

Note that the base frequency $\omega_b = 2\pi 60 = 376.8$ rad/s and, consequently, the constant power zone covers a speed ratio $\omega_{max1}/\omega_b = 501.77/376.8 = 1.33$. The corresponding currents, $i_{dk}$, $i_{qk}$ are obtained from

$$i_{qk}/i_{dk} = \frac{L_s}{L_{sc}} 20 \cdot \frac{0.08}{0.01} = 8 \tag{9.72}$$

$$\begin{aligned} \lambda_{smax} &= \frac{V_{sn}}{\omega_{max1}} = i_{dk}\sqrt{L_s^{\;2} + (L_{sc} i_{qk}/i_{dk})^2} \\ &= i_{dk}\sqrt{0.08^2 + (0.01 \cdot 8)^2} = 0.1128 \cdot i_{dk} \end{aligned} \tag{9.73}$$

$$i_{dk} = \frac{220\sqrt{2}}{\sqrt{3} \cdot 501.77} \cdot \frac{1}{0.1128} = 3.1679\,\text{A} \tag{9.74}$$

$$i_{qk} = 8 \cdot i_{dk} = 8 \cdot 3.1679 = 25.34\,\text{A} \tag{9.75}$$

For the base speed, the rated current $i_{sn}$ is

$$i_{sn} = \sqrt{i_{dn}^2 + i_{qn}^2} = \sqrt{5.948^2 + 17.929^2} = 18.89\,\text{A} \tag{9.76}$$

$$i_{sk} = \sqrt{i_{dk}^2 + i_{qk}^2} = \sqrt{3.1679^2 + 25.34^2} = 25.537\,\text{A} \tag{9.77}$$

So even this narrow constant power speed range (CPSR) ratio $\omega_{max1}/\omega_b = 1.33$ is obtained at the price of higher stator current which implies lower power factor and, perhaps, efficiency. Reducing $L_{sc}$ is a sure way to increase the value of $\omega_{max1}$ (9.71) and thus a wider CPSR is obtained. Leaving a voltage reserve at base speed or switching from a star to delta winding connection in the motor are two practical methods to widen the CPSR.

## 9.9 Impressing Voltage and Current Waveforms through PWM

As already mentioned, for VC, PWM is either closed loop, that is, a.c. current control driven (Figure 9.6), or open loop, that is, voltage waveform driven (Figure 9.10). PWM is thus a key part of VC with voltage-source inverters and has been given worldwide attention [15] with quite a few commercial implementation schemes.

### 9.9.1 Switching State Voltage Vectors

A PWM two-level three-phase voltage-source inverter (Figure 9.14) produces in the a.c. motor symmetrical rectangular voltage potentials $V_{ap}$, $V_{bp}$, $V_{cp}$ provided that the conducting PES triplet is on for 60 electrical degrees. This means six pulses per period or six switchings per period only.

The neutral potential $V_{np}$ is either positive or negative as two upper or lower PESs in the inverter lags are on. $V_{np}$ has three times the fundamental frequency and thus contains the triple harmonics that do not appear in the phase voltages $V_{an}$, $V_{bn}$, and $V_{cn}$. This is true also when the conduction angle of various PES triplets is less than 60° via pulse width modulation (PWM) performed to modify the fundamental of the motor phase voltages.

The maximum voltage fundamental $V_{1six\text{-}step}$ is obtained for six-pulse switching and the modulation index m is

$$m = \frac{V_1}{V_{1six\text{-}step}}; \quad 0 \le m \le 1 \tag{9.78}$$

where from Figure 9.15a

$$V_{1six\text{-}step} = \frac{2}{\pi} V_d \tag{9.79}$$

The ideal maximum modulation index is equal to unity. Various PWM schemes allow an $m_{max} < 1$ which represents an important performance criterion as the inverter maximum kVA depends on the maximum voltage at motor terminals.

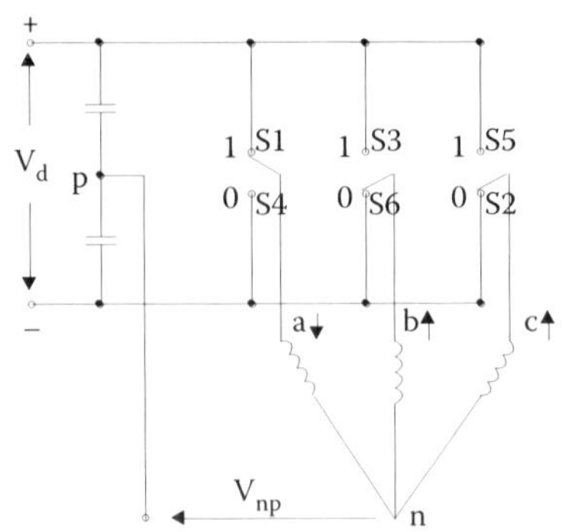


**FIGURE 9.14**
PWM voltage-source inverter: one switch per leg conducting at any time.

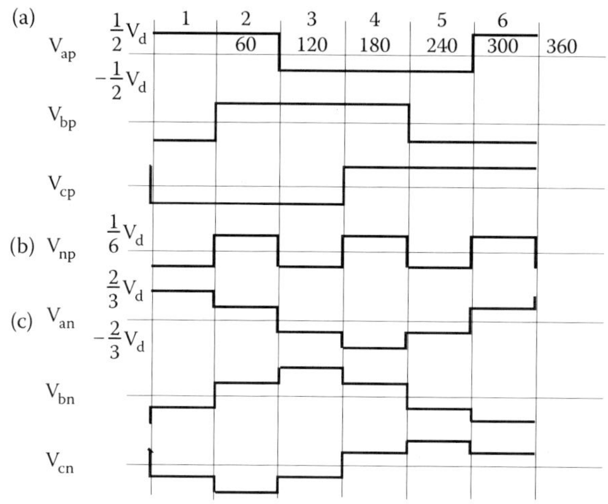


**FIGURE 9.15**
Voltage waveforms for six switchings per period: (a) voltage potentials at motor terminals, (b) neutral potential, and (c) phase voltages.

The PWM voltage-source inverter allows for six SCR triplets that produce nonzero voltage space-vectors (Figure 9.14): (100, 110, 010, 011, 001, 101) and two zero voltage space-vectors (111, 000).

We may use space-phasors to describe the six nonzero switching situations as

$$\overline{V}_s(t) = \frac{2}{3}(V_{an}(t) + V_{bn}(t)\cdot e^{j(2\pi/3)} + V_{cn}(t)\cdot e^{-j(2\pi/3)}) \tag{9.80}$$

with $V_{an}$, $V_{bn}$, $V_{cn}$ from Figure 9.15c, we obtain six space-phasors, 60° apart (Figure 9.16).

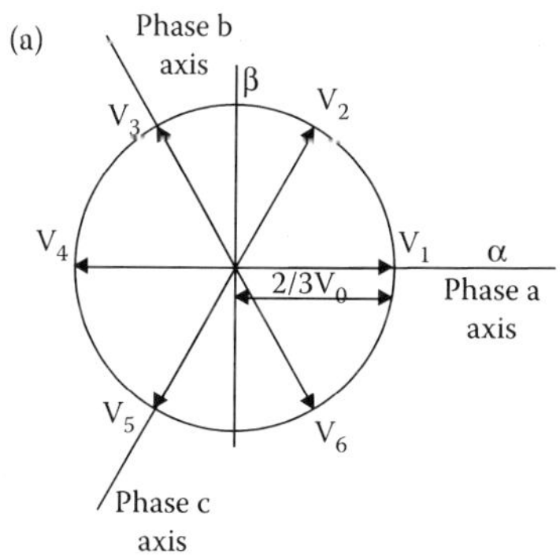


(b)

| $\overline{V}_s$ | $V_1$ | $V_2$ | $V_3$ | $V_4$ | $V_5$ | $V_6$ | $V_0$, $V_7$ |
|---|---|---|---|---|---|---|---|
| | 100 | 110 | 010 | 011 | 001 | 101 | 111, 000 |
| $V_{an}$ | $2/3V_0$ | $1/3V_0$ | $-1/3V_0$ | $-2/3V_0$ | $-1/3V_0$ | $1/3V_0$ | 0 |
| $V_{bn}$ | $-1/3V_0$ | $1/3V_0$ | $2/3V_0$ | $1/3V_0$ | $-1/3V_0$ | $-2/3V_0$ | 0 |
| $V_{cn}$ | $-1/3V_0$ | $-2/3V_0$ | $-1/3V_0$ | $1/3V_0$ | $2/3V_0$ | $1/3V_0$ | 0 |

**FIGURE 9.16**
(a) Voltage space-vectors, (b) the corresponding phase voltages.

Timing the eight voltage space-vectors $V_1, \ldots, V_8$ is, in fact, the art of PWM. *Impressing the voltage commands* required by the VC strategies may be done directly by *open-loop PWM*. *Impressing the current commands* through the same voltage space-vectors is done through *closed-loop PWM*. Among various PWM methods—treated extensively in the power electronics literature [15,16]—we deal here with the two open-loop and two closed-loop PWM strategies considered here most representative.

### 9.9.2 Open-Loop Space-Vector PWM

In space-vector PWM, the reference voltage space-vector of the motor is treated directly and not phase by phase. The reference voltage space-vector $V_1^*$ is sampled at a fixed clock frequency $2f_s$ (Figure 9.17a) being constructed through adequate timing of adjacent nonzero inverter voltage space-vectors $V_1$ to $V_6$ and the zero voltage space-vectors $V_0$, $V_7$ (Figure 9.17b)

$$2f_s(t_1V_i + t_2V_{i+1}) = V_s^*(t) \tag{9.81}$$

$$t_0 = \frac{1}{2f_s} - t_1 - t_2 \tag{9.82}$$

The respective timings $t_1$, $t_2$ are

$$t_1 = \frac{1}{2f_s}V_s^*(t)\frac{2\sqrt{3}}{\pi}\sin(60° - \alpha) \tag{9.83}$$

$$t_2 = \frac{1}{2f_s}V_s^*(t)\frac{2\sqrt{3}}{\pi}\sin\alpha \tag{9.84}$$

In fact, this technique produces an average of three voltage space-vectors $V_i$, $V_{i+1}$, and $V_0$ ($V_7$) over a subcycle $T = 1/2f_s$.

For the minimum number of commutations, with $V_1^*$ in the first sector, the switching sequence is

$$V_0(t_0/2)\ldots V_1(t_1)\ldots V_2(t_2)\ldots V_7(t_0/2)\ldots \tag{9.85}$$

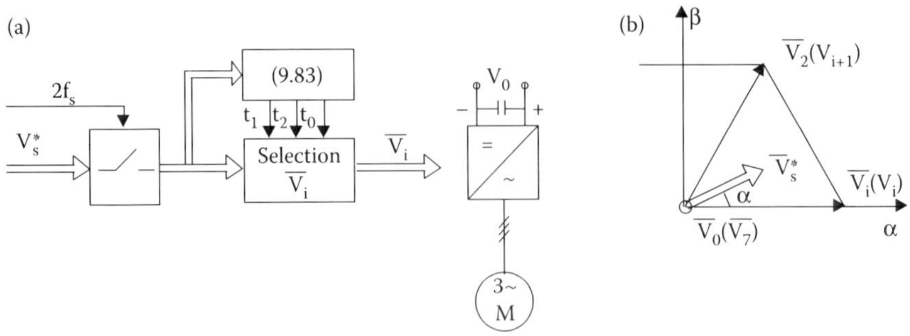


**FIGURE 9.17**
Open-loop space-vector PWM: (a) structural diagram; (b) voltage space-vector in the first sector.

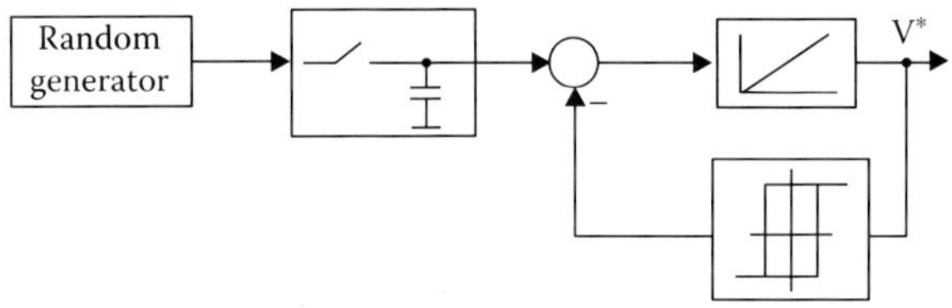


**FIGURE 9.18**
Random PWM principle.

in all odd subcycles (for all six sectors) and

$$V_7(t_0/2)\ldots V_2(t_2)\ldots V_1(t_1)\ldots V_0(t_0/2) \quad (9.86)$$

for all even subcycles of all sectors.

Other subsequences [17] might be imagined. For example,

$$V_0(t_0/3)\ldots V_1(t_1/3)\ldots V_2(t_2/3)\ldots V_2(t_2/3)\ldots V_1(t_1/3)\ldots V_0(t_0/3) \quad (9.87)$$

The loss factor is smaller in sequence (9.87) modified space-vector PWM than in Equations 9.85 and 9.86.

The space-vector PWM produces high performance but requires prediction (online computation) of the reference voltage space-vector V*(t).

Carrier-based (or fixed frequency) PWM exhibits selected harmonics especially around the carrier frequency and thus produces increased noise in the machine. Distributing the harmonics energy over a large frequency band reduces noise. Random PWM does just that [18] (Figure 9.18).

When the carrier signal reaches one of its peak values, its slope is reversed by a hysteresis block and a sample is taken from the random generator (Figure 9.18), which triggers an additional variation on the slope. This way the duration of subcycles is obtained while only the average switching frequency remains constant.

#### 9.9.2.1 Overmodulation

The zero voltage timing decreases with the increase of reference voltage. This means that the reference voltage $\overline{V}_s^*$ touches the outer hexagon of the inverter space-vectors $\overline{V}_{1,\ldots,6}$ (Figure 9.17b).

At this point, the modulation index reaches $m_{max1} \approx 0.9$, and the next step would be six-pulse operation. Special overmodulation techniques are required for a smooth transition up to $m_{max2} = 1$ [15]. To further improve performance, optimized open-loop PWM methods have been introduced, especially for low switching frequency (GTOs, GSITs) [15].

#### 9.9.2.2 Dead Time: Effect and Compensation

In order to prevent short-circuiting an inverter leg, there should be a lockout time $T_d$ between the turn-off of one PES and the turn-on of the next. $T_d$ should be larger than the maximum particle storage time of the PES, $T_{st}$. The effect of the lockout time $T_d$ is a distortion $\Delta V$ on the reference voltage $V_s^*$.

$$V_{av} = V_s^* - \Delta V; \quad \Delta V = \frac{T_d - T_{st}}{T_d} \, sig\bar{i}_s \tag{9.88}$$

The voltage distortion $\Delta V$ changes sign with current space-vector $sig\bar{i}_s$ function

$$sig\bar{i}_s = \frac{2}{3}[sign(i_a) + e^{j(2\pi/3)} sign(i_b) + e^{-j(2\pi/3)} sign(i_c)] \tag{9.89}$$

and is proportional to safety time $T_d - T_{st}$.

This voltage distortion is considered by the fact that the on-time of the upper bridge arm is shortened by $T_d - T_{st}$ for positive current and is increased by the same amount for negative sign of current.

If closed-loop (current controller) PWM is used, a compensator of dead time may not be required.

For open-loop PWM, dead time compensation is mandatory to avoid electromechanical instabilities caused by voltage distortion at low frequency (speed). As the delay is present every PWM cycle, the errors increase with speed (frequency) so dead time compensation is required at high speed.

Hardware or software compensation is used [16]. This is added to the pulse before the dead time generation for positive current and subtracted from the pulse for negative current.

### 9.9.3 Closed-Loop PWM

Closed-loop PWM involves, in general, current or flux closed-loop control and may be left nonoptimal or real-time optimized.

Hysteresis current control may be performed either with three independent (phase) current (a.c.) controllers (Figure 9.19) or through coordinated control of the current space-phasor error $\Delta\bar{i}_s$ in stator coordinates (a.c.).

When the phase current error $\Delta i_a = i_a^* - i_a > +h$, the upper inverter leg PES, (A+), is turned on, while, when $\Delta i_a < -h$, the lower leg PES (A–) is turned on. The same procedure is followed independently in phases b and c. Evidently, no zero voltage space-vector $V_0(V_7)$ may

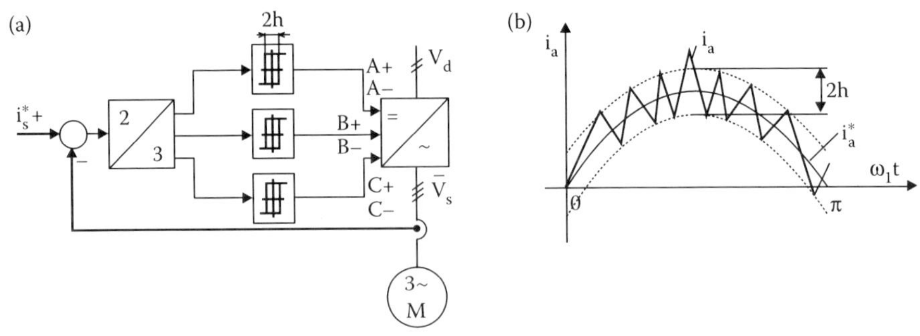


**FIGURE 9.19**
Independent hysteresis current control: (a) signal flow diagram, (b) phase current waveform.

be applied. Also, the switching frequency $f_s$ is variable, depending on the hysteresis band 2h, motor parameters, and speed. The value of h may be adjusted to maintain the average switching frequency between acceptable limits, to limit current harmonics, noise, etc. The absence of zero voltage vector requires high switching frequency at low fundamental frequency (speed)—low voltage amplitude—that is, low motor speeds; subharmonics may also occur.

In order to reduce the switching frequency and decrease current harmonics, an appropriate nonzero or zero voltage vector of the inverter may be applied, based on current phasor error $\Delta\bar{i}_s$ and its derivative $d\Delta\bar{i}_s/dt$ position corroborated with the e.m.f. vector $\bar{E}$ and the existing voltage vector $\bar{V}_i$ position in one of the six 60°-wide sectors.

A table of optimal switchings may thus be defined [19] based on the machine equation in stator coordinates:

$$L_{sc}\frac{d\Delta\bar{i}_s}{dt} \approx \bar{E} - \bar{V}_i; \quad \Delta\bar{i}_s = \bar{i}_s^* - \bar{i}_s \tag{9.90}$$

$$\bar{E} = \bar{E}_0 + L_{sc}\frac{d\bar{i}^*}{dt} + r_s\bar{i}_s^* \tag{9.91}$$

$\bar{E}_0$ is $\bar{E}$ when $\bar{i}_s = \bar{i}_s^*$.

Equation 9.90 shows that $d\Delta\bar{i}_s/dt$ is determined solely by the choice of $\bar{V}_i$. The position of $\bar{E}$ is one of the six sectors (Figure 9.20), which may be found knowing only the position of $\Delta\bar{i}_s$ and of the applied $\bar{V}_i$ (Figure 9.20a and b).

The new voltage vector to be applied to the inverter should provide the smallest $\Delta\bar{i}_s$ for steady state (to reduce current harmonics) or the largest $\Delta\bar{i}_s$ for transients (to produce fastest current [torque] response) (Figure 9.20c). Results [19] obtained on a 1.5-kW motor compare favorably with the case of independent hysteresis controllers [19]. Fast response is preserved while harmonics content and noise level are reduced.

However, a.c. controllers are shown to be load, motor parameter, and frequency (speed) dependent, while d.c. (d-q) current controllers are rather independent of frequency and cross-coupling effects [11]. The d.c. (synchronous or d-q) current controllers are better and may be implemented in stator (a.c.) coordinates by transforming their equation in PI form in flux coordinates [11]

$$\bar{V}_s^e = K\left(\tau_i + \frac{1}{s}\right)\left(\bar{i}_s^{e*} - \bar{i}_s^e\right) \tag{9.92}$$

through Park transformation $e^{-j\theta_1}$, to the final form

$$\bar{V}_s = \bar{x}_s + K\tau_i\left(\bar{i}_s^* - \bar{i}_s\right); \quad \frac{d\theta_1}{dt} = \omega_1 \tag{9.93}$$

$$s\bar{x}_s = K\left(\bar{i}_s^* - \bar{i}_s\right) + j\omega_1\bar{x}_s \tag{9.94}$$

These equations may be conveniently implemented, though, for cross-coupling compensation, the value of the flux speed $\omega_1$ has to be calculated [11,17].

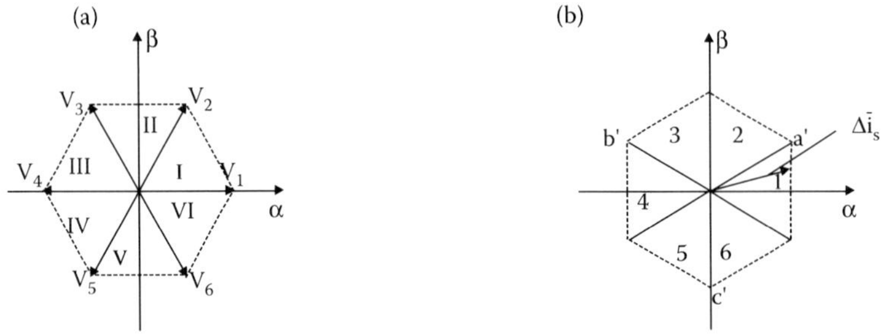


$$\begin{vmatrix} \Delta i'_a \\ \Delta i'_b \\ \Delta i'_c \end{vmatrix} = \frac{1}{\sqrt{3}} \begin{vmatrix} 1 & 0 & -1 \\ -1 & 1 & 0 \\ 0 & -1 & 1 \end{vmatrix} \bullet \begin{vmatrix} \Delta i_a \\ \Delta i_b \\ \Delta i_c \end{vmatrix}$$

(c)

| V(k) | $\Delta i'_a$ | $\Delta i'_b$ | $\Delta i'_c$ | E |
|---|---|---|---|---|
| 0,7 | 1 | 1 | 0 | I |
| | 0 | 1 | 0 | II |
| | 0 | 1 | 1 | III |
| | 0 | 0 | 1 | IV |
| | 1 | 0 | 1 | V |
| | 1 | 0 | 0 | VI |
| 1 | | 1 | | I |
| | | 0 | | VI |
| 2 | 0 | | | II |
| | 1 | | | I |
| 3 | | | 0 | II |
| | | | 1 | III |
| 4 | | 1 | | III |
| | | 0 | | IV |
| 5 | 0 | | | IV |
| | 1 | | | V |
| 6 | | | 1 | V |
| | | | 0 | VI |

| Region of E | Region of $\Delta i$ | | | | | |
|---|---|---|---|---|---|---|
| | 1 | 2 | 3 | 4 | 5 | 6 |
| I | $V_1$ | $V_2$ | $V_2$ | $V_0, V_7$ | $V_0, V_7$ | $V_1$ |
| II | $V_2$ | $V_2$ | $V_3$ | $V_3$ | $V_0, V_7$ | $V_0, V_7$ |
| III | $V_0, V_7$ | $V_3$ | $V_3$ | $V_4$ | $V_4$ | $V_0, V_7$ |
| IV | $V_0, V_7$ | $V_0, V_7$ | $V_4$ | $V_4$ | $V_5$ | $V_5$ |
| V | $V_6$ | $V_0, V_7$ | $V_0, V_7$ | $V_5$ | $V_5$ | $V_6$ |
| VI | $V_1$ | $V_1$ | $V_0, V_7$ | $V_0, V_7$ | $V_6$ | $V_6$ |

| $\Delta i$ | V(k) |
|---|---|
| 1 | $V_1$ |
| 2 | $V_2$ |
| 3 | $V_3$ |
| 4 | $V_4$ |
| 5 | $V_5$ |
| 6 | $V_6$ |

**FIGURE 9.20**
Current vector hysteresis control (a, b) $\bar{E}$ zone detection; (c) switching tables. (After Nabae, A., Ogasawara, S., Akagi, H., *IEEE Trans.*, IA-22, 697–701, 1986.)

NOTE: As shown later in this chapter, DFTC may be assimilated with d-q (synchronous) current and voltage closed-loop random PWM thus yielding high-grade performance both at low speeds and at high speeds (flux-weakening zone, $\omega_1 > \omega_b$).

Now that most knowledge for *indirect* VC has been presented, a complete numerical example is given below.

## 9.10 Indirect Vector A.C. Current Control: A Case Study in MATLAB–Simulink (Available Online)

We will present here the simulation of a feedforward (indirect) vector current control system for IMs. The example was implemented in MATLAB–Simulink simulation program. The motor model was integrated in two blocks, the first represents the *current and flux calculation module* in d-q axis, and the second represents the *torque, speed, and position computing module*.

Changing of motor parameters for different simulations is as simple as possible; by clicking on their blocks, a dialog box appears and you can change them by modifying their default values.

The current decoupling network was integrated in a single block. At its first input, you have to set the value of the flux psim*. You can modify the speed regulator (*PI regulator*) parameters (amplification and integration time constant, default values are $K_i = 10$, $T_i = 0.8$ s) to study the behavior of the system. We have included a *Commutation Frequency* block to satisfy the sampling criteria (to control the commutation frequency). The integration step can be modified from Simulink's *Simulation/Parameters*.

To find out the structure of each block presented above, unmask them (*Options/Unmask*). The *Park axis transformation* has been integrated in the Park transform block.

The block structure of the electric drive system is presented in Figures 9.21 through 9.23.

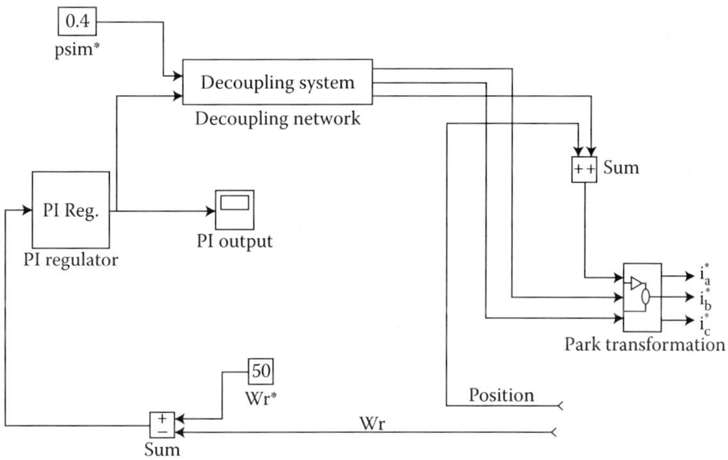


**FIGURE 9.21**
The indirect vector a.c. current control system for IMs.

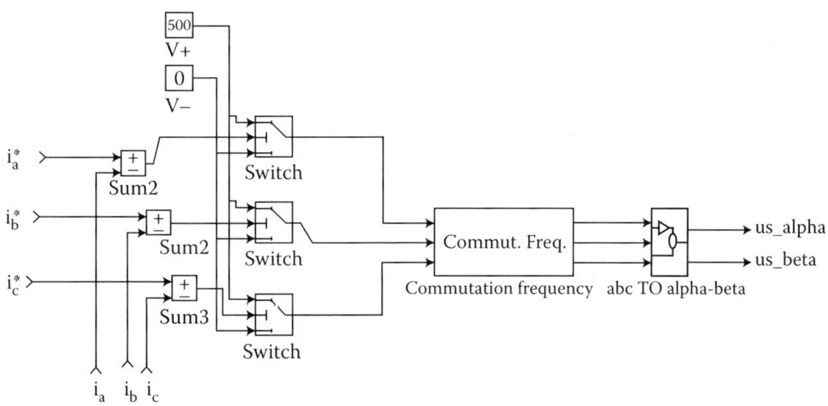


**FIGURE 9.22**
The a.c. current controllers.

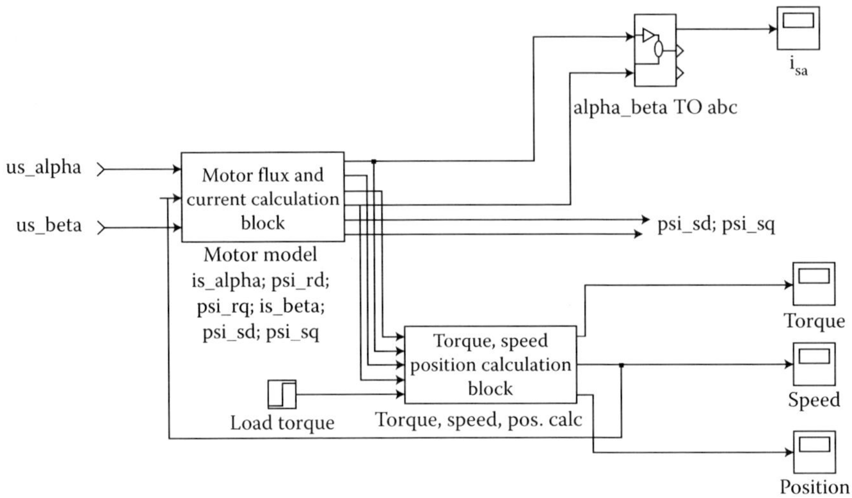


**FIGURE 9.23**
The motor space-phasor model.

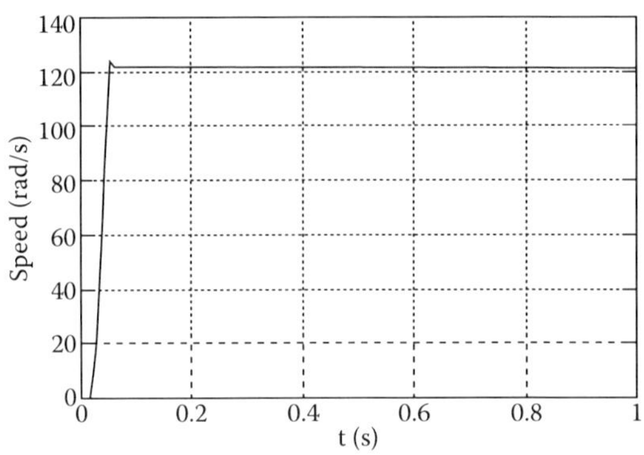


**FIGURE 9.24**
Speed transient response.

The motor used for this simulation has the following parameters: $P_n = 1100$ W, $V_{nf} = 220$ V, $2p = 4$, $r_s = 9.53\ \Omega$, $r_r = 5.619\ \Omega$, $L_{sc} = 0.136$ H, $L_r = 0.505$ H, $L_m = 0.447$ H, $J = 0.0026$ kg fm$^2$.

The following figures represent the speed, torque, current, and flux responses for *starting transients and a load torque applied at 0.4 s.*

Speed transients (Figure 9.24) are quick and smooth even immediately after load torque application at $t = 0.4$ s. Torque response (Figure 9.25) is fast and the torque pulsations are moderate for a chopping frequency $f_0 = 8$ kHz. Phase current waveform (Figure 9.26) also looks smooth with moderate harmonics content. The stator flux (Figure 9.27) and rotor flux (Figure 9.28) responses are somewhat similar, but the rotor flux transients are slower, as expected.

After starting from zero current and speed, the rotor flux settles to the reference value with very small transients when load torque occurs at $t = 0.4$ s.

We should mention that the parameters were fully tuned. The influence of parameter detuning could be investigated also. This is beyond the scope of this chapter.

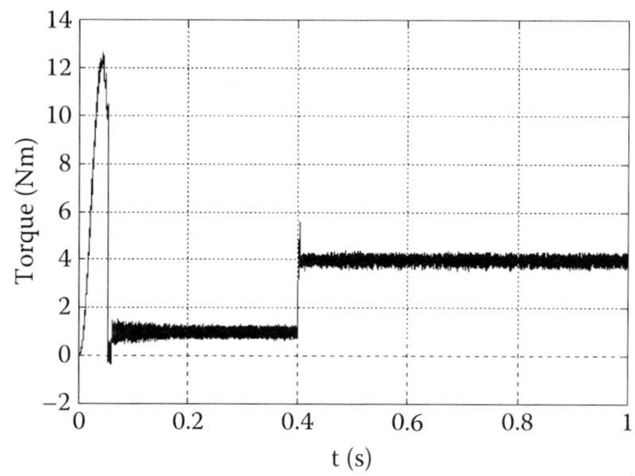


**FIGURE 9.25**
Torque response.

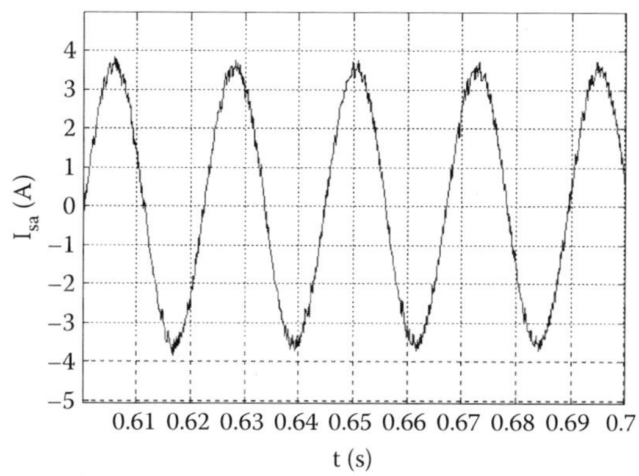


**FIGURE 9.26**
Phase current waveform under steady state.

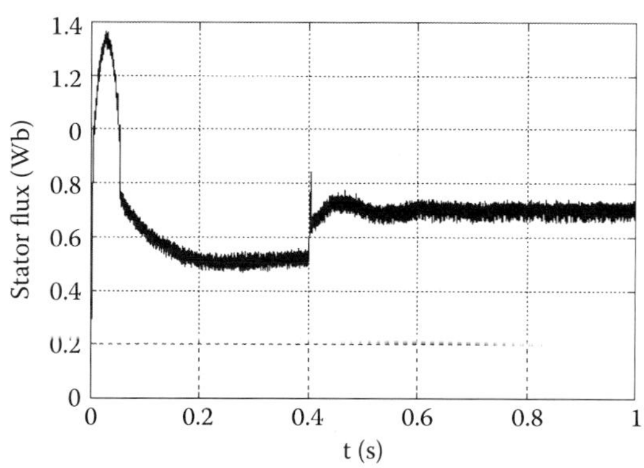


**FIGURE 9.27**
Stator flux amplitude.

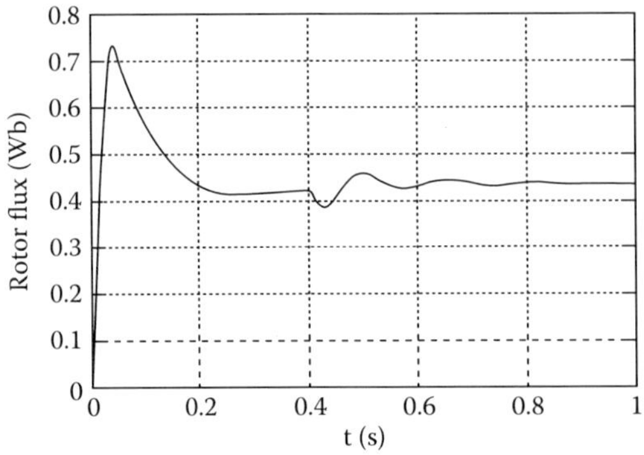


**FIGURE 9.28**
Rotor flux amplitude.

## 9.11 Indirect Vector Synchronous Current Control with Speed Sensor: A Case Study in MATLAB–Simulink (Available Online)

The simulation results of a VC system with IM based on d.c. current control (Figure 9.7) are now given. The simulation of this drive is implemented in MATLAB–Simulink. The motor model was integrated in two blocks, the first represents the *current and flux calculation module* in d-q axis (Figure 9.29), the second represents the *torque, speed, and position computing module* (Figure 9.30).

The commutation table was implemented in MATLAB as a function. Changing of motor parameters for different simulations is simple. By clicking on these blocks, a dialog box appears and you can change them by modifying their default values.

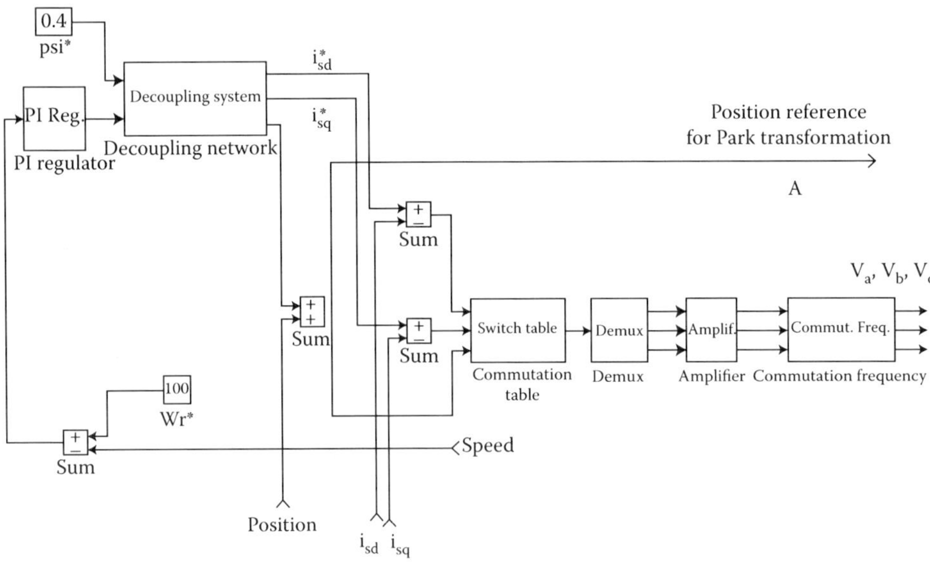


**FIGURE 9.29**
The indirect vector current control system.

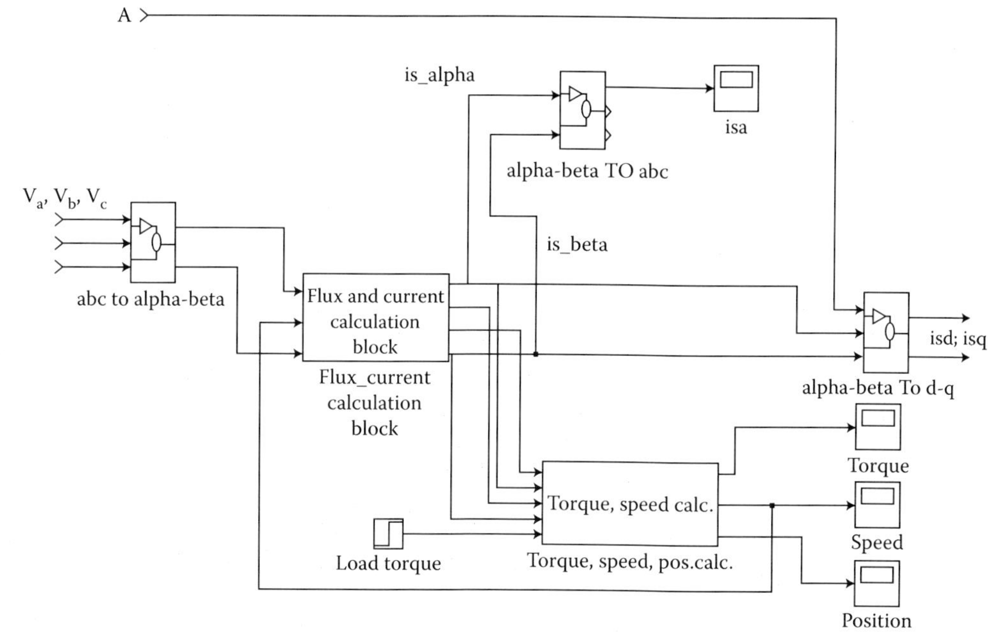


**FIGURE 9.30**
The motor space-phasor (d-q) model.

The current decoupling network was integrated in a single block. At its first input, you have to set the value of the flux $\lambda_r^*$. You can modify the speed regulator (*PI regulator*) parameters (amplification and integration time constant, default values are $K_i = 10$, $T_i = 0.8$ s) to study the behavior of the system. We included a *Commutation Frequency* block to satisfy the sampling criteria (to control the commutation frequency). The integration step (default: 50 μs) can be modified from Simulink's *Simulation/Parameters*.

To find out the structure of each block presented above, unmask it (*Options/Unmask*).

The errors in d-q currents $i_d^* - i_d$, $i_q^* - i_q$ may be translated directly into voltage vectors in the inverter in a kind of space-vector voltage PWM implemented through a commutation table. PI *d-q* current controllers have been used.

The motor used for this simulation has the following parameters: $P_n = 1100$ W, $V_{nf} = 220$ V, $2p = 4$, $r_s = 9.53\ \Omega$, $r_r = 5.619\ \Omega$, $L_{sc} = 0.136$ H, $L_r = 0.505$ H, $L_m = 0.447$ H, $J = 0.0026$ kg fm$^2$.

Figures 9.31 through 9.35 represent the speed, torque, current, and flux responses, for the *starting process and with load torque applied at 0.4 s*. The value of load torque is 4 Nm.

The speed response (Figure 9.31) is fast and smooth and so is the torque response with moderate torque pulsations (Figure 9.32). The current waveform has a small harmonics content (Figure 9.33) and the application of the load torque does not produce strong current transients, as expected. The stator and rotor flux linkage responses (Figures 9.34 and 9.35) are very similar to those of Section 9.10.

The listing of MATLAB file for switching table follows:

```
function [sys, x0] =tab3(t,x,u,flag,zindex,table)
%*****************************************
%Here starts the switching table lookup routine...
```

```
%******************************************
if flag == 3, Perform the desired table lookup
V0 = [0 0 0];
V1 = [1 0 0];
V2 = [1 1 0];
V3 = [0 1 0];
V4 = [0 1 1];
V5 = [0 0 1];
V6 = [1 0 1];
V7 = [1 1 1];

    sq = 0.866025404;
    x0 = u(1);
    y0 = u(2);
    z0 = u(3);
    if (x0 > 0), x0 = 1;end;
    if (x0 < 0), x0 = -1;end;
    if (x0 ==0), x0 = 0;end;
    if (y0 >=0), y0 = 1;end;
    if (y0 < 0), y0 = -1;end;
    if (x0 ==1)&(y0 ==1), xf = 1;end;
    if (x0 ==1)&(y0 ==-1), xf = 2;end;
    if (x0 ==-1)&(y0 ==1), xf = 3;end;
    if (x0 ==-1)&(y0 ==-1), xf = 4;end;
    if (x0 ==0),            xf = 5;end;
    if (cos(z0) > sq), zf = 1;end;
    if (cos(z0) <=sq)&(cos(z0) > 0)&(sin(z0) >=0), zf = 2;end;
    if (cos(z0) > -sq)&(cos(z0) <=0)&(sin(z0) >=0), zf = 3;end;
    if (cos(z0) < -sq), zf = 4;end;
    if (cos(z0) >=-sq)&(cos(z0) < 0)&(sin(z0) < 0), zf = 5;end;
    if (cos(z0) <=sq)&(cos(z0) >=0)&(sin(z0) < 0), zf = 6;end;
    r = table(xf, zf);
    if r ==0, sys = V0;end;
    if r ==1, sys = V1;end;
    if r ==2, sys = V2;end;
    if r ==3, sys = V3;end;
    if r ==4, sys = V4;end;
    if r ==5, sys = V5;end;
    if r ==6, sys = V6;end;
    if r ==7, sys = V7;end;
%***************************************
%Here ends the switching table lookup routine…
%***************************************
elseif flag == 0,
    % This part takes care of all initialization; it is used only once.
    x0 = [];
    % The system has no states, three outputs, and three inputs.
    sys = [0 0 3 3 0 1]';
else
    % Flags not considered here are treated as unimportant.
    % Output is set to [].
    sys = [];
end.
```

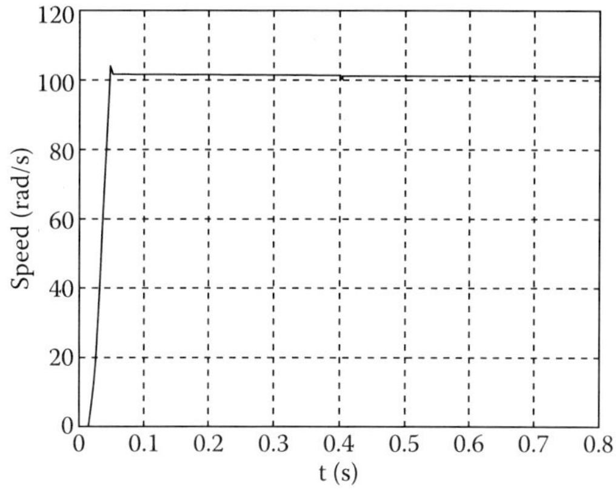


**FIGURE 9.31**
Speed transient response.

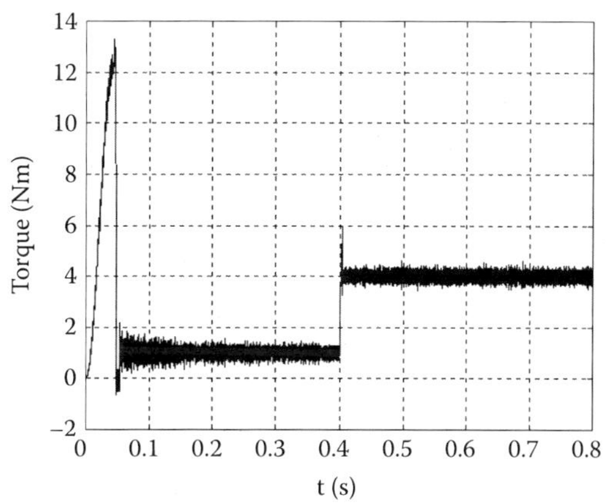


**FIGURE 9.32**
Torque response.

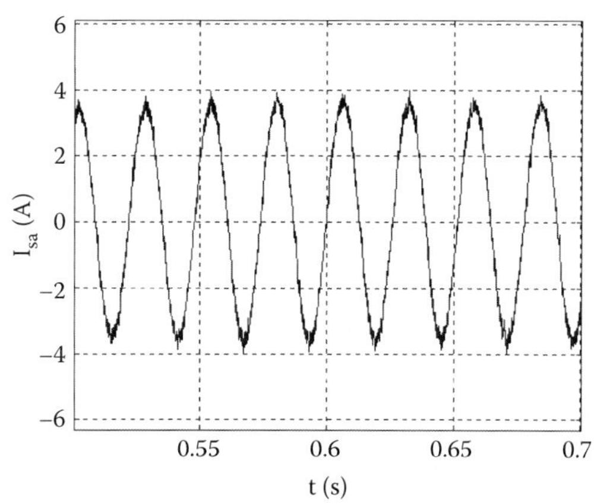


**FIGURE 9.33**
Phase current waveform.

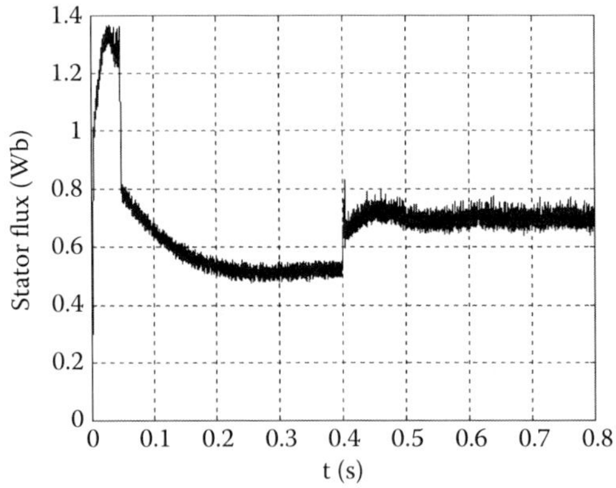


**FIGURE 9.34**
Stator flux amplitude.

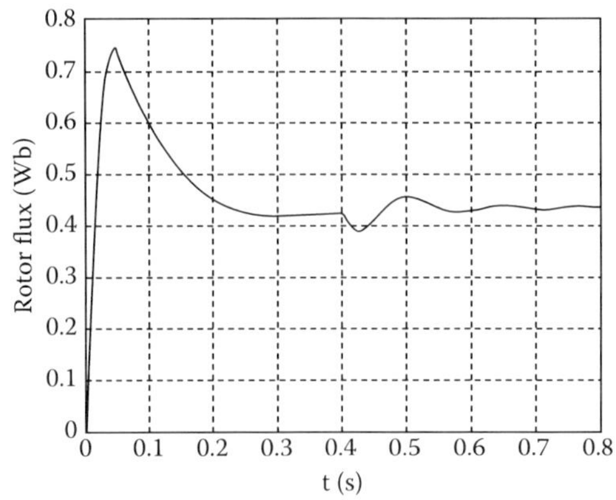


**FIGURE 9.35**
Rotor flux amplitude.

## 9.12 Flux Observers for Direct Vector Control with Motion Sensors

The motor stator or airgap flux space-phasor amplitude $\lambda_{ma}$ and its instantaneous position—$\theta_{er} + \gamma_a^*$—with respect to the stator phase a axis have to be computed online, based on measured motor voltages, currents, and, when available, rotor speed. The torque may be calculated from flux and current space-phasors and thus, once the flux is computed and stator currents measured, the torque problem is solved.

So flux observation is paramount for direct VC. We will present a few implementations of flux observers from simple to complex solutions.

Flux observers may be open loop or closed loop.

### 9.12.1 Open-Loop Flux Observers

Open-loop flux observers are based on the voltage model or on the current model. Voltage model makes use of stator voltage equation in stator coordinates (from Equation 9.32 with $\omega_1 = 0$)

$$\overline{V}_s^s = r_s \overline{i}_s^s + s\overline{\lambda}_s^s \tag{9.95}$$

From Equation 9.37 with $a = L_m/L_r$

$$\overline{\lambda}_r^s = \frac{L_r}{L_m}\left(\overline{\lambda}_s - L_{sc}\overline{i}_s\right) \tag{9.96}$$

Both stator flux, $\overline{\lambda}_s$, and rotor flux, $\overline{\lambda}_r$, space-phasors, may thus, in principle, be calculated based on $\overline{V}_s^s$ and $\overline{i}_s$ measured. The corresponding signal flow diagram is shown in Figure 9.36.

Unfortunately, the integration term and the variation of stator resistance with temperature make this observer hardly practical under 2 Hz without special corrections.

On the other hand, the current model for the rotor flux space-phasor is based on the rotor equation in *rotor coordinates* ($\omega_b = \omega_r$)

$$-r_r \overline{i}_s^r + (r_r + sL_r)\frac{\overline{\lambda}_r^r}{L_m} = 0 \tag{9.97}$$

Two coordinate transformations—one for current and the other for rotor flux—are required to produce results in stator coordinates. This time, the observer works even at *zero frequency* but is very sensitive to the detuning of parameters $L_m$ and $\tau_r$ due to temperature and magnetic saturation variation. Besides, it requires a rotor speed or position sensor. Parameter adaptation is a solution.

The corresponding signal flow diagram is shown in Figure 9.37.

More profitable, however, it seems to use closed-loop flux observers.

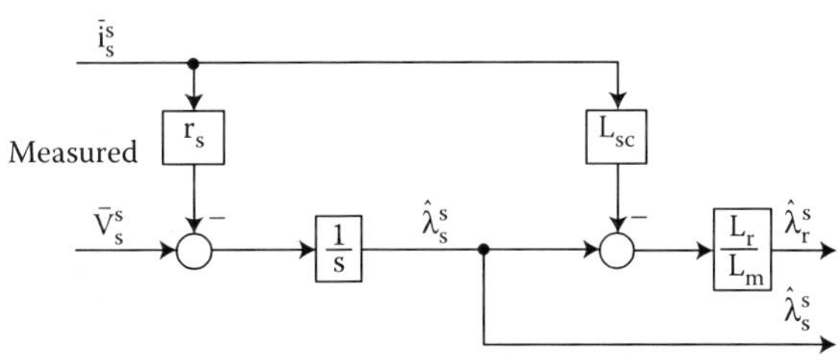


**FIGURE 9.36**
Voltage-model open-loop flux observer (stator coordinates).

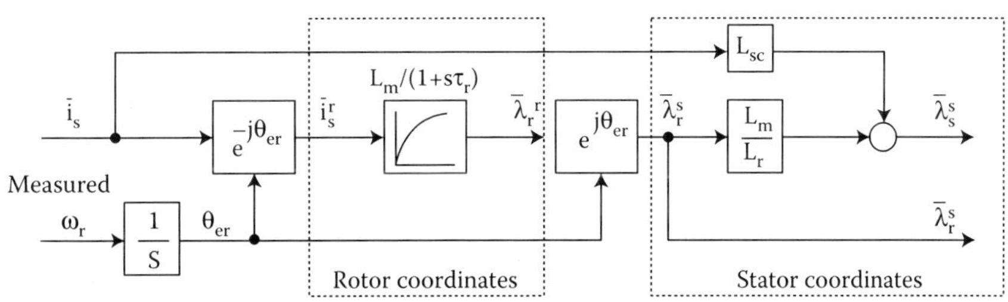


**FIGURE 9.37**
Current-model open-loop flux observer.

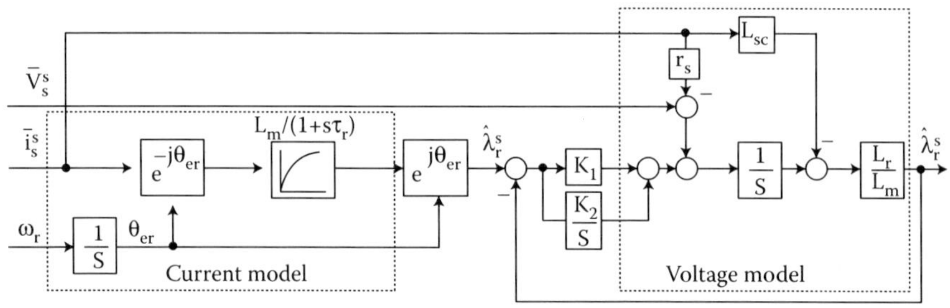


**FIGURE 9.38**
Closed-loop voltage and current model rotor flux observer.

### 9.12.2 Closed-Loop Flux Observers

We start by combining the voltage and current models to put together their advantages. At high speeds, the voltage model is better as the stator resistance influence is small while, at low speeds, the current model is better as it works even at zero frequency, during d.c. braking (Figure 9.38). The transition between the two models is decided by the two coefficients K1 and K2 such that at low frequency only the current model remains active while at high frequency the voltage model prevails.

Stator flux may simply be calculated from rotor flux using Equation 9.96. Values $K_1 = 33$ rad/s and $K_2 = 90$ (rad/s)$^2$ are practical. More details on the above flux observers are found in Reference 20. Many other flux observers have been proposed. Among them are the third flux (voltage) harmonic estimator [21], Gopinath observer [22], model reference adaptive, and Kalman filter observers.

They all require notable online computation effort and knowledge of IM parameters. Consequently, they seem more appropriate when used together with speed observers for sensorless IM drives as shown in Section 9.13.

## 9.13 Flux and Speed Observers in Motion Sensorless Drives

Sensorless drives are becoming predominant when only up to 100 to 1 speed control range is required even in fast torque response applications (1–5 ms for step-rated torque response). There is a rich literature on the subject, with quite a few proposed solutions [23,24] and some are already on markets worldwide. Essentially they are

- Without signal injection
- With signal injection

### 9.13.1 Performance Criteria

To assess the performance of various flux and speed observers for sensorless drives, the following performance criteria have become widely accepted:

- Steady-state error
- Torque response quickness

- Low-speed behavior (speed range)
- Sensitivity to noise and motor parameter detuning
- Dynamic robustness
- Complexity versus performance

### 9.13.2 A Classification of Speed Observers

The basic principles used for speed estimation (observation) may be classified as

A. Speed estimators
B. Model reference adaptive systems
C. Luenberger speed observers
D. Kalman filters
E. Rotor slot ripple

With the exception of rotor slot ripple, all the other methods imply the presence of flux observers to calculate the motor speed.

### 9.13.3 Speed Estimators

Speed estimators are in general based on the classical definition of rotor speed $\hat{\omega}_r$

$$\hat{\omega}_r = \hat{\omega}_1 - (S\omega_1) \tag{9.98}$$

where $\omega_1$ is the rotor flux vector instantaneous speed and $(S\omega_1)$ is the rotor flux slip speed. $\omega_1$ may be calculated in stator coordinates based on the formula

$$\hat{\omega}_1 = \frac{d}{dt}\left[\text{Arg}\left(\bar{\lambda}_r^s\right)\right]; \quad \bar{\lambda}_r^s = \lambda_{dr} + j\lambda_{qr} \tag{9.99}$$

or

$$\hat{\omega}_1 = \frac{\dot{\lambda}_{qr}^s \lambda_{dr}^s - \dot{\lambda}_{dr}^s \lambda_{qr}^s}{\left(\lambda_r^s\right)^2} \tag{9.100}$$

$\dot{\lambda}_{qr}^s, \lambda_{dr}^s, \dot{\lambda}_{dr}^s, \lambda_{qr}^s$ are to be determined from a flux observer (see Figure 9.38, for example). On the other hand, the slip frequency $(S\omega_1)$ (9.23) is

$$(S\omega_1) = \frac{(3/2)p(\lambda_{dr}i_{qs} - \lambda_{qr}i_{ds})L_m}{(3/2)p(\lambda_r^2/r_r)L_r} = \frac{(\lambda_{dr}i_{qs} - \lambda_{qr}i_{ds})L_m}{\lambda_r^2\tau_r} \tag{9.101}$$

Note that $(S\omega_1)$ is strongly dependent only on rotor resistance $r_r$ as $L_m/L_r$ is rather independent of magnetic saturation. Still, rotor resistance is to be corrected if good precision at low speed is required. This slip frequency value is valid both for steady state and transients and thus $\hat{\omega}_r$ is estimated quickly to allow fast torque response.

Such speed estimators may work even at 20 rpm although dynamic capacity of torque disturbance rejection at low speeds is limited.

This seems to be a problem with most speed observers. Asymmetrical Luenberger observers, with mechanical model, robust flux observer angle, and rotor resistance correction, have been proven dynamically robust down to 3 rpm with no signal injection.

### 9.13.4 Model Reference Adaptive Systems

Model reference adaptive systems (MRASs) are based on comparison of two estimators. One of them does not include speed and is called the *reference model*. The other, which contains speed, is the *adjustable model*. The error between the two is used to derive an adaption model that produces the estimated speed $\hat{\omega}_r$ for the adjustable model.

To eliminate the stator resistance influence, the airgap reactive power $q_m$ [25] may be the output of both models:

$$q_m = \bar{i}_s \otimes \left( \bar{V}_s - L_{sc} \frac{d\bar{i}_s^*}{dt} \right) \tag{9.102}$$

$$\hat{q}_m = L_m \left[ \hat{i}_m \bar{i}_s \hat{\omega}_r + \frac{1}{\tau_r} \left( \hat{i}_m \times \bar{i}_s \right) \right] \tag{9.103}$$

The rotor flux magnetization current $\hat{i}_m$ equation in stator coordinates is (Equation 9.4 with $\omega_1 = 0$)

$$\frac{d\hat{i}_m}{dt} = \hat{\omega}_r \otimes \hat{i}_m + \left( \bar{i}_s - \hat{i}_m \right) \frac{1}{\tau_r} \tag{9.104}$$

Now the speed adaptation mechanism is

$$\hat{\omega}_r = \left( K_p + \frac{K_i}{s} \right) (q_m - \hat{q}_m) \tag{9.105}$$

The signal flow diagram of the MRAS obtained is shown in Figure 9.39.

The effect of the rotor time constant $\tau_r$ variation persists and influences the speed estimation. However, if the speed estimator is used in conjunction with indirect vector current

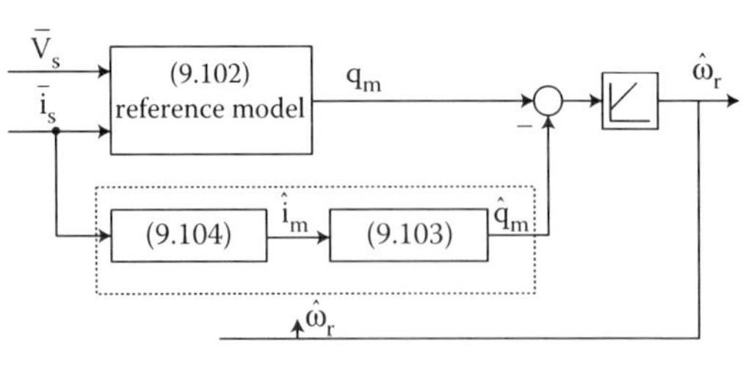


**FIGURE 9.39**
MRAS speed estimator based on airgap reactive power error.

control, at least rotor field orientation is maintained as the same (wrong!) $\tau_r$ also enters the slip frequency calculator.

The MRAS speed estimator does not contain integrals and thus works even at zero stator frequency (d.c. braking) (Figure 9.40a) and does not depend on stator resistance $r_s$. It works even at 20 rpm (Figure 9.40b) [25].

### 9.13.5 Luenberger Flux and Speed Observers

First the stator current and the rotor flux are calculated through a full-order Luenberger observer based on stator and rotor equations in stator coordinates

$$\frac{d}{dt}\begin{vmatrix}\bar{i}_s^s\\ \bar{\lambda}_r^s\end{vmatrix} = \begin{vmatrix}A_{11} & A_{12}\\ A_{21} & A_{22}\end{vmatrix}\cdot\begin{vmatrix}\bar{i}_s^s\\ \bar{\lambda}_r^s\end{vmatrix} + \begin{vmatrix}B_1\\ 0\end{vmatrix}\cdot|\bar{V}_s| \tag{9.106}$$

with

$$\bar{i}_s^s = [i_d \quad i_q]^T; \quad \bar{V}_s^s = [V_d \quad V_q]^T; \quad \bar{\lambda}_r^s = [\lambda_{dr} \quad \lambda_{qr}]^T \tag{9.107}$$

$A_{11}$, $A_{12}$, $A_{21}$, $A_{22}$ are defined in Equation 9.5 with $\omega_1 = 0$ and $a = L_m/L_r$

$$\begin{gathered}
A_{11} = -\frac{r_r}{L_{sc}}\left(1+\frac{L_m^2}{L_r^2}\right)\cdot I = a_{r11}\cdot I; \quad B_1 = \frac{1}{L_{sc}}\cdot I\\
A_{12} = \frac{L_m}{L_rL_{sc}}\left(\frac{1}{\tau_r}I - \hat{\omega}_r I'\right) = a_{r12}\cdot I + a_{i12}\cdot I'\\
A_{21} = \frac{L_m}{\tau_r}I = a_{r21}I; \quad A_{22} = \frac{1}{\tau_r}I + \hat{\omega}_r I'\\
I = \begin{vmatrix}1 & 0\\ 0 & 1\end{vmatrix}; \quad I' = \begin{vmatrix}0 & -1\\ 1 & 0\end{vmatrix}
\end{gathered} \tag{9.108}$$

with

$$\hat{x} = [i_{ds} \quad i_{qs} \quad \lambda_{dr} \quad \lambda_{qr}]^T \tag{9.109}$$

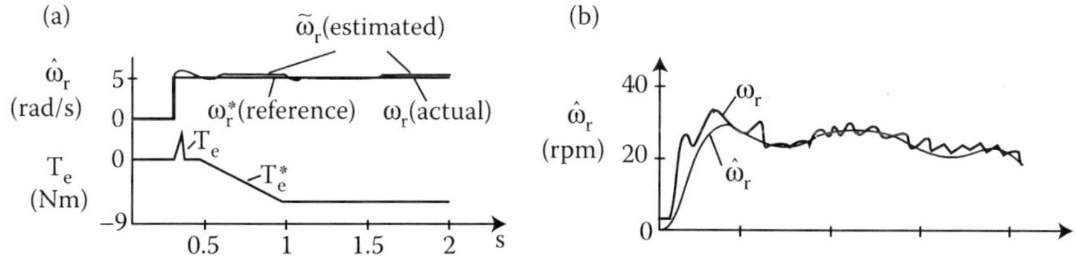


**FIGURE 9.40**
(a) Zero frequency, (b) low-speed operation of MRAS speed estimator.

The full-order Luenberger observer writes

$$\frac{d\hat{x}}{dt} = \hat{A}\hat{x} + B\hat{V}_s + G\left(\hat{i}_s - i_s\right) \tag{9.110}$$

The matrix G is chosen such that the observer is stable.

$$G = \begin{bmatrix} g_1 & g_2 & g_3 & g_4 \\ -g_2 & g_1 & -g_4 & g_3 \end{bmatrix}^T. \tag{9.111}$$

with

$$\begin{aligned} &g_1 = (K-1)(a_{r11} + a_{r22}); \quad g_2 = (K-1)a_{i22}; \quad g_4 = -c(K-1)a_{i12} \\ &g_3 = (K^2-1)(ca_{r11} + a_{r21}) - c(K-1)(a_{r11} + a_{r22}); \quad c = L_{sc} \\ &a_{r22} = 1/\tau_r; \quad a_{i22} = \hat{\omega}_r \end{aligned} \tag{9.112}$$

The speed estimator is based on rotor flux $\hat{\bar{\lambda}}_r^s$ and $\hat{\bar{i}}_s$ estimators

$$\hat{\omega}_r = \left(K_p + \frac{K_i}{s}\right) \text{Imag}\left[\left(\bar{i}_s - \hat{\bar{i}}_s\right)\hat{\bar{\lambda}}_r^s\right] \tag{9.113}$$

In essence, the speed estimator is based on some kind of torque error.

If the rotor resistance $r_r$ has to be estimated, an additional high-frequency current $i_{da}^*$ is added to the reference flux current $i_{ds}^*$. Then the rotor resistance may be estimated [26] as

$$\frac{d}{dt}\left(\frac{1}{\tau_r}\right) = -\hat{\lambda}_r\left[\left(\hat{i}_{ds} - i_{ds}^*\right) \cdot i_{da}^*\right] \tag{9.114}$$

Remarkable results have been obtained this way with minimum speed down to 30 rpm.

The idea of an additional high-frequency (10 times rated frequency) flux current may be used to determine both the rotor speed and rotor time constant $\tau_r$ [27].

Extended Kalman filters for speed and flux observers [28] also claim speed estimation at 20–25 rpm though they require considerable online computation time.

### 9.13.6 Rotor Slots Ripple Speed Estimators

The rotor slots ripple speed estimators are based on the fact that the rotor slotting openings cause stator voltage and current harmonics $\omega_{s1,2}$ related to rotor speed $\hat{\omega}_r$, the number of rotor slots $N_r$, and synchronous speed $\hat{\omega}_1$:

$$\hat{\omega}_{s1,2} = N_r\hat{\omega}_r \pm \hat{\omega}_1 \tag{9.115}$$

Band pass filters centered on the rotor slot harmonics $\hat{\omega}_{s1,2}$ are used to separate $\hat{\omega}_{s1,2}$ and thus calculate $\hat{\omega}_r$ from Equation 9.115. Various other methods have been proposed to obtain $\hat{\omega}_{s1,2}$ and improve the transient performance. The response tends to be rather slow and thus the method, though immune to machine parameters, is mostly favorable for a wide speed range, but for low dynamics (medium–high power) applications [29].

For an update on sensorless IM control with and without signal injection refer to Reference 30.

## 9.14 Direct Torque and Flux Control

DTFC is a commercial abbreviation for the so-called direct self-control proposed initially [2,3] for IMs fed from PWM voltage-source inverters and later generalized as torque vector control (TVC) in Reference 4 for all a.c. motor drives with voltage- or current-source inverters.

In fact, based on the stator flux vector amplitude and torque errors sign and relative value and the position of the stator flux vector in one of the six (12) sectors of a period, a certain voltage vector (or a combination of voltage vectors) is directly applied to the inverter with a certain average timing.

To sense the stator flux space-phasor and torque errors, we need to estimate the respective variables. So all types of flux (torque) estimators or speed observers good for direct VC are also good for DTFC. The basic configurations for direct VC and DTFC are shown in Figure 9.41.

As visible in Figure 9.41, DTFC is a kind of direct vector d.c. (synchronous) current control. Both strategies require flux and torque observers (also, of course, speed observers for motion sensorless control). However, direct torque and flux control (DTFC) has the right scale and thus no design of the PI d.c. current controllers ($i_{da}$, $i_{qa}$) is necessary. Also the vector rotator and the open-loop PWM are replaced with a table of switchings (TOS). Both these simplifications mean that DTFC is related to motor parameters only through the flux–torque–speed observers. Finally, DTFC works with stator flux and not with rotor flux.

It is true, however, that, in terms of dynamic properties, the two methods are somewhat similar though the flux dynamics is slower for the direct VC as the rotor flux to d axis current dynamics is, in general, neglected

$$i_d = (1 + s\tau_r)\frac{\lambda_r}{L_m} \tag{9.116}$$

Equation 9.116 may be added but this operation also implies a rotor time constant $\tau_r$ adaptation mechanism. So it seems clear that DTFC does "indirectly" direct VC in a simpler, more robust configuration, but still preserves high torque dynamics over a wide speed range.

Sliding mode or fuzzy logic may replace the two (three) positional hysteresis flux and torque controllers in Figure 9.41b. Knowing at any instant the stator flux amplitude and position, short source voltage sags (10–20 ms long) may be handled; thus providing for a ride through inherent capability.

These merits have prevailed so that DTFC has reached markets [33] in IM drives with motion sensors or sensorless, as a kind of universal (for all applications) IM drive. One drive with digital control does it all.

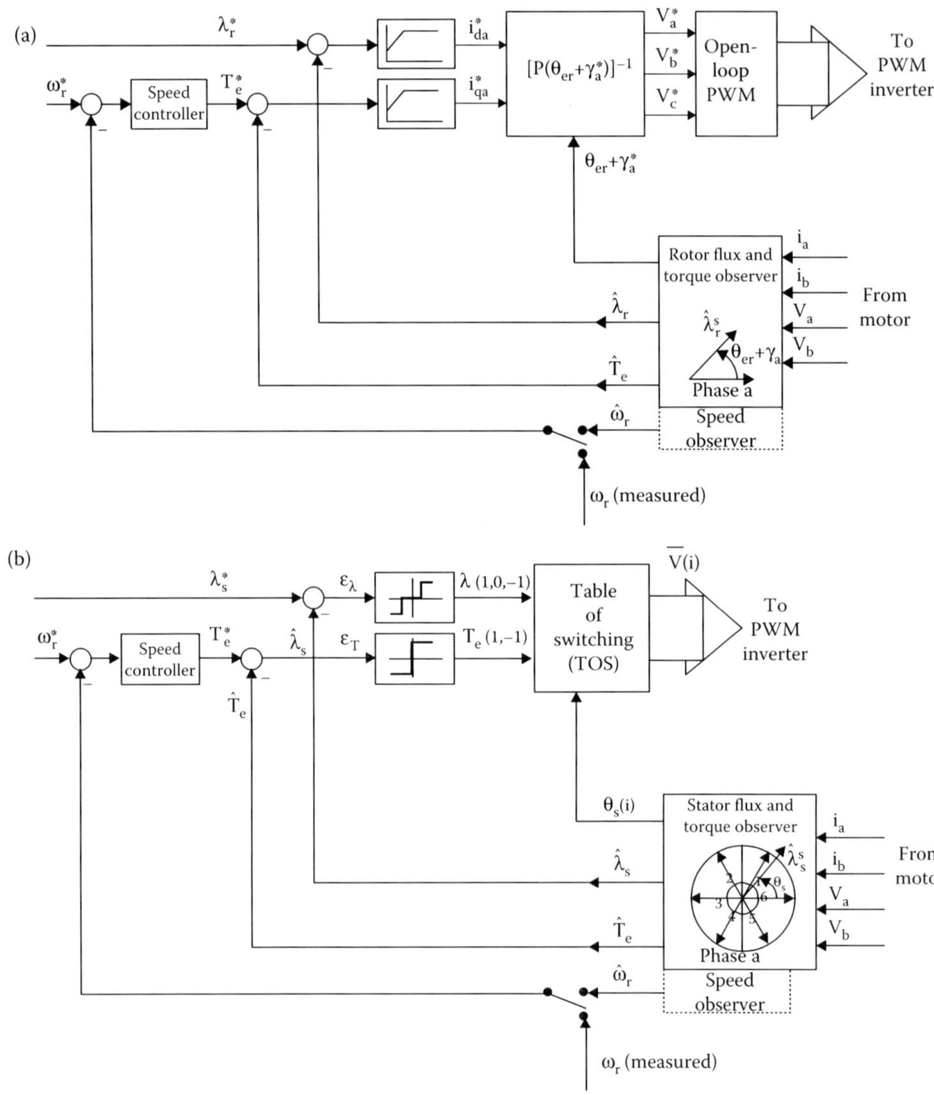


**FIGURE 9.41**
(a) Direct vector current control, (b) original DTFC control.

In what follows, we consider two aspects: DTFC principle and performance and DTFC sensorless IM drive—a case study.

### 9.14.1 DTFC Principle

Though Figure 9.41b uncovers the principle of DTFC, finding how the TOS is generated is the way to a successful operation.

Selection of the appropriate voltage vector in the inverter $\overline{V}_s(i)$ is based on a stator equation in stator coordinates

$$\frac{d\overline{\lambda}_s^s}{dt} = \overline{V}_s^s - r_s\overline{i}_s^s \quad (9.117)$$

By integration

$$\Delta\bar{\lambda}_s^s = \bar{\lambda}_s^s - \bar{\lambda}_{s0}^s = \int_0^{T_i}\left(\bar{V}_s^s - r_s\bar{i}_s^s\right)\cdot dt \approx \bar{V}_s(i)T_i \tag{9.118}$$

In essence the torque error $\varepsilon_T$ may be canceled by stator flux acceleration or deceleration. To reduce the flux errors, the flux trajectories will be driven along appropriate voltage vectors (9.118) that increase or decrease the flux amplitude.

When the flux amplitude has to be increased, a voltage vector phase shifted by an angle larger than 90° with respect to existing flux vector $\bar{\lambda}_{s0}^s$ is applied. In contrast, if flux amplitude has to be reduced, an angle less than 90° will be observed.

For the flux space-phasor in the first sector, the voltage vector selection is shown in Figure 9.42b.

The complete table of optimal switching, TOS, is shown in Table 9.2.

Either the hysteresis band will decide the timing of each voltage vector or, if the timing is constant, the switching frequency will be constant. The hysteresis band of the two controllers may be adapted to keep constant the average switching frequency.

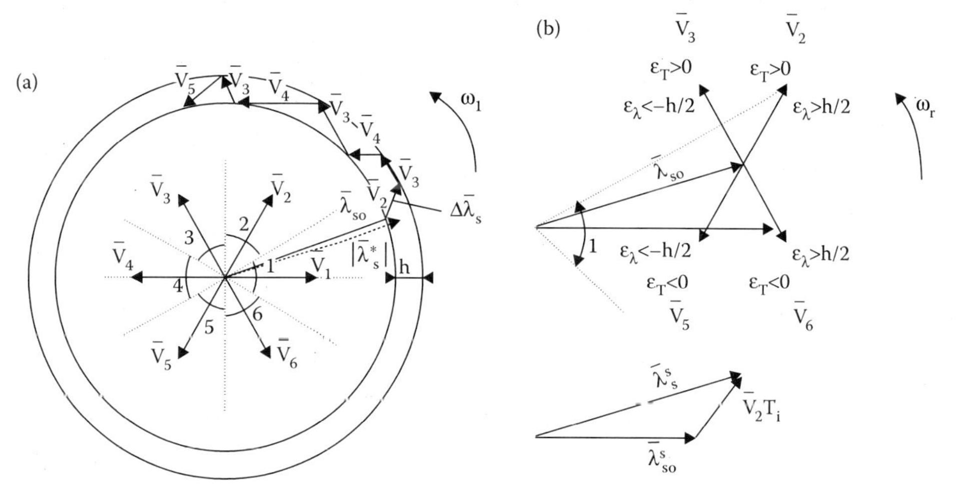

**FIGURE 9.42**
(a) Stator flux space-phasor trajectory, (b) selecting the adequate voltage vector in the first sector (–30° to +30°).

**TABLE 9.2**
Basic Voltage Vector Selection for DTFC

| $\theta_s(i)$ | | | | | | | |
|---|---|---|---|---|---|---|---|
| $\lambda$ | T | $\theta_s(1)$ | $\theta_s(2)$ | $\theta_s(3)$ | $\theta_s(4)$ | $\theta_s(5)$ | $\theta_s(6)$ |
| 1 | 1 | $V_2$ | $V_3$ | $V_4$ | $V_5$ | $V_6$ | $V_1$ |
| 1 | –1 | $V_6$ | $V_1$ | $V_2$ | $V_3$ | $V_4$ | $V_5$ |
| 0 | 1 | $V_0$ | $V_7$ | $V_0$ | $V_7$ | $V_0$ | $V_7$ |
| 0 | –1 | $V_0$ | $V_7$ | $V_0$ | $V_7$ | $V_0$ | $V_7$ |
| –1 | 1 | $V_3$ | $V_4$ | $V_5$ | $V_6$ | $V_1$ | $V_2$ |
| –1 | –1 | $V_5$ | $V_6$ | $V_1$ | $V_2$ | $V_3$ | $V_4$ |

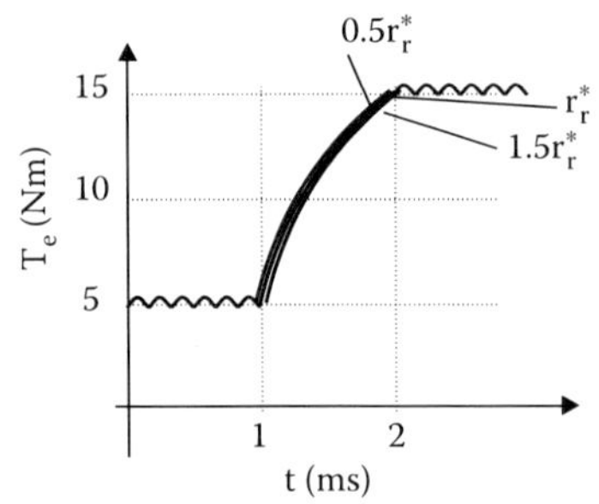


**FIGURE 9.43**
TVC torque response.

DTFC looks thus deceptively simple. Still, the stator flux has to be estimated and at least a combined voltage–current model (Figure 9.38) is necessary if acceptable performance down to 0.5 Hz is required.

Once the stator flux is known, the torque is simply

$$\hat{T}_e = \frac{3}{2} p \mathrm{Real}\left(j \bar{\lambda}_s^s \bar{i}_s^{s*}\right) \tag{9.119}$$

As expected, the torque response is quick (as in VC) but it is also rotor resistance independent, above 1–2 Hz (Figure 9.43) [2].

Note also that the torque pulsations are directly controlled and the core losses and noise may be controlled through the stator flux level selection.

For sensorless control both flux and speed observers, as presented in Section 9.13, are required; torque response times of 1–5 ms (as fast as for control with motion sensors) have been obtained for 100 to 1 speed control range and 0.1% of rated speed error [31–34]. Such sensorless DTFC drives are now commercially available [35].

## 9.15 DTFC Sensorless: A Case Study in MATLAB–Simulink (Available Online)

The simulation results of the original DTFC drive system for IMs [2] are presented. The example is implemented in MATLAB–Simulink. The motor model was integrated in two blocks, the first represents the current and flux calculation module in d-q axis (Figures 9.44 and 9.45); the second represents the torque, speed, and position computing module (Figure 9.46).

The commutation table was implemented as a MATLAB function to select the right voltage vector for the drive. The input values for this table are the two stator flux values from the motor model ($\lambda_{sd}$ and $\lambda_{sq}$), necessary to find out the position of the flux, the error values from comparing estimated torque (from the torque estimation block) with its reference value (the output of speed regulator), and, respectively, the calculated stator flux and its (constant) reference value.

The changing of motor parameters for different simulations is as simple as possible. By clicking on these blocks, a dialog box appears and you can change them by modifying their default values.

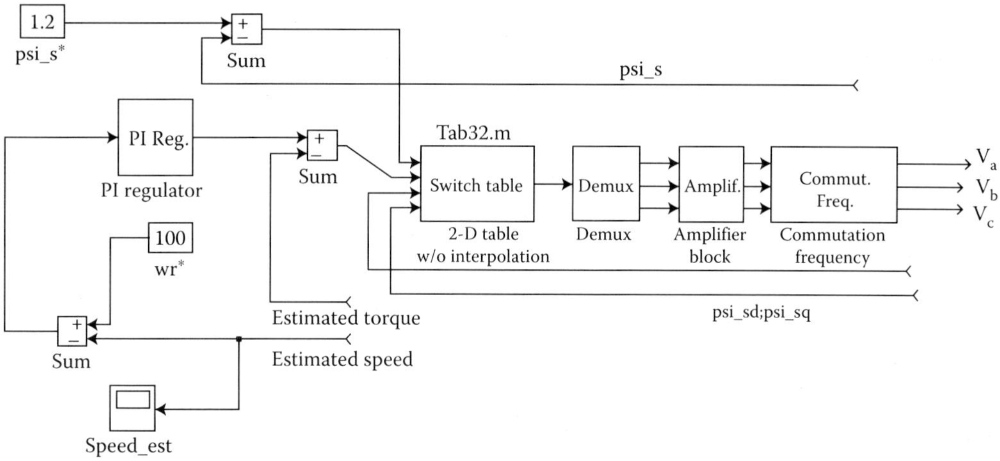


**FIGURE 9.44**
The DTFC system.

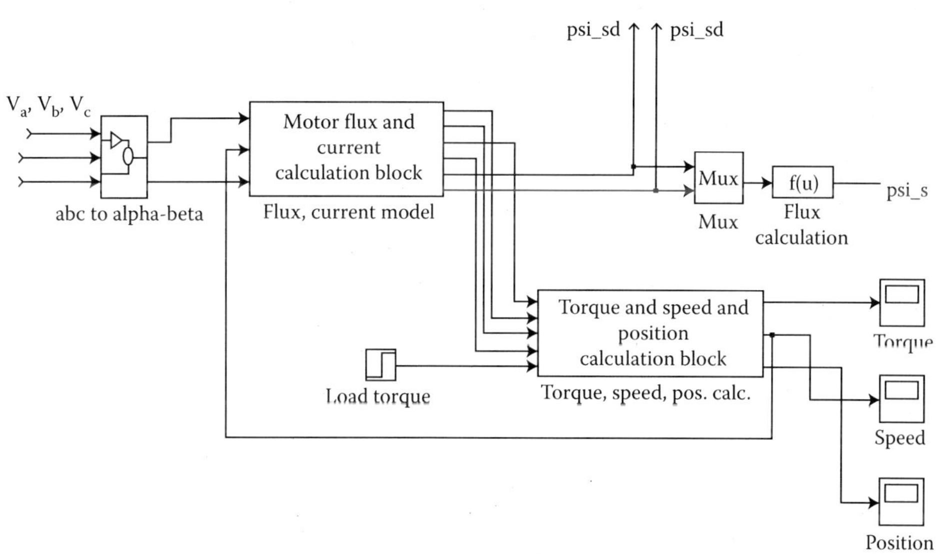


**FIGURE 9.45**
The IM model.

You can modify the speed regulator (*PI regulator*) parameters (amplification and integration time constant, default values are $K_i = 10$, $T_i = 0.8$ s) to study the behavior of the system. We included a *Commutation Frequency* block to satisfy the sampling criteria (to control the commutation/switching frequency). The integration step (default: 25 μs) can be modified from Simulink's *Simulation/Parameters*.

To find out the structure of each block presented above, unmask it (*Options/Unmask*).

The motor used for this simulation has the following parameters: $P_n = 1100$ W, $U_{nf} = 220$ V, $2p = 4$, $r_s = 9.53\ \Omega$, $r_r = 5.619\ \Omega$, $L_{sc} = 0.136$ H, $L_r = 0.505$ H, $L_m = 0.447$ H, $J = 0.0026$ kg fm$^2$.

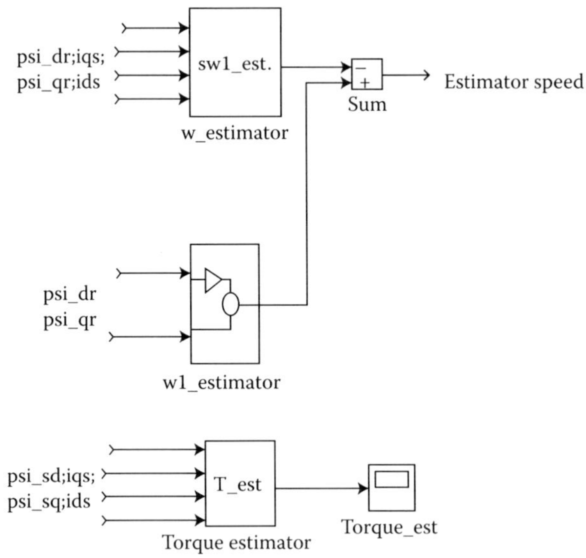


**FIGURE 9.46**
Speed and torque estimators.

The following figures represent the speed (Figure 9.47), current (Figure 9.48), torque (Figure 9.49), and flux responses (Figures 9.50 and 9.51) for the *starting process with load torque applied at 0.4 s*. The value of load torque is 8 Nm.

A combined voltage and current model speed observer was used. There is still a small steady-state error (reduction) in speed (Figure 9.47) after loading at 0.4 s.

The current waveform is slightly different from that obtained with indirect vector a.c. or d.c. current control.

The torque calculator (Figure 9.49) proves to be also satisfactory. The stator flux response (Figure 9.50) is quick and stable. The rotor flux (Figure 9.51) varies only during transients. Please remember that this time the stator flux amplitude is directly controlled.

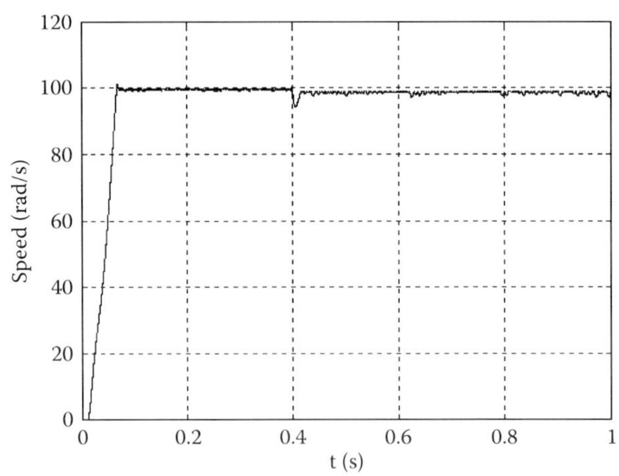


**FIGURE 9.47**
Speed transient response (measured and estimated).

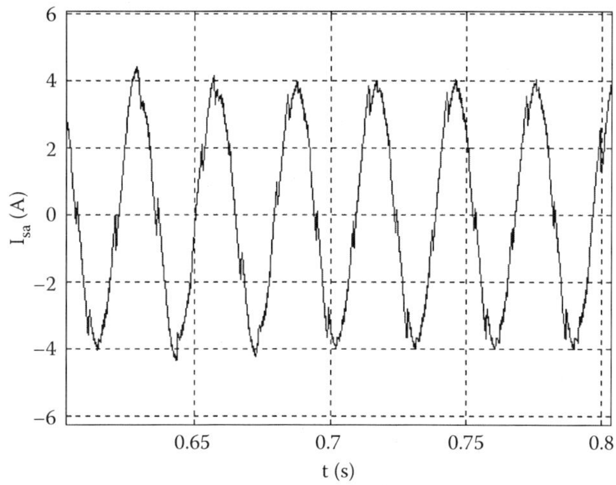


**FIGURE 9.48**
Phase current waveform (steady state).

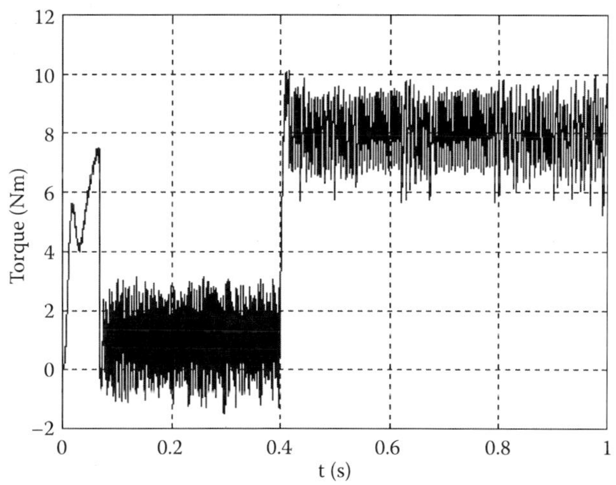


**FIGURE 9.49**
Torque response (measured and estimated).

The listing of the MATLAB file for switching table follows:

```
function [sys, x0]=tab3(t,x,u,flag,zindex,table)
% ****************************************
% Here starts the switching table lookup routine...
% ****************************************
if flag == 3,
% Perform the desired 2D table lookup

V0 = [0 0 0];V1 = [1 0 0];V2 = [1 1 0];V3 = [0 1 0];
V4 = [0 1 1];V5 = [0 0 1];V6 = [1 0 1];V7 = [1 1 1];
     x0 = u(1);
```

```
      y0 = u(2);
      z0 = u(3);
      z1 = u(4);
      if (x0 > 0.01), x0 = 1;end;
      if (x0 < -0.01), x0 = -1;end;
      if (x0 < 0.01)&(x0 > -0.01), x0 = 0;end;
      if (y0 >=0), y0 = 1;end;
      if (y0 < 0), y0 = -1;end;
      if (x0==1)&(y0==1), xf = 1;end;
      if (x0==1)&(y0==-1), xf = 2;end;
      if (x0==-1)&(y0==1), xf = 3;end;
      if (x0==-1)&(y0==-1), xf = 4;end;
      if (x0==0),             xf = 5;end;
      if (sign(z0)>=0)&(sign(z1)>=0)&(sign(sqrt(3)*abs(z1)-abs(z0))==
-1)&(sign(abs(z1)-sqrt(3)*abs(z0))==-1), zf = 1;end;
      if (sign(z0)>=0)&(sign(z1)>=0)&(sign(sqrt(3)*abs(z1)-
abs(z0))>=0)&(sign(abs(z1)-sqrt(3)*abs(z0))==-1), zf = 2;end;
      if (sign(z0)>=0)&(sign(z1)>=0)&(sign(sqrt(3)*abs(z1)-
abs(z0))>=0)&(sign(abs(z1)-sqrt(3)*abs(z0))>=0), zf = 2;end;
      if (sign(z0)==-1)&(sign(z1)>=0)&(sign(sqrt(3)*abs(z1)-
abs(z0))>=0)&(sign(abs(z1)-sqrt(3)*abs(z0))>=0), zf = 3;end;
      if (sign(z0)==-1)&(sign(z1)>=0)&(sign(sqrt(3)*abs(z1)-
abs(z0))>=0)&(sign(abs(z1)-sqrt(3)*abs(z0))==-1), zf = 3;end;
      if (sign(z0)==-1)&(sign(z1)>=0)&(sign(sqrt(3)*abs(z1)-
abs(z0))==-1)&(sign(abs(z1)-sqrt(3)*abs(z0))==-1),zf = 4;end;
      if (sign(z0)==-1)&(sign(z1)==-1)&(sign(sqrt(3)*abs(z1)-
abs(z0))==-1)&(sign(abs(z1)-sqrt(3)*abs(z0))==-1),zf = 4;end;
      if (sign(z0)==-1)&(sign(z1)==-1)&(sign(sqrt(3)*abs(z1)-
abs(z0))>=0)&(sign(abs(z1)-sqrt(3)*abs(z0))==-1),zf = 5;end;
      if (sign(z0)==-1)&(sign(z1)==-1)&(sign(sqrt(3)*abs(z1)-
abs(z0))>=0)&(sign(abs(z1)-sqrt(3)*abs(z0))>=0), zf = 5;end;
      if (sign(z0)>=0)&(sign(z1)==-1)&(sign(sqrt(3)*abs(z1)-
abs(z0))>=0)&(sign(abs(z1)-sqrt(3)*abs(z0))>=0), zf = 6;end;
      if (sign(z0) >=0)&(sign(z1)==-1)&(sign(sqrt(3)*abs(z1)-
abs(z0))>=0)&(sign(abs(z1)-sqrt(3)*abs(z0))==-1), zf = 6;end;
      if (sign(z0)>=0)&(sign(z1)==-1)&(sign(sqrt(3)*abs(z1)-
abs(z0))==-1)&(sign(abs(z1)-sqrt(3)*abs(z0)) ==-1),zf = 1;end;
      r = table(xf, zf);
      if r ==0, sys = V0;end;if r ==1, sys = V1;end;
      if r ==2, sys = V2;end;if r ==3, sys = V3;end;
      if r ==4, sys = V4;end;if r ==5, sys = V5;end;
      if r ==6, sys = V6;end;if r ==7, sys = V7;end;
   % Here ends the switching table lookup routine…
   % ****************************************
   elseif flag == 0,
        % This part takes care of all initialization; it is used only once.
        x0 = [];
        % The system has no states, three outputs, and four inputs.
        sys = [0 0 3 4 0 1]';
   else
        % Flags not considered here are treated as unimportant.
        % Output is set to [].
        sys = [];
end
```

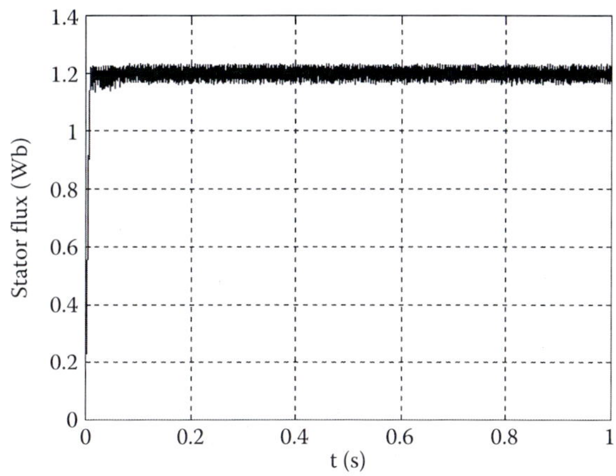


**FIGURE 9.50**
Stator flux amplitude.

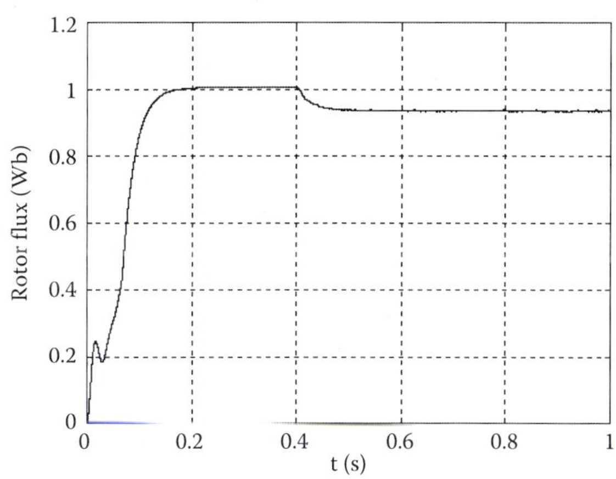


**FIGURE 9.51**
Rotor flux amplitude.

### 9.15.1 Sensorless DTFC with Space Vector Modulation

Notable torque and current pulsations are visible in Figures 9.48 and 9.49.

There is now a rich literature dealing with this problem [36–42].

In essence there are two main ways to reduce the torque ripple and current harmonics in DTFC-IM drives:

- Increasing the switching frequency in the PWM converter from 5–15 kHz to 20–40 kHz
- With limited switching frequency, the number of voltage vectors applied within 60° of stator flux vector evolution is increased

The latter type of method seems more practical with existing IGBT PWM voltage-source converters.

Open-loop space-vector modulation (SVM) (Section 9.9.2) may be used for limited switching frequency, to produce low torque ripple.

To apply open-loop SVM the reference (desired) voltage vector $\overline{V}_S^*$ in the stator has to be computed online.

Apart from the so-called stator flux VC [4] quite a variety of flux and torque controllers may be used to produce the reference voltage $\overline{V}_S^*$ required for open-loop SVM.

Linear variable structure [37] or hybrid (linear plus variable structure) torque and flux controllers have been proven practical for the scope [38,40].

The *sliding mode* (variable structure) surfaces $S_{\lambda s}$, $S_{Te}$ are defined based on stator flux and torque errors ($\varepsilon_{\lambda s}$, $\varepsilon_{Te}$):

$$S_{\lambda s} = \varepsilon_{\lambda s} + C_{\Psi s}\frac{d\varepsilon_{\lambda s}}{dt}$$
$$S_{Te} = \varepsilon_{Te} + C_{Te}\frac{d\varepsilon_{Te}}{dt} \quad (9.120)$$

The reference (command) d-q voltages in stator flux coordinates ($\theta_{\lambda s}$), $\overline{V}_{sd}^*$, $\overline{V}_{sq}^*$ are

$$\overline{V}_{sd}^* = \left(k_{p\lambda} + k_{i\lambda}\frac{1}{s}\right)\text{sgn}(S_{\lambda s})$$
$$\overline{V}_{sq}^* = \left(k_{pT} + k_{iT}\frac{1}{s}\right)\text{sgn}(S_{Te}) + \hat{\omega}_{\Psi s}\hat{\lambda}_s \quad (9.121)$$

where $\hat{\lambda}_s$, $\hat{\omega}_{\Psi s}$ are estimated stator flux speed and amplitude.

Park transformation to stator coordinates is operated to obtain the so-called relay with constant gains DTFC (Figure 9.52) [38,40].

As evident in Equation 9.121 and Figure 9.52:

- A Park transformation from stator flux to stator coordinates is required.
- E.m.f. compensation is applied also to improve torque response at high speeds.
- The low torque and current ripple are secured by the linear (PI) controller section.
- Robustness to torque, inertia, and machine parameter detuning is provided by the *sliding mode* controller section.

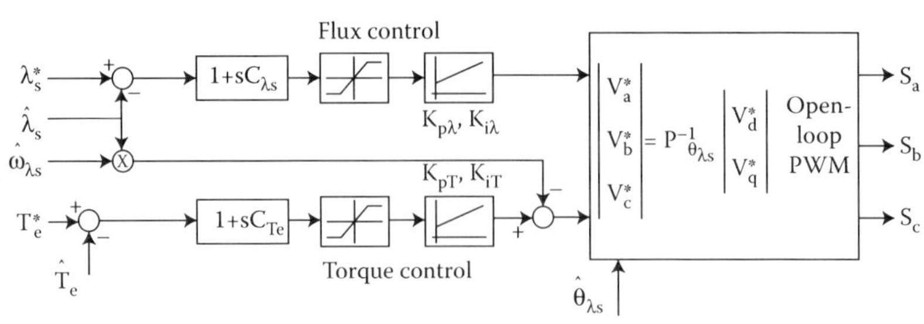


**FIGURE 9.52**
Linear + sliding mode SVM-DTFC scheme.

- To reduce chattering around the targeted variable values—under steady state—a boundary layer 2 h may be used to replace the sign function with a saturation function

$$\text{sat}(x) = \begin{cases} \text{sgn}\, x & \text{for } |x| > h \\ \dfrac{x}{h} & \text{for } |x| < h \end{cases} \tag{9.122}$$

Inside this layer the controller becomes high gain PI type and drives the error to zero. Outside the boundary layer sliding mode, robust and fast response of original DTFC is preserved.

- The state observer has to estimate the stator flux amplitude $\hat{\lambda}_s$, angle $\hat{\theta}_{\lambda s}$, electromagnetic torque $\hat{T}_e$, and stator flux vector speed $\hat{\omega}_{\lambda s}$.
- To perform speed closed-loop control, a rotor speed estimator is required.

A general voltage/current model, inherently motion sensorless, stator and rotor flux observer making use again of sliding mode concept is obtained simply by the stator equation in rotor flux coordinates [40]:

$$\frac{d\bar{\hat{\lambda}}_s}{dt} = -R_s\bar{i}_s + \bar{V}_s + k_1\,\text{sgn}(\bar{i}_s - \bar{\hat{i}}_s) + k_1'(\bar{i}_s - \bar{\hat{i}}_s) \tag{9.123}$$

$$\frac{d\bar{\hat{\lambda}}_r^r}{dt} \approx \frac{L_m R_r \bar{\hat{\lambda}}_s^r}{L_s L_{sc}} - \frac{R_r \bar{\hat{\lambda}}_r^r}{L_{sc}} + k_2\,\text{sgn}(\bar{i}_s^r - \bar{\hat{i}}_s^r) + k_2'(\bar{i}_s^r - \bar{\hat{i}}_s^r) \tag{9.124}$$

$$L_{sc} = L_s - L_m^2/L_r; \quad T_r = L_r/R_r$$

where $L_s$, $L_r$ are stator and rotor inductances, $R_s$, $R_r$ are stator and rotor resistances, $(\bar{i}_s, \bar{\hat{i}}_s)$ and $(\bar{i}_s^r, \bar{\hat{i}}_s^r)$ are the actual and estimated stator current vectors, respectively, in stator and rotor flux coordinates, respectively. As the stator and rotor resistances $R_s$, $R_r$ depend on temperature, they have to be corrected online. In general, $R_s = k_{sr}R_r$, so only the stator resistance has to be corrected.

The nonlinear term in *sgn* reflects the *sliding mode* contribution while also a linear (proportional, P) loop is added.

The current estimator is obtained from the stator and rotor flux definitions in stator coordinates as

$$\bar{\hat{i}}_s = \frac{L_r}{L_s L_{sc}}\bar{\hat{\lambda}}_s - \frac{L_m}{L_s L_{sc}}\bar{\hat{\lambda}}_r \tag{9.125}$$

The relationship between rotor flux and stator flux in stator coordinates is written as

$$\bar{\hat{\lambda}}_r = \frac{L_r}{L_m}\hat{\lambda}_s - L_{sc}\bar{i}_s; L_{sc} = L_s - \frac{L_m^2}{L_r} \approx L_{sl} + L_{rl} \tag{9.126}$$

Magnetic saturation plays a minor role in the stator current and rotor flux estimators above, because Lsc does vary only a few percent when the current is higher than 2–3 times rated current (for semiclosed or open slots in the motor) and the ratios Ls/Lm, Lr/Lm, and Lr/Ls are only mildly dependent on magnetic saturation.

The electromagnetic torque estimator is straightforward:

$$\hat{T}_e = \frac{3}{2} p \,\mathrm{Re}(j\bar{\hat{\lambda}}_s \hat{i}_s^*) \tag{9.127}$$

The structural diagram of the above hybrid (linear + SM) flux observer is shown in Figure 9.53.

For the speed estimation, an asymmetrical Luenberger (or equivalent PLL) speed observer may be adopted (Figure 9.54):

$$\frac{d}{dt}\begin{vmatrix}\hat{\alpha}\\ \hat{\omega}_r\\ \hat{T}_L\end{vmatrix} = \begin{vmatrix}0 & 1 & 0\\ 0 & 0 & -p/J\\ 0 & 0 & 0\end{vmatrix}\cdot\begin{vmatrix}\hat{\alpha}\\ \hat{\omega}_r\\ \hat{T}_L\end{vmatrix} + \begin{vmatrix}1 & 0\\ 0 & p/J\\ 0 & 0\end{vmatrix}\cdot\begin{vmatrix}\hat{\omega}_s\\ \hat{T}_e\end{vmatrix} + \begin{vmatrix}k_1\\ k_2\\ k_3\end{vmatrix}\cdot\varepsilon \tag{9.128}$$

$$\varepsilon = \sin(\hat{\theta}_{\lambda r} - \hat{\alpha}); \quad \hat{\omega}_s = S\omega = \hat{\omega}_{\lambda r} - \hat{\omega}_r \tag{9.129}$$

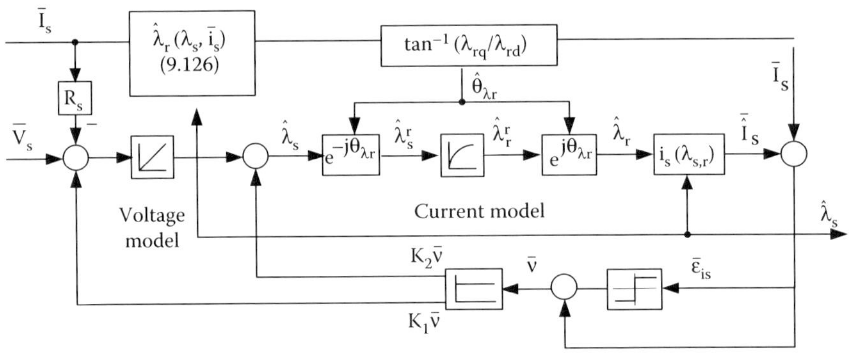


**FIGURE 9.53**
A generic hybrid (linear-SM) flux observer for IM. (After Lascu, C., Boldea, I., Blaabjerg, F., *IEEE Trans.*, IA-40(2), 582–590, 2004.)

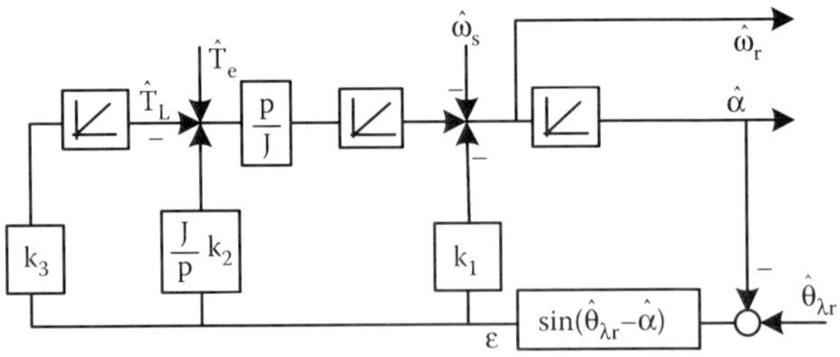


**FIGURE 9.54**
Asymmetrical Luenberger speed observer.

This observer roughly estimates the load torque $\hat{T}_L$, but estimates the rotor speed as well. $\hat{\omega}_s$ is the slip (S) speed of the rotor flux vector. Also, $\hat{\theta}_{\lambda r}$ is the estimated angle of rotor flux (Figure 9.53).

The error dynamics for this observer is simply

$$\frac{de}{dt} = (A - KC)e; \quad C = [1, 0, 0]; \quad K = [k_1\ k_2\ k_3]^T \tag{9.130}$$

where $|A|$ is the system matrix in Equation 9.128.

The eigen values of $|A| - |K||C|$ are the observer poles.

One negative real pole and two real-negative but complex poles should be provided for good dynamics.

The speed observer good transient behavior and small error are shown in Figure 9.55 for no parameters or inertia detuning [40].

The speed error immunity to large inertia change is high but rotor resistance correction is mandatory, to limit the speed error to 2–2.5 rpm at steady state.

The reduction of torque ripple at zero speed by SVM-DTFC with the above speed and flux observer is shown in Figure 9.56.

Fast torque response from standstill and larger PI gains in the flux and torque controller are evident in Figure 9.57 [40].

Steady 3 rpm speed control under load is shown in Figure 9.58.

The small current error (ripple) is also visible (Figure 9.58f).

The still visible torque ripple is also due to the brush friction in the d.c. brush PM machine that loads the drive by operating as motor in the opposite direction.

The above method is by no means the only one to work well without signal injection at very low speed (see, e.g., Reference 42 for VC at 5 rpm without signal injection).

The sensorless DTFC-SVM control of IMs proves to be capable of same performance as advanced VC in all respects for slightly lower online computation effort.

**NOTE:** Below 3 rpm, better speed and flux observers are required for motion sensorless control and perhaps the usage of signal injection may become mandatory [30].

New improvements relating to flux and speed observers for sensorless control of IMs, to operate stably during low speed regenerative braking have been reported more recently [31,33]. Besides, a single a.c. current sensor such sensorless drive was demonstrated [34].

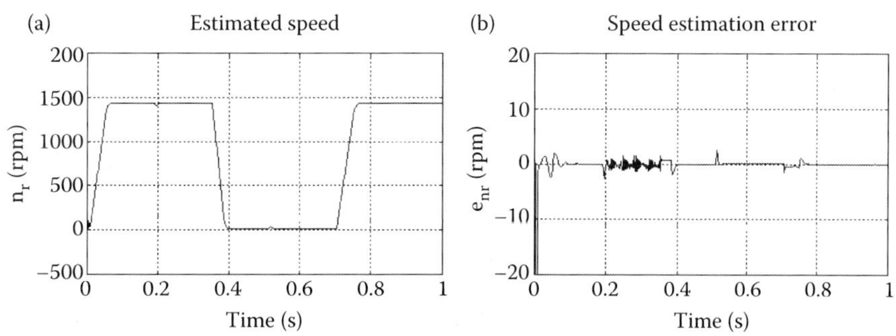


**FIGURE 9.55**
Asymmetrical Luenberger speed observer transient speed response (a) and error (b).

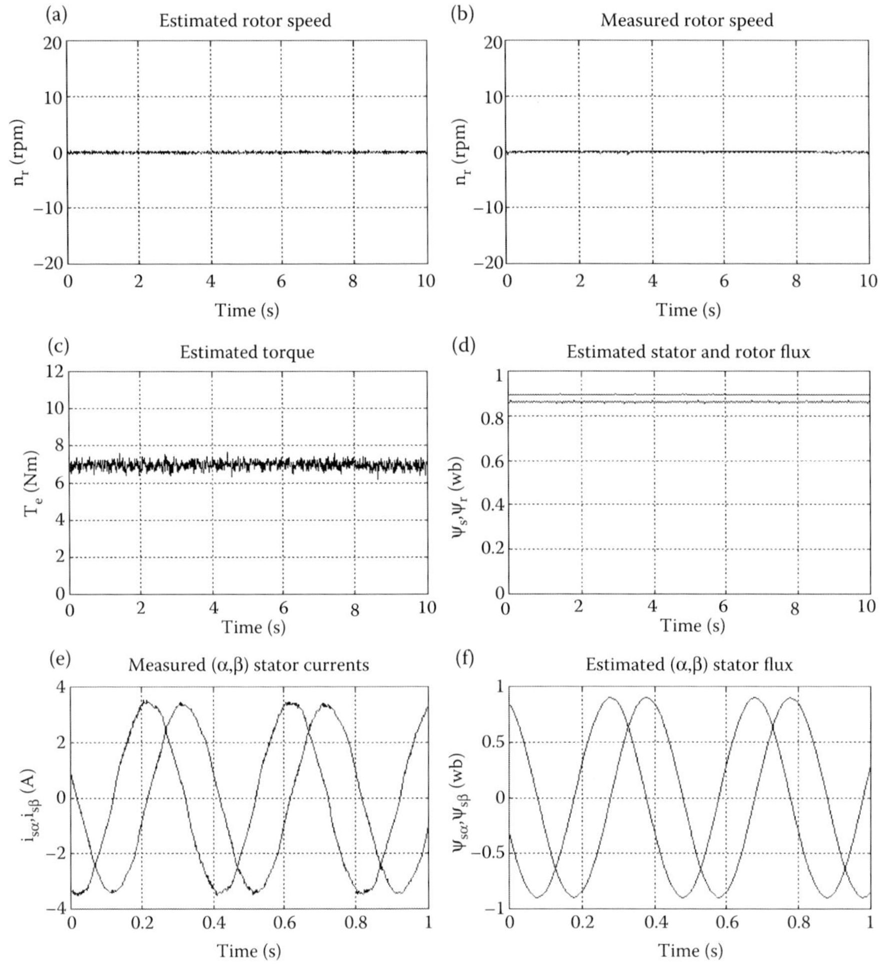


**FIGURE 9.56**
Measured and estimated variables at zero speed. (a) speed, (b) measured speed, (c) estimated torque, (d) estimated fluxes, (e,f) stator currents and flux wave forms. (After Lascu, C., Boldea, I., Blaabjerg, F., *IEEE Trans.*, IA-40(2), 582–590, 2004.)

## 9.16 Feedback Linearized Control

VC was invented to produce separate/decoupled flux and torque control as it is implicitly possible with d.c. brush motors.

It is also known that, with constant rotor flux, the torque is proportional to torque current and linear torque/speed characteristics may be obtained. Such an artificial decoupling and linearization of IM equations may, in principle, be done with some other nonlinear transformations. FLC [5–7] is such a method.

Still, Equations 9.106 through 9.108—in stator coordinates—with stator current and rotor flux d-q components, together with the equation of motion, are required.

The new output variables are the rotor flux squared $\Phi_1(x) = \lambda_r^2$ and the rotor speed $\Phi_2(x) = \omega_r$.

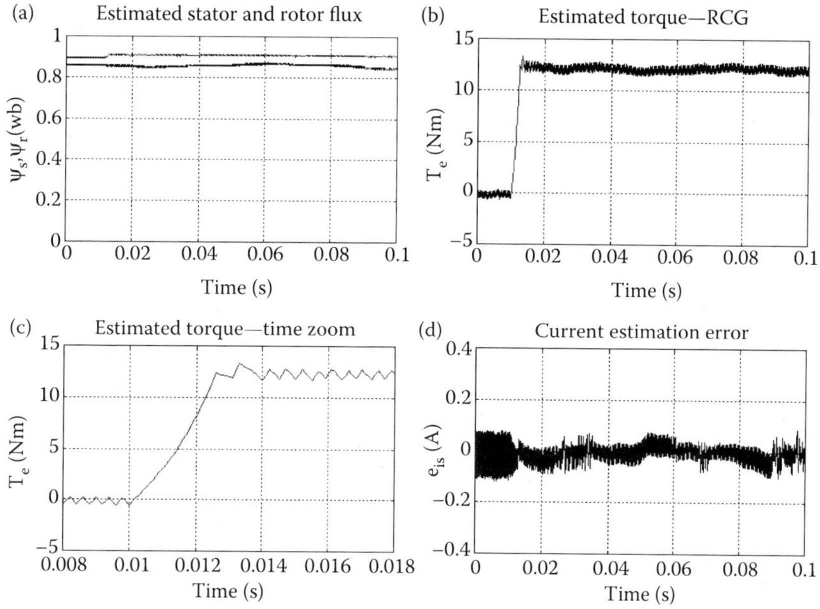


**FIGURE 9.57**
Torque transients from standstill: (a) stator and rotor flux, (b) torque, (c) torque—time zoom, and (d) current estimation error.

The basic signal flow diagram [7] is shown in Figure 9.59.

It should be noted that the computation effort is greater than for VC or DTFC and a thorough knowledge of motor parameters is necessary. Dynamic performance quite similar to both, advanced VC or DTFC has been obtained [35,36].

It seems too early to forecast the industrial prospects of FLC in competition with advanced VC or DTFC.

## 9.17 Predictive Control

Predictive control works on the principle of predicting the future behavior of the electric drive (along the controlled variable: current, or torque, or speed, or power) based on a model for each of the switching states in the PWM converter, and then selecting the optimal switching state according to a cost (objective) function (error) minimization. For the IM, the space-vector stator equation in stator coordinates is as shown [43]:

$$\bar{V}_s = R_s\bar{i}_s + L_s\frac{d\bar{i}_s}{dt} + L_r\frac{d\bar{i}_r}{dt} \tag{9.131}$$

The predictive model may simplify this equation by neglecting the rotor current variation ($d\bar{i}_r/dt$, for the sampling time):

$$\bar{V}_s \approx R_s\bar{i}_s + L_s\frac{d\bar{i}_s}{dt} \tag{9.132}$$

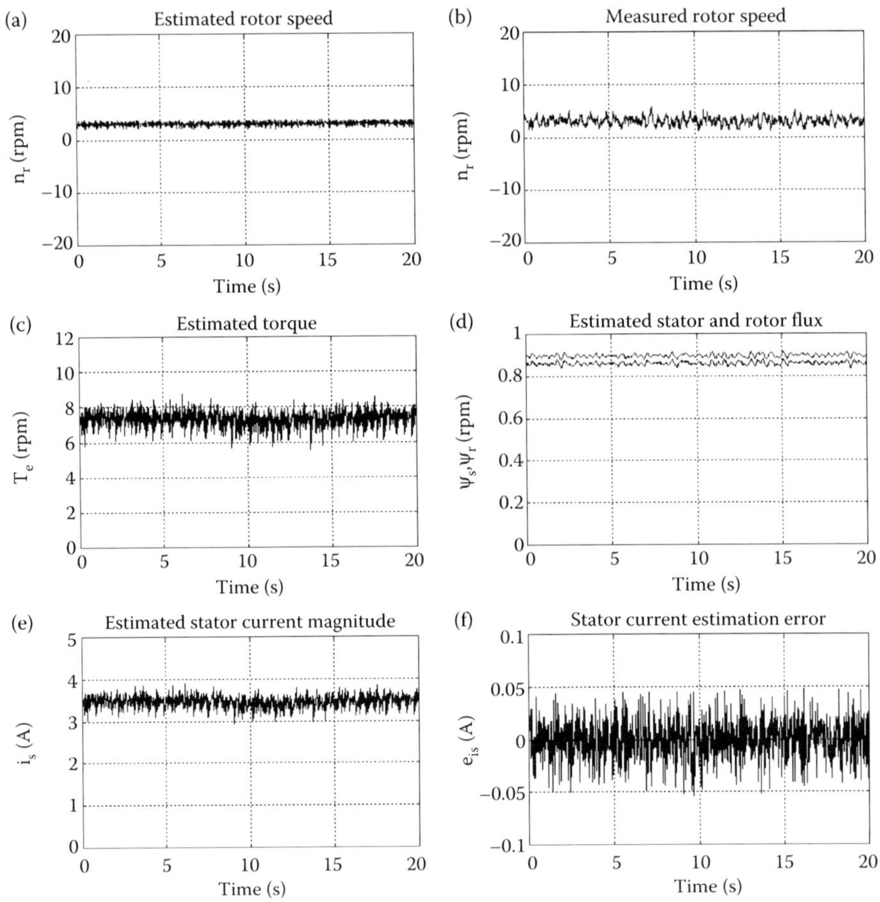


**FIGURE 9.58**
Very low speed (3 rpm) speed of DTFC-SVM IM drive with sliding mode observers and torque and flux controllers: (a,b) estimated and measured speed, (c,d) estimated torque and fluxes, (e,f) estimated stator current amplitude and error.

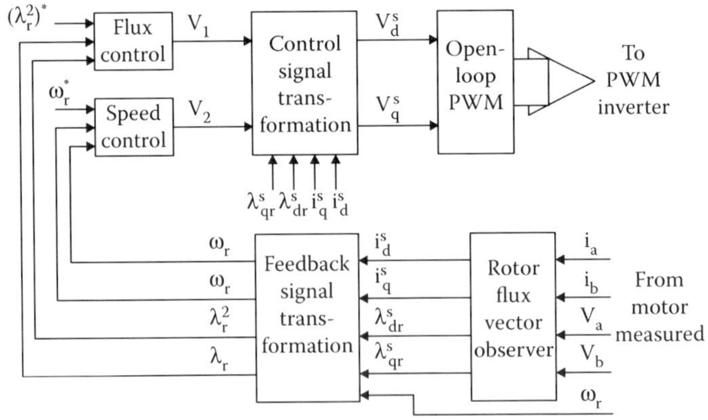


**FIGURE 9.59**
Feedback linearization control.

The predictive model (9.132) in digital form is

$$i_{(k+1)} = \left(1 - \frac{R_s}{L_s}T_s\right)\cdot \bar{i}_s(k) + \frac{1}{L_s}T_s \cdot \bar{V}_s(k); \quad \bar{V}_s(k) = S(k)\cdot V_{dc} \tag{9.133}$$

where $T_s$ is the sampling time and $\bar{V}_s(k)$ is the current–voltage vector of the PWM converter at time $kT_s$. S(k) is the switching vector for the used PWM converter. For a three-leg two-level voltage-source inverter, S(k) corresponds to the six nonzero and two zero voltage vectors:

$$\begin{aligned}
&(000) \quad \bar{V}_0 = 0,\\
&(100) \quad \bar{V}_1 = \frac{2}{3}V_{dc}\\
&(110) \quad \bar{V}_2 = \frac{1}{3}V_{dc} + j\frac{1}{\sqrt{3}}V_{dc}\\
&(010) \quad \bar{V}_3 = -\frac{1}{3}V_{dc} + j\frac{1}{\sqrt{3}}V_{dc}\\
&(011) \quad \bar{V}_4 = -\frac{2}{3}V_{dc}\\
&(001) \quad \bar{V}_5 = -\frac{1}{3}V_{dc} - j\frac{1}{\sqrt{3}}V_{dc}\\
&(101) \quad \bar{V}_6 = \frac{1}{3}V_{dc} - j\frac{1}{\sqrt{3}}V_{dc}\\
&(111) \quad \bar{V}_7 = 0
\end{aligned} \tag{9.134}$$

The cost function is related to the controlled variable(s) error(s) minimization (or another energy function minimization):

$$CF = \left(i_s^k - i_s(k+1)\right)^2 \tag{9.135}$$

for predictive current control.

$$CF = \left(T_e^* - T_e(k+1)\right)^2 + k_{\Psi T}\left(\Psi_s^* - \Psi_s(k+1)\right)^2 \tag{9.136}$$

for torque and flux control.

For speed control, the cost function may be of the form

$$CF = CF_{\omega\omega} + k_{\omega T}\cdot f(\Delta\omega_r)\cdot CF_{\omega T} + k_{con}\cdot F_{con} \tag{9.137}$$

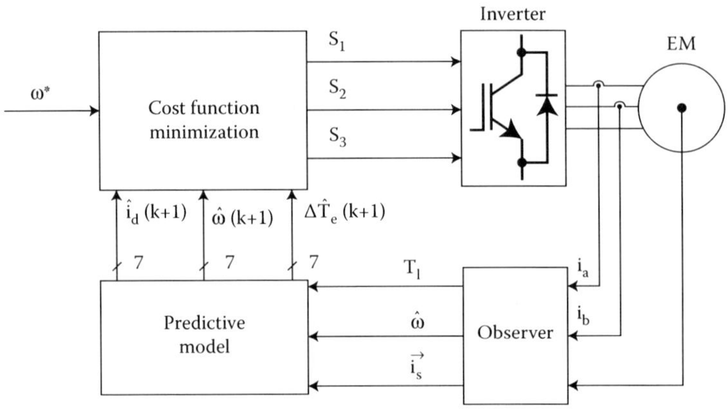


**FIGURE 9.60**
Predictive speed control of IM.

with

$$F_{ss} = \left(\omega_r^* - \omega_r(k+1)\right)^2 + k_{Te}\left(T_{load} - T_e(k+1)\right)^2 + k_d \cdot i_d^2 \quad (9.138)$$

By $f(\Delta\omega_r)$, the speed variation influence is limited and $CF_{ss}$ is a kind of steady-state (reference tracking) component of the cost function (CF); $k_{con}$ is related to the minimization of magnetization current ($i_d$) for the required torque (speed).

The $CF_{sT}$—related to speed transients—may be of the form

$$CF_{sT} = k_T'\left(\text{sign}(\Delta\omega_1)T_{e_{lim}} - T_e(k-1)\right)^2 + k_d' \cdot i_d^2 \quad (9.139)$$

For constant rotor flux the torque/speed $T_{elim}$ is the max torque allowed, while $k_d'$ relates to max torque per ampere (MTPA) conditions. Finally, the penalty component $F_{con}$ may be

$$F_{con} = \begin{cases} |\,i_s\,|^2 - i_{s\max}^2; & |\,i_s\,| > i_{s\max} \\ 0; & |\,i_s\,| < i_{s\max} \end{cases} \quad (9.140)$$

in which case it limits the peak current (allowed by the PWM converter).

All the above predictive controllers, replace only the closed-loop current or torque or speed regulators, while the state observer (for flux, torque, speed [in sensorless drives]) is still necessary.

A generic such predictive control system for speed control is shown in Figure 9.60.

The dynamic performance are similar to field orientation control (FOC), DTFC (without SVM), but it is claimed that the design of the control is simpler, though the online computation effort is still heavy. For more details, see Reference 43.

## 9.18 Scalar ($V_1/f_1$) Control

For pump and ventilator-like applications, the speed control range is only from 3 to 1 up to 10 to 1. Motion (speed) sensors are avoided in such drives.

Traditionally scalar, $V_1/f_1$, open-loop control has been used for such applications. In essence, the voltage amplitude $V_1$ and its frequency $f_1$ are related by

$$V_1^* = V_0^* + K_f f_1^* \tag{9.141}$$

$V_0^*$ is called voltage boost and is required to run the motor properly at low speeds.

The primary frequency is ramped as desired and, based on Equation 9.141, an open-loop PWM procedure is used to control the PWM inverter (Figure 9.61).

Ramping the frequency is performed slowly enough to maintain stability. Linearization of the IM equations around a steady-state point has backed experimental (industrial) findings of unstable open zones in the $i_{mr}(f_1)$ plane and ellipses of instability in the $V_1(f_1)$ plane [44] (Figure 9.62); $i_{mr}$ is the rotor flux current.

In general, $\omega_{1min}$ is [44]

$$\omega_{1min} > 2/\tau_r \tag{9.142}$$

where $\tau_r$ is the rotor time constant.

Also, it has been shown [44] that instabilities occur if the figure of merit $f_m$ is

$$f_m = \frac{\tau_m}{\tau_r} < 0.5 \tag{9.143}$$

where $\tau_m$ is the mechanical time constant.

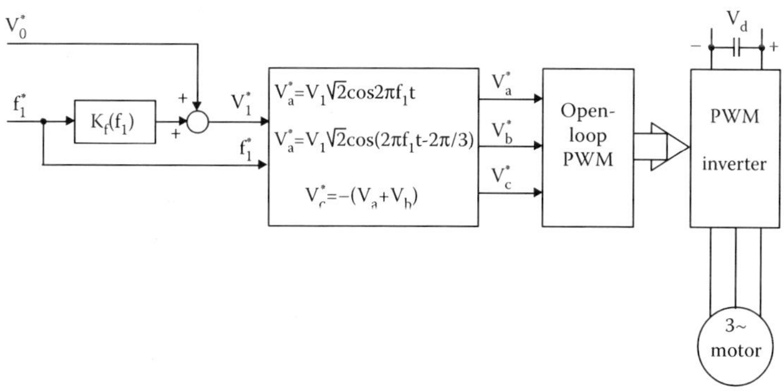


**FIGURE 9.61**
$V_1/f_1$ open-loop (scalar) control.

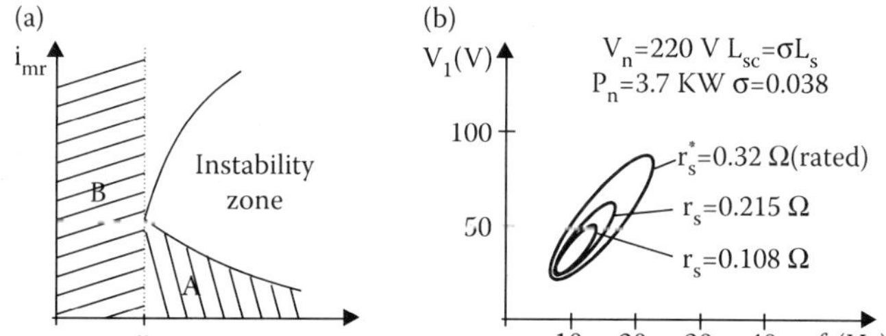


**FIGURE 9.62**
Instability zones for $V_1/f_1$ open-loop scalar control: (a) $i_{mr}/\omega_1$ plane, (b) $V_1/f_1$ plane.

The slope of the two asymptotes in Figure 9.62a are directly related to $\tau_r$ and $\tau_m$. The values of $f_m$ for a few 2-, 4-, 6-pole motors [44] are shown in Figure 9.63.

Two-pole motors tend to be unstable ($f_m < 0.5$) while 4- or 6-pole motors are less susceptible to instabilities. It should be mentioned that a higher short-circuit inductance $L_{sc}$ (in relative units) is beneficial from this point of view.

Besides the fact that open-loop scalar ($V_1/f_1$) drives are prone to instabilities, they are vulnerable to fast ramp accelerations or quick and (or) large torque perturbations. Moreover, increasing the load means a decrease in speed $\omega_r$:

$$\omega_r = \omega_1 - (S\omega_1) \tag{9.144}$$

For low speed (frequency), the reduction of speed with load becomes important and may not be tolerable. Compensating for the slip frequency $S\omega_1$ with load is a practical solution to this problem.

*Compensating the slip frequency* is attempted for steady state when, in fact, the rotor flux is constant in time and thus the torque $T_e$ (9.15) is

$$T_e = H(\omega_1 - \omega_r) = HS\omega_1 \tag{9.145}$$

with

$$H = \frac{3}{2}p\frac{\lambda_r^2}{r_r} \tag{9.146}$$

The motion equation is

$$\frac{J}{p}\frac{d\omega_r}{dt} = T_e - T_{load} \tag{9.147}$$

Assuming that the mechanical transients are slow, the rotor flux may be considered constant. The solution of $\omega_r$ for such slow transients is

$$\omega_r = \left(\omega_{r0} - \omega_1 + \frac{T_{load}}{H}\right)e^{-(t/\tau_m)} + \omega_1 - \frac{T_{load}}{H} \tag{9.148}$$

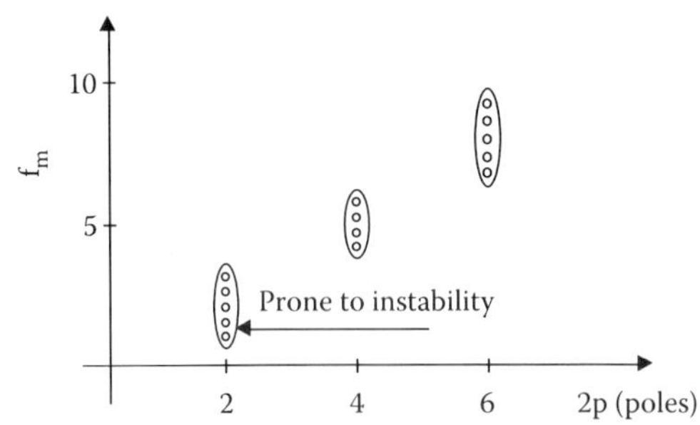


**FIGURE 9.63**
The figure of merit $f_m$ (9.143).

$$\tau_m = \frac{J}{pH} \tag{9.149}$$

For constant rotor flux, the torque speed curve is a straight line, so the stabilizing response in speed (9.148) is to be expected.

The variation of $r_r(\tau_r)$ with 25% produces, however, a 25% error in the slip frequency. For rated speed, this error is relatively low but at low speeds it remains a 25% speed error.

The principle of the slip frequency compensation method [45] consists of increasing the reference frequency $\omega_r^*$ by the estimated slip frequency $S\omega_1$ (Figure 9.64) to move, for given torque, from B to C and thus make $\omega_r$ independent of load.

The signal flow diagram (Figure 9.65) of such a scheme illustrates the estimation of the rotor flux from the voltage model.

The voltage model is adequate only above 2 Hz but this is acceptable in $V_1/f_1$ drives.

Equations 9.145 and 9.147 and Figure 9.65 suggest the simplified signal flow diagram of Figure 9.66.

The $\omega_r/\omega_r^*$ transfer function $G_0$ is simply (from Figure 9.66)

$$G_0 = \frac{sH(1+s\tau_2)}{J\tau_2^2 s^2 + (J + pH\tau_2 - JHK_1)s + pH} \tag{9.150}$$

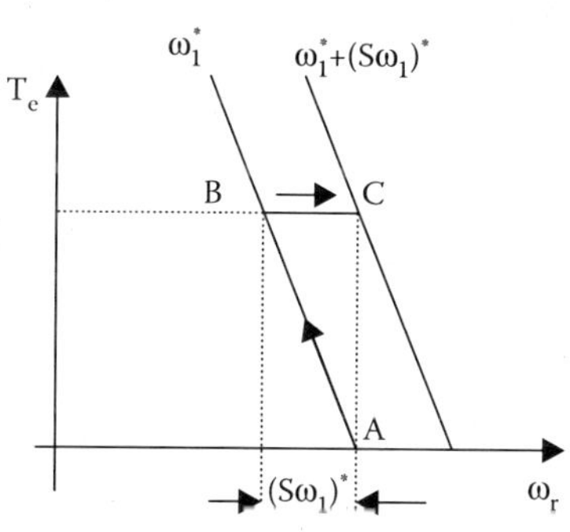


**FIGURE 9.64**
Principle of slip frequency compensation.

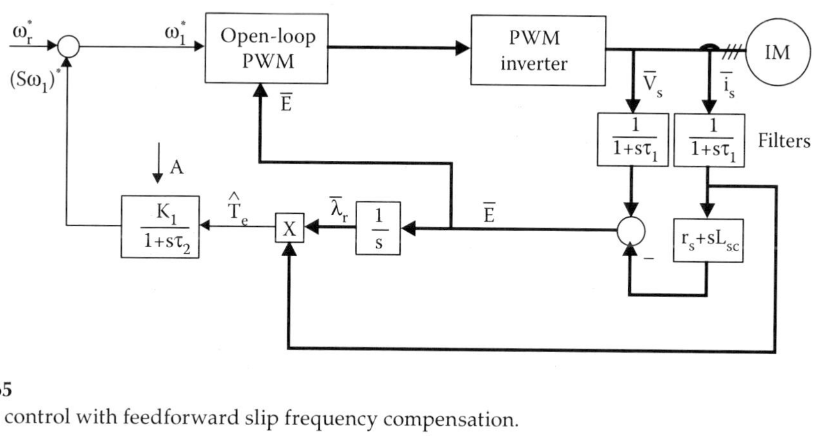


**FIGURE 9.65**
$V_1/f_1$ scalar control with feedforward slip frequency compensation.

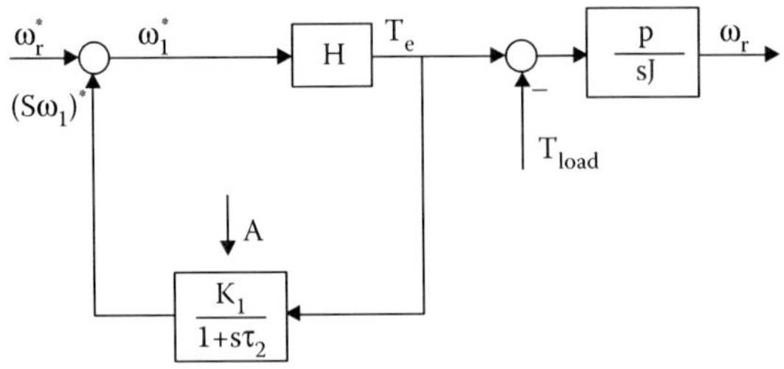


**FIGURE 9.66**
Signal flow diagram for slip frequency compensation.

$G_0$ offers a stable response only if

$$HK_1 < 1 + pH\tau_2/J \tag{9.151}$$

On the other hand, the transfer function $G_1$ between $\omega_r$ and $T_{load}$ is

$$G_1 = \frac{p(HK_1 - s\tau_2 - 1)}{J\tau_2^2 s^2 + (J + pH\tau_2 - JHK_1)s + pH} \tag{9.152}$$

Equation 9.152 suggests that for steady-state (s = 0 in Equation 9.152) $\Delta\omega_r$ for unit step load torque is

$$\Delta\omega_r = \frac{HK_1 - 1}{H} \tag{9.153}$$

So,

$$\begin{aligned} \Delta\omega_r &< 0; \quad \text{for } HK_1 \le 1 \\ \Delta\omega_r &= 0; \quad \text{for } HK_1 = 1 \\ \Delta\omega_r &> 0; \quad \text{for } HK_1 > 1 \end{aligned} \tag{9.154}$$

For full compensation $\Delta\omega_r = 0$ and thus $HK_1 = 1$ (point C in Figure 9.64). In this case, Equation 9.151 is automatically met. In special cases, at low speeds, $\Delta\omega_r < 0$ ($HK_1 \le 1$) in order to provide a stable operation in case of random friction torque perturbations. Good performance has been demonstrated down to 100 rpm [45].

NOTE: If flux weakening is to be performed (in order to reduce core loss), for low load levels, the gain $K_1$ in the filter in Figure 9.66 ($K_1 = 1/H$) has to be adapted based on the *a priori* known rotor flux level.

A good example is represented by drives where the load torque is a function of speed only and thus the value of $\lambda_r$ (and H) is solely a function of speed.

Again we mention that $V_1/f_1$ drives with slip compensation produce satisfactory precision in speed-sensorless control down to 100 rpm but they are adequate only for low dynamics applications (pumps, ventilators, and the like).

Speeds down to those corresponding to 1.2 Hz, full torque, have been obtained with more complete slip compensation or stabilizing loops methods (see Chapter 16).

## 9.19 Self-Commissioning

By self-commissioning we mean the parameter estimation and controller closed-loops calibration by the drive itself, when working on the initialization mode, at the site of application, before actual operation starts.

It is well understood that motor-converter parameters to be estimated depend on the particular control strategy adopted for the case.

As there are so many practical schemes for IM control we will treat here only a few basic principles of parameter estimation and controller calibration for a specific VC system [46–50].

Let us consider a direct vector current control scheme (Figure 9.67) with d.c. (synchronous) current controllers and current model open-loop rotor flux estimator.

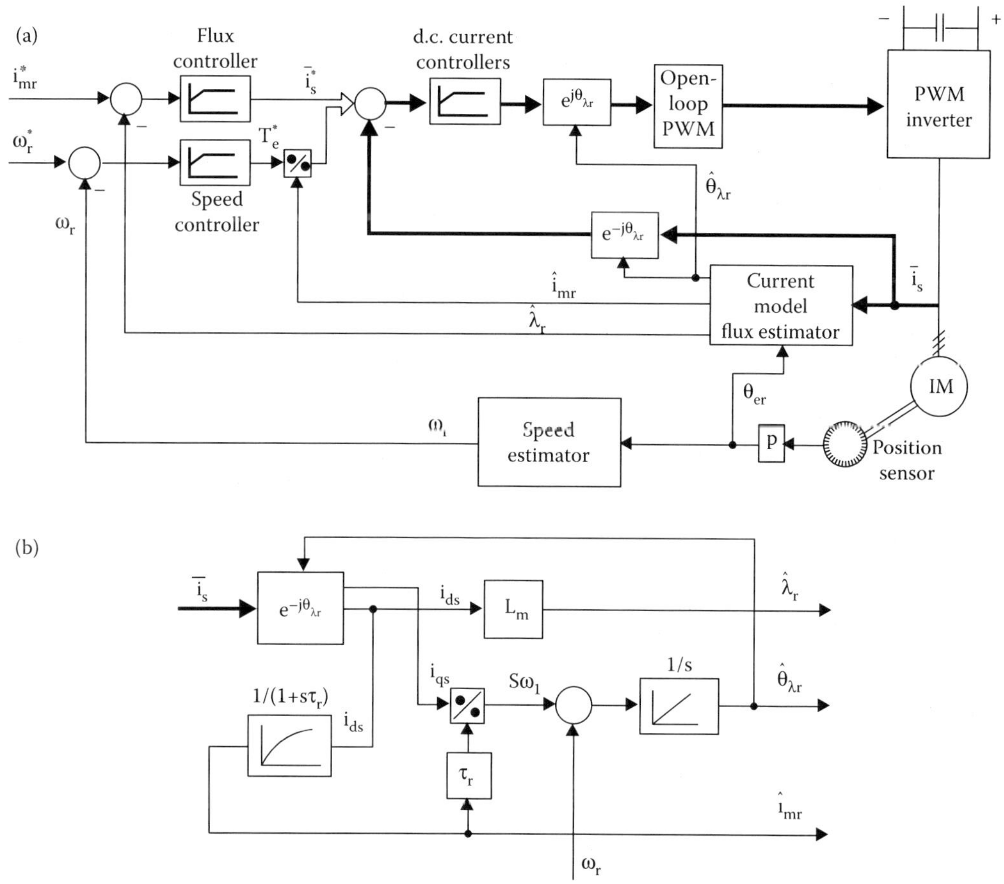


**FIGURE 9.67**
(a) Direct vector current control, (b) current model rotor flux estimator. (After Klambadkone, A.M., Holtz, J., *IEEE Trans.*, IE-38(5), 322–327, 1991.)

The apparent parameters are $L_m$ (main inductance) and the rotor time constant $\tau_r$. However, in order to calibrate the current, flux, and speed controllers:

- First, the stator resistance $r_s$ and the transient time constant $\tau' = L_{sc}/(r_s + r_r)$ have to be estimated and, based on these values, the current controllers are to be calibrated.
- Second, with the current controllers on, the rotor time constant $\tau_r$ is estimated and used in the rotor flux estimator (Figure 9.67b). Further, the flux controller is calibrated.
- Third, the reference rotor flux level and the corresponding magnetizing current $\hat{i}_{mr}$ are calculated.
- Fourth, through a no-load acceleration test, the mechanical time constant $\tau_m$ (9.149) is calculated.

Various methods to estimate the IM parameters $r_s$, $L_{sc}$, $L_m$, $\tau_r$, J may be classified into

- Step voltage response tests at standstill
- Frequency response tests at standstill
- Dynamic tests (nonzero speed)

Step voltage tests at standstill are rather easy to perform as the PWM may deliver voltage pulses for given time intervals. They still dominate the practice of self-commissioning programs.

However, recently, it has been shown that the step voltage response is influenced too much by the rotor cage skin effect, especially as the power of motor goes up. Consequently, the short-circuit inductance $L_{sc}$ is underestimated in comparison with the case of actual operation conditions [46].

Standstill frequency tests are proven to produce better results for a wide power range (1–500 kW) [46].

Here we still rely on step voltage responses at standstill [47]. The process of self-commissioning starts with introducing the motor specifications: rated voltage, rated current, rated frequency, and the number of pole pairs p.

The stator resistance is first calculated from a d.c. test knowing the d.c. link voltage and the modulation index $m_d$ (no voltage sensors are used in our case).

To account for the inverter dead time effect, measurements are taken for two different modulation indexes $m_{d1}$ and $m_{d2}$

$$r_s = \frac{V_1(m_{d1} - m_{d2})}{i_{s1} - i_{s2}} \tag{9.155}$$

Up to this point the current controllers are only grossly calibrated. The current levels are 50% and 100% of rated current and only a voltage vector is applied ($V_1$, for example). To estimate the motor transient time constant $\tau'$, the inverter is controlled through a binary port (with the modulator inhibited) for a few microseconds noting the time $t_{peak}$ and the peak current reached, $i_{peak}$ (Figure 9.68).

Due to the short time interval, the main flux does not occur and thus the motor equation is

$$\overline{V}_s = (r_s + r_r)\overline{i}_s + L_{sc}\frac{d\overline{i}_s}{dt} \tag{9.156}$$

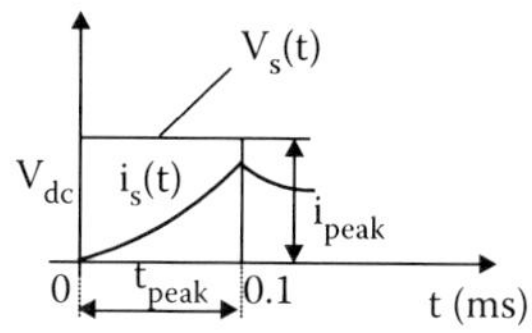


**FIGURE 9.68**
Step voltage response at standstill.

$$\tau' = \frac{L_{sc}}{r_s + r_r} = \frac{V_d t_{peak}}{(r_s + r_r) i_{peak}} \tag{9.157}$$

The average value of $\tau'$ from a few tests is computed.

Based on the above results, the current controllers (Figure 9.69) may be calibrated. $T_c$ is the sampling time.

Based on the optimal gain method [47], the PI controller constants are

$$V_{Pi} = \frac{\tau'}{2T_c V_d K_{inv}}; \quad T_i = \tau' \tag{9.158}$$

Further, we have to estimate the rotor time constant. We feed the inverter at standstill with d.c. current (applying $V_1$ with PWM) and then turn it off and record the stator voltage

$$\overline{V}_{s0}(t) = L_m \frac{d\bar{i}_r}{dt} \tag{9.159}$$

$$0 = \bar{i}_r + \tau_r \frac{d\bar{i}_r}{dt} \tag{9.160}$$

The rotor time constant $\tau_r$ is approximately

$$\tau_r \approx \frac{V_{s0}(t)}{V_{s0}(t_1) - V_{s0}(t_2)}(t_1 - t_2); \quad t_1 < t < t_2 \tag{9.161}$$

$V_{s0}$ has small values measured directly without attenuation. Once $\tau_r$ is known, the rated rotor flux magnetizing current $i^*_{ms}$ can be determined.

To do so, we use the drive in the $V_1/f_1$ open-loop mode at 10% of rated frequency and the motor no-load current is transformed into synchronous coordinates. The rotor time constant $\tau_r$ is changed until the $i_q$ current in synchronous coordinates is zero. The d axis current will be the rated magnetizing current $i^*_{mo}$ (provided $V_1/f_1$ ratio is the rated one) [47].

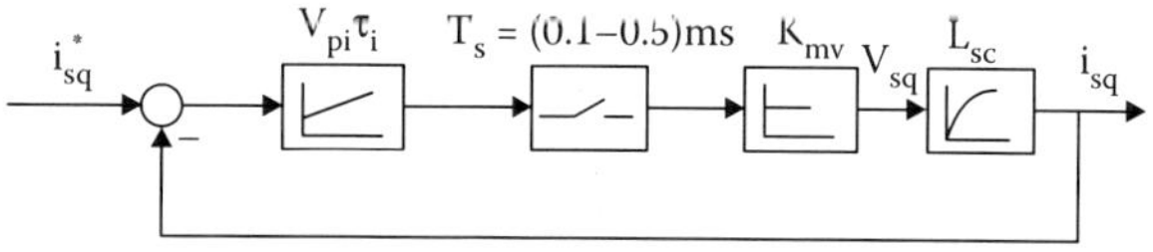


**FIGURE 9.69**
The d.c. (synchronous) current controller.

The same test is performed above the rated frequency for rated voltage to get the value of $i_{mr}(\omega_r)$.

Also, $L_m = \lambda_r/i_{mr}$ has to be found in the process. The flux controller (Figure 9.67) may be calibrated (Figure 9.70).

The mechanical time constant $\tau_m$ may be calculated from constant torque ($i_{mr}$, $i_q$ constant) rotor field orientation acceleration tests. Noting that the speed reached $\omega_a$ after the time $t_a$, the mechanical time constant is obtained from the motion equation

$$\frac{Jd\omega_r}{p_1 dt} = \frac{3}{2} p \frac{L_m^2}{L_r} i_{mr} i_q \tag{9.162}$$

$$J = \frac{3}{2} p^2 \frac{L_m^2}{L_r} i_{mr} i_q \frac{t_a}{\omega_a} \tag{9.163}$$

With $\tau_m$ from Equation 9.149 and J from Equation 9.163 we get

$$\tau_m = \frac{J}{pH} = \frac{1}{\tau_r} \frac{t_a}{\omega_a} \frac{i_q}{i_{mr}} \tag{9.164}$$

The speed controller may now be calibrated using standard methods.

The timing of self-commissioning operations may be summarized as in Figure 9.71 [47].

The whole process lasts about 60 s. After that the drive is ready for full performance operation. For sensorless operation slightly different techniques are required for self-commissioning.

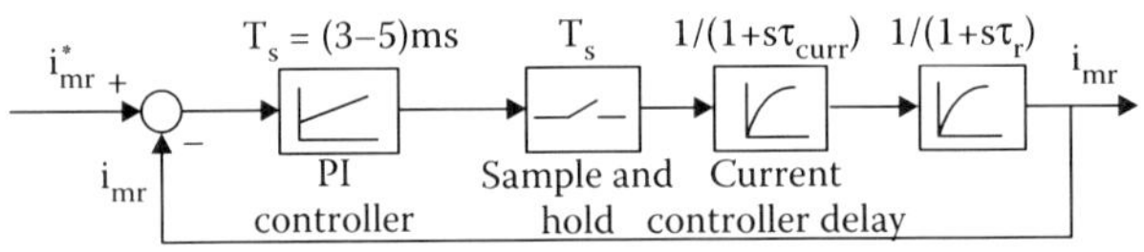


**FIGURE 9.70**
Rotor flux loop controller.

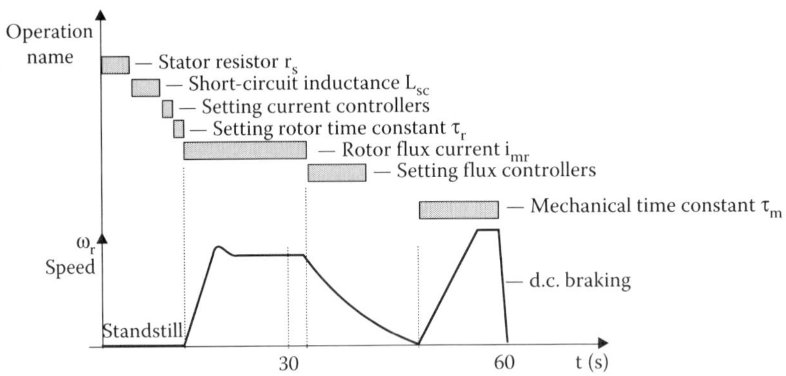


**FIGURE 9.71**
Self-commissioning timing.

In high-performance drives, the motor parameters are tuned through MRAC, online adaptation, or online estimators. These aspects are beyond the scope of this chapter [48,49]. Other intricate aspects of IM sensorless control are available in References 31 and 50.

## 9.20 Summary

IM drives may be used for torque, speed, or position control. This control may use motion sensors for high-performance response specifications or may be motion sensorless.

- Historically, $V_1/f_1$—coordinated voltage/frequency open-loop (scalar) control without motion sensors—has been used for moderate performance (up to 10/1 speed range).
- Fast torque dynamics, however, requires linearization of the torque speed curve and decoupled flux and torque control.
- VC, DTFC, or FLC may all provide for fast torque dynamics high performance in variable speed IMs.
- VC is performed with flux orientation and constant flux level so that fast torque control is obtained. Flux current and torque current controls have to be decoupled for the scope.
- Though general flux orientation is, in principle, possible, rotor flux orientation leads to simplified current decoupling network and stator flux orientation leads to simple voltage decoupling network.
- Motor parameter detuning influences on rotor flux orientation is severe in vector current control and, at least for torque control, rotor time constant correction is required.
- Flux orientation (vector) control may be achieved either indirectly (feedforward) or directly (feedback). In the latter case, flux estimators (observers) are required.
- The a.c. current controllers are surpassed by d.c. (synchronous, d-q) current controllers in terms of robustness to frequency and motor parameter detuning.
- Current control at low speeds and voltage control at high speeds are optimum in vector control (VC).
- Extending the torque/speed envelope is performed through flux weakening and has two regions above base speed $\omega_b$. One is the constant power zone when the d-q current (flux) angle (and slip frequency) varies up to the maximum torque/flux angle. In region two, the slip frequency is constant at $(S\omega_1)_k = 1/\tau_r$ but the power is not longer constant.
- Electric drives are current limited at low speeds and voltage (flux) limited at high speeds.
- Wide constant power zones ($\omega_{1max}/\omega_b > 2$) are obtained with low transient inductance $L_{sc}$ motors.
- Voltage or current control according to their command values is performed through PWM techniques in the voltage-source inverter. They may be open loop or closed loop.

- With open-loop PWM, the lockout time $T_d$ between the turn-off of one PES and the turn-on of the next along the same inverter leg has to be compensated to avoid voltage (current) distortion in low-frequency (speed) operation.
- In direct VC, closed-loop flux observers are required. Full-order or simplified observers are used. Up to 100 to 1 speed range, for fast torque response (1–5 ms for step-rated torque at zero speed) has been achieved with recent commercial drives.
- Direct torque and flux control (DTFC) means direct stator flux and torque error control to trigger a table of optimal switching in the PWM voltage-source inverter.
- DTFC is a kind of combined current and voltage VC in a simpler and more robust configuration but maintaining high steady state and transient performance.
- Scalar open-loop ($V_1/f_1$) control is prone to instability regions in the $V_1/f_1$ plane, especially with two-pole motors.
- Compensating the slip frequency in ($V_1/f_1$) drives is the typical solution to improve speed control precision at low speeds. Still a less than 15 to 1 speed control ratio is available with low torque dynamics.
- Self-commissioning means parameter estimation and subsequent closed-loops calibration by the drive itself at the site of application. It is an essential asset of modern electric drives [47]. Typically self-commissioning takes 1–2 min only. After that the drive is ready to run at full performance.

## 9.21 Problems

9.1 In a rotor flux orientation current decoupling network, calculate the torque and flux detuning effects for cases a and b.

a. $L_m/L_m^* = \beta = 0.812$. $S\omega_1\tau_r = 0.3$ and 0.5 for rated rotor time constant $\tau_r^* = 40\,\text{ms}$.

b. For a $\pm 50\%$ variation in the rotor resistance, $S\omega_1\tau_r = 0.5$ and no saturation influence ($\beta = 1$).

c. Based on the rotor flux detuning monotonic function (Figure 9.4) and the rotor flux error, using a rotor flux estimator, introduce a PI controller to correct the rotor time constant $\tau_r$ in the rotor flux orientation current decoupling network (Figure 9.4) and draw the corresponding signal flow diagram for this case.

9.2 A VC IM drive with rotor flux orientation operates in the flux-weakening mode. The magnetizing inductance $L_m$ of the motor varies linearly with the flux current $i_d$ in rotor flux coordinates:

$$L_m = 0.100\,\text{H}, \quad 0 < i_d \le 3\,\text{A};$$

$$L_m = 0.10 - 0.015 i_d/3, \quad \text{for } i_d \ge 1.5\,\text{A}$$

The leakage inductance $L_{ls} = L_{lr} = 0.005$ H, $r_r = r_s = 0.35\ \Omega$, rated phase voltage $V_{1n} = 120$ V (rms per phase), $p = 2$, $\omega_{1b} = 2\pi 60$ rad/s, $(S\omega_1)_{rated} = 3\pi$ rad/s.

Calculate

- The no-load phase current and the rated current
- The rated stator, rotor, and airgap flux motor amplitude
- Rated electromagnetic torque $T_{en}$
- Maximum torque at $2\omega_b$

9.3 As related to Section 9.17, an IM has the data p = 2 pole pairs, $\omega_1 = 2\pi 60$, $\lambda_r = 1$ Wb, $r_r = 0.4\ \Omega$, J = 0.1 kg $m^2$, $\tau_r = 0.1$ s, $S_n = 0.01$. For constant rotor flux and slow transients, calculate the speed response for a 10 Nm step increase in the load torque.

## References


1. F. Blaschke, The principle of field orientation as applied to the new transvector closed-loop control system for rotating field machines, *Siemens Rev.*, vol. 34, 1972, pp. 217–220. [In German.]
2. I. Takahashi, T. Noguchi, A new quick response and high efficiency control strategy of an induction motor, *Record of IEEE-IAS-1985, Annual Meeting*, pp. 496–502 (IEEExplore).
3. M. Depenbrock, Direct self-control (DSC) of inverter-fed induction machine, *IEEE Trans.*, vol. PE-3, no. 4, 1988, pp. 420–429.
4. I. Boldea, S.A. Nasar, Torque vector control (TVC) a class of fast and robust torque, speed and position digital controllers for electric drives, *EMPS*, vol. 15, 1988, pp. 135–148.
5. R. Marino, R. Peresada, P. Valigi, Adaptive partial feedback linearization of induction motors, *Proceedings of the 29th Conference on Decision and Control*, Honolulu, Hawaii, 1990, pp. 3313–3318 (IEEExplore).
6. M. Bodson, J. Chiarsson, R. Novotnak, High performance induction motor control via input-output linearization, *IEEE Control Syst.*, 1994, pp. 25–33.
7. D.L. Sobczuk, Nonlinear control of PWM inverter fed induction motor drives, *Record of ISIE '96*, Warsaw, Poland, vol. 2, pp. 958–962 (IEEExplore).
8. R. De Doncker, D.W. Novotny, The universal field oriented controller, *Record of IEEE-IAS, Annual Meeting*, 1988, pp. 450–456 (IEEExplore).
9. I. Boldea, S.A. Nasar, *Vector Control of A.C. Drives*, Chapter 3, CRC Press, Boca Raton, Florida, 1992.
10. R. Krishnan, F.C. Doran, Study of parameter sensitivity in high performance inverter-fed induction motor drives, *Record of IEEE-IAS, Annual Meeting*, 1984, pp. 510–524 (IEEExplore).
11. T.R. Rowan, R.J. Kerkman, A new synchronous current regulator and an analysis of current-regulated PWM inverters, *IEEE Trans.*, vol. IA-22, 1986, pp. 678–690.
12. S.H. Kim, S.K. Sul, Maximum torque control of an induction machine in the field weakening region, *IEEE Trans.*, vol. IA-31, no. 4, 1995, pp. 784–794.
13. R. Krishnan, Review of flux weakening in high performance vector controlled induction motor drives, *Record of ISIE'96*, Warsaw Poland, vol. 2, pp. 917–922 (IEEExplore).
14. H. Grotstollen, A. Bünte, Control of induction motor with orientation on rotor flux or on stator flux in a very wide field weakening region—Experimental results, *Record of ISIE'96*, Warsaw Poland, vol. 2, pp. 911–916 (IEEExplore).
15. J. Holtz, Pulsewidth modulation for electronic power conversion, *Proc. IEEE*, vol. 82, no. 8, 1994, pp. 1194–1214.
16. D. Legatte, R.J. Kerkman, Pulse-based dead-time compensation for PWM voltage inverters, *IEEE Trans.*, vol. IE-44, no. 2, 1997, pp. 191–197.
17. J. Holtz, E. Bube, Field oriented asynchronous PWM for high performance a.c. machine drives operating at low switching frequency, *IEEE Trans.*, vol. IA-27, no. 3, 1991, pp. 574–581.

18. A.M. Trzynadlowski, An overview of modern PWM techniques for three phase-voltage-controlled voltage-source inverters, *Record of ISIE-96*, Warsaw, Poland, vol. 1, pp. 25–39 (IEEExplore).
19. A. Nabae, S. Ogasawara, H. Akagi, A novel control scheme for current controlled PWM inverters, *IEEE Trans.*, vol. IA-22, 1986, pp. 697–701.
20. P.L. Jansen, R.D. Lorenz, D.W. Novotny, Observer based direct filed orientation and comparison of alternative methods, *IEEE Trans.*, vol. IA-30, no. 4, 1994, pp. 945–953.
21. L. Kwindler, J.C. Moreira, A. Testa, T.A. Lipo, Direct field orientation controller using the stator phase voltage third harmonics, *IEEE Trans.*, vol. IA-30, no. 2, 1994, pp. 441–447.
22. Y. Hori, T. Umeno, Implementation of robust flux observer based field orientation controller for induction machines, *Record of IEEE-IAS, Annual Meeting*, 1989, pp. 523–528 (IEEExplore).
23. J. Holtz, Methods for speed sensorless control of a.c. drives, *Record of IEEE-PCC-Yokohama*, 1993, pp. 415–420 (IEEExplore).
24. C. Has, A. Betini, L. Ferraris, G. Griva, F. Profumo, Comparison of different schemes without shaft sensors for field oriented control drives, *Record of IEEE-IECON*, 1994, pp. 1579–1588 (IEEExplore).
25. F.Z. Peng, T. Fukao, J.S. Lai, Low speed performance of robust speed identification using instantaneous reactive power for tacholess vector control of induction motors, *Record of IEEE-IAS, Annual Meeting*, 1994, pp. 509–514 (IEEExplore).
26. H. Kubota, K. Matsuse, Speed sensorless, field oriented control of induction motor with rotor resistance adaption, *IEEE Trans.*, vol. IA-30, no. 5, 1994, pp. 1219–1224.
27. S.I. Yong, J.W. Choi, S.K. Sul, Sensorless vector control of induction machine using high frequency current injection, *Record of IEEE-IAS, Annual Meeting*, 1994 (IEEExplore).
28. Y.R. Kim, S.K. Sul, M.H. Park, Speed sensorless control of an induction motor using an extended Kalman Filter, *Record of IEEE-IAS, Annual Meeting*, 1992, Part I, pp. 594–599 (IEEExplore).
29. K.D. Hurst, T.G. Habetler, G. Griva, F. Profumo, Speed sensorless field-oriented control of induction machines using current harmonics spectral estimation, *Record of IEEE-IAS, Annual Meeting*, 1994 (IEEExplore).
30. J. Holtz, Sensorless control of induction machines—with and without signal injection, Keynote address at *OPTIM*, 2002, Poiana Brasov, Romania (IEEExplore).
31. L. Harnefors, M. Hinkkanen, Complete stability of reduced-order and full-order observers for sensorless IM drives, *IEEE Trans.*, vol. IE-55, no. 3, 2008, pp. 1319–1329.
32. L. Harnefors, Globally stable speed-adaptive observers for sensorless induction motor drives, *IEEE Trans.*, vol. IE-54, no. 2, 2007, pp. 1243–1245.
33. K. Wang, W. Yao, K. Lee, Z. Lu, Regenerating mode stability improvements of combined voltage and current model flux observer in speed-sensorless IM control, *IEEE Trans.*, vol. IA-50, no. 4, 2014, pp. 2564–2573.
34. V. Verma, C. Chakraborty, S. Maiti, Y. Hori, Speed sensorless vector controlled IM drive using single current sensor, *IEEE Trans.*, vol. EC-28, no. 4, 2013, pp. 938–950.
35. P. Tiitinen, P. Pohyalainen, J. Lalu, The next generation motor control method-direct torque control, DTC, *Proceedings of the EPE Chapter Symposium*, Lausanne, Switzerland, 1994.
36. M.P. Kazmierkowski, Control philosophies of PWM inverter-fed induction motors, University of Aalborg lecture, February, 1997, DK.
37. C. Lascu, Direct torque control of sensorless IM drives, PhD thesis, 2002, University Politehnica Timisoara, Romania.
38. C. Lascu, I. Boldea, F. Blaabjerg, Variable-structure direct torque control—A class of fast and robust controllers for induction machine drives, *IEEE Trans.*, vol. IE-51, 2004, pp. 785–792.
39. D. Casadei, F. Profumo, G. Serra, A. Tani, FOC and DTC: Two viable schemes for induction motors torque control, *IEEE Trans.*, vol PE-17, 2002, pp. 779–787.
40. C. Lascu, I. Boldea, F. Blaabjerg, Direct torque control of sensorless induction motor drives—A sliding mode approach, *IEEE Trans.*, vol. IA-40, no. 2, 2004, pp. 582–590.
41. I. Boldea, A. Trica, Torque vector control (TVC) voltage-fed induction motor drives—Very low speed performance via sliding mode, *Record of ICEM*, 1990, vol. 3, pp. 1212–1217 (IEEExplore).

42. J. Holtz, J. Quan, Sensorless vector control of induction motors at very low speed using a nonlinear inverter model and parameters identification, *IEEE Trans.*, vol. IA-38, no. 4, 2002, pp. 1087–1095.
43. H. Abu-Rub, M. Malinowski, K. Al-Haddad, Editors, *Power Electronics for Renewable Energy Systems, Transportation and Industrial Applications*, Chapter 11, IEEE-Wiley Press, 2014.
44. R. Ueda, T. Sonada, K. Koga, M. Ichikawa, Stability analysis in induction motor driven V/f controlled general-purpose inverter, *IEEE Trans.*, vol. IA-28, no. 2, 1992, pp. 472–481.
45. K. Koga, R. Ueda, T. Sonada, Constitution of V/f control for reducing the steady-state error to zero in induction motor drive system, *IEEE Trans.*, vol. IA-28, no. 2, 1992, pp. 463–471.
46. R.J. Kerkman, J.D. Thunes, T.W. Rowan, D.W. Schlegel, A frequency based determination of transient inductance and rotor resistance for field commissioning purposes. *IEEE Trans.*, vol. IA-32, no. 3, 1996, pp. 577–584.
47. A.M. Klambadkone, J. Holtz, Vector controlled induction motor drive with a selfcommissioning scheme, *IEEE Trans.*, vol. IE-38, no. 5, 1991, pp. 322–327.
48. M. Ruff, A. Bünte, H. Grotstolen, A new selfcommissioning scheme for an asynchronous motor drive system, *Record of IEEE-IAS, Annual Meeting*, 1994, Part I, pp. 616–623 (IEEExplore).
49. R.J. Kerkman et al., A new flux and stator resistance identifier for a.c. drives, *IEEE Trans.*, vol. IA-32, no. 3, 1996, pp. 585–593.
50. J. Ruan, S. Wang, A prediction error self-commissioning scheme for parameter identification of induction motors in sensorless drives, *IEEE Trans.*, vol. EC-30, no. 1, 2015, pp. 384–393.

# 10

# Synchronous Motors for Drives

## 10.1 Introduction

Synchronous motors (SMs) are, in general, three-phase a.c. fed in the stator and d.c. (or PM) excited in the rotor. As the stator currents produce an m.m.f. traveling at the electric speed $\omega_1$

$$\omega_1 = 2\pi f_1 \tag{10.1}$$

the rotor m.m.f. (or PM) is fixed to the rotor. The rotor electrical speed $\omega_r$ is

$$\omega_r = \omega_1 = 2\pi np \tag{10.2}$$

in order to obtain two m.m.f. waves at standstill with each other. It is a known fact that only in this situation is a nonzero average torque per revolution obtained.

Yet, in an alternative interpretation, torque is produced when the magnetic coenergy in the machine $W_{co}$ varies with rotor position

$$T_e - \left(\frac{\partial W_{co}}{\partial \theta_r}\right)_{i_i=ct}; \quad \frac{d\theta_r}{dt} = \frac{\omega_r}{p} \tag{10.3}$$

where $\theta_r$ is the geometrical rotor position angle.

Thus a magnetically anisotropic (reluctance)—exciterless—rotor may also be used. In all cases the number of pole pairs is the same on the stator and on the rotor.

PM or reluctance rotors are preferred, in general, for low- (even medium-) power motors (up to 50–500 kW in general) and d.c. excited rotors are used for medium- and high-power (hundreds of kW, MW, and tens of MWs) applications.

As the SM speed is rigidly related to the stator frequency, only the development of PECs—variable voltage and frequency sources—has made the SM suitable for variable speed drives.

The higher efficiency, power density, and power levels per unit have thus become the main assets of variable speed SM drives.

## 10.2 Construction Aspects

As with any electrical motor, the SM has a stator and a rotor. The stator is made of a laminated core with uniform slotting. The stator slots accommodate, in general, a three-phase, single- or double-layer winding made like for IMs.

In general, the number of slots per pole and phase $q \geq 2$ and thus the coupling inductances vary sinusoidally with rotor position. However, in small-power PMSMs, concentrated windings may be used ($q = 1$ slot per pole per phase) when the sinusoidality is lost [1,2].

On the other hand, the SM rotors may be

- Active
  - Excited
  - With PMs
- Passive
  - Highly anisotropic (variable reluctance)

The excited and PM rotors may be with salient or nonsalient poles that are magnetically anisotropic or isotropic.

Also d.c. excited-rotor SMs may have a squirrel cage on the rotor to reduce the commutation inductance in current-source PEC variable speed drives.

Figure 10.1 shows salient pole SMs with d.c. excited and PM rotors, while Figure 10.2 exhibits nonsalient pole SMs.

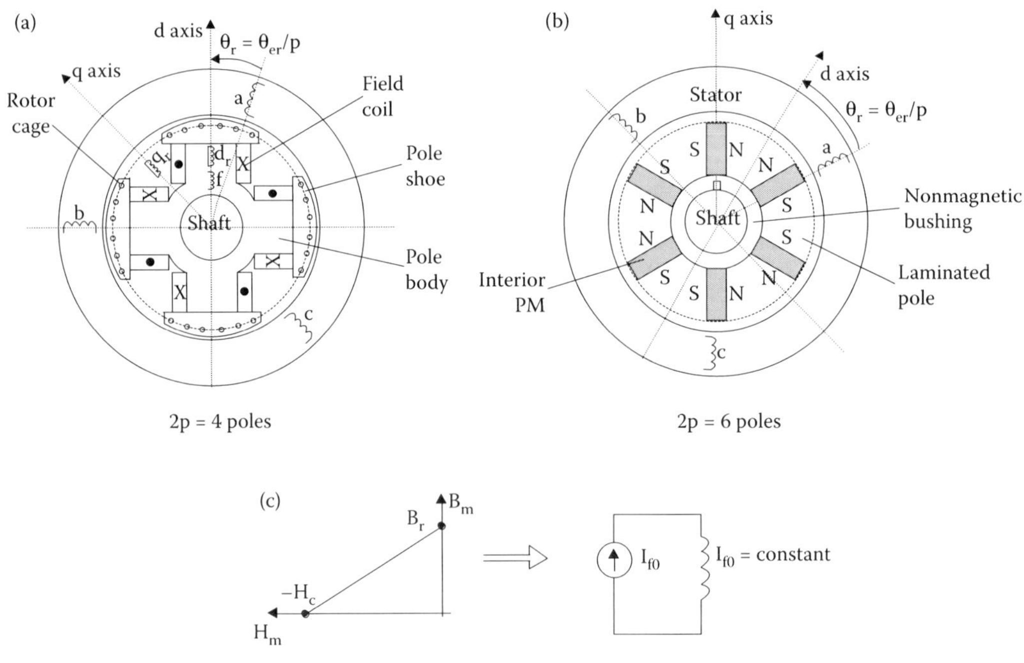


**FIGURE 10.1**
Salient pole active rotor SMs with (a) d.c. excited rotor, (b) interior PMs, (c) PM curve.

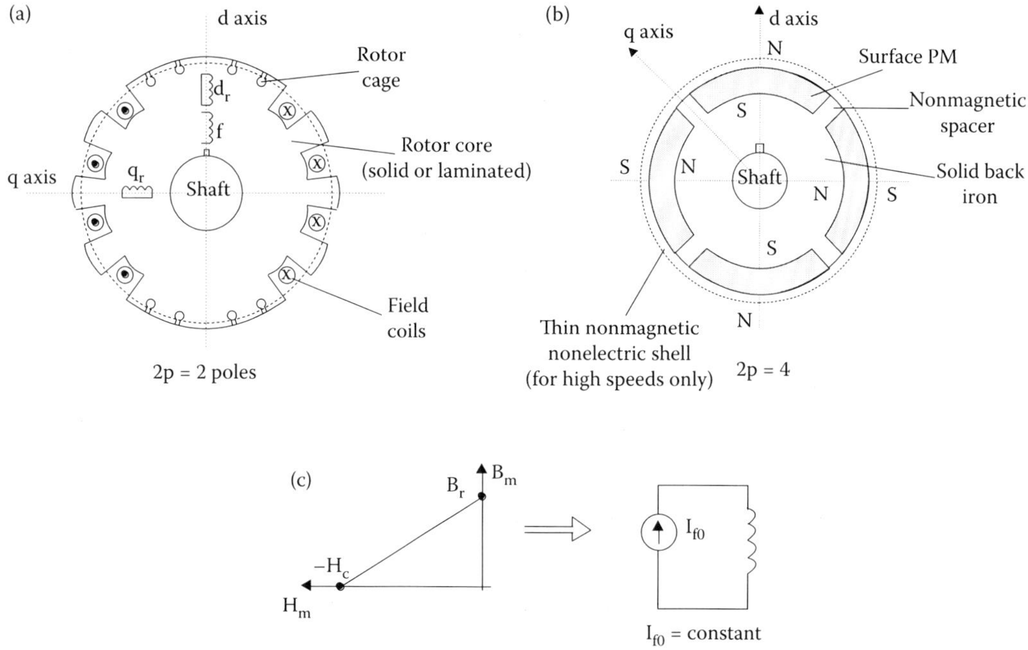


**FIGURE 10.2**
Nonsalient pole active rotors of SMs: (a) d.c. excited, (b) with surface PM, (c) PM B–H curve.

The high-energy PMs have a rather linear demagnetization characteristic in the second quadrant (Figure 10.1c) and thus such PMs may be considered as constant current ideal field coils or artificial superconducting coils at room temperature.

Once magnetized with a few high current pulses (milliseconds long), capable to produce (2–3) $H_c$ and (2–3) $B_r$ (see Figure 10.1), the magnetic energy stored in the PMs stays intact for years if accidental demagnetization through ultrahigh stator currents or overtemperature are avoided.

The energy required to magnetize the PMs is much smaller than the power losses in an equivalent field winding over the active life of the PMs (4–5 years or more). However, the cost of PMs is notably higher than the field winding costs. If the capitalized cost of losses is considered, however, the extra cost of PMs is paid back in less than 2–3 years.

While the nonsalient pole configuration is more suitable for high peripheral speeds for d.c. excited rotors, with PM rotors the opposite is true; that is, interior PMs are, in general, preferred for high speed as PM "protection" against large centrifugal and radial electromagnetic forces is implicitly provided. Radial—attraction—forces between the PM rotor and the stator core are exerted through the surface PMs and through the laminated poles (for the interior PM rotors). Note that all PMs are rather severely temperature limited (100°C for NdFeB, 150°C for $SmCo_5$, etc.) and thus, in adverse environments, the PMs constitute a liability.

Passive rotors are characterized by high magnetic anisotropy (or saliency). They may be made of conventional or of axial laminations (Figure 10.3a and b). Multiple flux barriers are provided to increase the saliency for the conventional rotor (Figure 10.3a). The same scope is served easily by alternating axial laminations with insulation layers (Figure 10.3b).

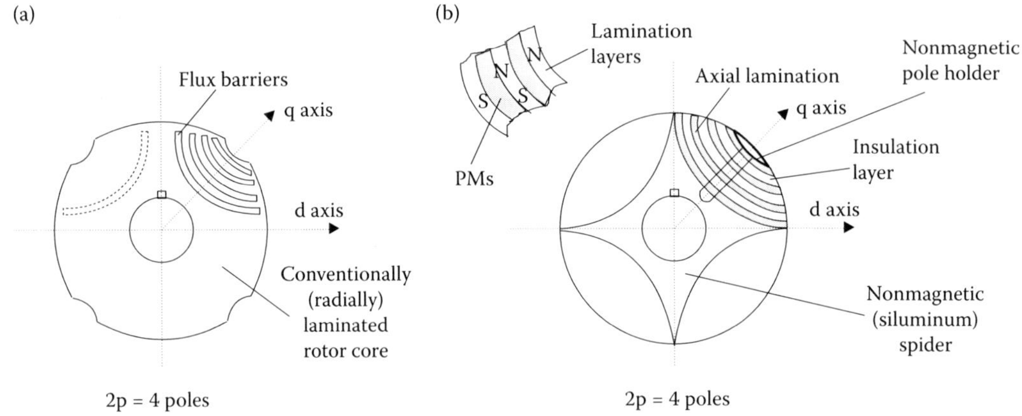


**FIGURE 10.3**
Passive-anisotropic (reluctance)-rotors for SMs: (a) multiple flux barrier rotor, (b) axially laminated-anisotropic (ALA) rotor.

Special measures need to be taken to reduce the rotor core harmonics losses in the axially laminated anisotropic rotor [3].

Saliency ratios (for rated conditions) above 10 to 1 have been obtained, thus pushing the performance of RSMs to the level of IMs [3]. Obviously, the absence of the rotor cage makes both PMSMs and RSMs fully dependent on PECs.

The insulation layers may be replaced by PMs along q axis (Figure 10.3b), in order to improve the torque density and the power factor. Also, in the presence of a rotor cage on both PM rotors or variable reluctance rotors, the same motors may be line-started and could operate at constant speed for constant voltage–frequency conditions.

## 10.3 Pulsating Torque

It is a known fact that an ideal SM—with sinusoidal m.m.f. and constant airgap—when fed with sinusoidal symmetric currents in the stator at frequency $\omega_1 = \omega_r$ ($\omega_r$—rotor speed)—produces a constant torque.

In reality, pulsating torques may occur due to

a. Stator (and rotor) slot openings
b. Magnetic saturation caused by flux harmonics
c. Current waveforms
d. PM field pulsations due to stator slot openings (cogging torque)

Items *a* to *c* cause the so-called electromagnetic pulsating torques while *d* causes the zero stator current or cogging torque.

Rotor pole (or PM) span correlation with stator slot openings, stator slot inclination, or PM pole inclination, fractional q (slots per pole and phase) and, finally, special current waveform shaping through PEC control are all methods to reduce these, basically reluctance, parasitic, torques [4] to less than 1% of rated torque. High-performance smooth torque drives capable of operating below 20 rpm in sensorless control or under 1 rpm with position sensor control are thus obtained.

NOTE: Pulsating torque investigation requires, in most cases, FEM analysis—two, quasi-two, or three-dimensional [4].

## 10.4 Phase Coordinate Model

The phase coordinate model is based on phase equations using stator and rotor circuits. For $q \geq 2$ (slots per pole per phase)—distributed windings—the inductance matrix contains sinusoidal terms. We will take the case of the salient pole rotor as the nonsalient pole rotor is a particular case of the former.

The phase inductance matrix [L] is (see Figures 10.1 and 10.2)

$$[L] = \begin{bmatrix} & a & b & c & f & d_r & q_r \\ a & L_{aa} & L_{ab} & L_{ac} & L_{af} & L_{ad_r} & L_{aq_r} \\ b & L_{ab} & L_{bb} & L_{bc} & L_{bf} & L_{bd_r} & L_{bq_r} \\ c & L_{ac} & L_{bc} & L_{cc} & L_{cf} & L_{cd_r} & L_{cq_r} \\ f & L_{af} & L_{bf} & L_{cf} & L_{ff} & L_{fd_r} & 0 \\ d_r & L_{ad_r} & L_{bd_r} & L_{cd_r} & L_{fd_r} & L_{d_rdr} & 0 \\ q_r & L_{aq_r} & L_{bq_r} & L_{cq_r} & 0 & 0 & L_{q_rq_r} \end{bmatrix} \tag{10.4}$$

For distributed windings ($q \geq 2$) all the stator self, mutual, and stator-rotor inductances are rotor position, $\theta_{er}$, dependent

$$\begin{aligned} L_{aa} &= L_{sl} + L_0 + L_2 \cos 2\theta_{er} \\ L_{bb} &= L_{sl} + L_0 + L_2 \cos\left(2\theta_{er} + \frac{2\pi}{3}\right) \\ L_{cc} &= L_{sl} + L_0 + L_2 \cos\left(2\theta_{er} - \frac{2\pi}{3}\right) \\ L_{bc} &= -L_0/2 + L_2 \cos 2\theta_{er} \\ L_{ac} &= -L_0/2 + L_2 \cos\left(2\theta_{er} + \frac{2\pi}{3}\right) \\ L_{ab} &= -L_0/2 + L_2 \cos\left(2\theta_{er} - \frac{2\pi}{3}\right) \end{aligned} \tag{10.5}$$

$$L_{ad_r} = L_{sd_r}\cos\theta_{er},\ L_{bd_r} = L_{sd_r}\cos\left(\theta_{er} - \frac{2\pi}{3}\right),$$

$$L_{cd_r} = L_{sd_r}\cos\left(\theta_{er} + \frac{2\pi}{3}\right)$$

$$L_{af} = L_{sf}\cos\theta_{er},\ L_{bf} = L_{sf}\cos\left(\theta_{er} - \frac{2\pi}{3}\right),$$

$$L_{cf} = L_{sf}\cos\left(\theta_{er} + \frac{2\pi}{3}\right)$$

$$L_{aq_r} = -L_{sq_r}\sin\theta_{er},\ L_{bq_r} = -L_{sq_r}\sin\left(\theta_{er} - \frac{2\pi}{3}\right),$$

$$L_{cq_r} = -L_{sq_r}\sin\left(\theta_{er} + \frac{2\pi}{3}\right)$$

Rotor inductances are evidently independent of rotor position as the saliency is on the rotor itself

$$\begin{aligned} L_{ff} &= L_{fl} + L_{fm} \\ L_{d_rd_r} &= L_{d_rl} + L_{d_rm},\ L_{q_rq_r} = L_{q_rl} + L_{q_rm} \end{aligned} \tag{10.6}$$

where $L_{fl}, L_{d_rl}, L_{q_rl}$ are leakage inductances while the others are related to main flux path.

On the other hand, for *concentrated* windings ($q = 1$), when nonsalient pole PM rotors are used, all stator inductances are independent of rotor position, the rotor cage is eliminated.

Only the motion-related inductances between the constant field-current PM equivalent circuit and the stator windings depend on rotor position.

$$[L]_{\substack{PM \\ q=1}} = \begin{bmatrix} L_s & L_{ab} & L_{ab} & L_{af}(\theta_{er}) \\ L_{ab} & L_s & L_{ab} & L_{bf}(\theta_{er}) \\ L_{ab} & L_{ab} & L_s & L_{cf}(\theta_{er}) \\ L_{af}(\theta_{er}) & L_{bf}(\theta_{er}) & L_{cf}(\theta_{er}) & 0 \end{bmatrix} \tag{10.7}$$

The stator self-inductance $L_s$ and the stator-stator mutual inductances $L_{ab}$ are all equal to each other but different from $-L_0/2$ (as they would become for nonsalient rotors [$L_2 = 0$]). To a first approximation, $L_{ab} = -L_0/3$ [1].

The voltage–current matrix equations in phase coordinates (stator coordinates for stator, rotor coordinates for rotor), for the PM machine with no rotor cage, are

$$[V] = [r]\cdot[i] + \frac{d[\lambda]}{dt} \tag{10.8}$$

with

$$[\lambda] = [L(\theta_{er})] \cdot [i] \tag{10.9}$$

$$[i] = [i_a, i_b, i_c, i_{f0}] \tag{10.10}$$

$$[r] = \text{Diag}\left[r_s, r_s, r_s, 0\right] \tag{10.11}$$

$$[\lambda] = [\lambda_a, \lambda_b, \lambda_c, 0] \tag{10.12}$$

Finally, the electromagnetic torque $T_e$ may be calculated from the coenergy derivative with respect to rotor position

$$T_e = \frac{dW_{co}}{d(\theta_{er})} = \frac{d}{d(\theta_{er})} \int_0^{[i]} [\lambda]\, d[i]^T \tag{10.13}$$

After neglecting magnetic saturation, we multiply Equation 10.8 by $[i]^T$

$$\begin{aligned} [i]^T \cdot [V] &= [r] \cdot [i] \cdot [i]^T + \frac{d}{dt}\left(\frac{1}{2}[L] \cdot [i] \cdot [i]^T\right) \\ &+ \frac{1}{2} \cdot [i]^T \cdot \frac{\partial}{\partial(\theta_{er})}[L(\theta_{er})] \cdot [i] \cdot \frac{d(\theta_{er})}{dt} \end{aligned} \tag{10.14}$$

The last term is the electromagnetic power $P_{elm}$

$$T_e = \frac{P_{elm} \cdot p}{\omega_r} = \frac{p}{2} \cdot [i]^T \cdot \frac{\partial}{\partial \theta_{er}}[L(\theta_{er})] \cdot [i] \tag{10.15}$$

The motion equations are

$$\frac{J}{p}\frac{d\omega_r}{dt} = T_e - T_{load}; \quad \frac{d(\theta_{er})}{dt} = \frac{\omega_r}{p} \tag{10.16}$$

As for the IM, for the general synchronous machine (with excitation and cage in the rotor) we end up with a set of eight nonlinear differential equations with time-varying coefficients (basically, inductances). As such, these equations are used for special cases for machines with some asymmetry or for unbalanced supply voltage operation.

Also, for the concentrated stator winding ($q = 1$) and the PM nonsalient cageless rotor, the phase variable model is the only one to be used as $L_{af}(\theta_{er})$, $L_{bf}(\theta_{er})$, and $L_{cf}(\theta_{er})$ are far from sinusoidal functions. In order to get rid of the inductance dependence on rotor position for $q \geq 2$, the space-phasor (d-q) model is used.

## 10.5 Space-Phasor (d-q) Model

Proceeding as for the IM, we may define the stator current space-phasor, $\bar{i}_s$ in stator coordinates

$$\bar{i}_s^s = \frac{2}{3}\cdot(i_a + a\cdot i_b + a^2\cdot i_c) \tag{10.17}$$

On the other hand, from Equation 10.9 with Equation 10.4, the phase a flux linkage $\lambda_a$ is

$$\lambda_a = L_{aa}i_a + L_{ab}i_b + L_{ac}i_c + L_{af}i_f + L_{ad_r}i_{d_r} + L_{aq_r}i_{q_r} \tag{10.18}$$

Making use of the inductance definition of Equation 10.5 we find

$$\begin{aligned}\lambda_a &= L_{sl}\,\mathrm{Re}(\bar{i}_s) + \frac{3}{2}L_0\,\mathrm{Re}(\bar{i}_s) + \frac{3}{2}L_2\,\mathrm{Re}\left(\bar{i}_s^{*}e^{2j\theta_{er}}\right)\\ &+L_{af}\,\mathrm{Re}\left(i_f^r e^{j\theta_{er}}\right) + L_{sd_r}\,\mathrm{Re}\left(i_{d_r}^r e^{j\theta_{er}}\right) - \mathrm{Re}\left(jL_{sq_r}^r i_{q_r} e^{j\theta_{er}}\right)\end{aligned} \tag{10.19}$$

The stator flux space-phasor $\bar{\lambda}_s$ in stator coordinates is

$$\bar{\lambda}_s^s = \frac{2}{3}(\lambda_a + a\lambda_b + a^2\lambda_c) \tag{10.20}$$

Making use of Equation 10.18 for $\lambda_b$, $\lambda_c$, Equation 10.20 yields

$$\begin{aligned}\bar{\lambda}_s^s &= L_{sl}\,\bar{i}_s^s + \frac{3}{2}L_0\,\bar{i}_s^s + \frac{3}{2}L_2\,\bar{i}_s^{s*}e^{2j\theta_{er}}\\ &+ L_{sf}i_f^r e^{j\theta_{er}} + L_{sd_r}i_{d_r}^r e^{j\theta_{er}} + L_{sq_r}ji_{q_r}^r e^{j\theta_{er}}\end{aligned} \tag{10.21}$$

Multiplying Equation 10.21 by $e^{-j\theta_{er}}$ we obtain

$$\begin{aligned}\bar{\lambda}_s^s e^{-j\theta_{er}} &= \left(L_{sl} + \frac{3}{2}L_0\right)\bar{i}_s^s e^{-j\theta_{er}} + \frac{3}{2}L_2\left(\bar{i}_s^s e^{-j\theta_{er}}\right)^{*}\\ &+ L_{sf}i_f^r + L_{sd_r}i_{d_r}^r + jL_{sq_r}i_{q_r}^r\end{aligned} \tag{10.22}$$

We should now note that

$$\bar{\lambda}_s = \bar{\lambda}_s^s e^{-j\theta_{er}};\quad \bar{i}_s = \bar{i}_s^s e^{-j\theta_{er}} \tag{10.23}$$

are space-phasors in rotor coordinates (aligned with rotor d axis).

With Equation 10.23, Equation 10.22 becomes

$$\bar{\lambda}_s = \left(L_{sl} + \frac{3}{2}L_0\right)\bar{i}_s + \frac{3}{2}L_2\bar{i}_s^* + L_{sf}i_f^r + L_{sd_r}i_{d_r}^r + jL_{sq_r}i_{qr}^r \tag{10.24}$$

Next, the stator-phase equations in stator coordinates are

$$\begin{aligned} r_s i_a - V_a &= -\frac{d\lambda_a}{dt} \\ r_s i_b - V_b &= -\frac{d\lambda_b}{dt} \\ r_s i_c - V_c &= -\frac{d\lambda_c}{dt} \end{aligned} \tag{10.25}$$

We may translate them in space-phasors as

$$r_s\bar{i}_s^s - \bar{V}_s^s = -\frac{d\bar{\lambda}_s^s}{dt} = -\frac{d}{dt}\left(\bar{\lambda}_s e^{j\theta_{er}}\right) = -e^{j\theta_{er}}\frac{d\bar{\lambda}_s}{dt} - j\omega_r\bar{\lambda}_s e^{j\theta_{er}} \tag{10.26}$$

The final form of Equation 10.26 is

$$r_s\bar{i}_s - \bar{V}_s = -\frac{d\bar{\lambda}_s}{dt} - j\omega_r\bar{\lambda}_s; \quad \omega_r = \frac{d\theta_{er}}{dt} \tag{10.27}$$

and

$$V_0 = r_s i_0 + L_{sl}\frac{di_0}{dt}; \quad i_0 = \frac{i_a + i_b + i_c}{3} \tag{10.28}$$

Equation 10.27 is, in fact, identical to that for the induction machine. Only the flux expression (10.24) is different.

$$\bar{\lambda}_s = \lambda_d + j\lambda_q; \ \lambda_d = L_{sl}i_d + \lambda_{dm}; \ \lambda_{dm} = L_{dm}i_d^r + L_{sf}i_f^r + L_{sd}i_{d_r}^r \tag{10.29}$$

$$\lambda_q = L_{sl}i_q + \lambda_{qm}; \quad \lambda_{qm} = L_{qm}i_q + L_{sq}i_{q_r}^r \tag{10.30}$$

with

$$\begin{aligned} L_{dm} &= \frac{3}{2}(L_0 + L_2) \\ L_{qm} &= \frac{3}{2}(L_0 - L_2) \end{aligned} \tag{10.31}$$

where $L_{dm}$ and $L_{qm}$ are called the d-q magnetizing inductances.

We may reduce the rotor to stator currents

$$\frac{i_f}{i_f^r} = \frac{L_{sf}}{L_{dm}} = K_f; \quad \frac{i_{d_r}}{i_{d_r}^r} = \frac{L_{sd}}{L_{dm}} = K_d; \quad \frac{i_{q_r}^r}{i_{d_r}^r} = \frac{L_{sq}}{L_{dm}} = K_q \tag{10.32}$$

$$\begin{aligned} &\lambda_{dm} = L_{dm} i_{dm}; \quad i_{dm} = i_d + i_f + i_{d_r} \\ &\lambda_{qm} = L_{qm} i_{qm}; \quad i_{qm} = i_q + i_{q_r}; \quad i_m = \sqrt{i_{dm}^2 + i_{qm}^2} \end{aligned} \tag{10.33}$$

Magnetic saturation may be determined by unique $\lambda_{dm}(i_m)$ and $\lambda_{qm}(i_m)$ functions, to be either calculated or measured [5].

The stator equation (10.27) in d-q coordinates becomes

$$V_d = r_s i_d + \frac{d\lambda_d}{dt} - \omega_r \lambda_q \tag{10.34}$$

$$V_q = r_s i_q + \frac{d\lambda_q}{dt} + \omega_r \lambda_d \tag{10.35}$$

Now we should add the rotor equations in rotor d-q coordinates (as along axes d and q, the rotor windings are not at all symmetric)

$$V_f = r_f i_f + \frac{d\lambda_f}{dt}; \quad \lambda_f = L_{fl} i_f + \lambda_{dm} \tag{10.36}$$

$$0 = r_{d_r} i_{d_r} + \frac{d\lambda_{d_r}}{dt}; \quad \lambda_{d_r} = L_{d_r l} i_{d_r} + \lambda_{dm} \tag{10.37}$$

$$0 = r_{q_r} i_{q_r} + \frac{d\lambda_{q_r}}{dt}; \quad \lambda_{q_r} = L_{q_r l} i_{q_r} + \lambda_{qm} \tag{10.38}$$

The torque, $T_e$, is

$$T_e = \frac{3}{2} p \,\mathrm{Re}\left(j\bar{\lambda}_s \bar{i}_s^*\right) = \frac{3}{2} p(\lambda_d i_q - \lambda_q i_d) \tag{10.39}$$

Finally, the d-q variables are related to the abc variables by the Park transformation (in rotor coordinates)

$$\bar{V}_s = V_d + jV_q = \frac{2}{3}(V_a + aV_b + a^2 V_c)e^{-j\theta_{er}}; \quad a = e^{j2\pi/3} \tag{10.40}$$

Note also that all rotor variables are reduced to the stator

$$V_f = V_f^r / K_f; \quad r_f = r_f^r / K_f^2; \quad L_{fl} = L_{fl}^r / K_f^2 \tag{10.41}$$

$$r_{dr} = r^{r}_{dr} / K_d^2; \quad L_{drl} = L^{r}_{drl} / K_d^2 \tag{10.42}$$

$$r_{qr} = r^{r}_{qr} / K_q^2; \quad L_{qrl} = L^{r}_{qrl} / K_q^2 \tag{10.43}$$

The motion equations (10.16) are to be added.

The d-q model is also of eighth order and basically nonlinear, but the coefficients are position (time) independent. The signal flow diagram of the d-q model is shown in Figure 10.4.

PM or reluctance rotor SMs used for variable speed lack the excitation winding and thus $i_f = \text{const.}$ (zero for RSM).

They also lack the damper cage ($\tau_{dr} = \tau'_{rd} = \tau_{qr} = \tau'_{qr} = 0$) (Figure 10.5).

The signal flow diagrams may be arranged into equivalent circuits (Figure 10.6).

For the PMSM (Figure 10.6), the field winding circuit is replaced by an ideal (constant) current source $i_{f0}$; $i_{f0} = 0$ for the RSM.

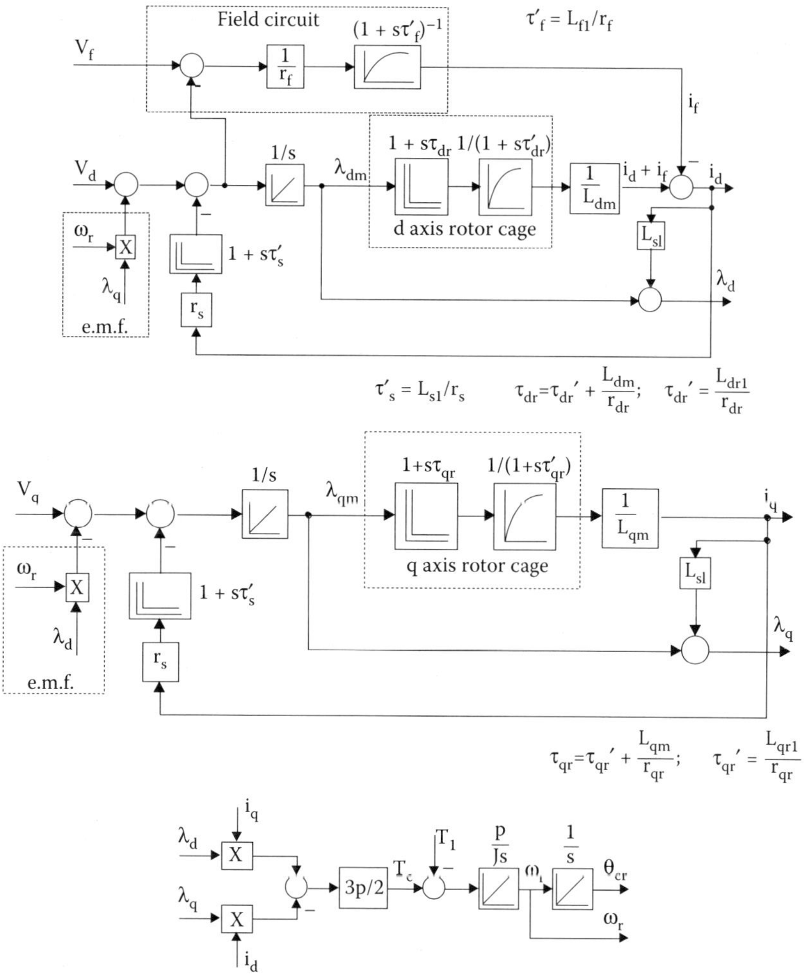


**FIGURE 10.4**
Signal flow diagram of SMs.

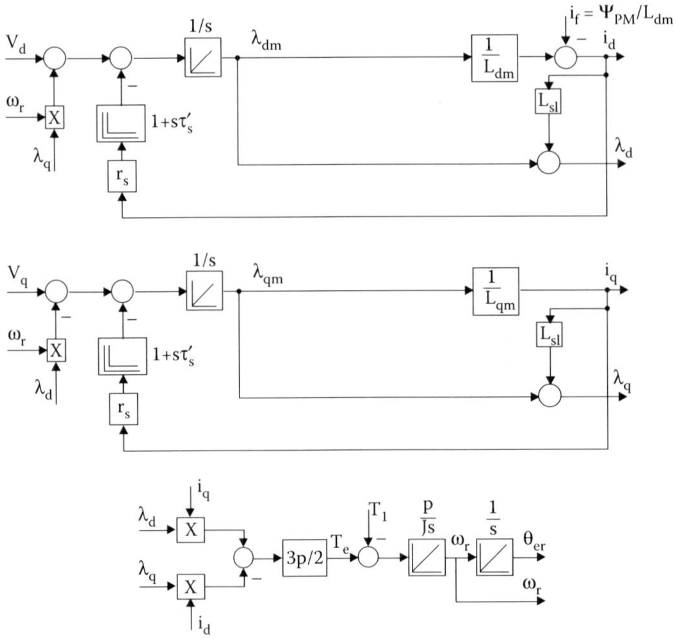


**FIGURE 10.5**
Flow signal diagram of PM and reluctance SMs (no rotor damper cage).

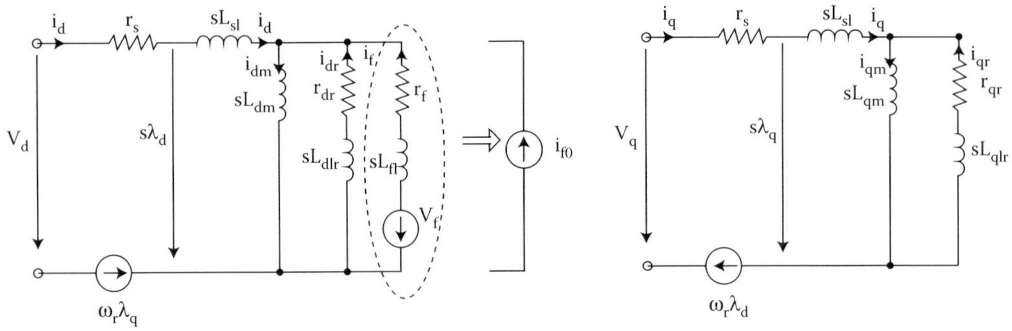


**FIGURE 10.6**
SM equivalent circuits.

Finally, for the cageless rotor, $r_{dr} = r_{qr} = \infty$ ($i_{dr} = i_{qr} = 0$). For steady state, simply $s = 0$ as we will demonstrate in Section 10.6.

## 10.6 Steady-State Operation

For steady state the stator-phase voltages are given by

$$V_{abc} = \sqrt{2}V\cos\left[\omega_1 t - (i-1)\frac{2\pi}{3}\right]; \quad i = 1,2,3 \tag{10.44}$$

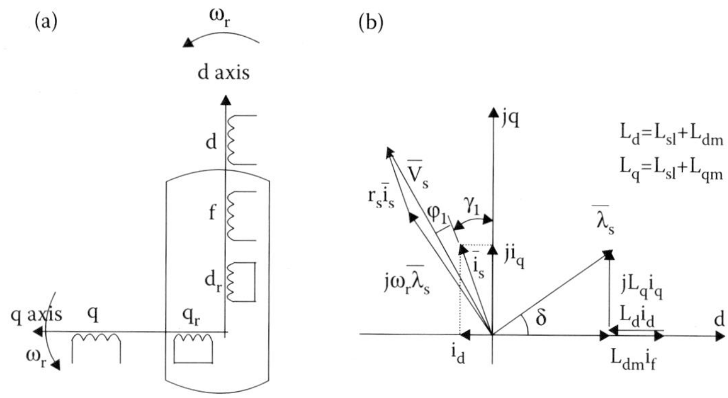


**FIGURE 10.7**
SM—steady-state—with excitation or PMs (a) d-q axis model, (b) space-phasor diagram.

where $\omega_1$ = const., and thus the rotor position $\theta_{er}$ is

$$\theta_{er} = \int \omega_r dt = \int \omega_1 dt = \omega_1 t + \theta_0 \tag{10.45}$$

Note that $\omega_r = \omega_{r0} = \omega_1$ for steady state.

Using Equation 10.40, the d-q voltage components $V_d$, $V_q$ in rotor coordinates are

$$\bar{V}_s = V_d + jV_q = V\sqrt{2}\cos\theta_0 - jV\sqrt{2}\sin\theta_0 \tag{10.46}$$

As $V_d$, $V_q$ are time independent so will be the currents in the stator $i_d$, $i_q$. Also d/dt = 0 (s = 0) and thus from Equation 10.38, $i_{d_r} = i_{q_r} = 0$ (zero damper cage currents) and $i_f = i_{f0} = V_f/r_f$ = const.

The space-phasor diagram (Figure 10.7) follows from Equations 10.34 and 10.35 which may be written in space-phasor form as the stator is symmetric electrically and magnetically.

$$\bar{V}_s = r_s\bar{i}_s + j\omega_r\bar{\lambda}_s; \quad \omega_r = \omega_1 \tag{10.47}$$

For steady state all variables in the space-phasor (or d-q) model are d.c. quantities. Space angles translate into time-phase shift angles between stator quantities (like $\varphi_1$—the power factor angle) in phase coordinates.

Using the d-q model for speed control is thus ideal, as steady-state means d.c.

## 10.7 To Vary Speed, Variable Frequency Is Mandatory

SMs keep a tight relationship between speed and frequency as the rotor m.m.f. (if any) is of d.c. (or PM) character

$$n = f_1/p; \quad \omega_r = \omega_1; \quad \Omega_r = \omega_r/p \tag{10.48}$$

Varying the number of pole pairs p is unusual so, in fact, to vary speed variable frequency is mandatory. How to vary frequency in relation to stator voltage $V_s$ and field current $i_f$ (or voltage $V_{ex}$)—if any—to obtain steady state and (or) transient performance suitable for various applications is the key to SM drives control.

While this problem will be dealt with in detail in the following chapters, the main principles will be illustrated through three numerical examples.

### EXAMPLE 10.1: UNITY POWER FACTOR AND CONSTANT STATOR FLUX

Let us consider an excited SM with the data: $V_n = 660$ V (line voltage, rms, star connection), $I_n = 500/\sqrt{2}$ (rms), $r_s = 0.016\ \Omega$, $L_d = 2L_q = 0.0056$ H, $L_{sl} = 0.1L_d$, $\omega_1 = 2\pi 60$ rad/s, $p = 2$ pole pairs.

Determine:

a. The steady-state space-phasor diagram for unity power factor ($\varphi_1 = 0$).
b. Calculate the values of $i_f$, $i_d$, $i_q$ for unity power factor and rated phase current, $I_n$.
c. Calculate the power angle $\delta$ and torque for $\varphi_1 = 0$.
d. Derive and plot the torque/speed curve for rated and half-rated stator flux for variable frequency and constant stator voltage.

**Solution**

a. The required space-phasor diagram comes from Figure 10.7b with $\varphi_1 = 0$.
b. From Figure 10.8 we have

$$V_s = r_s i_s + \omega_r \lambda_s \tag{10.49}$$

$$\bar{\lambda}_s = \lambda_d + j\lambda_q = L_d i_d + L_{dm} i_f + jL_q i_q \tag{10.50}$$

For no-load

$$V_{s0} = \omega_r L_{dm} i_f; \quad (i_d = i_q = 0); \quad \omega_r = \omega_1 \tag{10.51}$$

$$V_s = \frac{660\sqrt{2}}{\sqrt{3}} = 537.92\,\text{V}; \quad i_s = i_n\sqrt{2} = \frac{500}{\sqrt{2}}\sqrt{2} = 500\,\text{A} \tag{10.52}$$

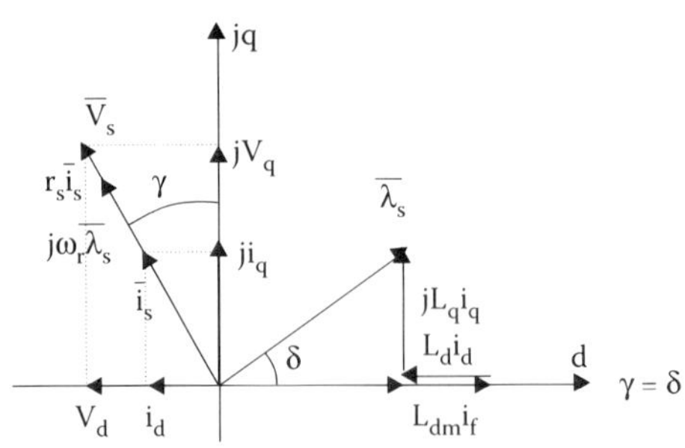


**FIGURE 10.8**
Space-phasor diagram for unity power factor.

From Equation 10.49 the rated stator flux $\lambda_s$ is

$$\lambda_s = \frac{V_s - r_s i_s}{\omega_r} = \frac{537.92 - 0.016 \cdot 500}{2\pi 60} = 1.406\,\text{Wb} \tag{10.53}$$

$$i_q = i_s \cos\delta \tag{10.54}$$

$$L_q i_q = \lambda_s \sin\delta \tag{10.55}$$

Thus

$$\lambda_s \tan\delta = L_q i_s \tag{10.56}$$

$$\tan\delta = \frac{L_q i_s}{\lambda_s} = \frac{0.0028 \cdot 500}{1.406} \approx 1 \Rightarrow \delta = 45° \tag{10.57}$$

$$-i_d = i_q = \frac{i_s}{\sqrt{2}} = \frac{500}{\sqrt{2}} = 354.61\,\text{A} \tag{10.58}$$

$$\lambda_d = \lambda_q = \frac{\lambda_s}{\sqrt{2}} = \frac{1.406}{\sqrt{2}} = 0.997\,\text{Wb} \tag{10.59}$$

$$\lambda_d = L_d i_d + L_{dm} i_f \tag{10.60}$$

The field current $i_f$ (reduced to the stator) is

$$i_f = \frac{\lambda_d - L_d i_d}{L_d - L_{sl}} = \frac{0.997 - 0.0056 \cdot (-354.61)}{0.0056(1 - 0.1)} = 591.82\,\text{A} \tag{10.61}$$

c. The electromagnetic torque $T_e$ is

$$\begin{aligned} T_e &= \frac{3}{2} p(\lambda_d i_q - \lambda_q i_d) \\ &= \frac{3}{2} \cdot 2 \cdot (0.997 \cdot 354.61 - 0.997 \cdot (-354.61)) = 2121\,\text{Nm} \end{aligned} \tag{10.62}$$

d. Note that for $\cos\varphi_1 = 1$ the torque expression (10.39) may be written as

$$T_e = \frac{3}{2} p \lambda_s i_s \tag{10.63}$$

The stator flux and current space-phasors are rectangular, as in a d.c. motor. Consequently, for constant stator flux $\lambda_s$, the torque is proportional to stator current.

Varying the frequency $\omega_1 = \omega_r$, such that $\lambda_s = \text{const.}$ and $\cos\varphi_1 = 1$ appears as an optimum way to speed control.

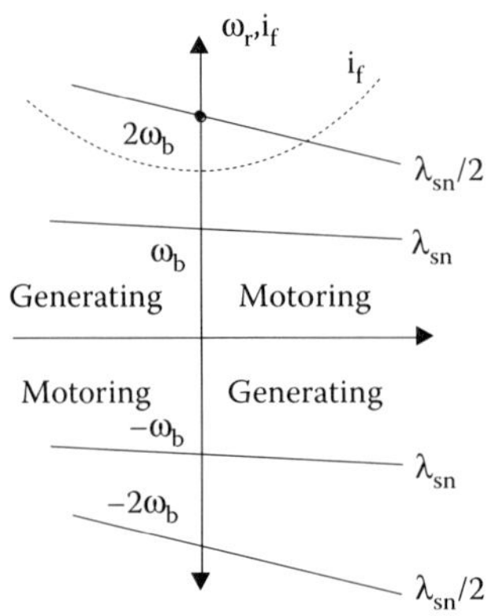


**FIGURE 10.9**
Torque/speed curves for constant flux and unity power factor.

The torque/speed curve may be derived from Equation 10.49 with Equation 10.63 by eliminating the stator current $i_s$

$$i_s = \frac{2T_e}{3p\lambda_s} \tag{10.64}$$

$$V_s = \omega_r\lambda_s + \frac{2r_sT_e}{3p\lambda_s} \tag{10.65}$$

$\omega_1 = \omega_r$, varying speed means, implicitly, that the frequency changes.

The torque/speed curve is a straight line as in a separately excited d.c. brush motor. The ideal no-load speed ($T_e = 0$) $\omega_{r0}$ is

$$\omega_{r0} = \frac{V_s}{\lambda_s} \tag{10.66}$$

For rated flux and full voltage, $V_{sn}$, the base speed $\omega_b$ is obtained. Above the base speed, $\omega_{r0}$ may be modified only through flux weakening as in a d.c. brush motor. Halving the flux doubles the ideal no-load speed $\omega_{r0}$. The torque/speed curves are illustrated in Figure 10.9.

Four-quadrant operation is implicit as negative frequency $\omega_1 = \omega_r$ means negative order ($a \to c \to b$ instead of $a \to b \to c$) of stator phases fed from a PEC.

Note that in order to maintain unity power factor, the field current has to be increased with torque (power) (Figure 10.9).

A tight coordination of stator and rotor currents control is required to maintain the above conditions, but once this is done—through vector control—ideal (linear) torque/speed curves are obtained.

Low-speed, large-power SMs for cement mills or the like, fed from cycloconverters or PWM voltage-source inverters, operate, in general, under such conditions (Chapter 14).

### EXAMPLE 10.2: LEADING POWER FACTOR AND CONSTANT STATOR FLUX

Consider an excited SM with the same data as in Example 10.1.

Determine

- The steady-state space-phasor diagram for a leading power factor angle $\varphi_1 = -12°$.
- Calculate the values of $i_f$, $i_d$, $i_q$ for rated current and $\varphi_1 = -12°$.
- Calculate the stator flux $\lambda_s$, its components $\lambda_d$ and $\lambda_q$ and the corresponding electromagnetic torque $T_e$.
- Calculate the torque/speed curve for rated and half-rated stator flux and $\varphi_1 = -12°$.

**Solution**

The space-phasor diagram of Figure 10.7b is redrawn with $\varphi_1 = -12°$ (leading) as in Figure 10.10.

From Figure 10.10 we may write approximately

$$V_s \approx r_s i_s \cos\varphi_1 + \omega_r \lambda_s \tag{10.67}$$

The torque $T_e$ is

$$T_e = \frac{3}{2} p \operatorname{Re}\left(j\bar{\lambda}_s \bar{i}_s^*\right) = \frac{3}{2} p \lambda_s i_s \cos(\gamma - \delta) \approx \frac{3}{2} p \lambda_s i_s \cos\phi_1 \tag{10.68}$$

$$\lambda_s \approx \frac{V_s - r_s i_s \cos\varphi_1}{\omega_r}$$

$$= \frac{537.92 - 500 \cdot 0.016 \cdot \cos(-12°)}{2\pi 60} = 1.4068\,\text{Wb} \tag{10.69}$$

The corresponding torque $T_e$ is

$$T_e = \frac{3}{2} \cdot 2 \cdot 1.4068 \cdot 500 \cdot 0.978 = 2063.83\,\text{N m} \tag{10.70}$$

$$i_q \approx i_s \cos(\delta + |\varphi_1|) \tag{10.71}$$

$$L_q i_q = \lambda_s \sin\delta \tag{10.72}$$

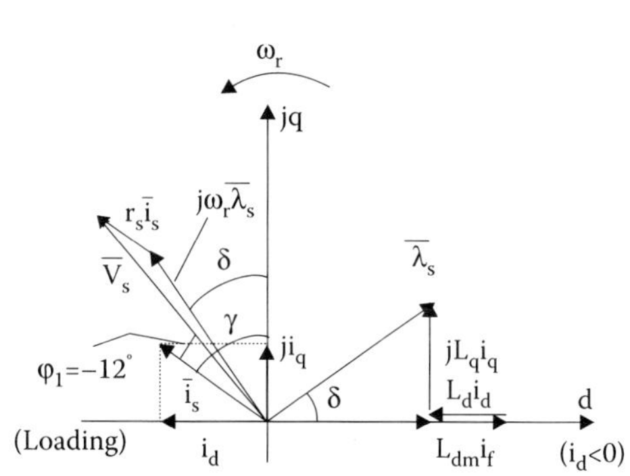


**FIGURE 10.10**
Leading power factor and constant stator flux.

Consequently,

$$\frac{\sin\delta}{\cos(\delta+|\varphi_1|)} = \frac{L_q i_s}{\lambda_s} = \frac{0.0028 \cdot 500}{1.4068} = 0.99 \quad (10.73)$$

$\delta = 39°$ (it was 45° for $\cos\varphi_1 = 1$ in Example 10.1)

Now

$$i_q = 500 \cdot \cos(39° + 12°) = 314.66\,\text{A} \quad (10.74)$$

$$i_d = -500 \cdot \sin(39° + 12°) = -388.57\,\text{A} \quad (10.75)$$

$$\lambda_q = L_q i_q = 0.0028 \cdot 314.66 = 0.881\,\text{Wb} \quad (10.76)$$

$$\lambda_d = \lambda_s \cos\delta = 1.4068 \cdot \cos 39° = 1.09388\,\text{Wb} \quad (10.77)$$

Finally, from Equation 10.58 the field current is obtained

$$i_f = \frac{\lambda_d - L_d i_d}{L_{dm}} = \frac{1.09388 - 0.0056 \cdot (-388.57)}{0.0056 \cdot 0.9} = 648.78\,\text{A} \quad (10.78)$$

This value has to be compared with $i_f = 591.82$ A for $\cos\varphi_1 = 1$ in Equation 10.61.

The torque/speed curve may be obtained from Equations 10.67 and 10.68 under the form

$$V_s = \omega_r \lambda_s + \frac{2 r_s T_e}{3 p \lambda_s} \quad (10.79)$$

Thus the torque/speed curve is quite similar to that of Equation 10.65 for $\cos\varphi_1 = 1$.

The main difference is that for the same stator current, the same stator voltage and speed, we obtain slightly less torque and we need a higher field current to provide for leading power factor angle $\varphi_1 < 0$. This value of $\varphi_1 = -12 < 0$ might be preserved for various loads if the field current is continuously adjusted. This type of coordinated frequency-stator flux-field current change method is currently used in high-speed high-power d.c. excited-rotor SM drives with current-source inverters. It will be detailed in the chapter dedicated to large-power drives (Chapter 14).

### EXAMPLE 10.3: PMSM OPERATION FOR CONSTANT $I_D$

A salient pole rotor PMSM has the data: $V_n = 180$ V (rms, line voltage, star connection of phases), $L_d = 0.4L_q = 0.05$ H, $r_s = 1\ \Omega$, no-load line voltage (rms) at $n_n = 1500$ rpm ($f_1 = 60$ Hz), $V_{on} = 180$ V (rms).

Determine:

- The PM flux linkage $\lambda_{PM}$.
- The torque/speed curve for $i_d = 0$, $i_d = \pm 5$ A and the ideal no-load speed in the three cases, torque for $i_q = 10$ A.
- Draw d-q equivalent circuits for steady-state and introduce the iron losses in the model.

**Solution**

The no-load voltage $V_{so}$ "translated" into space-vector terms, $V_{so}$

$$V_{so} = \frac{V_{on}\sqrt{2}}{\sqrt{3}} = \frac{180\sqrt{2}}{\sqrt{3}} = 146.7\,V \tag{10.80}$$

According to Equation 10.50

$$(\lambda_d)_{i_d=0} = L_{dm} i_{f0} = \lambda_{PM} \tag{10.81}$$

$$(\lambda_q)_{i_q=0} = 0 \tag{10.82}$$

$$V_{so} = \omega_r L_{dm} i_{f0} = \omega_r \lambda_{PM} \tag{10.83}$$

$$\lambda_{PM} = \frac{V_{s0}}{\omega_r} = \frac{146.70}{2\pi 60} = 0.389\,Wb \tag{10.84}$$

As

$$\lambda_d = L_d i_d + \lambda_{PM} \quad \lambda_q = L_q i_q \tag{10.85}$$

the electromagnetic torque $T_e$ is

$$T_e = \frac{3}{2} p(\lambda_d i_q - \lambda_q i_d) = \frac{3}{2} p[\lambda_{PM} + (L_d - L_q) i_d] i_q \tag{10.86}$$

The d-q voltage components ((10.34) and (10.35)), with d/dt = 0, are

$$\begin{aligned} V_d &= r_s i_d - \omega_r \lambda_q \\ V_q &= r_s i_q + \omega_r \lambda_d \end{aligned} \tag{10.87}$$

Now we replace $i_q$ from Equation 10.86 into Equation 10.87. Thus we have

$$\begin{aligned} V_d^2 + V_q^2 = V_s^2 &= r_s^2 \left(i_d^2 + i_q^2\right) + 2\omega_r r_s (\lambda_d i_q - \lambda_q i_d) \\ &+ \omega_r^2 \left[(L_d i_d + \lambda_{PM})^2 + (L_q i_q)^2\right] \end{aligned} \tag{10.88}$$

or

$$\begin{aligned} V_s^2 &= r_s^2 i_d^2 + \left(r_s^2 + \omega_r^2 L_q^2\right) \cdot \frac{4}{9} \frac{T_e^2}{p^2} \frac{1}{[\lambda_{PM} + (L_d - L_q) i_d]^2} \\ &+ \frac{4}{3} \frac{\omega_r}{p} r_s T_e + \omega_r{}^2 (L_d i_d + \lambda_{PM})^2 \end{aligned} \tag{10.89}$$

The ideal no-load ($T_e = 0$) speed $\omega_{r0}$ is

$$\omega_{r0} = \frac{\sqrt{V_s^2 - r_s^2 i_d^2}}{L_d i_d + \lambda_{PM}} \tag{10.90}$$

For $i_d = 0$

$$(\omega_{r0})_{i_d=0} = \frac{V_{sn}}{\lambda_{PM}} = \frac{180\sqrt{2}}{0.389\sqrt{3}} = 377.13\,\text{rad/s} \tag{10.91}$$

Also,

$$(\omega_{r0})_{i_d=-5A} = \frac{\sqrt{146.30^2 - (1.5)^2}}{(0.05\cdot(-5) + 0.389)} \approx 984\,\text{rad/s} \tag{10.92}$$

$$(\omega_{r0})_{i_d=5A} = \frac{\sqrt{146.30^2 - (1.5)^2}}{(0.05\cdot 5 + 0.389)} \approx 229\,\text{rad/s} \tag{10.93}$$

Note that $i_d < 0$ means a demagnetizing effect. However, the torque expression (10.86) shows that as $L_d < L_q$ only $i_d < 0$ produces a positive torque contribution.

A positive torque contribution for $i_d < 0$ is accompanied by a reduction in the stator flux as

$$(\lambda_s)_{\substack{i_q=0\\ i_d=-5A}} = \lambda_d = L_d i_d + \lambda_{PM} = -0.05\cdot 5 + 0.389 = 0.149\,\text{Wb} \tag{10.94}$$

$$(\lambda_s)_{\substack{i_q=0\\ i_d=5A}} = 0.05\cdot 5 + 0.389 = 0.639\,\text{Wb} \tag{10.95}$$

The torque for $i_d = 0, -5$ A, $+5$ A, $i_q = 10$ A is

$$(T_e)_{\substack{i_d=0\\ i_q=10A}} = \frac{3}{2} p \lambda_{PM} i_q = \frac{3}{2}\cdot 2\cdot 0.389\cdot 10 = 11.67\,\text{Nm} \tag{10.96}$$

$$(T_e)_{\substack{i_d=-5A\\ i_q=10A}} = \frac{3}{2}\cdot 2\cdot(0.389 + (0.05 - 0.125)\cdot(-5))\cdot 10 = 22.92\,\text{Nm} \tag{10.97}$$

$$(T_e)_{\substack{i_d=5A\\ i_q=10A}} = \frac{3}{2}\cdot 2\cdot(0.389 + (0.05 - 0.125)\cdot 5)\cdot 10 = 0.42\,\text{Nm} \tag{10.98}$$

Positive $i_d$ values are to be avoided as they increase the flux level and decrease the torque considerably.

The d-q equivalent circuits for steady-state ($s = 0$) reflect, in fact, Equations 10.87 (Figure 10.11).

Though the core losses are related to the main flux components we may assume them to be produced by the stator flux.

The core losses occurring in the stator are considered produced in a resistance $r_{Fe}$, to be determined through measurements.

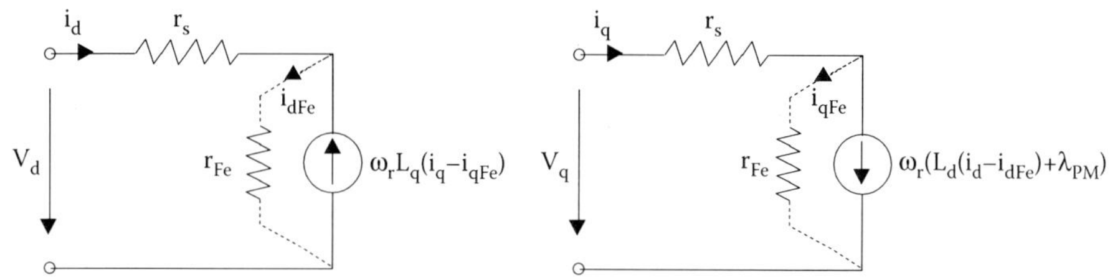


**FIGURE 10.11**
Steady-state d-q equivalent circuits of PMSM with core losses included.

Now, with core losses included, Equations 10.87 are no longer strictly valid but they may be modified simply based on the equivalent circuit. Introduced in parallel with e.m.f.s, $r_{Fe}$ should not vary dramatically with frequency.

### EXAMPLE 10.4: THE RSM STEADY STATE

A high-performance reluctance synchronous motor (RSM) has the data: $L_d = 1$ H, $L_q = 0.1$ H, $p = 2$, $r_s = 1\ \Omega$, $f_1 = 60$ Hz.

a. Determine the space-phasor diagram.
b. Determine the maximum torque per given stator current and for given stator flux.
c. Introduce a PM along axis q to reduce the q axis flux. Draw the new space-phasor diagram and discuss it. Calculate the torque for $i_d = 3$ A, $i_q = 15$ A without and with PM flux $\lambda_{PM} = -L_q i_q$.
d. Calculate the stator flux to produce the respective torques.

**Solution**

a. The space-phasor diagram is obtained simply from the general one (Figure 10.7) with $i_f = 0$ and $L_d \gg L_q$ (Figure 10.12).

The torque expression (10.86) with $i_f = 0$ and $L_d \gg L_q$ is

$$T_e = \frac{3}{2} p(L_d - L_q) i_d i_q \tag{10.99}$$

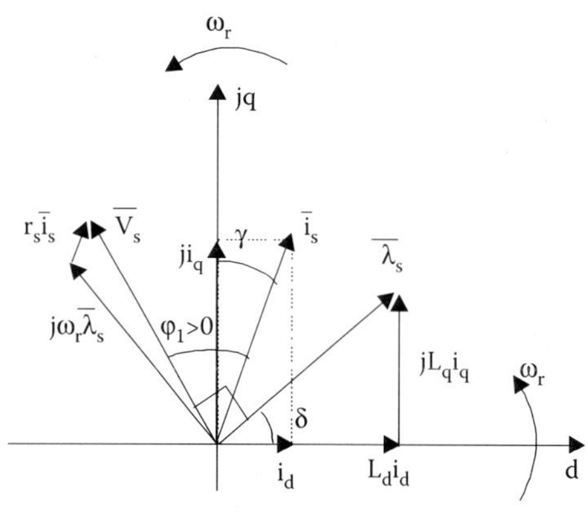


**FIGURE 10.12**
RSM—space-phasor diagram.

The voltage equations (10.87) become

$$V_d = r_s i_d - \omega_r L_q i_q$$
$$V_q = r_s i_q + \omega_r L_d i_d \tag{10.100}$$

The stator flux $\bar{\lambda}_s$ and stator current $\bar{i}_s$ are

$$\bar{\lambda}_s = \lambda_d + j\lambda_q = L_d i_d + jL_q i_q \tag{10.101}$$

$$\bar{i}_s = i_d + ji_q \tag{10.102}$$

b. The maximum torque for given stator current, $i_s$

$$i_s = \sqrt{i_d^2 + i_q^2} \tag{10.103}$$

may be obtained from Equations 10.99 and 10.103

$$\frac{\partial T_e}{\partial i_d} = 0 \tag{10.104}$$

After eliminating $i_q$ we get

$$i_{di} = i_{qi} = \frac{i_s}{\sqrt{2}}; \quad T_{ei} = \frac{3}{2}p(L_d - L_q)\frac{i_s^2}{2} \tag{10.105}$$

Consequently, $\gamma = 45°$ in Figure 10.12.

If $i_s = i_n$ (rated current), $i_{di} = i_{sn}/\sqrt{2}$, the machine is driven into heavy saturation as $i_{di}$ is too high in comparison with the rated no-load phase current $i_0/\sqrt{2}$, with the exception of very-low-power motors when $i_0/i_n \approx 0.707$. Though it leads to maximum torque per ampere it may be used for torques lower than 50%–60% of rated torque. This may be the case for low-speed operation in some variable speed drive applications.

On the other hand, for given stator flux (10.101), the maximum torque is obtained for

$$i_{dk} L_d = i_{qk} L_q = \frac{\lambda_s}{\sqrt{2}} \tag{10.106}$$

The maximum torque is

$$T_{ek} = \frac{3}{2}p(L_d - L_q)\frac{\lambda_s^2}{2L_d L_q} \tag{10.107}$$

This time the d-q flux angle is $\delta = 45°$ in Figure 10.12.

As it provides maximum torque per given flux, this is the limit condition at high speeds when the flux is limited (flux-weakening zone).

c. Introducing the PM in axis q

$$\lambda_q = L_q i_q - \lambda_{PMq}; \quad \lambda_d = L_d i_d \tag{10.108}$$

In Equation 10.108 $\lambda_{PMq} \geq L_q i_{qmax}$ to avoid PM demagnetization. The torque expression (10.86) becomes

$$i_{dk} = i_{di} = \frac{i_0}{\sqrt{2}}; \quad T_e = \frac{3}{2} p[(L_d - L_q) i_q + \lambda_{PM}] i_d \tag{10.109}$$

This time, to change the torque sign (for regenerative braking) the sign of $i_d$ (rather than $i_q$) has to be changed in variable speed drives. The space-phasor diagram of PM-RSM is shown in Figure 10.13.

Note that for motoring $\delta < 0$ (the flux vector is lagging [not leading] the d axis) and the power factor has been increased, together with torque.

Only low remnant flux density (moderate cost) PMs are required as $L_q i_{sn}$ is still a small flux ($L_d/L_q \gg 1$).

d. Let us now calculate the torque and stator flux for the two situations

$$(T_e)_{\lambda_{PMq}=0} = \frac{3}{2} \cdot p(L_d - L_q) i_d i_q = \frac{3}{2} \cdot 2 \cdot 0.1 \cdot \left(1 - \frac{1}{10} \cdot 3.15\right) = 12.15\,\text{Nm} \tag{10.110}$$

$$\lambda_s = \sqrt{L_d^2 i_d^2 + L_q^2 i_q^2} = \sqrt{0.1^2 \cdot 3^2 + \left(\frac{0.1}{10} \cdot 15\right)^2} = 0.335\,\text{Wb} \tag{10.111}$$

$$(T_e)_{\lambda_{PMq} \neq 0} = \frac{3}{2} \cdot 2 \cdot \left(0.1 \cdot \left(1 - \frac{1}{10}\right) \cdot 15 + \frac{0.1}{10} \cdot 15\right) \cdot 3 = 13.15\,\text{Nm} \tag{10.112}$$

$$\lambda_s = L_d i_d = 0.1 \cdot 3 = 0.3\,\text{Wb} \tag{10.113}$$

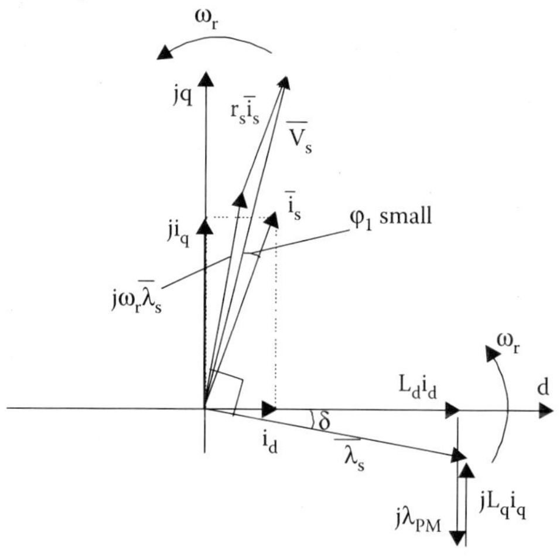


**FIGURE 10.13**
PM-SRM—space-phasor diagram.

So, the PMs in axis q produce almost a 10% torque increase for a 10% decrease in flux level.

The improvements are more spectacular in lower saliency machines ($L_d/L_q < 4$), which are easier to build.

## 10.8 Cogging Torque and Tooth-Wound PMSMs

PMSMs develop a pulsating torque at zero stator current due to the PM magnetic coenergy variation with rotor position in the presence of stator slot openings.

This zero current torque which has a zero average value per 360° is called the cogging torque.

Cogging torque stems from the PM airgap flux density variation due to stator slot openings and depends also on other factors such as

- PM shape and placement
- The number of stator slots Ns and rotor poles 2p

Figure 10.14 shows the PM airgap flux density obtained from finite element analysis (FEA) at the center of airgap for a 27-stator slot 6 rotor PM poles PMSM [6].

The cogging torque may be calculated, within FEA by the coenergy method or by the Maxwell stress tensor method.

There is also a relationship between the e.m.f. harmonics content and the cogging torque as evident in Table 10.1 for the same Ns = 27 slot, 2p = 6 pole PMSM and surface radial (SR), surface parallel (SP), and bread-loaf (BL)-shaped PMs (Figure 10.14).

While the e.m.f. fundamental and, consequently, the electromagnetic torque for given sinusoidal current would be the same for all three shapes, the cogging torque is notably smaller for the BL PMs. The e.m.f. waveform and cogging torque versus position curves depend also

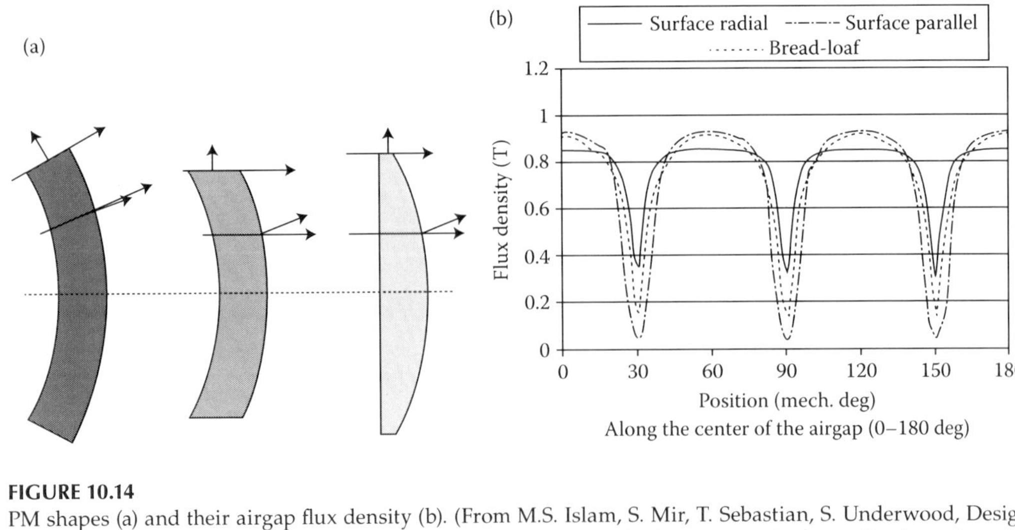


**FIGURE 10.14**
PM shapes (a) and their airgap flux density (b). (From M.S. Islam, S. Mir, T. Sebastian, S. Underwood, Design consideration of sinusoidal excited permanent magnet machines for low torque ripple applications, *Record of IEEE-IAS*, 2004, Seattle, Washington.)

**TABLE 10.1**

Cogging Torque and e.m.f. Harmonics

| PM Shape | $T_{coggpp}$ (mNm) | Line to Line e.m.f. at 1000 rpm (rms) 1st (V) | 5th (V) | 7th (V) |
|---|---|---|---|---|
| SR | 47.88 | 2.857 | 0.05604 | 0.012256 |
| SP | 15.63 | 2.8681 | 0.00648 | 0.008132 |
| BL | 5.562 | 2.8691 | 0.03459 | 0.004897 |

*Source:* Adapted from M.S. Islam, S. Mir, T. Sebastian, S. Underwood, Design consideration of sinusoidal excited permanent magnet machines for low torque ripple applications, *Record of IEEE-IAS*, 2004, Seattle, Washington.

on the number of slots per pole per phase q. For q = 1 ($N_s = 24$, $2p_1 = 8$ poles) the e.m.f., cogging torque, and electromagnetic torque for sinusoidal current are all shown in Figure 10.15.

It is evident that the number of cogging torque fundamental periods N is the lowest common multiplier of $N_s$ and 2p

$$N = LCM(N_s, 2p) \tag{10.114}$$

The larger N, the smaller the peak value of cogging torque. In the case in Figure 10.15, $N = N_s = 24$.

There are applications, such as automotive power steering, where the total torque pulsations should be well less than 1% of peak electromagnetic torque of the motor.

Torque pulsations include cogging torque but also refer to e.m.f. and current time harmonics; magnetic saturation of the magnetic circuit also plays an important role [6].

Among the many ways to reduce cogging torque we mention here

- A higher N = LCM ($N_s$, 2p)
- PM shape and span optimization (Figure 10.16)
- Rotor or stator skewing (the electromagnetic torque is also reduced)

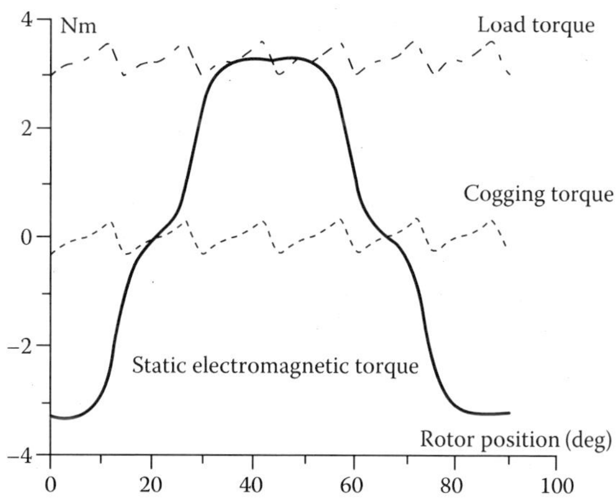


**FIGURE 10.15**

E.m.f., cogging torque, total torque (for $N_s = 24$, $2p_1 = 8$ poles) versus rotor position for sinusoidal current. (From Cross, J., Viarouge, P., *IEEE Trans.*, EC-17(2), 248–253, 2002.)

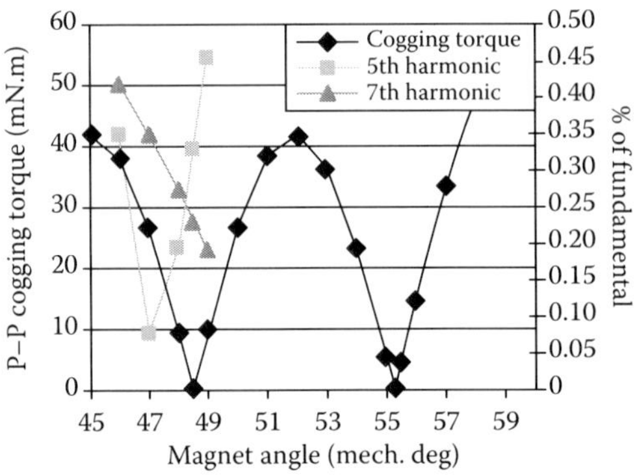


**FIGURE 10.16**
Cogging torque and line e.m.f. harmonics content with PM angle for surface PM rotor poles ($N_s = 24$, $2p = 8$).

There are two PM span angles in Figure 10.16 for which the cogging torque is zero; the e.m.f. harmonics are not zero for those situations, but also small.

The tooth-wound stator windings have been introduced to reduce stator cogging torque by reducing the end-connection length and to reduce the copper losses by increasing the value of $N = LCM\ (N_s, 2p)$. They may also be called fractionary windings as q (slot/pole/phase) <1.

They may be built in one layer or two layers (Figure 10.17).

The e.m.f. waveform and cogging torque of a 6/4 slot/pole combination PMSM is shown in Figure 10.18. The e.m.f. is almost sinusoidal. The complete geometrical data are given in Table 10.2 [9].

Despite the fact that the PM airgap flux density has a rather rectangular distribution in the airgap, the e.m.f. may be rather sinusoidal by geometrical optimization, either with surface PM or with interior PMs on the rotor [9].

It is also feasible to produce a trapezoidal e.m.f. waveform, in general, for unilayer windings with nonuniform slots [9,10].

To secure a large enough e.m.f. fundamental the winding factor $K_{w1}$ should be large enough (>0.866).

For fractionary q and $N_s/2pm = z/n$ slots/pole/phase, the space harmonics that occur with these windings are

$$\nu = \pm\frac{1}{n}(2mg + 2);\ g = 0, \pm1, \pm2, \pm3 \quad \text{for n an even number}$$

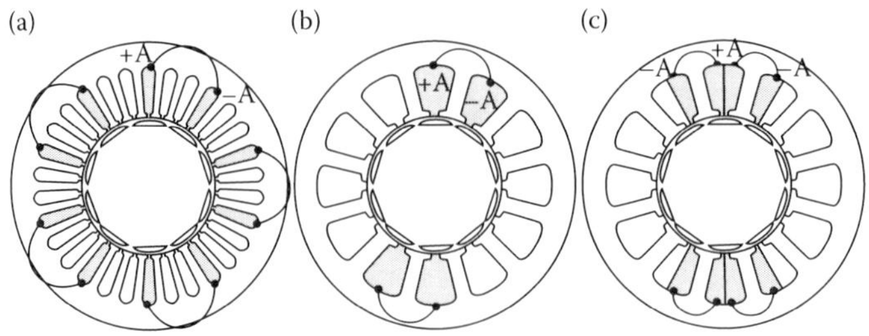


**FIGURE 10.17**
Windings for a $2p_1 = 10$ pole machine: (a) conventional: $N_s = 30$, $q = 1$, (b) one-layer tooth-wound: $N_s = 12$, $q = 0.4$, (c) two-layer tooth-wound, $N_s = 12$, $q = 0.4$.

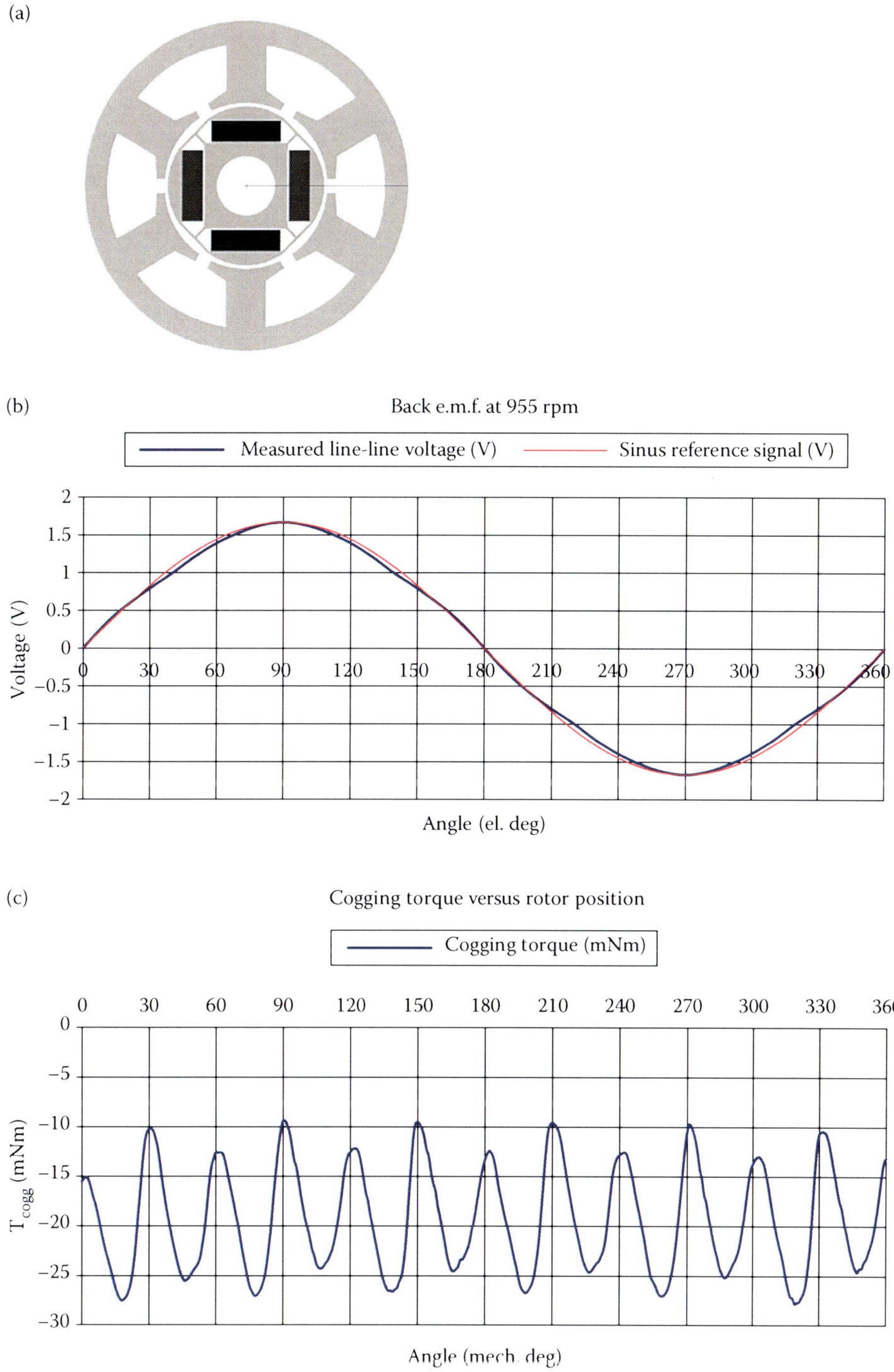


**FIGURE 10.18**
E.m.f. and cogging torque of a two-layer winding 6 slot/4 poles PMSM: (a) Cross section of IPMSM; (b) shape of the back e.m.f. at 955 rpm; and (c) cogging torque versus rotor position angle.

**TABLE 10.2**

Dimension and Properties of the Two-Layer Winding 6 Slot/4 Pole PMSM

| Parameter | Value | Unit |
|---|---|---|
| *Topology* | *Inner rotor IPMSM* | |
| Number of phases | 3 | – |
| Number of stator slots | 6 | – |
| | 4 | – |
| *Geometry* | | |
| Stator outer diameter | 56 | mm |
| Stator inner diameter | 28 | mm |
| Airgap (minimal) | 0.5 | mm |
| Stack length | 45 | mm |
| Magnet width | 12 | mm |
| Magnet height | 3.5 | mm |
| Winding | | |
| Nb. slots/pole/phase | 0.5 | – |
| Nb. winding layer | 2 | – |
| Nb. turns per phase | 20 | – |
| *Materials* | | |
| Core material | M800-50A | |
| Magnet type | NdFeB (1.2 T) | |

$$\nu = \pm\frac{1}{n}(2mg+1);\ g = 0, \pm1, \pm2, \pm3 \quad \text{for n an odd number}$$

Integer and fractionary (sub)harmonics may be present.

The m.m.f. space harmonics produce additional components in the leakage inductance of the tooth-wound PMSM, while subharmonics may also produce torque pulsations of low frequency.

Such double-layer windings are known to having lower (or no) subharmonics.

The fundamental winding factor for some main stator slot $N_s$, pole number $2p_1$ combinations for single-layer and double-layer slot-wound PMSMs are shown in Tables 10.3 and 10.4.

Consequently slot-wound PMSM, not only enjoy larger efficiency (because of shorter coil end-turns) but also may be controlled for either sinusoidal current or for rectangular current.

They are already applied in industry, from hard disks, to servodrives automotive wheel steering or low-speed high torque motor/generators.

NOTE: Other nonoverlapping coil winding PM brushless motor configurations have been proposed recently but have not reach the markets yet.

They are

- Transverse flux PM brushless machines (TFM) [11–13] (with a ring shape coil for a large number of poles).
- Claw pole stator composite magnetic core PM brushless machines [14] to cut the manufacturing costs in small motors.

**TABLE 10.3**

Winding Factor/Single-Layer Winding

| Q \ P | 2 | 4 | 6 | 8 | 10 | 12 | 14 | 16 |
|---|---|---|---|---|---|---|---|---|
| 3 | – | – | – | – | – | – | – | – |
| 6 | – | **0.866** | – | 0.866 | 0.500 | – | – | – |
| 9 | – | 0.736 | **0.667** | **0.960** | 0.960 | 0.667 | 0.218 | 0.177 |
| 12 | – | – | – | 0.866 | **0.966** | – | **0.966** | 0.866 |
| 15 | – | – | 0.247 | 0.383 | 0.866 | 0.808 | 0.957 | 0.957 |
| 18 | – | – | – | 0.473 | 0.676 | 0.866 | 0.844 | 0.960 |
| 21 | – | – | – | 0.248 | 0.397 | 0.622 | 0.866 | 0.793 |
| 24 | – | – | – | – | 0.430 | – | 0.561 | 0.866 |

*Source:* Adapted from F. Magnussen, C. Sadarangani, Winding factors and Joule losses of permanent magnet machines with concentrated windings, *Record of IEEE-IEMDC*, vol. 1, 2003, pp. 333–339.

**TABLE 10.4**

Winding Factor/Double-Layer Winding

| Q \ P | 2 | 4 | 6 | 8 | 10 | 12 | 14 | 16 |
|---|---|---|---|---|---|---|---|---|
| 3 | **0.866** | 0.866 | – | – | – | – | – | – |
| 6 | – | **0.866** | – | 0.866 | 0.500 | – | – | – |
| 9 | – | 0.617 | **0.866** | **0.945** | **0.945** | 0.764 | 0.473 | 0.175 |
| 12 | – | – | – | 0.866 | 0.933 | – | 0.933 | 0.866 |
| 15 | – | – | 0.481 | 0.621 | 0.866 | **0.906** | **0.951** | **0.951** |
| 18 | – | – | – | 0.543 | 0.647 | 0.866 | 0.902 | 0.931 |
| 21 | – | – | – | 0.468 | 0.565 | 0.521 | 0.866 | 0.851 |
| 24 | – | – | – | – | 0.463 | – | 0.760 | 0.866 |

*Source:* Adapted from F. Magnussen, C. Sadarangani, Winding factors and Joule losses of permanent magnet machines with concentrated windings, *Record of IEEE-IEMDC*, vol. 1, 2003, pp. 333–339.

- Flux reversal PM brushless machines (FRM) [15] with standard laminations and stator PMs (in general) to cut manufacturing costs in medium–large torque low-speed applications.
- Axial-airgap PMSMs have been introduced to increase torque/volume for various applications [16].

## 10.9 Single-Phase PMSM

For low-power (say below 500 W) home and automotive actuator applications, the single-phase PMSM is sometimes used in order to cut the costs of the power electronics converter for variable speed, by reducing the number of power switches.

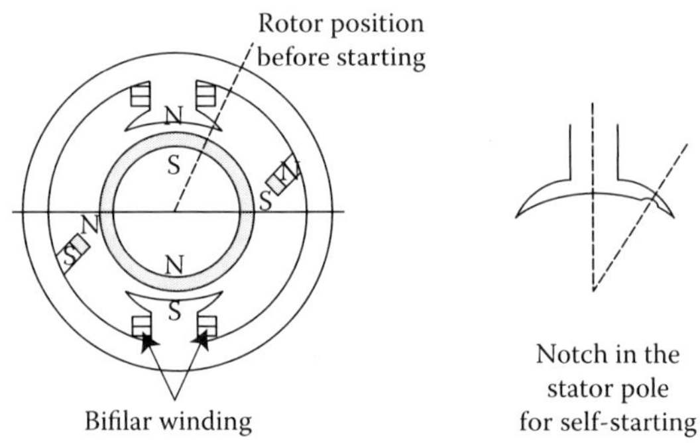


**FIGURE 10.19**
Single-phase PMSM with bifilar stator winding.

A typical cylindrical type single-phase PMSM is shown in Figure 10.19.

The stator magnetic core is made of silicon laminations (0.5 mm or less in thickness).

There are two semicoils for the positive polarity of current and other two semicoils for the negative polarity of current. The bifilar winding is thus obtained. Though it leads to an increase of copper losses it allows for a low-power switch count controller (Chapter 11).

The 2 slot/2 pole configuration leads to a 2-period cogging torque per revolution.

The machine will not start as the initial position sees the PMs aligned to the stator poles where the electromagnetic (interaction) torque is zero.

To produce a nonzero electromagnetic torque position a few solutions may be applied: a parking magnet pair is placed off-settled between the main poles (Figure 10.11) or a notch is placed asymmetrically on the stator poles toward the airgap; alternatively, a stepped airgap under the stator pole solution may be applied.

In all cases only motion in one direction is provided. But once the motor starts in one direction it may be braked and then moved eventually into opposite direction based on a Hall proximity sensor signal (Chapter 11), if the PEC allows for a.c. current in the phase coils.

## 10.10 Steady-State Performance of Single-Phase PMSM

At least for sinusoidal e.m.f., the single-phase PM brushless motor is, in fact, a synchronous machine. So all the equations already developed in this chapter apply here, but for a single stator electric circuit.

For steady state the phasor form of stator voltage equation is straightforward

$$\underline{V}_s = R_s\underline{I}_s + \underline{E}_s + j\omega L_s\underline{I}_s; \quad \underline{E}_s = j\omega_r\lambda_{PM} \tag{10.115}$$

With a sinusoidal e.m.f. $E_s$

$$E_s(t) = E_{s1}\cos(\omega_r t) \tag{10.116}$$

and $V_s$, $I_s$, are the stator voltage and current, $L_s$ is the stator inductances (independent of rotor position for surface PM rotors), and $\lambda_{PM}$ is the PM flux linkage in the stator phase.

For steady state

$$I_s(t) = I_{s1}\cos(\omega_r t - \gamma) \tag{10.117}$$

The electromagnetic (interaction) torque is

$$T_e = p\frac{E_s(t)I(t)}{\omega_r} = \frac{p}{\omega_r}\frac{E_{s1}I_{s1}(t)}{2}(\cos\gamma + \cos(2\omega_r t - \gamma)) \tag{10.118}$$

As for the single-phase IM or for the a.c. brush motor the electromagnetic torque pulsates at double stator current frequency (speed, here).

Now the cogging torque, with $N_s = 2$ slots, $2p = 2$ poles, has $N = LCM\ (2, 2) = 2$ periods

$$T_{cogg} \approx T_{cog\max}\cos(2\omega_r t - \gamma_{cog}) \tag{10.119}$$

It may be feasible that with such a cogging torque variation to secure $\gamma_{cog} = \gamma$ and, for a certain load torque

$$T_{cog\max} = -\frac{pE_{s1}I_{s1}}{2\omega_r} \tag{10.120}$$

This way the torque pulsations may be canceled (the large cogging torque (10.120) may be used for rotor parking at a proper angle to secure safe starting).

At lower load, pulsations occur into the total torque as their complete elimination is performed only for a chosen load.

The phasor diagram that corresponds to Equation 10.115 is shown in Figure 10.20.

The core losses $p_{iron}$ may be expressed as proportional to speed and stator flux squared:

$$p_{iron} \approx \frac{\omega_r^2\lambda_s^2}{R_{core}} \tag{10.121}$$

$R_{core}$ may be measured or calculated at the design stage.

In general the copper losses are larger than the iron losses and may be calculated afterwards and added to the efficiency formula.

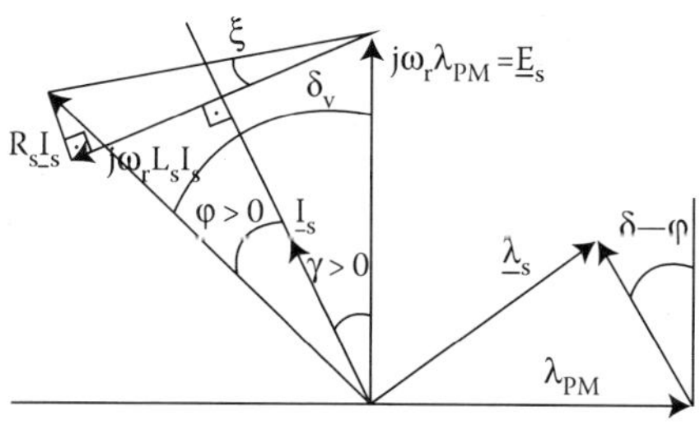


**FIGURE 10.20**
The steady-state phasor diagram of the single-phase PMSM.

To calculate the steady-state performance for given voltage power angle $\delta_v$, stator voltage $V_{s1}$, and $\omega_r$ ($E_s = \omega_r\lambda_{PM}$—given also) we obtain

$$I_s = \frac{\sqrt{V_s^2 + E_s^2 - 2V_sE_s\cos\delta_v}}{Z} \tag{10.122}$$

$$Z = \sqrt{R_s^2 + \omega_r^2L_s^2}; \quad \tan\xi = \frac{R_s}{\omega_rL_s}; \quad \cos(\phi + \xi) = \frac{E}{I \cdot Z}\sin\delta_v \tag{10.123}$$

$$\lambda_s^2 = \lambda_{PM}^2 + L_s^2i_s^2 - 2\lambda_{PM}L_si_s\cos(\delta_v - \phi) \tag{10.124}$$

The mechanical power $P_{mec}$ is

$$P_{mec} \approx V_sI_s\cos\phi - I_s^2R_s - \frac{\omega_r^2\lambda_s^2}{R_{core}} - p_{mec} \tag{10.125}$$

where $p_{mec}$ is mechanical losses.

Efficiency $\eta$ is

$$\eta = \frac{P_{mec}}{V_sI_s\cos\phi} \tag{10.126}$$

The electromagnetic torque $T_e$ versus power angle for a typical 150 W single-phase PMSM is shown in Figure 10.21 in p.u. for zero and nonzero stator resistance $R_s$.

It is evident that the stator resistance should not be neglected because otherwise the peak torque (at $\delta_v = 90°$) is grossly overestimated.

In a V/f scalar control variable speed drive the torque should not surpass the real peak torque, but better be notably smaller (by 30%–40%) to secure rather stable operation against moderate load torque perturbation. Typical efficiency and power factor curves for a 150 W motor [17] show mildly good performance (Figure 10.22).

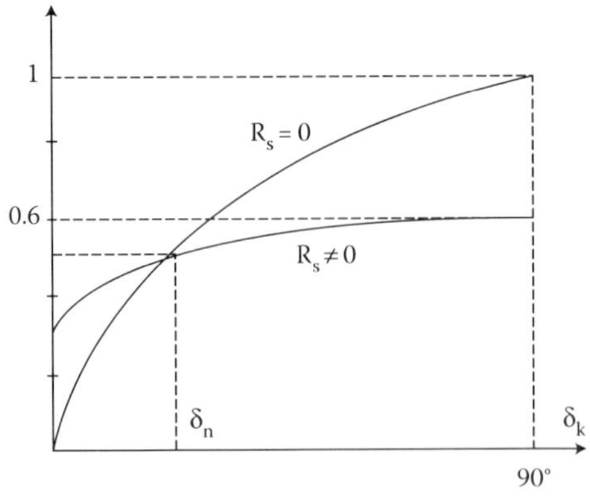


**FIGURE 10.21**
P.u. torque versus power (voltage) angle.

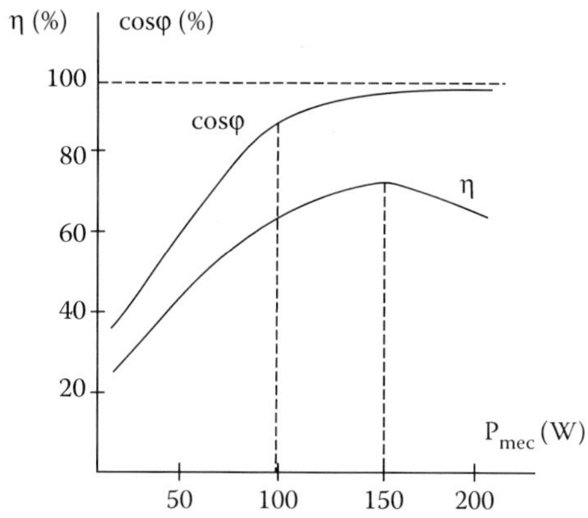


**FIGURE 10.22**
Efficiency and power factor versus output power $P_{mec}$ (W).

**EXAMPLE 10.5**

A single-phase PMSM is fed at $V_{sn} = 120$ V (rms) at the frequency $f_1 = 60$ Hz and has two rotor poles 2p = 2. At the corresponding speed, the no-load voltage $E_s$, when the motor is driven by a drive is: $E_s = 0.95\ V_{sn}$. The stator resistance $R_s = 5\ \Omega$ and the stator inductance $L_s = 0.05$ H. Calculate

a. The speed at $f_1 = 60$ Hz.
b. The stator current for $\delta_v = 0°, 45°, 90°$.
c. The power factor versus $\delta_v$.
d. The average torque for the voltage power angle $\delta_v = 0°, 45°, 90°$, with core and mechanical losses neglected.

**Solution**

a. The speed n is
$n = f_1/p = 60/1 - 60\ \text{rps} = 3600\ \text{rpm}$
b. The stator current from Equation 10.122 is

$$I_s = \frac{\sqrt{V_s^2 + E_s^2 - 2V_sE_s\cos\delta_v}}{Z}$$

$$= \frac{\sqrt{120^2 + (120\cdot 0.95)^2 - 2\cdot 120\cdot 120\cdot 0.95\cdot \cos\delta_v}}{Z}$$

$$= \begin{cases} 18/19.5 = 0.92\,\text{A}, & \delta_v = 0° \\ \dfrac{86.583}{19.5} = 4.44\,\text{A}, & \delta_v = 45° \\ \dfrac{157.49}{19.5} = 8.07, & \delta_v = 90° \end{cases}$$

with $Z = \sqrt{5^2 + (2\pi 60\cdot 0.05)^2} = 19.5\Omega$

c. The power factor angle φ comes from Equation 10.123 with

$$\xi = \tan^{-1}\frac{R_s}{\omega_r L_s} = \tan^{-1}\frac{5}{377\cdot 0.05} = 14.38°$$

$$\cos(\varphi+\xi)=\frac{E_s}{Z\cdot I_s}\sin\delta_v$$

For $\delta_v = 0$: $(\varphi + \xi) = 90°$, $\varphi = 90°–14.38° = 75.32°$

$\delta_v = 45°$: $\cos(\varphi + \xi) = 0.833$, $(\varphi + \xi) = 33.6°$, $\varphi = 33.6°–14.38° = 19.20°$

$\delta_v = 90°$: $\cos(\varphi + \xi) = 0.648$, $(\varphi + \xi) = 49.59°$, $\varphi = 49.59°–14.38° = 35.21°$

d. The electromagnetic torque $T_e$ is

$$T_e=\frac{pE_sI_s\cos(\delta_v-\phi)}{\omega_r}$$

$$=\begin{cases}\delta_v=0, T_e=\frac{1}{377}102\cdot 0.92\cdot\cos(0-75.62)=0.0618\,\text{Nm}\\ \delta_v=45°, T_e=\frac{1}{377}102\cdot 4.44\cdot\cos(45-19.20)=1.0815\,\text{Nm}\\ \delta_v=90°, T_e=\frac{1}{377}102\cdot 8.07\cdot\cos(90-35.21)=1.2589\,\text{Nm}\end{cases}$$

From $\delta_v = 45°$ to $\delta_v = 90°$ the torque increases by less than 25%, in part due to the stator resistance presence (the increase would be 30% with zero resistance).

## 10.11 Single-Phase PMSM FEM Modeling for Transients

FEM analysis of a 1-phase 4-pole PMSM with tapered airgap is summarized here [18] to illustrate the need of FEM to prepare torque/position, inductance/position, functions for the phase circuit model of such small motor drives.

### 10.11.1 Cogging Torque and Total Torque Analysis

Cogging torque is produced by the interaction between the PMs of the rotor and the stator slots at no load. It is used to assure the starting position of the rotor in which the axis of symmetry of the rotor and the stator magnetic field do not overlap. A tapered airgap was used here [18] (Figures 10.23 and 10.24).

A high as possible positive value and a small and smooth negative value for the cogging torque are desired (Figure 10.23), but the cogging torque cannot always be minimized.

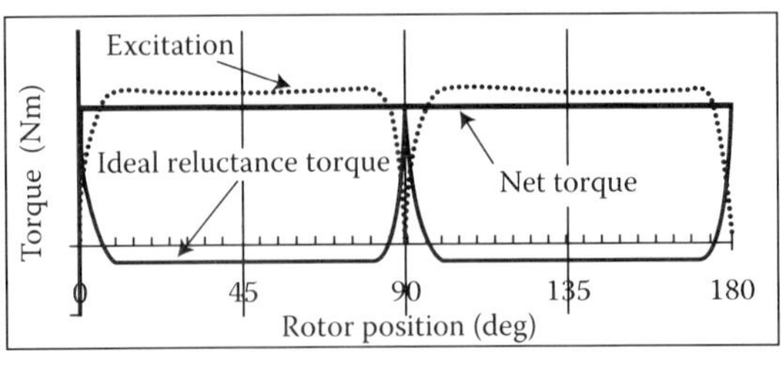


**FIGURE 10.23**
Ideal cogging torque waveform.

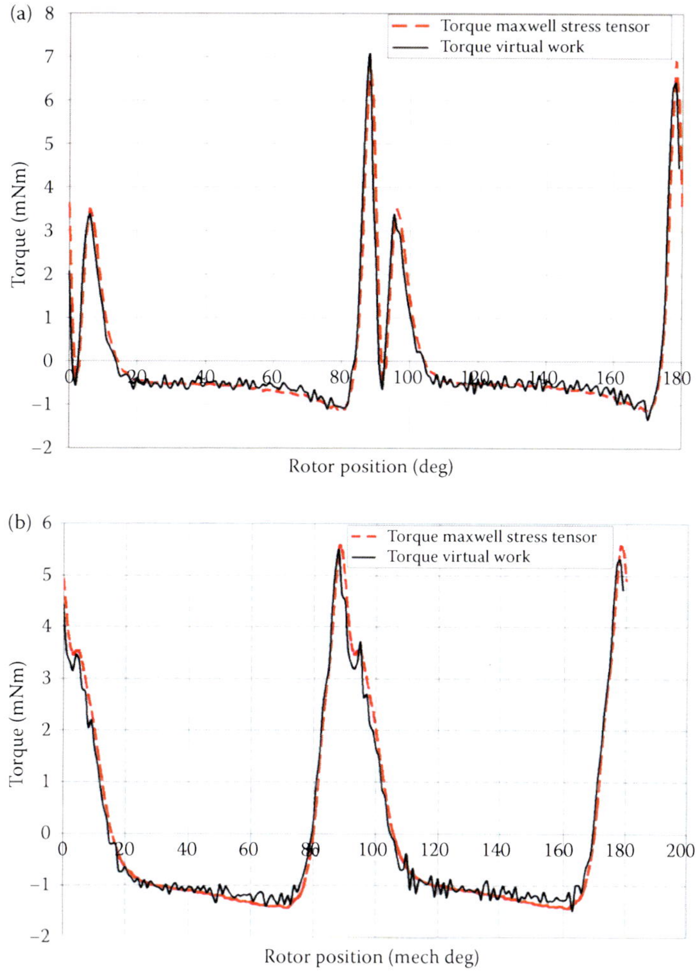


**FIGURE 10.24**
Cogging torque versus position: (a) radial magnetization, (b) modified magnetization pattern. (After L. Iepure, D. Iles-Klumpner, M. Risticevic, I. Boldea, Small blower PM single phase brushless d.c. motor drives FEM characterization with experiments, *Record of IEEE-IEMDC*, 2009, Miami, Florida (IEEExplore).)

The cogging torque is a very challenging problem and a proper calculation is needed for further advanced control tasks. It has been calculated by both Maxwell stress integration and from the rate of change of coenergy.

For fan and pump applications, the starting position is defined by $T_{cogg} = 0$ and $(dT_{cogg}/d\theta) < 0$.

Figures 10.25 and 10.26 show that clever airgap tapering leads to positive only total torque with rotor position; that is safe self-starting with increasing torque (Figure 10.27). However, the e.m.f. is nonsymmetric (Figure 10.28) and the inductance varies (slightly) with rotor position (Figure 10.28).

### 10.11.2 E.M.F., Inductance, and Measurements

Measurements (Figures 10.28 through 10.30) confirm the FEM calculations on the 20 mNm fan motor [18].

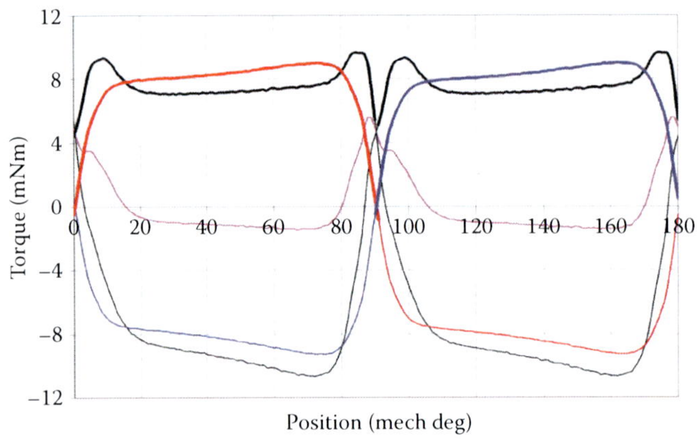


**FIGURE 10.25**
Torque components at positive and negative 1 A current value. (After L. Iepure, D. Iles-Klumpner, M. Risticevic, I. Boldea, Small blower PM single phase brushless d.c. motor drives FEM characterization with experiments, *Record of IEEE-IEMDC*, 2009, Miami, Florida (IEEExplore).)

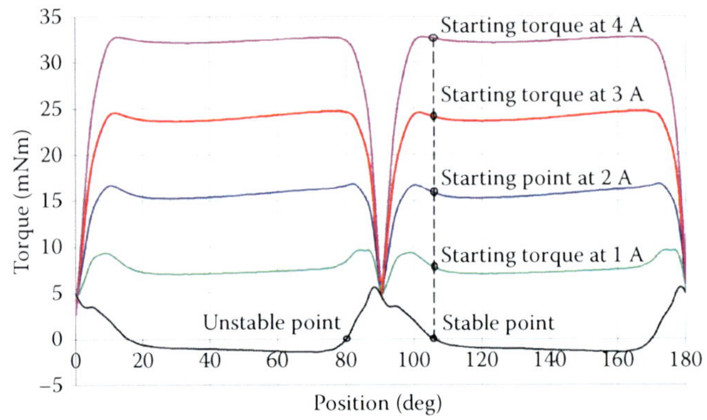


**FIGURE 10.26**
Detent position and starting torque explicative drawing.

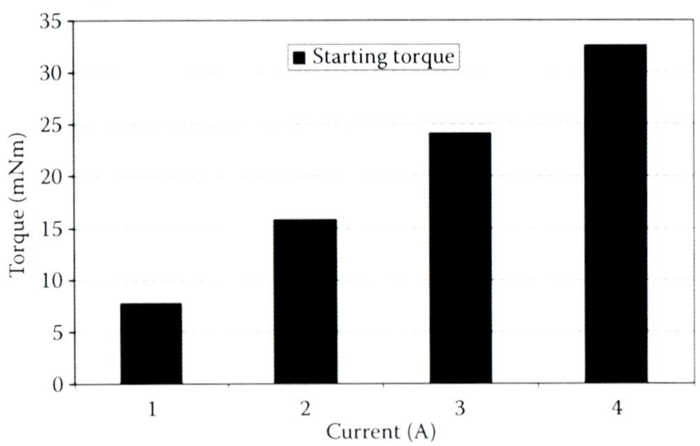


**FIGURE 10.27**
Starting torque for different currents.

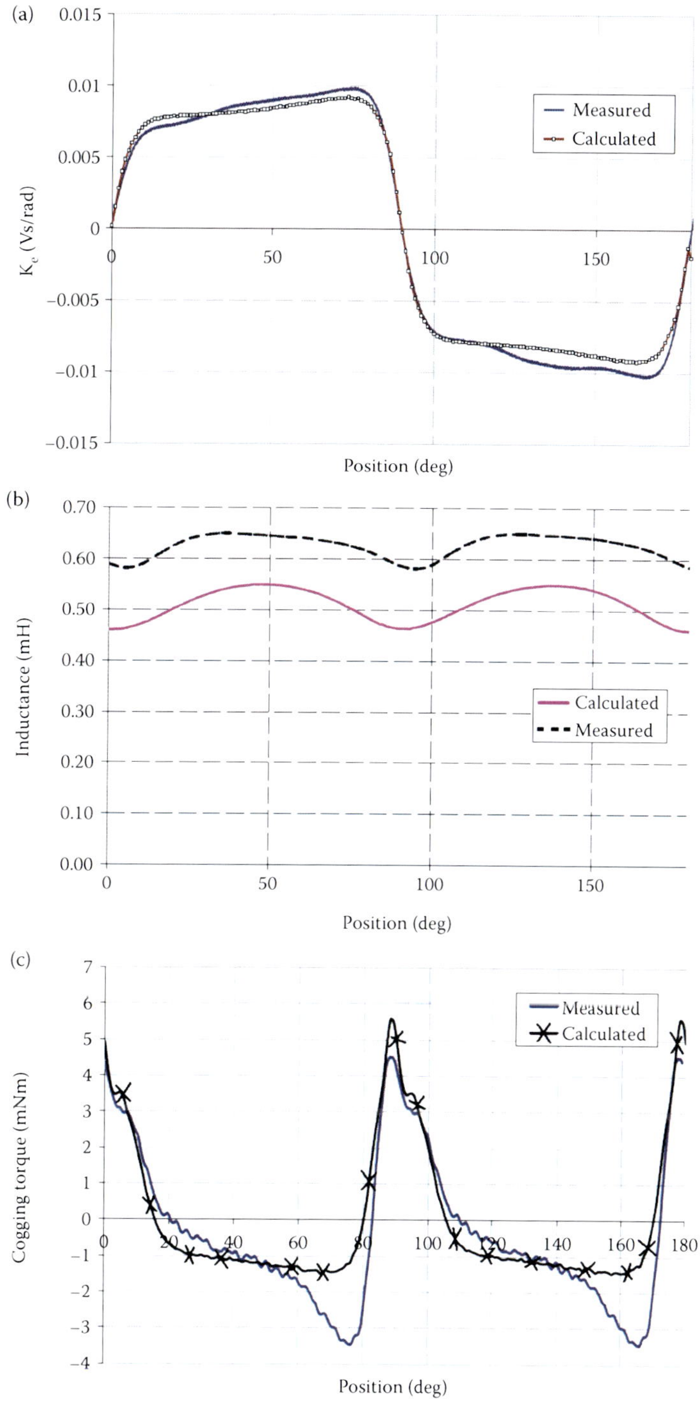


**FIGURE 10.28**
FEM results versus measurements: (a) BEMF, (b) phase inductance, (c) cogging torque. (After L. Iepure, D. Iles-Klumpner, M. Risticevic, I. Boldea, Small blower PM single phase brushless d.c. motor drives FEM characterization with experiments, *Record of IEEE-IEMDC*, 2009, Miami, Florida (IEEExplore).) *(Continued)*

(d)
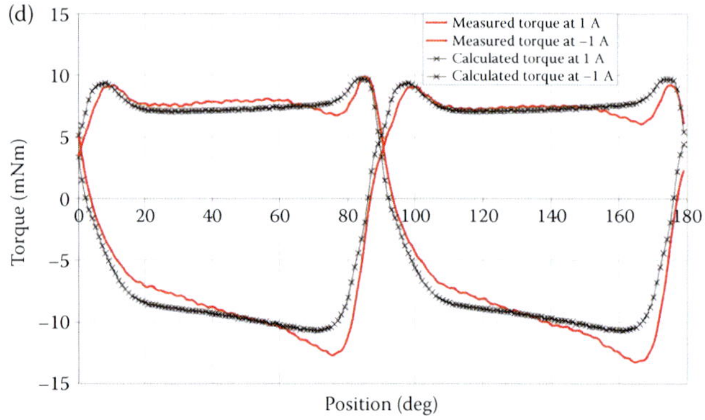


**FIGURE 10.28 (*Continued*)**
FEM results versus measurements: (d) total torque. (After L. Iepure, D. Iles-Klumpner, M. Risticevic, I. Boldea, Small blower PM single phase brushless d.c. motor drives FEM characterization with experiments, *Record of IEEE-IEMDC*, 2009, Miami, Florida (IEEExplore).)

(a)
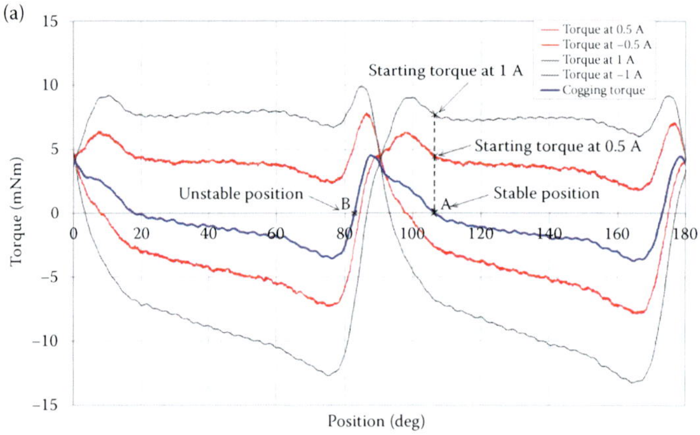


(b)
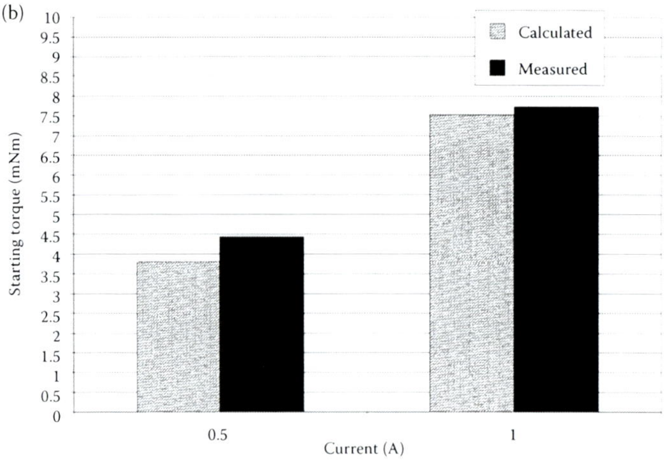


**FIGURE 10.29**
(a) FEM versus (b) measured starting torque for 0.5 and 1 A. (After L. Iepure, D. Iles-Klumpner, M. Risticevic, I. Boldea, Small blower PM single phase brushless d.c. motor drives FEM characterization with experiments, *Record of IEEE-IEMDC*, 2009, Miami, Florida (IEEExplore).)

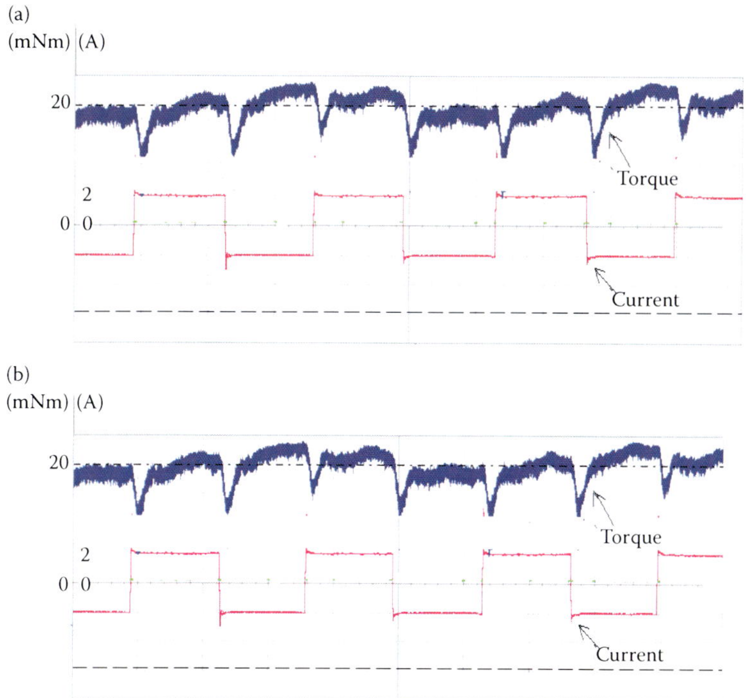


**FIGURE 10.30**
Dynamic torque measurement: (a) 110 rpm, (b) 220 rpm. (After L. Iepure, D. Iles-Klumpner, M. Risticevic, I. Boldea, Small blower PM single phase brushless d.c. motor drives FEM characterization with experiments, *Record of IEEE-IEMDC*, 2009, Miami, Florida (IEEExplore).)

## 10.12 Summary

- SMs are, in general, three-phase a.c. motors with d.c. excited or PM excited or variable reluctance rotors.
- Because the stator m.m.f. travels along the periphery with a speed $n_1 = f_1/p$ ($f_1$—stator frequency, p—pole pairs), the rotor will rotate at the same speed as only two m.m.f.s at relative standstill produce nonzero average torque.
- Changing the speed means changing frequency.
- Excited-rotor SMs require copper-rings and brushes to be connected to a d.c. controlled power supply. Alternatively, a rotary transformer with diodes on the rotor (secondary) side may provide for brushless energization of the excitation (field) windings.
- From all variable frequency-fed SMs only excited-rotor SMs, fed from current-source PECs (as shown in Chapter 14) have a squirrel cage on the rotor, to reduce the machine commutation inductance.
- Unity or leading power factor operation at variable speed and torque is possible only with excited-rotor SMs based on continuous field-current control, to nullify the input reactive power to the machine. For this case, if the stator flux is also constant, the torque/speed curve becomes linear, as for a d.c. brush motor with separate excitation. Flux weakening is performed above base speed.

- PM rotor SMs may be operated at variable speeds by their control in d-q orthogonal axes rotor coordinates, through adequate coordination of $i_d - i_q$ relationship with torque and speed requirements. Negative (demagnetizing) $i_d$ is required for positive reluctance torque as $L_d < L_q$.
- Multiple flux barrier rotors with conventional or axial laminations, eventually with PMs in axis q (of highest reluctance, $L_d > L_q$), have been shown to produce power factors and efficiencies equal or higher than those of induction machines in the same stator, for same torque.
- The d-q (space-phasor) model of SMs seems the most adequate vehicle for the design and analysis of variable speed drives with SMs.
- To reduce copper losses and frame size tooth-wound PMSMs have been introduced for various applications: from hard disks, through industrial servodrives, to active power steering.
- Various configurations of stator—PM double—saliency motors, which operate as PMSMs have been proposed recently in an effort to reduce losses and increase torque density and motor reliability [19].
- Single-phase PMSMs are recommended for low power, low system costs home appliance, or automotive electric actuation technologies [18,20].
- Chapter 11 will deal only with PM and reluctance SM drives—low and medium powers (up to hundreds of kW)—fed from PWM-IGBT converters. Large-power industrial drives with d.c. excited-rotor SMs will be treated in Chapter 14 as they require special PECs.

## 10.13 Problems

10.1 A PMSM with interior PMs ($L_d < L_q$) is applied a step voltage along the q axis, $\Delta V_q$, while $V_d$ = constant and the speed is constant. Using Equations 10.34 through 10.35 and Figure 10.5, determine the Laplace form of the current $\Delta i_d$, $\Delta i_q$, and torque responses.

10.2 A PMSM with surface PMs and $q \geq 2$ has the data: $L_d = L_q = 0.01$ H, $r_s = 0.5\ \Omega$, $p = 2$ (pole pairs), rated current (rms, phase current, 10 A), no-load voltage (at $n_0 = 1800$ rpm), $V_0 = 80$ V (rms, phase voltage).

Calculate

a. The PM flux linkage in the d-q model $\lambda_{PM}$.

b. The torque at rated current with $i_d = 0$ and the corresponding rated terminal voltage for n = 1800 rpm.

c. The ideal no-load speed for $i_q = 0$ and $i_d$ corresponding to rated current.

d. The torque at rated current and voltage for n′ = 3600 rpm.

10.3 An RSM has the data: rated voltage $V_n = 120$ V (rms, phase voltage), $\omega_{1n} = 120$ rad/s, $i_{sn} = 20$ A (rms, phase current), and the inductances $l_d$, $l_q$ in relative units: $l_d = 3$ p.u., $l_q = 0.3$ p.u., $r_s = 0.05$ p.u., $p = 2$ pole pairs.

Calculate

a. $L_d$, $L_q$, $r_s$ in $\Omega$ if $l(p.u.) = L(V_n/I_n\omega_{1n})^{-1}$.

b. For $i_d/i_q = \sqrt{L_q/L_d}$ and rated current and voltage determine the base speed $\omega_b$, torque, input power, and power factor angle $\varphi_1$.

c. Add a PM in axis q (acting against $i_q$) such that the no-load voltage, at base speed $\omega_b$, $e_0 = 0.15$ p.u. and calculate the ideal no-load speed for $i_q = i_{sn}$.

## References


1. T.J.E. Miller, *Brushless Permanent Magnet and Reluctance Motor Drives*, Oxford: Oxford University Press, 1989.
2. S.A. Nasar, I. Boldea, L. Unnewehr, *Permanent Magnet, Reluctance and Self-Synchronous Motors*, CRC Press, Florida, 1993.
3. I. Boldea, *Reluctance Synchronous Machines and Drives*, Oxford: Oxford University Press, 1996.
4. R.P. Deodhar, D.A. Staton, T.M. Jahns, T.J.E. Miller, Prediction of cogging torque using the flux-m.m.f. diagram technique, *IEEE Trans.*, vol. IA-32, no. 3, 1996, pp. 569–576.
5. I. Boldea, S.A. Nasar, Unified treatment of core losses and saturation in orthogonal axis model of electric machines, *Proc. IEE*, vol. 134, Part B, 1987, pp. 355–363.
6. M.S. Islam, S. Mir, T. Sebastian, S. Underwood, Design consideration of sinusoidal excited permanent magnet machines for low torque ripple applications, *Record of IEEE-IAS*, 2004, Seattle, Washington (IEEExplore).
7. J. Cross, P. Viarouge, Synthesis of high performance PM motors with concentrated windings, *IEEE Trans.*, vol. EC-17, no. 2, 2002, pp. 248–253.
8. F. Magnussen, C. Sadarangani, Winding factors and Joule losses of permanent magnet machines with concentrated windings, *Record of IEEE-IEMDC*, vol. 1, 2003, pp. 333–339 (IEEExplore).
9. D. Iles-Klumpner, I. Boldea, Optimization design of an interior permanent magnet synchronous motor for an automotive active steering system, *Record of OPTIM*, vol. 2, 2004, pp. 129–134, Brasov, Romania.
10. D. Ishak, Z.Q. Zhu, D. Howe, Permanent magnet brushless machines with unequal tooth widths and similar slot and pole numbers, *Record of IEEE-IAS*, 2004, Seattle, Washington (IEEExplore).
11. H. Weh, H. May, Achievable force densities for permanent magnet excited machines in new configurations, *Record of ICEM*, vol. 3, 1986, pp. 1107–1111, Munchen, Germany.
12. E. Henneberger, I.A. Viorel, *Variable Reluctance Electrical Machines*, Shaker Verlag, Aachen, 2001.
13. I. Luo, S. Huang, S. Chen, T. Lipo, Design and experiments of a novel axial circumferential current permanent magnet (AFCC) machine with radial airgap, *Record of IEEE-IAS*, 2001 (IEEExplore).
14. I. Cross, P. Viarouge, New structure of polyphase claw pole machines, *IEEE Trans.*, vol. IA–40, no. 1, 2004, pp. 113–120.
15. I. Boldea, I. Zhang, S.A. Nasar, Theoretical characterization of flux reversal machine in low speed servodrives—The pole—PM configuration, *IEEE Trans.*, vol. IA 38, no. 6, 2002, pp. 1549–1557.
16. M. Cerchio, G. Griva, F. Profumo, A. Tenconi, "Plastic" axial flux machines: Design and prototyping of a multidisk PM synchronous motor for aircraft applications, *Record of ICEMS*, 2004, Je Ju Island, Korea (IEEExplore).

17. V. Ostovic, Performance comparison of U core and round stator single phase PM motors for pump applications, *IEEE Trans.*, vol. IA-38, no. 2, 2002, pp. 476–482.
18. L. Iepure, D. Iles-Klumpner, M. Risticevic, I. Boldea, Small blower PM single phase brushless d.c. motor drives FEM characterization with experiments, *Record of IEEE-IEMDC*, 2009, Miami, Florida (IEEExplore).
19. Z.Q. Zhu, J.T. Chen, Advanced switched flux PM brushless machines, *IEEE Trans*, vol. MAG-46, no. 6, 2010, pp. 1447–1453.
20. A. Isfanuti, L. Tutelea, F. Kalluf, I. Boldea, A novel design of a stator—Ferrite PM single phase doubly salient small motor: FEM characterization and controlled dynamics, *Record of OPTIM*, 2014, pp. 284–290 (IEEExplore).

# 11

# *PM and Reluctance Synchronous Motor Drives*

## 11.1 Introduction

PMSMs and RSMs are associated with PWM voltage-source inverters in variable speed drives.

While PMSM drives now enjoy worldwide markets, RSM drives have a small share of the market in the low-power region, though recently high saliency (and performance) RSMs with q axis PMs have been successfully introduced for wide constant power speed range applications (such as spindle drives).

High-performance applications are, in general, provided with PMSM drives. RSM drives may be used for general applications as the costs of the latter are lower than those of the former and, in general, close to the costs of similar IM drives.

In what follows, we will deal first with PMSM drives and after that, with RSM drives. Constant power operation will be discussed toward the end of the chapter for both types of SMs. Control of single-phase PMSM ends this chapter.

## 11.2 PMSM Drives: Classifications

Basically, we may distinguish three ways to classify PMSM drives: with respect to current waveform, voltage–frequency correlation, and motion sensor presence.

From the point of view of current waveform we distinguish

- Rectangular current control (Figure 11.1a)—the so-called brushless d.c. motor drive—q = 1 slot per pole per phase, surface PM rotor
- Sinusoidal current control (Figure 11.1b)—the so-called brushless a.c. drive—q ≥ 2 slots per pole per phase

From the point of view of motion sensor presence, there are

- Drives with motion sensors
- Drives without motion sensors (sensorless) [1]

Finally, sinusoidal current (brushless a.c.) drives may have

- Scalar (V/f) control—a damper cage on the rotor is beneficial
- Vector control (VC): current or current and voltage
- Direct torque and flux control (DTFC)

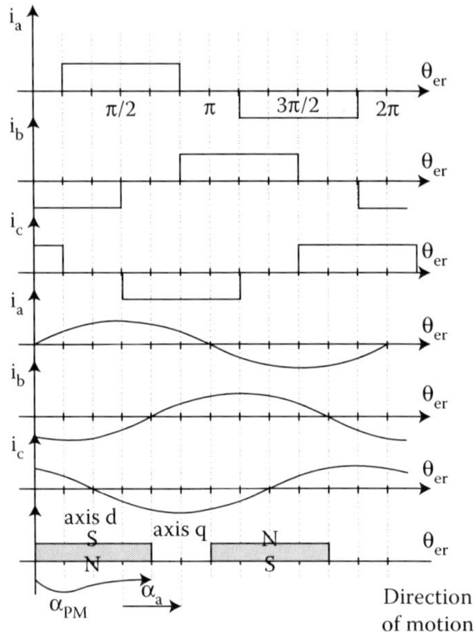


**FIGURE 11.1**
Rectangular or sinusoidal current control: $\alpha_a$— advance angle.

*The stator current waveforms—rectangular or sinusoidal (Figure 11.1)—have to be synchronized with the rotor position.*

This may be done by turning on and off the SCRs in the PWM inverter in relation to the rotor position $\theta_{er}$.

In general, to compensate for the e.m.f. increase with speed and for the machine inductance delay in current variation, the phase commutation is advanced by an angle $\alpha_a$ which should increase with speed. Current waveforming is obtained through current controllers.

Rectangular current control is preferred when the PM e.m.f. is nonsinusoidal (trapezoidal), $q \le 1$, concentrated coil stator windings—to reduce torque pulsations and take advantage of a simpler position (proximity) sensor or estimator.

Scalar control (V/f) is related to sinusoidal current control without motion sensors (sensorless) (Figure 11.2).

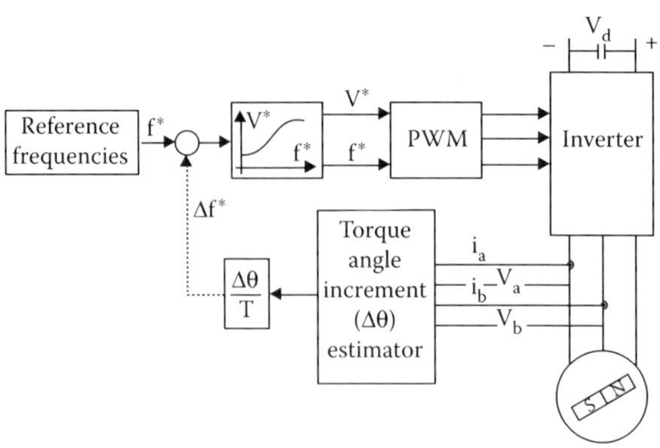


**FIGURE 11.2**
V/f (scalar) control for PMSM (and for RSM) with torque angle gradient compensation.

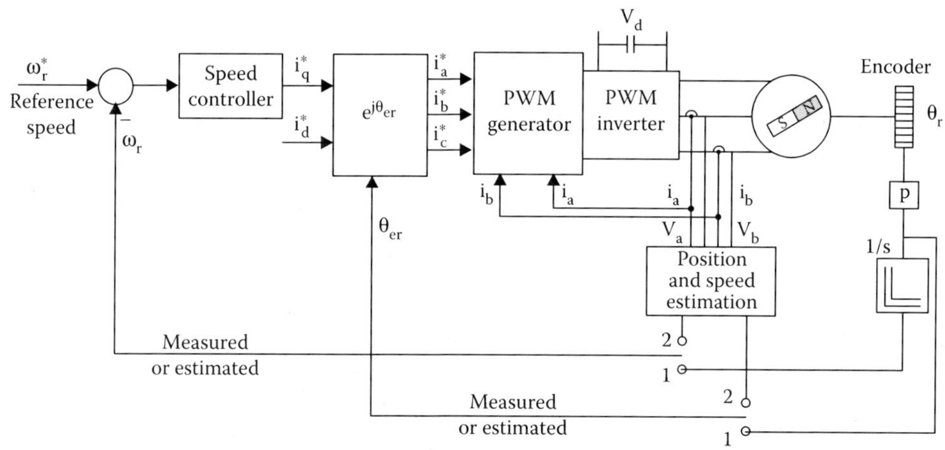


**FIGURE 11.3**
Basic VC of PMSM (or for RSM): 1—with motion sensor; 2—sensorless.

The torque angle increment Δθ is estimated and the reference frequency is increased by Δf* to compensate for the torque variation and keep the motor currents in synchronism with the rotor during transients. A rotor cage is useful to damp the oscillations as ramping the frequency is limited.

Low dynamics applications may take advantage of such simplified solutions. For faster dynamics applications, VC is used (Figure 11.3).

The rotor position and speed are either measured or estimated (for sensorless drives) and used as speed feedback ($\omega_r$) and vector rotator ($\theta_{er}$) to generate reference phase currents. Closed-loop or open-loop PWM is used to "construct" the current (or voltage) waveforms locked into synchronism with the rotor. The PMSM is controlled along the d-q model in rotor coordinates, which corresponds to d.c. for steady state (Chapter 10).

Correlating $i_d^*$ with $i_q^*$ is a matter of optimization, according to some criterion. VC is considerably more complicated in comparison with V/f control but superior dynamic performance is obtained (quicker torque control, in essence).

To simplify the motor control, the DTFC for IMs has been extended to PMSM (and to RSMs) as TVC in Reference 2.

The direct stator flux and torque control leads to a table of voltage switching (voltage vector sequence). Vector rotation has been dropped but flux and torque observers are required. While speed is observed, rotor position estimation is not required in sensorless driving (Figure 11.4).

Again, fast flux and torque control may be obtained even in sensorless driving.

Rectangular current control and sinusoidal current control (through VC or DTFC) are going to be detailed in what follows for motion sensor and sensorless driving [1–9].

## 11.3 Rectangular Current Control (Brushless D.C. Motor Drives)

Rectangular current control is applied to nonsinusoidal (trapezoidal) PM-e.m.f. waveform, typical for concentrated coil stator windings (three slots per pole: q = 1). To reduce torque

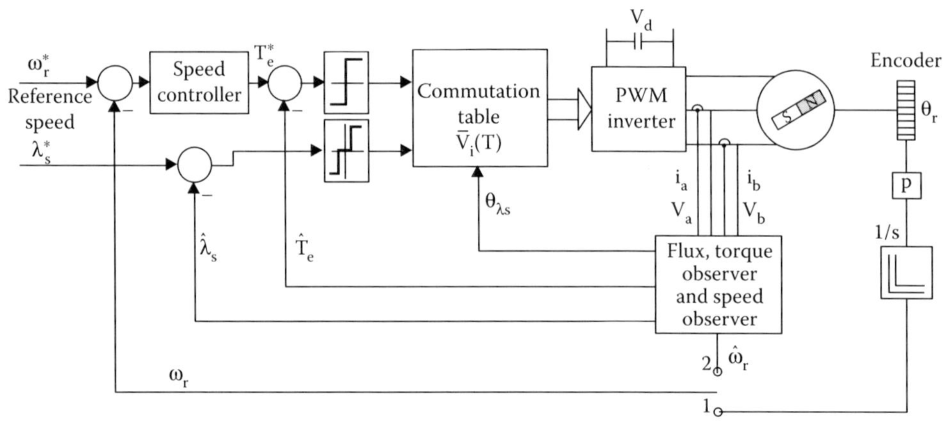


**FIGURE 11.4**
DTFC of PMSMs (and RSMs).

pulsations, rectangular current is needed. With rectangular current, however, the reluctance torque production is not very efficient and thus nonsalient pole (surface PM) rotors are preferred.

The motor phase inductance $L_a$ is [9]

$$L_a = L_{sl} + L_{ag} \tag{11.1}$$

$$L_{ag} \approx \frac{\mu_0 W_1^2 L\tau}{2pg_e}; \quad g_e \approx K_c g + \frac{h_{PM}}{\mu_{rec}} \tag{11.2}$$

where $W_1$ is turns per phase, L is the stack length, $\tau$ is the pole pitch, g is the airgap, $K_c$ is the Carter coefficient, $h_{PM}$ is the PM radial thickness, p is the pole pairs, and $L_{sl}$ is the leakage inductance.

The mutual inductance between phases, $L_{ab}$, is

$$L_{ab} \approx -\frac{L_{ag}}{3} \tag{11.3}$$

Thus the cycling inductance $L_s$ (per phase) for two-phase conduction is

$$L_s = L_a - L_{ab} = L_{sl} + \frac{4}{3}L_{ag} \tag{11.4}$$

### 11.3.1 Ideal Brushless D.C. Motor Waveforms

In principle, the surface PM extends over an angle $\alpha_{PM}$ less than 180° (which represents the pole pitch, Figure 11.1). The two limits of $\alpha_{PM}$ are $2\pi/3$ and $\pi$. Let us suppose that the PM produces a rectangular airgap flux distribution over $\alpha_{PM} = \pi$ (180°) (Figure 11.5a). The stator phase m.m.f. is supposed to be rectangular, a case corresponding to $q = 1$ (three slots per pole). Consequently, the PM flux linkage in the stator winding $\lambda_{PM}(\theta_{er})$ varies linearly with

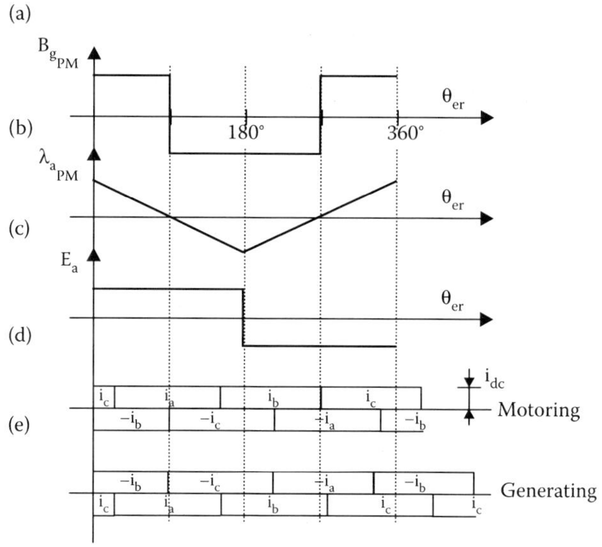


**FIGURE 11.5**
Ideal waveforms for BLDC (brushless d.c. motor): (a) PM airgap flux density; (b) PM flux per phase a; (c) e.m.f. in phase a; (d) ideal currents for motoring; and (e) ideal currents for generating.

rotor position (Figure 11.5b). Finally the phase e.m.f. $E_a$ is rectangular with respect to rotor position (Figure 11.5c).

For zero advance angle $\alpha_a = 0$ (Figure 11.1)—the phase currents are in phase with the e.m.f. ($i_a$, $E_a$) for motoring (Figure 11.5d).

At any instant, due to assumed instantaneous current commutation, only two phases are in conduction. Consequently, only two-thirds of the PMs are utilized. So the back core flux is 50% larger than necessary. On the other hand, if the currents extend over 180° (with three phases conducting at any time) for 120° wide PMs, it would mean 50% more than necessary copper losses for a given torque.

For 120° wide magnets, a delta connection is used, while 180° wide magnets require a star connection. To reduce the fringing (leakage) flux between neighboring magnets their span is 150°–160°.

Returning to the linear phase flux linkage variation with rotor position (Figure 11.5b), we have

$$\lambda_{aPM}(\theta_{er}) = \left(1 - \frac{2}{\pi}\theta_{er}\right)\lambda_{PM} \tag{11.5}$$

The maximum flux linkage per phase $\lambda_{PM}$ is

$$\lambda_{PM} = W_1 B_{gPM} \tau L \tag{11.6}$$

The phase e.m.f., $E_a$, is

$$E_a = -\frac{d\lambda_{aPM}(\theta_{er})}{d\theta_{er}} \cdot \frac{d\theta_{er}}{dt} = \frac{2}{\pi}\lambda_{PM}\omega_r \tag{11.7}$$

where $\omega_r$ is the electrical angular speed.

As two phases conduct at any time the ideal torque $T_e$ is constant.

$$T_e = 2E_a i_{dc}\frac{p}{\omega_r} = \frac{4}{\pi}\lambda_{PM} p i_{dc} \tag{11.8}$$

Between any two current instantaneous commutations the phase current is constant ($i_d = i_{dc}$) and thus the voltage equation is

$$V_d = 2r_s i_{dc} + 2E_a = 2r_s i_{dc} + \frac{4}{\pi}\lambda_{PM}\omega_r \tag{11.9}$$

From Equations 11.9 and 11.8 we obtain

$$\omega_r = \omega_{r0}\left(1 - \frac{T_e}{T_{esc}}\right) \tag{11.10}$$

with

$$\omega_{r0} = \frac{\pi}{4}\frac{V_d}{\lambda_{PM}};\quad T_{esc} = \frac{4}{\pi}\lambda_{PM} p i_{sc};\quad i_{sc} = \frac{V_d}{2r_s} \tag{11.11}$$

The ideal torque speed curve is linear (Figure 11.6) as for a d.c. brush PM motor.

The speed may be reduced and reversed by reducing the level and changing the polarity of d.c. voltage through supplying each motor phase by proper commutation in the PWM inverter (Figure 11.6). The current level is reduced by chopping.

The value of maximum flux linkage in a phase ($\lambda_{PM}$) may be reduced by advancing the current in the phase by the angle $\alpha_a \neq 0$. On the other hand, for an advancing angle $\alpha_a = \pi$ the electromagnetic power becomes negative and thus regenerative braking mode is obtained (Figure 11.5e).

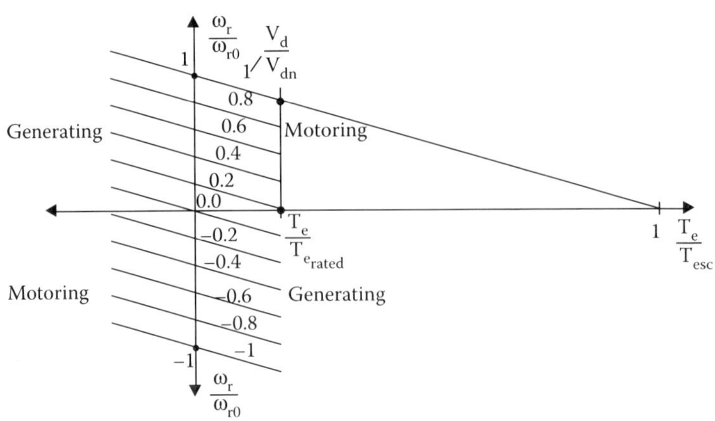


**FIGURE 11.6**
Ideal torque speed curves of BLDC.

### 11.3.2 Rectangular Current Control System

In general, a rectangular current control system contains the BLDC motor, the PWM inverter, the speed and current controllers, the position (speed) sensors (or estimators, for sensorless control), and the current sensors (Figure 11.7).

The currents sequence, produced through inverter adequate control, with 120° current waveforms as in Figure 11.8 also shows the position of the six elements of the proximity sensors with respect to the axis of phase a for a zero advance angle $\alpha_a = 0$.

The location of proximity sensors P (a+, a−, b+, b−, c+, c−) is situated 90° (electrical) behind the pertinent phase with respect to the direction of motion.

With two phases conducting, the stator active m.m.f. is on from 60° to 120° with respect to the rotor position. The ideal current vector (Figure 11.8) also jumps 60° for any phase commutation in the inverter. Each phase is on 120° out of 180°, for the 120° conducting strategy.

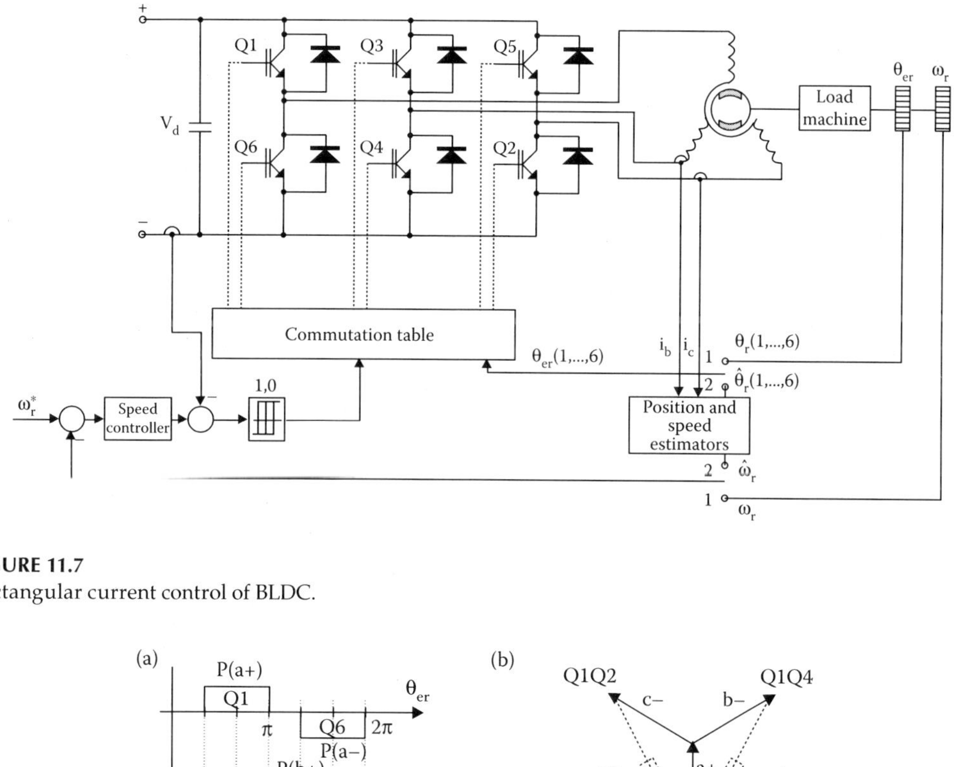


**FIGURE 11.7**
Rectangular current control of BLDC.

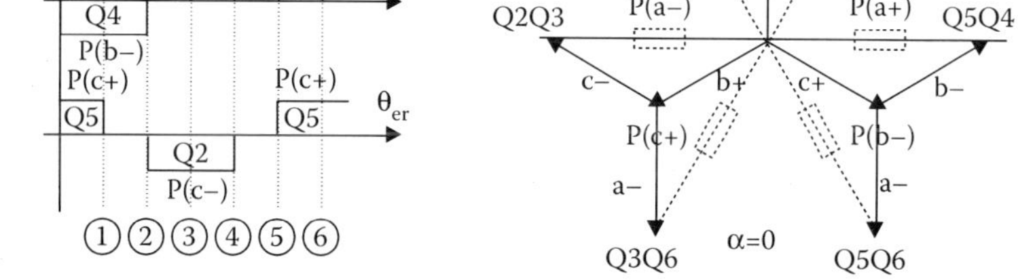


**FIGURE 11.8**
(a) Current sequencing; (b) phase connection.

To reverse the speed, the addresses (IGBTs) of the proximity sensor elements action are shifted by 180° (P(a+) → P(a–); P(b+) → P(b–); P(c+) → P(c–)). The proximity sensor has been located for zero advance angle to provide similar performance for direct and reverse motion. However, through electronic means, the advance angle may be increased as speed increases to reduce the peak PM flux in the stator phase and thus produce more torque, for limited voltage, at high speeds.

Using similar hardware we may also provide for 180° conduction conditions, at high speeds, when all three phases conduct at any time.

### 11.3.3 Hysteresis Current Controller

The d.c. current control requires only one current sensor, placed in the d.c. link, to regulate the current level. The distribution of the current through the six groups of two phases at a time is triggered by the proximity position sensor.

For 180° conduction, with three phases working at a time, a single current sensor in the d.c. link may not fully "represent" the three phase currents, especially if the dead time in the inverter is not properly compensated.

Quite a few current controllers may be applied for the scope. We prefer here the hysteresis controller as it allows a quick understanding of motor–inverter interactions [3]. Once the current in a phase is initiated (as triggered by the proximity sensor) it increases until it reaches the adopted maximum value $i_{max}$. In that moment, the phase is turned off until the current decreases to $i_{min}$. The duration of on and off times, $t_{on}$ and $t_{off}$, is determined based on the hysteresis band: ($i_{max}$–$i_{min}$) (Figure 11.9).

Let us consider the motor equations for the on ($t_{on}$) and off ($t_{off}$) intervals with Q1 and Q4 and D1 D4 in conduction (Figure 11.10), that is a + b conduction.

During the on-time, Figure 11.10a, the a + b equation is

$$V_d = 2r_s i + 2L_s \frac{di}{dt} + E_a - E_b \tag{11.12}$$

$$E_a - E_b \approx 2E_a = E_0 = \text{const.} \tag{11.13}$$

Note that, if the turn-on is advanced by $\alpha_a$, during part of the on-time $E_a$–$E_b$ = 0 and thus a faster current increase is possible.

The solution of Equation 11.12 is

$$i(t) = \frac{V_d - E_0}{2r_s}(1 - e^{-t(r_s/L_s)}) + i_{min} e^{-t(r_s/L_s)} \tag{11.14}$$

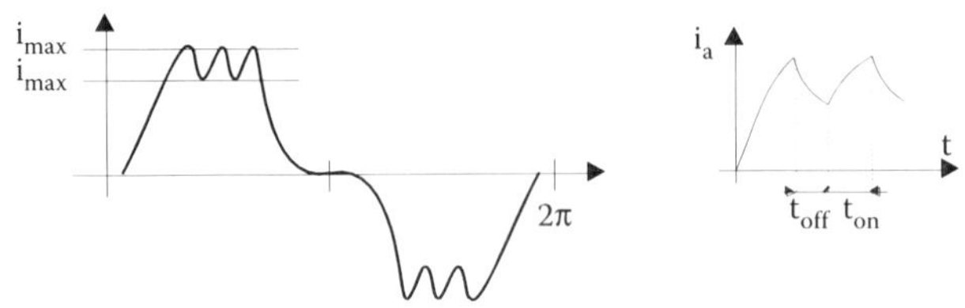


FIGURE 11.9
Current chopping.

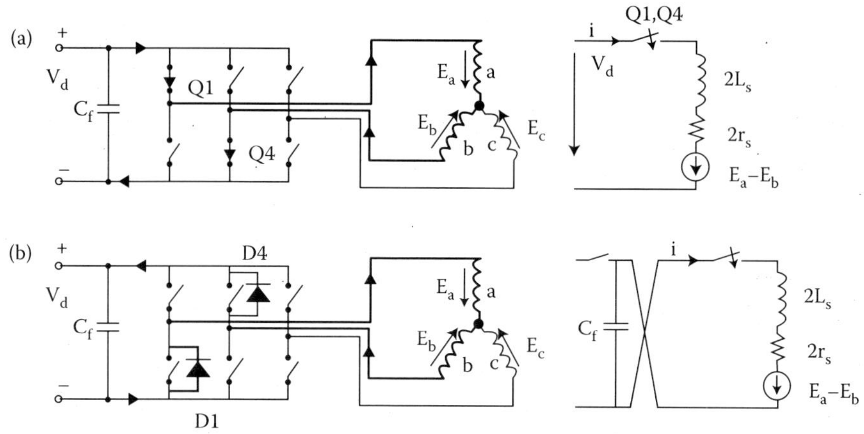


**FIGURE 11.10**
Conduction of phases a and b: (a) on-time; (b) off-time.

To allow for current rising $V_d > E_0$. Above a certain speed $V_d < E_0$ and thus current chopping is not feasible anymore. The current waveform contains, in this case, a single on–off pulse, triggered by the proximity sensor (estimator).

During the off-time (diodes D1 and D4 conducting, in Figure 11.10b) the voltage equation is

$$0 = 2r_s i + 2L_s \frac{di}{dt} + (E_a - E_b) + \frac{1}{C_f}\int i dt + V_{c0} \quad (11.15)$$

where $V_{c0}$ is the capacitor voltage at the end of the on-time, or at the beginning of the off-time.

The solution of Equation 11.15 with $t' = t - t_{on}$ is

$$i(t) = -\frac{V_{c0} + E_0}{2\omega L_s} e^{-\alpha_1 t'} \sin \omega t' - i_{max} \frac{\omega_0}{\omega} e^{-\alpha_1 t'} \sin(\omega t' - \phi) \quad (11.16)$$

with

$$\omega_0 = \sqrt{\frac{1}{2C_f L_s}}; \quad \alpha_1 = \frac{r_s}{2L_s}; \quad \omega = \sqrt{\omega_0^{\,2} - \alpha_1^{\,2}}; \quad \phi = \tan^{-1}\frac{\omega}{\alpha_1} \quad (11.17)$$

The torque $T_e(t)$ expression is

$$T_e(t) = \frac{(E_a - E_b)i(t)}{\omega_r / p} \quad (11.18)$$

So if the e.m.f. is constant in time the electromagnetic torque reproduces the current pulsations between $i_{min}$ and $i_{max}$.

**EXAMPLE 11.1**

A BLDC PM motor is fed through a PWM inverter from a 300 V d.c. source ($V_d = 300$ V). Rectangular current control is performed at an electrical speed $\omega_r = 2\pi 10$ rad/s. The no-load line voltage at $\omega_r$ is $E_0 = 48$ V = const. The cyclic inductance $L_s = 0.5$ mH, $r_s = 0.1\ \Omega$, and the stator winding has q = 1 slot per pole per phase and two poles (2p = 2).

The filter capacitor $C_f = 10$ mF, the current chopping frequency $f_c = 1.25$ kHz, and $t_{on}/t_{off} = 5/3$.

Determine

a. The minimum and maximum values of current ($i_{min}$, $i_{max}$) during current chopping
b. The torque expression and plot it

**Solution**

a. To find $i_{min}$ and $i_{max}$ we have to use Equation 11.14 for $i(t) = i_{max}$ and $t = t_{on}$ and Equation 11.16 for $i(t) = i_{min}$ and $t' = t_{off}$.

$$i_{max} = \frac{(300-48)}{2\cdot 0.1}(1-e^{-0.5\cdot 10^{-3}\cdot 0.1/(0.5\cdot 10^{-3})}) + i_{min}e^{-0.5\cdot 10^{-3}\cdot 0.1/(0.5\cdot 10^{-3})} \tag{11.19}$$

Note that

$$t_{on} = 0.5\text{ ms} \quad \text{and} \quad t_{off} = 0.3\text{ ms} \tag{11.20}$$

Also from Equation 11.17

$$\omega_0^2 = \frac{1}{2\cdot 0.5\cdot 10^{-3}\cdot 10^{-2}} = 10^5; \quad \alpha_1 = \frac{0.1}{2\cdot 0.5\cdot 10^{-3}} \tag{11.21}$$

$$\omega = \sqrt{\omega_0^2 - \alpha_1^2} = \sqrt{10^5 - 100^2} = 300\text{ rad/s} \tag{11.22}$$

$$\phi = \tan^{-1}\left(\frac{300}{100}\right) = 71.56^\circ \tag{11.23}$$

$$\begin{aligned} i_{min} = &-\frac{(300+48)}{2\cdot 300\cdot 0.5\cdot 10^{-3}}\cdot e^{-100\cdot 0.3\cdot 10^{-3}}\cdot \sin(300\cdot 0.3\cdot 10^{-3}) - \\ &- i_{max}\cdot \frac{316.22}{300}e^{-100\cdot 0.3\cdot 10^{-3}} \sin 9300\cdot 0.3\cdot 10^{-3} - 71.56\cdot \pi/180) \end{aligned} \tag{11.24}$$

From Equations 11.19 and 11.24 we may calculate $I_{max}$ and $I_{min}$:

$$I_{max} = 186.77\text{ A} \quad I_{min} = 73.90\text{ A} \tag{11.25}$$

The average current is

$$I_0 = \frac{(I_{max} + I_{min})}{2} \approx 130\text{ A}$$

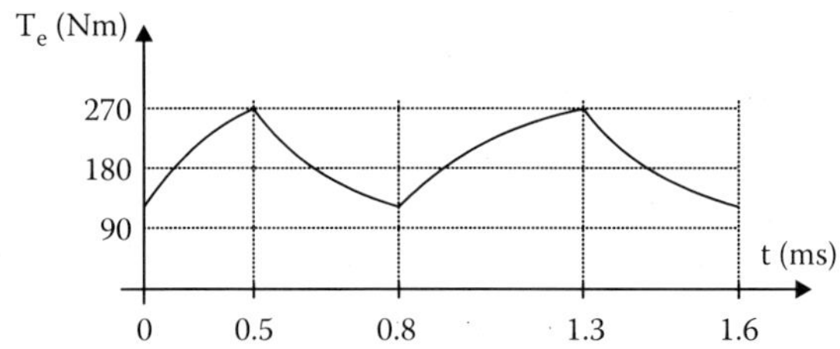


**FIGURE 11.11**
Torque pulsations due to current chopping.

b. The average torque $T_{av}$ is

$$T_{av} = \frac{2E_0I_0}{\omega_r/p} = \frac{2 \cdot 48 \cdot 130}{2 \cdot \pi \cdot (10/1)} = 198.72 \text{ Nm} \tag{11.26}$$

The instantaneous torque (11.18) includes current pulsations (as from Equations 11.14 and 11.16, Figure 11.11).

It should be noted that the chopping frequency is low for the chosen speed ($E_0 \ll V_d$) and thus the current and torque pulsations are large.

Increasing the chopping frequency will reduce these pulsations. To keep the current error band ($I_{max}$–$I_{min}$) within reasonable limits, the chopping frequency should vary with the speed (higher at lower speeds and lower at medium speeds). Though the high-frequency torque pulsations due to current chopping are not followed by the motor speed, due to the much larger mechanical time constant, they produce flux density pulsations and, thus, notable additional core and copper losses (only the average current $I_0$ is, in fact, useful).

### 11.3.4 Practical Performance

So far the phase commutation transients—current overlapping—have been neglected. They, however, introduce notable torque pulsations at $6\omega_r$ frequency (Figure 11.12), much lower than those due to current chopping. To account for them, complete simulation or testing is required [4].

There are also some spikes in the conducting phase when the other two phases commute (points A and B on Figure 11.12).

Also not seen from Figures 11.11 and 11.12 is the cogging torque produced at zero current by the slot openings in the presence of rotor PMs. Special measures are required to reduce the cogging torque to less than 2%–5% of rated torque for high-performance drives.

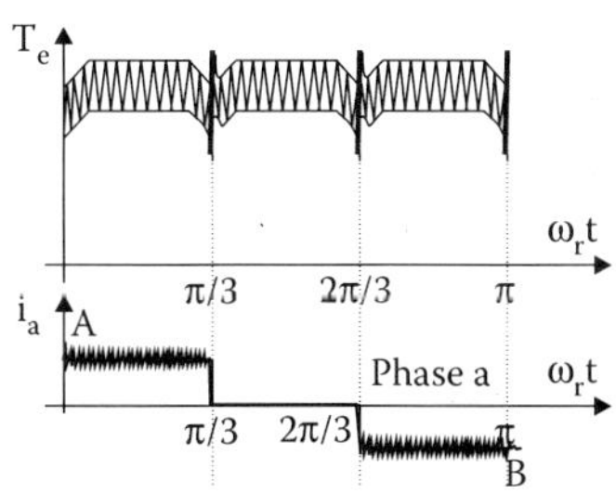


**FIGURE 11.12**
Torque pulsations due to phase commutation.

While at low speeds current chopping is feasible, at high speeds only one current pulse remains (Figure 11.13). The current controller gets saturated and the required current is not reached.

As the advance angle is zero ($\alpha_a = 0$), there is a delay in "installing" the current and thus, as the e.m.f. is "in phase" with the reference current, a further reduction in torque occurs.

### 11.3.5 Extending the Torque/Speed Domain

Extending the torque/speed domain may be obtained (for a given drive) by advancing the phase commutation time by an angle $\alpha_a$ dependent on speed. This phase advancing allows fast current rise before the "occurrence" of the e.m.f. (assuming a PM span angle $\alpha_{PM} < 150°–160°$). An approximate way to estimate the advance angle required $\alpha_a$, for 120° conduction, may be based on linear current rise [5] to the value I

$$(\alpha_a)_{120^0} = \omega_r \frac{L_s I}{V_{dc}}; \quad \omega_r = 2pn\pi \tag{11.27}$$

where n is the rotor speed in rps.

Torque at even higher speeds may be obtained by switching from 120° to 180° current conduction (three phases working at any time). The current waveform changes, especially with advancing angle (Figure 11.14).

This time the e.m.f. is considered trapezoidal, that is close to reality.

The advancing angle $\alpha_a$ may be, for high speeds, calculated assuming sinusoidal e.m.f. [5] and current variation

$$(\alpha_a)_{180^\circ} = \tan^{-1}\left(\frac{\omega_r L_s}{r_s}\right) \tag{11.28}$$

It has been demonstrated [4,5] that 120° conduction is profitable at low to base speeds while 180° conduction with advancing angle is profitable for high speeds (Figure 11.15).

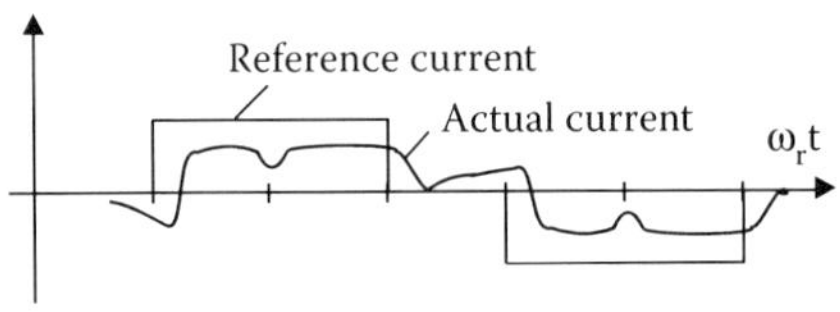


**FIGURE 11.13**
Current waveform at high speeds.

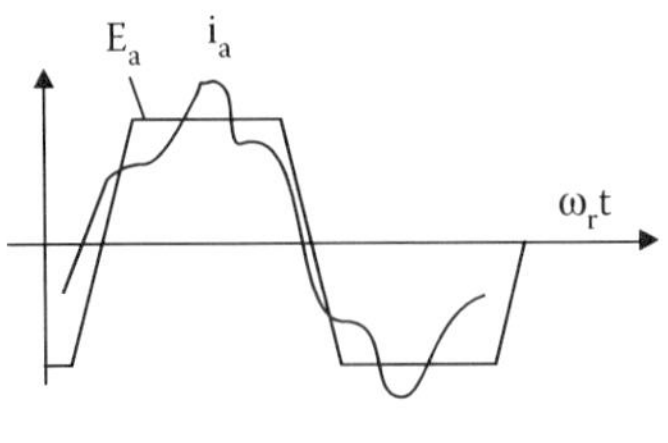


**FIGURE 11.14**
180° conducting with advancing angle at high speed.

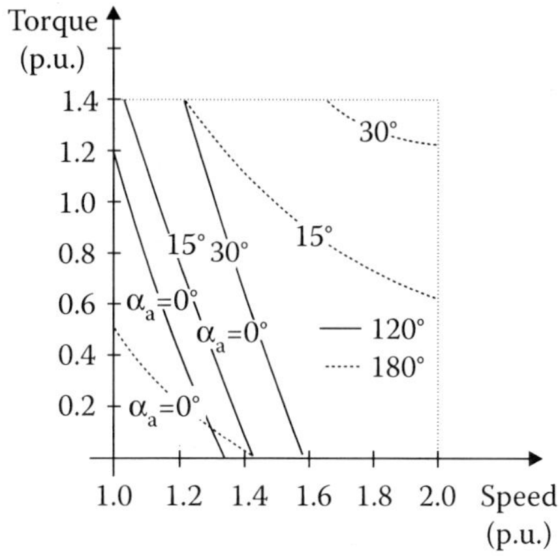


**FIGURE 11.15**
Torque speed curves for various advancing angle $\alpha_a$.

A smooth transition between 120° and 180° conduction is required to fully exploit the torque/speed capabilities of brushless d.c. PM motor drives. However, for 180° conduction, the summation of phase currents is not zero and thus, at least, the winding null point should be tied to the twin series capacitor midpoint in the d.c. voltage link.

**EXAMPLE 11.2: DIGITAL SIMULATION: A BRUSHLESS D.C. PM MOTOR DRIVE IN MATLAB/SIMULINK (AVAILABLE ONLINE)**

We will present here the simulation results on a PM brushless D.C. PM motor drive (BLDC). The motor equations (see Section 10.6, Chapter 10) are

$$L_s \cdot s \begin{vmatrix} i_a \\ i_b \\ i_c \end{vmatrix} = \begin{vmatrix} 1 & 0 & 0 \\ 0 & 1 & 0 \\ 0 & 0 & 1 \end{vmatrix} \cdot \begin{vmatrix} V_{an} \\ V_{bn} \\ V_{cn} \end{vmatrix} - \begin{vmatrix} R_s & 0 & 0 \\ 0 & R_s & 0 \\ 0 & 0 & R_s \end{vmatrix} \cdot \begin{vmatrix} i_a \\ i_b \\ i_c \end{vmatrix} - \begin{vmatrix} e_a(\theta_{er}) \\ e_b(\theta_{er}) \\ e_c(\theta_{er}) \end{vmatrix} \tag{11.29}$$

with

$$L_s = L_a + M \tag{11.30}$$

where $L_a$, M are self and mutual inductances. Introducing the null point of d.c. link (0), the phase voltages are

$$\begin{aligned} V_a &= V_{a0} - V_{n0} \\ V_b &= V_{b0} - V_{n0} \\ V_c &= V_{c0} - V_{n0} \end{aligned} \tag{11.31}$$

For star connection of phases

$$V_{n0} = \frac{[(V_{a0} + V_{b0} + V_{c0}) - (e_a + e_b + e_c)]}{3} \tag{11.32}$$

Here n represents the star connection point of stator windings. $V_{a0}$, $V_{b0}$, $V_{c0}$ could easily be related to the d.c. link voltage and inverter switching state.

The simulation of this drive was implemented in MATLAB–Simulink. The motor model was integrated in a block (PM_SM).

The changing of motor parameters for different simulations is as simple as possible. After clicking on this block, a dialog box appears and you can change them by modifying their default values.

The drive system consists of a PI speed controller ($K_i = 20$, $T_i = 0.05$ s), a reference current calculation block, hysteresis controllers, and motor blocks. The study examines system behavior for starting, load perturbation, and speed reversal. The motor operates at desired speed and the phase currents are regulated within a hysteresis band around the reference currents (as functions of rotor position).

The integration step (50 µs) can be modified from Simulink's *Simulation/Parameters*.

To find out the structure of each block presented above, unmask it (*Options/Unmask*). Each masked block contains a short help screen describing that block (inputs/outputs/parameters).

The block structure of the electric drive system is presented in Figure 11.16 and the motor block diagram in Figure 11.17.

The drive and motor used for this simulation have the following parameters: $V_{dc} = 220$ V, $2p = 2$, $R_s = 1\ \Omega$, $L_s = 0.02$ H, $M = -0.006667$ H, $J = 0.005$ kg m$^2$, $K = 0.763$, hb (hysteresis band) = 0.2.

The following figures represent the speed (Figure 11.18), torque (Figure 11.19) and current (Figure 11.20) responses, and e.m.f. waveform (Figure 11.21), for a *starting process, loading (6 Nm) at 0.2 s, unloading at 0.4 s, reversal at 0.5 s, and loading (6 Nm) again at 0.8 s.*

Details of part of the program are given in Listing 1 for easy reference:

*Listing 1:*

```
function [sys, x0]=bldc0(t,x,u,flag,K)
if flag==3,
%Inputs: position, speedcontroller output/K (reference current
amplitude);
sq3=1.732050808;
pos=u(1);
ir=u(2);
sq=0.866025;
per=fix(pos/2/pi)*360;
%Sector selection from 0 degrees (0-60;60-120...);
if (sin(pos)<sq3/2)&(cos(pos)>0.5)&(sin(pos)>=0),
iar=ir;ibr=-ir;icr=0;fa=1;fb=-1;fc=((per-pos*180/pi)/30+1);end;
if (sin(pos)>sq3/2)&(abs(cos(pos))<0.5),
iar=ir;ibr=0;icr=-ir;fa=1;fb=((pos*180/pi-60-per)/30-1);fc=-1;end;
if (sin(pos)<sq3/2)&(cos(pos)<-0.5)&(sin(pos)>=0),
iar=0;ibr=ir;icr=-ir;fa=((per+120-pos*180/
pi)/30+1);fb=1;fc=-1;end;
if (sin(pos)>-sq3/2)&(cos(pos)<-0.5)&(sin(pos)<0),
iar=-ir;ibr=ir;icr=0;fa=-1;fb=1;fc=((pos*180/pi-180-per)/30-1);end;
if (sin(pos)<-sq3/2)&(abs(cos(pos))<=0.5),
iar=-ir;ibr=0;icr=ir;fa=-1;fb=((per+240-pos*180/
pi)/30+1);fc=1;end;
if (sin(pos)>-sq3/2)&(cos(pos)>0.5)&(sin(pos)<0),
iar=0;ibr=-ir;icr=ir;fa=((pos*180/pi-300-per)/30-1);fb=-1;fc=1;end;
ean=K*fa;
ebn=K*fb;
ecn=K*fc;
sys=[iar ibr icr ean ebn ecn];
elseif flag==0,
  x0=[];
  sys=[0 0 6 2 0 1]';
else sys=[];
end
```

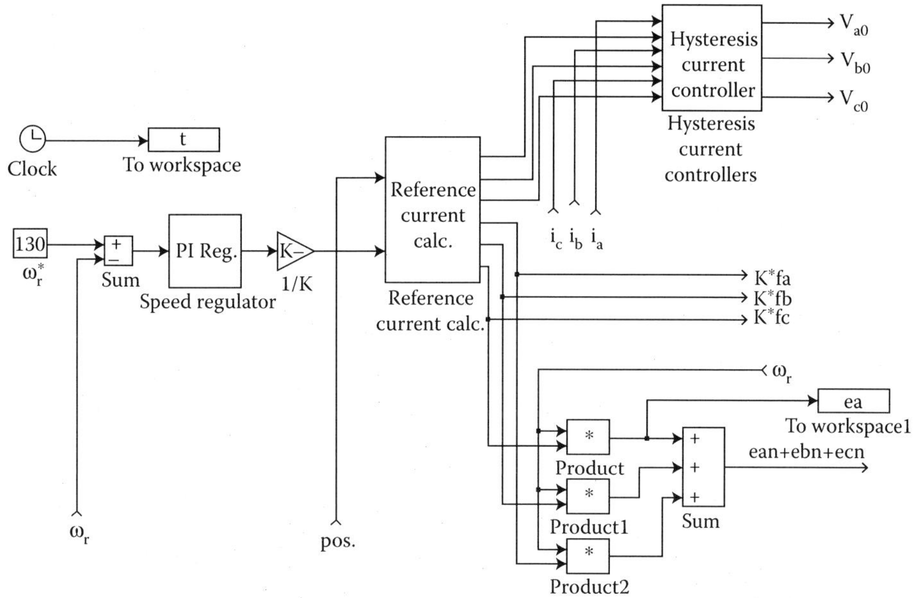


**FIGURE 11.16**
The BLDC rectangular current-drive controller.

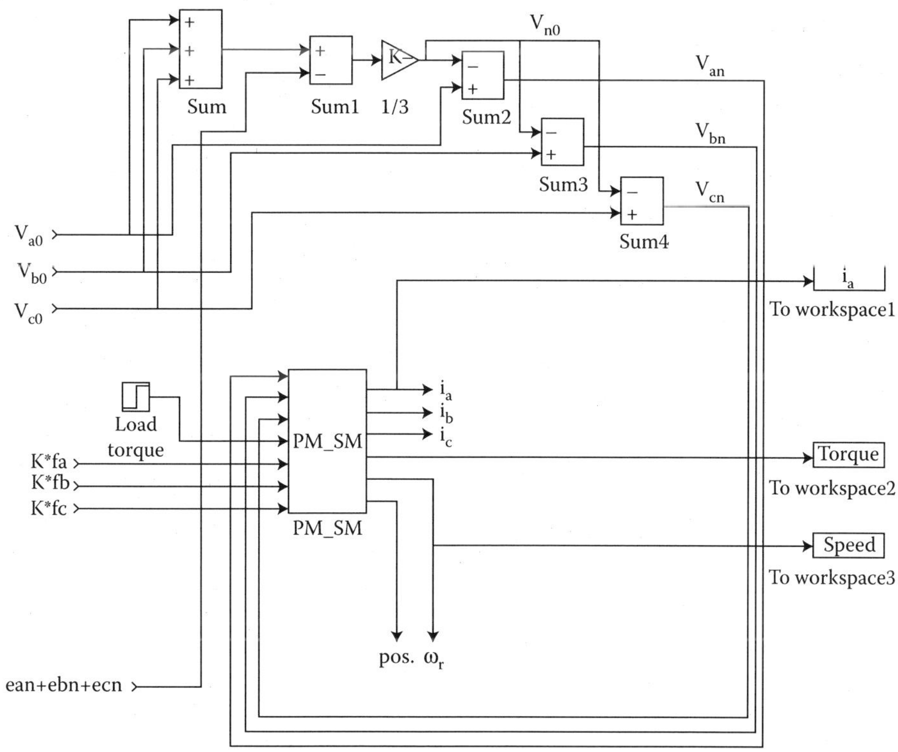


**FIGURE 11.17**
The BLDC motor block diagram.

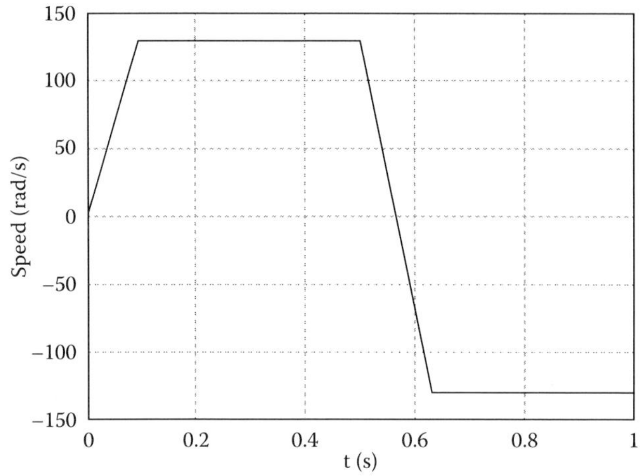


**FIGURE 11.18**
Speed transient response.

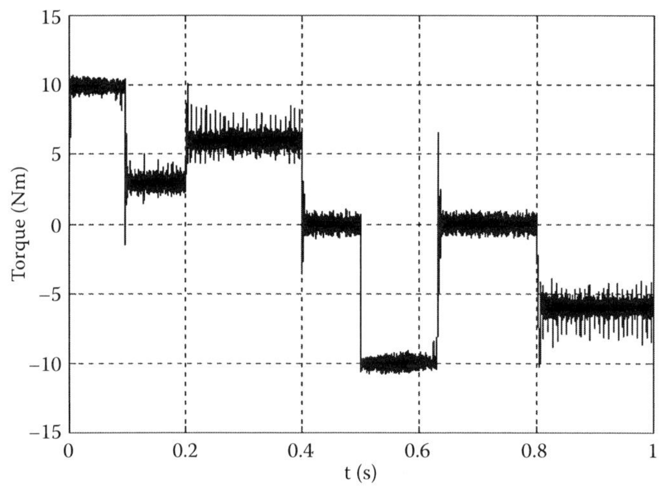


**FIGURE 11.19**
Torque response.

## 11.4 Vector (Sinusoidal) Control

The vector (sinusoidal) control is applied both for surface PM and interior PM rotors and distributed stator windings ($q \geq 2$); and also to tooth-wound PMSMs with sinusoidal e.m.f.

According to Chapter 10, the torque $T_e$ expression is

$$T_e = \frac{3}{2}p\left[\lambda_{PM}i_q + (L_d - L_q)i_d i_q\right] \tag{11.33}$$

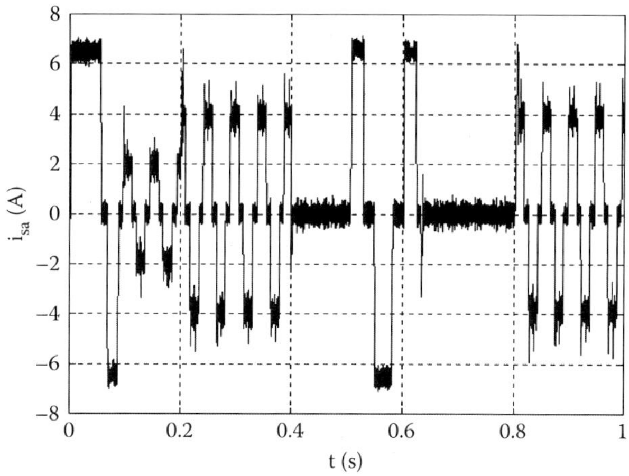


**FIGURE 11.20**
Current waveform ($i_a$).

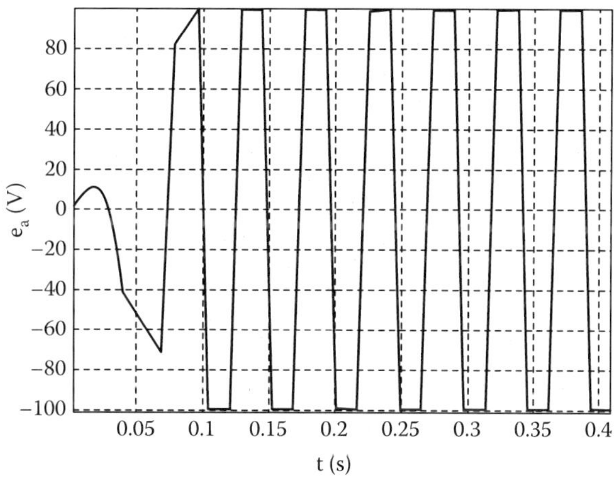


**FIGURE 11.21**
Induced voltage waveform ($e_a$).

The space vector equation, Chapter 10, is

$$\overline{V}_s = r_s \overline{i}_s + \frac{d\overline{\lambda}_s}{dt} + j\omega_r \overline{\lambda}_s \tag{11.34}$$

$$\overline{V}_s = V_d + jV_q; \quad \overline{i}_s = i_d + ji_q; \quad \overline{\lambda}_s = \lambda_d + j\lambda_q \tag{11.35}$$

$$\lambda_d = \lambda_{PM} + L_d i_d; \quad \lambda_q = L_q i_q \tag{11.36}$$

In general, $L_d \leq L_q$ and thus the second (reactive) torque component is positive only if $i_d \leq 0$. For $i_d \leq 0$ the PM flux is diminished by the $i_d$ m.m.f. (Figure 11.22), but the PM always prevails to avoid PM demagnetization.

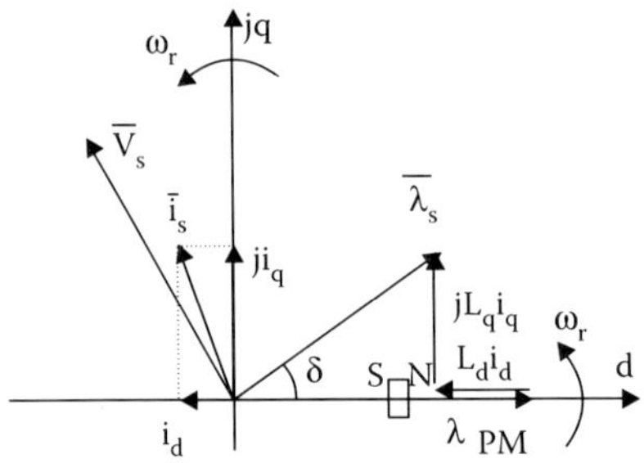


**FIGURE 11.22**
Space vector diagram of PMSM with negative $i_d$ ($i_d < 0$).

For VC, in general, the rotor does not have a damper cage and thus there is no rotor circuit, and, as a direct consequence, no current decoupling network is necessary, in contrast to the IM case.

In VC the PMSM is controlled in d-q rotor coordinates and then the reference d-q currents (or voltages) are transformed into stator coordinates, through the vector rotator (inverse Park transformation), to be then realized by PWM in the inverter.

### 11.4.1 Optimum $i_d$–$i_q$ Relationships

Since the current decoupling network is missing and the drive command variable is the reference torque $T_e^*$, as required by a speed (speed and position) controller, we should now choose $i_d^*$ and $i_q^*$ from the torque equation (11.33). Evidently we need one more equation. This additional information may be obtained through an optimization criterion such as maximum torque per current, maximum torque per flux, maximum efficiency, etc.

As at low speeds the drive is current limited and above base speed $\omega_b$, it is flux limited, we may use these two criteria combined for a high-performance drive. The maximum torque/ampere criterion is applied by using Equation 11.34 and

$$i_s^* = \sqrt{i_d^{*2} + i_1^{*2}} \tag{11.37}$$

$$T_e^* = \frac{3}{2}p\left[\lambda_{PM} + (L_d - L_q)i_d^*\right]\sqrt{i_s^{*2} - i_d^{*2}} \tag{11.38}$$

to find

$$i_{di}^2 + i_{di}\frac{\lambda_{PM}}{L_d - L_q} - i_{qi}^2 = 0 \tag{11.39}$$

For $L_d = L_q$, starting with Equation 11.33, it follows that

$$(i_{di})_{L_d = L_q} = 0 \tag{11.40}$$

On the other hand, for maximum torque per flux

$$\lambda_s^* = \sqrt{(\lambda_{PM} + L_d i_d)^2 + (L_q i_q)^2} \tag{11.41}$$

Proceeding as above we obtain a new relationship between $i_{d\lambda}$ and $i_{q\lambda}$

$$(\lambda_{PM} + L_d i_{d\lambda})^2(2L_d - L_q) - L_d L_q i_{d\lambda}(\lambda_{PM} + L_d i_{d\lambda}) + (L_d - L_q)\lambda_s^{*2} = 0 \quad (11.42)$$

where

$$\lambda_s^* \approx \frac{V_s^*}{\omega_r} \quad (11.43)$$

For $L_d = L_q$, Equation 11.41 becomes (starting all over with Equation 11.33)

$$(i_{d\lambda})_{L_d = L_q} = -\frac{\lambda_{PM}}{L_d} \quad (11.44)$$

Equation 11.44 signifies the complete cancelation of PM flux. The PM is still not completely demagnetized as part of the $L_d i_d$ is leakage flux which does not flow through the PM. Also note that constant current $i_s$ means a circle in the $i_d$–$i_q$ plane and constant stator flux $\lambda_s$ means an ellipse.

We may now represent Equations 11.38, 11.40 through 11.42 as in Figure 11.23.

So, for each value of reference torque $T_e^*$, according to one of the two optimization criteria, unique values of $i_d^*$ and $i_q^*$ are obtained, provided the current limit $i_s^*$ and the flux limit $\lambda_s^*$ are not surpassed.

Note that the flux limit is related to speed $\omega_r$ (through Equation 11.43). So, in fact, the torque $T_e^*$ is limited with respect to stator current and flux (speed) (Figure 11.24a).

$$i_d^* = f_{d_{i,\lambda}}(T_e^*) \quad (11.45)$$

$$i_q^* = f_{q_{i,\lambda}}(T_e^*) \quad (11.46)$$

$T_e^*$ limit is dependent on speed (Figure 11.24b).

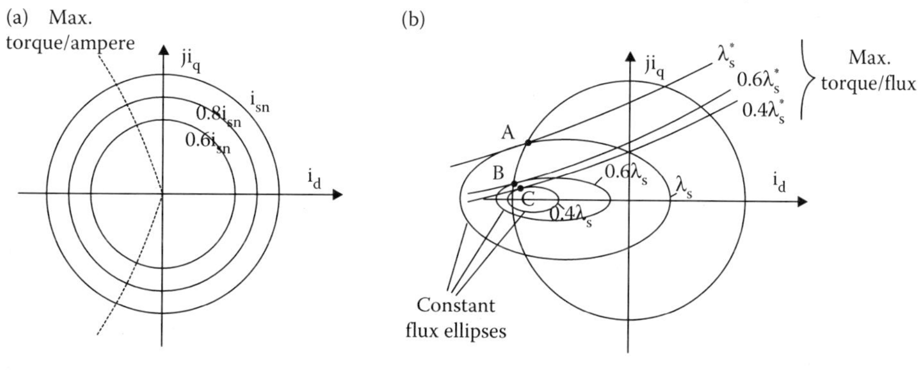


**FIGURE 11.23**
$i_d$–$i_q$ (a) for given current $i_s^*$, (b) for given flux $\lambda_s^*$.

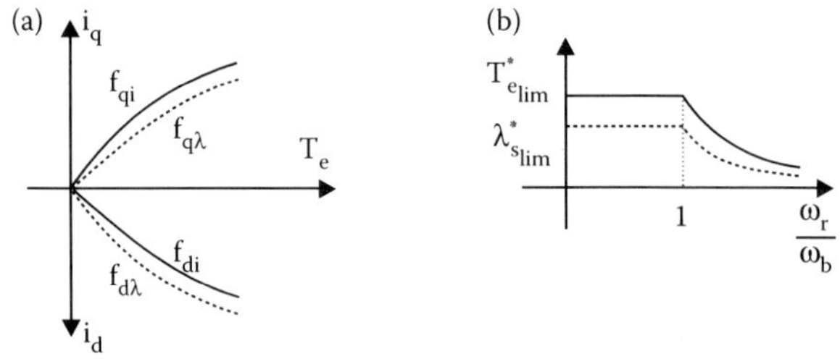


**FIGURE 11.24**
(a) $i_d$–$i_q$ optimum relationship and (b) torque and flux limits with speed.

**EXAMPLE 11.3**

A PMSM with $V_{sn} = 120\sqrt{2}$ V, $i_{sn} = 20$ A, $l_d = 0.6$ p.u., no-load voltage $e_0 = 0.7$ p.u., $l_q = 1.2$ p.u., $n_n = 15$ rps, $p = 1$ (pole pairs), $r_s = 0$, is driven at a current of 10 A up to base speed $n_b$, according to the criterion of maximum torque per current.

Calculate the torque $T_{ei}$ up to base speed and the base speed $\omega_b$.

**Solution**

First, we have to calculate $l_d$, $l_q$, $e_0$ in absolute units $L_d$, $L_q$, $E_0$

$$L_d = l_d \cdot \frac{V_{sn}}{i_{sn} \cdot \omega_n} = \frac{0.6 \cdot 120\sqrt{2}}{2\pi 15 \cdot 20} \approx 0.0539 \text{ H} \tag{11.47}$$

$$L_q = l_q \cdot \frac{V_{sn}}{i_{sn} \cdot \omega_n} = \frac{1.2 \cdot 120\sqrt{2}}{2\pi 15 \cdot 20} \approx 0.1077 \text{ H} \tag{11.48}$$

$$E_0 = e_0 \cdot V_{sn} = 0.7 \cdot 120\sqrt{2} = 118.44 \text{ V} \tag{11.49}$$

The PM flux $\lambda_{PM}$ is

$$\lambda_{PM} = \frac{E_0}{\omega_b} = \frac{118.44}{2\pi 15} = 1.257 \text{ Wb} \tag{11.50}$$

Making use of Equation 11.39 with Equation 11.50

$$2{i_{di}}^2 + i_{di} \cdot \frac{1.257}{0.0539 - 0.1077} - 10^2 = 0 \tag{11.51}$$

$$i_{di} = -3.645 \text{ A} \tag{11.52}$$

$$i_{qi} = \sqrt{100 - 3.645^2} = 9.312 \text{ A} \tag{11.53}$$

and the torque

$$T_{ei} = \frac{3}{2}p[\lambda_{PM} + (L_d - L_q)i_d]\cdot i_q$$

$$= \frac{3}{2}\cdot 1\cdot[1.257 + (0.0539 - 0.1077)(-3.645)]\cdot 9.312 = 20.3\ \text{Nm} \tag{11.54}$$

Now we have to find the maximum speed for which this torque may be produced

$$\lambda_{di} = \lambda_{PM} + L_d i_d = 1.257 + 0.0539\cdot(-3.645) = 1.06\ \text{Wb}$$

$$\lambda_{qi} = L_q i_q = 0.1077\cdot 9.312 = 1.00\ \text{Wb} \tag{11.55}$$

$$\lambda_{si} = \sqrt{\lambda_{di}^2 + \lambda_{qi}^2} = \sqrt{1.06^2 + 1.00^2} = 1.457 \tag{11.56}$$

Consequently, for rated voltage (maximum inverter voltage), the base speed $\omega_b$ is

$$\omega_b = \frac{V_{sn}}{\lambda_{si}} = \frac{120\sqrt{2}}{1.457} = 116.108\ \text{rad/s} \tag{11.57}$$

$$n_b = \frac{\omega_b}{2\pi p} = \frac{116.108}{2\pi 1} = 18.454\ \text{rps} \approx 1110\ \text{rpm} \tag{11.58}$$

In a similar way we may proceed above base speed using the maximum torque/flux criterion.

Note: The value of $T_{ei}$, if provided for up to base speed, does not leave any extra voltage for acceleration (at base speed).

This is only a small item that sheds light on the intricacies of a high-performance drive design.

### 11.4.2 Indirect Vector Current Control

Indirect vector current control means, in fact, making use of precalculated $f_{di,\lambda}$ and $f_{qi,\lambda}$ (11.45 and 11.46), to produce the reference d-q currents $i_d^*$, $i_q^*$. Then, with the Park transformation, the reference phase current controllers are used to produce PWM in the inverter. A rotor position sensor is required for position and speed feedback (Figure 11.25).

As for the IMs, the a.c. controllers may be replaced by d.c. current controllers (in rotor coordinates) to improve the performance at high speeds especially. Though, in principle, direct VC is possible, it is hardly practical unless the drive is sensorless.

### 11.4.3 Indirect Voltage and Current Vector Control

As expected, the vector current control does not account for the e.m.f. effect of slowing down the current transients with increasing speed. This problem may be solved by using the stator voltage equation for voltage decoupling

$$V_d^* = (r_s + sL_d)i_d^* - \omega_r L_q i_q^* \tag{11.59}$$

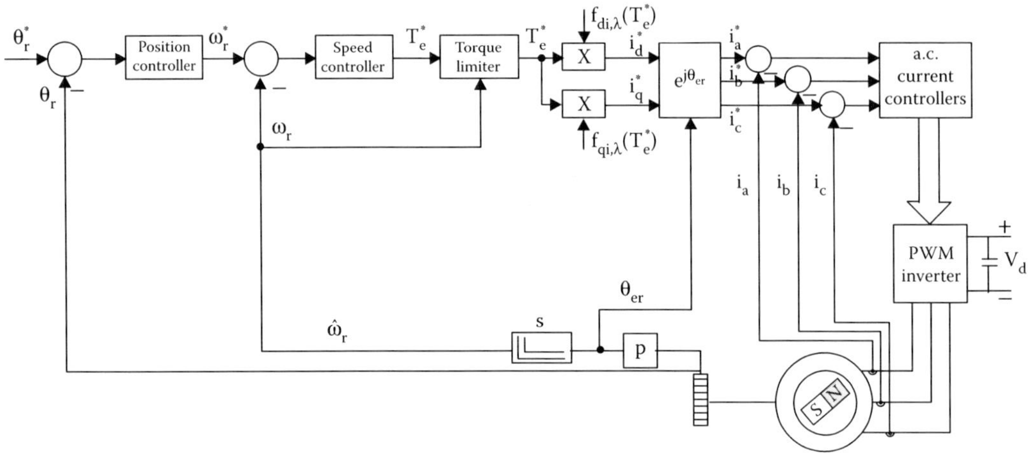


**FIGURE 11.25**
Indirect vector current control of PMSM.

$$V_q^* = (r_s + sL_q)i_q^* + \omega_r(\lambda_{PM} + L_d i_d^*) \tag{11.60}$$

The d.c. current controllers may replace the first terms in Equations 11.59 and 11.60 and thus only the motion-induced voltages are feedforwarded. Then PWM is performed open-loop, to produce the phase voltages at the motor terminals, through the PWM inverter (Figure 11.26).

As expected, the open-loop (voltage) PWM has to observe the voltage limit

$$(V_d^*)^2 + (V_q^*)^2 \le V_{sn}^2 \tag{11.61}$$

At low speeds the current controllers prevail, while at high speeds the voltage decoupler takes over. Results obtained with such a method [6], based only on $f_{di}$, $f_{qi}$ (maximum

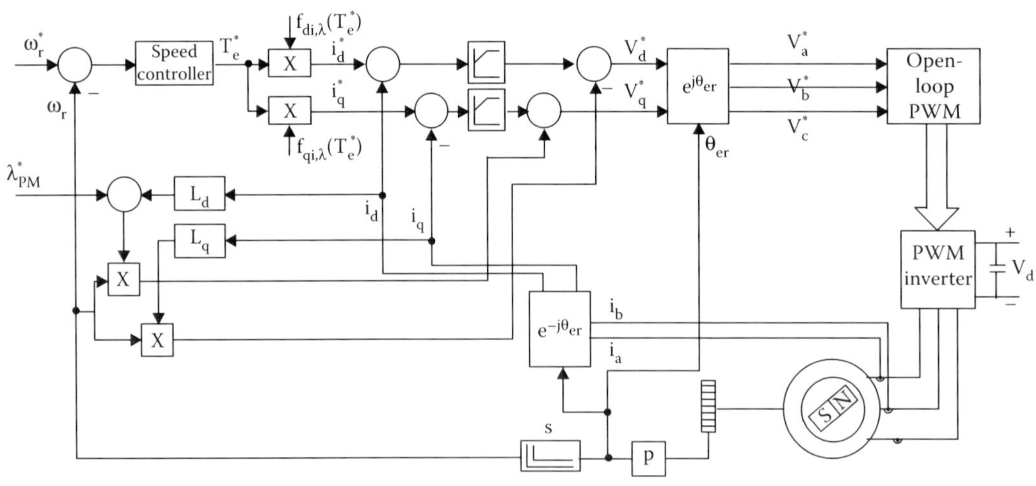


**FIGURE 11.26**
Indirect voltage and current VC of PMSM with d-q (d.c.) current controllers.

torque per current criterion) proved a remarkable enlargement of the torque/speed envelope (Figure 11.27), for an interior PMSM.

NOTE: The e.m.f. compensation in indirect current VC may be solved by desaturating the $i_q$ controller based on d axis controller saturation for a given interval of time [7]. Similar results with those in Figure 11.27 have been thus obtained with a notably smaller online computation effort.

### 11.4.4 Fast Response PMSM Drives: Surface PM Rotor Motors with Predictive Control

Interior PMSM drives have been, so far, treated for VC. As $L_q$ is rather high in such motors, at least $i_q$ response is rather slow in comparison with IM drives. However, for surface PM rotor PMSMs, $L_d = L_q = L_s$ is small and thus fast current (and torque) response may be obtained.

The current increment $\Delta i_s$ is

$$\Delta\bar{i}_s(t) \approx \frac{\bar{V}_s - \bar{E}}{L_s}\Delta t$$
$$\Delta\bar{i}_s = \bar{i}_s^* - \bar{i}_s \tag{11.62}$$

So the current controller error, $\Delta\bar{i}_s$, with the e.m.f. space vector $\bar{E}$ estimated, is used to calculate the required (predicted) stator voltage vector which, then, may be open-loop pulse width modulated to control the drive.

This classical method [8] has produced spectacular results with position sensors (encoders) (Figure 11.28).

The low speed/time linearity [8] is obtained also by using special measures to reduce all torque pulsations to less than 1.5% of rated torque.

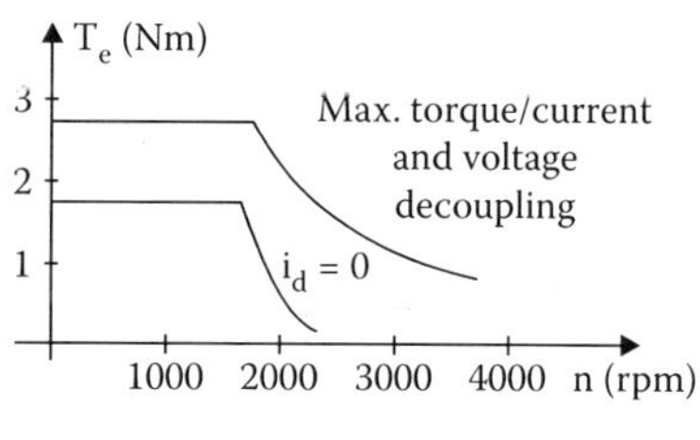


**FIGURE 11.27**
Torque/speed envelope for $i_d = 0$ and for voltage and current VC.

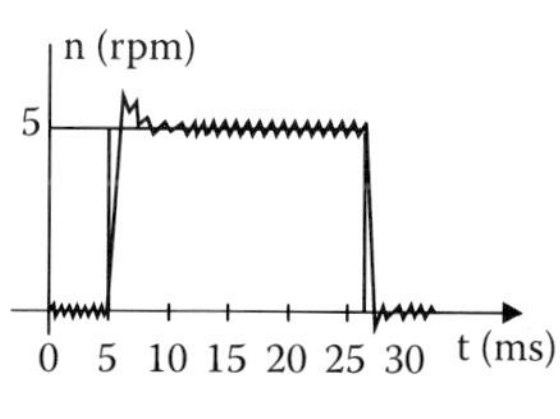


**FIGURE 11.28**
Speed responses at low speed.

## EXAMPLE 11.4: DIGITAL SIMULATION: INDIRECT VECTOR A.C. CURRENT CONTROL IN MATLAB/SIMULINK (AVAILABLE ONLINE)

For the PMSM motor in Example 11.3, making use of the indirect vector a.c. current control system of Figure 11.25, perform digital simulations.

The simulation of this drive was implemented in MATLAB–Simulink. The motor model was integrated in a block (PM_SM) (Figures 11.29 through 11.31).

The changing of motor parameters for different simulations is as simple as possible. By clicking on this block, a dialog box appears and you can change the parameters by modifying their default values.

The reference values for the current components ($i_d^*$ and $i_q^*$) have been obtained from the maximum torque/current criterion (11.39). You can modify the speed regulator (*PI regulator*) parameters (amplification and integration time constant) to study the behavior of the system. A *Commutation Frequency* block has been included to satisfy the sampling criteria (to control the commutation frequency). The integration step can be modified from Simulink's *Simulation/Parameters*.

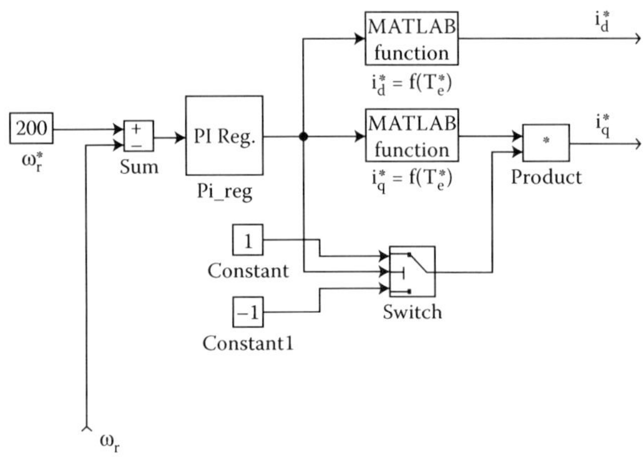


**FIGURE 11.29**
$i_d^*$, $i_q^*$ referencers.

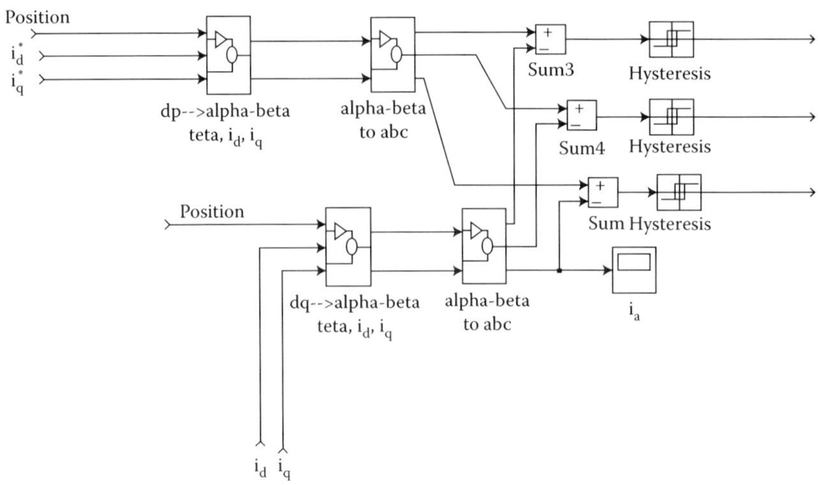


**FIGURE 11.30**
A.c. current controllers.

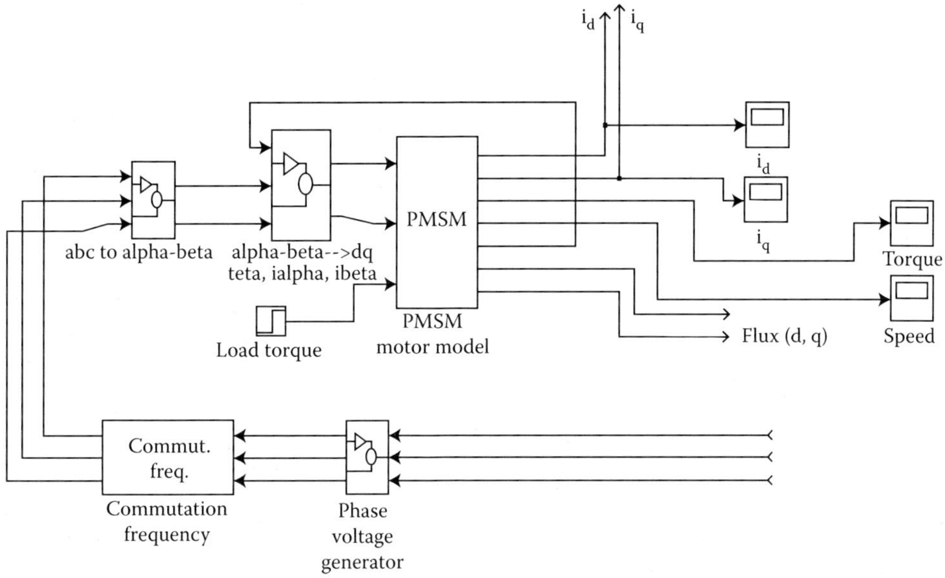


**FIGURE 11.31**
PMSM block diagram.

To find out the structure of each block presented above, unmask them (*Options/Unmask*). The *abc_to_alpha-beta* and *alpha-beta_to_dq* blocks produce the coordinate transform.

The block structure of the electric drive system is presented in Figures 11.29 through 11.31.

The motor used for this simulation has the following parameters: $P_n = 900$ W, $U_n = 220$ V, $2p = 4$, $n = 1700$ rpm; $I_n = 3$ A, $\lambda_{PM} = 0.272$ Wb, $R_s = 4.3\ \Omega$, $L_d = 0.027$ H, $L_q = 0.067$ H, $J = 0.00179$ kg m$^2$.

The following figures represent the speed (Figure 11.32), torque (Figure 11.33), and current waveform (Figure 11.34), for a *starting process and load torque (2 Nm) applied at 0.4 s.*

The d-q reference current relationships to torque for maximum torque per current are given in Figure 11.35.

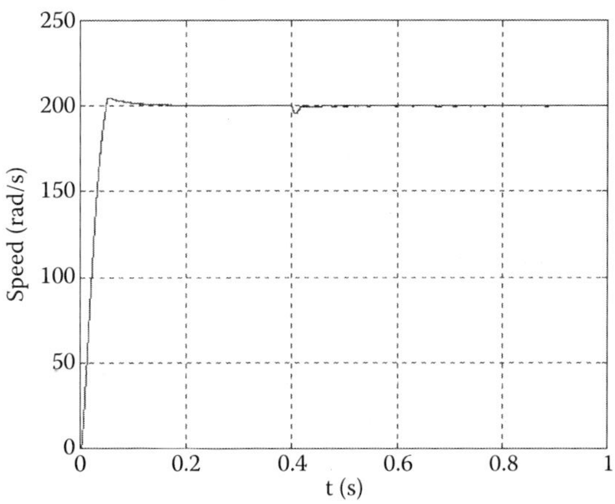


**FIGURE 11.32**
Speed transient response.

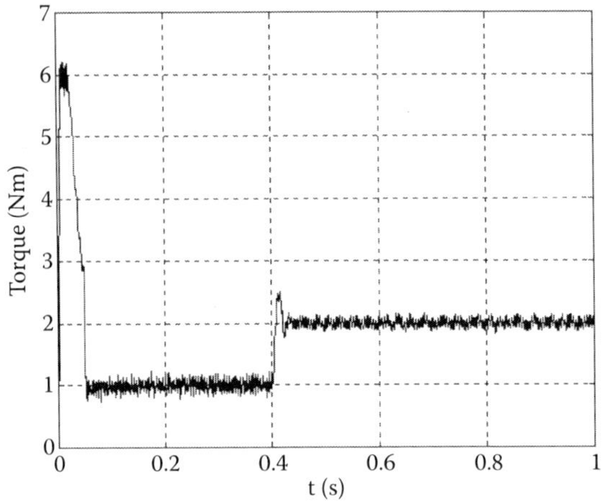


**FIGURE 11.33**
Torque response.

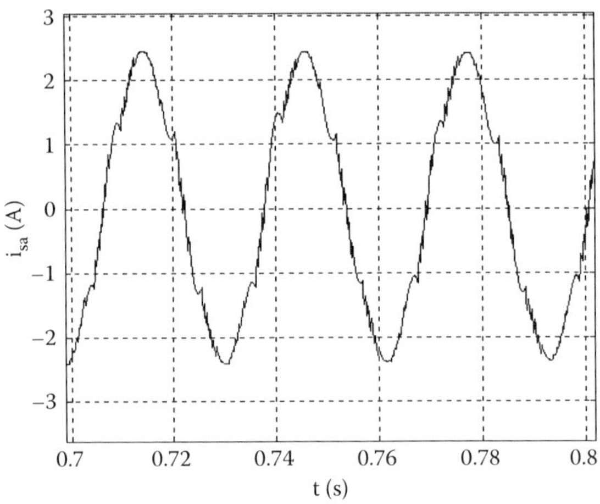


**FIGURE 11.34**
Current waveform ($i_a$) under steady state.

## EXAMPLE 11.5: DIGITAL SIMULATION: INDIRECT VECTOR D.C. CURRENT CONTROL OF PMSM IN MATLAB/SIMULINK (AVAILABLE ONLINE)

The simulation of this drive (Figure 11.36) was implemented in MATLAB–Simulink simulation program. The motor model was integrated in the PM_SM block (Figure 11.37).

Changing of motor parameters for different simulations is as simple as possible. By clicking on the block, a dialog box appears and you can change the parameters by modifying their default values.

The reference values for the current components ($i_d^*$ and $i_q^*$) have been obtained from the maximum torque/current criterion (11.38). This part of simulation has been developed as a MATLAB routine, Listing 2 and Figure 11.38.

You can modify the speed regulator (*PI regulator*) parameters (amplification and integration time constant) to study the behavior of the system. A *Commutation Frequency* block was introduced to satisfy the sampling criteria (to control the commutation frequency). The integration step can be modified from Simulink's *Simulation/Parameters*.

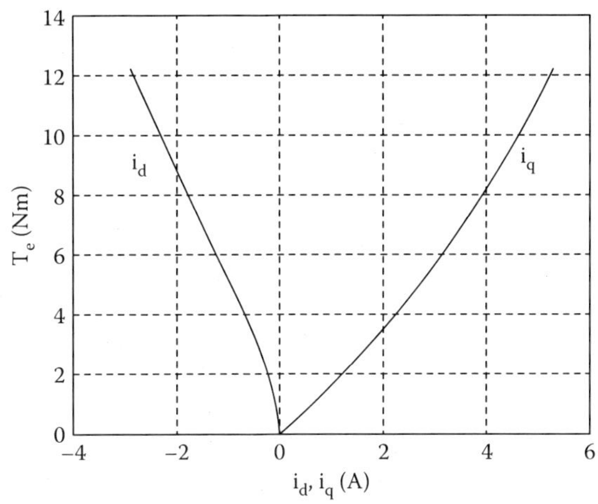


**FIGURE 11.35**
The d-q currents versus torque.

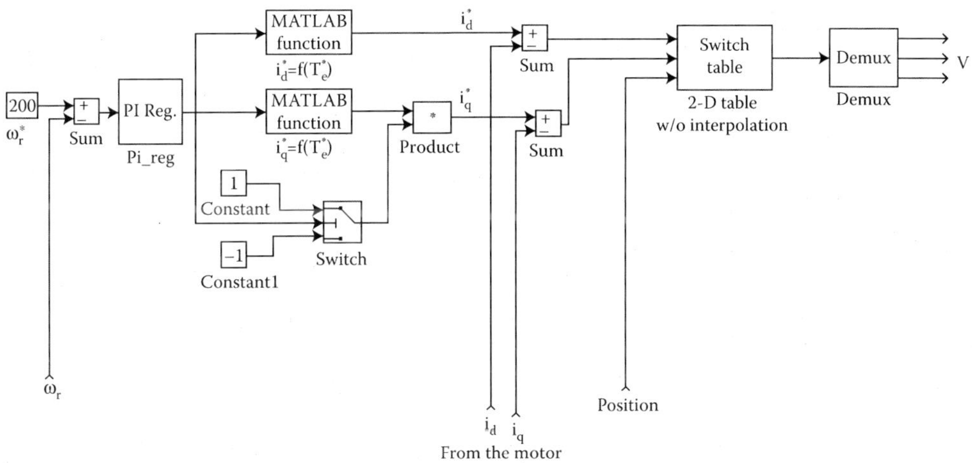


**FIGURE 11.36**
The indirect vector d.c. current controllers.

To find out the structure of each block presented above, unmask them (*Options/Unmask*). The *abc_to_alpha-beta* and *alpha-beta_to_dq* blocks produce the coordinate transform.

The motor used for this simulation has the following parameters: $P_n = 900$ W, $U_n = 220$ V, $2p = 4$, $n = 1700$ rpm; $I_n = 3$ A, $\lambda_{PM} = 0.272$ Wb, $R_s = 4.3\ \Omega$, $L_d = 0.027$ H, $L_q = 0.067$ H, $J = 0.000179$ kg m$^2$.

The following figures represent the speed (Figure 11.39), torque (Figure 11.40), and d-q current responses (Figures 11.41 and 11.42), for a *starting process and load torque (2 Nm) applied at 0.4 s.*

*Listing 2:*

```
% This program creates the table for id* = f(Te*) and iq* = f(Te*).
% Variable definition
  clear;
  lPM = 0.272;
  Ld = 0.027;
```

```
  Lq = 0.067;
  p1 = 4;
  Isn = 6;
k = 1;
for is = 0:0.1:Isn
    p = [2 lPM/(Ld-Lq) -is*is];
    R = roots(p);
    if (R(1) < R(2)), id = R(1);
    else
     id = R(2);
    end
    iq1 = sqrt(is*is-id*id);
    iq2 = -iq1;
    Te1 = 1.5*p1*(lPM + (Ld-Lq)*id)*iq1;
    Te2 = -Te1;
    Vid(k) = id;
    Viq1(k) = iq1;
    Viq2(k) = iq2;
    VTe1(k) = Te1;
    VTe2(k) = Te2;
    i = is;
    k = k + 1;
end
M1 = [VTe1;Vid]';
M21 = [VTe1;Viq1';
M22 = [VTe2;Viq2';
plot(Vid,VTe1,'-',Viq1,VTe1,'-g')
xlabel('id,iq')
ylabel('Te')
title('Variation of Id & Iq with Te')
```

As seen from Figure 11.36 the d-q current error drives a switch table to produce the desired (adequate) voltage vector in the inverter; this explains the larger speed, torque, and d-q currents ripple.

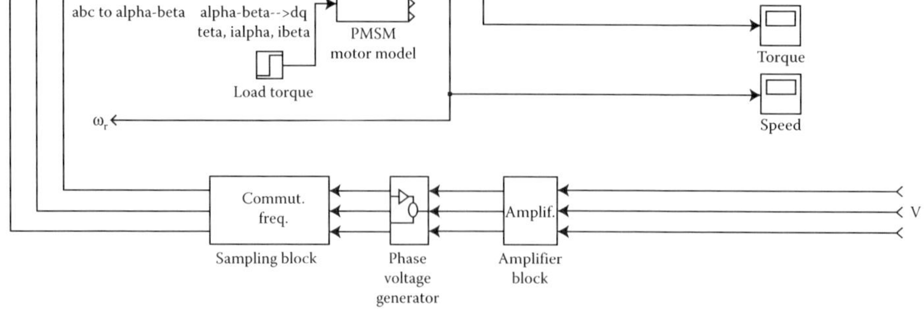


**FIGURE 11.37**
PMSM model.

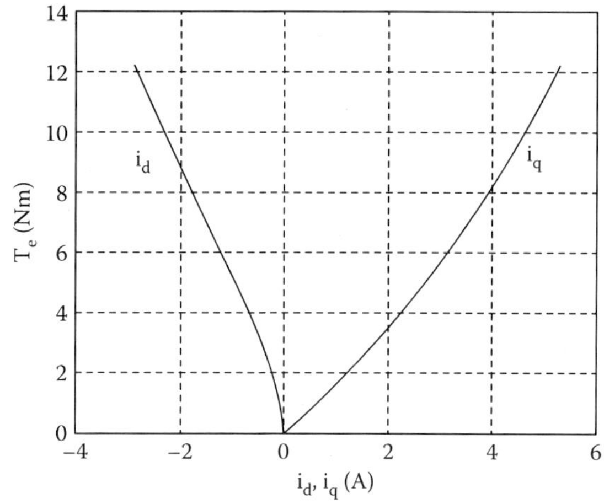


**FIGURE 11.38**
$i_d$–$i_q$ reference currents versus torque for maximum torque/current criterion.

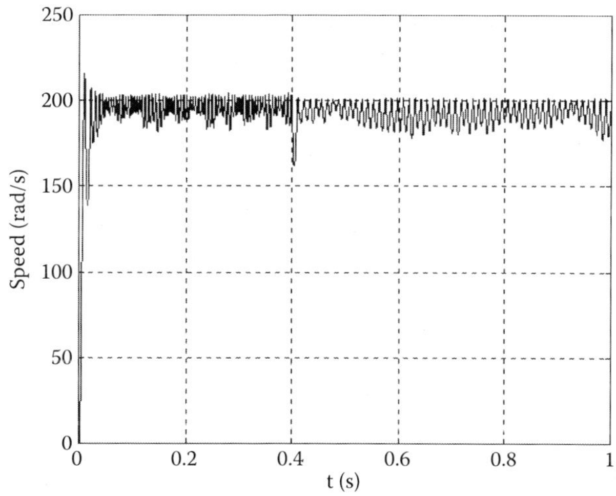


**FIGURE 11.39**
Speed transient response.

## 11.5 DTFC of PMSMs

DTFC, as presented for the IM, implies direct torque and stator flux control through direct triggering of one (or a combination of) voltage vector(s) in the PWM voltage-source inverter that supplies the PMSM.

A general configuration of a DTFC system for PMSMs is shown in Figure 11.43.

The table of switchings (TOS) is obtained based on the principle that the stator flux space vector variation goes in the direction of the applied voltage space vector

$$\bar{\lambda}_s \approx \bar{\lambda}_{s0} + \bar{V}(i) \cdot T \tag{11.63}$$

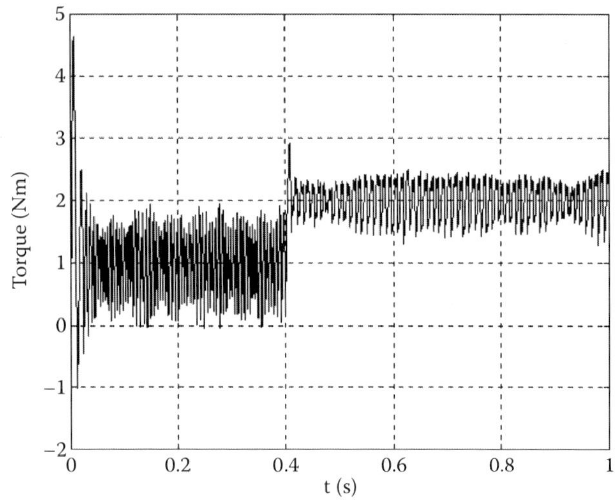


**FIGURE 11.40**
Torque response.

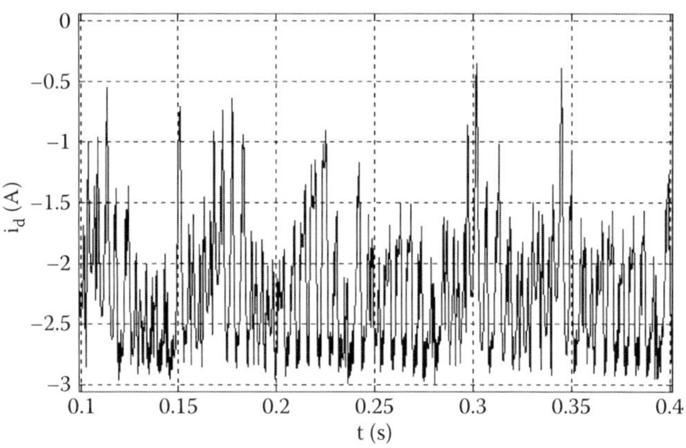


**FIGURE 11.41**
Current waveform ($i_d$) under steady-state.

Also, to increase torque, we have to advance the flux vector in the direction of motion. Flux increasing means an angle above 90° between $\overline{V}(i)$ and $\overline{\lambda}_{s0}$. Consequently, the adequate voltage vector is also chosen based on the 60° wide sector ($\theta_i$) where the initial flux vector is located (Figure 11.44).

The complete table of switching (TOS) is the same as for the IM (Chapter 9, Table 9.2). The TOS is "visited" with a certain frequency and thus constant switching frequency is required.

Applying all voltage vectors for a given (constant) time interval $T = 1/f_c$ may not be the most adequate solution for low speeds where short nonzero voltage vectors and long zero vectors are required.

The $T_{on}$ time of nonzero voltage vectors may thus be made speed dependent, provided $T_{on} < T$ and $T_{on} > T_{min}$ (minimum conduction time allowable). On the other hand, if

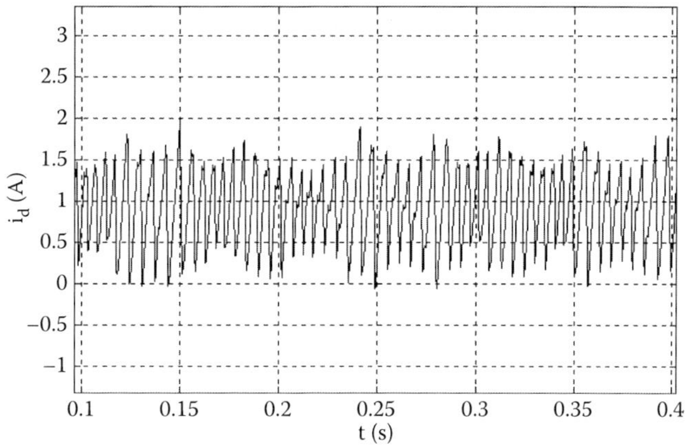


**FIGURE 11.42**
Current waveform ($i_q$) under steady-state.

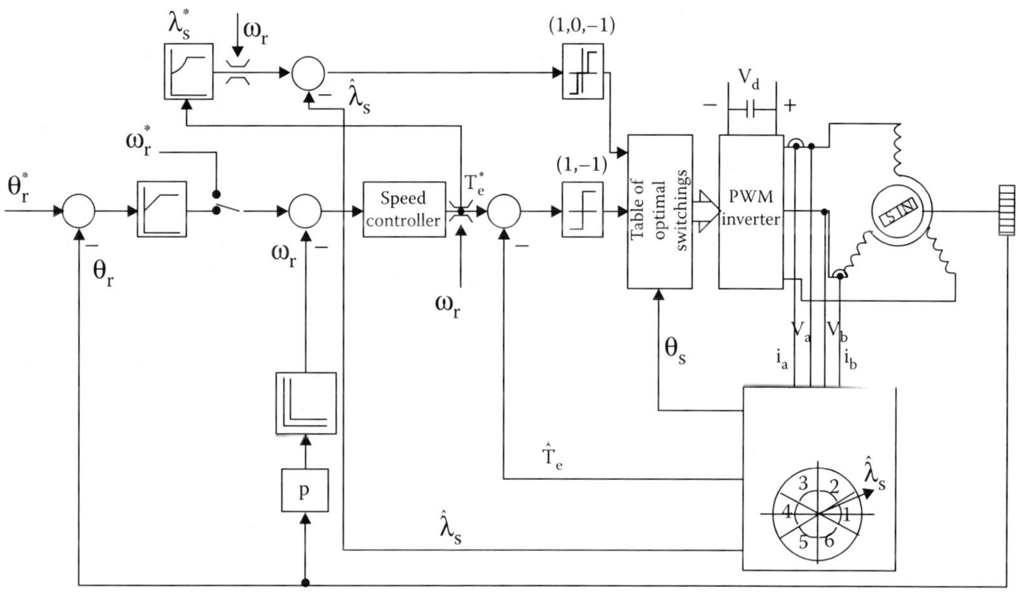


**FIGURE 11.43**
DTFC for PMSMs.

the "exact" position of stator flux $\hat{\theta}_s$ is estimated, the required voltage vector $\overline{V}_s^*$ may be predicted

$$\overline{V}_s^* = r_s \overline{i}_s^* + \Delta\overline{\lambda}_s \tag{11.64}$$

This time the space vector flux error (not only the amplitude error) is required. Once $\overline{V}_s^*$ is computed, space vector modulation techniques may be applied to calculate the timings $t_1$, $t_2$ of the adjacent voltage vectors V(i), V(i + 1), and of the zero voltage vector $V_0(V_7)$

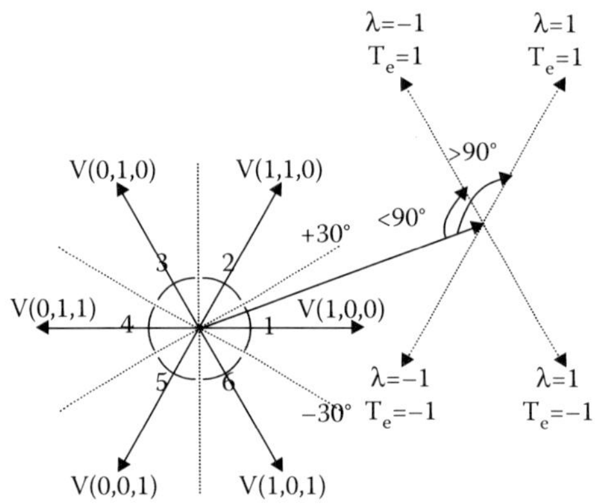


**FIGURE 11.44**
Voltage vector selection for the first sector.

$$\overline{V}_s^* = \overline{V}(i)\cdot t_1 + \overline{V}(i+1)\cdot t_2 + V_0(0)\cdot t_0 \tag{11.65}$$

$$t_0 + t_1 + t_2 = T \tag{11.66}$$

Smoother torque and flux control is obtained this way for given commutation frequency. The flux/torque relationship (Figure 11.43) is derived directly from the maximum torque/current and maximum torque/flux criteria.

Also the maximum reference flux and torque are limited, depending on speed, to keep the current and voltage within available limits.

The key component of DTFC is represented by the stator flux and torque observer (Figure 11.45) which represents the "price" for its control simplicity and robustness.

### 11.5.1 Stator Flux and Torque Observer

It is, in general, assumed that the stator currents $i_a$, $i_b$ and voltages $V_a$, $V_b$ (or the d.c. link voltage $V_d$) are measured. Also, an encoder provides both rotor position $\theta_r$ and speed feedback

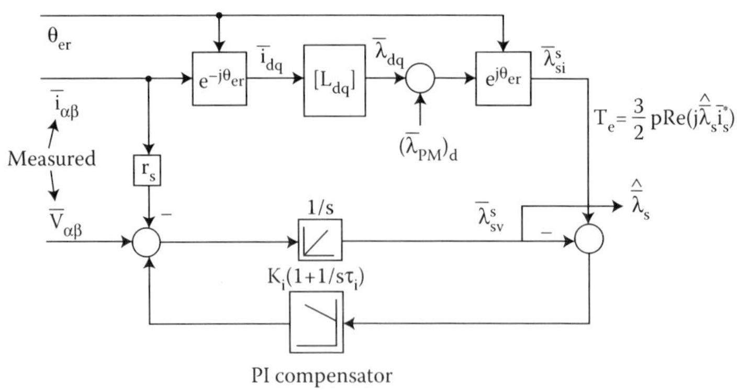


**FIGURE 11.45**
Voltage and current stator flux observer.

in all high-performance drives [9]. The observers considered for IMs may be applied here also. Among the real, practical ones we mention the voltage–current, full-order observers, Kalman filters, and MRAS closed-loop configurations.

The voltage–current closed-loop observer is presented here in some detail.

The voltage model in stator coordinates is

$$\frac{d\overline{\lambda}_{sv}^{s}}{dt} = \overline{V}_s - r_s\overline{i}_s \tag{11.67}$$

while the current model, in rotor coordinates, is

$$\overline{\lambda}_{dq} = \overline{\lambda}_{si}^{r} = L_d i_d + \lambda_{PM} + jL_q i_q; \quad \overline{i}_{dq} = \overline{i}_s^{r} = i_d + ji_q \tag{11.68}$$

The current in rotor coordinates $\overline{\lambda}_{si}^{r}$ is

$$\overline{i}_s^{r} = \overline{i}_s^{s} e^{-j\theta_{er}} \tag{11.69}$$

The stator flux $\overline{\lambda}_{si}^{r}$ is transformed back to stator coordinates as

$$\overline{\lambda}_{si}^{s} = \overline{\lambda}_{si}^{r} e^{j\theta_{er}} \tag{11.70}$$

It is evident from the above that the voltage model is plagued by the stator resistance error and the integration drift at low frequencies.

The current model is influenced by parameter detuning (due to magnetic saturation) and by the position error, but it works from zero speed (frequency). A combination of the two with a PI compensator is designed such that the current model is predominant at low speeds, while the voltage model takes over at high speeds (Figure 11.45).

The PI compensator provides for this discrimination. The frequency band for the transition is provided by an adequate choice of $K_i$ and $\tau_i$ based on the given observer poles $\omega_1$ and $\omega_2$ [10], real and negative

$$K_i = -(\omega_1 + \omega_2); \quad \tau_i = \frac{|K_i|}{\omega_1 \omega_2} \tag{11.71}$$

Typical values of $\omega_1$ and $\omega_2$ are given by $\omega_1 = (3–10)$ rad/s and $\omega_2 = (3–10)|\omega_1|$.

Low speed and torque response for such a DTFC drive with sliding mode speed controller and $L_d = 4.1$ mH, $L_q = 8.2$ mH, $\lambda_{PM} = 0.2$ Wb, $r_s = 0.6\ \Omega$, $J = 0.005$ kg m$^2$, $p = 4$ pole pairs, $f_c = 10$ kHz, $V_d = 200$ V, $T_{emax} = 12$ Nm are given in Figure 11.46 [10]. The flux and torque hysteresis bands are 0.5% and 1%, respectively.

Fast torque control is demonstrated, as expected. The absence of speed overshoot is due to the use of a sliding mode speed controller.

The torque is calculated easily from flux and current (Figure 11.45). Providing for low- and high-speed good performance, DTFC is a strong contender for vector voltage and current control as it is simpler and more robust.

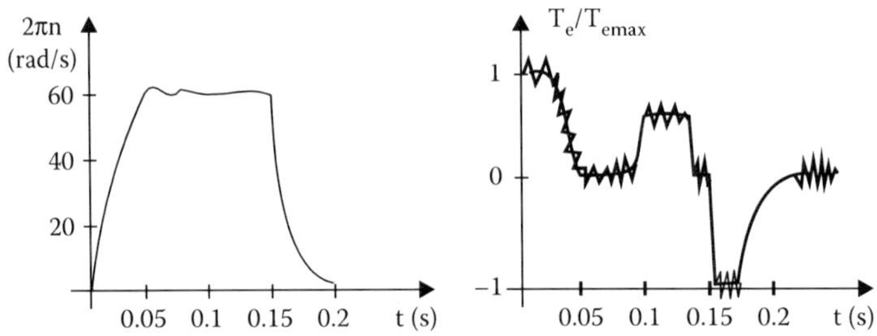


**FIGURE 11.46**
Low speed and torque transients with DTFC of PMSM. (After I. Boldea, S.A. Nasar, *Vector Control of a.c. Drives*, Chapter 10, CRC Press, Boca Raton, Florida, 1992.).

## 11.6 Sensorless Control of PMSMs

Sensorless control is targeted so far for speed-controlled drives where a less than 100 to 1 speed control range is required. It reduces hardware costs and improves mechanical reliability. High-performance drives—for more than 100 to 1 speed range and for precision positioning—still require motion sensor feedback. Unfortunately, VC requires, even for speed control only, rotor position information.

As for IM sensorless, control may be open loop or closed loop. Open loop (V/f) runs into problems of stability and is adequate for low dynamics applications that require a maximum of 10 to 1 speed control range with low precision. Such drives need a damper cage on the rotor for improved stability (Figure 11.2).

Feedforward torque compensation may be added to slightly improve the dynamic response. On the other hand, closed-loop sensorless PMSM or BLDC motor drives require rotor speed and position observers to produce a 100 to 1 speed control range with fast dynamics. Closed-loop sensorless control is approached in what follows.

BLDC-PM motor drives (with surface PM rotors, trapezoidal e.m.f., and rectangular current control) require special proximity (position) estimators. A classification of such estimators includes zero crossing of open phase e.m.f., the back e.m.f. integration approach, third harmonic method, and conduction time of diodes in the PWM inverter.

A collection of papers is available in Reference [1, Section 2]. Only 10/1 speed control range is usually claimed with BLDC-PM motor drives as, especially, speed estimation from position estimation is coarse at low speeds. On the other hand, vector-controlled PMSM use, for position and speed estimation, voltage and current or full-order observers, Kalman filters or MRAS.

As DTFC requires a flux and torque observer, for sensorless control only speed estimation is additionally required. Note that DTFC does not require explicit rotor position for speed control while for VC it does (for vector rotators). However, implicit rotor position is required for speed calculation.

An example of a speed calculator is obtained with respect to Figure 11.47.

$$\hat{\omega}_r = \frac{d\hat{\theta}_{\lambda s}}{dt} - \frac{d\hat{\delta}}{dt}$$

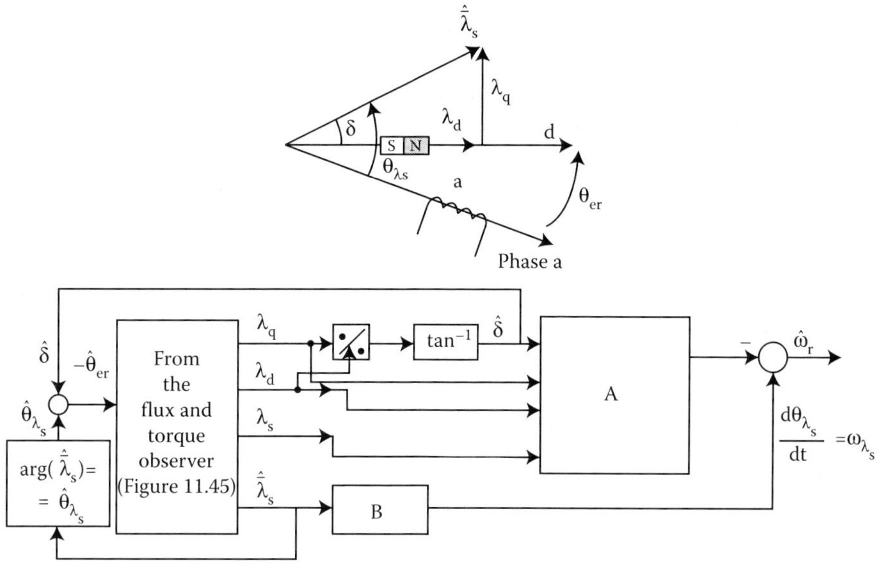


**FIGURE 11.47**
Rotor speed estimator for DTFC control of PMSMs.

$$\frac{d\delta}{dt} = \frac{\lambda_d(k-1)\lambda_q(k) - \lambda_q(k-1)\lambda_d(k)}{\lambda_s^2 T} \tag{11.72}$$

$$\frac{d\theta_{\lambda_s}}{dt} = \frac{\lambda_{s\alpha}(k-1)\lambda_{s\beta}(k) - \lambda_{s\beta}(k-1)\lambda_{s\alpha}(k)}{\lambda_s^2 T}$$

As the d-q flux angle δ varies only in the first and fourth quadrants ($\lambda_d \geq 0$, to avoid PM demagnetization), its estimation is straightforward. The derivatives required to estimate rotor speed $\hat{\omega}_r$ are done based on sin α and cos α derivative properties in discrete form with T as the sampling time (Figure 11.47).

Note that the speed estimator does not use inertia and, consequently, the motion equation has not been used for the scope. The latter may be used either to identify the load perturbation torque or somehow to correct the speed estimated from electrical equations, so useful during speed transients.

### 11.6.1 Initial Rotor Position Detection

Most position and speed observers are not capable of detecting the initial rotor position. Special starting methods or initial position identification, without moving the rotor to a known position, are required for safe, stable starting. Initial rotor identification may be performed by the drive itself through sending short voltage vector signals $V_1$, $V_3$, $V_5$ for a few microseconds until the phase currents reach a certain limit. From the time required for the current rising in the three cases, in case of interior PM rotors, the rather exact initial rotor position may be obtained based on the sinusoidal inductance-position dependence. On the other hand, for surface-PM rotor, additionally, negative m.m.f. is required for each position ($V_2$, $V_4$, $V_6$) to find out in which 60°-wide sector the rotor is placed based on the slight magnetic saturation (for positive m.m.f.) and desaturation of PMs (for negative m.m.f.) [11].

Alternatively, it is to find the rotor position sector (within 60°) by comparing the phase voltage decay through the freewheeling diodes after the turn off of a nonzero voltage vector.

DTFC sensorless tends to self-start from any position after a short lived oscillation as a certain voltage vector is applied until the flux surpasses the reference value for the first time.

For electric drives with nonhesitant start from any rotor position and very low speed (below 10 rpm) high-performance (low speed error and fast torque response) *signal injection methods* have been introduced for rotor position estimation. Initial position estimation is included.

They are based on tracking the machine natural or saturation-produced saliency.

In advanced solutions special saliency tracking observers with online magnet polarity discrimination are applied.

Rotating carrier-signal injection or a.c. carrier-signal injection at a given frequency, larger than the rated fundamental frequency of the drive, is used for the scope, while the current response to these signals is checked for the components with rotor position information [12].

Precision of a few mechanical degrees has been obtained this way for the initial rotor position estimation.

Above a certain low speed the injection voltage signal has to be eliminated and a different, fundamental model-based rotor position estimation is required. Switching from one position observer to the other should be smooth and nonhesitant [13,14,15].

Making use of special PWM voltage sequences to analyze the current transient response and thus finding the initial rotor position and the rotor position at low speeds is another way to handle this problem.

Sensorless PMSM drives with 100 to 1 speed control range, speed precision of 1% of rated speed, and fast torque response (milliseconds) are now close to reaching markets and thus much progress on the subject is expected in the near future.

### 11.6.2 Wide Speed Range Sensorless Control of PM-RSM via "Active Flux:" A Case Study

PM-RSM refers to a ferrite PM-assisted (low PM flux in axis q) RSM with notable saliency (in general $L_d/L_q > 2.5$). Most of the torque is produced by saliency while ferrite PMs produce some torque (less than 30% of rated torque) but improve the power factor and efficiency of the motor and thus reduce the converter kVA.

PM-RSM is also an interior PMSM with high saliency. The term IPMSM (with PMs in axis d) is reserved for strong PMs low saliency motors when most of the torque is produced through PM field. The PM-RSM (even IPMSM) is used in wide CPSR such as traction (HEVs [hybrid electric vehicles], EVs, etc.).

A peak torque, speed, envelope is defined by the application. At low speed the current has to be limited (to protect the PWM converter) while at high speeds the d.c. link voltage limits the fundamental voltage to the motor and thus the max-speed for given power.

In general a base speed $\omega_b$ is defined for which full power at full voltage is obtained. Then $\omega_{max}/\omega_b$ ratio at CPSR defines the speed range for constant power. Ratios $\omega_{max}/\omega_b = 3–3.5$ are needed in traction drives and may be obtained through PM-RSMs without excessive oversizing the motor or the PWM converter.

Here, among the many motion-sensorless control strategies we choose one [16,17] based on the concept of "active flux" which simplifies not only the "position and speed observer"

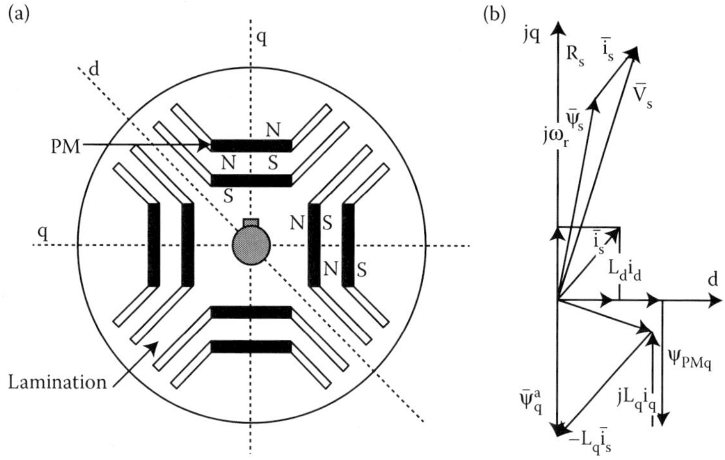


**FIGURE 11.48**
PM-RSM: (a) rotor cross section; (b) vector diagram.

but also the equations for maximum torque/current (MTPA), required whenever the max. voltage ceiling is not reached (at low speeds and high torque and at higher speeds but lower torque than the torque/speed envelope).

#### *11.6.2.1 PM-RSM Mathematical Model*

The mathematical model of a saturated PM-RSM with PM in the q axis (Figure 11.48), in the dq synchronous reference frame is given by

$$\bar{V}_s = R_s \cdot \bar{i}_s + \frac{d\bar{\psi}_s}{dt} + j\omega_r \cdot \bar{\psi}_s$$

$$\bar{\Psi}_q^a = \bar{\Psi}_s - L_d \bar{i}_s \tag{11.73}$$

$$\bar{\psi}_s = L_d(i_d) \cdot i_d + j(L_q \cdot i_q - \psi_{PMq})$$

$$T_e = 1.5p[\Psi_{PMq} + (L_d(i_d) - L_q) \cdot i_q] \cdot i_d = 1.5p \cdot \Psi_q^a \cdot i_d$$

where $\bar{V}_s$, $\bar{i}_s$, $\bar{\psi}_s$ are the stator voltage, current, and flux vector, respectively, $R_s$ is the stator resistance, $L_d$, $L_q$ are the dq axes inductances, $\Psi_{PMq}$ is the PM flux in the q axis, $T_e$ is the electromagnetic torque, $\omega_r$ is the electrical rotor speed, p is the number of pole pairs, and $\bar{\Psi}_q^a$ is the "active-flux" vector [16].

#### *11.6.2.2 A Practical Wide-Speed Range Control for PM-RSM*

A practical wide-speed range sensorless control system for PM-RSM is presented in Figure 11.49. The hereby torque current referencing strategy for optimal current pair $(i_d^*, i_q^*)$ used to obtain near-maximum available torque in wide speed range, with reference to torque, limited by maximum current constraint, is an alternative to the wide-speed range standard VC strategies of PMSM, but with much less computation effort. The maximum torque

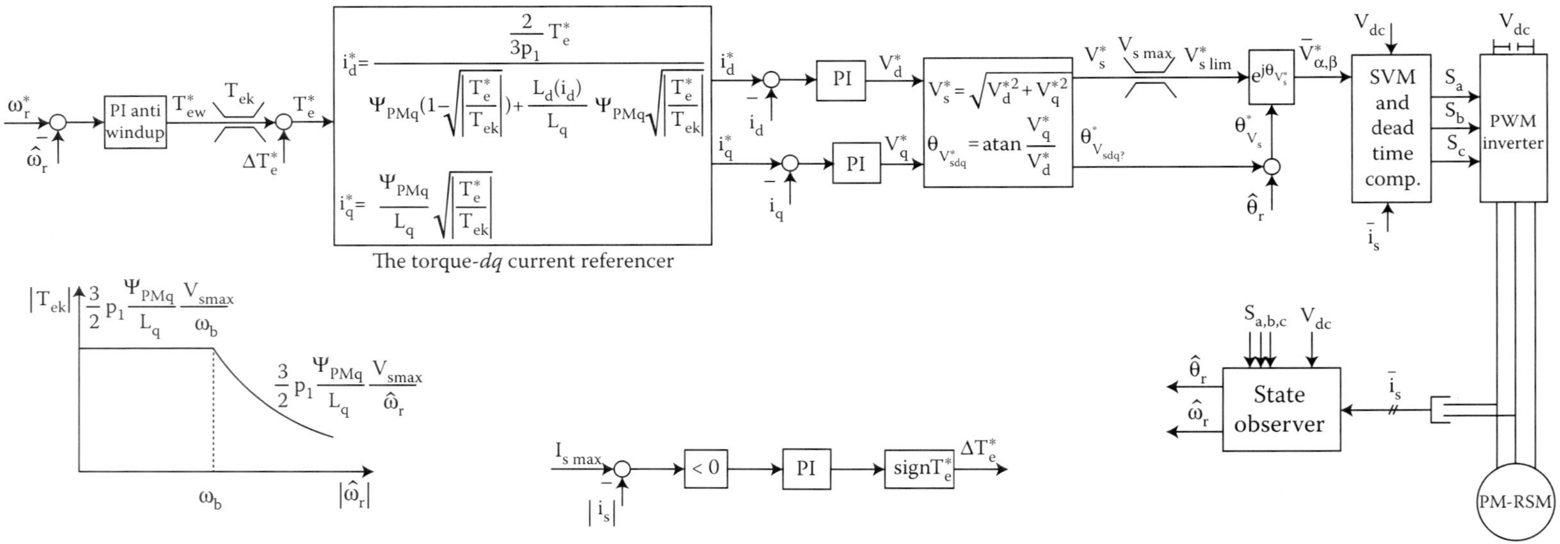


**FIGURE 11.49**
Sensorless control system in wide speed range with torque-dq current reference calculator and torque limitation.

in Equation 11.73 can be obtained when the q axis flux is zero ($\Psi_q = 0$), that is, for the maximum q axis current $i_{qk} = \lambda_{PMq}/L_q$. In this case, the torque $T_e = 1.5pL_{d(id)}i_d i_{qk}$ and the voltage vector $\bar{V}_s \approx j\omega_r L_d i_d$. Thus the equivalent PM-RSM looks like a d.c. machine. The maximum torque ($T_{ek}$) as a function of speed $\omega_r$ is obtained from Equation 11.73 as

$$\begin{aligned} T_{ek} &= T_{e\max b} \approx \frac{3}{2} p \cdot \frac{\Psi_{PMq}}{L_q} \cdot \frac{V_{s\max}}{\omega_b}, \quad \text{for } |\omega_r| \le \omega_b \\ T_{ek} &= T_{e\max b}\left|\frac{\omega_b}{\omega_r}\right|, \quad \text{for } |\omega_r| > \omega_b \end{aligned} \tag{11.74}$$

If the required reference torque $T_e^* < T_{ek}(\omega_r)$, then the current reference $i_q^*(T_e^*)$ is reduced from $i_{qk}$ by the square root of torque reference, in order to improve efficiency, by "sliding" toward the maximum torque/current control:

$$i_q^* = \frac{\psi_{PMq}}{L_q}\sqrt{\left|\frac{T_e^*}{T_{ek}}\right|} \tag{11.75}$$

In this case, by using Equation 11.73, the current reference $i_d^*$ ($T_e^*$) is given by

$$i_d^* = \frac{-2|T_e^*|/(3p)}{|\Psi_q^{as}|} \tag{11.76}$$

which replaces the need for the maximum torque per current family of curves $i_d^*(T_e^*)$, $i_q^*(T_e^*)$, that require *a priori* calculations or elaborated experiments.

This "square-root" simplification is acceptable for $L_d/L_q > 2.5$ and weak PM flux $\Psi_{PMq}\omega_r/V_{sn} < 0.3$. Otherwise, for a regular IPMSM (with low saliency ($L_q > L_d$), PMs in axis d, by manipulating the active flux, now in axis d, $\psi_d^a$ we obtain new robust expressions for the $i_d^*$, $i_q^*$ for MTPA:

$$i_d^* = -|i_s|\sqrt{\frac{|\Psi_d^a| - \Psi_{PMd}}{2|\Psi_d^a| - \Psi_{PMd}}}; \quad i_q^* = -\frac{+(2/3p_1)T_e^*}{|\Psi_d^a|} \tag{11.77}$$

In what follows we concentrate only on PM-RSM.

When the current limit is surpassed due to parameter mismatch, an additional torque reference correction $\Delta T_e^*$ is added to maintain the current below the maximum limit $I_{smax}$ (Figure 11.49). Also, when the voltage limit is surpassed (again only due to parameter mismatch), the stator voltage is limited to $V_{smax}$, but its angle remains given by $V_d^*$ and $V_q^*$; this way the parameter detuning is handled with little effort. Only the magnetization curve $L_d(i_d)$, the constant value of $L_q$, and the PM flux $\psi_{PMq}$ are required for control calibration.

#### *11.6.2.3 State Observer for Sensorless Control*

Rotor position and speed observers are needed to provide for sensorless-motion control. As prolonged very low-speed operation is not considered here, only a model-based active-flux state observer is used.

#### *11.6.2.3.1 "Active-Flux" Observer*

This observer is based on the "active-flux" concept developed in Reference 16 for all a.c. drives. For PM-RSM case, it leads to the estimation of both active-flux amplitude $\hat{\psi}_q^{as}$ and angle $\hat{\theta}_{\psi_q^{as}}$ that corresponds to the negative q axis (Figure 11.48b). Therefore, the electrical rotor position ($\theta_r$) is specified by $\hat{\theta}_r = \hat{\theta}_{\psi_q^{as}} + \pi/2$.

The "active-flux" vector $\bar{\psi}_q^{as}$ observer in stator coordinates is given by

$$\bar{\psi}_q^a = \int (\bar{V}_s^s - R_s \bar{i}_s^s + \bar{V}_{comp})\, dt - L_d(i_s^s)\bar{i}_s^s \tag{11.78}$$

Again, the "active-flux" vector $\bar{\psi}_q^a$ is oriented along the negative q axis and renders the IPMSM as a nonsalient pole a.c. machine with $L_d(i_s)$ as its inductance along both d and q axes [18]. The implementation scheme of the "active-flux" observer (Figure 11.50) is based on Equation 11.78 and consists of stator flux observer in stator coordinates from which the term $L_d(i_s^s)\cdot\bar{i}_s^s$ is subtracted.

The stator flux observer is an adaptive observer type combining the flux estimator based on the voltage model given by Equation 11.73 in stator coordinates, with $\omega_r = 0$ ($\hat{\bar{\psi}}_s^s$ in Figure 11.50—reference model—with the flux estimator based on the current model given by Equation 11.73 in rotor coordinates [$\hat{\bar{\psi}}_{si}^s$ in Figure 11.50—adaptive model—which has the estimated rotor position $\hat{\theta}_r$ as the adaptive variable obtained from the estimated active-flux vector $\hat{\bar{\psi}}_q^a$]). The stator flux error between the current model and the voltage model is employed by a PI compensator ($k_{pc} + k_{ic}/s$) that acts as $V_{comp}$ at the pure integrator level of the voltage model. This structure eliminates the negative effect of the input d.c. offset present in the stator current measurement chain and of errors in the integrator initial condition.

The observer has the output $\bar{\psi}_s^s$—the estimated stator flux—and two inputs: the e.m.f. voltage $\bar{e} = \bar{V}_s^s - R_s\bar{i}_s^s$ and $\bar{\psi}_{si}^s$—the stator flux estimation based on the current model.

The associated scalar transfer functions are, respectively,

$$\frac{\hat{\psi}_s^s}{e} = \frac{s}{s^2 + k_{pc}s + k_{ic}}; \quad \frac{\hat{\psi}_s^s}{\hat{\psi}_{si}^s} = \frac{k_{pc}s + k_{ic}}{s^2 + k_{pc}s + k_{ic}} \tag{11.79}$$

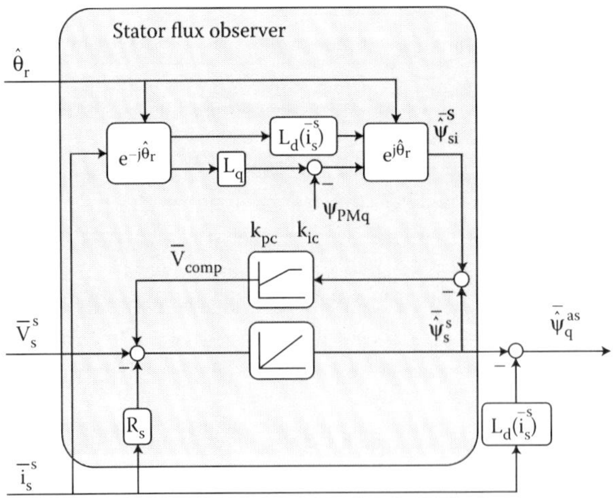


**FIGURE 11.50**
Active flux observer.

where $\hat{\psi}_s^s$, $\hat{\psi}_{si}^s$, and e are the Laplace transforms of stator fluxes and e.m.f. voltage, respectively. They characterize the observer frequency behavior, which depends on the chosen poles $\omega_1, \omega_2$ ($\omega_1 \le \omega_2$) as follows. At low speeds ($\omega_r < \omega_1$), the stator flux estimation based on the current model prevails, and at medium-high speeds ($\omega_r > \omega_2$), the voltage model prevails, with a smooth transition between them.

The PI parameters are designed by the pole placement method:

$$k_{pc} = \omega_1 + \omega_2; \quad k_{ic} = \omega_1 \omega_2 \tag{11.80}$$

For the chosen poles $\omega_1 = \omega_2 = 2$ rad/s, the PI gains are $k_{pc} = 4$ and $k_{ic} = 4$.

*11.6.2.3.2 Phase-Locked Loop State Observer*

A phase-locked loop (PLL) tracking observer shown in Figure 11.51 is used for rotor position and speed estimations, having as input the estimated active-flux components ($\hat{\psi}_{q\alpha}^{as}$, $\hat{\psi}_{q\beta}^{as}$). The PLL acts like a zero-phase filter and provides a smooth estimation.

The PLL extracts the rotor position $\hat{\theta}_r$ and speed $\hat{\omega}_r$ estimations from the phase of the active-flux vector $\overline{\psi}_q^{as}$ in stator reference frame, with the PLL position error given by

$$\varepsilon = \hat{\psi}_q^a \Delta\theta_r = \hat{\psi}_{q\beta}^a \cos\left(\frac{\hat{\theta}_r - \pi}{2}\right) - \hat{\psi}_{q\alpha}^a \sin\left(\frac{\hat{\theta}_r - \pi}{2}\right) \tag{11.81}$$

To eliminate the quite large speed error during start up transients, a feedforward term (Equation 11.82) is added to the estimated speed in the PLL closed loop after the PI compensator:

$$\Delta\hat{\omega}_r = \frac{p_1 h \hat{T}_e}{J} \tag{11.82}$$

where $\hat{T}_e$ is the estimated electromagnetic torque, h is the sampling period, and J is the inertia of the rotating system. The estimated torque $\hat{T}_e$ is straightforward once the flux observer is implemented.

The PLL parameter values for PI compensator ($k_p$, $k_i$) are specified in Table 11.1. A first-order low-pass filter (LPF) with a time constant $T_{iF} = 0.001$ s is applied on the estimated speed to reduce noise.

Experimentally, the PLL state observer tends to be better for high-speed operation.

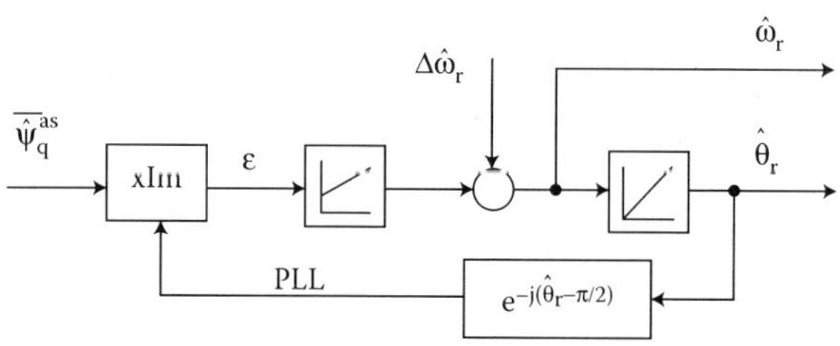


**FIGURE 11.51**
PLL state observer.

**TABLE 11.1**
PLL Observer Parameters (Figure 11.51)
$H_{PI} = k_p(1 + k_i/s)$

| Speed | $k_p$ | $k_i$ |
|---|---|---|
| 1 rpm | 10,000 | 100 |
| 1000 rpm | 5,000 | 300 |
| 6000 rpm | 10,000 | 400 |

#### *11.6.2.4 Digital Simulation Results*

Digital simulations of the just developed sensorless VC strategy are implemented in MATLAB/Simulink package for the PM-RSM model with parameters specified in Table 11.2.

The following digital simulations are performed at the inverter d.c. voltage of 48 V d.c. to check the sensorless control in large speed range, having sample results presented in Figures 11.52 and 11.53: startup to 1 rpm at $t = 0.5$ s followed by a full step load torque at $t = 3$ s (Figure 11.52); startup to 6000 rpm followed by a step load torque (of 30% rated torque) at $t = 4$ s (Figure 11.53).

Figure 11.52 illustrates the transients for step startup to the very low speed of 1 rpm with full step load at $t = 3$ s. Figure 11.52a shows a very good speed response for 1 rpm step speed reference, without load torque. Figure 11.52b shows a quick speed recovery at full load torque and a small speed ripple in the estimated speed. The position estimation during torque transients seems very reliable (Figure 11.52c) and, because the modulo $2\pi$ operator is used, a discontinuity in rotor position appears at 3 s. The speed transients that occur in the interval $t = 3$–$3.4$ s are typical for sudden loading and show a momentary speed and position reversal in the first instant. After that, the speed controller reacts and then, the speed returns to its reference of 1 rpm in about 0.4 s.

Note that the motor operates with constant torque (Figure 11.52e) because the command speed (1 rpm) is well below the base speed. Figure 11.52f shows that the stator flux

**TABLE 11.2**
Parameters of the PM-RSM Prototype

| Number of Pole Pairs ($p_1$) | 2 |
|---|---|
| Rated power | 750 W |
| Rated speed | 1500 rpm |
| Rated frequency | 50 Hz |
| Rated torque | 6.3 Nm |
| Rated phase to phase voltage | 22 V(rms) |
| Rated phase current | 50 A(rms) |
| Stator resistance per phase ($R_{s-}$) | 0.037 Ω |
| Stator resistance per phase including wires resistance ($R_s$) | 0.065 Ω |
| d-axis inductance ($L_d(i_d)$) (unsaturated) | 2.5 mH |
| q axis inductance ($L_q$) = const. | 0.5 mH |
| Permanent magnet flux ($\Psi_{PMq}$) | 0.011 Wb |
| Inertia of the rotating system (J) | $16 \times 10^{-3}$ kg m$^2$ |
| Viscous friction coefficient (B) | $10^{-4}$ Nm s/rad |

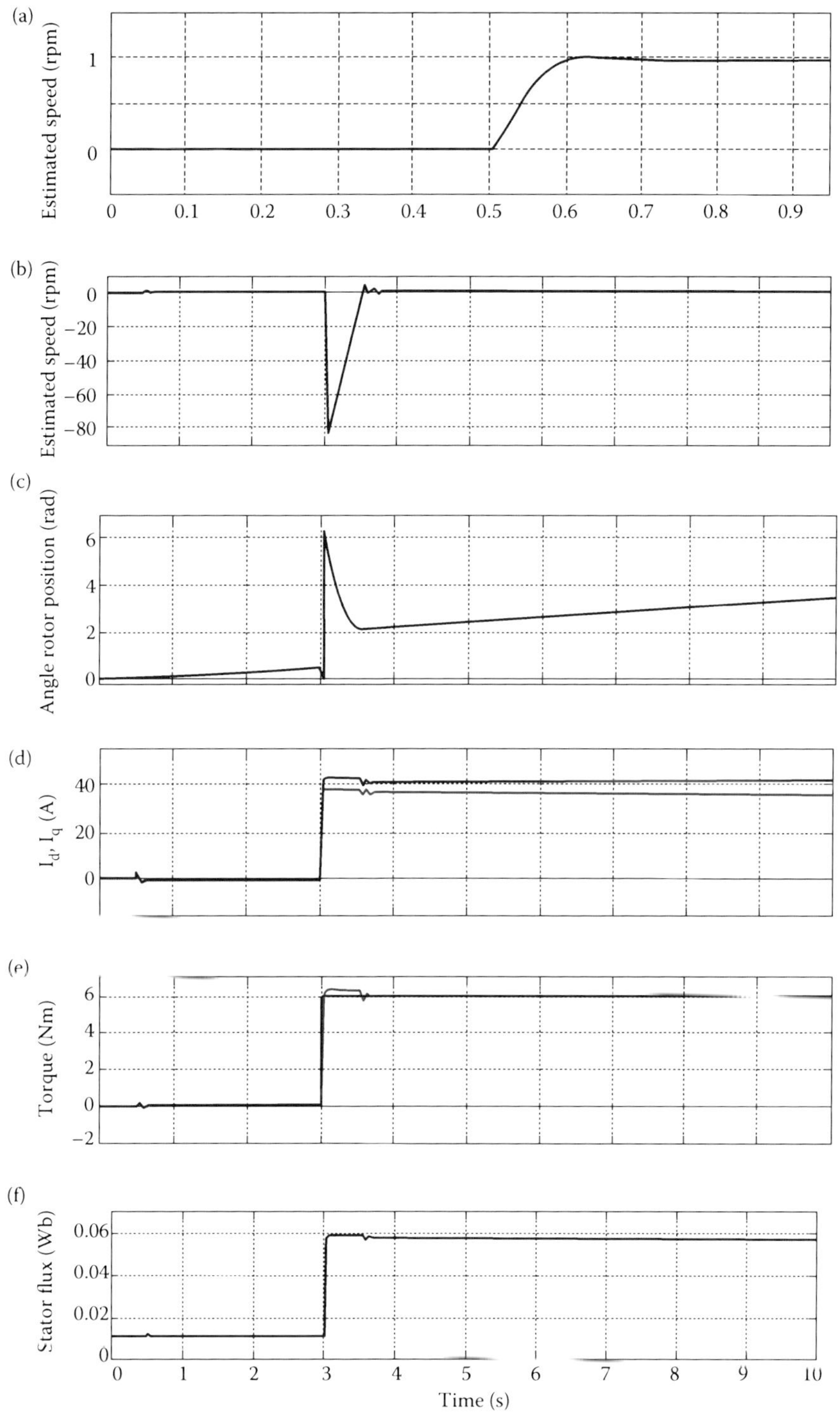


**FIGURE 11.52**
Startup to very low speed of 1 rpm step at 0.5 s followed by a full step load torque at t = 3 s: (a) zoom of estimated speed during 0–1 s; (b) estimated speed; (c) estimated rotor position; (d) $I_q$, $I_d$ (blue) currents; (e) load torque reference and estimated torque; (f) estimated stator flux.

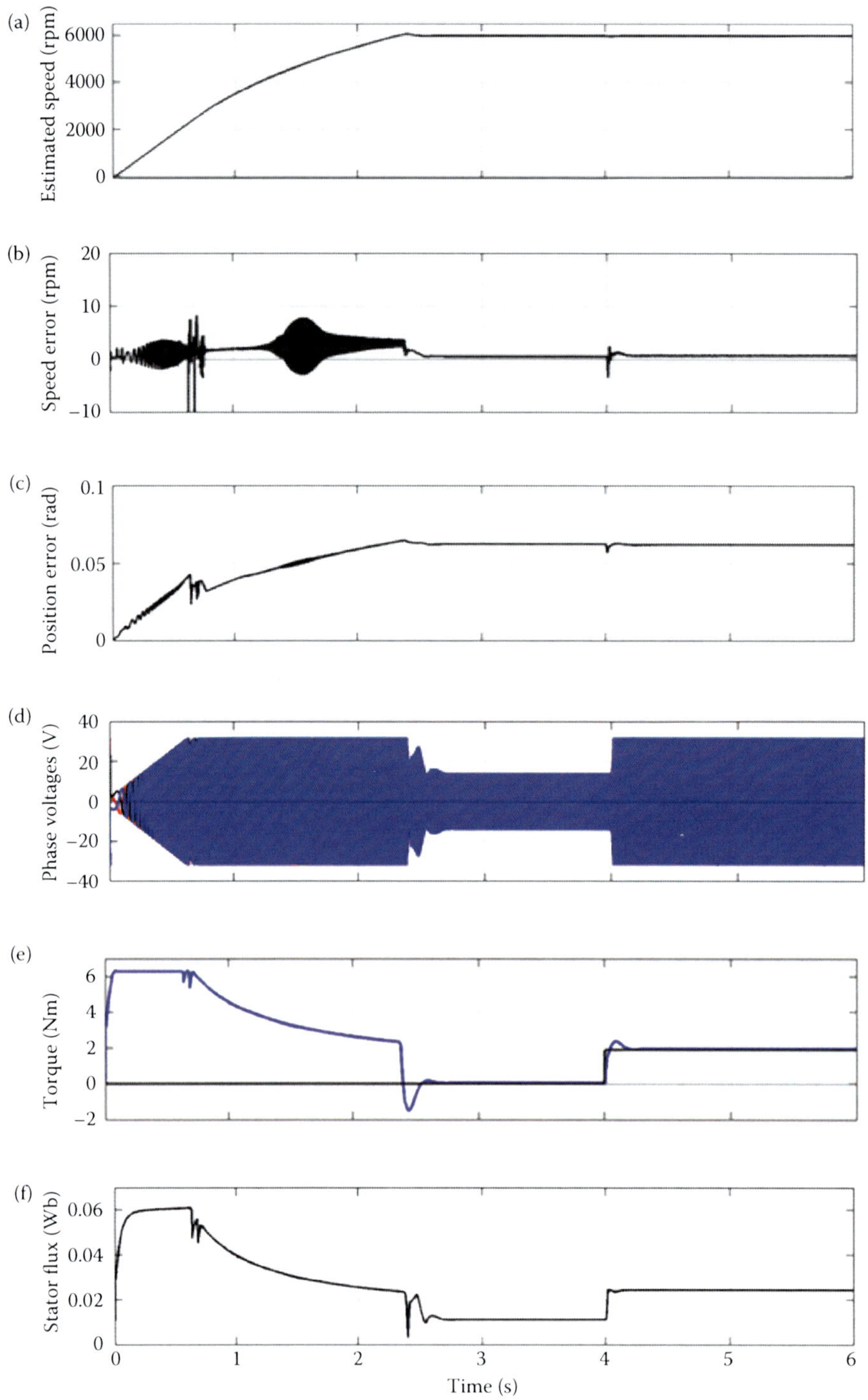


**FIGURE 11.53**
Startup to 6000 rpm (above base speed) followed by a step load torque (of 30% rated torque) at t = 4 s: (a) estimated speed; (b) error between encoder and estimated speed; (c) error between actual and estimated angle rotor position; (d) phase voltages; (e) load torque reference and estimated torque; (f) estimated stator flux.

increases at flux reference value only during torque transients in order to produce the demanded torque; otherwise, it is maintained close to the PM flux (low) value.

Figure 11.53 illustrates the transients for step startup to 6000 rpm (above base speed) with 30% full step load at t = 4 s. Figure 11.53a and e shows well-behaved speed and torque transients. In Figure 11.53c, the position error is small (0.065 rad), which means that the estimation of rotor position is good, even when the speed reference or load disturbance changes quickly. The stator voltage reduction (Figure 11.53d) after which the speed reaches steady state, for low load, clearly shows the optimum torque/current concept capability to reduce losses even at high speed, but also for low torque, when the machine uses the entire available voltage and current. Figure 11.53f shows how, once the speed reaches the speed reference of 6000 rpm on no-load, the flux is decreased until it achieves almost PM flux (low) value.

But when the load occurs the stator flux increases again, even at 6000 rpm.

As the magnetization curve $L_d(i_s)$ is considered, to some low extent, only the stator resistance error influences the "active-flux" angle, that is, the rotor position estimation. The stator resistance online correction was not done here, but it should be done for better low-speed response of the drive

In short, Figure 11.53 clearly shows the potential of the introduced strategy to handle a wide speed control range with stator flux level changing pertinently with speed and torque, with low core losses under steady state, and full available torque for fast dynamics.

#### 11.6.2.5 Experimental Work and Results

The experimental platform setup (Figure 11.54) contains the following main parts: a PM-RSM (parameters shown in Table 11.2), a 300 A/48 V d.c. Sauer-Danfoss MOSFET inverter (Sauer-Danfoss Inc., Nordborg, Denmark), a dSpace 1103 real-time control system (dSpace GmbH, Paderborn, Germany), and an induction motor with direct torque control (IM-DTFC) drive as load. *All experiments have been done at low inverter d.c. voltage of 12 V d.c., that is, 25% rated voltage.*

Experimental results using the developed control system (Figure 11.49), in sensorless operation at the lowest achieved speed reversal of ±50 rpm, with small speed ripple, and without load, are presented in Figure 11.55. The lowest (50 rpm) speed looks high

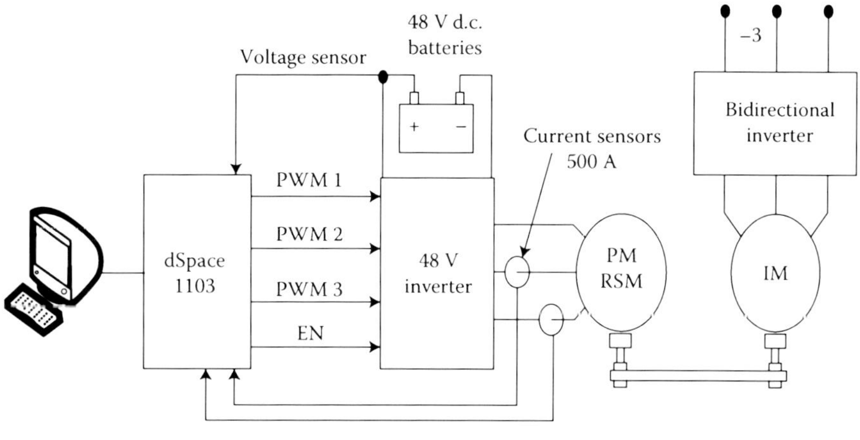


**FIGURE 11.54**
Experimental platform setup.

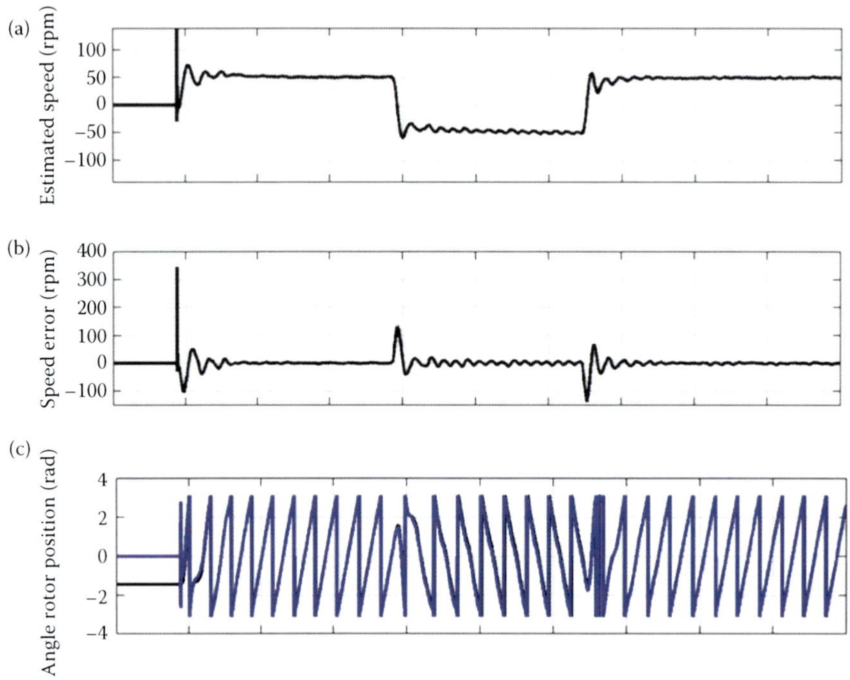


**FIGURE 11.55**
Sensorless speed reversal at low speed of ±50 rpm, no load, at inverter d.c. voltage of 12 V d.c.: (a) estimated speed; (b) error between encoder and estimated speed; (c) actual and estimated rotor position angle.

in comparison with the 1 rpm speed from digital simulations (Figure 11.52). Numerous causes lead to this situation; we consider here: mechanical problems with the drive-load pulsations at low speeds, motor cogging torque, inverter nonlinearities at 12 V d.c. (only partly compensated), parameter detuning. For general (1/100 speed ratio) drives, the results may be considered acceptable, though more efforts are due for further improvements in the state observer.

The speed dynamics during low-speed reversal is good, excepting a single spike during startup (Figure 11.55a) due to the initial position error in the state observer. The speed estimation error in steady state is near zero with low-speed ripple, but with important dynamic error at speed reversal (Figure 11.55b). The rotor position dynamic estimation properly follows the actual rotor position (Figure 11.55c).

For the developed sensorless control system with relatively low rated stator voltage, it was mandatory to compensate the voltage drops on the power devices and the inverter dead time. Even so, the estimated rotor position angle is not very smooth; but it has acceptable pulsations (Figure 11.55c).

Sensorless operating in the flux weakening by developing maximum available torque for a certain speed, employing the new torque-dq current reference calculator, is demonstrated by experiments for two high-speed startup operations to 1000 rpm, above the base speed of 375 rpm, taking into account that the inverter d.c. voltage is 25% rated voltage.

Using a single battery of 12 V d.c. instead of 48 V d.c. for inverter, the speed of 1000 rpm would correspond to a speed of 4000 rpm in the case of using all four batteries.

Figure 11.56 illustrates the experimental results for step start to 1000 rpm (above the base speed), no load, followed at 12.5 s by a 30% full step load torque, that is, the maximum

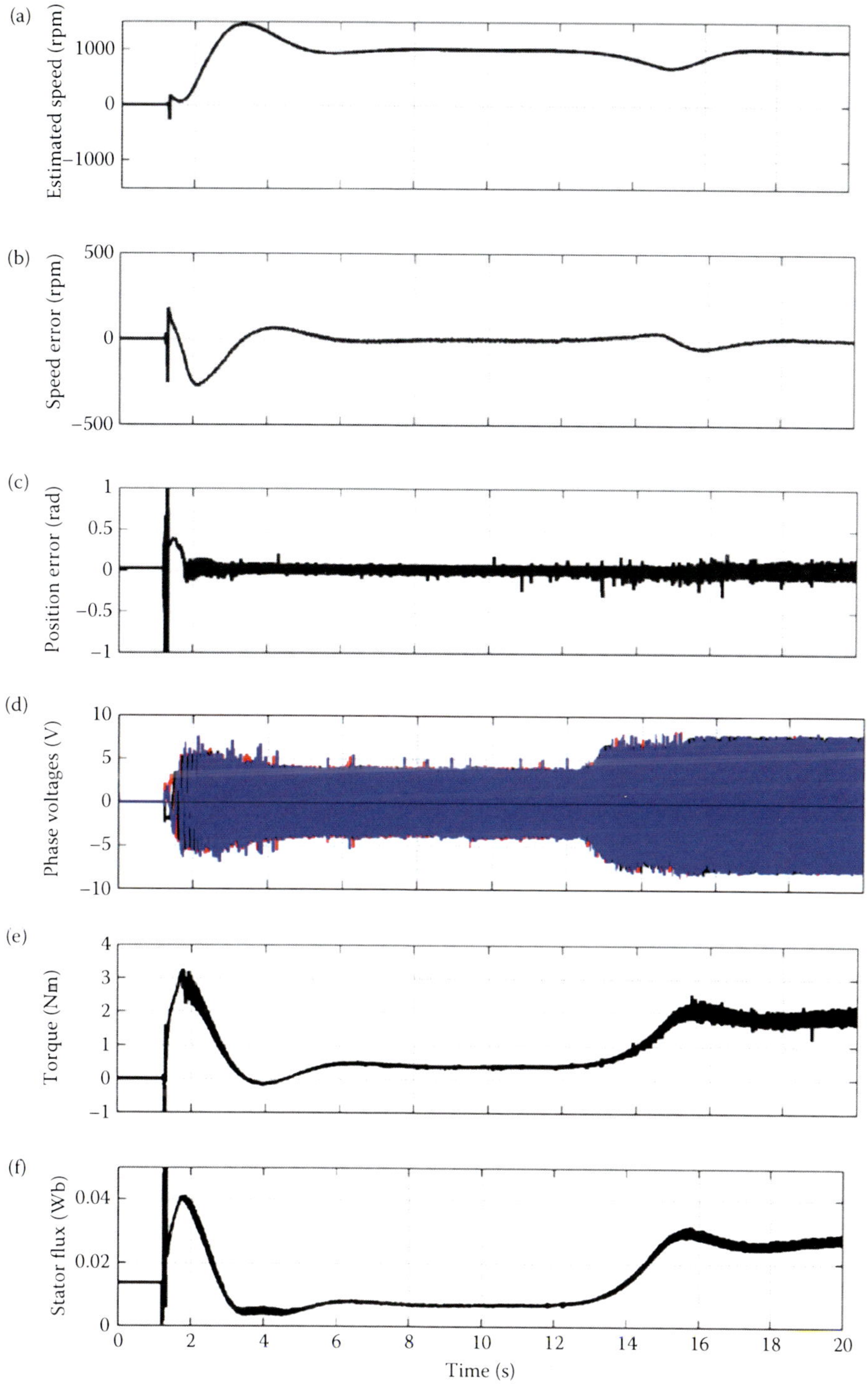


**FIGURE 11.56**
Start up to 1000 rpm (above base speed) at inverter d.c. voltage of 12 V d.c., followed by a 30% full step load torque at 12.5 s: (a) estimated speed; (b) error between actual and estimated speed; (c) error between actual and estimated angle rotor position; (d) phase voltages; (e) estimated torque; (f) estimated stator flux.

available torque at this speed for inverter d.c. voltage of 12 V d.c. At startup, a small negative polarity in the estimated speed waveform occurs, then it quickly recovers and the speed achieves the target in about 2 s (Figure 11.56a), having a large overshoot. The speed estimation error tends to zero in steady state (Figure 11.56b) and the average position estimation error tends also to zero with maximum error below 0.25 rad (Figure 11.56c). When operating at medium-high speed, the rotor position estimation error is lower than when operating at low speed, because now the control system is less sensitive to parameter mismatch and to inverter nonlinearities.

After 12.5 s during step loading, the phase voltage attains the maximum available voltage (8 V) provided by inverter at 1000 rpm for maximum available torque of 2 Nm (Figure 11.56d), and therefore, the estimated torque and flux ripples increase in this zone practically without voltage reserve. Figure 11.56e shows the estimated torque, where the maximum starting torque is quite high (3 Nm) at lower speed, and then maximum torque is reduced to 2 Nm at the high speed of 1000 rpm where the motor is operating in the constant power region. This is in accordance with the simulation results from Figure 11.53e for speed above base speed. The estimated stator flux remains at a low value (approximately PM flux) during steady-state low torque (0.4 Nm), and increases during loading in order to produce the demanded torque (Figure 11.56f).

To show the operation in the extreme extended high-speed region, the speed command was step changed from 0 to 3000 rpm, no load, using again a single battery of 12 V d.c. In the case of using all four batteries, this speed of 3000 rpm would correspond to 12,000 rpm, which would be 8 times the base speed (CPSR $\approx$ 8). The results for 3000 rpm acceleration are not given here for lack of space [17].

In summary, the developed control strategy, sensorless, operating in a very wide speed range, experimentally proves its capacity to "slide" smoothly from approximately maximum torque/current (good efficiency) at all torques to the full voltage utilization when the maximum torque/speed envelope is required.

## 11.7 RSM Drives

RSMs are characterized by the absence of rotor currents and, in rotor configurations with multiple flux barriers (or anisotropic axially laminated [ALA], Figure 11.57a and b), high $L_d/L_q$ rates provide for the torque density (Nm/kg) power factor and efficiency comparable or slightly superior to IMs. For $L_d/L_q > 8$, the RSM is fully competitive with the IM. The absence of rotor currents, especially when an encoder is available for rotor position feedback, makes the RSM drive control notably simpler than that for IMs.

As ideally the core losses in the rotor are zero (in reality some harmonics flux loss currents exist), the RSM may work safely at low speeds and high torque, provided the stator is properly cooled. On the other hand, the slightly fragile rotor and small airgap limits the peripheral rotor speeds to, probably, 50–60 m/s. So RSM might be considered from 100 W up to 1 MW. Recently multiflux-barrier rotor RSM drives in the tens to hundreds of kW have become commercial, with 3%–4% more efficiency and 7%–8% less power factor than IM drives of similar output.

Naturally, for conventionally laminated anisotropic rotors (Figure 11.48c), $L_d/L_q < 3$, low power factors (about 0.45–0.50) at low power at higher peripheral speeds are acceptable, due to their simplicity/low cost.

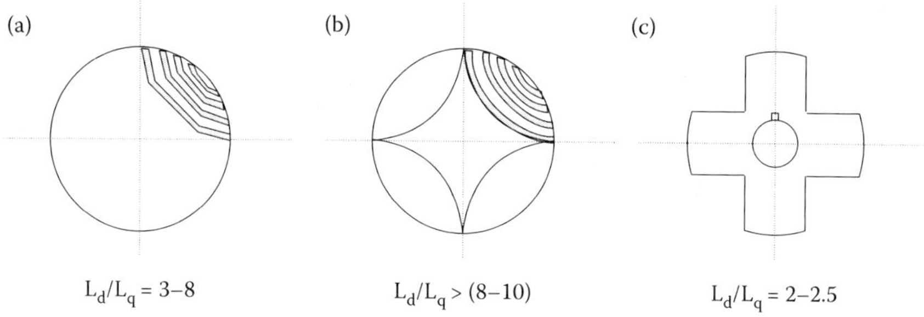


**FIGURE 11.57**
RSM rotors: (a) with multiple flux barriers; (b) ALA; (c) conventional.

### 11.7.1 RSM Vector Control Principles

The RSM may be considered as a nonlinear system with two entries; the stator flux $\bar{\lambda}_s$ and the torque $T_e$ (Chapter 10, Equations 10.116 and 10.117)

$$\bar{\lambda}_s = L_d i_d + j(L_q i_q - \lambda_{PMq})$$
$$T_e = \frac{3}{2} p[\lambda_{PM_q} + (L_d - L_q) i_q] i_d \tag{11.83}$$

A PM in axis q (of lower inductance) has been included.

Note that the PM flux is opposite to and higher than any $L_q i_q$ contribution: $\lambda_{PMq} \geq L_q i_q$.

Also, $i_q$ should not change sign if positive reluctance torque in the presence of PMs is desired. To switch from positive to negative torque, the sign of $i_d$ is changed. Conversely, if $\lambda_{PMq} = 0$, in general, to change the torque sign, the polarity of $i_q$ is changed. As $L_d > L_q$, $i_d$ transients are slower than $i_q$ transients, so faster torque response is obtained for $i_q$ control.

As there are no rotor currents, no current decoupling network is required. Instead, it is paramount to find $i_d$, $i_q$ relationships to torque. This is done based on optimization performance criteria, as already explained for PMSMs earlier in this chapter.

Practical criteria would be maximum torque/current at low speeds, maximum torque/flux at high speeds, and maximum power factor or efficiency.

Three basic criteria are discussed separately for the "pure" RSM ($\lambda_{PMq} = 0$).

For maximum torque/current, using Equation 11.83 we obtain

$$i_{di} = i_{qi} = \frac{i_s}{\sqrt{2}}$$
$$T_{ei} = \frac{3}{2} p(L_d - L_q) i_{di}^2 \tag{11.84}$$

As $i_{di} < 0.5 i_n \cdot \sqrt{2}$ ($i_n$ rated phase current, rms) to avoid heavy magnetic saturation, about half of the rated torque may be expected when using this criterion.

The maximum power factor criterion ($r_s = 0$) is based on the definition

$$\tan \varphi_1 = \frac{L_d i_d^2/2 + L_q i_q^2/2}{(L_d - L_q) i_d i_q} \tag{11.85}$$

Finally, we find (for d tan $\varphi_1/d(i_d/iq) = 0$)

$$\left(\frac{i_d}{i_q}\right)_{\varphi_{1\min}} = \sqrt{\frac{L_q}{L_d}} \tag{11.86}$$

$$T_{e\varphi_1} = \frac{3}{2}p(L_d - L_q)i_{di}^2\sqrt{\frac{L_d}{L_q}} \tag{11.87}$$

$$(\cos\varphi_1)_{max} = \frac{1 - L_q/L_d}{1 + L_q/L_d} \tag{11.88}$$

For maximum torque per flux

$$\frac{i_{d\lambda}}{i_{q\lambda}} = \frac{L_q}{L_d}; \quad \lambda_{d\lambda} = \lambda_{q\lambda} = \frac{\lambda_s}{\sqrt{2}} \tag{11.89}$$

Again, for $i_{d\lambda} = i_{di}$

$$T_{e\lambda} = \frac{3}{2}p(L_d - L_q)i_{di}^2\frac{L_d}{L_q} \tag{11.90}$$

or

$$T_{e\lambda} = \frac{3}{2}p(L_d - L_q)\frac{\lambda_s^2}{2L_dL_q} \tag{11.91}$$

The torque obtained with the criterion of maximum torque/flux, corresponds to lowest power factor (lower than 0.707). The base speed $\omega_b$, defined for maximum torque available, is thus

$$\omega_b = \frac{V_{sn}}{L_d i_{di}\sqrt{2}} \tag{11.92}$$

For a more detailed analysis of steady-state and transient torque capabilities of RSM, see Reference 19.

In essence, for servo drives without constant power (flux weakening) zone, and fast speed response, the criterion of maximum power factor may be used for steady state. For fast transients, the maximum torque/flux approach is adequate. As expected, the PWM inverter has to be rated for the rather high currents required for fast torque (speed) transients.

Trajectories of $i_d$–$i_q$ currents and $\lambda_d$–$\lambda_q$ fluxes for the three main optimal control strategies are shown in Figure 11.58.

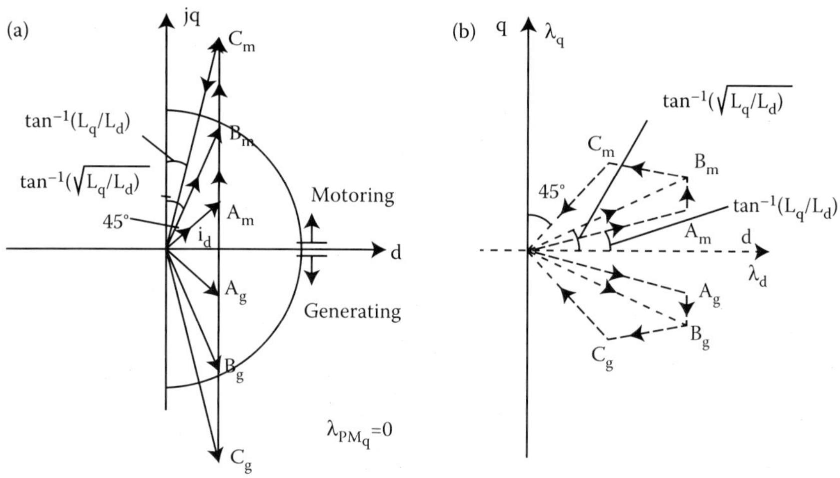


**FIGURE 11.58**
(a) Current and (b) flux trajectories with speed increasing.

It has been shown that the influence of magnetic saturation (of $L_d$ mainly) is small for maximum power factor control [20].

On the contrary, for maximum torque/flux control, the influence of magnetic saturation on performance is notable. For small torques and below base speed, the control may follow the d-q current trajectory $0A_mB_m$ while above base speed, for constant power control, the trajectory $0B_mC_m$ is to be followed. Consequently, the $i_d$, $i_q$ angle $\gamma$

$$\gamma = \tan^{-1}\frac{i_d}{i_q} \tag{11.93}$$

should be made variable with torque (and speed).

The same conditions may be met by working with flux linkage d-q angles (Figure 11.58b). In general, for a wide ($\omega_{max}/\omega_b > 2$) constant power speed zone, PMs in axis q are required [21]. For low speeds vector current control is more adequate while above base speed, vector voltage control is better.

A combination of the two is, perhaps, the practical solution for wide speed range control. Vector current control [22,23], eventually combined with voltage decoupling in d-q orientation or in stator flux orientation [24], has been proposed. They are very similar to the corresponding methods developed for PMSMs. In what follows, we will only illustrate the VC of RSM through a vector d-q current control system with $i_d$ constant up to base speed $\omega_b$ and speed-dependent above $\omega_b$ [25]. We will also treat DTFC [26].

### 11.7.2 Indirect Vector Current Control of RSM

The indirect vector current control scheme (Figure 11.59) is similar to that for the PMSMs.

Either speed or position control may be performed. The reference magnetization current $i_d^*$ is constant up to base speed $\omega_b$ and decreases inversely with speed above $\omega_b$.

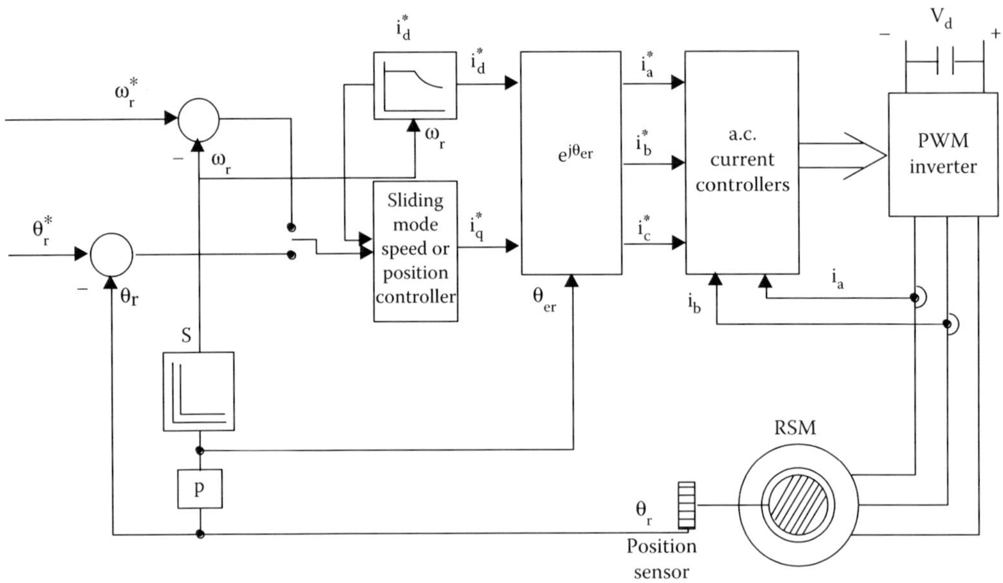


**FIGURE 11.59**
Indirect vector current control of RSM.

Let us introduce some details on the sliding mode speed controller simulation and test results. The sliding mode speed controller output produces the reference torque current $i_q^*$ and the simplest control law is

$$\begin{cases} i_q^* = +i_{qk} & \text{for } s_\omega > \omega_h \quad \text{or} \quad \left(|s_\omega| < \omega_h \text{ but } \dot{s}_\omega < 0\right) \\ i_q^* = -i_{qk} & \text{for } s_\omega \le -\omega_h \quad \text{or} \quad \left(|s_\omega| < \omega_h \text{ but } \dot{s}_\omega > 0\right) \end{cases} \tag{11.94}$$

where $\omega_h$ is the hysteresis band of the controller and $s_\omega$ is the sliding mode functional

$$s_\omega = K_{p\omega}(\omega_r^* - \omega_r) - K_{d\omega}\frac{d\omega_r}{dt} \tag{11.95}$$

Besides meeting the conditions of touching the straight line $s_\omega = 0$ (in the state-space plane) we need to know $K_{p\omega}$ and $K_{d\omega}$. To do so, we admit that $i_q$ varies linearly in time and the targeted speed $\omega_r^*$ is reached after $s_\omega$ changes sign at least once (Figure 11.60).

The ratio $K_\omega = K_{d\omega}/K_{p\omega}$ is to be determined first.

If, for $t = t_0$, $s_\omega = 0$, for no-load, Equation 11.95 yields

$$\omega_r^* - \omega_r(t_0) = K_\omega \frac{d\omega_r}{dt} = K_\omega \frac{p}{J} T_e(t_0) \tag{11.96}$$

where $T_e$ is the torque, J is inertia, and p is pole pairs.

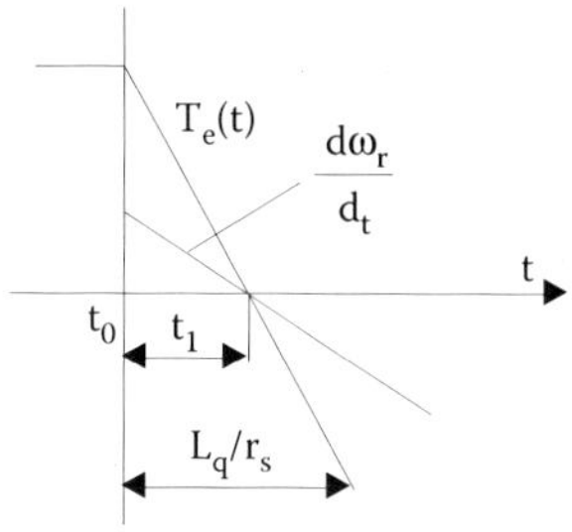


**FIGURE 11.60**
Ideal transients (constant $i_d$, linear variation of $i_q$).

If $i_q$ varies linearly, the torque $T_e$ does the same thing ($i_d$ = const.) and, from $t_0$ to $t_0 + t_1$ (Figure 11.60), we have

$$\frac{d\omega_r}{dt} = \frac{p}{J} T_e(t_0) \cdot \left[1 - \frac{t - t_0}{t_1}\right] \tag{11.97}$$

From Equation 11.97, through integration, we have

$$\omega_r^* - \omega_r(t_0) = \frac{p}{2J} t_1 T_e(t_0) \tag{11.98}$$

Comparing with Equation 11.96 yields

$$K_\omega = \frac{t_1}{2} \tag{11.99}$$

A good approximation could be $t_1 = L_q/2r_s$ and, thus,

$$K_\omega = \frac{L_q}{4r_s} \tag{11.100}$$

By limiting the steady-state error we may find the second relation, to calculate $K_{p\omega}$ and $K_{d\omega}$ and thus the sliding mode controller is fully designed.

In a similar way, but for linear speed deceleration during braking, an sliding mode position controller may be designed. Its constants would be $K_{p\theta}$ and $K_{d\theta}$ [25].

For an RSM with data $2p = 6$, $L_q = 24$ mH, $L_d = 88$ mH for $i_d^* = 3$ A, $r_s = 0.8\ \Omega$, $J = 0.0157$ kg m$^2$, $V_0 = 80$ V, $f_c = 5$ kHz, for $i_{dm}{}^* = 3$ A and $i_{qm}{}^* = 5$ A, $K_{p\omega} = 5$, $K_\omega = 7.5 \times 10^{-3}$ s, $K_{d\omega} = 0.01$ s, $\omega_h = 5$ rad/s, $J_{max} = 5$ J, $K_{p\theta} = 5$, $K_{d\theta} = 0.75$, $\theta_h = 3\pi/50$ rad, the digital simulation and test results are shown for comparison in Figures 11.61 and 11.62.

Note that the acceleration decreases above 500 rpm as stepwise $i_d$ decrease with speed is performed (Figure 11.62a).

As the position sliding mode controller is designed for 5J it is no wonder that the responses for J and 5J do not differ notably (Figure 11.63). This kind of robustness, with the elimination of the overshoot and with fast response, is an intrinsic merit of sliding mode controllers.

**NOTE:** A complete simulation application is not given here as it may be treated as a particular case of the previous examples for PMSMs.

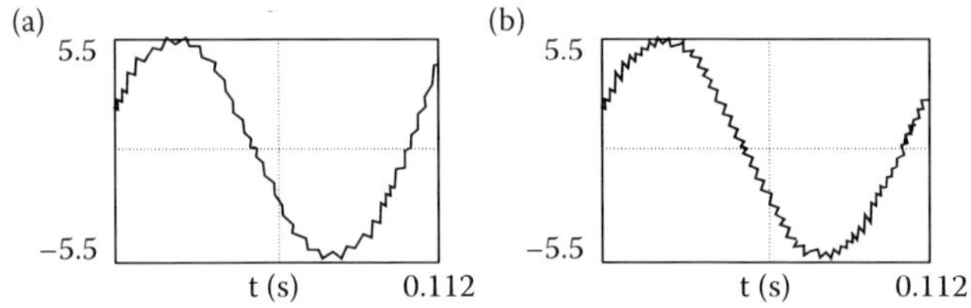


**FIGURE 11.61**
Steady-state phase current: (a) digital simulation; (b) test results.

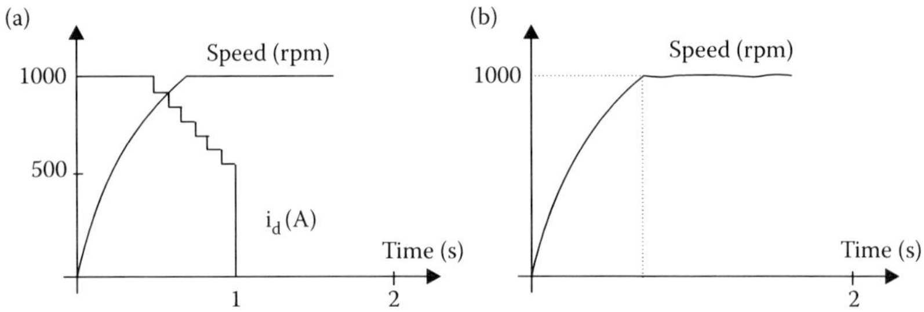


**FIGURE 11.62**
Step speed responses: (a) digital simulation; (b) test results.

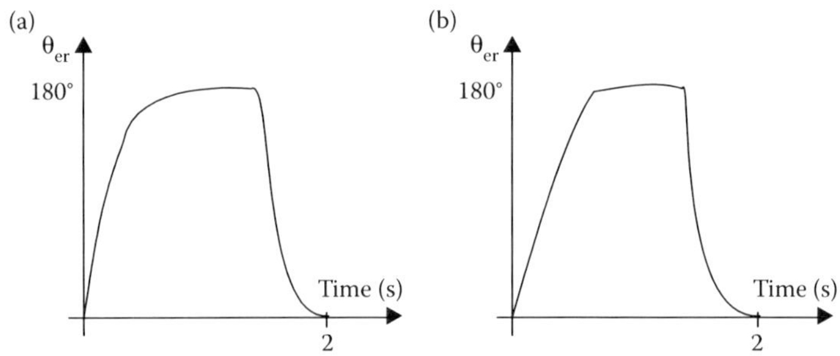


**FIGURE 11.63**
Positioning responses (calculated): (a) rated inertia J; (b) 5J.

### 11.7.3 DTFC of RSM

The DTFC of RSM is similar to that of PMSM. The basic configuration is shown in Figure 11.64.

The flux and torque observers are paramount to DTFC. A speed observer (estimator) is required, in addition, for sensorless control (Figure 11.64). The rationale and table of optimal switchings developed for PMSMs remain valid here (Section 11.5). Only the flux and torque observer bear some peculiarities in the sense that the PM flux is dropped from axis d (Figure 11.45) and eventually subtracted from axis q (if a PM along axis q is added). Also $L_d > L_q$.

Digital simulation results [26] with DTFC are now presented for a 1.5-kW, 2-pole motor with $L_d = 140.77$ mH, $L_q = 7.366$ mH, $r_s = 0.955\ \Omega$, $J = 2.5 \times 10^{-3}$ kg m$^2$, switching frequency $f_c = 15$ kHz in Figure 11.66. The time constant is $\tau_s$ in the speed sliding mode controller functional $s_\omega$

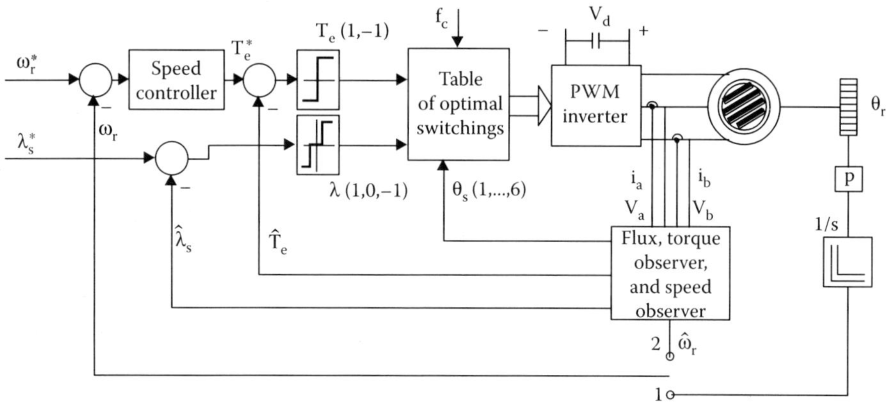


**FIGURE 11.64**
DTFC of RSM: 1—with motion sensor; 2—sensorless.

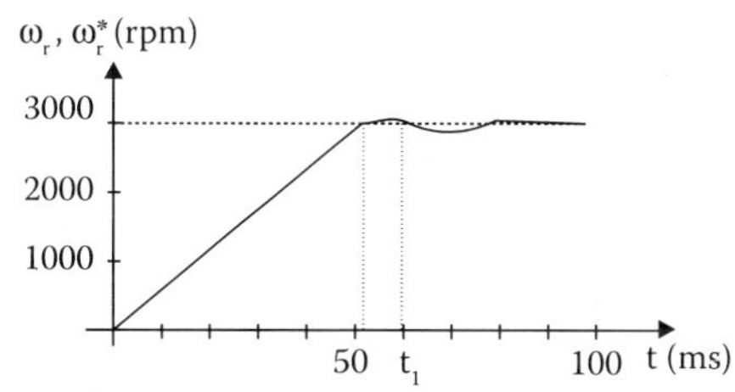


**FIGURE 11.65**
Speed step response with step load 3 Nm at $t_1 = 60$ ms.

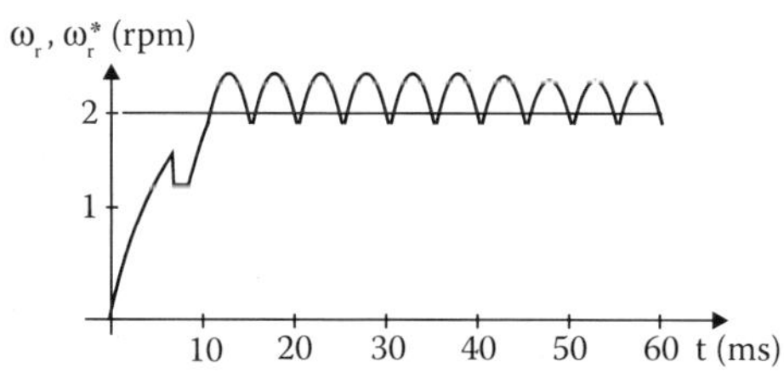


**FIGURE 11.66**
Low speed step response.

$$s_\omega = (\omega_r^* - \omega_r) - \tau_s \frac{d\omega_r}{dt} \tag{11.101}$$

$$\tau_s = 1.2\,\text{ms}$$

Fast speed response without overshoot is evident both for low and high speeds (Figures 11.65 and 11.66).

### 11.7.4 Sensorless Control of RSM: A Case Study

The RSMs can be today considered as a reliable alternative to induction, synchronous, and switched reluctance machines in the medium performance drive applications due to the rotor structures based on multiple flux barriers. Shortcomings like low power factor and low torque density vanish for ratios $L_d/L_q > 5$. Other key aspects such as the following are important strengths of RSMs: the ability to operate at high speeds in high-temperature environments (due to its simple and robust mechanical structure), the absence of rotor Joule losses, and the reduced iron losses in the rotor.

A large variety of control strategies for RSM are present in the literature, such as the following: DTC control of axially laminated anisotropic (ALA) rotor with good results, artificial neural network-based control, predictive torque control, feedback linearization based, and current sensorless field-oriented control (FOC). Superior results to V/f control compared to I–f open-loop control are reported.

In order to reduce the costs and increase the robustness of the electric drive system, rotor position estimation techniques have been widely investigated. The first and most popular method for rotor position estimation for RSM is based on the back electromotive force (e.m.f.). This method, along with the INFORM method, is developed for an RSM in order to achieve wide-speed-range operation. By combining an e.m.f.-based method and a signal injection method, a position sensorless control of a transverse-laminated RSM, which takes into account the effects of cross saturation, is presented here. Extended e.m.f. techniques with online parameter identification were investigated in.

A second method used to detect the rotor position is based on signal injection. In order to extract the RSM rotor saliency, high-frequency signals having low magnitude are superimposed to the fundamental component signals. The injecting signals can be a rotating vector in the stationary frame, pulses in the stationary frame, rotating vector in the synchronous frame, and pulsating vectors in the synchronous frame. Other methods use only the ripples created by the inverter switching in the phase currents, or the induced third harmonic voltage.

Here we present a hybrid motion sensorless control based on two estimators: e.m.f. tracking estimator based on the fundamental model and a saliency tracking estimator based on signal injection. Because the first one is speed dependent, the estimators based on it fail at zero and very low speeds.

Since the signal injection is relatively straightforward in VC schemes, high-frequency-injection methods are added for sensorless control at very low speeds. So, the second estimator is based on a well-known demodulation scheme of the position information contained by the negative sequence component of the high-frequency current obtained by injecting a rotating voltage vector in the stationary frame.

### *11.7.4.1 ALA-RSM Model and Parameters*

The ALA-RSM is credited with high saliency (high performance), and thus, despite its apparent manufacturing difficulties, it is investigated here. The space vector model of RSM in the rotor reference frame is

$$\bar{V}_s^r = \bar{i}_s^r \cdot R_s + \frac{d\bar{\lambda}_s^r}{dt} + j\cdot\omega_{er}\cdot\bar{\lambda}_s^r \tag{11.102}$$

$$\bar{\lambda}_s^r = \lambda_d + j\lambda_q = i_d(L_{dm} + L_{s\sigma}) + j\cdot i_q(L_{qm} + L_{s\sigma}) \tag{11.103}$$

where $\bar{V}_s^r = V_d + jV_q$ denotes the stator voltage; $\bar{i}_s^r = i_d + ji_q$ the stator current vector; $\bar{\lambda}_s^r$ the stator flux vector; $R_s$ the stator phase resistance; $L_{s\sigma}$, $L_{dm}$, and $L_{qm}$ denote the leakage, d axis, and q axis magnetizing inductances; and $\omega_{er}$ is the rotor electrical speed. The RSM electrical parameters were determined after performing d.c. current decay tests with the rotor aligned along the d and q axes, respectively, at standstill. The electrical diagram of the test is shown in Figure 11.67. As can be seen from Figure 11.67, the current decay occurs in phases B and C, which are actually connected in series, while in phase A, different values of constant current $i_{cc}$ are injected.

The d-q fluxes were calculated using Equation 11.104

$$\lambda_{d,q} = \frac{\left(2 \cdot R_s \cdot \int idt + \int V_D dt\right)}{2} \tag{11.104}$$

where $R_s$ denotes the phase resistance, i the decaying current, and $V_D$ the diode voltage drop. The $\lambda_d$–$i_d$ and $\lambda_q$–$i_q$ dependencies are shown in Figure 11.68a and b, respectively.

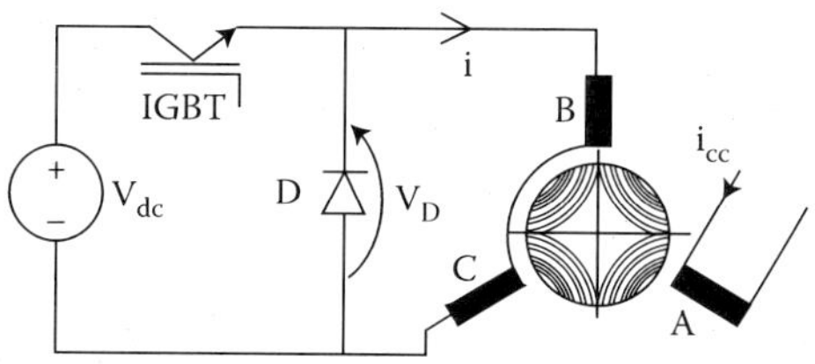


**FIGURE 11.67**
D.c. current decay standstill test arrangement for an ALA-RSM with four poles.

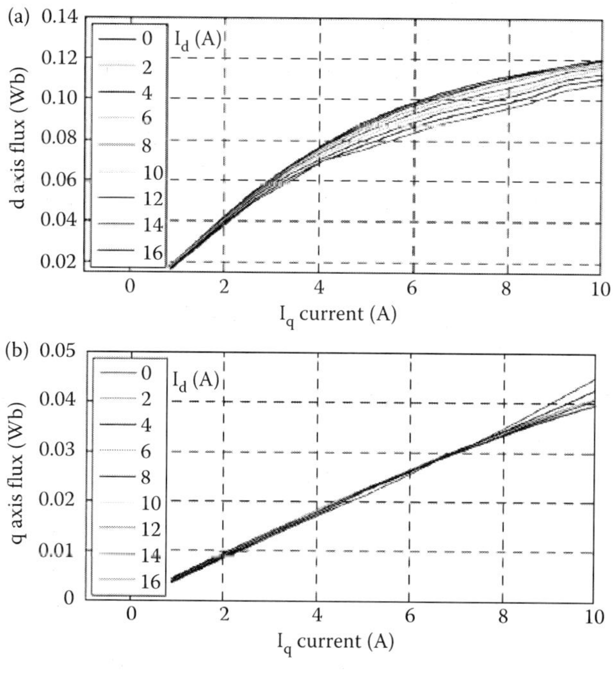


**FIGURE 11.68**
Flux linkage/current curves of ALA-RSM: (a) $\lambda d - i_d$ and (b) $\lambda q - i_q$.

The cross-coupling saturation is visible mainly in d axis, while $L_q$ is rather constant for an ALA rotor, as shown in Figure 11.68b. This is only an approximation as

$$\frac{\partial \lambda_d}{\partial i_q} = \frac{\partial \lambda_q}{\partial i_d} \tag{11.105}$$

The other parameters of the ALA-RSM are listed in Table 11.3.

#### 11.7.4.2 Active-Flux-Based Speed and Position Estimator

The "active-flux" concept [16] or torque producing flux turns the RSM into a fictitious non-salient pole machine with $L_q$ as its inductance. The active flux is defined as the flux that multiplies the current $i_q$ in the expression of the electromagnetic torque $T_e$

$$T_e = \left(\frac{3}{2}\right) \cdot p \cdot (L_d - L_q) \cdot i_d \cdot i_q = \left(\frac{3}{2}\right) \cdot p \cdot \lambda_d^a \cdot i_q \tag{11.106}$$

with $L_d \gg L_q$ for RSM. In Equation 11.106, $\lambda_d^a$ represents the maximum value of active flux per phase. By definition, the active-flux vector observer, in stator reference frame, $\bar{\lambda}_d^{as}$, is

$$\bar{\lambda}_d^{as} = \bar{\lambda}_s^s - L_q \cdot \bar{i}_s^s \tag{11.107}$$

The equivalent synchronous inductance $L_q$ in Equation 11.107 is the q axis inductance.

The active-flux axis falls along the d axis, and thus, its position with respect to a stator reference frame is equal to the rotor electrical position $\theta_{er}$

$$\theta_{\lambda_d^{as}} = a\tan\left(\frac{\lambda_{d\beta}^{as}}{\lambda_{d\alpha}^{as}}\right) = \theta_{er} \tag{11.108}$$

where $\lambda_{d\beta}^{as}$ and $\lambda_{d\alpha}^{as}$ are the two orthogonal components of the active flux in stator coordinates. Cross-coupling saturation (Figure 11.68) may create a quantifiable error in estimated rotor position by Equation 11.108.

**TABLE 11.3**

ALA-RSM Parameters

| Parameter | Value |
|---|---|
| Rated power | 157 W |
| Number of pole pairs | $p = 2$ |
| Stator phase resistance | $R = 0.61\ \Omega$ |
| Drive d.c. link voltage | $V_{dc} = 42$ V |
| Rated frequency | $f_n = 50$ Hz |
| Rated speed | $n_n = 1500$ rpm |
| Rated phase current | 5.2 A |
| Inertia of rotor | $J = 1 \times 10^{-3}$ kg m$^2$ |
| Viscous friction coefficient | $B = 0.7 \times 10^{-3}$ Nm s/rad |

Introducing the definition of the active flux (11.107) in Equation 11.102, we obtain active-flux (rotor) coordinates

$$\bar{V}_s^r = R_s \cdot \bar{i}_s^r + (s + j\omega_{er})L_q \cdot \bar{i}_s^r + (s + j\omega_{er}) \cdot \bar{\lambda}_d^a \tag{11.109}$$

It is evident that Equation 11.109 transforms the RSM into a nonsalient pole equivalent machine. For a better understanding of the concept, the steady-state vector diagram is shown in Figure 11.69a.

In the "active-flux" control, the RSM behaves as if it has a nonsalient rotor with torque current $i_q$ and inductance $L_q$. The active-flux-based speed and position estimator is shown in Figure 11.69b and is active for the medium- and high-speed range. The observer, in stator coordinates, comprises both the voltage model (11.110); superscript "su," and the current model (11.111); superscript "si"

$$\bar{\lambda}_s^{su} = \int (\bar{V}_s^s - R_s \cdot \bar{i}_s^s + \bar{V}_{comp})dt \tag{11.110}$$

$$\bar{\lambda}_s^{si} = (L_d i_d + jL_q i_q) \cdot e^{j\hat{\theta}_{er}} \tag{11.111}$$

where $\hat{\theta}_{er}$ represents the estimated electrical rotor position.

The difference between the two models is fed into a PI compensator which produces the compensation voltage $V_{comp}$ needed to handle the parameter estimation errors, their changes during operation, offsets in the measured variables within the motor drive and

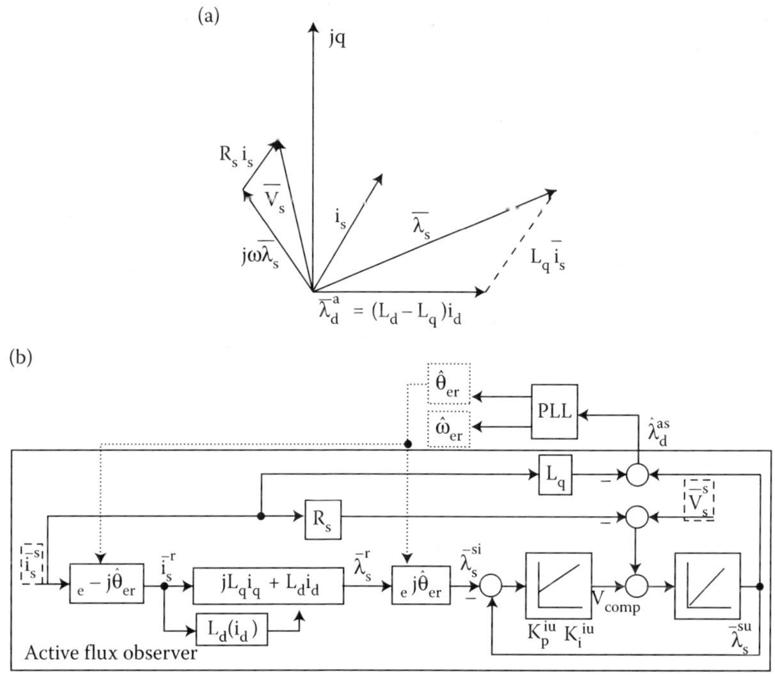


**FIGURE 11.69**
(a) Vector diagram of ALA-RSM. (b) Speed and position estimator based on the active-flux observer. (After Agarlita, S., Boldea, I., Blaabjerg, F., *IEEE Trans.*, IA-48(6), 1931–1939, 2012.)

quantization errors of the digital implementation. All those issues can cause drift of the integration. In this way, the voltage model is adapted through the PI compensator:

$$\bar{V}_{comp} = \left( \frac{K_p^{ui} + K_i^{ui}}{s} \right) \cdot (\bar{\lambda}_s^{su} - \bar{\lambda}_s^{si}) \tag{11.112}$$

The coefficients $K_p^{ui} = 18$ and $K_i^{ui} = 80$ have been chosen such that, for frequencies below 10 Hz of the fundamental model, the current model stands alone ($\bar{\lambda}_s^{si}$), and, for higher frequencies, the voltage model prevails ($\bar{\lambda}_s^{su}$). By combining Equations 11.110 and 11.112, the estimated flux in stator reference frame becomes

$$\bar{\lambda}_s^{su} = \frac{s}{s^2 + k_p^{ui} s + k_i^{su}} (\bar{V}_s^s - R_s \bar{i}_s^s) + \frac{s \cdot k_p^{ui} + k_i^{ui}}{s^2 + k_p^{ui} s + k_i^{su}} \bar{\lambda}_s^{si} \tag{11.113}$$

The term that multiplies the voltage model is a bandpass filter behaving as an integrator for frequencies above 10 Hz, and the one that multiplies the current model is a second-order LPF. The active flux is simply obtained from the stator flux vector and the measured stator current (11.107).

In order to extract the phase and velocity of the active flux, a Luenberger-style speed and position observer is used, as shown in Figure 11.70. This observer is nothing but a PLL enhanced with the mechanical acceleration component obtained from an estimated mechanical equation needed to improve the estimation during speed transients. The input error is the imaginary part of the product:

$$\begin{aligned} \mathrm{Im}(\hat{\bar{\lambda}}_d^{as} \cdot \bar{\lambda}_1) &= \hat{\lambda}_{d\beta}^{as} \cdot \cos(\hat{\theta}_{er}) - \hat{\lambda}_{d\alpha}^{as} \cdot \sin(\hat{\theta}_{er}) \\ &= \left|\bar{\lambda}_d^a\right| \cdot \left[ \sin\theta_{\bar{\lambda}_d^a} \cdot \cos(\hat{\theta}_{er}) - \cos\theta_{\bar{\lambda}_d^a} \cdot \sin(\hat{\theta}_{er}) \right] = \left|\bar{\lambda}_d^a\right| \cdot \sin\Delta\theta_{er} \end{aligned} \tag{11.114}$$

where $\bar{\lambda}_1 = 1 \cdot e^{-j\hat{\theta}_{er}} = \cos(\hat{\theta}_{er}) - j \cdot \sin(\hat{\theta}_{er})$. The observer gains are the following: $K1 = 100$, $K2 = 5000$, and $K3 = -15$. The gains have been chosen such that the damping of the observer is 0.707 ($B = 2 \times 10^{-3}$ N m s/rad, $J = 1.3 \times 10^{-3}$ kg m$^2$, and load included).

#### *11.7.4.3 High Frequency Signal Injection (HFSI)-Based Speed and Position Estimator*

A high-frequency voltage injection is used in the stationary reference frame by superimposing a carrier-signal voltage vector rotating at the carrier-signal frequency $f_c = 500$ Hz and amplitude $V_c$ that varies with speed. The injection is made on top of the fundamental

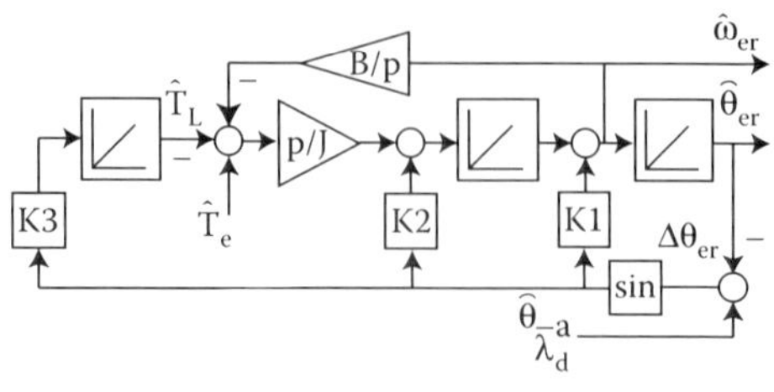


**FIGURE 11.70**
PLL-based position and speed observer.

excitation (Figure 11.71), where $V_d^*$ and $V_q^*$ are the prescribed voltages, in rotor reference frame, corresponding to d and q axis, respectively, $V_{dc}$–d.c. link voltage and $D^*_{a,b,c}$ prescribed duty cycles. LPF and BSF stand for low-pass filter and band stop filter, respectively. The decision not to use a BPF on the HFSI channel, as many classical demodulation schemes do, is argued in Sector 11.7.4.5. The stator inductance of a RSM can be modeled in the rotor reference frame by the following matrix:

$$L_s^r = \begin{bmatrix} L_d & 0 \\ 0 & L_q \end{bmatrix} \tag{11.115}$$

where $L_d \neq L_q$ are position invariant quantities. By transforming the machine equations

$$\begin{bmatrix} V_d \\ V_q \end{bmatrix} = R_s \begin{bmatrix} i_d \\ i_q \end{bmatrix} + \begin{bmatrix} L_d & 0 \\ 0 & L_q \end{bmatrix} s \begin{bmatrix} i_d \\ i_q \end{bmatrix} + \hat{\omega}_{er} \begin{bmatrix} 0 & -L_q \\ L_d & 0 \end{bmatrix} \begin{bmatrix} i_d \\ i_q \end{bmatrix} \tag{11.116}$$

to the stationary reference frame, the position-dependent and cross-coupling inductances are obtained as

$$\begin{aligned} \bar{V}_s^s &= R \cdot \bar{i}_s^s + s \cdot \left( \sum L \cdot \bar{i}_s^s + \Delta L \cdot e^{j2\hat{\theta}_{er}} \cdot \bar{i}_s^{s*} \right) \\ &\quad + j \cdot \hat{\omega}_{er} \cdot \left( \sum L \cdot \bar{i}_s^s + \Delta L \cdot e^{j2\theta_{er}} \cdot \bar{i}_s^{s*} \right) \end{aligned} \tag{11.117}$$

where $\Sigma L = (L_d + L_q)/2$, $\Delta L = (L_d - L_q)/2$ are stator average and differential high-frequency inductances, $\bar{i}_s^{s*}$ is the complex conjugate stator current, and s is the Laplace operator. The motion-induced voltage component along with the voltage drop on the stator resistance, in Equation 11.117, can be neglected for $\omega_c \gg \omega_{er}$. At high frequencies, the machine equation is

$$\bar{V}_s^{s\_c} = s \cdot \left( \sum L \cdot \bar{i}_s^{s\_c} + \Delta L \cdot e^{j2\theta_{er}} \cdot \bar{i}_s^{s\_c*} \right) \tag{11.118}$$

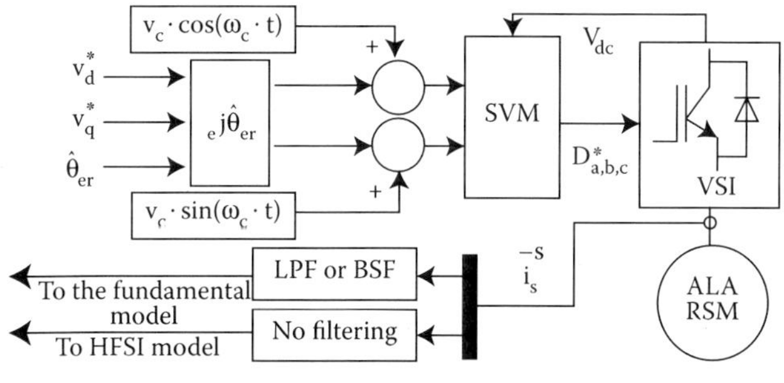


**FIGURE 11.71**
Injection of high-frequency voltage in the stationary frame.

For $\bar{V}_s^{s\_c} = V_c \cdot e^{j\omega_c t}$, representing the injected high-frequency voltage in stator reference frame, the resulting high-frequency current, in the same reference frame, will have the following form:

$$i_s^{s\_c} = i_{pc} \cdot e^{j(\omega_c t + \varphi_1)} + i_{nc} \cdot e^{j(-\omega_c t + 2\theta_{er} + \varphi_2)} \quad (11.119)$$

The interaction between the injected carrier voltage and the sinusoidally distributed machine saliency, having the period equal to a pole pitch, will produce a carrier-signal current containing a positive ($i_{pc}$) and a negative sequence component ($i_{nc}$). Only the negative one contains spatial information and will be processed to extract the rotor speed and position. The implemented demodulation is based only on a LPF used in the negative sequence carrier reference frame, as shown in Figure 11.72.

The magnitude of the negative sequence component determines the values of the $K_p$ (80) and $K_I$ (4000), the two PI constants inside the PLL in Figure 11.72, which has the structure presented in Figure 11.70. The magnitude is dependent upon the amplitude of the injected voltage vector and the differential stator high-frequency inductance. A small magnitude of the negative component means high values for both constants and, consequently, high sensitivity to noise. An important feature of the HFSI observer is that its steady-state tracking ability does not depend on motor parameters.

The inverter nonlinearities are also addressed: dead time ($T_d$), turn-on ($T_{on}$), and turn-off ($T_{off}$) time, and voltage drop on the switching power devices (threshold: $u_{D0}$, $u_{T0}$). The added duty cycle error has the form presented in Equation 11.120. The compensation is made in a feedforward manner

$$\Delta = \left( \frac{T_d + T_{on} - T_{off}}{T_s} + \frac{u_{D0} + u_{T0}}{2V_{dc}} \right) \cdot \text{sign}(i_f) \quad (11.120)$$

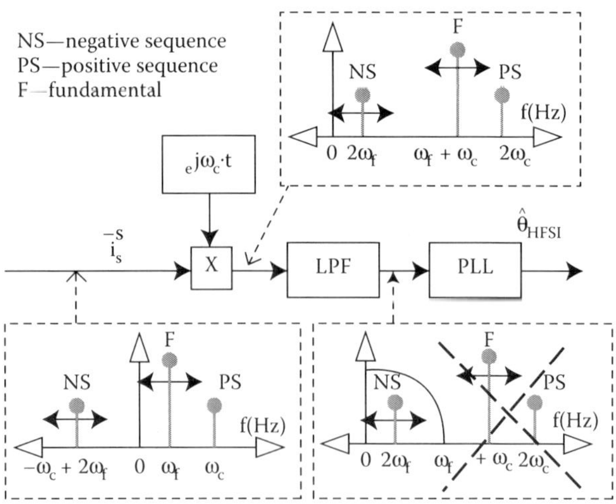


**FIGURE 11.72**
Demodulation technique of the HFSI currents. (After Agarlita, S., Boldea, I., Blaabjerg, F., *IEEE Trans.*, IA-48(6), 1931–1939, 2012.)

where $T_s$ is the PWM period, $V_{dc}$ is the d.c. link voltage, and $i_f$ represents the current of the compensated phase. The ohmic voltage drop caused by the on-state dynamic resistance of the transistor and diode was added to the stator resistance.

#### *11.7.4.4 Sensor Fusion Position Estimator*

The active-flux-based speed and position estimator (AFE) is fed with the fundamental excitation components of the machine, $V_s$ and $I_s$, and experiences estimation problems at low and very low speed. To improve the position estimation in this region, the difference between the two estimators is considered an error and is fed into a PI compensator which will add to the AFE the needed difference to stabilize its position estimation. The proposed sensor fusion structure is shown in Figure 11.73. Moreover, if $K_p$ and $K_i$ are reduced down to zero, as the rotor speed increases, the correction will vanish. When the correction vanishes completely, at 3.3 Hz (100 rpm, 2 pole pairs), the current model, present in the AFE, continues to assist the voltage model until 10 Hz, from where the latter will remain alone. As experiments demonstrate, this type of fusion structure, combining a high-frequency method, the current model, and the voltage model, offers high robustness over a wide speed range.

At the same time the high-frequency voltage amplitude is decreased gradually with speed, from 3 to 0 V, similarly to the PI controller coefficients. The need to stop the injection at high speeds is justified by the following shortcomings: additional core losses, limitation of the maximum available voltage vector, motor vibrations, acoustical noise, etc.

#### *11.7.4.5 Motion Sensorless Control of ALA-RSM*

The adopted configuration of the motion sensorless VC system for RSM is shown in Figure 11.74. The reference current $i_q^*$ is obtained from the speed control loop, which employs a PI controller that has as input the speed error

$$i_q^* = K_p^{\omega}\left(\frac{1+K_i^{\omega}}{s}\right)(\omega_r^* - \hat{\omega}_{er}) \tag{11.121}$$

The PI gains $K_p^{\omega} = 0.028$, $K_i^{\omega} = 29$ are selected for high enough dynamic response to avoid interference between the speed loop and the RSM current loops, but still secure good rotor

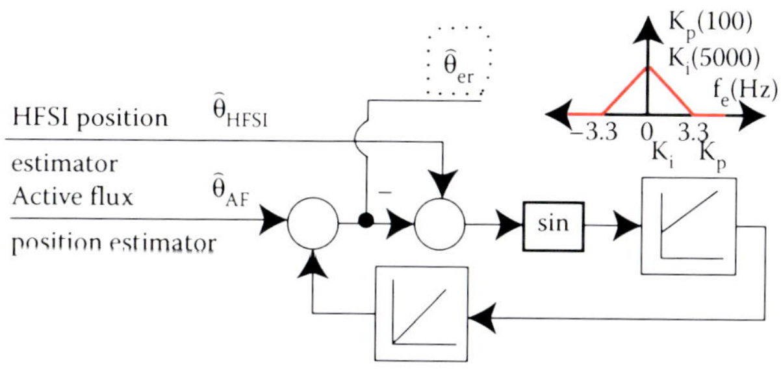


**FIGURE 11.73**
Sensor fusion structure.

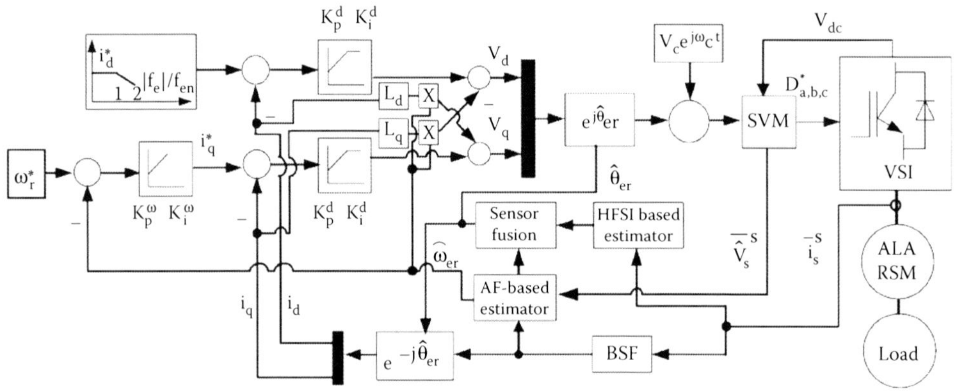


**FIGURE 11.74**
Complete motion sensorless control scheme of ALA-RSM with position and speed estimation.

speed dynamics. The current controllers are implemented in the rotor reference frame and produce the reference voltage vector components:

$$
\begin{aligned}
V_d^* &= K_p^d\left(\frac{1+K_i^d}{s}\right)(i_d^* - i_d) - \hat{\omega}_{er}\cdot L_q\cdot i_q \\
V_q^* &= K_p^q\left(\frac{1+K_i^q}{s}\right)(i_q^* - i_q) + \hat{\omega}_{er}\cdot L_d\cdot i_d
\end{aligned}
\tag{11.122}
$$

where the PI gains are $K_p^d = 12$ and $K_i^d = 83$, and, respectively $K_p^q = 3.55$ and $K_i^q = 384$. The characteristic equation of the equivalent electrical system for q axis containing the motion-induced term, H(s), is

$$
1 + G(s)\cdot H(s) = 1 + \frac{1}{R + sL_q}\cdot\frac{k_T\cdot\lambda_d\cdot p}{B_t + sJ_t} \tag{11.123}
$$

where $k_T = 1.5\cdot p(L_d - L_q)i^*_d$ is the torque constant of RSM, and $B_t$ and $J_t$ are the viscous friction coefficient and the inertia of the motor and load. The zero of the current controller, for each axis, was chosen to neutralize the pole of the corresponding electrical equation. The proportional gain is then found based on the magnitude optimum criterion (damping ratio $\zeta = 0.707$).

By considering all the transfer functions of the control system components: control algorithm (estimators), inner q axis current loop (characteristic equation in Equation 11.123), zero-order hold, VSI (voltage-source inverter), and current sensors, along with the corresponding compensator, the speed closed-loop system is found. By using the bilinear transformation (sampling time: h = 2*e−4) the discrete-time model is obtained.

The presence of all the poles of the speed control system within the unity circle demonstrates the system stability, as shown in Figure 11.75. The dominant complex poles are located such that 0.707 damping factor is obtained. The gain margin and phase margin are 13.2 dB and 63.4°, respectively.

Finally, the q current controller bandwidth is 310 Hz, obtainable only if the LPF for the fundamental currents (Figure 11.71) is replaced with a narrow Band Stop Filter centered at 500 Hz. If a stop band of 20 Hz is chosen the attenuation is under 3 dB until 490 Hz, while

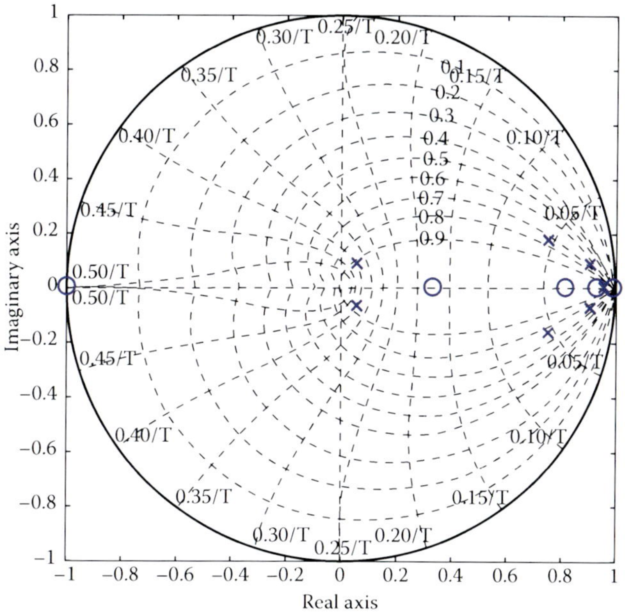


FIGURE 11.75
Zero-pole location in the z-plane of the speed loop.

the phase lag is almost zero until 390 Hz. Such a filter will provide good rejection of high-frequency components without disturbing the 310 Hz bandwidth of the current loop.

For the HFSI estimator input, recent publications have shown that a BPF (present in the classical demodulation schemes) needed to reject the fundamental component is redundant and will impair the dynamic response of the estimator. The bandwidth of the speed controller is 56 Hz, under pure active-flux control, but in the low-speed region it can be reduced mainly due to the LPF needed to reject the positive component and the fundamental of current in the HFSI estimator. It is not the case here because the chosen corner frequency is twice the speed controller bandwidth.

#### 11.7.4.6 Experimental Results

The above-described algorithms were implemented in MATLAB/Simulink package and executed on a dSpace platform, using 200 μs sampling rate. The test rig is shown in Figure 11.76.

The test rig contains: two U405—FC 302/1.5 kW Danfoss inverters fed with two autotransformers to adapt the d.c. link voltage, the dSpace 1103 Platform, current sensors (LEM LA 55-P), one incremental encoder (2RH type-4096 pulses/revolution), and the RSM with ALA rotor and the load.

The load torque is safely produced by an indirect current vector-controlled IM drive, down to zero speed. In what follows zero and very low-speed operation (1, 5 rpm), rated speed operation, 1500 and 3000 rpm results will be presented. Test results for zero speed operation with 1 Nm load torque are illustrated in Figure 11.77 and for 1 and 5 rpm with 0.5 Nm loading in Figures 11.78 and 11.79.

During the following test, the machine travels to 1500 rpm, suffers a sudden loading with 1 Nm, followed by an unloading action. A reversing reference is given to reach −1500 rpm, followed by a similar torque disturbance (Figure 11.80a and b). As one can

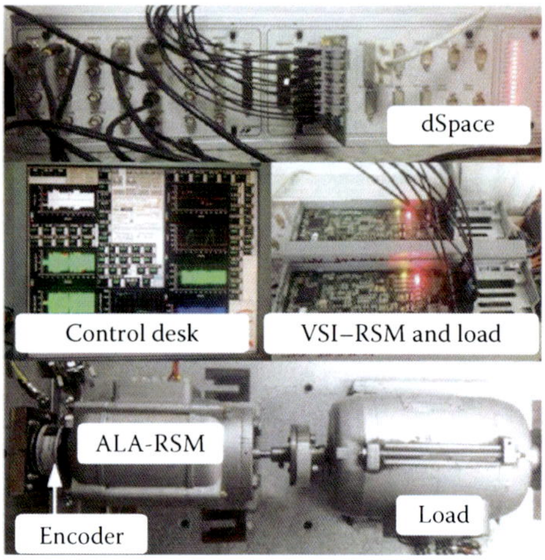


**FIGURE 11.76**
Photograph of the complete test rig.

observe in Figure 11.80c four different position estimations are illustrated: $\Theta_{AF}$ is the position obtained from AFE, $\Theta_{enc}$ is the position from encoder, $\Theta_{HFSI}$ is the position estimated by HFE, and $\Theta_{EI}$ is the position used in all coordinate transformations within the control. $\Theta_{EI}$, being $\Theta_{AF}$ corrected by $\Theta_{HFSI}$ in the low-speed region, is located very near $\Theta_{HFSI}$ for speeds below 100 rpm. After the rotor speed passes this threshold value, $\Theta_{EI}$ becomes identical to $\Theta_{AF}$ because the correction stops.

Next, 3000 rpm operation followed by a reversing action is illustrated in Figure 11.81a–d. The loading test with 0.2 Nm at 3000 rpm is illustrated in Figure 11.81e–g.

#### 11.7.4.7 Discussion

As it can be seen from the test results, using HFSI for rotor position estimation to correct the active-flux-based position estimator, through a new sensor fusion structure, safe operation at zero speed, 1 and 5 rpm with torque perturbations is obtained, as shown in Figures 11.77 through 11.79. Acceleration to 1500 rpm, Figure 11.80, and deceleration to zero shows stable and rather quick response.

Even during speed notable dynamics the speed estimation error is small. Quick and stable acceleration/deceleration to ± 3000 rpm with load perturbation (Figure 11.81) proves the robustness of the proposed hybrid, motion sensorless, control.

A 0–1500 rpm speed control has been shown on this 42 V d.c. RSM drive, but the simple reduction of active flux with speed over 1500 rpm did extend the speed range to 3000 rpm. Even more speed may be obtained by further decreasing $i_d$ (flux weakening).

## 11.8 High-Frequency (Speed) PMSM Drives

By high frequency we mean fundamental frequency above 500 Hz when the 10–15 kHz switching frequency of the IGBTs becomes less than desired to produce a close to sinusoidal current waveform in the PMSM. Also the online computation cycle—which has to be kept within 5° (electrical)—becomes too small for complex online computation controls.

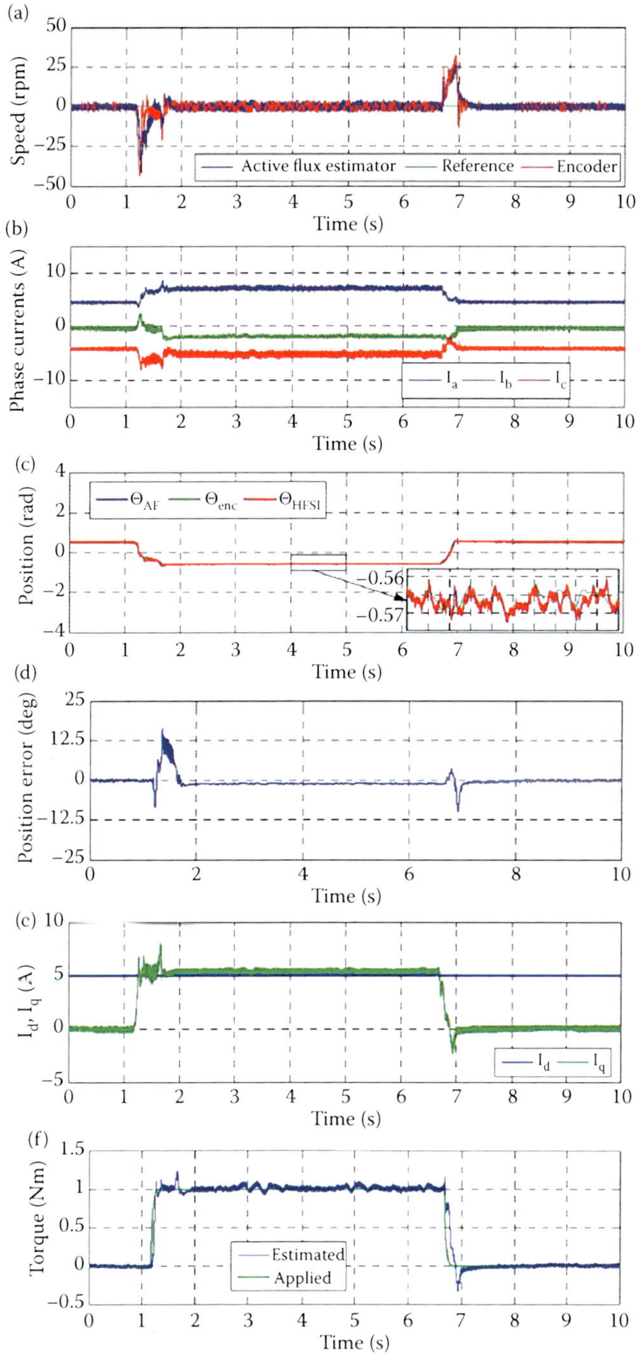


**FIGURE 11.77**
Zero speed operation with 1 Nm torque step loading: (a) speed, (b) phase currents, (c) electric position, (d) position error, (e) synchronous currents, (f) estimated and applied torque. (After Agarlita, S., Boldea, I., Blaabjerg, F., *IEEE Trans.*, IA-48(6), 1931–1939, 2012.)

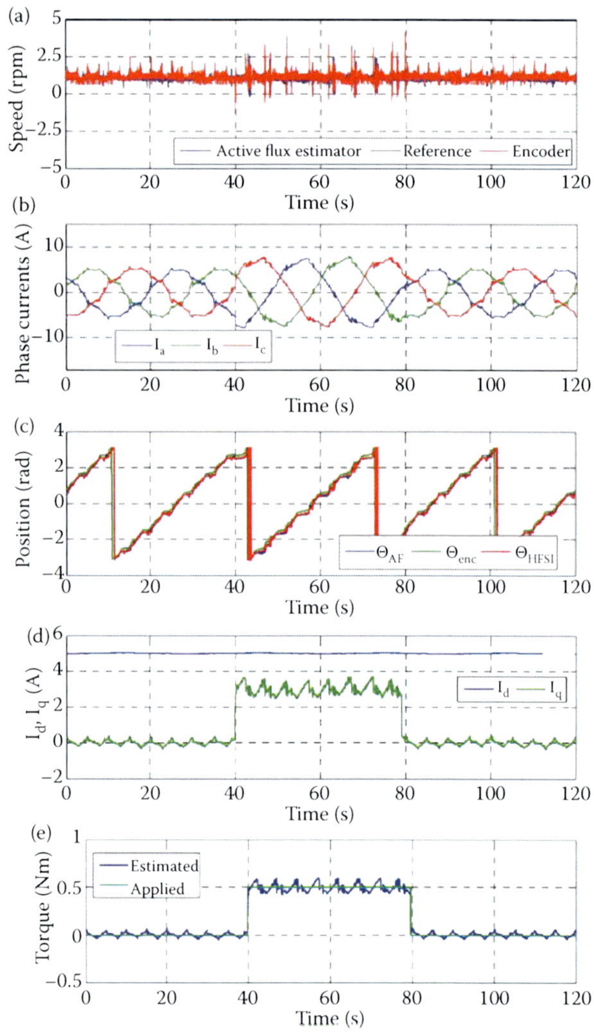


**FIGURE 11.78**
1 rpm, 0.5 Nm torque step loading: (a) speed, (b) phase currents, (c) electric theta, (d) $I_d$, $I_q$ currents, (e) reference and estimated torque. (After Agarlita, S., Boldea, I., Blaabjerg, F., *IEEE Trans.*, IA-48(6), 1931–1939, 2012.)

Applications for high-frequency (high speed) PMSM drives refer to dental drillers and compressors or gas turbines. The latter have reached 150 kW at 70,000 rpm and 3 MW at 15,000 rpm. Even at 15,000 rpm, because the large power machine size has to be reduced further by using a large number of poles, the fundamental frequency goes up to 1–1.5 kHz.

There are three main ways to control high-frequency (speed) PMSM:

- Rectangular current control with variable d.c. link voltage
- V/f scalar control with stabilizing loops
- VC

All strategies are motion sensorless, at most they may use three low cost Hall proximity sensors, which is standard for rectangular current control.

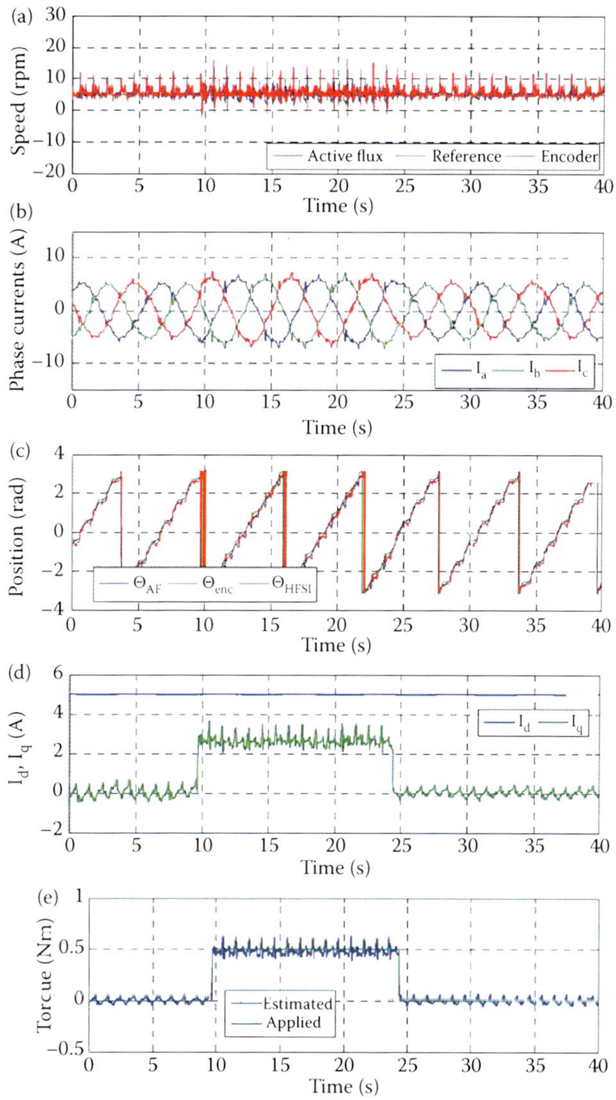


**FIGURE 11.79**
5 rpm with 0.5 Nm load, (a) speed, (b) phase currents, (c) electrical theta, (d) $I_d$ and $I_q$ currents, (e) reference and estimated torque. (After Agarlita, S., Boldea, I., Blaabjerg, F., *IEEE Trans.*, IA-48(6), 1931–1939, 2012.)

### 11.8.1 Rectangular Current Control

Due to the problems with rising delay in high-speed (frequency) drives, for rectangular current control, the d.c. link voltage may be made variable (increasing) with speed (Figure 11.82) [28].

The d.c. link output voltage $V_0$ of the step-down d.c.–d.c. converter is simply

$$V_0 = dV_{dc} \tag{11.124}$$

where $V_{dc}$ is the input d.c. voltage and d is the duty ratio.

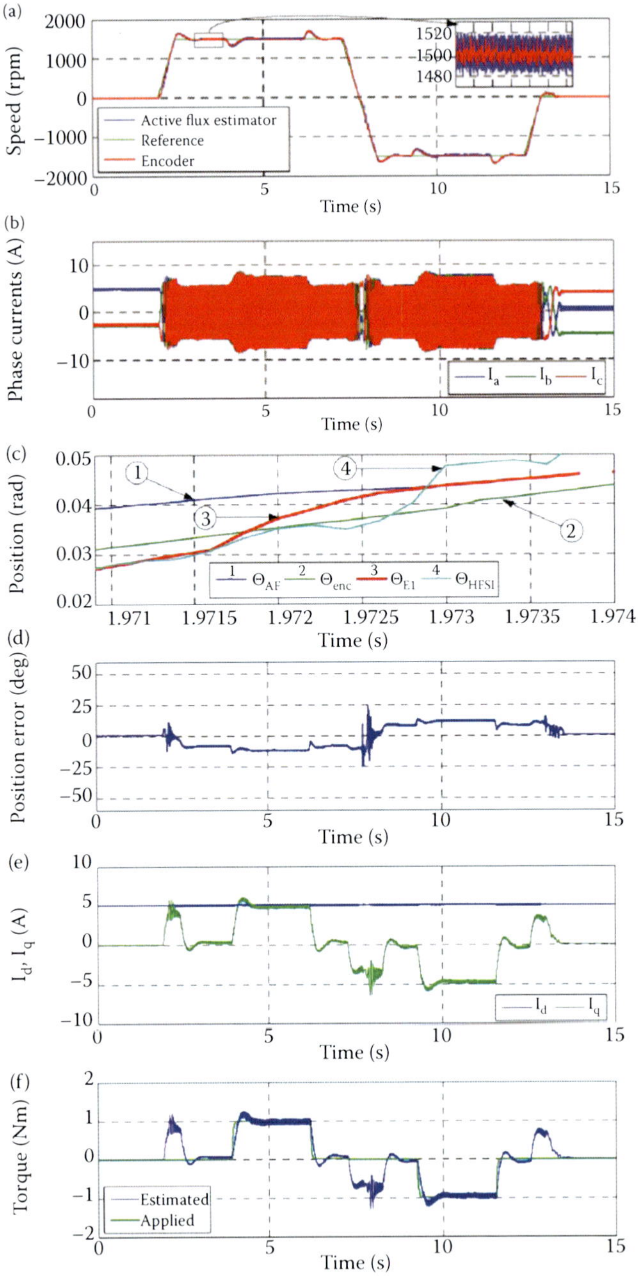


**FIGURE 11.80**
From 0 to 1500 rpm to –1500 rpm with 1 Nm load: (a) speed, (b) phase currents, (c) electrical rotor position transition, (d) position error, (e) synchronous currents, (f) applied and estimated torque. (After Agarlita, S., Boldea, I., Blaabjerg, F., *IEEE Trans.*, IA-48(6), 1931–1939, 2012.)

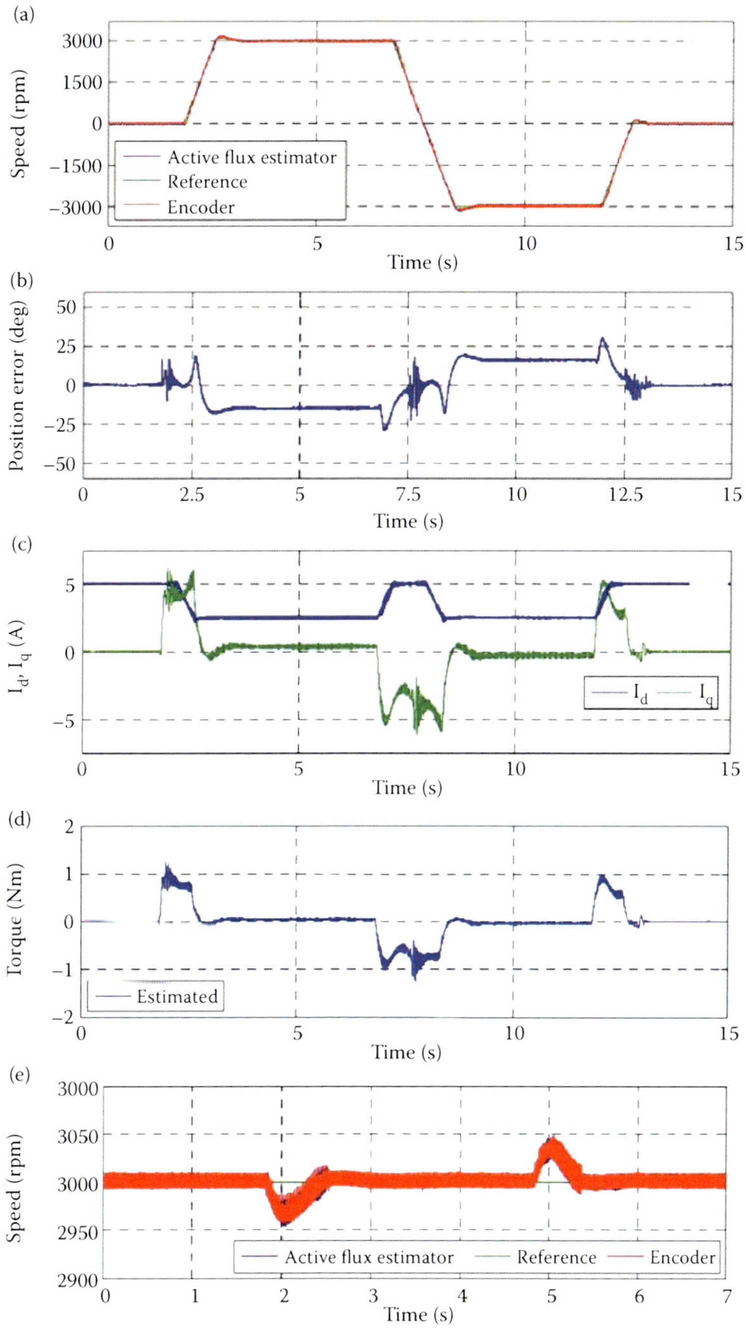


**FIGURE 11.81**
Speed reversal from 3000 to −3000 rpm: (a) speed, (b) position error, (c) $I_d$ and $I_q$ currents, (d) applied and estimated torque; loading at 3000 rpm, (e) speed. (After Agarlita, S., Boldea, I., Blaabjerg, F., *IEEE Trans*., IA-48(6), 1931–1939, 2012.)

*(Continued)*

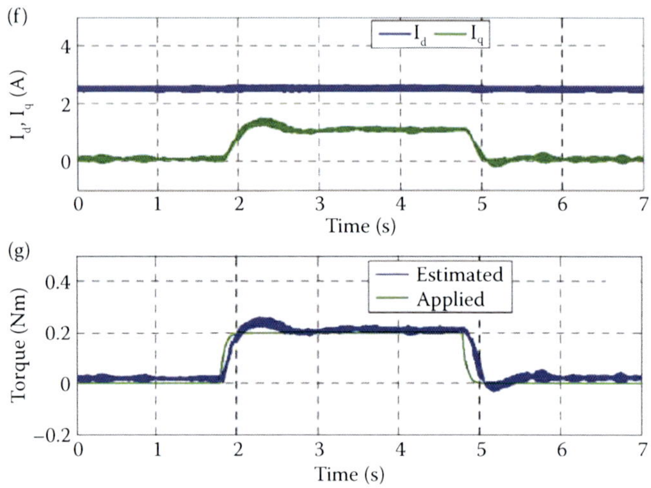


**FIGURE 11.81** ***(Continued)***
Speed reversal from 3000 to −3000 rpm: (f) $I_d$ and $I_q$ currents, (g) applied and estimated torque. (After Agarlita, S., Boldea, I., Blaabjerg, F., *IEEE Trans.*, IA-48(6), 1931–1939, 2012.)

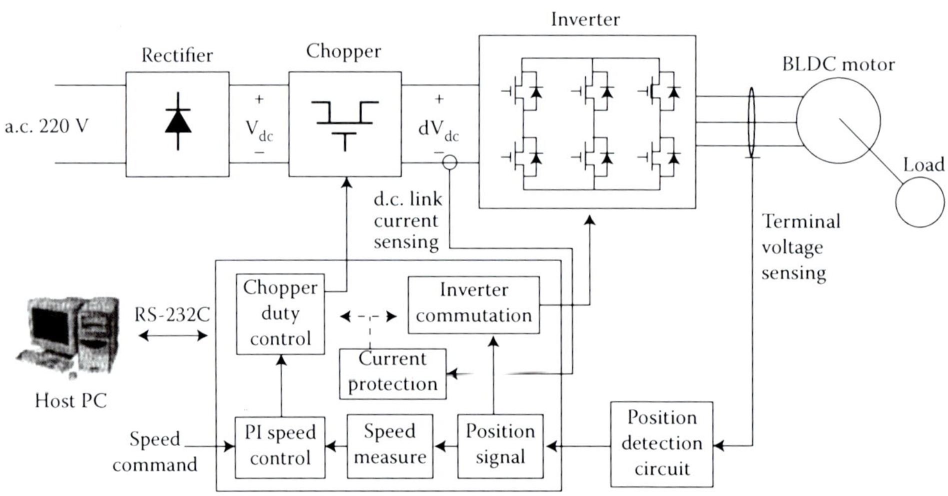


**FIGURE 11.82**
Variable d.c. link voltage PWM converter.

The value of d may be made dependent on speed error $\omega_r^* - \omega_r$

$$d = (\omega_r^* - \omega_r) \cdot \left( K_p + \frac{K_i}{s} \right) \quad (11.125)$$

This is a PI controller.

The speed $\omega_r$ has to be calculated (estimated). If Hall sensors are used, $\omega_r$ may be approximately calculated from the discrete information on rotor position, after proper filtration.

In discrete form Equation 11.125 becomes

$$\begin{aligned} &d(k) = d(k-1) + K_p[\varepsilon(k) - \varepsilon(k-1)] + K_i T_s \varepsilon(k-1) \\ &0 \le d(k) \le 1 \end{aligned} \quad (11.126)$$

where $T_s$ is the sampling time.

It follows that the speed is regulated through the d.c. link voltage $V_0$. Consequently, only the commutation of phases every electrical 60° is provided through the PWM converter.

As a result, the commutation losses in the latter are reduced drastically, at the price of the losses in the d.c.–d.c. converter.

Reference 28 shows the beneficial effect of the d.c.–d.c. converter with current control in the d.c. link upon the current pulse waveform of current control system (Figure 11.83).

The sharp rise and fall of rectangular current at 50,000 rpm (820 Hz) is a clear indication of very good performance.

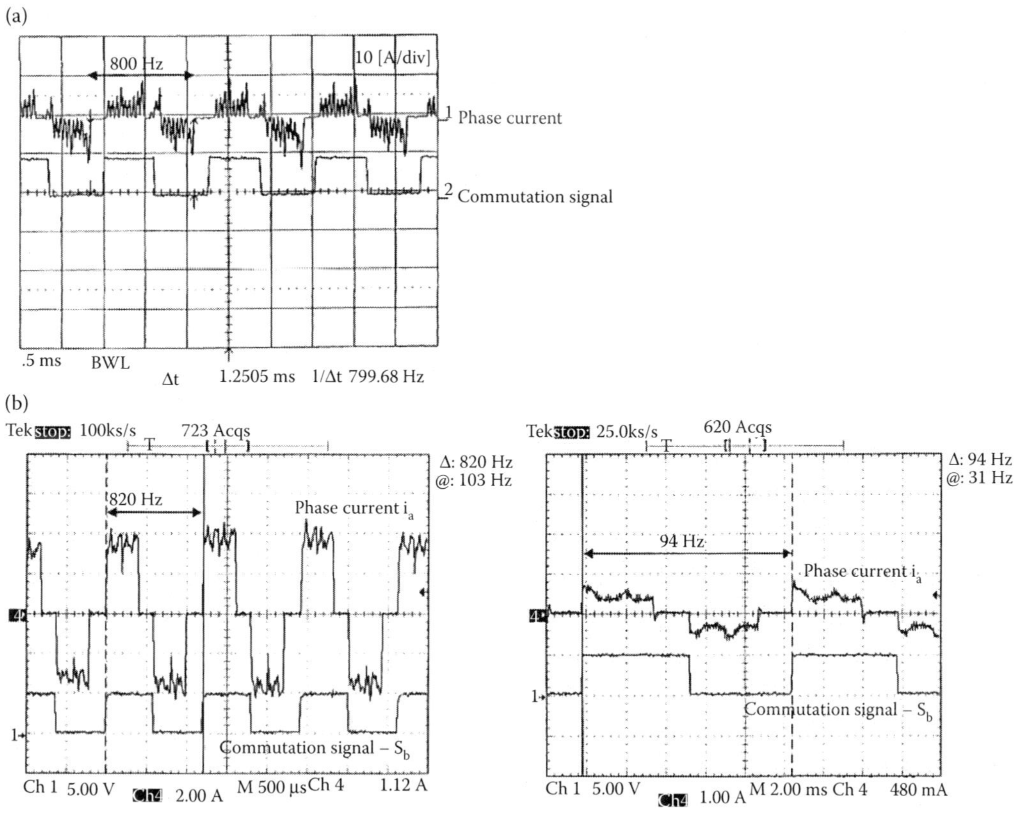


**FIGURE 11.83**
PWM converter control of PMSM: (a) without d.c.–d.c. converter; (b) with d.c.–d.c. converter.

### 11.8.2 V/f Control with Stabilizing Loops

As the fundamental frequency increases, the online available computation time decreases. A simplified control scheme is required. Fortunately the transient torque response quickness is not so demanding in high-speed drives.

V/f open control is rather simple but it needs a stabilizing loop. It could be a current stabilizing loop or an active power variation damping that would modify the reference frequency f* in a feedforwarding manner [29].

Alternatively, it is possible to assembly a dual stabilizing loop:

- One of reference voltage amplitude, to control the stator flux amplitude
- One on reference voltage phase angle, to closed-loop control the current to PM flux linkage vector angle $\gamma$ (Figure 11.84)

$$\Delta V = (\lambda_s^* - \lambda_s)\cdot\left(K_{p\lambda} + \frac{K_{i\lambda}}{s}\right)$$
$$\Delta\gamma = (\gamma^* - \gamma)\cdot\left(K_{p\gamma} + \frac{K_{i\gamma}}{s}\right) \tag{11.127}$$

The maximum torque per current for surface-PM rotor ($L_d = L_q$) occurs for $\gamma^* = \pi/2$ (e.m.f. in phase with current).

The PM produced flux linkage $\bar{\lambda}_{PM}$ is straightforward:

$$\bar{\lambda}_{PM} = -L_s\bar{I}_s + \bar{\lambda}_s \tag{11.128}$$

But the stator flux linkage $\bar{\lambda}_s$ is

$$\bar{\lambda}_s = \int(\bar{V}_s - R_s\bar{I}_s)\cdot dt \tag{11.129}$$

Equations 11.128 and 11.129 represent the PM flux and stator flux estimators. As the speed is large the voltage model suffices while the integral offset has to be eliminated by filtering.

Typical digital simulations with such a scheme at 80,000 rpm with a small PMSM with step load at t = 12 s are shown in Figure 11.85.

The introduction of the power angle control stabilizes the drive while the flux loop control just improves the energy conversion ratio.

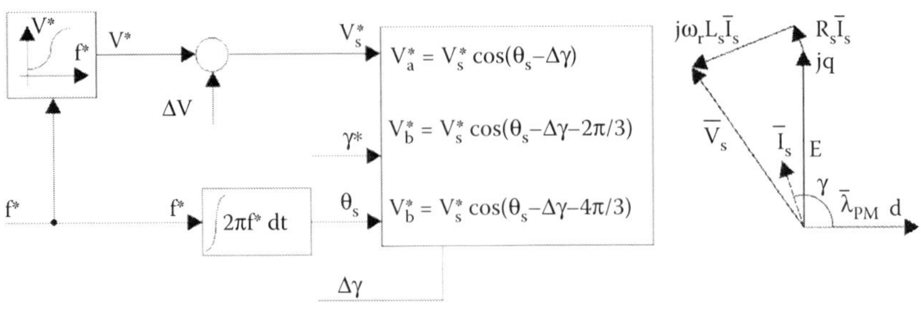


**FIGURE 11.84**
V/f control with flux and torque angle stabilizing loops.

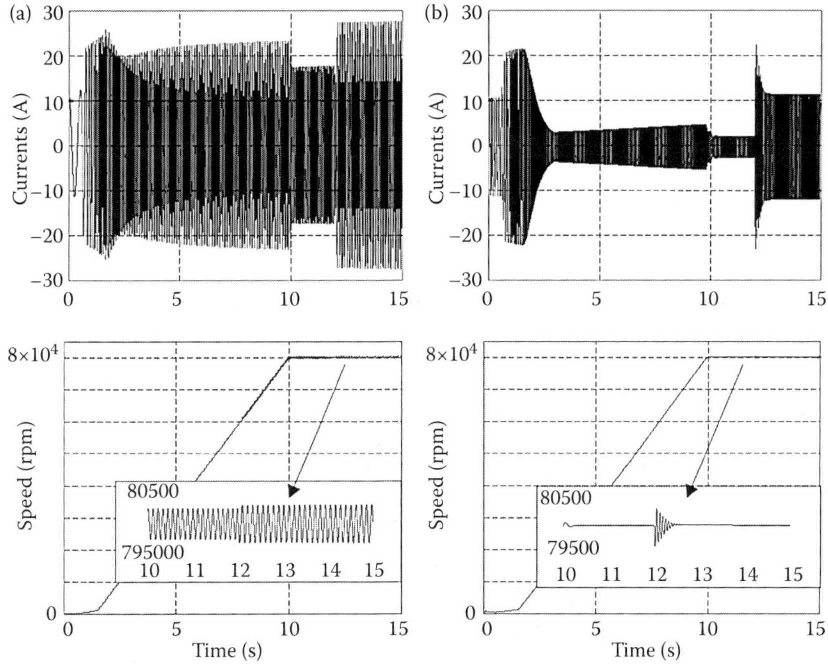


**FIGURE 11.85**
Steady-state performance of V/f control of a PMSM at 80,000 rpm: (a) without stabilizing loops, (b) with stabilizing loops.

### 11.8.3 Vector Control

If speed and rotor position angle $\theta_{er}$ may be observed by using simplified observers, even VC may be applied up to 70,000–100,000 rpm.

In essence a combined voltage–current VC e.m.f. compensation is used in Reference 30. Sample results are shown in Figure 11.86.

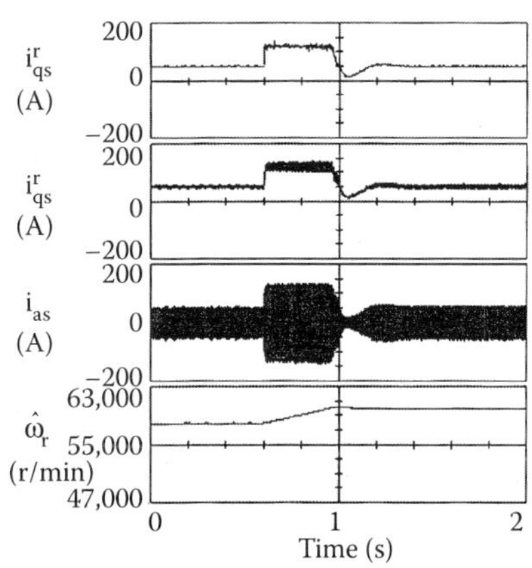


**FIGURE 11.86**
Acceleration characteristics of the PMSM with sensorless VC scheme (from 58,000 to 60,000 rpm). (After Bae, B.-H., Sul, S.-K., Kwon, J.-H., Byeon, Ji-.S., *IEEE Trans.*, IA-39(3), 811–818, 2003.)

A problem with high-frequency (speed) PMSM drives is the fact that the synchronous inductance $L_s$ tends to be too small and thus the current ripple tends to be large with limited switching frequency.

The addition of external inductors in series with all phases leads to a reduction of current ripple [30]. Also it is feasible to increase the slot leakage inductance by filling only the bottom part of the semiclosed slots of the stator. Finally, for the surface-PM rotors at high speeds, a resin should be placed above a 0.5–0.7-mm-thick copper shield to reduce the rotor iron losses produced by the stator space and time harmonics. Such a copper shield also attenuates the current ripple. More elaborated recent V/f (or I–f) scalar control schemes are assessed in Reference 31 and in Chapter 16.

## 11.9 Single-Phase PMSM Control

Low-power electric drives are applied mainly for home appliances and on automobiles. Air purifiers, air conditioning heater blowers, room temperature sensors, vacuum pumps, electric remote control mirrors, automatic speed adjustment pumps, retractable head lamps, radiator cooling fans, oil cooling fans, throttle controls, sun roofs, variable shock absorbers, auto door locks, fuel pumps, rear wipers, windscreen wipers are all typical automobile small-power electric drive applications.

PM brushless motors are a solution for these low-power drives. As three phases PMSM drive control has already been treated earlier in this chapter, here only typical single-phase PMSM drive control is introduced.

A lower SRC count power electronics control, typical for a single-phase PMSM, is maybe the way for a lower global cost drive systems below 50 and 100 W. The radial or axial airgap single-phase PMSM with safe-starting rotor parking provision (Chapter 10) with a small number of stator coils is typical for such a low-power drive.

### 11.9.1 Typical Converter Topologies

The typical power converter for such a drive is composed of a diode rectifier and a four active switch single-phase PWM inverter (Figure 11.87a).

A lower count (two) switch PWM converter with split d.c. capacitor with voltage doubler and half bridge diode rectifier is shown in Figure 11.87b.

The voltage doubler (through $L_d$) is required to preserve the d.c. voltage of the typical four-switch PWM converter (Figure 11.87a). The same converter may be used for a bidirectional motion drive with capacitor two-phase PMSM (Figure 11.87b′).

To simplify the drivers sources (by eliminating the voltage shifting) for the bifilar single-phase PMSM, two active switches $T_1$ and $T_2$ in the lower leg are connected in series with the two parts of the a.c. winding (one of them serves one polarity) (Figure 11.87c). In addition, a third switch T3 with a $C_d$-dump capacitor and a high-frequency transformer is used to channel the freewheeling energy when $T_1$ and $T_2$ are off, such that to restore the input a.c. current sinusoidality. $C_d$ is designed at twice the rated d.c. voltage to provide fast phase turn off.

Another option of lower cost converter (Figure 11.87d) makes use of two thyristors ($T_1$, $T_2$) that provide phase commutation, while a single fast switch ($T_3$) d.c.–d.c. converter

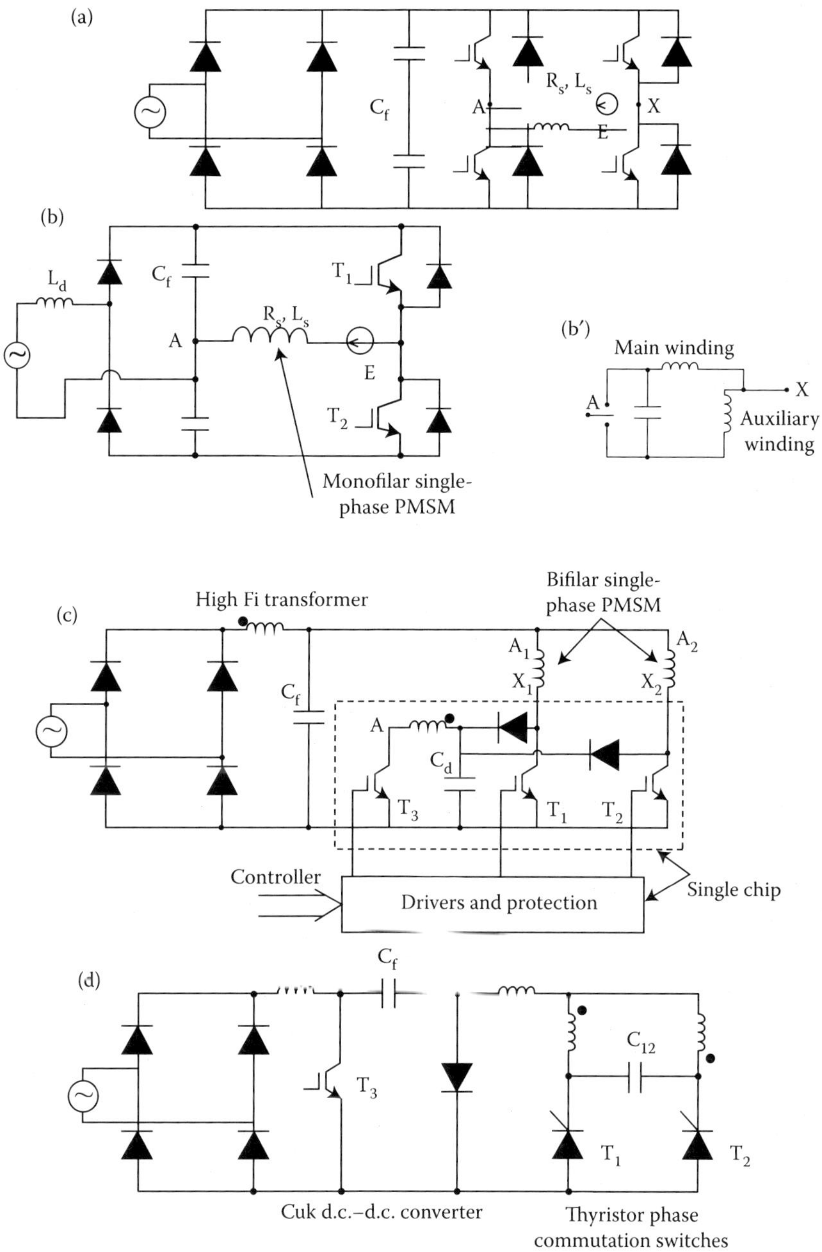


**FIGURE 11.87**
Representative power electronics converters for single-phase PMSM: (a) with diode rectifier and four switch PWM inverter single phase; (b) with half diode bridge, voltage doubler and two switch PWM single-phase converter (two-phase capacitor PMSM connection- (b′)); (c) with diode rectifier, bifilar single-phase winding, and two active switches PWM converter with $C_{dump}$ and high-frequency transformer for sinusoidal a.c. source current, (d) with diode rectifier, Cuk chopper, and thyristor phase commutator.

may be used also to restore the sinusoidality of the a.c. source input current according to today's standards (IEC 1000–3-2 for class D equipment, for example).

### 11.9.2 State-Space Model

Before addressing the problem of open or closed-loop speed control through power electronics, the state-space model of the single-phase PMSM is required:

$$I_sR_s - V_s = -L_s\frac{dI_s}{dt} - \omega_r\lambda_{PM}(\theta_{er})$$

$$\frac{J}{p}\frac{d\omega_r}{dt} = T_e + T_{cogging} - T_{load} \quad (11.130)$$

$$T_e = p\lambda_{PM}(\theta_{er})I_s(t); \frac{d\theta_{er}}{dt} = \omega_r$$

The cogging torque $T_{cogging}(\theta_{er})$ which is the zero current torque, has to be specified but also it has to provide for a parking position from which safe starting can be effective.

A FEM investigation of a two pole single-phase PMSM with gradually variable airgap under the stator poles and surface PM rotor poles is shown in Figure 11.88.

It is evident that if the rotor parks in positions A, B there is enough interaction torque ($T_e$) and time to move the rotor in a single direction (from left to right) with a positive voltage pulse.

### 11.9.3 Control Strategies

One or two low-cost Hall sensors may be placed on the stator to detect the rotor proximity. One of them would be good to sense the zero crossing e.m.f. E (or maximum PM flux linkage) in the stator.

After a starting sequence, flux $\lambda_{PM}(\theta_{er})$ may be estimated:

$$\lambda_{PM}(\theta_{er}) = \int (V_s - R_sI_s)\cdot dt - L_sI_s \quad (11.131)$$

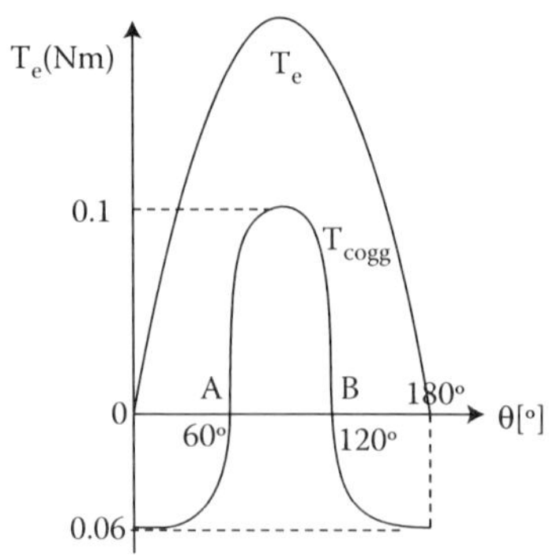


**FIGURE 11.88**
Cogging torque $T_{cogg}$ versus position and interaction torque $T_e$ (for constant current) with continuously increasing airgap under the stator poles.

This way the rotor position $\hat{\theta}_{er}$ and then the speed may be estimated. The Hall sensors may be used to operate corrections to the rotor position estimator.

A typical control system for sinusoidal or trapezoidal current in phase with the e.m.f. is shown in Figure 11.89.

Based on the PM flux estimation the cos $\theta_{er}$, cos $3\theta_{er}$ terms are found, but the stator current should be phase shifted by 90°, and thus contain terms in sin $\theta_{er}$, sin $3\theta_{er}$ which may be simply obtained. The d.c.–d.c. converter uses a single current closed-loop controller, to produce the required current amplitude and waveform. The thyristors provide the phase commutation at zero current crossing as each of them serves half the winding.

To simplify the control it is possible to build a single phase I/f open-loop speed control by carefully ramping the speed to maintain synchronism. A stabilizing loop may be used to dump speed oscillations brought by torque perturbations.

### 11.9.4 Single-Phase PMSM Drive: A Case Study in MATLAB/Simulink (Available Online)

For a sinusoidal current control system with the rotor position available from an encoder, digital performance simulation on a single-phase PMSM with the data:

$V_n = 200$ V
$f_n = 50$ Hz
$P_n = 200$ W
$n_n = 3000$ rpm
$2p = 2$

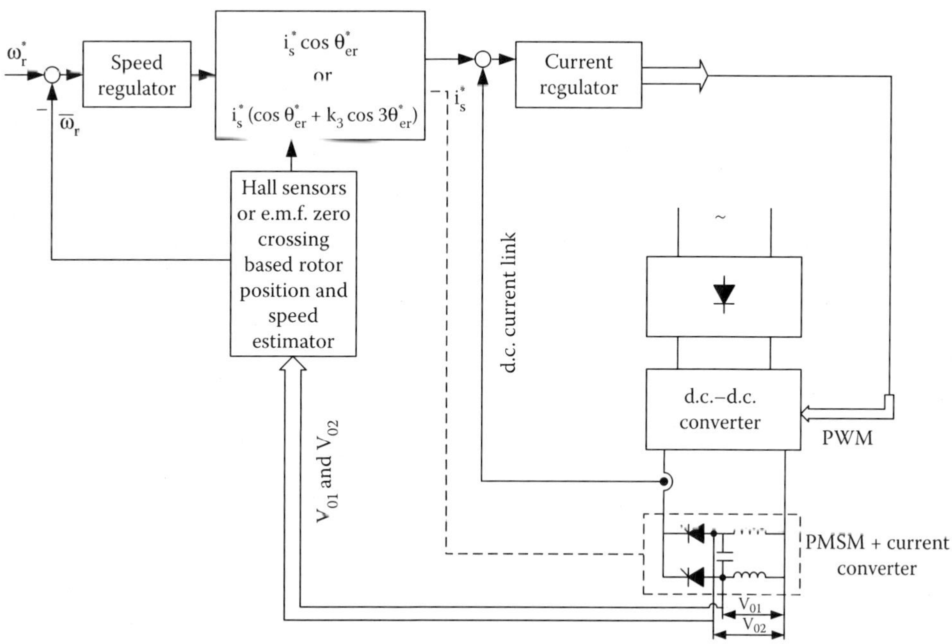


**FIGURE 11.89**
Generic sinusoidal or trapezoidal VC (e.m.f. and current in phase) of a bifilar winding single-phase PMSM.

$$\eta_n = 0.8$$
$$\cos\varphi_n = 0.8$$
$$E_n = 0.8\ V_n$$
$$J = 10^{-3}\ \text{kg m}^2$$
$$T_{cogg} \approx T_{cog\max}\cos(2\theta_{er} + \gamma)$$
$$\frac{T_{cog\max}}{T_{en}} \approx 0.3$$

has been run for the full count converter (Figure 11.87a) and for the two thyristors + one fast-switch converter (Figure 11.87d).

The computer programs written in MATLAB/Simulink comprise the motor, the converter with sensors and the control system. Transients during acceleration with 0.2 Nm step load at 1 s show both control systems capable of acceptable mechanical performance (Figure 11.90).

On the other hand, the steady-state no load and step load transients in torque and current show notable differences between the two solutions (Figure 11.91).

The presence of the $C_{12}$ capacitor between the bifilar windings (Figure 11.87d) leads to adequate polarity currents in both bifilar windings all the time.

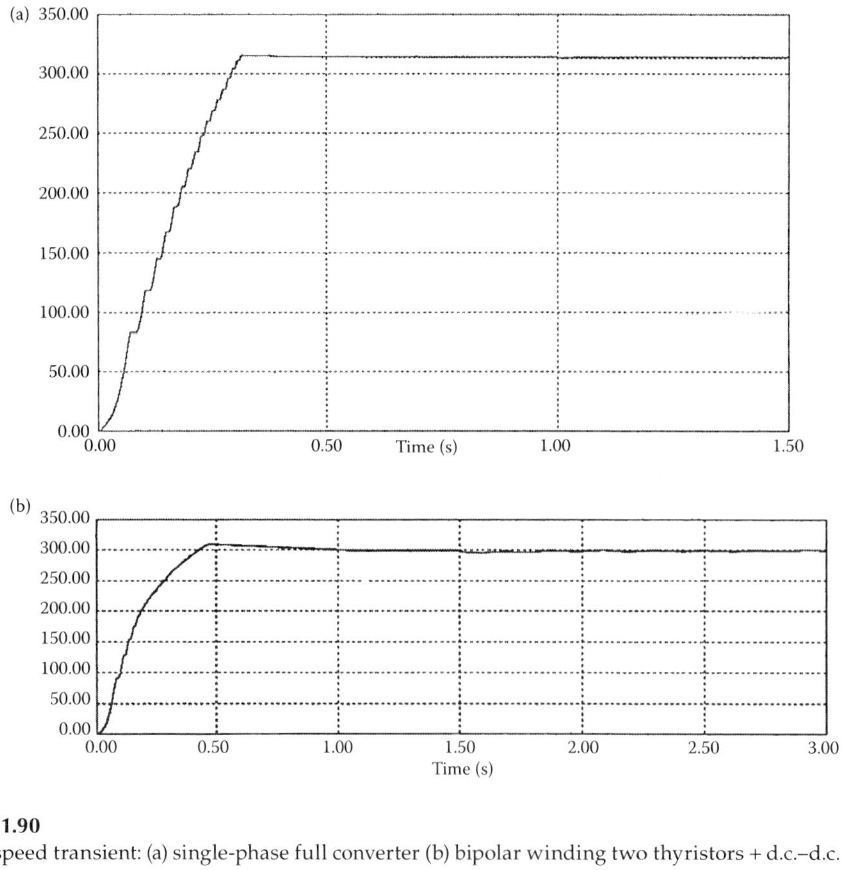


**FIGURE 11.90**
Start-up speed transient: (a) single-phase full converter (b) bipolar winding two thyristors + d.c.–d.c. converter.

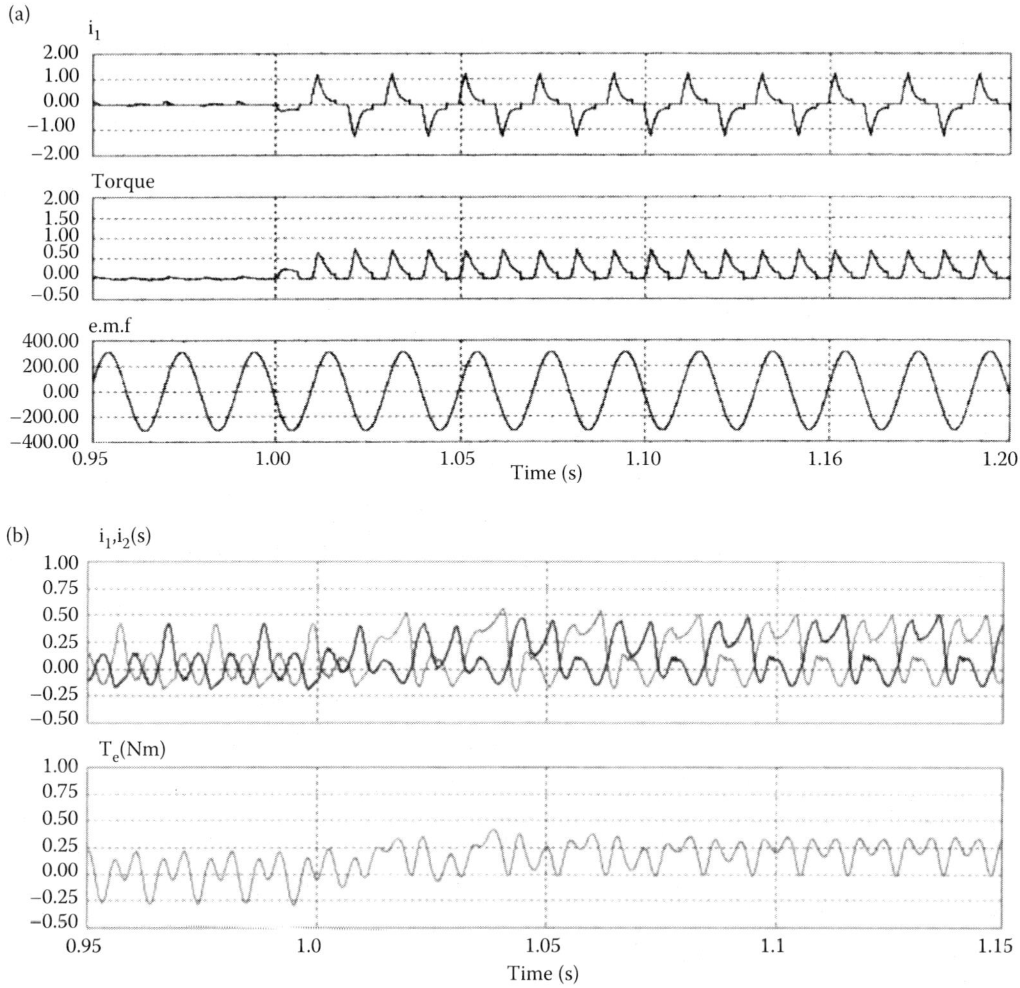


**FIGURE 11.91**
Machine currents, torque, and e.m.f. with load at 3000 rpm: (a) the single-phase full converter, (b) bipolar winding two thyristors + d.c.–d.c. converter.

This way the disadvantage of partial use of copper is attenuated and, also, the torque pulsations are reduced notably for the case of two thyristor + d.c.–d.c. converter.

In terms of total system costs, the additional passive components in the two thyristor + d.c.–d.c. converter seem to offset the reduction by three the count of fast switches (IGBTs) of the four-switch PWM converter but the filter capacitor $C_f$ (Figure 11.87d) is also reduced in the former.

However, the torque pulsations are reduced (peak torque reduced from 0.6 to 0.35 Nm for an average torque of 0.2 Nm—Figure 11.91b). Also the phase current is reduced from 1.1 A (Figure 11.91a) to 0.55 A (Figure 11.91b).

The a.c. source current waveform is about the same for both cases.

Finally, in the two thyristors + d.c.–d.c. converter case the current sensor and its control are placed in the d.c. link.

## 11.10 Summary

- PM or reluctance rotor a.c. motors are used in high-performance drives with PWM voltage-source inverters. The higher efficiency and torque density of these motors are their main merits.
- PM a.c. motors may be controlled with rectangular (trapezoidal) current or with sinusoidal current.
- Sinusoidal current control is performed on PMSMs with distributed windings ($q \geq 2$ slots/pole/phase) or in tooth-wound coil ($q < 1$) windings with sinusoidal e.m.f. and with RSMs (reluctance SMs) without or with PMs in axis q ($L_d > L_q$).
- Tooth-wound PMSMs with sinusoidal or trapezoidal e.m.f. lead to lower motor costs and copper losses.
- Rectangular current control (for PM a.c. motors with $q = 1$, called brushless d.c. motors) tends to be simpler as proximity position sensors suffice, if a speed sensor is available. For low speed, 120° current conduction per polarity per phase is adequate; for high speeds, 180° current conduction is better in terms of available torque.
- PMSM motor drives with motion sensors and VC are particularly suitable for high-performance position or wide speed range (from 100 to 1 to 10,000 to 1) applications. Fast torque response is obtained only with surface magnets on the rotor.
- Special optimization $i_d/i_q$ ($\lambda_d/\lambda_q$) relationships to torque are required to obtain high-performance for steady state and transients over wide speed (torque) range.
- RSMs tend to retain the good performance of PMSMs at lower costs and for similar control systems.
- Speed control ratios up to 100 to 1 in sensorless drives have been obtained. Involved flux, torque, speed, and position observers are needed for sensorless VC.
- DTFC may be used for PMSMs and RSMs for high performance in simpler and highly robust control systems, both with motion sensors and sensorless.
- Wide constant power speed range has been demonstrated with PMSMs or RSMs with PMs.
- The unity power level constant power speed range $K_{cpr}$ is [6,18]

$$K_{cpr} \approx 0.7\omega_c - 1$$

$$\omega_c = \frac{1}{e_0 - x_d} \tag{11.132}$$

where $\omega_c$ is the critical (zero torque) speed in relative units (per base speed $\omega_b$), $e_0$ is the PM (no-load) voltage in relative units (per rated (maximum) voltage), and $x_d$ is the d axis reactance in relative units.

If we consider that for RSM the d axis is along the low permeance path as for the PMSMs (change of places of d and q axis) $\omega_c$ is valid also for RSM with PMs.

A theoretically infinite constant power zone (11.132) is obtained for

$$e_0 - x_d = 0 \tag{11.133}$$

however, at lower levels of power.

- Condition (11.133) is met in conventional PMSMs with $e_0 > 0.5$, in general, to retain some acceptable power factor and with smaller $e_0$ ($e_0 < 0.3$) for RSMs with low costs PMs (in axis q) and good power factor.
- Constant power zone requirements are in some conflict with high torque requirements in steady state and transients below base speed $\omega_b$.
- Super high speed PMSM drives are characterized by fundamental frequency above 500 Hz, up to 2.4 kHz. In such cases, with the limited switching frequency of IGBT converters, synchronous current VC or V/f control with stabilizing loop(s) and open-loop PWM are to be preferred. Also the machine inductance has to be increased to reduce phase current time harmonics. Simplified control is imposed by the limited online computation cycle that should correspond to a few electrical degrees, in order to maintain tight control.
- At low power levels (below 1 kW in general), beside three phase, single-phase PMSM (with self-starting rotor parking provision) with high dynamic performance, including regenerative braking, drives may become practical in terms of system costs [32].
- New solutions of control for sensorless PMSM drives continue to be introduced [33–35].
- Home appliance and automotive actuators are such typical low-power electric drives.
- Reduced active power switch count converter systems are required to reduce costs at low power.
- The two thyristors with d.c.–d.c. converter (with one IGBT) and bifilar winding single-phase PMSM drives, investigated in this chapter through digital simulations, are believed to be a strong example of a practical lower cost solution.
- Motion sensorless V/f or I/f control with stabilizing loops are to be used for the scope to reduce the costs of DSP and of control electronics in small-power electric drives.

## 11.11 Problems

11.1 A brushless d.c. motor has the data: peak PM flux per phase $\lambda_{PM} = 1.25$ Wb, $p = 2$ pole pairs, stator resistance $r_s = 2\ \Omega$, inverter d.c. input voltage $V_d = 500$ V. For zero advancing angle and rectangular current $i_{dc} = 10$ A,

a. Determine the airgap torque with two phases conducting at any time.

b. Determine the ideal no-load speed.

c. Draw the speed versus torque curve thus obtained.

11.2 For a PMSM with interior PM-rotor with the data $L_d = 0.06$ H, $L_q = 0.10$ H, $V_{sn} = 300$ V, $p = 2$, $r_s \approx 0$, and $I_{sn} = 10$ A, $\lambda_{PM} = 1$ Wb, determine the following:

a. The current components $I_d$, $I_q$, and the torque at base speed $\omega_b$ (voltage $V_{sn}$ and current $I_{sn}$) for maximum torque/current criterion.

b. The stator flux level, $\lambda_s^*$, $i_d$, $i_q$, and the torque and power for maximum torque/flux condition (11.42) for $3\omega_b$.

11.3 For the indirect vector current control of RSM presented in Section 11.7.2 write a computer program in PSPICE or MATLAB–Simulink and reproduce the results given there.

11.4 For the DTFC of RSM presented in Section 11.7.3, write a computer program in MATLAB–Simulink or PSPICE and reproduce the results given there, adding other output data.

## References

1. K. Rajashekara, A. Kawamura, K. Matsuse (editors), *Sensorless Control of A.C. Motor Drives*, Section 2, IEEE Press, 1996, pp. 259–379.
2. I. Boldea, S.A. Nasar, Torque vector control (TVC)—A class of fast and robust torque, speed and position digital controllers for electric drives, *EMPS*, vol. 15, 1988, pp. 135–148.
3. B.V. Murty, Fast response reversible brushless d.c. drive with regenerative braking, *Record of IEEE-IAS-1984 Annual Meeting*, pp. 475–480 (IEEExplore).
4. T.M. Jahns, Torque production in permanent magnet synchronous motor drives with rectangular current excitation, *IEEE Trans.*, vol. IA-20, no. 4, 1984, pp. 803–813.
5. S.K. Safi, P.P. Acarnley, A.G. Jack, Analysis and simulation of the high-speed torque performance of brushless d.c. motor drives, *IEE Proc.*, vol. EPA-142, no. 3, 1995, pp. 191–200.
6. S. Morimoto, M. Sonoda, Y. Takeda, Wide speed operation of interior permanent magnet synchronous motors with high performance current regulator, *IEEE Trans.*, vol. IA-30, no. 4, 1994, pp. 920–925.
7. T. Jahns, Flux weakening regime operation of an interior permanent magnet synchronous motor drive, *IEEE Trans.*, vol. IA-23, 1987, pp. 681–687.
8. G. Pfaff et al., Design and experimental results of a brushless a.c. drive, *IEEE Trans.*, vol. IA-20, no. 2, 1984, pp. 814–821.
9. I. Boldea, S.A. Nasar, *Vector Control of AC Drives*, Chapter 10, CRC Press, Boca Raton, Florida, 1992.
10. G.D. Andreescu, Robust direct torque vector control system with stator flux observer for PM-SM drives, *Record of OPTIM-1996*, vol. 5, Brasov, Romania.
11. L. Cardoletti, A. Cassat, M. Jufer, Sensorless position and speed control of a brushless d.c. motor from start-up to nominal speed, *EPE J.*, vol. 2, no. 1, 1988, pp. 25–34.
12. H. Kim, K.K. Huh, R.D. Lorenz, T.M. Jahns, A novel method for initial rotor position estimation of IPM synchronous machine drives, *IEEE Trans.*, vol. IA-40, no. 5, 2004, pp. 1369–1378.
13. Y. Yeong, R.D. Lorenz, T.M. Jahns, S. Sul, Initial rotor position estimation of IPM motor, *Record of IEEE-IEMDC*, 2003, pp. 1218–1223 (IEEExplore).
14. S.Shinnaka, New mirror-phase vector control for sensorless drive of PMSM with pole saliency, *IEEE Trans.*, vol. IA-40, no. 2, 2004, pp. 599–606.
15. M. Linke, R. Kennel, J. Holtz, Sensorless speed and position control of synchronous machines using alternating carrier signal injection, *Record of IEEE-IEMDC*, 2003, vol. 2, pp. 1211–1217 (IEEExplore).
16. I. Boldea, M.C. Paicu, G.D. Andreescu, Active flux concept for motion-sensorless unified a.c. drives, *IEEE Trans.*, vol. PE-23, no. 5, 2008, pp. 2612–2618.
17. M.C. Ancuti, L. Tutelea, G.D. Andreescu, F. Blaabjerg, C. Lascu, I. Boldea, Practical wide range sensorless control system for PMSM drives via active flux model, *EPCS J.*, vol. 42, 2014, pp. 91–102.
18. S. Morimoto, M. Sonoda, Y. Takeda, Inverter driven synchronous motors for constant power, *IEEE-IA Mag.*, vol. 2, no. 6, 1996, pp. 18–24.
19. I. Boldea, *Reluctance Synchronous Machines and Drives*, Oxford University Press, Oxford, 1996.

20. R.E. Betz, Theoretical aspects in control of synchronous reluctance machines, *Proc. IEE*, vol. B-139, no. 4, 1992, pp. 355–364.
21. A. Fratta, A. Vagati, F. Villata, Design criteria of an IPM machine suitable for field weakening operation, *Record of ICEM, 1990*, Part III, Cambridge, Massachusetts, 1990, pp. 1059–1065.
22. L. Xu, X. Xu, T.A. Lipo, D.W. Novotny, Vector control of a reluctance synchronous motor including saturation and iron losses, *IEEE Trans.*, vol. IA-27, no. 5, 1991, pp. 977–985.
23. I. Boldea, Z.X. Fu, S.A. Nasar, Digital simulation of a vector controlled ALA rotor synchronous motor servodrive, *EMPS*, vol. 19, 1991, pp. 415–424.
24. A. Fratta, A. Vagati, F. Villata, *Permanent Magnet Assisted Synchronous Reluctance Drives for Constant Power Applications: Comparative Analysis and Control Requirements*, PCIM Europe, Nurnberg, Germany, 1992, pp. 187–203.
25. I. Boldea, N. Munteanu, S.A. Nasar, Robust low cost implementation of vector control for reluctance synchronous machines, *IEE Proc. EPA*, vol. 141, no. 1, 1994, pp. 1–6.
26. I. Boldea, Z. Fu, S.A. Nasar, Torque vector control (TVC) of ALA rotor reluctance synchronous motors, *EMPS*, vol. 19, 1991, pp. 381–398.
27. S. Agarlita, I. Boldea, F. Blaabjerg, High-frequency-injection-assisted "active-flux"-based sensorless vector control of reluctance synchronous motors, with experiments from zero speed, *IEEE Trans.*, vol. IA-48, no. 6, 2012, pp. 1931–1939.
28. K.-H. Kim, M.J. Youn, DSP-based high speed sensorless control of a brushless d.c. motor using d.c. link voltage control, *EMPS J.*, vol. 30, no. 2, 2002, pp. 889–906.
29. P.D. Chandana Perera, F. Blaabjerg, J.K. Pedersen, P. Thorgersen, A sensorless, stable V/f control method for PMSM drives, *IEEE Trans.*, vol. IA-39, no. 3, 2003, pp. 783–791.
30. B.-H. Bae, S-.K. Sul, J-.H. Kwon, Ji-.S. Byeon, Implementation of sensorless vector control of superhigh speed PMSM for turbo-compresor, *IEEE Trans.*, vol. IA–39, no. 3, 2003, pp. 811–818.
31. I. Boldea, L. Tutelea, A. Moldovan (Popa), Scalar V/f and I–f control of ac. motor drives: An overview. *Record of ACEMP/OPTIM/Electromotion*, 2015, Side, Turkey (on IEEExplore).
32. E. Iepure, I. Boldea, F. Blaabjerg, Hybrid I-f starting and observer based sensorless control of single phase BLDC-PM motor drives, *IEEE Ind. Electron. Trans.*, vol. 59, no. 9, 2012, pp. 3436–3444.
33. C. Lascu, I. Boldea, F. Blaabjerg, Direct torque control via feedback linearization for PMSM drives, *Record of OPTIM 2012* (IEEExplore).
34. C. Lascu, I. Boldea, F. Blaabjerg, Super-twisting sliding mode control of torque and flux in PMSM drives, *Record of IEEE IECON*, 2013, Vienna, Austria.
35. C. Lascu, G.D. Andreescu, Self-commissioning of electrical parameters for PMSM in sensorless drives, *Record of ACEMP/OPTIM/Electromotion*, 2015, Side, Turkey (IEEExplore).

# 12

# *Switched Reluctance Motor Drives*

## 12.1 Introduction

SRMs are doubly salient, singly excited electric motors with passive (windingless) rotors. Their concentrated coil phases are turned on sequentially, to produce torque, through d.c. voltage pulses which result in unipolar controlled current.

Due to their simple and rugged topology, SRMs [1–5] have been given considerable attention in the last two decades in the hope of producing alternative low-grade and high-grade brushless motor electric drives at a lower cost and equivalent performance when compared with a.c. (induction or synchronous) motor drives.

At the time of this writing the SRM drive enjoys only an incipient market but a more aggressive penetration of worldwide markets is expected in the near future. Power ratings of SRMs range from a few watts to practically MW units for low-speed control range low dynamics, as well as high-grade (servo) applications, especially in thermally and chemically harsh environments.

Given the topological peculiarities of the SRM, a few details are considered in order. Also, the operation of SRM with an active phase only at a time in a three-phase stator, in the presence of magnetic saturation, with open-loop (voltage PWM) or closed-loop (current control) PWM, requires elaborate digital simulation methods.

The basic strategies for low-grade and high-grade speed and position control with and without motion sensors are also treated in some detail. Some numerical examples are provided to facilitate quick access to magnitudes for a realistic assessment of performance.

## 12.2 Construction and Functional Aspects

SRMs are made of laminated stator and rotor cores with $N_s = 2mq$ poles on the stator and $N_r$ poles on the rotor. The number of phases is m and each phase is made of concentrated coils placed on 2q stator poles.

Most favored configurations—amongst many more options—are the 6/4 three-phase and the 8/6 four-phase SRMs (Figure 12.1a and b).

These two configurations correspond to q = 1 (one pair of stator poles [and coils] per phase) but q may be equal to 2, 3 when, for the three-phase machine, we obtain 12/8 or 18/12 topologies applied either for low-speed high torque direct drives or for high-speed motor-generator systems for aircraft [2]. The stator and rotor pole angles $\beta_s$ and $\beta_r$ are, in general, almost equal to each other, to avoid zero torque zones

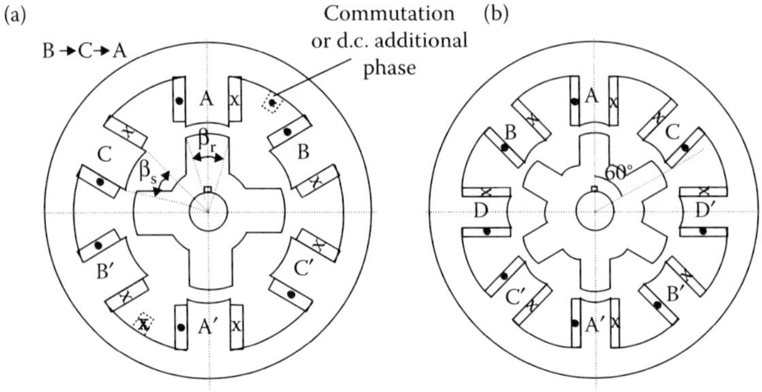


**FIGURE 12.1**
Representative SRM configurations: (a) 6 slots 14 poles for 3 phases and (b) 8 slots 16 poles (3 phases).

The symmetry of the magnetic circuit leads to almost zero mutual flux linkage in the SRM phases, even under saturated conditions. This means that the SRM may work with $m^{-1}$ phases, since no induced voltage or current will appear in the short-circuited phase. Hence the SRM is more fault tolerant than any a.c. motor where the interaction between phases is at the core of their principle of operation (some tooth-wound PMSMs [q < 1] enjoy also zero mutual inductances between phases). The self-inductance of each phase alone thus plays the key role in torque production.

In the absence of magnetic saturation, the phase self-inductance varies linearly with rotor position, while, in presence of saturation, the respective dependence is nonlinear (Figure 12.2).

If the phase flux linkage λ is calculated and plotted versus current for the various rotor positions, the $\lambda(\theta_r,i)$ curve family is obtained (Figure 12.3).

The influence of magnetic saturation is evident from Figure 12.3 and is a practical reality in well-designed SRMs, as will be shown later in this section.

The instantaneous torque $T_e(i)$ per phase may be obtained through the known coenergy, $W_{mc}(\theta_r)$, formula

$$T_e(i) = \left(\frac{\partial W_{mc}(\theta_r)}{\partial \theta_r}\right)_{i=\text{cons.}}; \quad W_{mc} = \int_0^i \lambda(\theta_r, i)\, di \tag{12.1}$$

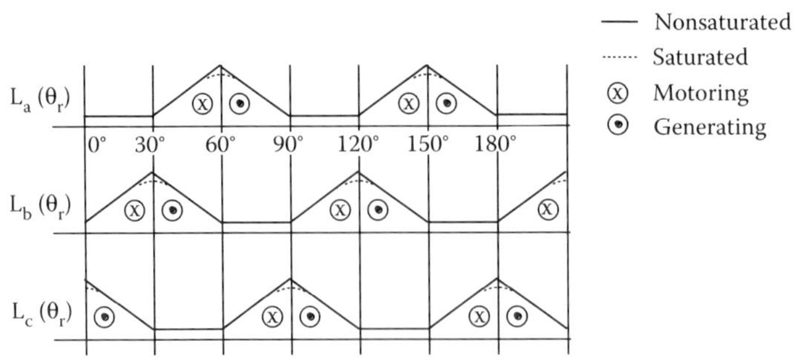


**FIGURE 12.2**
The phase inductances and the operation modes of three-phase 6/4 SRM.

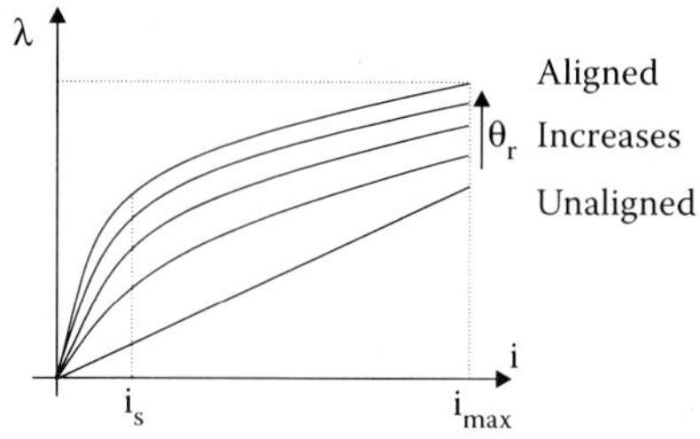


**FIGURE 12.3**
Flux/current/position curve family.

Equation 12.1 demonstrates the necessity of knowing, through calculations or test, the family of curves $\lambda(\theta_r,i)$.

The total instantaneous torque is

$$T_e = \sum_{i=1}^{m} T_e(i) \tag{12.2}$$

Only in the absence of saturation, the instantaneous torque is

$$T_e = \sum_{i=1}^{m} \frac{1}{2} i_i^{\,2} \frac{\partial \lambda_i(\theta_r)}{\partial \theta_r} \tag{12.3}$$

Ideally, a phase is turned on when rotor poles, along the direction of motion, lay between neighboring stator poles $\theta_0 = 0$ (Figure 12.4), for the motoring operation mode of the respective phase. Only one voltage pulse is applied for a conduction (dwell) angle $\theta_w = \theta_c - \theta_{on}$ in

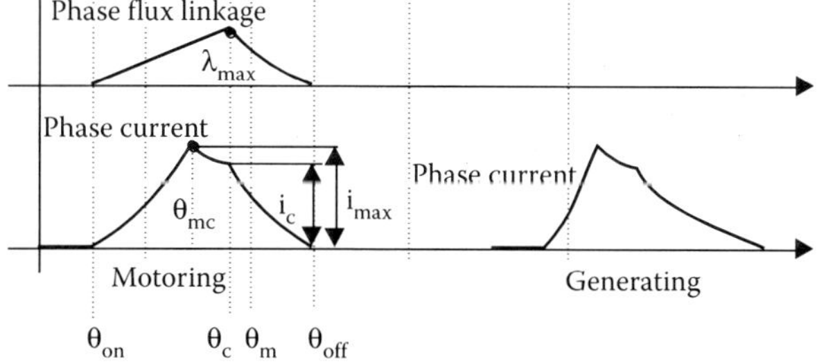


**FIGURE 12.4**
Phase inductance, voltage, flux linkage, and current.

Figure 12.4. During this period, neglecting the resistive voltage drop, the maximum phase flux linkage $\lambda_{max}$, for constant speed $\omega_r$, is

$$\lambda_{max} = \int_0^t V_d dt = V_d \frac{\theta_w}{\omega_r} \tag{12.4}$$

The maximum value of $\theta_w$, for $\theta_{on} = 0$ (zero advance angle), is given by motor design

$$\theta_{w\,max} = \theta_m = \frac{\pi}{N_r} \tag{12.5}$$

The base speed $\omega_b$ corresponds to $\theta_{wmax}$ and single-voltage pulse $V_d$ with maximum flux linkage $\lambda_{max}$, which is dependent on machine design and on the level of saturation. Thus, to increase the base speed, we have to saturate the magnetic circuit of that machine.

For higher speeds $\omega_r$ (above $\omega_b$), the value of $\theta_w$ may be slightly reduced but definitely, the maximum $\lambda_{max}$ has to be reduced. This is called flux weakening. Above base speed $\omega_b$, also, the turn-on angle $\theta_{on}$ may be advanced to reach the available maximum flux $\lambda_{max}$ at a smaller angle $\theta_c$ and thus allow the current to reach its maximum (at $\theta_{mc}$) sooner and at a higher level and thus produce more torque. As a consequence, the speed/torque envelope may be enlarged. On the other hand, the turn-off process of a phase starts at $\theta_c \leq \theta_m$ and terminates at $\theta_{off}$ in the "generating" zone.

The smaller the angle $\theta_{off}$–$\theta_m$, the smaller the negative torque "contribution" of the phase going off. In reality, at $\theta_r = \theta_m$ (aligned position) if the current $i_m$ is already less than (25%–30%) of the peak current, the negative (generating) torque influence becomes small.

Once one phase is turned off at $\theta_c$, another one is turned on, eventually to contribute positive torque so as to lower the total torque pulsations caused by the reduction of torque in the phase going off.

It is now evident that the entire magnetic energy of each phase is "pumped" in and out for each conduction cycle. There are $mN_r$ cycles per mechanical revolution. A part of this energy is passed over to the incoming phase through the PEC and the rest to the d.c. bus filtering capacitor of PEC. Below base speed $\omega_b$ the current is limited (and controlled) through PWM (Figure 12.5).

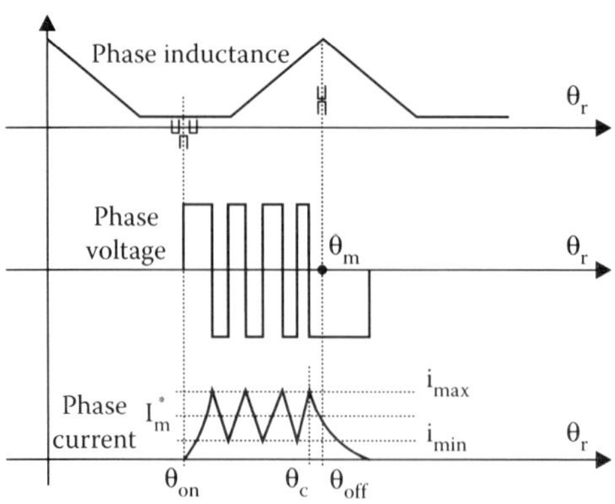


**FIGURE 12.5**
PWM below base speed.

It should be noted that the interval of conduction is prolonged close to $\theta_m$ where the phase inductance is maximum. As mentioned above, at high speeds the phase turn-on angle $\theta_{on}$ is advanced and so is the turn-off angle $\theta_c$.

## 12.3 Average Torque and Energy Conversion Ratio

From the energy cycle point of view, based on the family of curves $\lambda(\theta_r, i)$, the two situations presented in Figures 12.4 and 12.5 (single-voltage pulse) and, respectively, current chopping, are shown in Figure 12.6a–c.

The average torque $T_{eav}$ is proportional to the hatched area in Figure 12.6a and b, $W_{mec}$. Consequently, for constant speed, single-phase conduction, with m phases and $N_r$ rotor poles, the SRM average torque $T_{eav}$ is

$$T_{eav} = \frac{W_{mec} m N_r}{\theta_c - \theta_m} \tag{12.6}$$

In parallel, an energy conversion ratio, E.C. (kW/kVA), is defined as

$$\text{E.C.} = \frac{\text{Area of } \overline{\text{OABCO}}}{\text{Area of } \overline{\text{OABDO}}} = \frac{W_{mec}}{W_{mec} + W_r} \tag{12.7}$$

As the energy returned during diode conduction, $W_r$, is reduced by magnetic saturation, the E.C. ratio increases with saturation. Consequently, the kVA rating of the PEC (which depends on $W_{mec} + W_r$) is reduced by magnetic saturation. The E.C. ratio is only 0.5 for linear flux/current curves—in the absence of magnetic saturation—but it could reach values of 0.65–0.67 with magnetic saturation. Also, to keep the $W_{mec}$ high, in comparison with the linear case, the airgap is reduced such that adequate saturation is obtained at the

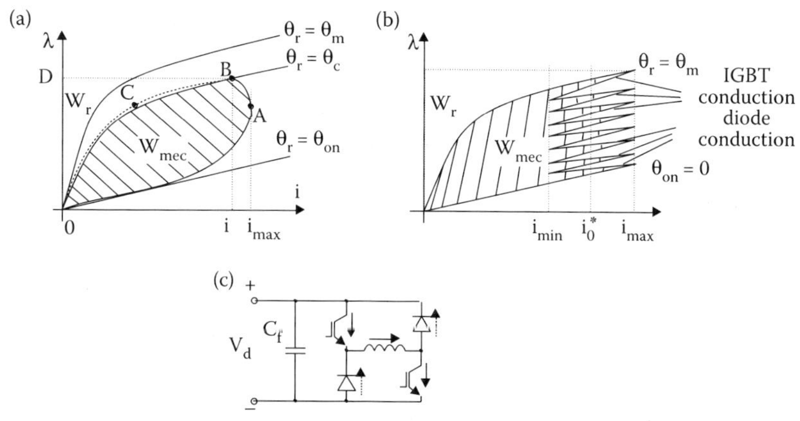


**FIGURE 12.6**
Energy cycle per phase: (a) for high speeds (one voltage pulse, Figure 12.4); (b) for low speeds (current chopping, Figure 12.5); (c) phase PEC for unipolar current per phase.

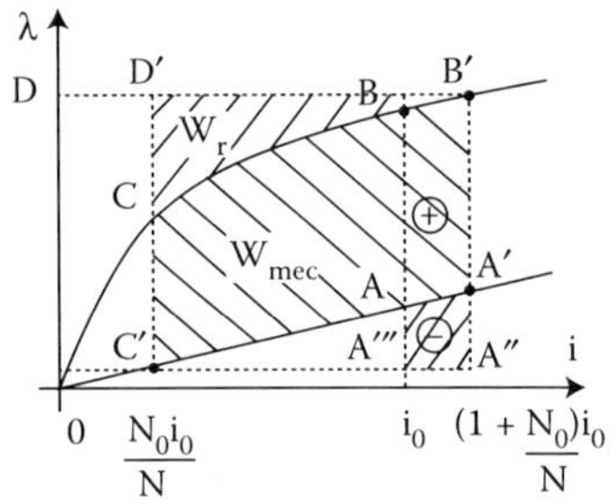


**FIGURE 12.7**
Energy cycle with d.c. premagnetization.

same level of currents for most rotor positions. This way the level of average torque can be maintained.

Note that the maximum flux per phase and the maximum inductance are reduced by magnetic saturation. Current commutation becomes faster, while the base speed for a given voltage is also increased.

Further increase in the E.C. ratio may be obtained through d.c. premagnetization by a diametrical coil (Figure 12.1a) in series with all phases [3].

The energy cycle is moved to the right (Figure 12.7) with the equivalent current $i_0N_0/N$ ($N_0$—turns in the premagnetization coil, N–turns in the phase coil). The mechanical energy is increased by the area AA′B ′B minus area AA′A″A ‴, while the energy retrieved during phase demagnetization, $W_r$, is reduced by the area OCD′D. Consequently, the energy conversion ratio E.C. is notably increased (from 0.64 to 0.74, [3]) at the price of the losses in the premagnetization coil (19% additional copper losses, [3]).

## 12.4 Peak kW/kVA Ratio

The peak kW/kVA ratio [4] is

$$\text{peak kW/kVA} = \frac{\beta_s N_r Q}{8\pi} \quad (12.8)$$

where $\beta_s$ is the stator pole pitch ratio ($\alpha_s = 0.4–0.5$), and Q is

$$Q \approx C\left(2 \quad \frac{C}{C_s}\right) \quad (12.9)$$

where C is the ratio between the conduction angle below the stator pole and the stator pole angle $\beta_s$.

In general, C = 1 at zero speed and decreases down to C = 0.65 at base speed.

$$C_s = \frac{\lambda_u - 1}{\lambda_u \sigma - 1};\ \lambda_u = \frac{L_a^u}{L_u} \approx (6 \div 10);\ \sigma = \frac{L_a^s}{L_a^u} \approx (0.3 \div 0.4) \quad (12.10)$$

where $L_u$ is the unaligned inductance, and $L_a^u$ and $L_a^s$ are the aligned unsaturated and saturated values of phase inductance, respectively.

The peak apparent power, S, of switches in the converter is

$$S_1 = 2mV_d I_{peak} \tag{12.11}$$

where $V_d$ is the d.c. source voltage and $I_{peak}$ is the peak current value.

For an inverter-fed IM drive the peak kW/kVA is

$$(\text{peak KW/KVA})_{IM} = \frac{3}{\pi}\frac{V_d I \cdot PF}{K(6V_d I)} = \frac{3}{\pi}\frac{PF}{6K} \tag{12.12}$$

where K is the ratio between the peak current $I_{peak}$ and the peak value of the current fundamental in IMs when inverter-fed (for the 6 pulse mode K = 1.1–1.15); and PF is the power factor.

**EXAMPLE 12.1**

For a 6/4 SRM with $\sigma = L_a^s / L_a^u = 0.4$, $l_u = L_a^u / L_u = 6$, $C = 1$, $\beta_s = 0.4$, calculate the peak kW/kVA ratio.

**Solution**

As $\sigma = 0.4$ and $\lambda_u = 6$, the coefficient $C_s$ (from 12.10) is

$$C_s = \frac{\lambda_u - 1}{\lambda_u \sigma - 1} = \frac{6-1}{6 \cdot 0.4 - 1} = \frac{5}{1.4} = 3.57 \tag{12.13}$$

Further on, from Equation 12.9, Q is

$$Q = C\left(2 - \frac{C}{C_s}\right) = 1\left(2 - \frac{1}{3.57}\right) = 1.72 \tag{12.14}$$

Consequently, from Equation 12.8

$$\text{kW/kVA} = \frac{\beta_s N_r Q}{8\pi} = \frac{0.4 \cdot 4 \cdot 1.72}{8\pi} = 0.1095 \tag{12.15}$$

For an induction motor with PF = 0.85 and K = 1.12

$$(\text{kW/kVA})_{IM} = \frac{3}{\pi}\frac{0.85}{6 \cdot 1.12} = 0.1208 \tag{12.16}$$

So, in terms of the PWM inverter rating, the IM (with rather good power factor) is only 10%–20% better than the PEC for the equivalent SRM, which is supposed to produce the same power at the same efficiency.

## 12.5 Commutation Windings

The turn-off process angle (time) $\theta_c - \theta_{off}$ has to be limited. In general, $\theta_{off} - \theta_c < \beta_s$ ($\beta_s$ is the stator pole angle) is not sufficient, as a strong negative (generating) torque may occur.

Conversely, in the generator mode, the turn-on process has to be advanced within motoring regime (positive inductance slope), but not too much, in order to avoid strong motoring torque influences.

For both cases and all speeds $\theta_{off} - \theta_c < 3\text{–}4°$. Through advanced saturation (reduced airgap) and the advancing of turn-on and turn-off angles $\theta_{on}$ and $\theta_{off}$, the turn-off interval angle $\theta_{off} - \theta_m$ may be kept (up to a certain speed) below high limits at the price of higher torque pulsations.

If the diametrical coil (proposed for demagnetization) (Figure 12.1a) is turned on whenever an active phase is de-energized, the magnetic energy will be passed to this commutation phase, partly. After the current is zero in the going-off phase, the commutation winding is turned off. Notably smaller turn-off times are obtained this way (Figure 12.8) [6].

The effect of the commutation current turn-off process is evident but we have to consider that an additional coil with an additional phase in the PEC and the corresponding losses make the price for the improvement thus obtained high. The energy conversion rate EC is also increased by a few percent. Finally, the interaction of the commutation winding with, say, a faulty (short-circuited) phase notably reduces the fault tolerance claims of conventional SRM. Other distributed winding arrangements for SRMs have been proposed [7,8] for increasing the torque density (Nm/kg of rotor), but they all seem to impede on the fault tolerance claims of conventional SRMs, while the efficiency is not notably increased either.

In what follows we will deal with modeling, digital simulation, and the control of the basic 6/4 and 8/6 SRMs as they are most representative.

## 12.6 SRM Modeling

The mathematical model of SRM is highly nonlinear due to magnetic saturation influence on the $\lambda(\theta_r, i)$ curve family, but it allows for phase-torque superposition as the interaction between phases is minimal.

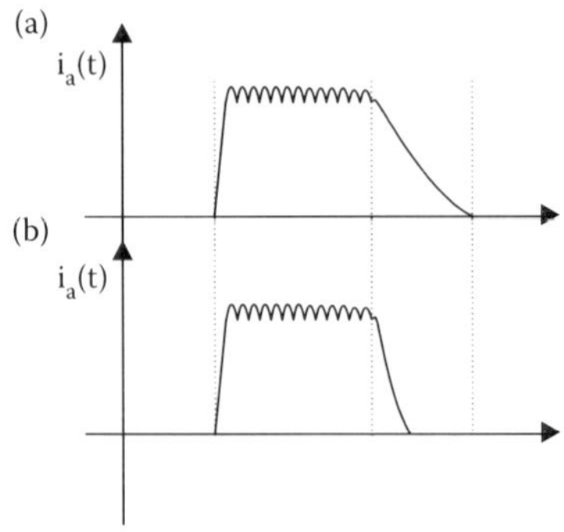


**FIGURE 12.8**
Current profile: (a) conventional, (b) with commutation coil (winding).

As SRM has double saliency, stator (phase) coordinates are mandatory.

The phase equations are

$$V_{a,b,c,d} = r_s i_{a,b,c,d} + \frac{d\lambda_{a,b,c,d}(\theta_r, i_{a,b,c,d})}{dt} \tag{12.17}$$

with the family of curves $\lambda_{a,b,c,d}(\theta_r, i_{a,b,c,d})$ required for one phase only (as the periodicity is $\pi/N_s$). These curves may be obtained either through theory or through tests. Analytical or finite element methods are used for the scope. Accounting for magnetic saturation and airgap flux fringing is mandatory in all cases.

The motion equations are

$$J\frac{d\omega_r}{dt} = T_e - T_{load}; \quad \frac{d\theta_r}{dt} = \omega_r \tag{12.18}$$

with

$$T_e = \sum_{a,b,c,d} T_{ea,b,c,d}; \quad T_{ea,b,c,d} = \frac{\partial}{\partial\theta_r}\int_0^{i_{a,b,c,d}} \lambda_{a,b,c,d}(\theta_r, i_{a,b,c,d})di_{a,b,c,d} \tag{12.19}$$

Let us use the subscript i for one dominating phase.

Equation 12.16 may be written as

$$V_i = r_s i_i + \frac{\partial\lambda_i}{\partial i_i}\frac{di_i}{dt} + \frac{\partial\lambda_i}{\partial\theta_r}\frac{d\theta_r}{dt} \tag{12.20}$$

Denoting $\partial\lambda i/\partial i_i$ as the transient inductance $L_t$:

$$L_t(\theta_r, i_i) = \frac{\partial\lambda_i(\theta_r, i_i)}{\partial i_i} \tag{12.21}$$

The last term in Equation 12.20 represents the pseudo-back e.m.f. $E_i$

$$E_i = \frac{\partial\lambda_i}{\partial\theta_r}\cdot\omega_r \tag{12.22}$$

So Equation 12.20 becomes

$$V_i = r_s i_i + L_t(\theta_r, i_i)\frac{di_i}{dt} + E_i(\omega_r, \theta_r, i_i) \tag{12.23}$$

An equivalent circuit with time-dependent parameters may be defined based on Equation 12.23 (Figure 12.9).

The core losses are represented by the variable resistances in parallel with the e.m.f. $E_i$, based on the assumption that only the main flux produces core losses. Core losses occur

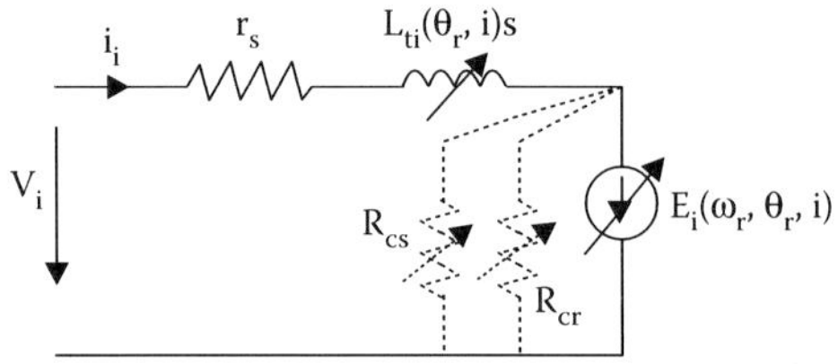


**FIGURE 12.9**
Equivalent circuit of SRM with core losses accounted for.

both in the stator and in the rotor core as this machine does not operate on the traveling field principle [9,10].

Especially in high-speed applications (above 6000 rpm), core loss has to be considered not only for efficiency calculations but also in the transient current response assessment.

NOTE: Only for the linear case (no magnetic saturation) the instantaneous torque $T_e(t)$ is

$$T_e(t) = \sum \frac{1}{2} \frac{E_i(\omega_r, \theta_r, i_i) i_i}{\omega_r} \tag{12.24}$$

As usual, heavy magnetic saturation is present, $E_i(\omega_r, \theta_r, i_i)$ as in Equation 12.22 is a pseudo e.m.f. as it also includes a part related to the magnetic energy storage. Consequently, the torque is to be calculated only from the coenergy formula (12.19).

For more details on this aspect, see Reference 11.

## 12.7 Flux–Current–Position Curve Fitting

For digital simulations and control purposes the $\lambda_i(\theta_r, i)$ curves family has to be known. The safest way is to use measurements at standstill or with the machine in rotation. Finite element calculations represent the second best approach. Once this is done what still remains is to determine the inverse function $i(\lambda_i, \theta_r)$ and, eventually, $\theta_r(\lambda_i, i_i)$. There are two main ways to invert the $\lambda_i(\theta_r, i)$ to find either $i(\lambda_i, \theta_r)$ or $\theta_r(\lambda_i, i_i)$.

One way is to use analytical functions (polynomials or exponentials) [12,13]. The second way is to use direct approximations, for example, fuzzy logic [14] or other curve fitting methods. For digital simulations the computation time is not important. In the case of control torque calculation, $T_{ei}(\lambda_i, \theta_r, i_i)$, or position estimation, $\theta_r(\lambda_i, i_i)$, is done on line.

Using exponential approximations [12,13], the phase flux linkage $\lambda_i(\theta_r, i)$ is

$$\lambda(\theta_r, i) = a_1(\theta_r)(1 - e^{-a_2(\theta_r) i}) + a_3(\theta_r) i \tag{12.25}$$

The periodicity of $\lambda$ is built in the $a_{1,2,3}(\theta_r)$ functions expressed as Fourier series

$$a_m = \sum_{k=0}^{\infty} A_{mk} \cos(k\alpha\theta_r) \tag{12.26}$$

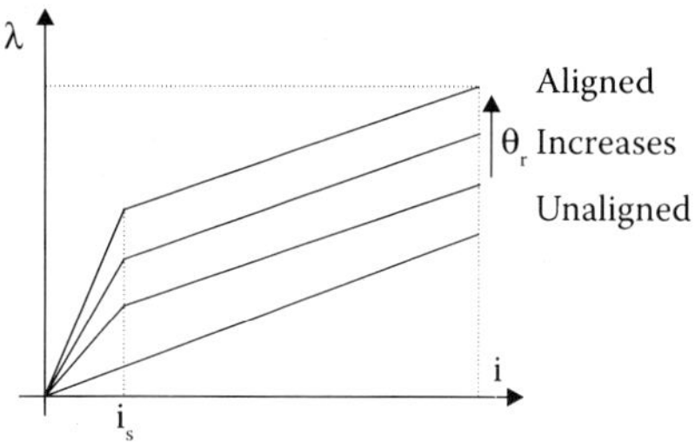


**FIGURE 12.10**
Linear $\lambda(\theta_r,i)$ approximations.

Here $\alpha = 4$ for a 6/4 machine and $\alpha = 6$ for an 8/6 machine. $A_{mk}$ is the Fourier coefficient of kth order in $a_m$.

Other rather cumbersome analytical functions of current, based on linear position dependence approximations, are also feasible [15].

For transients, even linear approximations will do (Figure 12.10).

It is supposed that magnetic saturation occurs at the constant current level $i_s$, irrespective of rotor position. As the airgap is small, in heavily saturated motors, the local overlapping pole zone may saturate quickly. Thus the flux linkage varies linearly with rotor position. It is known, however, that close to the aligned position such an approximation model seems less adequate

$$\lambda = i\left(L_u + \frac{K_s(\theta_r - \theta_0)}{i_s}\right); \quad i \le i_s \tag{12.27}$$

$$\lambda = L_u i + K_s(\theta_r - \theta_0); \quad i \ge i_s \tag{12.28}$$

where $L_u$ is the unaligned inductance value; $K_s$ is the only coefficient to be found from the family of $\lambda(\theta_r,i)$ curves—eventually one intermediate position information suffices; $\theta_0$ is the position angle of the pole entry end; and $\theta_{rmax} = \theta_m$ is the pole exit end position angle.

## 12.8 SRM Drives

SRM drives may be classified with respect to a few criteria:

- With motion (position, speed) sensors
- Without motion sensors (sensorless)

They also may be

- General—for low dynamics applications (moderate speed range and costs)
- High grade (performance)—for servos

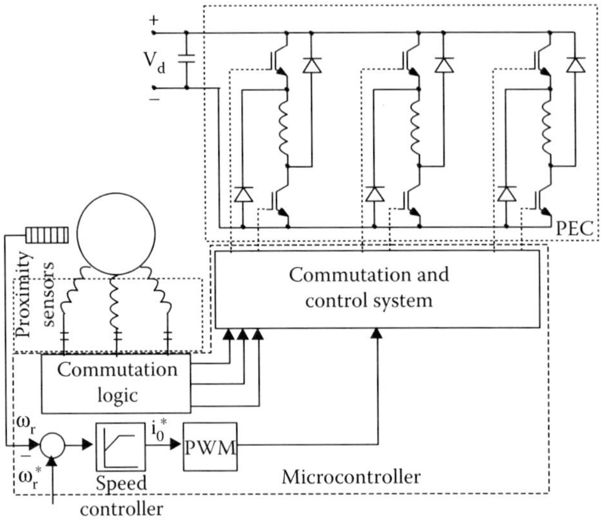


**FIGURE 12.11**
Basic (historical) SRM drive.

They differ in complexity, costs, and performance. Performance is defined by the energy conversion ratio, speed control range and precision, and quickness of torque response (Figure 12.11).

Servo applications require precise speed or position control and fast torque response (in the milliseconds range). Historically, for low-cost SRM drives, proximity (Hall) position sensors have been used to trigger the turn-off conducting phase and at the same time, the turn-on of the incoming phase. For constant speed only, even the position angle between two proximity sensors signals may be estimated.

Consequently, advancing the turn-on and turn-off angle may be performed in such drives only for steady-state or low dynamics.

Also, starting under load poses problems because, from the three (four) proximity sensor signals, we may only infer what phase has to be turned on but not exactly the initial position. A separate speed signal is necessary for safe starting under load.

In the following section we will discuss an up-to-date general drive, a high-grade drive with precision (encoder) position feedback, and an advanced motion sensorless drive.

## 12.9 General-Purpose Drive with Position Sensor

For applications with wide speed range but moderate energy conversion or dynamic performance, a precision position sensor is used to commutate the phases at speeds down to a few rpm. No speed sensor is used (Figure 12.12).

The core of the general-purpose SRM drive in Figure 12.12 is the dependence of the turn-on $\theta_{on}$ and turn-off $\theta_{off}$ angles (for motoring and generating) on speed. A linear dependence is the obvious choice.

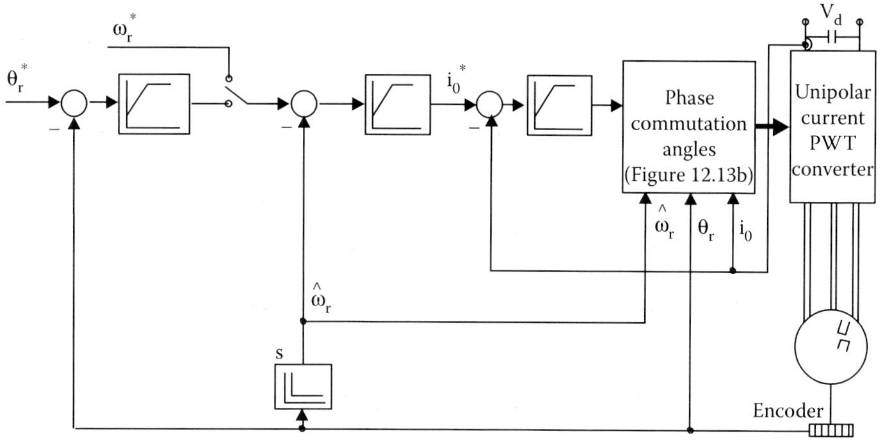


**FIGURE 12.12**
General-purpose SRM drive with encoder.

We may distinguish three regions:

- The low speed—constant torque—zone: $\theta_{on}$ = cons., $\theta_{off}$ = cons., $\omega_r < \omega_b$
- The constant power zone, $T_e\omega_r$ = cons.: $\theta_{on}$ and $\theta_{off}$ decrease with speed $\omega_r > \omega_b$
- The $T_e\omega_r^2$ = cons. zone, above $\omega_{m1}$ (Figure 12.13a)

The above control philosophy assumes flat top currents below base speed, obtained through closed-loop (current control) PWM and single-voltage pulses above base speed when (and if) current control is no longer feasible due to lack of voltage reserve (the e.m.f. overcomes the d.c. link voltage $V_d$).

As is easy to see, no complicated data on machine flux/current/position curves is necessary as $(\theta_{on})_{min}$ and $(\theta_{off})_{min}$ are to be obtained through trial and error. For more details on such a drive, see Reference 16.

The drive could produce moderate performance over a wide speed range but quick torque response optimum energy conversion or maximum torque/speed envelope is not guaranteed. However, there are many applications where the general-purpose SRM drive suffices.

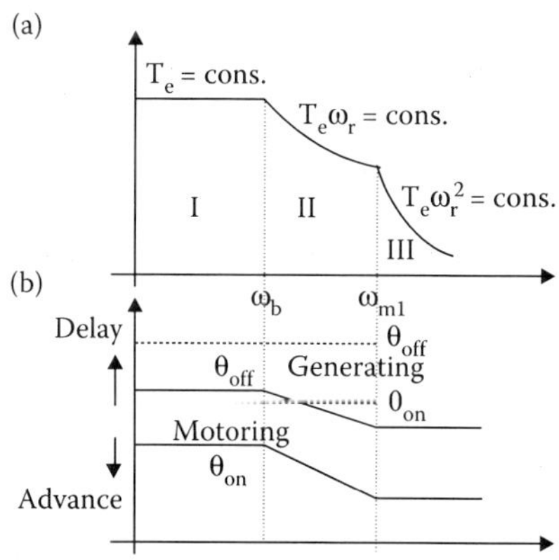


**FIGURE 12.13**
(a) Torque/speed zones, (b) turn-on and turn-off angles $\theta_{on}$ and $\theta_{off}$ for motoring and generating.

### EXAMPLE 12.2: SRM DRIVE SIMULATION BY MATLAB/ SIMULINK (AVAILABLE ONLINE)

A 6/4 three-phase SRM has the following data: $\beta_s = \beta_r = 30°$ (stator and rotor pole angle), $J = 0.002$ kg m$^2$, maximum current $i_{omax} = 10$ A, d.c. voltage $V_d = 300$ V, $r_s = 1.5\ \Omega$, maximum flux (aligned position) $\lambda_{max} = 0.8$ Wb, minimum flux $\lambda_{min} = 0.16$ Wb, linear flux/current curves (12.27 and 12.28) are supposed, with $i_s = 2$ A (Figure 12.10).

A position sensor with 1024 pulses per revolution is available.

Calculate

- The unaligned inductance, $K_s$ coefficient in Equations 12.27 and 12.28 and the maximum average torque available for $i_{max} = 10$ A at zero speed.
- Using MATLAB–Simulink run digital simulations on a general-purpose SRM drive (Figures 12.14 through 12.17) and choose the turn-on and turn-off angles to explore starting transients, step load, and step-speed responses.

#### Solution

At zero speed the conduction angle $\theta_{rmax}-\theta_0 = \beta_s = 30° = \pi/6$ with $\theta_0 = 0$, for the aligned position $\theta_{rmax} = 30°$.

In the unaligned position $\theta_r = \theta_0$ and from Equation 12.27

$$\lambda_{min} = i_{max} L_u \tag{12.29}$$

So the unaligned inductance $L_u$ is

$$L_u = \frac{\lambda_{min}}{i_{max}} = \frac{0.16}{10} = 0.016\,\text{H} \tag{12.30}$$

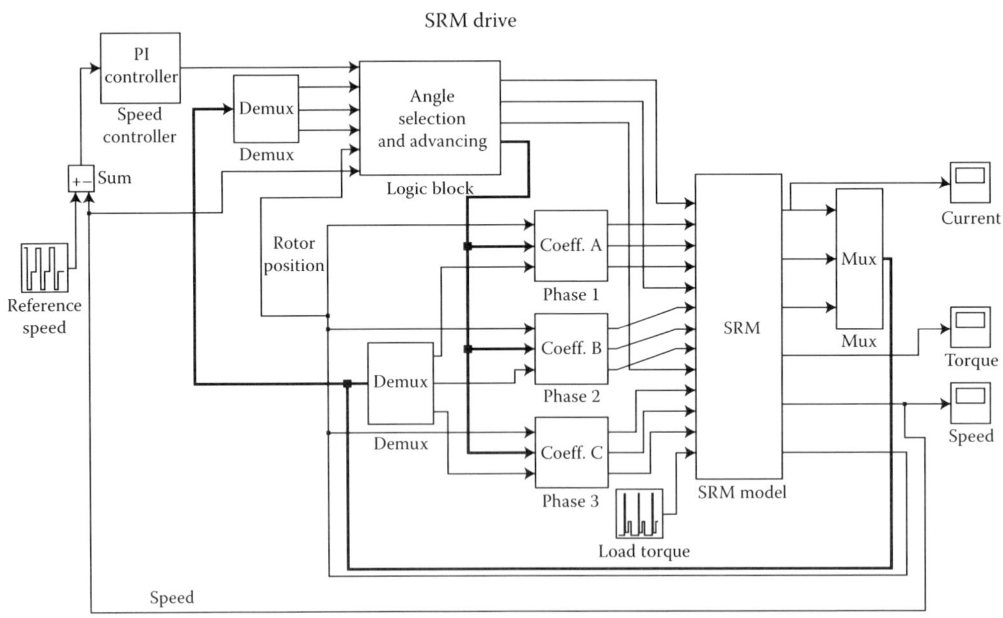


**FIGURE 12.14**
Block diagram of simulated SRM drive.

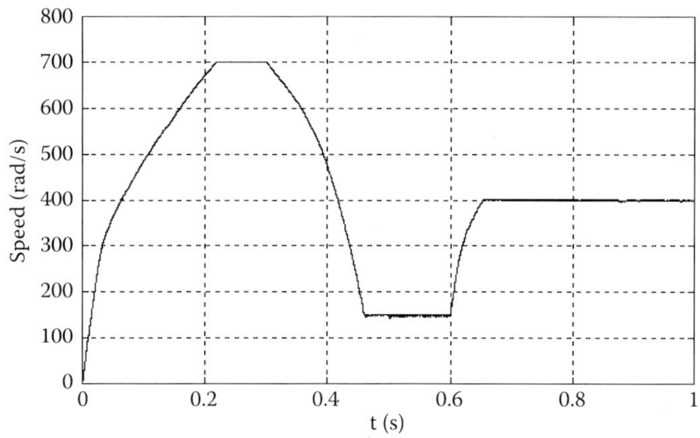


**FIGURE 12.15**
Speed response.

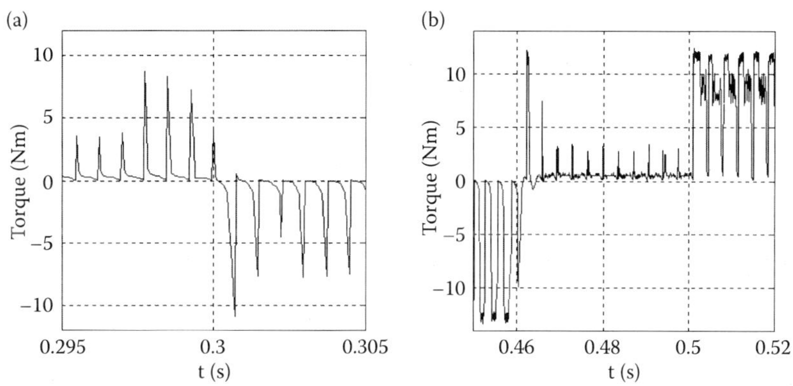


**FIGURE 12.16**
Torque response: (a) At 700 rad/s, (b) for deceleration, at 150 rad/s and acceleration to 400 rad/s (Figure 12.15).

Also, from Equation 12.28 $\lambda_{max}$ is

$$\lambda_{max} = i_{max} L_u + K_s \beta_s \tag{12.31}$$

$$K_s = \frac{0.8 - 0.16}{\pi/6} = 1.223 \text{ Wb/rad} \tag{12.32}$$

MATLAB/Simulink simulation results on a SRM motor drive with PEC are presented. The motor model was integrated in a block (SRM in Figure 12.14).

Changing motor parameters is done by clicking on this block. A dialog box appears and you may change them by modifying their default values.

The drive system consists of PI speed controller ($K_i = 10$, $T_i = 0.05$s), motor block (voltage equations for each phase, motion equations (12.18 through 12.20), angle selection

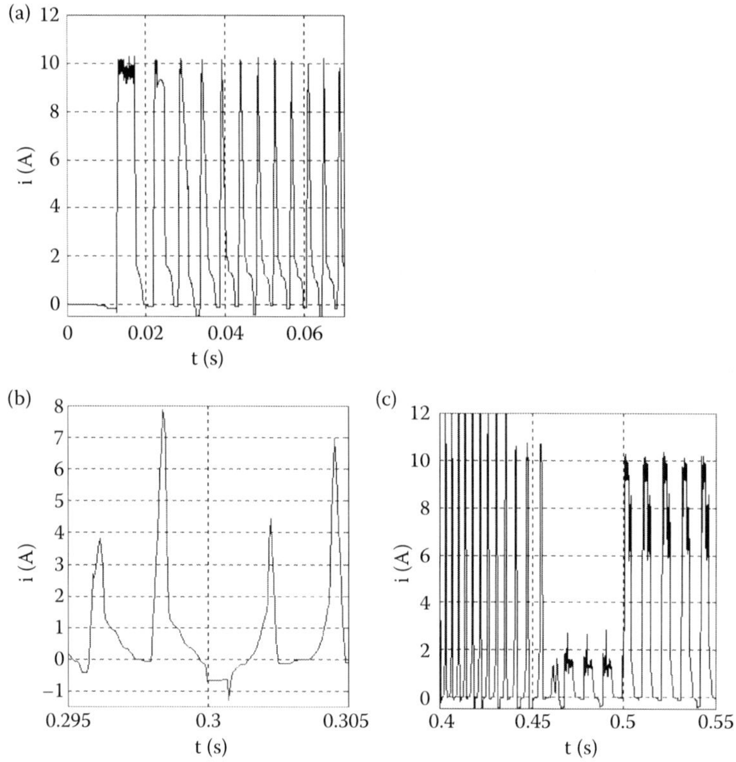


**FIGURE 12.17**
Current waveforms: (a) During acceleration, (b) at 700 rad/s, and (c) during acceleration, at 150 rad, and acceleration to 400 rpm (Figure 12.6).

and advancing block, and three blocks (A, B, C) for matrix calculation [as functions of $\theta_r$ and phase current]: Equations 12.33 through 12.35).

$$\frac{\partial\lambda_i}{\partial i_i} = \begin{bmatrix} L_u & L_u \\ L_u + \frac{K_s}{i_s}(\theta_r - \theta_0) & L_u \\ L_u + \frac{K_s}{i_s}(\theta_0 + \pi/3 - \theta_r) & L_u \end{bmatrix} \tag{12.33}$$

$$\frac{\partial\lambda_i}{\partial\theta_r} = \begin{bmatrix} 0 & 0 \\ \frac{K_s}{i_s} i_i & K_s \\ -\frac{K_s}{i_s} i_i & -K_s \end{bmatrix} \tag{12.34}$$

$$\frac{\partial\left(\int \lambda_i di_i\right)}{\partial\theta_r} = \begin{bmatrix} 0 & 0 \\ \frac{K_s}{2i_s} i_i^2 & K_s i_i \\ -\frac{K_s}{2i_s} i_i^2 & -K_s i_i \end{bmatrix} \quad (12.35)$$

Each line of these matrices represents a case function of phase inductance for three operation modes (see Figure 12.2) based on Equations 12.27 and 12.28. The columns represent the two cases for $i \le i_s$ and $i \ge i_s$.

The angle selection and advancing block produce the voltage PWM between $\theta_{on}$ and $\theta_c$ angles and, respectively, negative voltage supply until $\theta_{off}$ (where the current for the working phase becomes zero); see Figure 12.5. In this block the firing angle is selected for each phase as a function of rotor position information.

The study examines the system's behavior for starting, load perturbation, motoring, and generating of the SRM drive.

The integration step (50 μs) can be modified from Simulink's *Simulation/Parameters*.

To find out the structure of each block presented above, unmask them (*Options/Unmask*). Each masked block contains a short help screen describing that block (inputs/outputs/parameters).

The block diagram of the electric drive system is presented in Figure 12.14.

The following figures represent the speed (Figure 12.15), torque (Figure 12.16) responses, and current waveforms (Figure 12.17) for speed step response from 0 to 700 rad/s (advancing angle is 5° after 500 rad/s). When changing the speed reference at 0.3 s (from 700 to 150 rad/s, the machine is working as a generator until it reaches the reference speed); load torque is applied at 0.5 s (8 Nm); unloading is done at 0.6 s; and, at the same time, the speed reference is changed to 400 rad/s and loading is done again at 0.8 s (3 Nm).

## 12.10 High-Grade (Servo) Drives

High-grade (servo) drives have a strong capability for position, speed, or torque control, characterized by high-energy conversion ratios, precision, robustness, and quickness of torque control and rather wide speed range control. Traditionally, d.c. brush, brushless (PMSM), or (recently) advanced control IM drives are used as servos.

The strong nonlinearity in the flux/current/position and torque/current/position curves makes, in principle, the SRM apparently less attractive for servos. However, the motor simplicity and ruggedness in such applications may not be overlooked.

An attempt to build a SRM servo drive may start with a revisitation of torque production in the SRM.

### 12.10.1 Torque Sharing by Current Shaping

In servos, torque pulsations should be reduced to less than 2% of rated torque if smooth (precise) torque control is to be obtained. If only one phase can produce torque (positive inductance slope) at any time, chances to reduce the torque pulsations are low, especially if winding losses are to be limited.

A four-phase (8/6) SRM has two phases capable of producing torque at the same time, while, in a three-phase machine, only one phase can produce torque at any time. Consequently, four phases seem necessary for servo performance. In this case, single-phase torque production will be alternated with two-phase torque action for low losses or low-torque pulsations. The total torque $T_e$ is the sum of the torques per phase $T_{ei}$ (Figure 12.18)

$$T_e = \sum_{i=1}^{m} T_{ei}(i_i, \theta_{ri}); \quad i = 1,...,m; \; m = 4 \tag{12.36}$$

The torque response time is basically equal to the time to reach the maximum flux level in the machine and it is in the order of milliseconds for the kW power range.

How to turn on and turn off various phases and profile the current waveforms to minimize torque ripple is not a problem with a unique solution.

A second condition is required. To determine the flux/torque relationships for SRMs we may apply two optimization criteria for torque sharing:

- Maximum torque per unit winding losses (below base speed)
- Maximum torque per unit admissible flux level (above base speed)

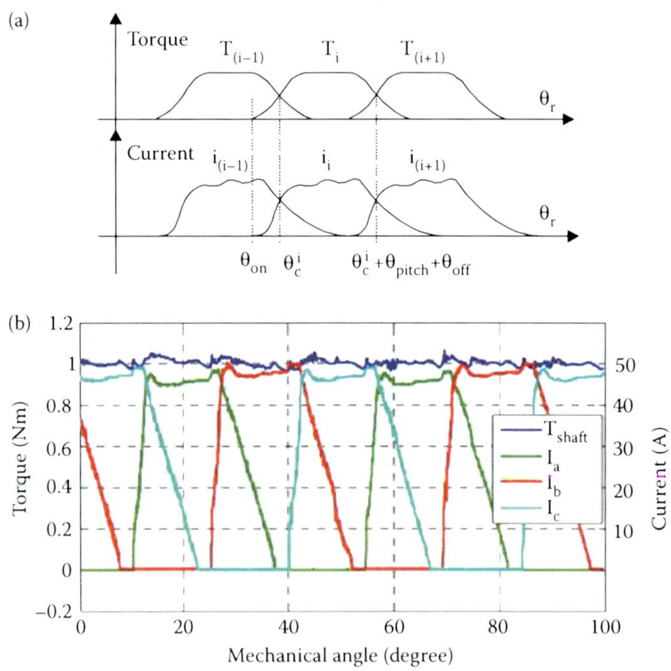


**FIGURE 12.18**
(a) Phase-torque components and phase current shaping, (b) sample result for a 12/8 three-phase SRM. (From R.B. Inderka, R.W. De Doncker, Simple average torque estimation for control of switched reluctance machines, *Record of EPE-2000*, vol. 5, 2000, pp. 176–181.)

For both criteria the crucial problem is when to switch from one phase to the next. In Reference 17 the switch over angle $\theta_c^i$ (for the first criterion) and $\theta_c^\lambda$ (for the second) are chosen so that in these conditions the values of torque are equal to each other (half the reference torque: $T_e^*/2$) and so are the currents (Figure 12.18), and the flux levels.

The $\theta_{on}$ angle is calculated to allow the current rise required in the machine characteristics. Before $\theta_c^i$ ($\theta_c^\lambda$), the phase with stronger torque was in charge, while after that the incoming phase predominates in torque production.

The problem is how to find the current versus position profiles for various speeds and reference torques. This task definitely requires a thorough knowledge of the $\lambda(i,\theta_r)$ curves which have to be inverted to produce $i_i(\omega_r, T_i, \theta_r)$ functions. Various mathematical approximations of $\lambda_i(i_i,\theta_r)$ functions determined through tests may be used for the scope [17]. Here is an example

$$\lambda(i,\theta_k) = a_0(\theta_k)\tan^{-1}(a_1 i) + a_2(\theta_k)i \tag{12.37}$$

where $\theta_k$ are discrete rotor positions.

From the coenergy formula the corresponding torque is computed. The torque sharing functions may be implemented in a feedforward manner in the sense that four reference currents and current controllers make sure that the reference currents are closely tracked (Figure 12.19).

To store all the waveforms for various speeds would require much memory and thus, in Reference 17, only the waveforms for maximum speed are stored. For lower speeds the phase voltage (flux) available is not surpassed.

It may be seen from Figure 12.19 that the switch over angle $\theta_c$ ($\theta_c^i$ at low speeds and $\theta_c^\lambda$ at high speeds) has to be precalculated based on the optimization criteria mentioned above. In general, $\theta_c^i > \theta_c^\lambda$, as expected.

Torque pulsations of less than 4% from 0.5 to 5000 rpm and good positioning performance have been reported in References 17 and 18. Also, it has been shown that the maximum torque/loss criterion produces more power at low speeds while the maximum torque/flux enlarges the CPSR.

What has still not been solved satisfactorily is the self-commissioning mode when the drive itself, before calibration, should gather and process all the data for the $\lambda(\theta_r,i)$ curve family within one minute as is done currently for a.c. motor drives.

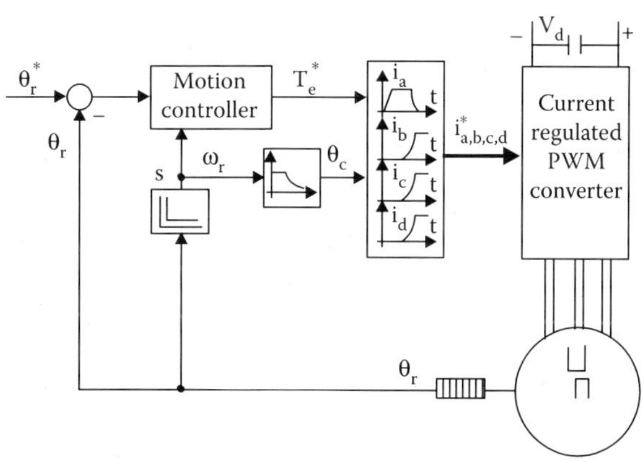


**FIGURE 12.19**
High-grade SRM drive with feedforward torque sharing optimization.

## 12.11 Sensorless SRM Drives

Sensorless SRM drives may be open loop or closed loop with respect to speed (or position) control. Open-loop sensorless driving implies fixed dwell angle $\theta_w = (\theta_c - \theta_{on})$ control with synchronization as is done in stepper motors [18] (Figure 12.20). Increasing the dwell angle tends to increase the stability, but with lower efficiency. To solve this problem the dwell angle is increased with d.c. link current (load in general). Though simple, such a method has stability problems, produces low dynamics, and allows for a limited speed control range.

Consequently, open-loop sensorless drive solutions seem today of some historical importance but hardly practical.

Closed-loop sensorless SRM drives have been proposed to compete with the, by now, commercial sensorless a.c. drives capable of at least 100 to 1 speed range and torque response in the milliseconds range, with good speed precision (less than 0.2%–0.3% of rated speed).

Competitive closed-loop sensorless SRM drives should be high grade and, in addition, capable of producing, through estimation (or observation) the rotor position $\hat{\theta}_r$ and speed $\hat{\omega}_r$ with good precision, robustness, and on time (fast). Among the closed-loop strategies for position and speed estimation we mention the incremental inductance estimator in the active phase, diagnostic pulse injection in the activated or passive phase, mutual inductance estimation, observers, etc. For a collection of papers on these solutions see Reference 19, Section 4.

To the best of our knowledge, no high-performance closed-loop sensorless SRM drive has reached the markets as of the time of writing. We will dwell only a little on two promising solutions: the combined voltage–current model-based [20] and the fuzzy logic (rule-based) observers [14].

## 12.12 Voltage–Current Model-Based Position Speed Observer

Basically, the phase voltages and currents are measured with wideband frequency sensors with optoisolation and analog–digital fast (2 μs or so) converters. In any case, the $\lambda(\theta_r, i)$

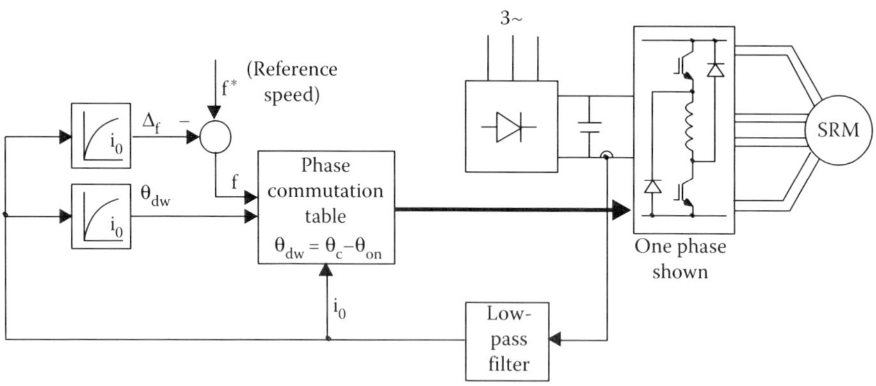


**FIGURE 12.20**
Basic open-loop sensorless SRM drive.

family of curves has to be *a priori* measured or FEM calculated and curve fitted through derivable functions.

The voltage model is represented by the stator phase equations

$$\frac{d\lambda_i}{dt} = V_i - r_s i_i \tag{12.38}$$

This model may be used to estimate the flux, but, for low speeds, the model is erratic due to integration drift, voltage and current measurements noise, and stator resistance variation with temperature.

From the $\lambda(\theta_r,i)$ curves, with current measured and $\hat{\theta}_r$ estimated in a previous time step, we may estimate back the current $\hat{i}_i$ with the current error $\Delta i_i(k)$. A position correction $\Delta\theta_r(k)$ is operated to drive this error to zero. Thus the voltage and current models have already been used together once [20]

$$\Delta\theta_i(k) = -\left(\frac{\partial\lambda_i}{\partial i_i}\right)\Delta i_i(k) / \frac{\partial\lambda_i}{\partial\theta_r(k)} \tag{12.39}$$

The three position corrections (from all three-phases) may be averaged (or weighted) to give

$$\Delta\theta_r = \frac{(\Delta\theta_a + \Delta\theta_b + \Delta\theta_c + \cdots)}{m} \tag{12.40}$$

Thus the estimated position $\theta^e(k)$ is

$$\theta^e(k) = \theta_p(k) + \Delta\theta_r \tag{12.41}$$

where $\theta_p(k)$ is the value estimated in the previous time step.

Finally, the rotor position is predicted using its estimated values at three time steps K–2, K–1, K and a quadratic prediction [20]

$$\theta_p(k) = 3\theta^e(k) - 3\theta^e(k-1) + \theta^e(k-2) \tag{12.42}$$

Now, with a new position $\theta^e(k)$ and the predicted flux values from the voltage model, once again new currents are estimated and new current errors $\Delta' i_i$ appear. Flux corrections $\Delta\lambda_i(k)$, based on $\lambda(\theta_r,i)$ curves, are

$$\Delta\lambda_i = \frac{\partial\lambda_i}{\partial i_i}\Delta' i_i \tag{12.43}$$

$$\Delta\lambda_i(k) = \lambda_i(k) + \Delta\lambda_i \tag{12.44}$$

The flux correction reduces the integrator drift. Mechanical load parameters are not required and the magnetic saturation is included through the $\lambda(\theta_r,i)$ curves. Still, the speed estimator (from position) is to be added. It has to be noted that a great number of calculations are required. A TMS310C31 DSP may provide for a 100 μs computing (and control) cycle. Good results, for position control only, have been reported, down to 30 rpm, in Reference 20.

The control system for the high-grade SRM drive has to be added to the voltage–current model position (and speed) observer (Figure 12.21), to produce a high-performance SRM drive.

Fuzzy logic may also be used to realize a voltage–current observer similar to that presented in this section.

In essence, the $\lambda(\theta_r,i)$ curve family is approximated (fitted) through the fuzzy logic approach, avoiding analytical approximations as above. Digital simulations have shown [14] such a system capable of position estimation with an error less than 0.4°.

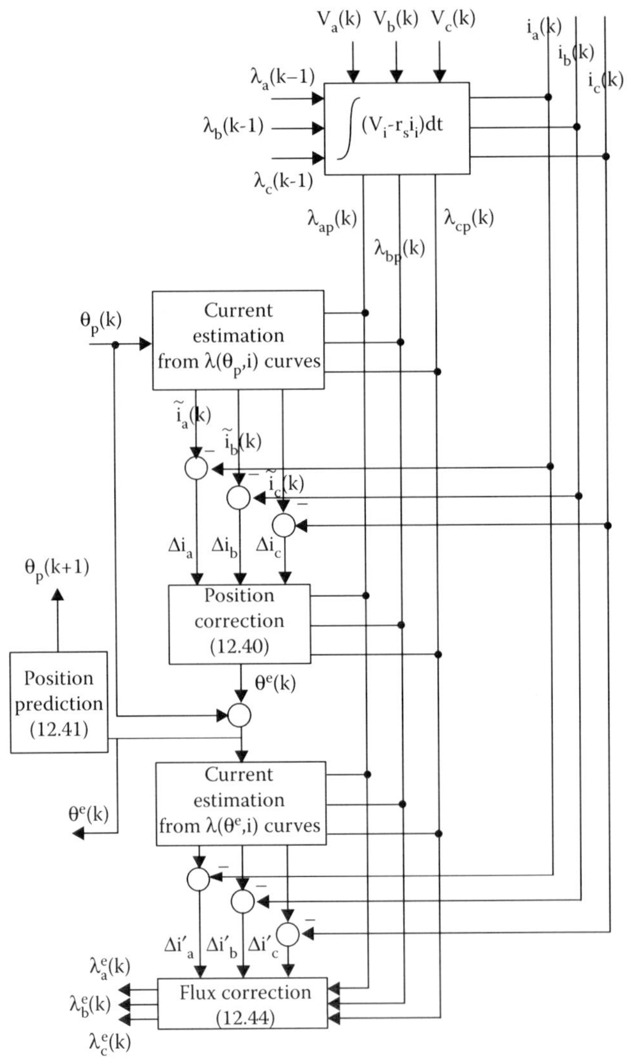


**FIGURE 12.21**
Voltage–current model position observer.

Alternatively, sliding-mode rotor position and speed estimators based on given $L_s(\lambda_s, \theta_r)$ curves, with flux estimator, have been also proven to produce good motion sensorless control [21] for a 5:1 constant power range, by switching from maximum torque/current to maximum torque/flux optimal turn on $\theta_{on}$ and turn off $\theta_{off}$ phase angle estimation. High torque dynamic response has been also demonstrated with direct average torque control and online correction of off-line reference current and reference $\theta^*_{on}$ and $\theta^*_{off}$, based on PI torque error output [22].

For more on position sensorless control, see References 23 through 25.

The first sensorless 100 to 1 speed range (and 0.3% of rated speed error and quick torque response) SRM drive is apparently still due.

## 12.13 Single-Phase SRM Control

The single-phase SRM with equal count of stator and rotor poles (2/2 or 4/4) and self-starting parking position provision seems very attractive for home appliances or automotive actuators below 1 Nm of torque because of the possibility to utilize only one active power switch; that is, a lower cost drive is obtained.

In addition, with 6/6, 4/4, and even 2/2 stator/rotor pole count, the single-phase SRM may be designed in some conditions for more torque density than a three-phase SRM, which is characterized, anyway, by one phase acting at any time.

The self-starting parking position may be obtained by

- Parking PMs on top of special poles (Figure 12.22a).
- Step airgap under the stator poles $g_2/g_1 = 1.5–2.5$ (Figure 12.22b) or small cam rotor shape.
- Magnetic saturation of roughly one-half of the rotor poles with the other half providing a double airgap $g_2 \sim 2g_1$ ($g_1$ airgap under the saturable zone—Figure 12.22c) [26].
- With an additional starting phase that is automatically turned on when the active phase turns off, some additional torque is produced by the auxiliary phase even during operation at nonzero speed (Figure 12.22d) [27, 28].

Basically all methods of providing the self-starting position allow for unidirectional motion from standstill.

However once the motor starts in one direction, in the presence of two Hall sensors (in general), it is feasible to brake the rotor regeneratively and then rotate it in the reverse direction of motion. This way bidirectional (or four quadrant) motion is claimed with single-phase SRM [28].

A basic power electronics converter for the single-phase SRM without and with starting phase is shown in Figure 12.23.

The single-phase SRM main winding A works only during about half the time as it is only then that the torque is unidirectional (positive or negative).

However, the saturated rotor half-pole with stepped airgap configuration (Figure 12.22c) claims that the motoring cycle (negative torque) may be wider than the generating one (Figure 12.24).

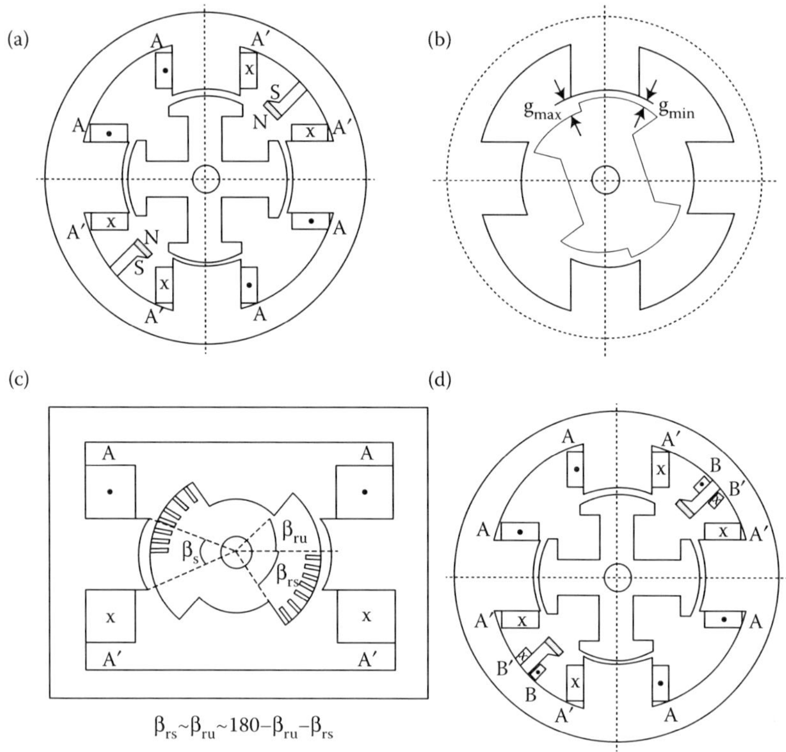


**FIGURE 12.22**
Self-starting parking position methods: (a) parking magnet, (b) stepped airgap, (c) saturated half rotor pole plus stepped airgap, (d) energy-drain starting phase.

There are two zero torque positions, at 75° and, respectively, 110° besides the 0° and 180° ones. Moreover, for low current, the torque between 75° and 110° is positive, while for large current it is negative (motoring).

If the rotor is positioned for $\theta_r = 75°$, then, with a small current, the rotor is moved to $\theta_r = 110°$ in the opposite direction of desired motion.

Further on, with high current (at 110° rotor position) the large negative torque should be able to move the rotor to the left till $\theta_r = 0$, when again negative torque is available, to continue motor acceleration.

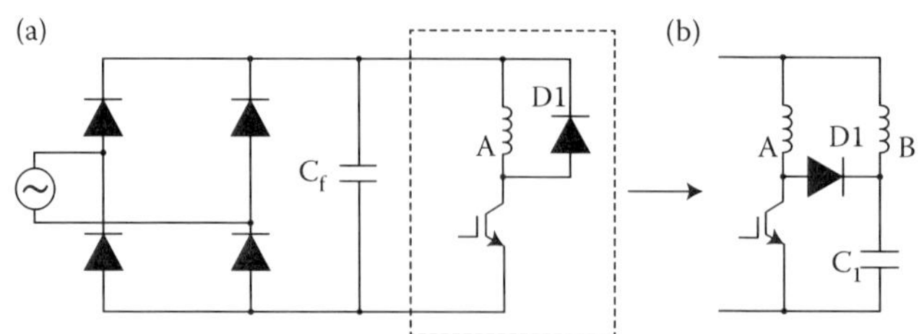


**FIGURE 12.23**
Single IGBT (MOSFET) converter: (a) for single-phase SRM, (b) for single-phase SRM with energy-drain starting phase B.

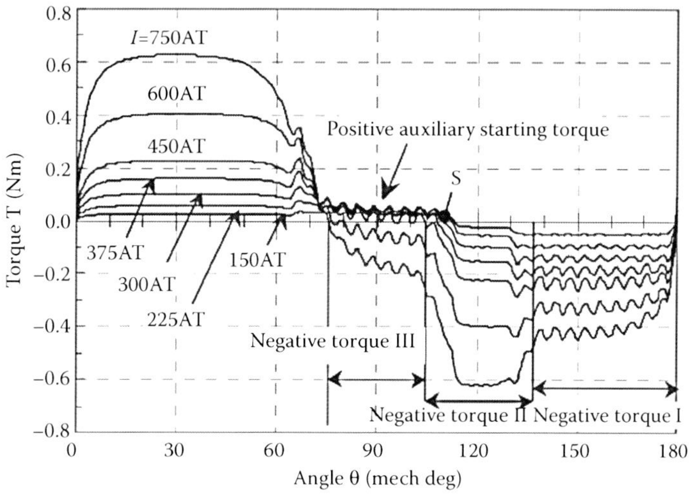


**FIGURE 12.24**
Torque cycle of single-phase SRM of Figure 12.22c.

Playing with magnetic saturation zone (made of slots and teeth on the rotor half-pole), the starting torque at low current may be increased to 10% of rated torque, to handle easy starts. Heavy starts are, apparently, hardly possible.

On the other hand, with the starting phase B present (Figure 12.22d), if the rotor is initially aligned to the main phase A poles, a current pulse does not move the rotor, but, when the phase A is turned off, the energy is drained through the starting phase B, tending to align the rotor to its axis. This way, when the second voltage pulse is applied to the main phase, the latter should provide for motor self-starting from the starting phase axis toward the main phase axis. Once the motor starts in the preferred direction of motion, and two Hall proximity sensors are available, the motor may be regeneratively braked and then controlled (on the fly) to accelerate in the opposite direction of motion (Figure 12.25). Four-quadrant operation is thus claimed [28].

A nonhesitant heavy (direct) bidirectional motor start does not seem possible even with this solution.

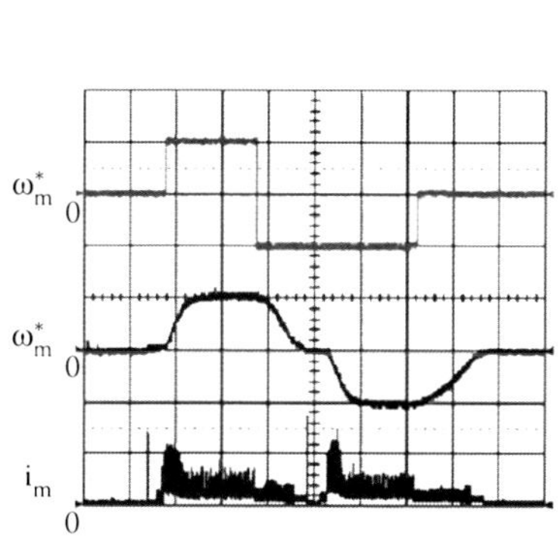


**FIGURE 12.25**
Four quadrant operation 15,000 rpm/div and 5 s/div.

## 12.14 Recent Reluctance Motor Drives

Switched reluctance machines, in their standard multiphase configurations are characterized by

a. Doubly salient (stator and rotor) magnetic cores with $N_s$ and $N_r$ salient poles
b. No rotor winding or PMs
c. Stator tooth-coil windings connected in *m* phases
d. The unipolar stator currents are intact with rotor position in discontinuous mode or at high speeds in continuous modes
e. Motoring (when phase inductance increases with rotor position) and generating (when phase inductance decreases with rotor position) modes are available

In an effort to increase torque density (in Nm/kg or Nm/L) the standard SRM in the three-phase Ns = 12 and Nr = 10 poles has been optimally designed to produce 15 Nm at 45 Nm/L and 92%–93% measured efficiency at 1200 rpm with max speed of 6000 rpm (at same 50 kW) at about same for 93% efficiency in HEV applications [29–31].

Torque pulsations at high speeds and large radial forces vibration and noise in standard SRM have also been reasons to look for further improvements, maintaining somehow the rotor ruggedness and the stator simplicity and ease of fabrication. Among them we treat here the following recent results to stir further scientific inquires:

- The double stator (DS) SRM with segmented rotor
- The d.c. + a.c. fed stator SRM machines

### 12.14.1 DS-SRM with Segmented Rotor

The DS-SRM with segmented rotor and radial airgap cross section is shown in Figure 12.26 [32]. The coils of the dual stator coils are diametrical (not tooth-wound) and their m.m.f.s are opposite (counter series connected).

**FIGURE 12.26**
DS-SRM with radial airgap.

This way the flux lines are bent tangentially, reducing the radial forces (that is, reducing vibration and noise), while the machine inductance is reduced also with its derivative with rotor position $(dLi(\theta_{er})/d\theta_{er})$ increased for higher torque density.

The segments of the rotor are mounted in a siluminium frame (10 time larger electric resistivity than aluminum: $6*10^{-8}\ \Omega m$), with a radial double sided gap of up to 1 mm between the two stators and rotor. The rotor is mechanically a bit frail and the additional siluminium losses in its structure constitutes a thermal bottleneck in design.

A 100 kW, 2000 rpm DS-SRM with a long stator stack (250 mm) would weigh 90 kg of active materials at an efficiency above 0.92, with good vibration/noise performance for current densities below 10–12 A/mm$^2$. This is quite remarkable progress. To increase the rotor mechanical ruggedness and reduce machine fabrication costs, an axial airgap topology with large maximum diameter length can be imagined.

### 12.14.2 D.C. + A.C. Fed Double Saliency Machine Drives

In an effort to feed a.c. (sinusoidal current) in the stator windings from a rather standard inverter and possibly increase SRM torque density, at similar efficiency, but better inverter voltage utilization for large CPSR and lower torque pulsations vibration and noise, basically tooth-wound or two-slot pitch span [33] alternate d.c. and a.c. coil stator winding double saliency machine drives have been proposed. The VC and DTFC sensorless control is similar to that of d.c. excited SMs. The rotor may be built with salient poles or with segmented poles embedded in a nonmagnetic structure [34].

We introduce here the two-slot pitch span coil d.c. + a.c. winding double saliency wound field flux switching machine (WFFSM), as it seems to produce competitive results for HEV and EV propulsion [33] in high-speed topologies with high gear ratio (max. machine speed 20,000 rpm, 123 kW, 3.5 kW/kg, max torque 213 Nm). The topology for 24 stator open slots (teeth) and 10 rotor salient poles is shown in Figure 12.27.

The optimal design performance at sinusoidal a.c. current is shown in Figure 12.28a–c (power factor, torque and power versus speed, and efficiency versus speed at critical operation points).

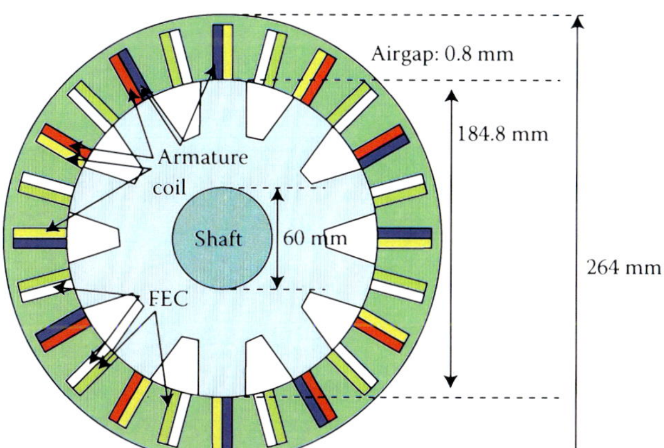


**FIGURE 12.27**
Wound (d.c.) field flux switching machine (WFFSM).

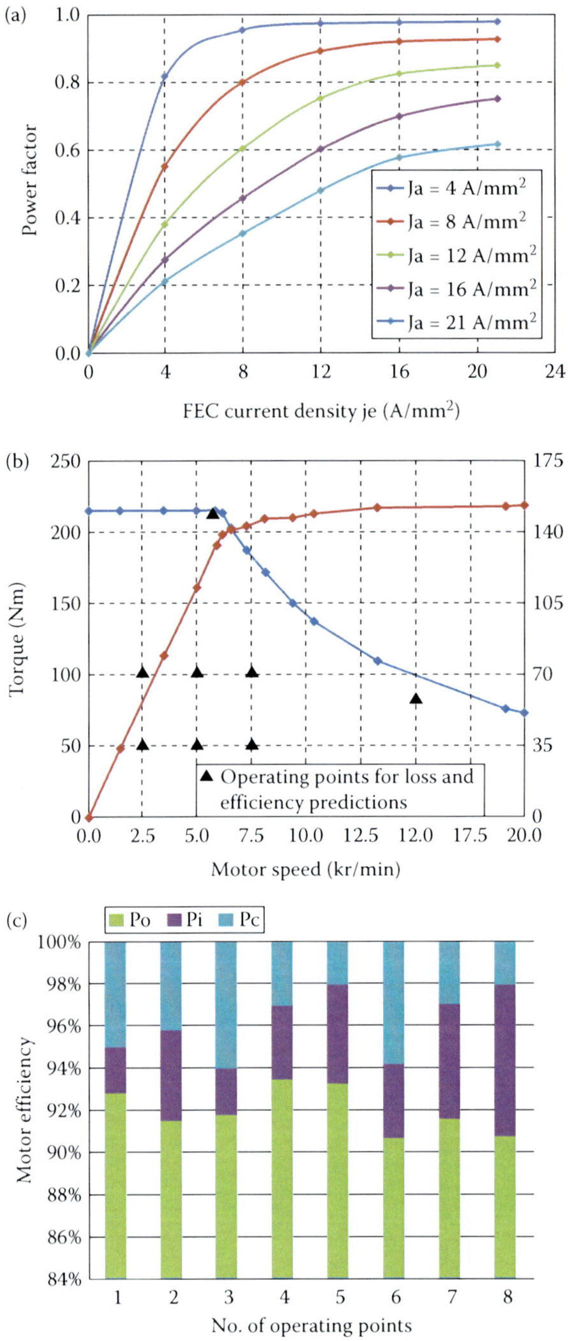


**FIGURE 12.28**
(a) Power factor; (b) torque and power/speed; (c) efficiency versus speed at critical operation points. (After E.B. Sulaiman, T. Kosaka, N.Matsui, Design study and experimental analysis of wound field flux switching motor for HEV applications, *Record of ICEM*, 2012, pp. 1269–1275.)

The results in Figure 12.28 warrant remarks such as

- The machine is claimed to deliver higher power density than the IPMSM for same geometrical envelope at similar (good) efficiency for the entire power/speed range. The higher power density is obtained at higher speed (for higher ratio transmission), however.
- The machine initial and fabrication costs are notably smaller than for IPMSMs.
- For the maximum torque (at 213 Nm), the machine is heavily saturated and the equivalent power factor is only 0.6 (as in SRMs).
- Consequently the inverter kVA (cost) is increased in comparison with IPMSMs drives.
- Also, the fundamental maximum frequency in the investigated case study is large for sinusoidal current control (3.33 kHz!), which means high switching frequency (above 20 kH) in the inverter; a multilevel inverter may be the economical solution here.

NOTE: Other d.c. + a.c. tooth-wound coils stator SRMs have been proposed recently as diode rectified d.c. output controlled autonomous auxiliary generators [34,35].

## 12.15 Summary

- SRM is a single- or multiphase doubly salient simple and rugged electric machine with a passive (windingless) rotor.
- The SRM phases are fed sequentially with voltage pulses through a sort of PWM.
- Phase commutation is rotor position dependent.
- The torque has to be calculated from the coenergy derivative method, due to magnetic saturation.
- In general, the motor phases have almost zero interaction and thus SRM is claimed to be fault tolerant.
- The SRM is a motor totally dependent on the unipolar current two-quadrant multiphase chopper.
- One or, at most, two phases may conduct at a time.
- Torque pulsations are reduced by torque sharing and current profiling, when a precision position sensor (estimator) is required.
- While low-grade—general-purpose—SRM drives are not difficult to control, high-grade (servo) SRM drives require notable online (and off-line) computation efforts [29,30].
- The flux/current/position curve family is required for high-grade (servo) SRM drives with motion sensors or sensorless.
- These curves are either FEM calculated or measured at standstill or in running (dynamic) tests. They are also curve fitted through analytical functions or mapped by a fuzzy logic approach.

- SRMs are simple and rugged, may withstand thermally harsh environments (like avionics, metallurgy, etc.) and thus, despite their more complicated control (for high-grade performance), they seem to have a future. An incipient market has surfaced and strong marketing strides seem close.
- Single-controlled power switch single-phase SRM drives at low power seem feasible apparently for four-quadrant operation in light start residential and automotive applications.
- Recent advanced designs [31–33] and control have apparently put the SRM on a fully competitive basis to other (IPMSM) propulsion systems for HEVs and EVs.
- Doubly salient (SRM like) machines with stator PMs (flux switching) or (and) with additional d.c. excitation have been investigated recently for better performance [33–35], but have not yet reached industrial stage.

## 12.16 Problems

12.1 A 6/4 three-phase SRM has the data: maximum current $i_{omax} = 10$ A, unaligned inductance $L_u = 10$ mH, aligned inductance $L_a^s = 6L_u$, flux/current/position curves are considered piecewise linear (Figure 12.29). The stator pole width angle $\beta_s = 30°$. The interpole width angle is $\beta_i = 30°$. The d.c. link voltage $V_d = 300$ V for the unipolar current PWM converter. Determine

a. The maximum flux linkage.

b. The base speed $\omega_b$ for zero advance angle ($\theta_0 = 0°$).

c. The maximum average torque at zero speed.

d. Torque versus position for constant current $i_{av} = 5$ A.

e. For $\theta_c = -5°$ and $\theta_{off} = 22°$ calculate the maximum flux level in the phase at $2\omega_b$; determine the corresponding current from flux/current/position curves in Figure 12.29.

12.2 For the SRM of problem 12.1 and constant $\theta_{on} = \theta_0 = 0°$ and $\theta_{off} = 24°$ use Simulink to investigate the starting transients from zero to base speed with a load of 50% of maximum average torque at zero speed and the controller in Example 12.2.

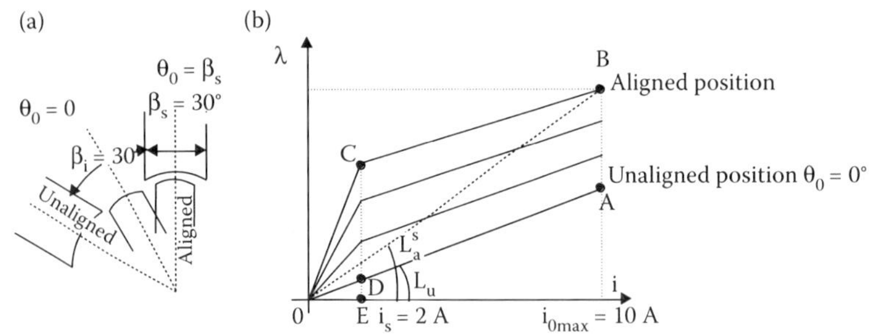


**FIGURE 12.29**
(a) Unaligned and aligned position and (b) linear piecewise flux/current curve.

## References

1. T.J.E. Miller, *Switched Reluctance Motors and Their Control*, Oxford University Press, Oxford, 1993.
2. C. Ferreira, E. Richter, About channel independence for multichannel switching reluctance generating systems, *Record of IEEE-IAS, Annual Meeting*, 1996, pp. 816–821 (IEEExplore).
3. D.A. Philips, Switched reluctance drive, new aspects, *IEEE Trans.*, vol. PE-5, no. 4, 1990, pp. 454–458.
4. T.J.E. Miller, Converter volt-ampere requirements of the switched reluctance motor drive, *IEEE Trans.*, vol. IA-21, no. 5, 1985, pp. 1136–1144.
5. W.F. Ray, P.L. Lawrenson et al., High performance switched reluctance brushless drives, *IEEE Trans.*, vol. IA-22, no. 4, 1986, pp. 722–730.
6. F. Liang, Y. Liao, T.A. Lipo, A new variable reluctance motor utilizing an auxiliary commutation winding, *IEEE Trans.*, 1993.
7. H.Y. Li, Y. Zhao, T.A. Lipo, F. Liang, A doubly salient doubly excited variable reluctance motor, *Record of IEEE-IAS, Annual Meeting*, 1992 (IEEExplore).
8. B.C. Mecrow, New winding configurations of doubly salient reluctance machines, *Record of IEEE-IAS, Annual Meeting*, Part I, 1992, pp. 249–256 (IEEExplore).
9. P. Materu, R. Krishnan, Estimation of switched reluctance motor losses, *Record of IEEE-IAS, Annual Meeting*, 1988, vol. I, pp. 176–187 (IEEExplore).
10. Y. Hayashi, T.J.E. Miller, A new approach to calculating core losses in the SRM, *Record of IEEE-IAS, Annual Meeting*, 1994, vol. I, pp. 322–328 (IEEExplore).
11. V.V. Athani, V.N. Walivadeker, Equivalent circuit for switched reluctance motor, *EMPS*, vol. 22, no. 4, 1994, pp. 533–543.
12. I.A. Byrne, J.B. O'Dwyer, Saturable variable reluctance machine simulation using exponential functions, *Proceedings of International Conference on Stepping Motors and Systems*, July 1976, University of Leeds, pp. 11–16.
13. D. Torrey, An experimentally verified variable reluctance machine model implemented in the Saber™ circuit simulation, *EMPS J.*, vol. 23, 1995.
14. A. Cheok, N. Ertugrul, A model-free fuzzy-logic-based rotor position sensorless switched reluctance motor drive, *Record of IEEE-IAS, Annual Meeting*, 1996, vol. I, pp. 76–83 (IEEExplore).
15. A. Radun, Design considerations for the switched reluctance motor, *Record of IEEE-IAS, Annual Meeting*, 1994, vol. I, pp. 290–297 (IEEExplore).
16. B.K. Bose, T.J.E. Miller, Microcomputer control of switched reluctance motor, *Record of IEEE-IAS, Annual Meeting*, 1985, pp. 542–547 (IEEExplore).
17. P.C. Kjaer, J.J. Gribble, T.J.E. Miller, High grade control of switched reluctance machines, *Record of IEEE-IAS, Annual Meeting*, 1996, vol. I, pp. 92–100 (IEEExplore).
18. J.T. Buss, M. Eshani, T.J.E. Miller, Robust torque control of SRM without a shaft-position sensor, *IEEE Trans.*, vol. IA-33, no. 3, 1996, pp. 212–216.
19. K. Rajashekara, A. Kawamura, K. Matsuse (editors), *Sensorless Control of AC Drives*, IEEE Press, 1996, pp. 433–485.
20. P.P. Acarnley, C.D. French, IH. Al-Bahadly, *Position Estimation in Switched-Reluctance Drives*, Record of EPE-95, Seville, Spain, pp. 3765–3770.
21. M.S. Islam, M.M. Anwar, I. Husain, A sensorless wide-speed range SRM drive with optimally designed critical rotor angles, *Record of IEEE-IAS-2000 Annual Meeting* (IEEExplore).
22. R.B. Inderca, R.W. De Doncker, High dynamic direct torque control for switched reluctance drives, *Record of IEEE-IAS-2001 Annual Meeting* (IEEExplore).
23. M. Eshani, B. Fahimi, Position sensorless control of switched reluctance motor drives, *IEEE Trans.*, vol. IE-49, no. 1, 2002, pp. 40–48.
24. B. Fahimi, A Emadi, R.B. Sepe, Position sensorless control, *IEEE-IAS Appl. Mag.*, vol. 10, 2004, pp. 40–47.

25. R. Krishnan, *Switched Reluctance Motor Drives, Book*, CRC Press, Boca Raton, Florida, 2002.
26. T. Higuchi, J.O. Fiedler, R.W. De Doncker, On the design of a single-phase switched reluctance motor, *Record of IEEE-IEMDC*, 2003, pp. 561–567 (IEEExplore).
27. R. Krishnan, A.M. Staley, K. Sitapati, A novel single phase switched reluctance motor drive, *Record of IEEE-IECON*, 2001, pp. 1488–1493 (IEEExplore).
28. R. Krishnan, S.T. Park, K. Ha, Theory and operation of a four quadrant SRM drive with a single controllable switch—The lowest cost four quadrant brushless motor drive, *Record of IEEE-IAS-2004 Annual Meeting* (IEEExplore).
29. R.B. Inderka, R.W. De Doncker, Simple average torque estimation for control of switched reluctance machines, *Record of EPE*, vol. 5, 2000, pp. 176–181.
30. R.B. Inderka, M. Menne, R.W. De Doncker, Control of SRM drives for electric vehicle applications, *IEEE Trans.*, vol. IE-49, no. 1, 2002, pp. 48–53.
31. K. Kiyota, A. Chiba, Design of switched reluctance motor competitive to 60 kW IPMSM in third generation hybrid electric vehicle, *IEEE Trans.*, vol. IA-48, no. 6, 2012, pp. 2303–2309.
32. M. Abbasian, M. Moallem, B. Fahimi, Double stator switched reluctance machines (DS-SRM) fundamentals and magnetic force analysis, *IEEE Trans.*, vol. EC-25, no. 3, 2010, pp. 589–597.
33. E.B. Sulaiman, T. Kosaka, N. Matsui, Design study and experimental analysis of wound field flux switching motor for HEV applications, *Record of ICEM*, 2012, pp. 1269–1275.
34. A. Zulu, B.C. Mecrow, M. Armstrong, Permanent-magnet flux-switching synchronous motor employing a segmental rotor, *IEEE Trans.*, vol. IA-48, no. 6, 2012, pp. 2259–2267.
35. L. Yu, Z. Zhang, Z. Chen, Y. Yan, Analysis and verification of the doubly salient brushless dc. generator for automobile auxiliary power unit application, *IEEE Trans.*, vol. IE-61, no. 12, 2014, pp. 6655–6663.

# 13

# *Practical Issues with PWM Converter Motor Drives*

## 13.1 Introduction

IM, PMSM, RSM, and SRM are brushless multiphase motors fed through PWM voltage-source converters (PECs) with bipolar and, respectively (for SRM), unipolar current capability. In all these drives, the line side converter is (still), in general, a diode rectifier with a d.c. link filter capacitor.

Also, the motors are fed through a cable of notable length (sometimes up to some hundreds of meters), with ultrafast voltage pulses (1–3 μs or less in most cases). What are the effects of such drives on the motor itself and on the environment?

Let us enumerate a few of them:

- Additional motor losses due to current and flux harmonics.
- Current harmonics injected in the a.c. power supply (supply filter is necessary).
- Electromagnetic interference due to the rather high switching frequency (up to 20 kHz with IGBTs, in the hundreds of kW power range per unit).
- High-frequency leakage currents' influence on motor current control and on circuit breakers.
- Overvoltages along the stator coils first turns due to wave reflections of the PWM converter (special filters are required); these overvoltages are magnified by long cables.
- The steep-front voltage pulses produce both electromagnetic and electrostatic stray currents (voltages) in the bearings, causing wear unless special measures are taken.

All these issues are called practical as they occur in industry where these drives are used every day for many hours.

## 13.2 Basic PWM Converter Drive

A basic PWM inverter drive with a.c. motors (or with SRMs) is made of a PEC, a power cable, and the motor (Figure 13.1).

As variable speed drives are applied in various industries, the environment (thermal or chemical), the distance between the motor and the PEC, and the load cycle vary widely with

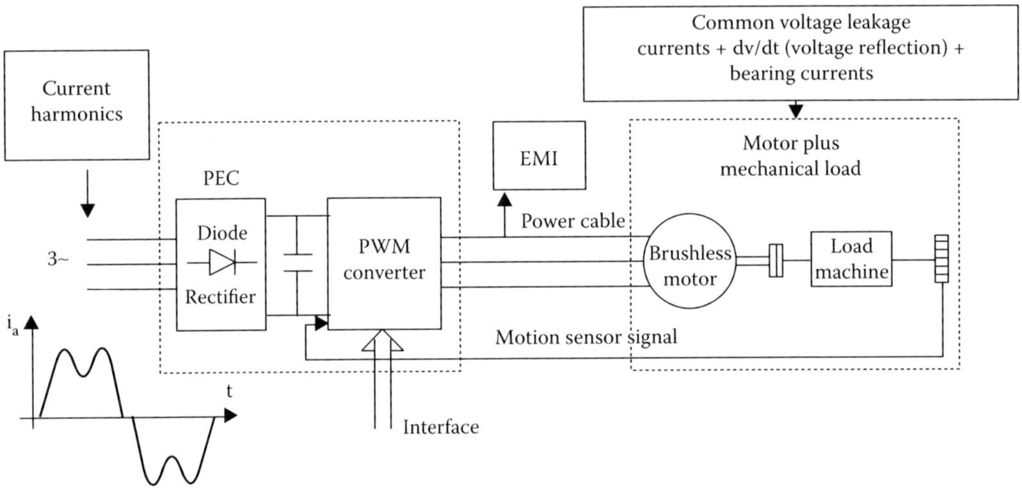


**FIGURE 13.1**
Basic PWM converter drive.

the application. Variable speed drives may be introduced as new systems (the PEC and the motor) or the old motor (IM or SM)—designed for sine wave supply—is not changed but rerated for the PEC supply and for the new speed control range.

Current harmonics in the power supply in the presence of a diode rectifier, long cable effects on voltage reflection (and dV/dt) at motor terminals, bearing currents and leakage currents, are all consequences of the short rise time of voltage pulses in the PWM converter.

## 13.3 Line Current Harmonics

Increasing the use of PWM converter drives for pumps, blowers, fans, and other industrial applications may lead to a notable infusion of current harmonics in the industrial systems.

A few questions arise:

- Are harmonics filters needed?
- What type of filters (5th, 7th, 11th, 13th, or combinations of these) are required?
- How to design (size) such filters?

Six-pulse PWM converters with diode rectifiers on the power system side draw harmonic currents of orders 5, 7, 11, 13, and 17. In principle, each PEC has its own distinct harmonic signature.

A harmonic frequency (HF) factor for current is defined in Reference 1:

$$HF = \frac{\sqrt{\sum_{5,7,11,\ldots} v^2 I_v^2}}{I_1} \quad (13.1)$$

The HF depends on the fact that the PWM converter is fed from a single-phase or three-phase supply, with or without line reactors, with or without an isolation transformer.

HF varies in the interval 2%–4% for three-phase supplies and up to 12% for single-phase a.c. supplies, for powers up to 400 kW or more and IGBTs.

The HF factor is related to the total harmonic distortion (THD) of voltage by

$$THD = HF \cdot \frac{\text{Drive kVA}}{\text{SC kVA}} \cdot 100\% \tag{13.2}$$

where drive kVA is the drive rated kVA and SC kVA is the short-circuit kVA of the power system at the point of drive connection.

In general, a less than 5% THD is required (IEEE-519-1992 and IEC 1000-3-2(3)-61000 standard). If this constraint is met, no line filter is needed. However, in general, this is not the case and a filter is necessary.

Harmonics filters consist of several sections of single (or double) tuned series L–C resonant circuits (Figure 13.2).

Separate filters for each drive above 15 kW are preferred, but aggregate filters for multidrive nodes may also be used, especially if not all drives are working at full load most of the time.

Slightly below the resonance frequency (0.95), series LC filters behave like a capacitor and their design for this situation avoids the possibility of parallel resonance due to component parameter change because of temperature or aging.

The resonance frequency of a single tuned filter $f_r$ is

$$f_r = \frac{1}{2\pi\sqrt{LC}}; \quad \alpha = \frac{f_r}{f_n} \tag{13.3}$$

where $\alpha$ is the detuning factor and $f_n$ is the nominal resonance frequency of the filter $f_n = \nu f_1$.

The filter impedance for $\nu$th harmonic, $\underline{Z}_f(\nu)$, is

$$\underline{Z}_f(\nu) = R_f + j\left[2\pi f_1 \nu L - \frac{1}{(2\pi f_1 \nu C)}\right] \tag{13.4}$$

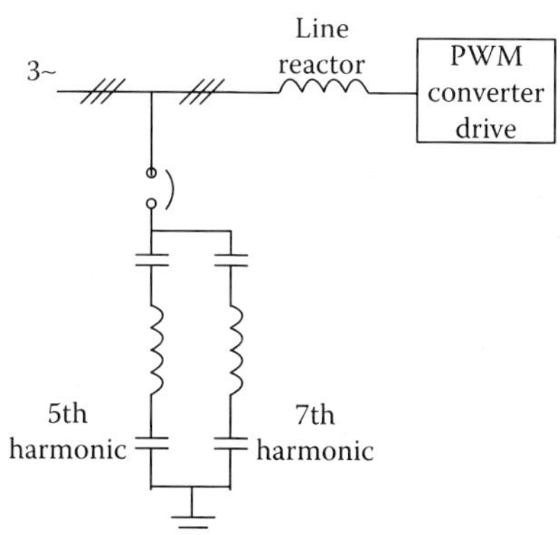


**FIGURE 13.2**
Line filters.

The filter capacitor reactive power kVAF (three-phase) for 60 Hz is

$$\text{kVAF} = 2\pi f_1 10^{-3} C V_{LL}^2 = 0.3777 C V_{LL}^2 \tag{13.5}$$

where $L_{VV}$ is the line voltage (rms) and C is in farad. The filter capacitor kVA is about (25%–30%) of drive kVA rating. A harmonic filter attenuates all harmonic voltages at the point of connection but maximum attenuation occurs for frequencies close to its resonance frequency.

The attenuation factor $a_v(h)$ is

$$a_v(h) = \frac{V(h)}{V_f(h)} \tag{13.6}$$

$$V(h) = \%I(h) \cdot h \cdot \left(\frac{\text{Drive kVA}}{\text{SC kVA}}\right) \tag{13.7}$$

where V(h) is the h harmonics voltage without the filter and $V_f(h)$ is with the νth-order filter.

Neglecting the filter resistance

$$a_v(h) = 1 + \frac{(\nu\alpha)^2}{1-(\nu\alpha/h)^2}\frac{\text{kVAF}}{\text{SC kVA}} \tag{13.8}$$

In general, $a_v(h)$ should be higher than 1.0 to provide some attenuation.

The THD has to be limited for the voltage harmonics, making use of V(h) of Equation 13.7 for finding $V_f(h)$.

The filter is designed such that the THD is less than 5%. The allowed current THD may increase with the short-circuit ratio $R_{sc}$ at the point of connection [2] from 5% for ($R_{sc} \le 20$) to 20% for $R_{sc} > 1000$

$$R_{sc} = \frac{\text{utility maximum short-circuit current}}{\text{maximum demand fundamental current}} \tag{13.9}$$

For a 5th-order harmonic current filter, the drive current waveform changes notably (Figure 13.3d).

The peak current is increased but continuous current conduction is obtained by eliminating the 5th current harmonic.

Designing adequate line filters for a local power industrial system with more than one PWM converter drive is a rather complex problem whose solution also depends on the power source impedance, through the short-circuit current ratio $R_{sc}$ or through the drive kVA/SC kVA.

As the relative power of PWM converter drives is increasing and the power quality standards are more and more challenging, it seems that the trend is to provide each drive with its line filter to comply with some strict regulations.

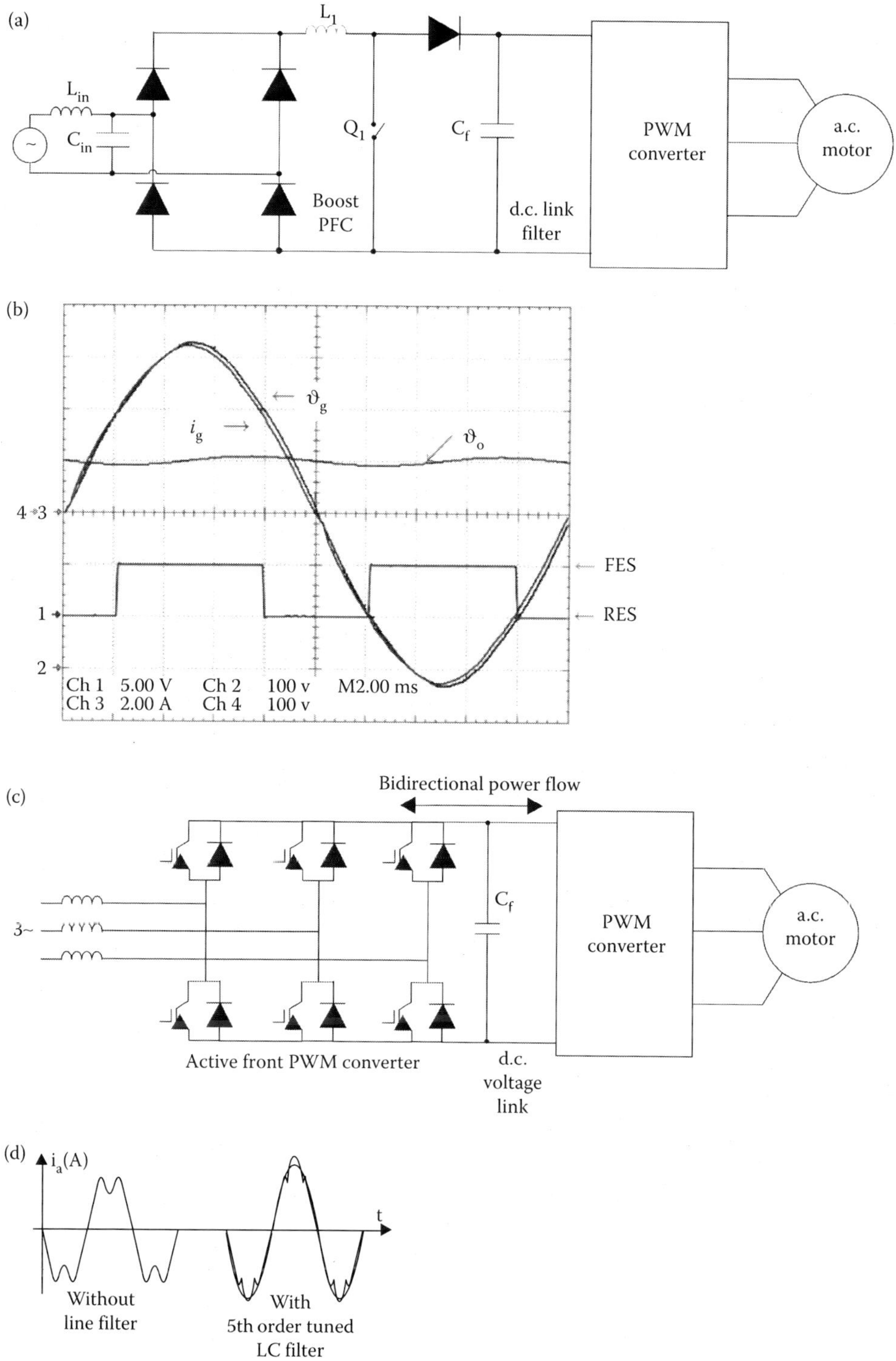


**FIGURE 13.3**
The boost PFC: (a) the configuration, (b) the obtained input voltage and current, (c) the active front rectifier, (d) 5th harmonic filtered in input current.

However, common point of connection filter additions may be made in "hot" PWM converter-drive zones with weak local power supplies.

In PWM converters we included both, the voltage-source inverters for a.c. motor drives, and the multiphase choppers for SRM drives.

Line voltage instabilities in the presence of line filters with some fast (DTFC) control systems of a.c. motor drives [3] and with d.c. chopper-fed motor drives [4] have been reported. This aspect has to be taken care of, when the drive with incorporated line filter is designed and built.

**EXAMPLE 13.1: LINE FILTER SIZING**

Let us consider that a PWM converter IM drive of 100 kVA is connected to a local power system with a short-circuit power SC kVA = 1600 kVA. The line current 5th, 7th, and 11th harmonics are $i_5 = 0.15I_1$, $I_7 = 0.03I_1$, $I_{11} = 0.01I_1$; line frequency $f_1 = 60$ Hz, line voltage $V_{LL} = 440$ V (rms).

Calculate

- The line current harmonic factor HF
- The total distortion factor THD for the voltage
- The filter capacitor C tuned to the 5th harmonic with $\alpha = 0.95$ (detuning factor), and the corresponding inductance $L_f$
- The voltage attenuation factors $a_5(h)$ for the 5th, 7th, and 11th harmonics
- The voltage total harmonic distortion factor THD with the filter off and on

**Solution**

According to Equation 13.1, HF is

$$HF = \frac{\sqrt{\sum_{5,7,11} v^2 I_v^2}}{I_1} = \sqrt{(5 \cdot 0.15)^2 + (7 \cdot 0.03)^2 + (11 \cdot 0.01)^2} = 0.7865 \tag{13.10}$$

From Equation 13.2, the THD voltage is

$$THD = HS \cdot \frac{\text{Drive kVA}}{\text{SC kVA}} \cdot 100\% = 0.7865 \cdot \frac{100}{1000} \cdot 100\% = 7.865\% \tag{13.11}$$

As THD is higher than the 5% recommended for drive kVA/SC kVA = 1/10, a line filter is required.

The filter is tuned to the 5th harmonic with $\alpha = 0.95$, so, from Equation 13.3

$$LC = \frac{1}{4\pi^2 (f_n \alpha v)^2} = \frac{1}{4\pi^2 (60 \cdot 0.95 \cdot 5)^2} = \frac{0.1308 \cdot 10^{-6}}{25} s^2 \tag{13.12}$$

Also, the capacitor kVA is

$$kVAF = 0.3 \cdot \text{Drive kVA} = 0.3 \cdot 100 = 30 \text{ kVA} \tag{13.13}$$

From Equation 13.5, C is

$$C = \frac{kVAF}{0.377 \cdot V_{LL}^2} = \frac{30}{0.377 \cdot 440^2} = 411 \cdot 10^{-6} \text{ F} = 411 \,\mu\text{F} \tag{13.14}$$

The harmonic voltage V(h) (with filter off) in the power source (in %) is found from Equation 13.7:

$$\begin{aligned} V(5) &= \%(I(5))\cdot 5\cdot\frac{100}{1000} = 15\cdot 5\cdot\frac{100}{1000} = 7.5\% \\ V(5) &= \%(I(7))\cdot 7\cdot\frac{100}{1000} = 3\cdot 7\cdot\frac{100}{1000} = 2.1\% \\ V(5) &= \%(I(11))\cdot 11\cdot\frac{100}{1000} = 1\cdot 11\cdot\frac{100}{1000} = 1.1\% \end{aligned} \tag{13.15}$$

We may calculate the voltage THD before filtering

$$\begin{aligned} \text{THD}(\%) &= \sqrt{(V(5))^2 + (V(7))^2 + (V(11))^2} \\ &= \sqrt{(7.5)^2 + (2.1)^2 + (1.1)^2} = 7.8657\% \end{aligned} \tag{13.16}$$

The voltage attenuation factors are

$$\begin{aligned} a_5(5) &= 1 + \frac{(5\cdot 0.95)^2}{1-(5\cdot 0.95/5)^2}\frac{0.3}{10} = 7.9423 \\ a_5(7) &= 1 + \frac{(5\cdot 0.95)^2}{1-(5\cdot 0.95/7)^2}\frac{0.3}{10} = 2.2545 \\ a_5(11) &= 1 + \frac{(5\cdot 0.95)^2}{1-(5\cdot 0.95/11)^2}\frac{0.3}{10} = 1.8320 \end{aligned} \tag{13.17}$$

Based on the definition of $a_v(h)$, Equation 13.6, we may calculate $V_f(5)$, $V_f(7)$, $V_f(11)$, and thus, from Equation 13.1, the new voltage, $THD_f$ (with the filter on) is

$$\begin{aligned} \text{THD}_f &= \sqrt{(V(5)/a_5(5))^2 + (V(7)/a_5(7))^2 + (V(11)/a_5(11))^2} \\ &= \sqrt{(7.5/6.942)^2 + (2.1/1.284)^2 + (1.1/0.862)^2} = 1.456\% \end{aligned} \tag{13.18}$$

The beneficial effect of the 5th-order harmonic filter is only partly evident: a reduction of voltage THD from 7.865% to 1.456%. Also, the short-circuit power of the local grid was unusually small. An increase in the SC kVA of the local grid would improve the results.

Apart from passive input filters, hybrid (active–passive) filters may be used to compensate the input current and voltage harmonics and eventually deal with short voltage sags in the a.c. source.

The boost power factor corrector (PFC) is a typical such device (Figure 13.3a), applied for single-phase a.c. source [5,6].

The boost PFC converter operates in continuous conduction mode. This way, with a proper control, the sinusoidal input current is reconstructed in phase with the a.c. source voltage (Figure 13.3b) [5].

Also for input current filtering and unity power factor, the active front end rectifier-inverter (Figure 13.3c) may replace the diode rectifier plus the passive input filter [7,8,9].

The active front end PWM converter provides quite a few functions:

- Stabilizes the a.c. link voltage "against" limited overload or against a.c. source short-lived voltage sags (three phase, two phase, single phase).
- Produces rather sinusoidal current input with limited current, controllable, power factor angle.
- Provides for fully bidirectional power flow, so necessary for fast braking drives (such as in elevators).

## 13.4 Long Motor Power Cables: Voltage Reflection and Attenuation

In many new and retrofit industrial applications, the PWM converter and motor must be placed in separate locations, and thus long motor cables are required. The high rate of voltage rise (dV/dt) of up to 6000 V/μs, typical for IGBT inverters, has adverse effects on motor insulation and produces electrostatic-caused currents in the bearings. The distributed nature of the long cable L–C parameters results in overvoltages which further stress the motor insulation. In addition, voltage reflection of the long cable is dependent on inverter pulse rise time ($t_r$ = 0.1 to 5 μs) and on the cable length which behaves like a transmission line.

PWM voltage pulses travel along the power cable at about half light speed ($U^*$ = 150–200 m/μs) and, if the pulses take longer than one-third the rise time to travel from converter to motor, full reflection occurs; the voltage pulse level is doubled at motor terminals.

Let us consider a finite-length cable with an infinite dV/dt voltage wave (pulse) traveling from converter to motor. The motor represents a high impedance with respect to such fast voltage pulses. The motor behaves like an equivalent uncharged capacitor in this case. Upon arrival, the incident wave is reflected (the capacitor charges). The reflected plus incident wave doubles the voltage at motor terminals (2E). The line is thus charged at 2E while at the sending end, the inverter output voltage, is still E. Hence a –E (negative reflection) wave now travels from inverter to motor and thus, finally, the motor voltage becomes again E after three "runs" along the cable length (Figure 13.4). This is why it is said that, if the pulses take longer than one-third of rise time to travel from inverter to motor, the reflected voltage is 2E at the motor terminals.

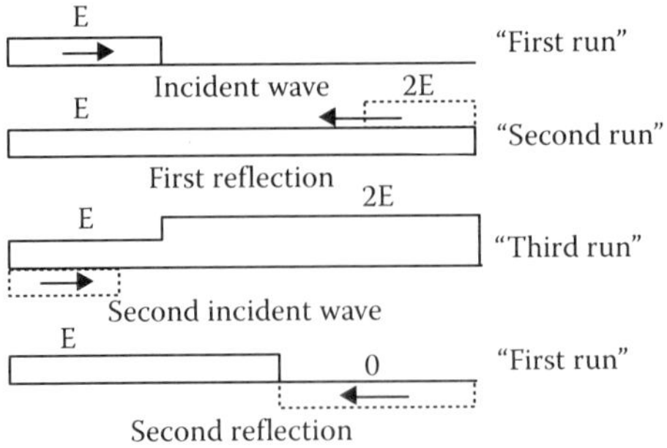


**FIGURE 13.4**
Reflection wave cycle.

We may thus calculate the ideal peak line to line voltage $V_{LL}$ at the motor terminals as [10,11]

$$V_{LL} = \begin{cases} \dfrac{3l_c \cdot E}{U^* \cdot t_r} \cdot \Gamma_m + E; & \text{for } t_{travel} < t_r/3 \\ E \cdot \Gamma_m + E; & \text{for } t_{travel} > t_r/3 \end{cases} \tag{13.19}$$

where $\Gamma_m$ is the motor realistic reflection coefficient derived from transmission line theory

$$\Gamma_m = \frac{Z_{motor} - Z_{0c}}{Z_{motor} + Z_{0c}} \tag{13.20}$$

where $Z_{motor}$ is the motor equivalent impedance and $Z_{0c}$ is the surge impedance of the cable (of length $l_c$)

$$Z_{0c} = \sqrt{\frac{L_c}{C_c}} \tag{13.21}$$

A similar reflection coefficient $\Gamma_c$ may be defined on the PWM converter output side. In the qualitative analysis in Figure 13.4, $Z_{motor} = \infty$ and, thus, $\Gamma_m = 1$. In reality, the motor impedance is more than 10–100 times larger than $Z_{0c}$ and thus $\Gamma_m \cong 1$.

For $\Gamma_m = 1$, to reduce the overvoltage to almost zero, $t_{travel} < t_r/3$ and thus

$$\frac{3l_c \cdot E}{U^* \cdot t_r} \cdot \Gamma_m \le 0.2E \tag{13.22}$$

Equation 13.22 provides a constraint condition to calculate the practical cable length $l_c$. If we replace the 0.2 in Equation 13.22 by 1.0, we will get the critical cable length $l_{cr}$ above which virtual voltage doubling occurs. For E = 650 V d.c. bus (480 V a.c. system), $\Gamma_m = 0.9$, $U^* = 165$ m/μs, from Equation 13.22 we obtain

$$\frac{3l_c \cdot E}{U^* \cdot t_r} \cdot \Gamma_m = E \tag{13.23}$$

$$l_c = \frac{U^* \cdot t_r}{3\Gamma_m} = \frac{165}{3 \cdot 0.9} \cdot t_r = (61.11 \cdot t_r(\mu s)) \text{ meters} \tag{13.24}$$

Five times shorter cables in comparison with the critical length $l_c$ (Figure 13.5) are required to reduce the motor overvoltage to 20% (13.22). This latter condition may not be met in many applications.

For long cables, the solution is to reduce the reflection coefficients at motor side $\Gamma_m(Z_m)$ (13.20) and on converter output side $\Gamma_c(Z_c)$. This may be done by adding a filter at motor terminals, for example, to reduce $Z_m$ (Figure 13.6). Consequently, a LPF at converter output will increase the voltage pulse rise time which has the effect of allowing cables longer than $l_c$ (Figure 13.6b).

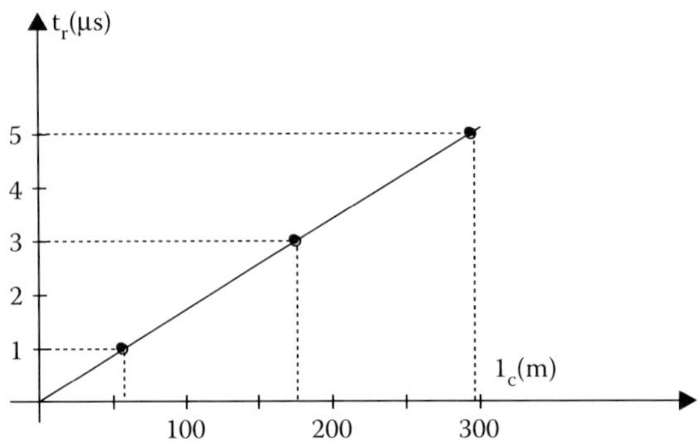


**FIGURE 13.5**
Voltage rising time $t_r$ versus critical cable length.

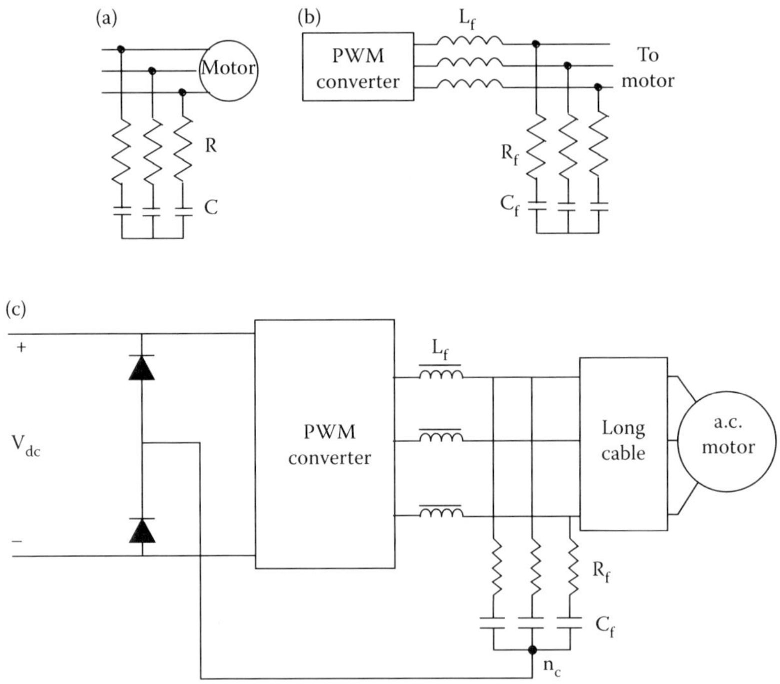


**FIGURE 13.6**
First-order filters: (a) at motor terminals, (b) at PWM converter output, (c) LRC filter with diode clamp.

According to Reference 11, the total voltage across the R–C motor terminal filter, $E_r$, is

$$E_r = 2E\left(1 - \frac{Z_{0c}}{R + Z_{0c}} e^{-(t/(R+Z_{0c})C)}\right); \quad R = Z_{0c} \tag{13.25}$$

To design the filter again, we may limit $E_r$ to $E_r = 1.2E_0$ (20% overvoltage) for $t = t_r$ (rising time):

$$1.2E = 2E\left(1 - \frac{Z_{0c}}{R + Z_{0c}} e^{-(t_r/2Z_{0c}C)}\right) \tag{13.26}$$

Thus, we may calculate the filter capacitor C while the resistance $R = Z_{0c}$ ($Z_{0c}$ is the cable surge impedance).

On the other hand, the influence of inverter output LPF is materialized by a voltage delay

$$V(t) = E\left(1 - e^{-(t_r/\tau)}\right) \tag{13.27}$$

$$\tau = \sqrt{L_f C_f} \tag{13.28}$$

where τ is the filter time constant. Evidently,

$$\tau \geq t_r \tag{13.29}$$

The capacitor has to be chosen as $C_f > l_c \cdot 10^{-10}$ (F): $L_f$ is obtained from Equation 13.28 with τ imposed and the resistance $R_f$ is calculated assuming an overdamped circuit

$$R_f \geq \sqrt{\frac{4L_f}{C_f}} \tag{13.30}$$

Typical results with a motor terminal filter and inverter output filter are given in Figure 13.7a and b, respectively.

NOTE: Increase in the rise time of the terminal voltage may impede the control system performance in the motor parameter identification (for tuning) mode. Care must be exercised when designing the drive to account for this aspect.

Recently, overvoltages (above the doubling effect of ideal reflection) of up to three times the d.c. link voltage E have been reported in connection with some PWM patterns which, once modified adequately by pulse elimination techniques, lead to lowering the ideal overvoltage to twice the value of E [12].

A rather low-cost solution to reduce drastically motor terminal overvoltage due to high dV/dt and their propagation along motor cable (as shown earlier in this paragraph) adds to the differential LRC filter (Figure 13.6c) two diodes in the d.c. link whose middle point is connected to the capacitors null point (Figure 13.6c) [13].

The LC filter still operates in the differential mode to reduce dV/dt but the diode-clamp occurs whenever an overvoltage due to ringing is present. When the capacitor null point $n_c$ voltage becomes larger than $\pm V_{dc}/2$ the corresponding diode (upper or lower) starts conducting; this way the d.c. link voltage is clamped. For a 2.2 kW drive, $L_f = 2.6$ mH, $C_f = 1.4$ μF would be suitable [13].

The simulated motor terminal voltage at 8 Hz fundamental frequency with $f_{sw} = 4.5$ kHz switching frequency (Figure 13.7c and d) for the LRC filter without and with diode-clamping shows a 15%–20% further reduction of motor terminal overvoltage with the diode-clamp. Experiments fully confirmed these results.

Medium and large IMs still experience failures in their windings due to voltage sources (high dV/dt and overvoltage) [14]. Attributing these failures to winding insulation weakness or to cable length and type (fully shielded or only bundled) or to PWM strategy in the converter is not always straightforward and a comprehensive study is required to find the actual cause.

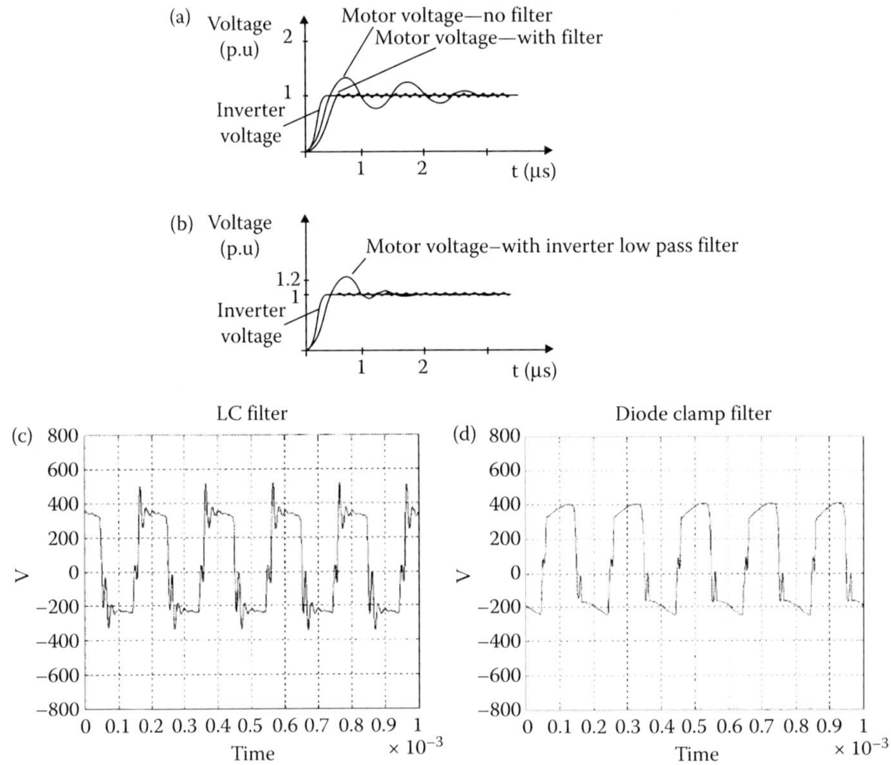


**FIGURE 13.7**
Motor terminal voltage: (a) with motor terminal RC filter, (b) with inverter LPF, (c) 2.2 kW induction motor with LPF $L_f = 2.6$ mH, $C_f = 1.4$ μF, (d) 2.2 kW induction motor with LPF $L_f = 2.6$ mH, $C_f = 1.4$ μF, and diode clamp.

Adding an LC filter with diode clamp looks like a fairly practical solution to the problem, especially when the existing motor is kept and only the inverter drive is added anew.

More on this subject is found in References 15 and 16.

## 13.5 Motor Model for Ultrahigh Frequency

For fast-rising voltage pulses produced by the PWM at the motor terminals, the motor behaves like a complex net of self, and mutual inductances and capacitors between winding turns $C_s$ and between them and earth $C_p$ (series parallel) (Figure 13.8).

Besides the stator model (Figure 13.7), the rotor windings (if any) have to be added. Owing to the presence of parallel capacitors $C_p$ and $C_{p1,2}$ the voltage pulse at the motor terminals (a, b, c), in the first nanoseconds is not distributed uniformly along the winding length. Thus, even if neutral n of the stator windings is isolated, the first coils of windings have to withstand more than 60%–70% of the voltage pulse. Consequently, they are heavily stressed and their insulation may wear out quickly. Increasing the voltage rise time at motor terminals reduces this electrostatic stress and thus even existing line-start motors may work with PWM converter drives for a long time.

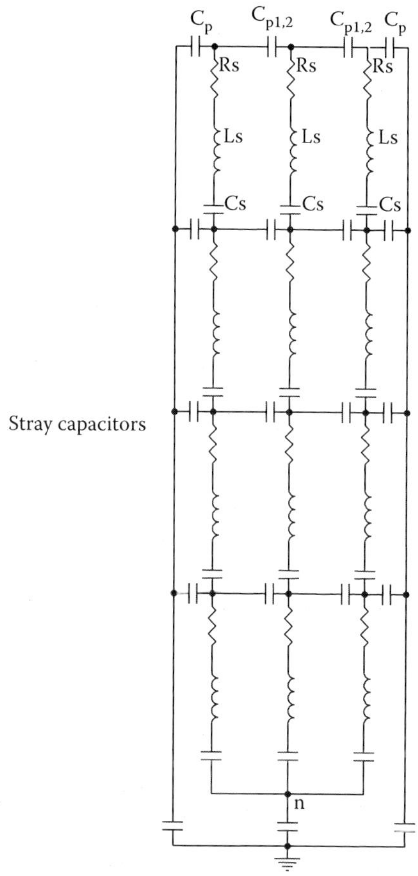


**FIGURE 13.8**
Motor RLC equivalent circuit for superhigh frequencies (stator only considered).

It is also possible to use special thin insulation for motor windings and/or use random windings to distribute the initial voltage pulse with high dV/dt randomly through the winding.

## 13.6 Common Mode Voltage: Motor Model and Consequences

The structure of a PWM inverter drive is shown in Figure 13.9.

Assume that the zero sequence impedance of the load (motor) is $Z_0$. The zero sequence voltage $V_0$ is

$$V_0 = \frac{V_{an} + V_{bn} + V_{cn}}{3} = \frac{V_a + V_b + V_c}{3} - V_n \tag{13.31}$$

and

$$i_0 = \frac{i_a + i_b + i_c}{3} \tag{13.32}$$

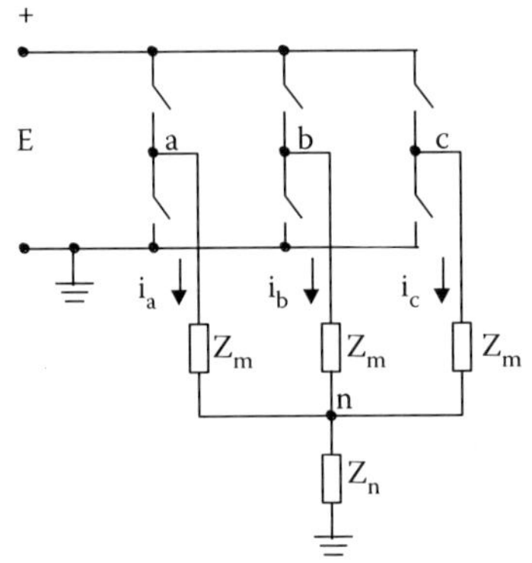


**FIGURE 13.9**
PWM inverter with motor and common mode impedance $Z_n$ (isolated null point).

The relationship between $V_0$ and $i_0$ is

$$V_0 = Z_0 i_0 \tag{13.33}$$

Consequently, common mode current $i_n$ and common mode voltage at neutral point $V_n$ are

$$i_n = i_a + i_b + i_c = \frac{3}{Z_0 + 3Z_n} \frac{V_a + V_b + V_c}{3} \tag{13.34}$$

$$V_n = \frac{3Z_n}{Z_0 + 3Z_n} \frac{V_a + V_b + V_c}{3} \tag{13.35}$$

As known, the common current mode is decoupled from the differential current mode (the balanced phase impedance $Z_m$ is not involved).

Consequently, the equivalent common mode input voltage $V_{0in}$ is

$$V_{0in} = \frac{V_a + V_b + V_c}{3} \tag{13.36}$$

and the machine impedance is as in Equation 13.35 and Figure 13.10

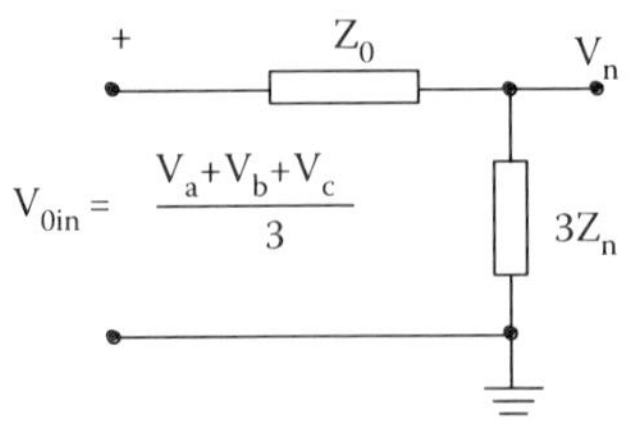


**FIGURE 13.10**
Motor model for common mode input voltage.

For each commutation sequence in the converter, the common mode input voltage $V_{0in}$ varies approximately with E/3 approximately (E—d.c. link voltage). To the quick common voltage pulses, the motor behaves like stray capacitors between stator windings and stator laminations in parallel with capacitors through the airgap to the shaft and in parallel with capacitors from the null point through bearings to the common grounding point (Figure 13.11) [17,18].

Let us note that the power cable, common mode choke (to be detailed later), and line reactors may be represented as common voltage series and parallel impedances $Z_s$, $Z_p$ between the inverter and the motor.

Also, the $3Z_n$ impedance may be replaced by capacitors and by the equivalent circuit of the voltage of the rotor shaft to ground $V_{rg}$ (Figure 13.11).

The bearing is represented by the bearing resistance ($R_b$), in series with a capacitor $C_b$, which is in parallel with a nonlinear impedance $Z_{fb}$ representing the intermittent shorting of capacitance $C_b$ through bearing film breaking by the voltage pulses.

$C_{sr}$: equivalent capacitor from stator to rotor

$C_{sf}$: equivalent capacitor between stator and frame

$C_{rf}$: equivalent capacitor between rotor laminations and frame

It now becomes evident that any voltage pulse from the inverter, when a switching in the inverter takes place, also produces a common mode voltage $V_{0in}$ of E/3. This common voltage pulse results in a high common mode leakage current $i_n$. If large, this common mode (leakage) current can influence the null protection system. Reducing $i_n$ means implicitly to reduce the shaft voltage $V_{rg}$ and the noncirculating bearing current $i_b$ [18]. Though $i_b \ll i_n$, it manages to deteriorate the bearings through the breakdown of lubricant film.

We will first deal with the stator common mode (leakage) current $i_n$ which, in some cases, may be as large as the rated current.

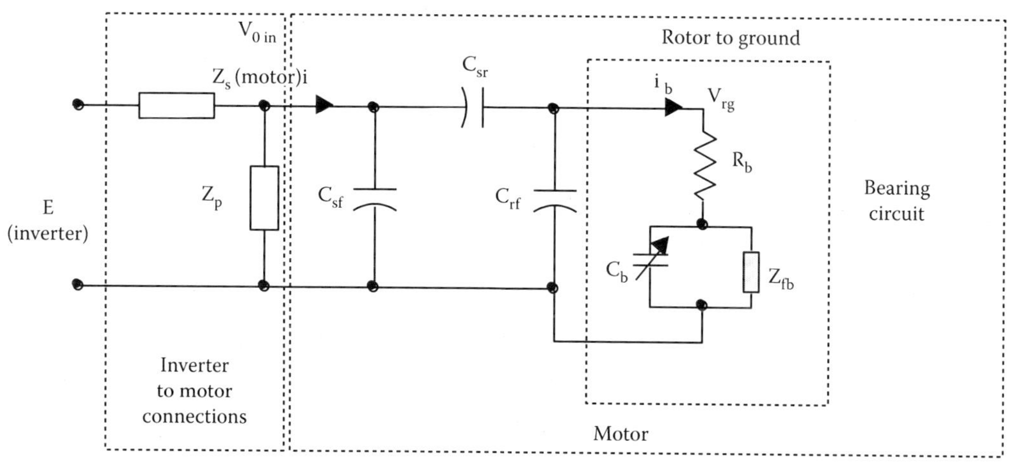


**FIGURE 13.11**
Common mode motor model for high frequencies.

## 13.7 Common Mode (Leakage) Stator Current Reduction

As we are, for the time being, interested only in the leakage current $i_l(t)$, we may lump the motor and cable through a series RLC circuit whose components may be determined experimentally (Figure 13.12) by applying a pulse from the converter and measuring the leakage current $i_l$, which looks like that in Figure 13.12b.

The leakage current response resembles that of an RLC circuit. L is related to the cable inductance, C is the stray capacitance between windings and frame, and R is related to the motor only.

After adding a common mode choke $L_c$ at motor terminals (Figure 13.13a), we obtain an equivalent circuit as in Figure 13.13b.

The peak leakage current is notably reduced (Figure 13.13b) by the presence of the common mode choke but the rms value is still not drastically reduced.

Placing a secondary coil (connected to a resistor $R_t$) on the common choke core [19] (Figure 13.14a) has proved to drastically reduce both peak and rms leakage current values (Figure 13.14b).

The common mode trafo with secondary resistor $R_t$ ($L_l$ is the leakage inductance, $L_{mag}$ is the magnetizing inductance) reduces the main flux in comparison with the common mode choke. Consequently, the ferrite core size may be reduced notably [19].

Drastically reducing the rms value of the leakage current leads to lowering the risk of incorrect operation of the residual current-operated circuit breakers. Implicitly, it reduces, to some extent, the shaft voltage $V_{rg}$ and the bearing currents. It appears, however, that bearing currents still have to be further reduced to avoid premature bearing damage through lubrication film electrostatic breakdown.

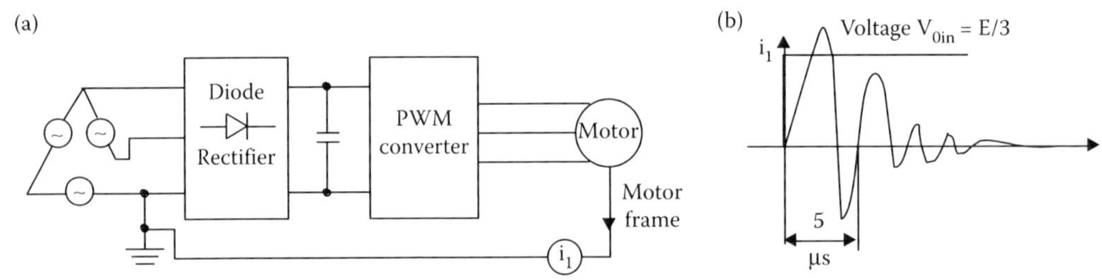


**FIGURE 13.12**
Measuring the leakage current: (a) test circuit, (b) current response to step voltage.

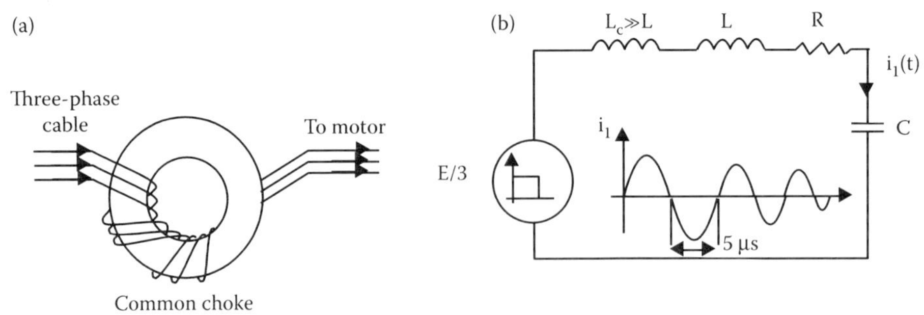


**FIGURE 13.13**
Common mode choke: (a) equivalent circuit, (b) leakage current $i_l(t)$.

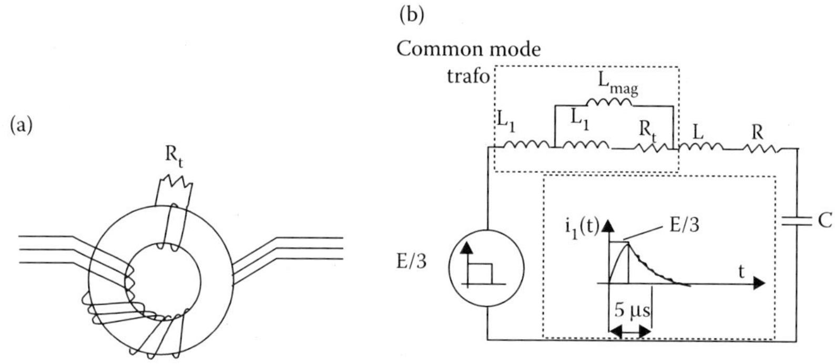


**FIGURE 13.14**
(a) Common mode trafo, (b) equivalent circuit leakage current response.

## 13.8 Circulating Bearing Currents

Noncirculating common mode bearing currents occur, as shown in Section 13.8, as a result of common mode voltage pulses (Figure 13.11).

Recently, an additional bearing current—the circulating component—due to the electrostatic current leaks to the laminated core, due to stator coil sides currents, along the machine stack length: $2\Delta i_a$, $2\Delta i_b$, $2\Delta i_c$, has been discovered (Figure 13.15a [18]).

These unbalanced longitudinal currents produce a circular net flux that is responsible for an additional, circulating, frame-bearing shaft current (Figure 13.15b).

## 13.9 Reducing the Bearing Currents

Examining the bearing failure in some PWM converter drives after only a few months of operation, bearing fluting (induced by electrical discharge machining [EDM]) has been noted. Bearing fluting manifests itself by the appearance of transverse grooves or pits in the bearing race. The shaft voltage $V_{rg}$ (Figure 13.11) is a strong indication for the presence of bearing currents.

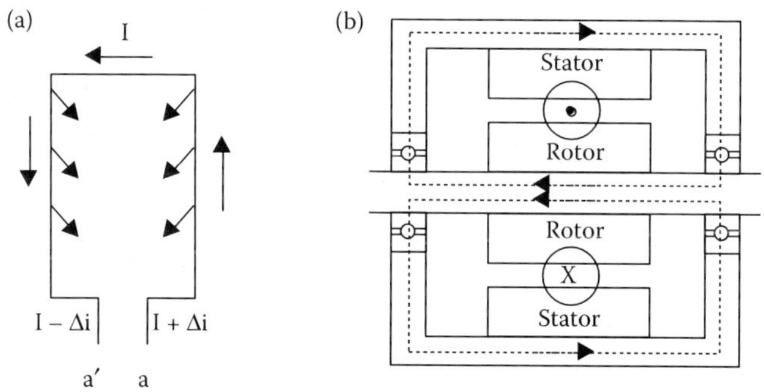


**FIGURE 13.15**
(a) Current unbalance, (b) net flux and bearing circulating current.

EDM may occur if the electrical field in the thin lubricant film is in excess of 15 V (peak)/μm which, for films between 0.2 to 2 μm, means 3–30$V_{pk}$ of shaft voltage, $V_{rg}$.

Among the measures proposed to reduce the shaft voltage $V_{rg}$ and, implicitly, bearing wear we mention here

- Outer-race insulation layer (Figure 13.16a).
- Dielectric-metallic Faraday airgap or complete foil (Figure 13.16b)—or paint along full stator length.
- Copper-plated slot stick covers (Figure 13.16c).
- One insulated bearing and shaft grounding brush constitutes a good motor grounding (to high frequency) from motor to mechanical load and from motor to inverter (drive) (Figure 13.16d) [14].

In fact, all methods introduce a small capacitor in series with the bearing circuit (Figure 13.11), and this is expected to ultimately reduce the bearing currents. The bearing insulation spacer seems to be a must. Only when stator and winding shielding is added—either through conductive foil or paint—is the shaft voltage reduced to less than 5% of its initial value (of the conventional machine) to values of 1–1.2 V(peak). Also, the stator temperature does not change notably due to these conductive shields (in which eddy currents are induced) and thus the methods proposed seem safe to apply [17]. Finally, when common mode chokes are used (see Section 13.8), the shaft voltage was initially around 30 V. After

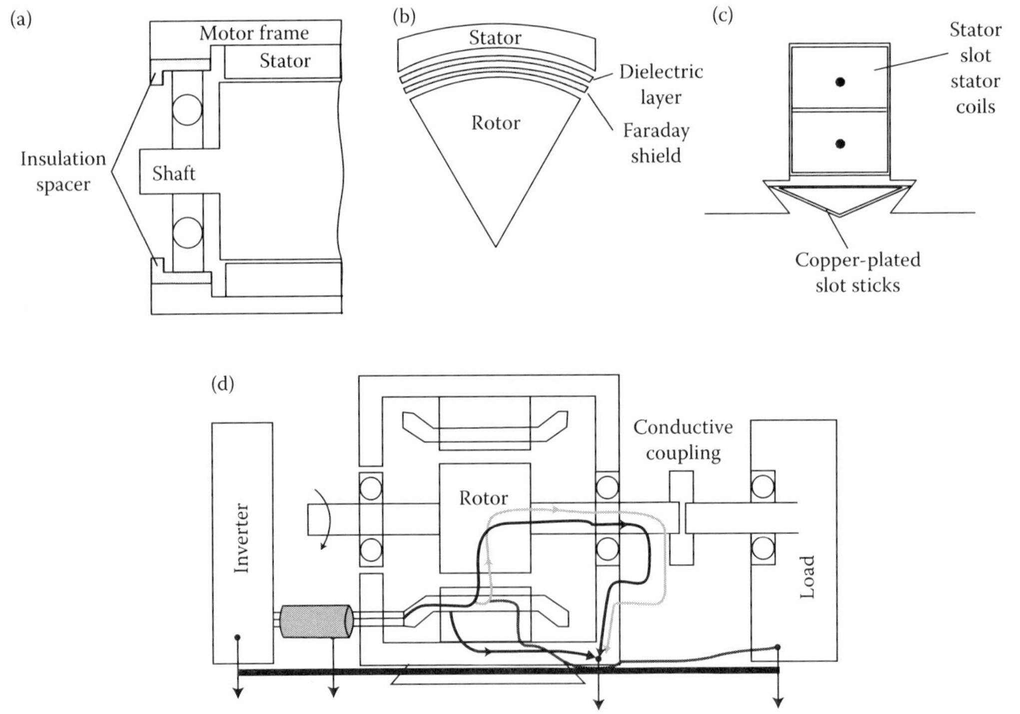


**FIGURE 13.16**
Methods to reduce bearing currents: (a) bearing insulation spacer, (b) conduction foil or paint, (c) conduction slot sticks, (d) one insulated bearing and shaft grounding brush.

using the full Faraday shield (Figure 13.16b) the shaft voltage was further reduced to less than 1.5 V [17].

It is yet to be seen, if the common mode trafo is capable of reducing enough the danger of bearing wear occurrence so as to eliminate or at least reduce the demand for Faraday shields.

## 13.10 Electromagnetic Interference

The step change in voltage caused by fast switching IGBTs produces high-frequency common mode (Section 13.7) and normal mode currents after each switching sequence in the PWM inverter, due to the parasitic stray capacitors of electric motors.

The oscillatory currents with frequencies from 100 kHz to several MHz radiate electromagnetic interference (EMI) fields (noises) in the environment, influencing various electronic devices such as AM radio receivers, medical equipment, etc.

A motor model, including stray capacitors and the corresponding common mode and normal mode current paths, is shown in Figure 13.17.

As expected, using a shielded three-core cable (with the shield used as the grounding wire) generally reduces the EMI noise to the limits prescribed through international standards (less than 40 dB μV/m).

Alternatively, making use of a common mode transformer (Section 13.7) and $R_F L_F$ normal mode filters (Figure 13.18) for the normal mode currents, produces acceptable EMI noise limits with a conventional power cable made through bundling the three feeding wires and the ground wire [20].

## 13.11 Audible Noise

The audible noise, in fact a part of the EMI spectrum (15 Hz to about 18 kHz), is a main performance index of an electric drive. Audible noise standards, measures to reduce the

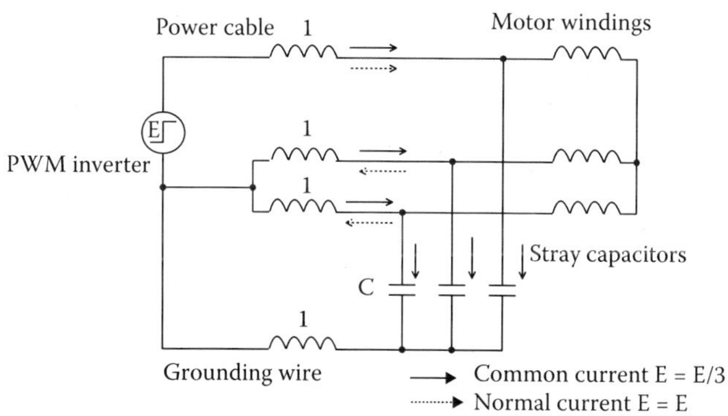


**FIGURE 13.17**
A.c. motor model with stray capacitors.

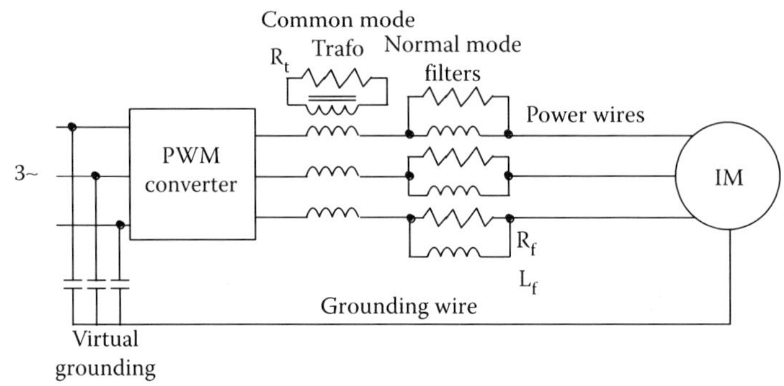


**FIGURE 13.18**
EMI reduction through common mode transformer and natural mode filters mounted at inverter output terminals.

audible noise in PWM electric drives, and methods to measure it (ISO 3740-46 standard) constitute a subject of wide interest and debate [21]. In general, the PWM supply in electric drives increases the level of noise related to grid connected electric motors.

Approaches to reduce PWM excited noise include: using an LC filter at motor terminals, changing the switching frequency, operation above 18 kHz, use of random PWM [22], or making the PWM drive play music with its noise.

Standards for noise emission of induction motors, for example, are ISO 1680, IEC-60034-9, NEMA M61-12.49.

## 13.12 Losses in PWM Converter Drives

Losses in PWM converter drives occur both in the motor and in the PEC. The PEC produces current and flux harmonics in the motor and thus additional winding losses and core losses are present in PWM converter drives [23]. The PWM converter losses are of conduction and commutation (switching) types [24].

Though there is a rich literature related to loss modeling in motors and PWM converters, experimental findings are still crucial in view of the complexity of the phenomena. As the switching frequency increases, the PEC losses increase while the motor losses decrease only slightly.

Total drive losses only slightly increase when switching frequency varies from 2 to 10 kHz in a 2-kW motor, for example. Higher-power units mean lower switching frequencies; also, it has been shown that if the motor efficiency is high so is PEC efficiency.

A maximum efficiency drive is obtained if the flux increases with torque and there are quite a few control methods to implement it. For vector or DTFC control, adequate flux/torque reference correlation solves the problem rather satisfactorily. However, the speed response in the first few milliseconds after a step torque reference increase occurs is rather slow, until the flux in the motor reaches the higher (according to torque reference) levels.

## 13.13 Summary

The introduction of a large scale of PWM converter drives with line-side diode rectifiers for powers now up to 500 kW per unit and more has produced considerable progress in energy usage and productivity increases through variable speed.

- However, there are some side effects related to the ultrafast rise time voltage pulses produced by the PWM converter, with switching frequency up to 20 kHz, and sent to the motor often through long power cables, as imposed by application.
- EMI is related to conductive (through power source) and electromagnetic voltage radiations at frequencies from 100 kHz to MHz and produces communication (radio, etc.) damage around or crosstalk in neighboring digital equipment. This aspect is treated extensively in power electronics books although it is by now a field in itself.
- Line current harmonics of 5th, 7th, 11th, and so on order occur. Limited by demanding international or national standards, they are defined by the THD factor of harmonic voltages produced in the power source. Passive line filters may solve the problem, but care must be exercised to check if filter voltage instabilities do not occur in presence of high-performance PWM converter drives.
- If power cables between PWM converter and motor are longer than a critical $l_c$, which corresponds to one-third of voltage pulse rise time $t_r$ ($t_r$ = 0.1–5 μs) travel at about half light speed, the voltage is doubled, by reflection, at the motor terminal. This fast high-voltage pulse may damage the motor insulation as it stresses initially only the first coils of the stator windings. So it has to be reduced—for cable length $l > l_c$—by inverter-side LPFs or by motor terminal filters.
- Increasing the voltage rise time might impede some self commissioning strategies for motor parameter identification based on step voltage response standstill tests, which are in use for some commercial drives.
- Some PWM strategies may lead to tripling the inverter voltage pulse at motor terminals, but special pulse elimination methods may reduce them to the classical double voltage, taken care of as explained above.
- Common mode (zero sequence) voltage pulses occur in the machine and they produce electrostatic leakage currents in the motor—to the frame structure. For such high frequencies a distributed capacitor model of the machine is required. The motor bearing appears also as an elaborate circuit in this case. The common mode stator leakage current may reach the level of the rated current and may impede the action of residual current circuit breakers and produce notable shaft voltage and bearing currents.
- Common mode chokes and common mode transformers are proposed to reduce the leakage current, shaft voltage, and bearing currents. Apparently they are not enough to secure a long life for bearings.
- Current bearing existence is related to rather high shaft voltages (in the order of 5–30 V peak values); reducing this voltage to 1–2 V would reduce the chances for bearing damage caused by electrostatic breakdown of the bearings' 2-μm-thick

film lubricant. This is done through Faraday complete shields (paints) placed along the stator bore and winding end connections, or similar methods.

- The future? It seems to be either to improve existing solutions for curing the side effects of PWM converters or to produce close to sinusoidal input–output converters where these problems are inherently solved at the price of a more sophisticated converter-motor control and higher costs (see References 25–30).

## 13.14 Problems

13.1. *The line filters*: For a PWM converter drive of drive kVA = 25 kVA, connected to a grid point with a short-circuit capacity SC kVA = 500 kVA, the measured line current harmonics are $i_5 = 25\%$, $i_7 = 5\%$, $i_{11} = 1.5\%$, line harmonics; $f_1 = 60$ Hz and line voltage $V_{LL} = 220$ V (rms).
Calculate

a. The line current harmonic factor HF
b. The filter capacitor C and inductance L tuned to the 5th harmonic with a detuning factor $\alpha = 0.95$
c. The voltage total harmonic distortion factor $THD_f$ after the filter is introduced

13.2 *The voltage reflection*: A PWM converter drive with a rated kVA of 10 kVA for maximum phase voltage $V_n = 120$ V (rms) at 60 Hz has the motor supplied through a power cable with a surge impedance $Z_0 = Z_n/10$ where $Z_n$ is the rated load impedance of the motor. The IGBTs in the PEC have a rising time $t_r = 0.5$ μs. The electromagnetic wave speed along the cable is $U^* = 160$ m/μs.
Calculate

a. The reflection coefficient $\Gamma_m$ at motor terminals and motor load impedance $Z_n$ for which the reflected voltage is only 20% of d.c. link voltage
b. The cable critical length $l_c$
c. For a 20% reflected wave, the required RC filter located at motor terminals

## References

1. S.M. Peeran, C.W.P. Cascadden, Application, design and specification of harmonic filters for variable frequency drives, *IEEE Trans.*, vol. IA-31, no. 4, 1995, pp. 841–847.
2. J.W. Gray, F.J. Haydock, Industrial power quality considerations when installing adjustable speed drive systems, *IEEE Trans.*, vol. 32, no. 3, 1996, pp. 646–652.
3. A.M. Walczyna, K. Hasse, R. Czarnecki, Input filter stability of drives fed from voltage inverters controlled by direct flux and torque control methods, *Proc. IEE*, vol. EPA-143, no. 5, 1996, pp. 396–401.
4. B. Mollit, J. Allan, Stability characteristics of a constant power chopper controller for traction drives, *Proc. IEE*, vol. B, no. 1(3), 1978, pp. 100–104.

5. D.M. Van de Sype, K. De Gusseme, A.P. Van den Bossche, J.A.A. Melkebeek, Sampling algorithm for digitally controlled boost PFC converters, *IEEE Trans.*, vol. PE-19, no. 3, 2004, pp. 649–657.
6. I. Zhang. F.C. Lee, M.M. Iovanovic, An improved CCM single phase PFC converter with a low frequency auxiliary switch, *IEEE Trans.*, vol. PE-18, no. 1, 2003, pp. 44–50.
7. S. Chattopadhyay, V. Ramanarayanan, Digital implementation of a line current shapping algorithm for three phase high power factor boost rectifier without voltage input sensing, *IEEE Trans.*, vol. PE-19, no. 3, 2004, pp. 709–721.
8. M. Kazmierkovski, R. Krishnan, F. Blaabjerg (editors), *Control in Power Electronics*, Academic Press, New York, 2003.
9. K. Stockman, M. Didden, F. D'Hulster, R. Belmans, Bag the sags, *IEEE-IA Mag.*, vol. 10, no. 5, 2004, pp. 59–65.
10. A. von Jouanne, D.A. Rendusora, P.M. Enjeti, J.W. Gray, Filtering techniques to minimize the effect of long motor leads on PWM inverter fed ac motor drive systems, *IEEE Trans.*, vol. IA-32, 1996, pp. 919–925.
11. A. von Jouanne, P. Enjeti, W. Gray, Application issues for PWM adjustable speed ac motor drives, *IEEE-IA Mag.*, vol. 2, no. 5, 1996, pp. 10–18.
12. R. Kerkman, D. Leggate, G. Skibinski, Interaction of drive modulation and cable parameters on ac motor transients, *Record of IEEE-IAS, Annual meeting*, vol. 1, 1996, pp. 143–152 (IEEExplore).
13. N. Hanigovszki, J. Poulsen, F. Blaabjerg, Novel output filter topology to reduce motor overvoltage, *IEEE Trans.*, vol. IA-40, no. 3, 2004, pp. 845–852.
14. M. Fenger, S.R. Campbell, J. Pedersen, Motor winding problems caused by inverter drives, *IEEE-IA Mag.*, vol. 9, no. 4, 2003, pp. 22–31.
15. L. Wei, Z. Liu; G.L. Skibinski, Investigation of voltage stresses inside adjustable speed drives, *IEEE Trans. Ind. Appl.*, vol. 49, no. 1, 2013, pp. 100–108.
16. V. Dzhankhotov, J. Pyrhonen, Passive LC Filters design considerations for motor applications, *IEEE Trans. Ind. Appl.*, vol. 60, no. 10, 2013, pp. 4253–4259.
17. D. Busse et al., An evaluation of electrostatic shielded induction motor: A solution for rotor shaft voltage build-up and bearing current, *IEEE Trans. Ind. Appl.*, vol. I, pp. 610–617.
18. S. Chen, T.A. Lipo, D.W. Novotny, Circulating type bearing current in inverter drives, *IEEE Trans. Ind. Appl.*, vol. 1, pp. 162–167.
19. S. Ogasawara, H. Akagi, Modeling and damping of high frequency leakage current in PWM inverter fed AC motor drive systems, *IEEE Trans.*, vol. IA-32, no. 5, 1996, pp. 1105–1114.
20. S. Ogasawara, H. Ayano, H. Akagi, Measurement and reduction of EMI radiated by a PWM inverter-fed AC motor drive system, *IEEE Trans.*, vol. IA-33, no. 4, 1997, pp. 1019–1026.
21. P. Enjeti, F. Blaabjerg, J.K. Pedersen, *Adjustable Speed AC Motor Drives, Application Workbook, EPE*, Trondheim, Norway, 1997.
22. F. Blaabjerg, J.K. Pedersen, L. Oesterguard, R.L. Kirlin, A.M. Trzynadlowski, S. Logowski, Optimized and nonoptimized random modulation techniques for VSI drives, *Proc. EPE-95*, vol. 1, 1995, pp. 19–26.
23. A. Boglietti, P. Ferraris, M. Lazzani, M. Pastorelli, Energetic behavior of induction motors fed by inverter supply, *Record of IEEE-IAS, Annual Meeting*, 1993, pp. 331–335.
24. F. Blaabjerg, U. Jaeger, S. Munk-Nielsen, J.K. Pedersen, Power losses in PWM-VSI inverter using NPT or PT IGBT devices, *IEEE Trans.*, vol. IE-10, no. 3, 1995, pp. 358–367.
25. K. K-F Yaen, H. Shu – H Chung, A low loss RL-plus-C filter for overvoltage suppression in inverter fed drive system with long motor cable, *IEEE Trans.*, vol. PE-30, no. 4, 2015, pp. 2167–2181.
26. I. Abrahamsen, J.K. Pedersen, F. Blaaberg, State of the art of optimal efficiency control of induction motor drives, *Proc. PEMC-96*, vol. 2, 1996, pp. 163–170.
27. F. Abrahamsen, F. Blaabjerg, J.K. Pedersen, On the energy optimized control of standard and high-efficiency induction motors in CT and HVAC applications, *Record of IEEE-IAS, Annual Meeting*, 1997, vol. 1, pp. 612–628.

28. S. Chattopadhyay, V. Ramanarayanan, V. Iayashankan, A predictive switching modulator for current mode control of high power factor boost rectifier, *IEEE Trans.*, vol. PE-18, no. 1, 2003, pp. 114–123.
29. D. Graovac, T. Hoffman, A. Haltmair, A transfer function approach to common mode filter optimization in PWM inverter supplied motor drives, *IEEE Trans.*, vol. EC-26, no. 1, 2011, pp. 93–101.
30. A. Muetze, V. Niskanen, I. Ahola, On radio frequency based detection of high frequency bearing current flow, *IEEE Trans.*, vol. IA-50, no. 4, 2014, pp. 2592–2601.

# 14

# *Large-Power Drives*

## 14.1 Power and Speed Limits: Moving Up

Large-power drives are defined as those drives beyond the reach of low-voltage (up to 660 V a.c. source) PWM voltage-source inverters (VSIs) with IGBTs (and high switching frequency). The IGBT-inverter power limit per unit has increased steadily in the last decade to about 2 MW (at the time of writing) by paralleling four 500 kW units. With respect to the voltage level, we have noted the introduction of 3–5 IGBT power cells (diode rectifier plus single-phase PWM inverter) in series per phase-properly insulated from each other and to ground to supply 1 MW (and more) a.c. motors at line voltages up to 4.5 kV (rms) [1].

So powers above 1–2 MW, where GTO or thyristor power electronic converters are a must, may be termed as large powers. High powers per unit have been achieved with thyristor rectifier current-source inverter (CSI) SMs: 100 MW for speeds up to 3000 rpm, 30 MW at 6000 rpm, 3 MW at 18,000 rpm. In low-speed applications, such as cement mills (up to 20 rpm, 5–6 Hz) and up to 11 MW power [2], thyristor cycloconverter-SM drives are predominant.

Recently, three-level GTO VSIs (at 15 MW, 60 Hz, and 6 kV line voltage [rms]) for SM drives have been introduced for steel main rolling mills [3]. It seems that this new breakthrough will drastically change the large-power drives technologies spectrum. For limited speed control (±20% around rated speed) doubly fed IM drives with a step down transformer-cycloconverter supplying the rotor circuit have been built for powers up to for pumped-storage 400 MW hydropower plants. In such applications, the motoring is used during off-peak energy consumption hours for pumped storage and generating is done during peak hours [4]. Variable speed is useful to reduce energy consumption during motoring and extract energy at maximum available hydroturbine efficiency during generating mode.

The rotor converter power in wound-rotor IM drives is proportional to speed control ratio (20%) and so are the costs. Resistive starting for motoring may still be required.

NOTE: By now, electric or diesel–electric multimotor propulsion systems for powers up to 6 MW are provided with two-level GTO (IGBT in the future) VSIs and IMs or with cycloconverters and induction (or SMs) or with thyristor-controlled rectifier current inverters. For ship electric propulsion, cycloconverters-SM drives at low speed prevail. Note that all these solutions are similar to those mentioned above for large-power drives.

A recent newcomer to the field is the diode rectifier VSI with IGCT for powers up to 5 MW at 4.16 kV. Finally, a multilevel active front bidirectional PWM converter SM drive at 50 MW/unit and at 60 KV (with cable made high-voltage stator winding) has been dedicated recently in a gas compressor submarine application. The six main large-power drive technologies as presented above are summarized in Table 14.1.

**TABLE 14.1**

Large-Power Drives

| Converter Type | Motor Type | Power Range | Speed Range | Power Quality | Power Flow |
|---|---|---|---|---|---|
| 3 level GTO rectifier VSI (a) with GTOs (b) with GCTs (c) with IGBTs | Synchronous (without rotor damper winding) | 10 MVA(15 MVA/1 minute) | 6 Hz | Unity PF, low line harmonics, high efficiency through regenerative GTO snubbers | Bidirectional |
| Thyristor cycloconverter | Synchronous (without rotor damper winding) | 10 MW | 6 Hz | Rather low PF | Bidirectional |
| Rectifier CSI (a) With thyristors (b) With GTOs (c) With IGCTs | Synchronous (with rotor damper winding) induction synchronous (with rotor damper winding) | 30 MW<br>100 MW<br>3 MW<br>1–4 MW | 120 Hz (6000 rpm)<br>60 Hz (3600 rpm)<br>300 Hz (18,000 rpm)<br>Up to 300 Hz | Low PF and high line current harmonics at low speed unity PF and low line current harmonics | Bidirectional |
| Step down transformer and cycloconverter in the rotor circuit | Doubly fed IM | Up to 100–200 MVA | Frequency: ±20 Hz variable, in the rotor, 50 (60) Hz constant in the stator | Rather low PF though stator active and reactive power control is possible | Bidirectional |
| Back-to-back PWM voltage converter | Doubly fed IM | Up to 10 MW or more | 60 Hz | Active and reactive power control | Bidirectional |
| Diode rectifier CSI with insulated gate commutated thyristors | Cage rotor IMs | 5 MW<br>4 KV | 3000 rpm | Close to unity source-side PF: it needs strong input current filter (transformer) | Unidirectional |

## 14.2 Voltage-Source Converter SM Drives

Both three-, five-, seven-level GTO (gate-commutated thyristor [GCT], IGBT) inverters and the cycloconverters are, in fact, voltage-source type PECs with load current control and they drive the d.c.-excited rotor SM at the unity PF, above 5% of rated speed. Being both a voltage-source type, the SM rotor may be cageless which is a notable simplification with sizeable motor costs reduction.

Also, the VC system of cycloconverter is similar to that of PWM-IGBT converter. There are major differences related to the fundamental output frequency limit, which is $f_1/3$ for the standard cycloconverter.

In contrast, the VSI is limited in frequency only by the switching frequency in the converter which, for large-power GTOs, is, today, around 300 Hz, while 500–600 Hz for IGBTs and GCTs [5,6].

Also, the cycloconverter is notably less expensive but the PF, especially at low motor frequencies (well below the rated one), is rather low.

A bidirectional three-level GTO inverter system is shown in Figure 14.1.

To reduce the losses in the GTOs a regenerative snubber circuit topology is used. Space vector voltage PWM techniques are applied to improve the motor current waveform, taking advantage of the three-level voltage available in the d.c. link voltage circuit.

Unity PF control for the motor side and on the source side may be performed through d.c. link voltage (filter capacitor voltage) control.

A high-power cycloconverter SM drive configuration is shown in Figure 14.2.

Each phase is fed from a double rectifier bridge—one for each current polarity—to produce operation in four quadrants (positive and negative output voltage and current) with phase angle control. The output voltage waveform is "carved" from a sequence of adequate (same polarity, highest value) sections of the three-phase voltage waveforms of the input source and constant frequency $f_1$ (Figure 14.3). This is why the output frequency is a fraction of the input frequency.

The motor current waveform is rather close to a sinusoid, while the voltage waveform (consequently the motor flux) has harmonics whose order is related to the ratio between the output frequency $f_2$ and input frequency $f_1$. Theoretically, $f_{2max} = f_1/2$. Unfortunately, the input currents are closer to rectangular waveforms (Figure 14.4) and are not even fully symmetric.

A lagging PF is obtained as the thyristors are line (input source) commutated.

To decouple the thyristor commutation process of three phases, the leakage inductance of the source-side transformer has to be concentrated mostly in its secondary.

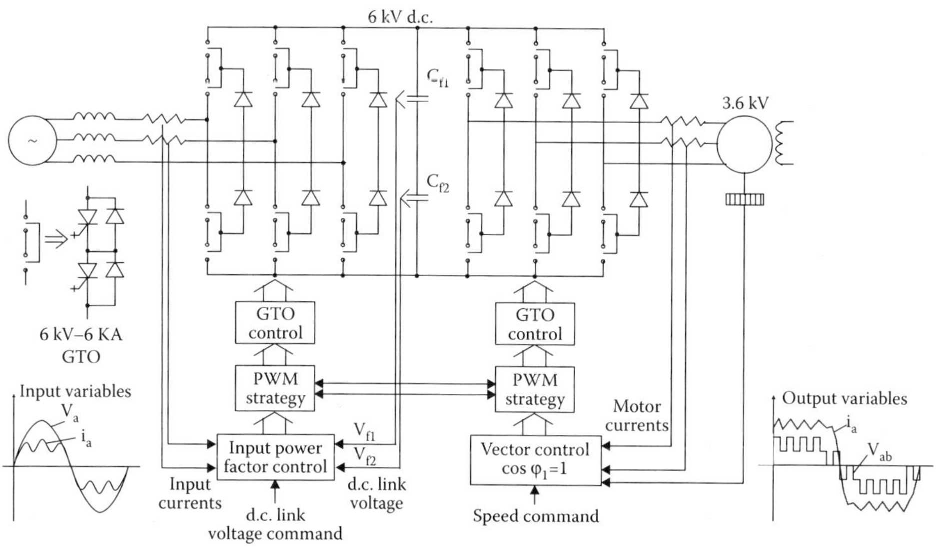


**FIGURE 14.1**
A 3-level GTO inverter-SM drive.

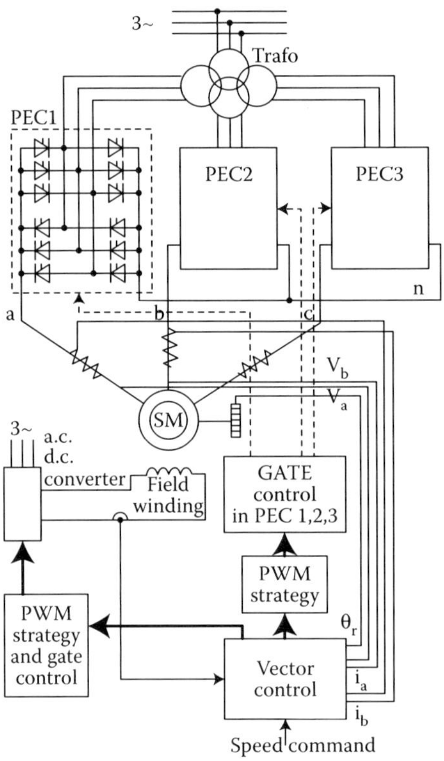


**FIGURE 14.2**
A cycloconverter SM drive.

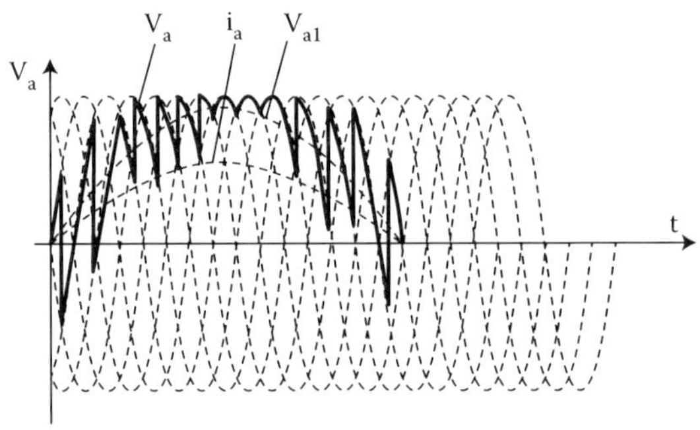


**FIGURE 14.3**
Output voltage $V_a$ and ideal current $i_a$ of the cycloconverter for unity PF operation of SM.

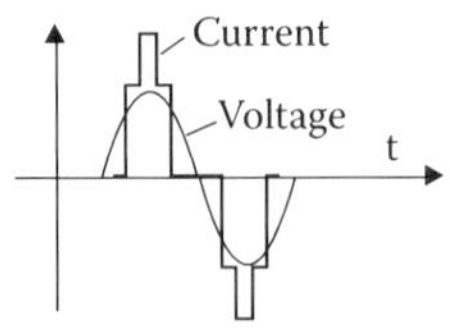


**FIGURE 14.4**
Input phase voltage and current of cycloconverters.

The line current harmonics and the input lagging PF (lower at lower motor speeds) are serious demerits of the otherwise rather simple and rugged cycloconverter configuration.

Now since the cycloconverter is also a voltage source, as the three-level GTO inverter system is, VC of both is similar, though, in general, the first one is applied for low-frequency drives ($f_2 < 6$ Hz) and the second one for up to 60 Hz (or more, as required).

## 14.3 High-Power SCRs

Large-power electric drives are characterized by medium voltage (3–6(11) kV line voltage RMS) and power in the MW to tens of MW range (Table 14.1).

For low fundamental frequency (less than 15 Hz) large electric motor drives, two-level VSIs with a few SCRs (GTOs, GCTs, IGBTs) in series may be used. But for fundamental frequency $f_{1n} \geq 50$ Hz multiple (3, 5, 7) level voltage-source converters are used to produce low-enough THD voltage and current output for the limited switching frequency of large-power GTOs, GCTs, and IGBTs (Figure 14.5a–c).

The GTO thyristor can be turned on by a positive current and turned off by a negative gate current (IG). They are currently fabricated up to 6 kV/4 kA per unit in asymmetric structure (good for voltage-source d.c.–a.c. [and a.c.–d.c.] PWM converters) and in symmetric structure (with reverse blocking capability), good for current-source PWM d.c.–a.c. (and a.c.–d.c.) converters. The GTO has a sign on-state current density and high blocking voltage but it is bulky, needs expensive turn-off snubbers due to low $dV_T/dt$, has high switching losses and requires a turn-on snubber to limit $di_T/dt$. These features explain why lately the gate-commutated thyristor (GCT or IGCT) is used to replace the GTO in large-power converters.

The GCT wafer is much thinner than the GTO wafer and thus on state power losses are reduced; it has a very low gate inductance (<5 nH) and thus lacks any snubber circuit. The

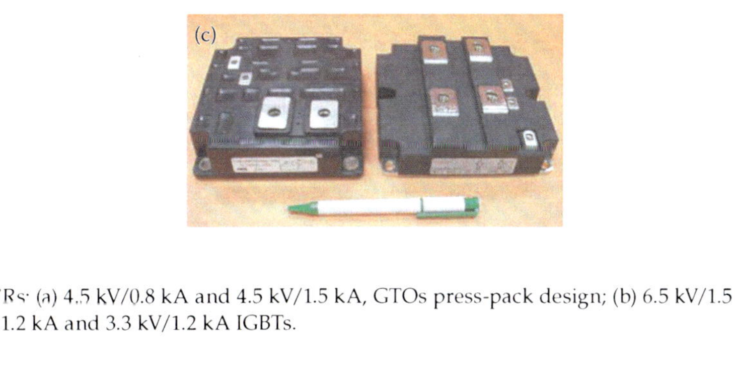


**FIGURE 14.5**
High power SCRs: (a) 4.5 kV/0.8 kA and 4.5 kV/1.5 kA, GTOs press-pack design; (b) 6.5 kV/1.5 kA symmetrical GCT; (c) 1.7 kV/1.2 kA and 3.3 kV/1.2 kA IGBTs.

rate of current change raises from 40 A/μs in the GTO to 3000 A/μs in the IGCT; the gate drive is provided with a 20–30 V d.c. power source. Again, they are made asymmetric reverse-conducting and symmetric. The asymmetric configuration is suitable for voltage source converter while the symmetric one is suitable for PWM current-source converters. The IGBT is a voltage-controlled device turned on by a +15 V gate voltage and turned off when the gate voltage is made zero (in fact a negative gate voltage is kept under turn-off mode, to increase immunity to noise).

When fully turned on and off the IGBT does not need any gate current, but it needs a few amperes gate current during switching transients due to the gate-emitter capacitance. Modular and press-back designs are available. Snubberless operation, high switching speed are implemented by its capability to control the collector current by the gate voltage, providing for reliable short-circuit protection and active control of dV/dt and of overvoltage at turn-off. Press-pack IGBTs allow direct series connection so needed in medium-voltage PWM converters. (For more details, see Reference 15.)

Representative multilevel VSI-fed a.c. drives (Figure 14.1) contain a multipulse diode (or SCR) rectifier and a multilevel VSI.

Their PWM techniques constitute generalizations of those used for 6 switch rectifier-6 switch (two level) inverter systems. But now the SVM methodology offers more voltage vectors which allow not only a low THD for input and output voltages and currents, but also a means to control the machine neutral voltage, for machine insulation protection.

There are three main multilevel voltage-source converters:

- Cascaded H-bridge type (Figure 14.6a)
- Diode-clamped type (Figure 14.6b)
- Flying-capacitor type (Figure 14.6c)

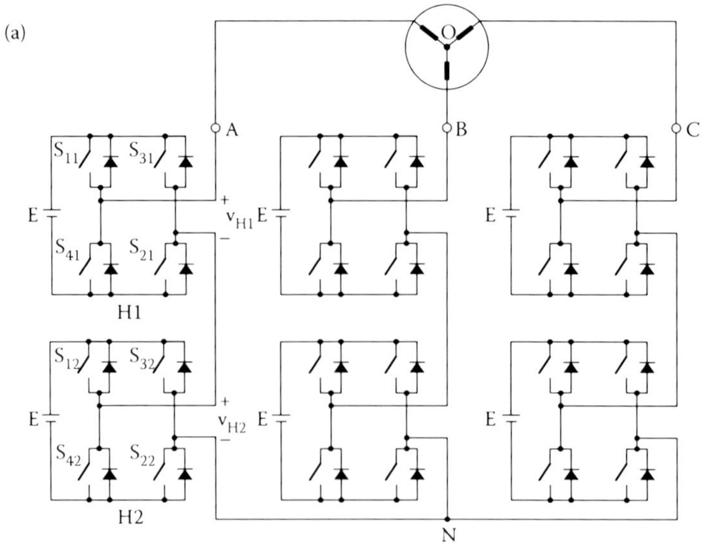


**FIGURE 14.6**
Typical multilevel VSIs. (a) Five-level cascaded H-bridge type. (*Continued*)

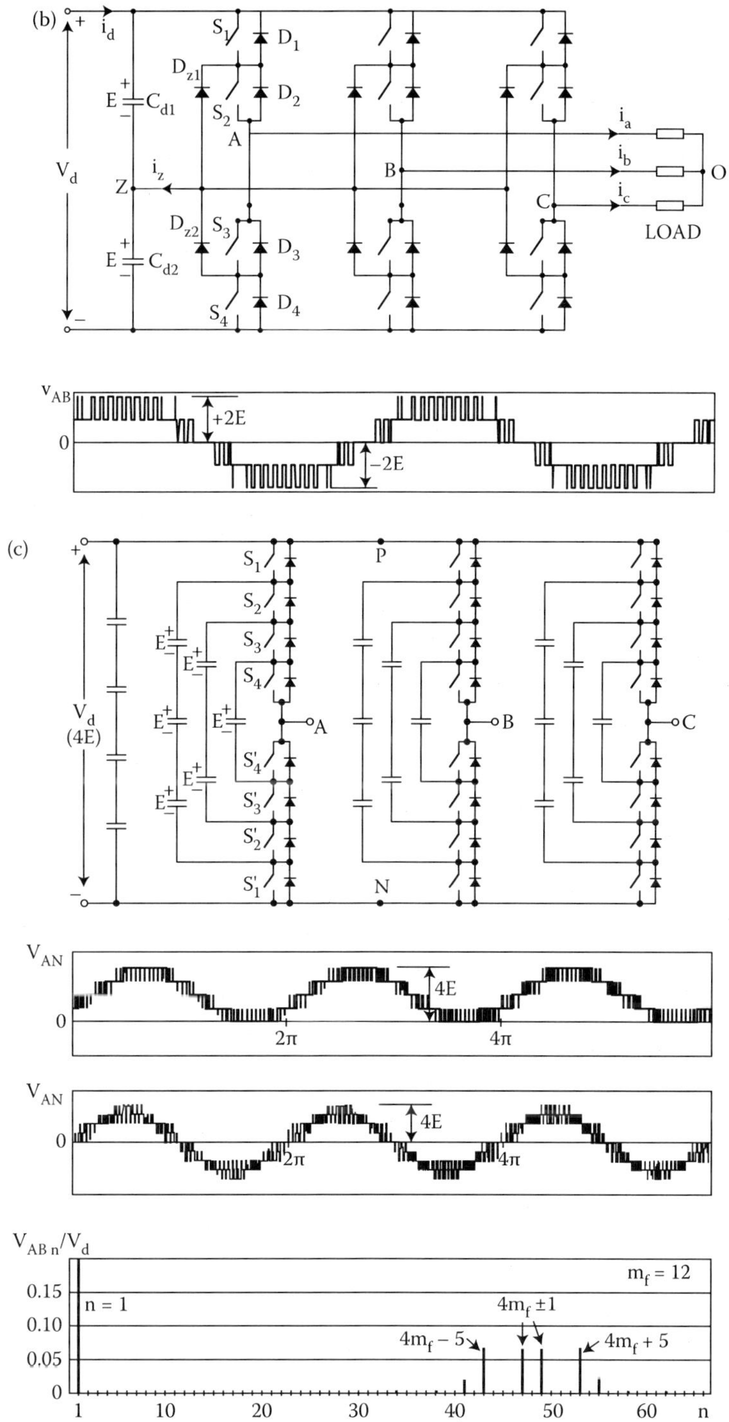


**FIGURE 14.6 (*Continued*)**
Typical multilevel VSIs. (b) Three-level NPC (neutral point clamped) type; (c) five-level flying capacitor type (the equivalent inverter switching frequency is 4 times the device switching frequency [say 720 Hz]).

The usage of H-bridge single-phase (4 switch) inverter as the building cell of multilevel voltage-source converters comes as natural and thus it has become very popular, but it needs quite a few isolated d.c. supplies, one for each H-bridge and needs a high component count while providing a modular structure (easy to repair), lower voltage THD and dV/dt and higher voltage operation without switching devices in series (which would require voltage balancing measures).

In contrast, the diode-clamped configuration employs clamping diodes and cascaded d.c. capacitors to produce a.c. voltages with multiple levels. Only the three-level (neutral point clamped) solution has been applied frequently in high-power medium-voltage drives [6]. The NPC (neutral point clamped) inverter keeps the reduced dV/dt and a.c. voltage THD without power switches in series (6000 V devices allow for a drive rated at 4160 V).

Each power switch in the NPC inverter experiences only 50% of total d.c. voltage during commutation while static voltage equalization is realized by proper design without additional components. Still, there are additional clamping diodes and possible deviation of neutral point voltage.

The main limitations of flying capacitor multilevel inverters refer to the large number of d.c. capacitors with the charge circuits and complex capacitor voltage balancing. This last drawback has limited drastically the usage of flying capacitor multilevel inverters in high-power drives.

For more on the PWM in multilevel converters, see Reference 15.

## 14.4 Vector Control in Voltage Source Converter D.C.-Excited SM Drives

In essence, VC of SMs is performed either in d–q rotor coordinates or in stator flux coordinates. The d–q model equations for a salient pole cageless rotor SM is (Chapter 10)

$$\bar{V}_s = r_s \bar{i}_s + \frac{d\bar{\lambda}_s}{dt} + j\omega_r \bar{\lambda}_s \tag{14.1}$$

$$\bar{\lambda}_s = \lambda_d + j\lambda_q;\ \ \bar{i}_s = i_d + ji_q;\ \ \bar{V}_s = V_d + jV_q \tag{14.2}$$

$$\lambda_d = L_{sl} i_d + L_{dm}(i_d + i_F);\ \ L_d = L_{dm} + L_{sl} \tag{14.3}$$

$$\lambda_q = L_q i_d;\ \ L_q = L_{qm} + L_{sl} \tag{14.4}$$

$$\frac{d\lambda_F}{dt} = V_f - r_f i_F \tag{14.5}$$

$$T_e = \frac{3}{2} p\left(\lambda_d i_q - \lambda_q i_d\right) \tag{14.6}$$

$$\frac{J}{p}\frac{d\omega_r}{dt} = T_e - T_{load} \tag{14.7}$$

For steady-state and unity PF the space vector diagram (from Equations 14.1 through 14.4) is as shown in Figure 14.7.

During the VC process the stator current space vector $\bar{i}_s$ differs from the reference space vector $\bar{i}_s^*$. Consequently, the actual current $\bar{i}_s$ may be written in flux (M, T) coordinates (denoted as (x, y) here)

$$\bar{i}_s = i_x + ji_y \tag{14.8}$$

Evidently, for unity PF (zero angle, $\varphi_1$, between $\bar{V}_s^*$ and $\bar{i}_s^*$) $i_M$ should be zero. In reality, during transients at least, as the field current $i_F$ correction is rather slow (to maintain constant stator flux), the unity PF condition is abandoned for a short time and a nonzero value of $i_M$ is referenced.

How to calculate, approximately, a reference flux current $i_M^*$ is not simple unless the motor has a nonsalient rotor: $L_d = L_q$. Let us calculate the torque from Equations 14.3, 14.4, and 14.6

$$T_e = \frac{3}{2}p\left[L_{dm}i_F + \left(L_d - L_q\right)\cdot i_d\right]\cdot i_q \tag{14.9}$$

As $L_d \geq L_q$ and $i_d < 0$ (demagnetizing effect, for unity PF) the reluctance torque (small as it may be) will be negative. *So it would be adequate to use a cageless nonsalient pole rotor SM.* This is possible in 2(4)-pole rotors used in high-speed applications, but it seems still possible for a 5–6 Hz, 4(6) multipole rotor, 11 MW SM where the rotor pole pitch is 0.66 m.

The stator flux may be written as

$$\lambda_s^* = L_{dm}i_M^* \tag{14.10}$$

where

$$i_M^* \approx i_F\cos\delta + i_x^*\cdot\frac{L_d}{L_{dm}} \tag{14.11}$$

is strictly valid for $L_d = L_q$.

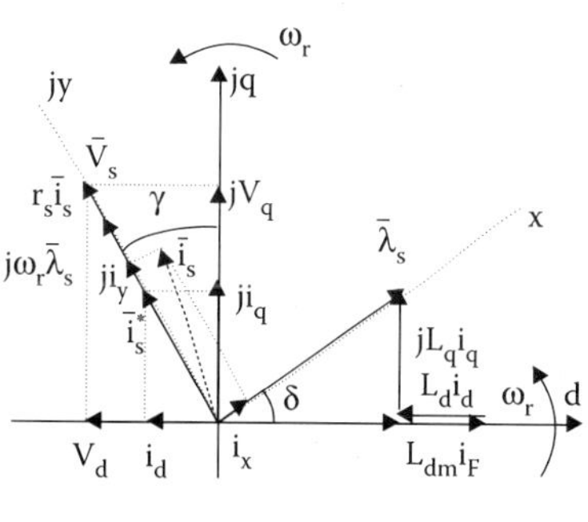


**FIGURE 14.7**
Space vector diagram of SM at steady-state and unity PF ($\varphi_1 = 0$).

For $L_d = L_q$ the torque from Equation 14.6 is

$$T_e = \frac{3}{2} p\lambda_s i_T \tag{14.12}$$

with

$$i_T^* = \frac{T_e^*}{3/2p\lambda_s^*} = i_y^* \tag{14.13}$$

From Equation 14.11 we may calculate

$$i_x^* = (i_M^* - i_F \cos\delta)\frac{L_{dm}}{L_d} \tag{14.14}$$

The stator flux has to be controlled directly through the field current control. But first the stator flux has to be estimated.

We again resort to the combined voltage (in stator coordinates) and current model (in rotor coordinates)

$$\hat{\bar{\lambda}}_{sv^s} = \int \left(\bar{V}_s - r_s \bar{i}_s\right) \cdot dt + \bar{\lambda}_{s0} \tag{14.15}$$

$$\hat{\bar{\lambda}}_{si^r} = L_{sl} i_d + L_{dm}(i_d + i_F) + jL_q i_q \tag{14.16}$$

$$\hat{\bar{\lambda}}_s = \hat{\bar{\lambda}}_{sv} . \frac{T}{1+sT_c} + \hat{\bar{\lambda}}_{si^r} e^{j\theta_{er}} \frac{1}{1+sT_c} \tag{14.17}$$

For low frequencies, the current model prevails while at high frequency, the voltage model takes over. If only short periods of low-speed operation are required, the current model might be avoided, by using instead the reference flux $\lambda_s^*$ in stator coordinates

$$\hat{\bar{\lambda}}_s = \hat{\bar{\lambda}}_{sv} \cdot \frac{T_c}{1+sT_c} + \lambda_s^* e^{j\theta_{er}} \frac{1}{1+sT_c} \tag{14.18}$$

$$\sin(\theta_{er} + \delta) = \frac{\operatorname{Im ag} \hat{\bar{\lambda}}_s}{\left|\hat{\bar{\lambda}}_s\right|}; \quad \cos(\theta_{er} + \delta) = \frac{\operatorname{Re} \hat{\bar{\lambda}}_s}{\left|\hat{\bar{\lambda}}_s\right|} \tag{14.19}$$

with $\theta_{er}$ measured, $\delta$ may be found (Figure 14.7).

Finally, the vector current control in rotor flux coordinates is shown in Figure 14.8.

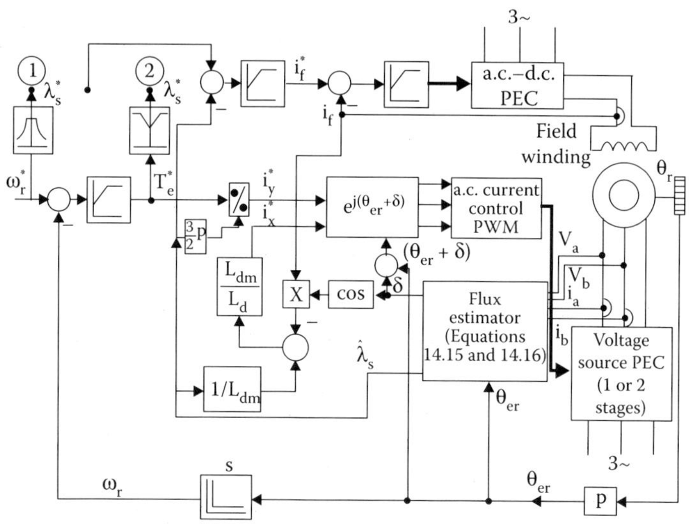


**FIGURE 14.8**
Vector current control of SM fed from voltage source type (one or two stages) PECs and unity PF.

The kind of direct vector current control system shown in Figure 14.8 "moves" most of parameter dependence problems into the flux estimator but relies heavily on the rotor position feedback (a resolver is used generally).

Voltage decoupling is not considered but it could be handled as for PMSM drives when d.c. current ($i_x$, $i_y$) controllers are used and the motion-induced voltage is added only along axis y

$$V_s^* = 0 \tag{14.20}$$

$$V_y^* = \omega_r \lambda_s \tag{14.21}$$

In this case, open-loop (voltage) PWM is performed (Figure 14.9).

The voltage decoupler with d.c. current controllers is known to give better performance around and above rated frequency (speed) and to be less sensitive to motor parameter

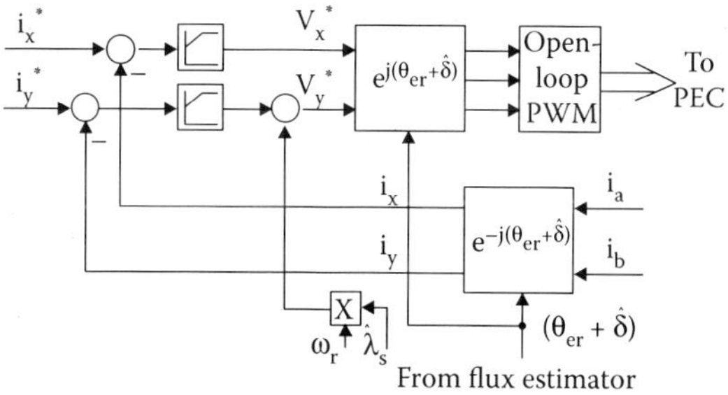


**FIGURE 14.9**
Voltage decoupler in stator flux coordinates and d.c. current controllers.

detuning. Flux control as done in Figure 14.8 proved to provide fast response in MW power motors due to the delay compensation by nonzero transient flux current reference $i_x^*$.

Also, fast speed control and reversal has been proved practical with vector current control.

In drives where the load is dependent only on speed the reference flux switch is on position 1 (Figure 14.8), while it is on position 2 where the torque may vary at any speed. The flux level has to increase with torque to preserve the unity PF with variable load.

Response frequency bandwidth of more than 600 rad/s in torque and speed control has been reported for an 11 MW drive with a three-level GTO inverter system [3]. Similar results have been obtained with a 2500 kW cycloconverter SM drive [7].

Besides vector current control with voltage decoupler, DTFC for unity PF may be performed to obtain a simpler and more robust control, while preserving quick response.

## 14.5 DTFC of D.C.-Excited SM Drives

Vector current control, as discussed in the previous paragraph, performs direct flux control but it does not do direct torque control. Also it uses current vector rotators.

The stator flux estimator equations (14.15 through 14.19) could easily be augmented to also perform torque estimation (Figure 14.10).

For the unity PF, the angle between stator flux and current is 90°. Also, a reactive torque RT is defined as

$$RT = Q_1/\omega_1 \tag{14.22}$$

where $Q_1$ is the reactive power and $\omega_1$ the primary frequency.

The "reactive torque" RT may be estimated from Equation 14.15 and works well for steady state and above 5% of rated speed. To make sure that the PF is unity, the RT should be driven to zero through adding or subtracting from reference field current (Figure 14.11). Finally, we should remember that the torque and flux error and stator flux position in one of the 60° wide sectors solely lead to a voltage vector (or a combination of) in the PECs, to form the table of switching (TOS) (and timings) as in any DTFC system. The TOS remains the same as for IMs or PMSMs. Consequently, the DTFC system is as shown in Figure 14.11.

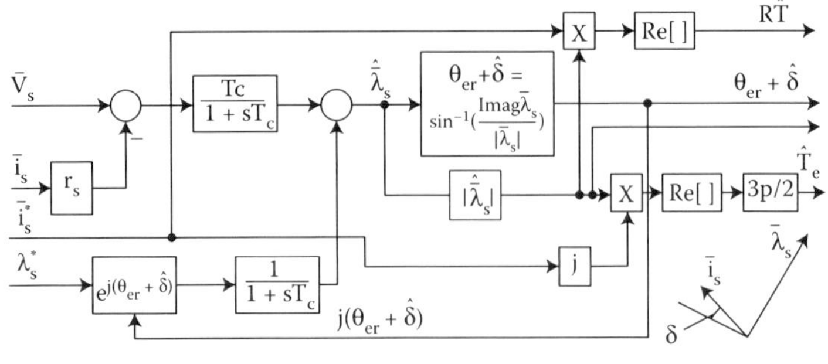


**FIGURE 14.10**
Flux and torque and "reactive torque" observer.

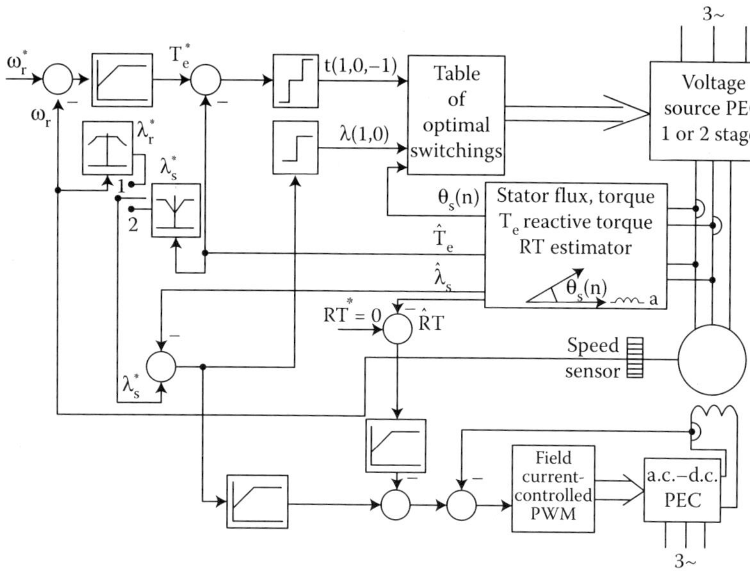


**FIGURE 14.11**
Generic DTFC of voltage source PWM converter SM drives.

A prompt start up process requires the flux to be present in the machine. To do so, first we initiate the field current control while the stator is fed through a zero voltage vector until the stator flux $\lambda_s$ surpasses the reference flux for the first time. After that, the motor may start.

## 14.6 Sensorless Control of a D.C.-Excited SM via "Active Flux:" A Case Study

In wide constant power speed range drives, the need for overall initial costs, dominated by PWM converter costs, tends to favor the d.c.-excited SM moderate [8]. The main reason is the possibility to magnetize the machine from the rotor in d.c., via a d.c.–d.c. converter with 5% peak power rating and thus, save up to a net 25% of PWM inverter peak kVA ($i_d = 0$) and costs. The lower the d.c. link voltage, the more substantial is the cost reduction, since the PWM inverter/ machine costs is reduced.

The possibilities to operate at unity PF, or maximum efficiency through adequate field current control, are also assets of this solution. Finally, the field current reduction in a low magnetic saliency rotor SM produces a large flux-weakening (constant power) speed range. The price to pay is the presence and costs of mechanical brushes, which, however, have to transfer to the rotor only up to 5% of the machine rated power.

VC of a d.c.-excited SM drives has been applied after 1970 in cement rolling mills and icebreakers with direct drives up to 10 MW and 5.5 Hz. Interest in the VC of d.c.-excited SM drives was revived recently for automotive applications at low-medium power level, together with DTFC. The main reason is, again, large constant power speed range for minimum inverter peak kVA.

In all these cases, to the best of our knowledge, a rotor position sensor was used for position feedback. While the cost of such a position sensor in a medium or large-power drive is not prohibitive, its reliability is, and the collateral money loss due to production stoppage for position sensor repair or replacement is the problem. This is the main reason why in this investigation a wide speed range motion-sensorless control system for d.c.-excited SM drives via VC is introduced and documented.

Because the rotor magnetic saliency of a salient pole d.c.-excited SM machine (such as shown in Figure 14.12) is small or even absent, and strongly dependent of magnetic saturation level, all signal injection or fundamental model magnetic saliency-based rotor position and speed observers, so similar with IMs and PMSMs have been disregarded. Instead, use is made of the so called "active flux" concept, which models a salient pole rotor a.c. machine by a nonsalient pole equivalent model. For the d.c.-excited SM, the active flux vector is aligned to the rotor position and thus, the rotor position and speed observation becomes a notably simpler task.

### 14.6.1 Active Flux Model of SM

The "active flux" concept (Chapter 11, Reference 25), defined as the "torque producing flux" $\Psi_d^a$ in the torque ($T_e$) formula of the d-q model in rotor coordinates (without superscript), is briefly particularized here for d.c.-excited SM:

$$T_e = 1.5p \cdot \psi_d^a \cdot i_q \tag{14.23}$$

$$\psi_d^a = L_{md} \cdot i_F + (L_d - L_q) \cdot i_d, \quad L_d > L_q \tag{14.24}$$

where p is the number of pole pair, $i_d$, $i_q$ are the d-q stator currents, and $i_F$ is the field current reduced to stator.

$L_d$, $L_q$ are the d-q inductances, $L_{md}$ is the d axis magnetization inductance $L_{md} = L_d - L_{sl}$, $L_{sl}$ is the stator leakage inductance.

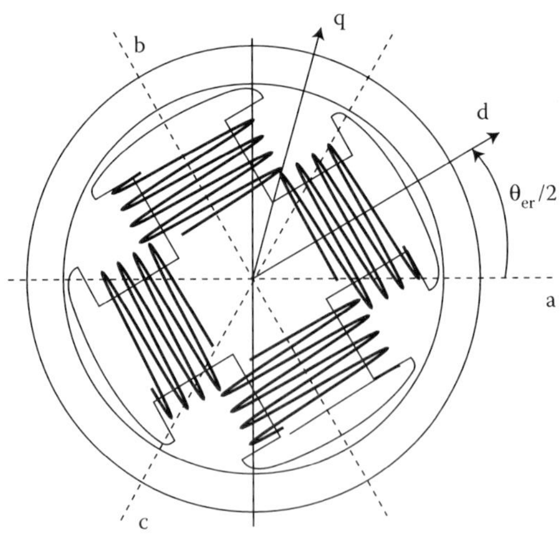


**FIGURE 14.12**
Four pole d.c.-excited synchronous machine.

The active flux vector $\bar{\psi}_d^a$ is aligned to the d axis, that is, its angle is equal with the electrical rotor position angle $\theta_{\psi_d^a} = \theta_{er}$.

The d-q (space-phasor) model equations of SM, in rotor coordinates, are

$$\bar{V}_s = R_s i_s + d\psi_s/dt + j\omega_r \psi_s \tag{14.25}$$

$$\bar{\psi}_s = L_{md} i_F + L_d i_d + jL_q i_q \tag{14.26}$$

where $\bar{V}_s$, $\bar{i}_s$, and $\bar{\psi}_s$ are the stator voltage, current, and flux vector, respectively, $\omega_r$ is the electrical rotor speed and $R_s$ is the stator resistance.

From Equations 14.24 and 14.26:

$$\bar{\psi}_d^a = \bar{\psi}_s - L_q \bar{i}_s \tag{14.27}$$

To clarify the concept quickly, the steady-state vector diagram of SM at cos $\varphi = 1$ is shown in Figure 14.13.

Now, with Eqauation 14.27 in 14.25 and d/dt → s (Laplace operator)

$$\bar{V}_s = R_s \bar{i}_s + (s + j\omega_r) L_q \bar{i}_s + (s + j\omega_r) \bar{\psi}_d^a \tag{14.28}$$

Equation 14.28 refers to a completely nonsalient rotor pole machine model with its inductance $L_q$, and with a new flux vector $\bar{\psi}_d^a$ aligned to the rotor position. This is the essence of the "active flux" model.

## 14.6.2 State Observer

### *14.6.2.1 Active Flux Observer*

Equations 14.25 and 14.27 lead straight to the "active flux" vector observer if stator coordinates (with superscript s) are used

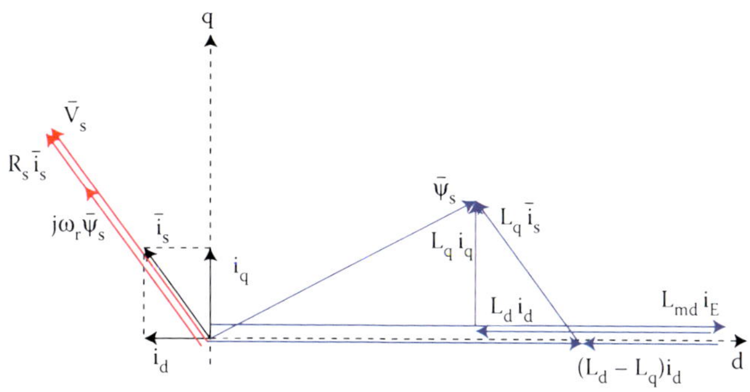


**FIGURE 14.13**
Steady state vector diagram of SM d-q model at cos $\varphi = 1$.

$$\bar{\psi}_s^s = \int (\bar{V}_s^s - R_s \bar{i}_s^s + \bar{V}_{comp})dt \qquad (14.29)$$

$$\bar{\psi}_d^{as} = \bar{\psi}_s^s - L_q(i_m)\bar{i}_s^s, \quad \bar{i}_m = i_d + i_F + ji_q \qquad (14.30)$$

where $\bar{V}_{comp}$ is a total compensator vector at the integrator input (14.29), which takes care of integral d.c. offset, error in initial rotor position identification, stator resistance variation with temperature, inverter nonlinearities, to make the flux observer applicable to wide speed range. $L_q$ depends on the total magnetization current $i_m$.

As $\bar{\psi}_d^{as}$ is aligned to rotor position $\theta_{er}$:

$$\bar{\psi}_d^{as} = \psi_d^a \cos\theta_{er} + j\psi_d^a \sin\theta_{er} = \psi_{d\alpha}^a + j\psi_{d\beta}^a \qquad (14.31)$$

### *14.6.2.2 Rotor Position and Speed Estimator*

Once the "active flux" observer 14.29 through 14.31 is applied, its components in stator coordinates αβ ($\psi_{d\alpha}^a$, $\psi_{d\beta}^a$) become available. The rotor position estimation, required for vector rotator operators in the VC system and in the current–flux model, is computed as

$$\theta_{er} = a\tan(\hat{\psi}_{d\beta}^a / \hat{\psi}_{d\alpha}^a) \qquad (14.32)$$

The rotor speed digital estimator is straightforward

$$\hat{\omega}_r = \frac{d\hat{\theta}_{er}}{dt} = \frac{\psi_{d\alpha}^a(k-1)\psi_{d\beta}^a(k) - \psi_{d\beta}^a(k-1)\psi_{d\alpha}^a(k)}{h(\psi_d^a)^2} \qquad (14.33)$$

where h is the sampling time. A LPF is needed, which involves a small delay in response as price paid for smoother speed response. The estimated speed $\hat{\omega}_r$ is needed for the speed feedback loop and for d-q axes voltage decoupling (e.m.f. compensation).

### *14.6.2.3 Complete State Observer*

Based on Sections 14.6.2.1 and 14.6.2.2, a complete state observer for the proposed motion-sensorless control system is shown in Figure 14.14. In this scheme, the stator flux observer $\Psi_s^s$ is based on a combination of the open-loop current–flux model in rotor reference (14.26) and the closed-loop voltage flux model in stator reference (14.29) with a PI compensator on the error between the current and voltage flux models ($\bar{\psi}_{si}^s$, $\bar{\psi}_s^s$). The PI compensator frequency behavior provides for the stator flux observer a smooth speed transition, between the current model, dominant at low speeds, sensitive to $L_q$ and $L_d$, and the voltage model, dominant at medium-high speeds, and sensitive to $R_s$ and stator voltage estimation $\bar{V}_s^s$, especially at low speeds.

Compensating for the stator resistance is not so important in this case, since the current model dominates the stator flux observer at low speeds. The crosscoupling saturation is included in the current model. The PI compensator eliminates the integrator offset and initial condition errors in the voltage model.

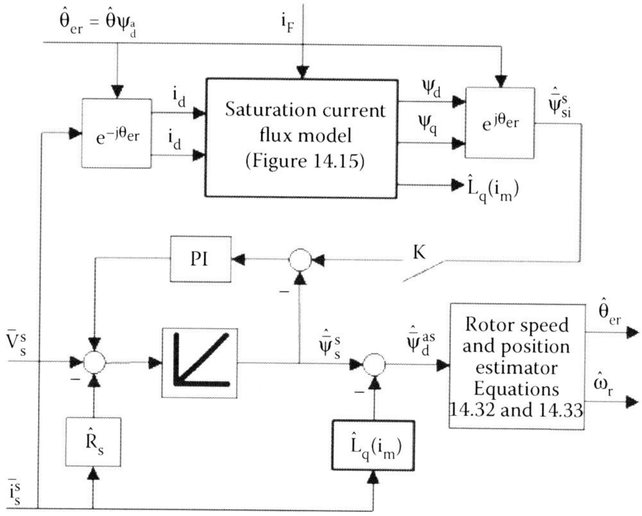


**FIGURE 14.14**
Complete proposed state observer for sensorless control of SM.

### 14.6.3 Magnetic Saturation Model and Its Identification

The magnetic saturation varies notably in the SM, especially for wide constant power speed range drives and thus its consideration in the state observer and in the control system is mandatory. There is a crosscoupling magnetic saturation effect between the d and q axes, which is a very involved phenomenon, with quite a few approximations proposed already to account for it. The solution with unique (but different) magnetization saturation curves along the d and q axes $\psi_d^*(i_m)$, $\psi_q^*(i_m)$, dependent solely on the resultant magnetization current $i_m$ (14.30) is used. By definition

$$\psi_d^*(i_m) = \psi_d \frac{i_m}{i_d + i_F}; \quad \psi_q^*(i_m) = \psi_q \frac{i_m}{i_q} \tag{14.34}$$

$$L_d = \frac{\psi_d^*}{i_m}; \quad L_q = \frac{\psi_q^*}{i_m}; \quad i_m = \sqrt{(i_d + i_F)^2 + i_q^2} \tag{14.35}$$

The identification procedure of $\psi_d^*(i_m)$ and $\psi_q^*(i_m)$ magnetization curves during self-commissioning is described in the following:

1. The machine operates at (constant) high speed in torque close-loop by slowly rising $i_s(i_d, i_q)$ and $i_F$ reference values.

   The measured variables are $\bar{i}_s^s(i_a, i_b)$, $i_F$, and $\theta_{er}$.
2. The voltage flux model with PI compensator from the state observer (Figure 14.14, K—open) provides for the stator flux vector $\bar{\psi}_s^s$ estimation. Then, the stator flux components $(\psi_d, \psi_q)$ and the stator current components $(i_d, i_q)$ in rotor coordinates are calculated

$$\psi_d + j\psi_q = \bar{\psi}_s^s e^{-j\theta_{er}}, \quad i_d + ji_q = \bar{i}_s^s(i_a, i_b)e^{-j\theta_{er}} \tag{14.36}$$

3. The resulting crosscoupling saturation curves used in current–flux model are computed according to Equations 14.34 and 14.35, as illustrated in Figure 14.15. Following this procedure, the experimental identification results for the crosscoupling saturation curves $\psi_d^*(i_m)$, $\psi_q^*(i_m)$ as function of the resultant magnetization current $i_m$ are illustrated in Figure 14.16.

NOTE: It has been proved by FEM on large SMs that the unique d, q magnetic saturation curves model is reliable for a wide range of saturation levels and combinations ($i_d$, $i_q$, $i_F$), except for the case of under-excited SM when a bifurcation behavior may occur.

Let us also notice the difference between $\psi_d^*(i_m)$, $\psi_q^*(i_m)$ curves and $\psi_d$, $\psi_q$ and also the definition of $L_d$, $L_q$ in Equations 14.34 and 14.35 which, if not used as they are, may lead to wrong results and conclusions.

Now it may be assumed that, if the crosscoupling saturation current–flux model is to be called upon at all speeds to yield $L_q(i_m)$ in Equation 14.30, the rotor position estimation $\hat{\theta}_{er}$ is used at all speeds for speed estimation and for Park operators in VC.

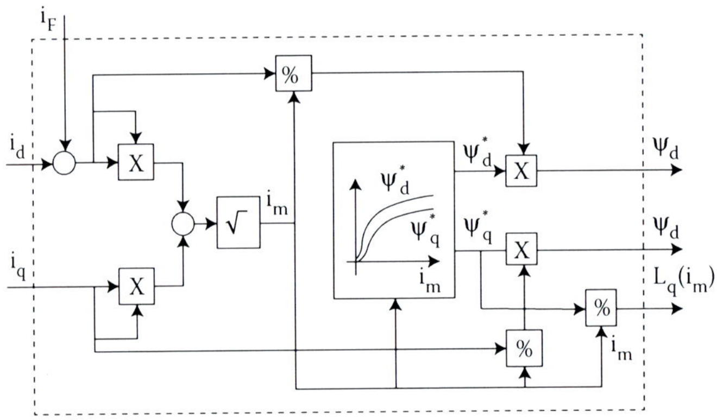

**FIGURE 14.15**
Crosscoupling saturation current–flux model of SM (used in Figure 14.14).

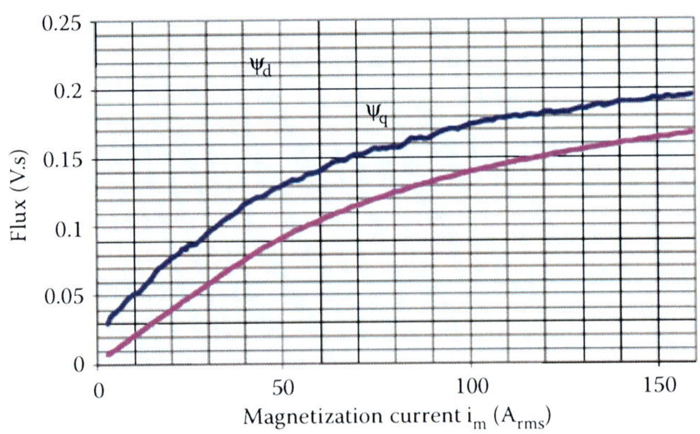


**FIGURE 14.16**
Magnetization saturation curves along d axis (upper) and q axis (lower) $\psi_d^*(i_m)$, $\psi_q^*(i_m)$ versus the magnetization current $i_m$ (experimental results).

Since, at high speeds, the flux weakening is inherent, $\psi_d^*(i_m)$, $\psi_q^*(i_m)$ curves degenerate into straight lines with $L_q$ = const., and thus the $\hat{\theta}_{er}$ observer errors become less important.

### 14.6.4 Initial Rotor Position Identification

The initial rotor position identification is essential in VC for safe starting. The procedure consists of the following steps:

1. The stator is short-circuited by a zero voltage vector applied to the PWM inverter.
2. A current ramp command is applied to the field circuit.
3. The induced stator currents $i_a(t)$, $i_b(t)$ are recorded.

Approximately, during such a sudden short-circuit mode, the stator current vector is opposite to the field current vector that is aligned to the rotor position. Therefore, the initial rotor position identification is given by

$$\hat{\theta}_{er0} \approx -\tan^{-1}(i_\beta / i_\alpha) \tag{14.37}$$

where

$$i_\alpha = i_a, \quad i_\beta = (i_a + 2i_b)/\sqrt{3}, \quad i_s = \sqrt{i_\alpha^2 + i_\beta^2}$$

The DSP implementation of the initial position searching procedure is based on the injection in the rotor winding of a current ramp command from zero to $2I_{R\_rated}$ in about 25 ms. During this transient, the stator currents are regularly sampled and the stator current vector in the stator reference frame is computed. The components of the maximum magnitude of stator current vector acquired during this transient are used for calculation of $\hat{\theta}_{er0}$ as in Equation 14.37.

It was proven that the average error in the initial rotor position estimation (done automatically before starting) was less than 5° electrical degrees, which is more than satisfactory for safe starting, even under heavy load.

### 14.6.5 Motion-Sensorless Vector Control System

The following main features of the proposed motion-sensorless VC system for d.c.-excited SMs, shown in Figure 14.17, are considered.

- The active flux-based state observer (Figure 14.14) is used to estimate the rotor position $\hat{\theta}_{er0}$ and speed $\hat{\omega}_r$ in wide speed range, without relying on machine saliency that is anyway magnetic saturation dependent.
- The current VC system employs an optimized algorithm for $i_d^*$, $i_q^*$, $i_F^*$ current references depending on $i_s^*$ (computed from $T_e^*$) and $\omega_r$. It manages both
    1. Maximum torque/current (approximately $i_d = 0$ control since the rotor saliency is small) below base speed, or for small torque loads
    2. Up to maximum torque/speed envelope available unity PF, with least commissioning efforts

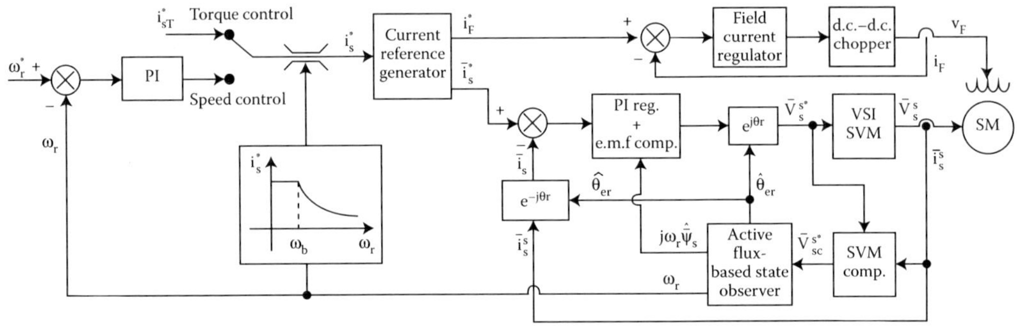


**FIGURE 14.17**
Motion-sensorless VC system for SM.

- An *a priori* calculated $i_F^*$ versus $i_s^*(i_F^* \sim i_s^*)$ reference curve for maximum torque/copper losses (minimum losses) is introduced.
- The torque reference $T_e^*$ is limited by an *a priori* known maximum torque/speed envelope (available at cos $\varphi = 1$).
- The $i_d$, $i_q$ current closed loops utilize PI controllers and e.m.f. decoupling compensation.
- The initial rotor position estimation is obtained by self-commissioning using the procedure from Section 14.6.4.
- There are two possible control schemes: torque control with $i_s^*$ as reference, or speed control with a PI anti-windup speed regulator.
- The VSI switching signals are generated by the SVM block, which employs the dead time and inverter nonlinearities compensations, ensuring reduced torque and current pulsations. This compensation realizes suitable stator voltage estimation $\bar{V}_{sc}^s$ for accurate stator flux estimation (14.29), especially at very low speeds where the stator voltage vector is very small.

However, in the absence of a rotor cage, there is strong interaction between field and stator circuits which still produces some errors in the proposed (or any) state observer.

### 14.6.6 Implementation on Test Platform and Experimental Results

#### *14.6.6.1 Test Platform Setup*

The proposed motion-sensorless control system for d.c.-excited SM is validated on an experimental drive.

Pictures in Figure 14.18 illustrate the experimental three-phase d.c.-excited SM torque controlled, directly coupled through a speed and torque meter to the load machine (IM), which is speed controlled over four quadrants by an independent converter.

The d.c.-excited SM used in these tests was specifically designed to be used in battery powered electric vehicles. Its mechanical and electrical specifications are given in Table 14.2.

The power electronic converter was designed for supplying the d.c.-excited SM at low voltage. All the data of the inverter and the parameters of the control system are given

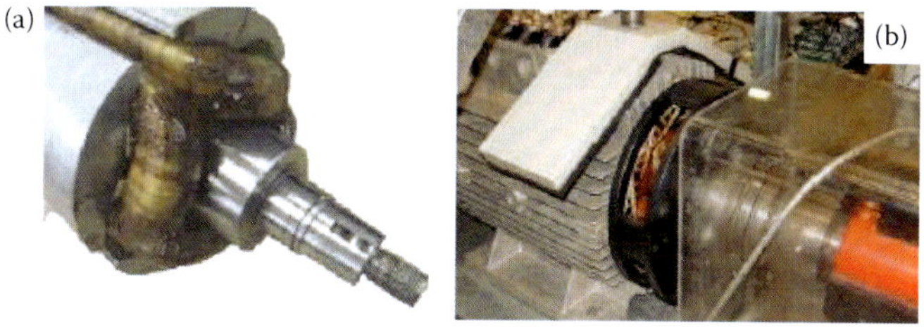

**FIGURE 14.18**
Pictures of the experimental platform: (a) exploded view of the rotor of the d.c.-excited SM machine; (b) d.c.-exited SM machine placed on the test bench.

in Table 14.3. The inverter characteristics which mainly affect the performance of the proposed motion-sensorless control technique are

1. The use of open-loop stator current transducer with poor linearity and sensitivity characteristics.
2. The use of parallel connection of Mosfets (24 per switch) for the inverter power stage introduces uncompensated voltage drop across the components, particularly relevant at low speed (note that the rated voltage of the machine is 54 V). Only compensation of dead time effects during branch switch over (BSO) has been implemented.

During testing, the actual rotor position and speed are provided by an incremental encoder with 1024 pulses per revolution, only for comparisons.

#### 14.6.6.2 Experimental Results for Motion-Sensorless Operation

Figure 14.19 shows the initial rotor position searching and a full load start up from 0 to 80 rpm. At the beginning of the transient, traces 3 and 4 illustrate the effects of the current ramp injected into the field winding on a short-circuited stator phase. After the initial position searching procedure, the rotor current is regulated at its rated value and the machine is kept magnetized for a prompt torque response. During the start-up transient, the matching between the estimated and actual speed is satisfactory.

**TABLE 14.2**
Data of the Experimental D.C.-Excited Synchronous Machine

| | |
|---|---|
| Rated stator voltage | 54 V |
| Rated stator current | 110 A |
| Pole pairs | 2 |
| Rated torque | 70 Nm |
| Rated speed | 1010 rpm |
| Efficiency | 95% |
| Rated field current | 7.5 A |
| Rated field current reduced to stator | 118 A |
| Field winding resistance $R_f$ | 1.8 Ω |
| Stator phase resistance $R_s$ | 12 mΩ |

**TABLE 14.3**

Parameters of the Experimental PEC and Control System

| | |
|---|---|
| *Supply System* | |
| Type | Lead acid battery bank |
| Rated voltage and capacity | 80 V; 400 Ah |
| *Stator Inverter* | |
| Inverter type | Three phase, two level |
| Technology | MOSFET 24 in parallel per switch |
| Rated power | 14 kVA |
| Rated output current | 140 $A_{RMS}$ |
| Maximum output current | 600 $A_{RMS}$ |
| Switching frequency | 6 kHz |
| Switching dead time | 3 μs |
| Modulation strategy | Symmetric SVM |
| *Rotor Chopper* | |
| Type | Buck, I quadrant |
| Rotor chopper rated current | 9 A |
| Rotor chopper max current | 30 A |
| Current control type | Digital hysteric regulator |
| Mean switching frequency | 10 kHz |
| *Inverter Control System* | |
| DSP | TMS320F2812—150 MHz |
| Sample time | 167 μs |
| Current transducers type | Open loop Hall effect |
| Rated current | 1000 $A_{pk}$ |
| Current transducer sensitivity | 7 A |
| Current transducer linearity | 1.5% |
| *Pi Regulators Parameters* | |
| Stator current regulator | $K_P = 0.25$ V/A |
| | $K_I = 1.00$ V/(A.s) |
| Flux estimator ($\bar{\hat{\psi}}_{si}^{s}$ $\bar{\hat{\psi}}_{s}^{s}$) error comp. | $K_P = 4.00$ 1/s |
| | $K_I = 1.5$ 1/s$^2$ |

Figure 14.20 illustrates the transient behavior, at no load, for step reference speed reversal from +13 rpm to –13 rpm and back to 13 rpm. The speed estimation (Figure 14.20) is rather satisfactory taking into account the characteristics of the experimental set up. At very low speed, major problems are the compensation of inverter nonlinearities and the reduction of stator and field winding interactions.

Figure 14.21 shows the transient responses, at 15 rpm, for step variation of the load torque: from 100% to 30% and back to 100%. The estimated rotor position (trace 1) follows the measured one (trace 2) with some delay. When torque perturbations are applied, this delay is inherent for any position and speed observer-based solely on current–flux models and VC machine equations.

Figure 14.22 illustrates the steady-state operation at 800 rpm and 100% load torque. The estimated speed shows a relatively low ripple around the measured speed.

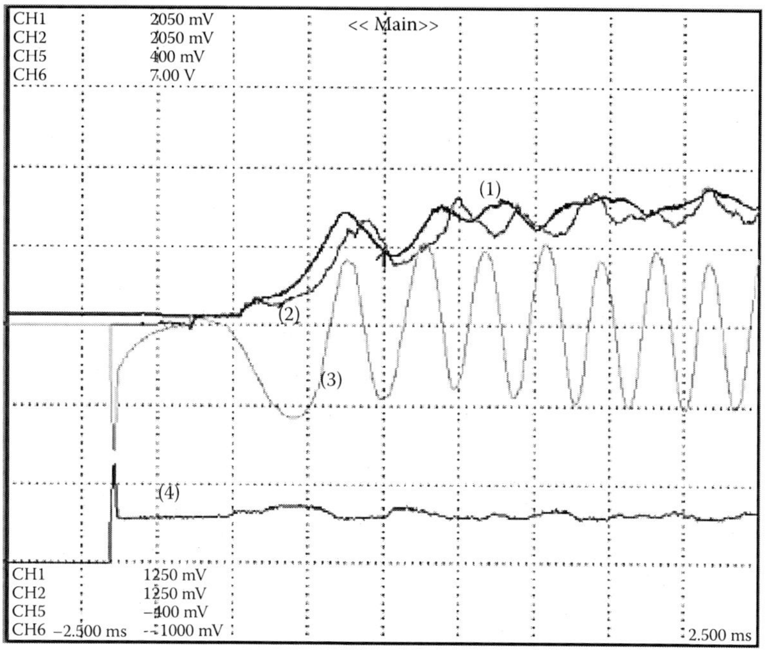


**FIGURE 14.19**
Initial rotor position searching and start up from 0 to 80 rpm, 100% load. (1) estimated and (2) measured rotor speed (10 rad/s/div), (3) phase stator current (100 A/div), (4) rotor current (10 A/div), time scale (500 ms/div). (After I. Boldea et al., Active flux based sensorless vector control of a DC excited SM drive, *Record of IEEE-ECCE*, 2009, pp. 2496–2503.)

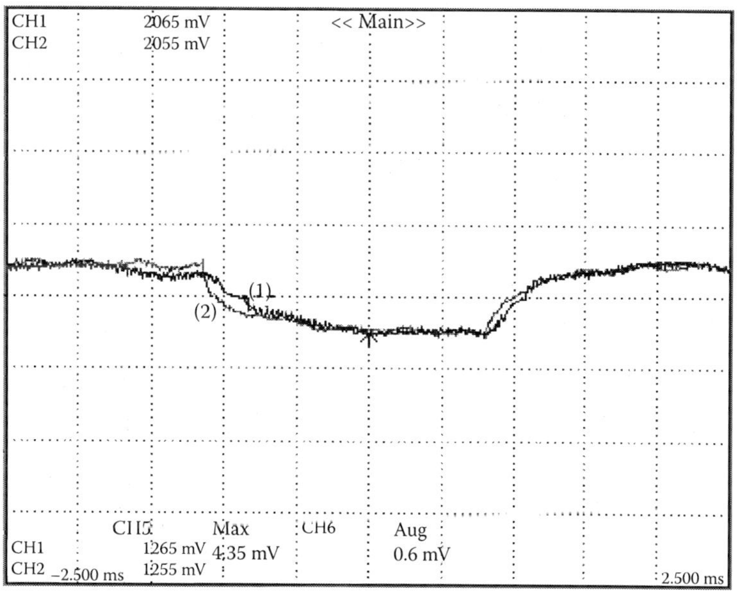


**FIGURE 14.20**
Step reference speed reversals from +13 to −13 rpm and back to +13 rpm, no load: (1) estimated and (2) measured rotor speed. Speed scale (5 rad/s/div), time scale (500 ms/div). (After I. Boldea et al., Active flux based sensorless vector control of a DC excited SM drive, *Record of IEEE-ECCE*, 2009, pp. 2496–2503.)

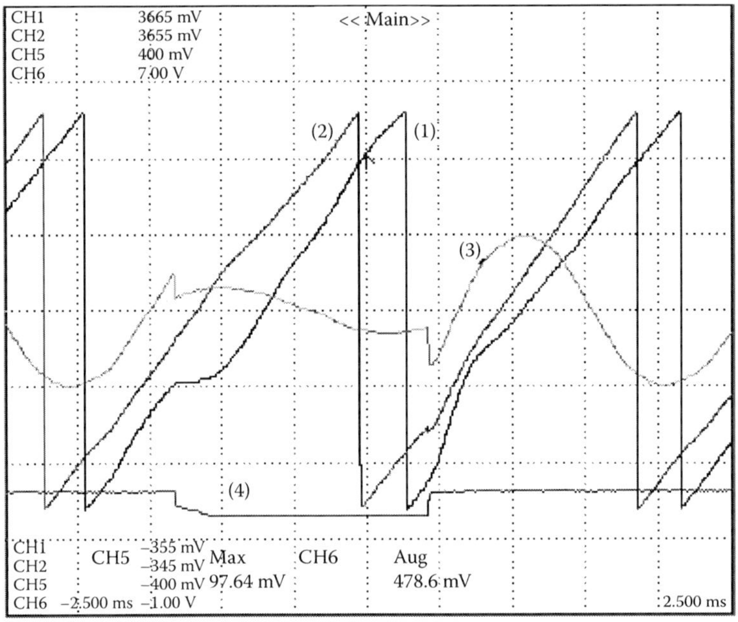


**FIGURE 14.21**
Step variations of the load torque at 15 rpm: (1) estimated and (2) measured rotor position (1.2 rad/div), (3) phase stator current (100 A/div), (4) rotor current (10 A/div), time scale (500 ms/div). (After I. Boldea et al., Active flux based sensorless vector control of a DC excited SM drive, *Record of IEEE-ECCE*, 2009, pp. 2496–2503.)

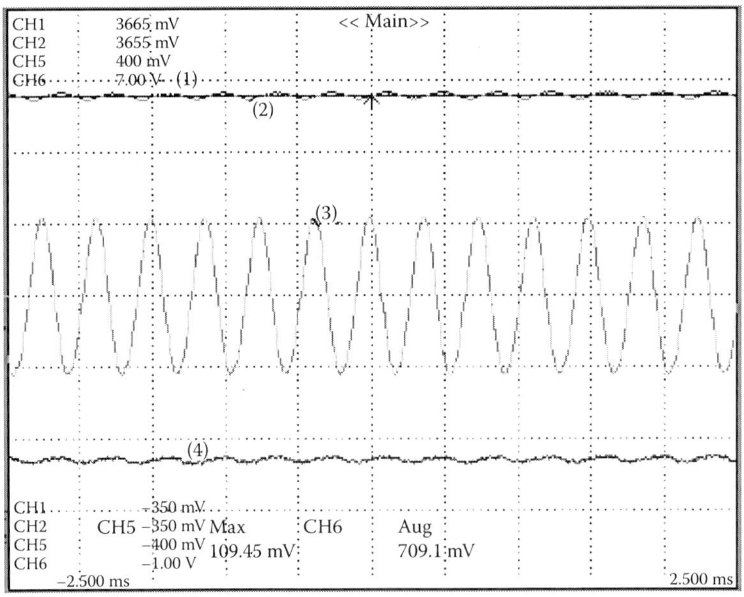


**FIGURE 14.22**
Steady state operation at 800 rpm and 100% load torque: {100%→30%→100%} (1) estimated and (2) measured rotor speed (30 rad/s/div), (3) phase stator current (100 A/div), (4) rotor current (10 A/div), time scale (50 ms/div). (After I. Boldea et al., Active flux based sensorless vector control of a DC excited SM drive, *Record of IEEE-ECCE*, 2009, pp. 2496–2503.)

Figure 14.23 shows the system behavior during a start-up transient to 800 rpm, at 100% friction load torque present from zero speed. The load is characterized by a high inertia to simulate the mass of an industrial fork-lift electric vehicle. The overload torque required during start-up is achieved by overloading both stator and field windings as shown by trace (4) for the stator current and trace (5) for the field current. Both stator and rotor current are characterized by smooth variations, without overshoot, which are specific required features for traction applications.

### 14.6.7 Discussion

This case study introduced an "active-flux" model based motion-sensorless VC system for d.c.-excited SMs in wide speed range. The developed sensorless control includes the following features:

- The active-flux vector is aligned always to the rotor position and thus the rotor position and speed estimator are simplified. The active flux is easy obtained from the estimated stator flux vector minus $L_q i_s$ vector with $L_q(i_m)$ depending on total magnetization current.
- A stator flux observer in wide speed range is developed based on combined current–voltage flux models, including magnetization curves with crosscoupling

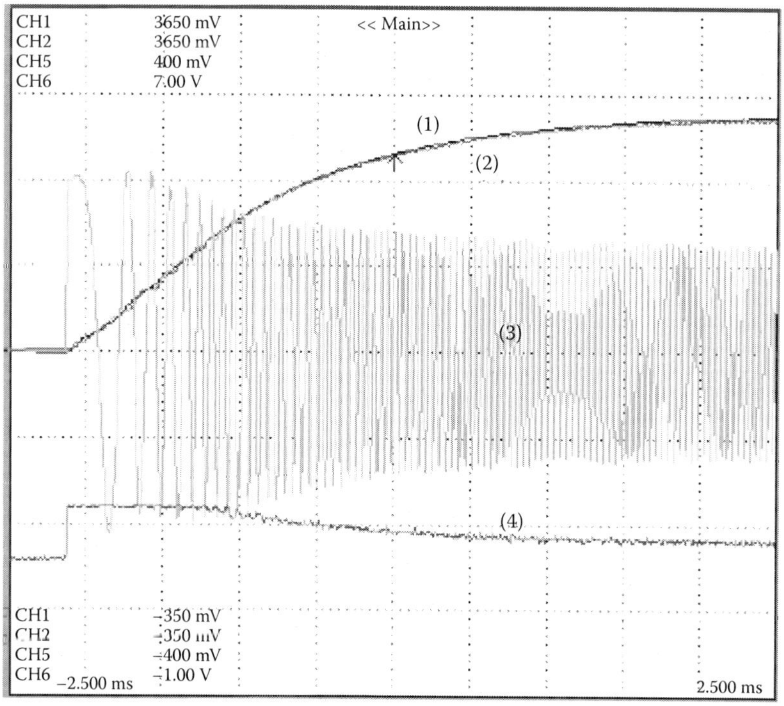


**FIGURE 14.23**
Start up from 0 rpm to 800 rpm, 100% load: (1) estimated and (2) measured rotor speed (30 rad/s/div), (3) phase stator current (200 A/div), (4) rotor current (10 A/div), time scale (500 ms/div). (After I. Boldea et al., Active flux based sensorless vector control of a DC excited SM drive, *Record of IEEE-ECCE*, 2009, pp. 2496–2503.)

saturation in the current model. A PI speed depending compensator provides a smooth transition between the current and voltage model "active-flux" observer. The input d.c.-offset and initial condition errors at the voltage-model integrator level are eliminated.

- A novel commissioning procedure to determine the crosscoupling magnetic saturation curves, employing only the voltage model (without PI compensator) is proposed.
- The initial rotor position at standstill is identified by a new method based on processing the stator current response to a quick ramp variation of the d.c. field current, with short-circuited stator.
- The current VC system employs an optimized algorithm for the calculation of $i_d^* i_q^*$, $i_F^*$ current references from $i_s^*$(computed from the torque command $T_e^*$), which allows maximum torque/current and unity PF to be achieved depending on speed range.
- The proposed sensorless control system, for a d.c.-excited SM with 80 V d.c. batteries used in traction applications, has been implemented on a DSP platform and has been experimentally validated by plenty of significant tests. The test results show good performance in a wide speed range starting from 15 rpm, with heavy load perturbation rejection, and rather quick torque response.
- Further improvements in low-speed performance, down to a few rpm could be expected in the near future, by refining sensitivity of current transducers and compensating voltage drop across switches.
- The concept can be extended to higher power VC SM drives.

## 14.7 Large Motor Drives: Working Less Time per Day Is Better

Large-power drives do not work, in general, 24 h a day, but many times they come close to this figure. Let us suppose that a cement mill has a 5 MW ball mill drive that works about 8000 h/year, that is, about 22 h/day (average) for 365 days and uses 45,000 MW h of energy.

A 16 MW drive would work for 4000 h to consume about 44,000 MW h (not much less). However, the 16 MW drive may work only off-peak hours.

The energy tariffs are high for 6 h (0.12$/kW h), normal for 8 h (0.09$/kW h), and low for 10 h (0.04$/kW h).

As the high-power drive works an average of 11 h/day/365 days (or 12 h/day excluding national holidays), it may take advantage of low tariff for 10 h and work only 2 h at normal tariff.

The lower-power drive, however, has to work at least 4 peak tariff hours. Though the energy consumption is about the same, the cost of energy may be reduced from $3,700,000 to $1,700,000 per year, that is, more than 50%.

As investment costs are not double for doubling the power (60% more) and the maintenance costs are about the same, the revenue time is favorable for the higher-power drive working less time per day.

## 14.8 Rectifier CSI-SM Drives: The Basic Scheme

Standard high-speed (50(60) Hz or more) large-power variable speed SM drives, traditionally above only 750 kW, use frequently rectifier current-source inverters (CSIs). Load commutation of thyristors is required. The SMs are capable of leading PF angle, $\varphi_1 = -(6° - 8°)$, and thus, above 5% of rated speed, load commutation is feasible.

Both the phase-delay rectifier (PDR) and the CSI make use of thyristors and thus they may be built at low costs.

To reduce torque pulsations at low speed, PWM (current notches) of the otherwise rectangular (120° wide) currents may be used.

Here we introduce the basic scheme (Figure 14.24), steady-state equations, and performance with load commutation and an advanced control scheme which provides constant leading PF with load and speed response without overshooting.

The basic scheme of the rectifier CSI-SM drive in Figure 14.24 uses an indirect vector current control that contains:

- A fully controlled (phase-delay) rectifier
- A current choke ($r_f$, $L_f$)
- A CSI-load commutated
- A rotor position sensor ($\gamma_0$) or a terminal voltage zero crossing sensor ($\gamma$)
- Advance angle ($\gamma$ and $\gamma_0$ increasing with load ($I_{dc}$)) controller for constant leading PF angle and safe commutation (limited overlapping angle u)

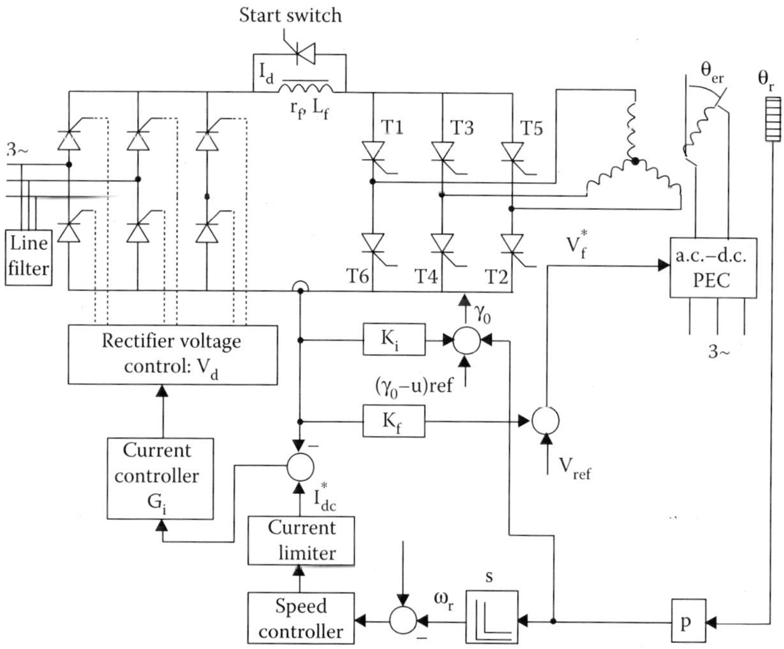


**FIGURE 14.24**
Basic scheme of rectifier CSI-SM drive.

- Speed controller with current limiter
- A d.c. current controller
- An a.c.–d.c. PEC for the field winding

The CSI provides only the synchronization between the rotor position and stator current 120°-wide blocks through the advance angle $\gamma_0(\gamma)$.

The controlled rectifier modifies the d.c. link (and stator) current level, that is the torque, in order to perform speed control.

The d.c. link voltage produced by the rectifier $V_d$ is positive for motoring and negative for generating. Through proper changing of the inverter firing sequence, regenerative braking and speed reversal are obtained. Sufficient e.m.f. (field current motion-induced stator voltage) to turn off the CSI thyristors is available only above 5% of rated speed.

## 14.9 Rectifier CSI-SM Drive: Steady State with Load Commutation

Let us consider that the machine is running above 5% of rated speed. Therefore, sufficient e.m.f. is provided by the field current to produce safe load commutation. By commutation, we mean the turning on and off process of current in the motor phases. Ideally, the d.c. link current is constant in time (because of a high filter inductance $L_f$). Thus the inverter distributes the d.c. link current as 120° current blocks of alternate polarity between the stator phases in pairs. With instantaneous (ideal) commutation, the ideal currents are as shown in Figure 14.25.

As the SM contains inductances, a sudden change in phase currents, as implied by instantaneous commutation (Figure 14.25) is, in fact, impossible.

Intuitively, it follows that the machine will impose some exponential (or almost linear) variation of currents throughout the commutation process. So the actual currents are trapezoidal rather than rectangular.

As the commutation process is rather fast—in comparison with the current period—the machine behaves approximately according to the voltage behind the subtransient inductance principle (Figure 14.26).

The voltage behind the subtransient inductance L″, $V_1$, is, in fact, the terminal voltage fundamental if phase resistance is neglected. The no-load (e.m.f.) voltage $E_1$ is also sinusoidal. The space vector (or phasor) diagram may be used for the current fundamental $I_1$ for

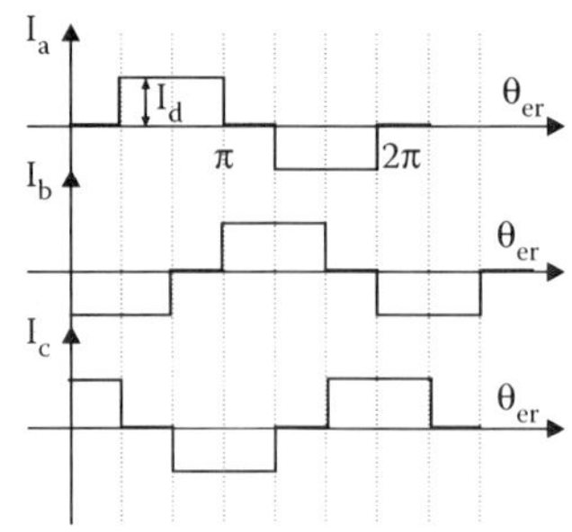


**FIGURE 14.25**
Ideal stator current waveforms.

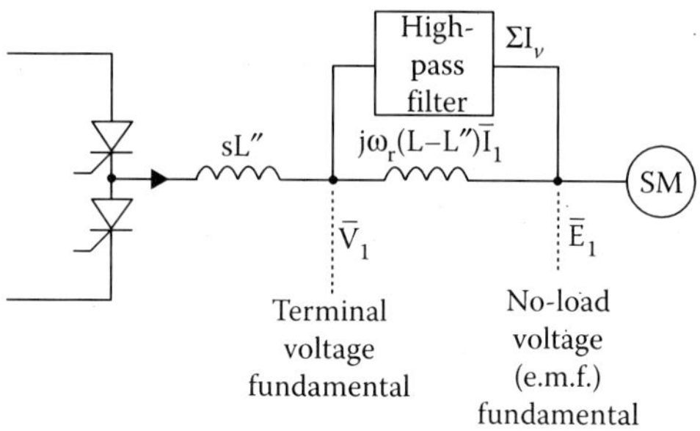


**FIGURE 14.26**
Equivalent circuit under mechanical steady state.

L–L″ with L as synchronous inductance and L″ the subtransient inductance along the d-q axes (Figure 14.27) [9].

$$V_1 \cos\delta_1 = E_1 - \omega_r(L_d - L_d'')I_1 \sin(\delta_1 - \varphi_1) \tag{14.38}$$

$$V_1 \sin\delta_1 = \omega_r(L_q - L_q'')I_1 \cos(\delta_1 - \varphi_1) \tag{14.39}$$

In general, the machine may have salient poles ($L_d \neq L_q$) and, definitely, damper windings ($L_d'' \ll L_d$, $L_q'' \ll L_q$). The currents induced in the damper windings due to the nonsinusoidal character of stator currents are neglected here.

During commutation intervals the machine reacts with the subtransient inductance. An average of d–q subtransient inductances $L_d''$ and $L_q''$ is considered to represent the so-called commutation inductance $L_c$ given by

$$L'' = L_c \approx \frac{1}{2}(L_d'' + L_q'') \tag{14.40}$$

### 14.9.1 Commutation and Steady-State Equations

To study the commutation process let us suppose that, at time zero, phases a and c are conducting (Figure 14.28) ($T_1T_2$ in conduction)

$$\text{at } t = 0 \;\; I_a = +I_d, \;\; I_c = -I_d \;\; \text{and} \;\; I_b = 0 \tag{14.41}$$

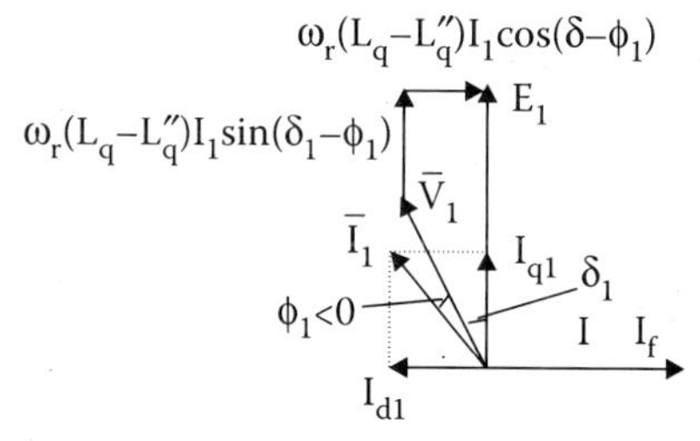


**FIGURE 14.27**
Space vector (or phasor) diagram for the voltage behind subtransient inductance.

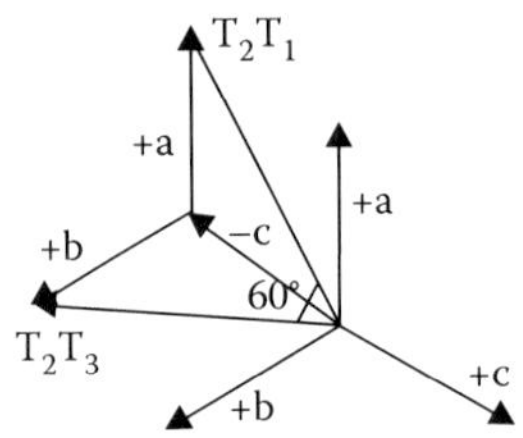


**FIGURE 14.28**
Stator m.m.f. (60° jump from $T_1T_2$ to $T_2T_3$ conduction).

The advance angle control requires a jump of 60° counterclockwise in the stator current space vector. That is, leave −c conducting and switch phase +a for +b with $T_2T_3$ (after $T_1T_2$) on.

During this commutation process phases +a and +b are both conducting (in parallel) with phase −c continuing conduction (Figure 14.29).

The loop made of phases a, b in parallel (Figure 14.29) has the equation

$$L_c \frac{di_a}{dt} + V_{a1} = L_c \frac{di_b}{dt} + V_{b1} \tag{14.42}$$

$$i_d = i_a + i_b \tag{14.43}$$

At the end of the commutation interval ($t = t_c$), phase b is conducting

$$i_a = 0, \quad i_b = I_d \quad \text{for } t = t_c \tag{14.44}$$

The fundamental voltages $V_{a1}$ and $V_{b1}$, represent, as explained above, voltages behind subtransient inductances and are thus considered sinusoidal in time.

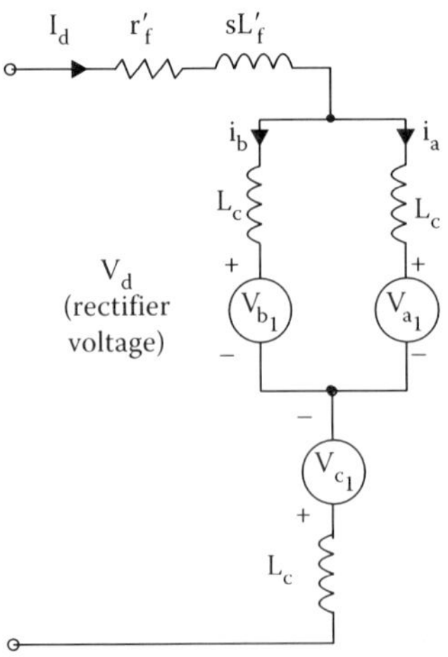


**FIGURE 14.29**
Equivalent circuit for commutation.

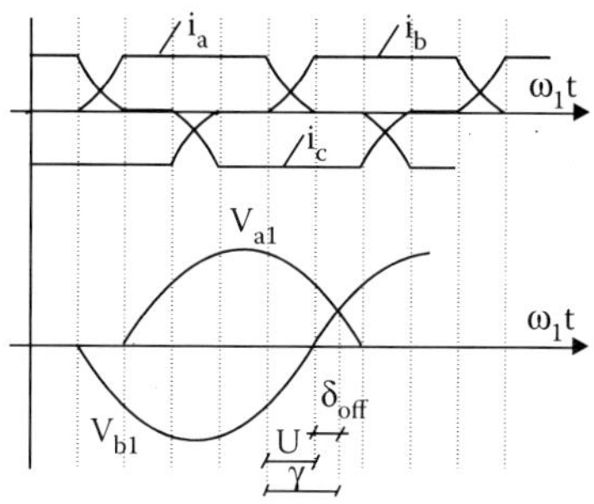


**FIGURE 14.30**
The commutation process.

Eliminating $i_b$ from Equations 14.42 and 14.43 yields

$$2L_c\frac{di_a}{dt} = -(V_{a1} - V_{b1}) \tag{14.45}$$

For successful commutation, the current in phase a should decrease to zero during the commutation interval ($di_a/dt < 0$). Consequently, the line voltage $V_{ab1}$ should necessarily be positive when the current switches from phase +a to phase +b (Figure 14.30).

If the commutation process is triggered sufficiently before $V_{ab1}$ goes to zero, from positive values, the phase a current $i_a$ will be driven to zero.

Before $V_{ab1}$ goes to zero, a time interval $t_{off}$ ($\delta_{off}/\omega_r$) is required with $i_a = 0$ to allow the recombination of charges in the thyristor $T_1$.

According to Figure 14.30 the phasing of $V_{ab1}$ is

$$V_{ab1} = V_{a1} - V_{b1} = -V_1\sqrt{6}\sin(\omega_r t - \gamma);\ \ 0 < \omega_r t < \gamma \tag{14.46}$$

with $V_1$ the rms value of phase voltage fundamental. Integrating Equation 14.46 from t = 0 to $t_c = u/\omega_r$, we obtain

$$V_1\sqrt{6}(\cos(\gamma - u) - \cos\gamma) = 2L_c I_d \omega_r \tag{14.47}$$

The angle u corresponds to the overlapping time of currents $i_a$ and $i_b$. For $I_d = 0$, u = 0 and γ should be greater than u for successful commutation. Ideally, $\gamma - u = \delta_{off}$ should be kept constant so both γ and u should increase with current.

It is now clear (as $\gamma_{max} = 60°$) that the maximum current commutated safely is inversely proportional to the commutation inductance $L_c$. The lower $L_c$ the better, *so a damper winding in the rotor is necessary*.

The rms phase current fundamental is related to the d.c. current $I_d$ by

$$I_1 \approx \frac{\sqrt{6}}{\pi}\frac{\sin(u/2)}{(u/2)}I_d \approx \frac{\sqrt{6}}{\pi}I_d \tag{14.48}$$

and was obtained by using the trapezoidal current shape of Figure 14.30. The PF angle $\phi_1$ between $V_1$ and $I_1$ (Figure 14.31) is approximately [9]

$$\phi_1 \approx \gamma - u/2 \tag{14.49}$$

The inverter voltage $V_I$ is made of line voltage segments disturbed only during the commutation process

$$V_I(t) = V_{a1}(t) - V_{c1}(t);\quad 0 \le \omega_r t \le (\pi/3 - u) \tag{14.50}$$

$$V_I(t) = V_{a1}(t) - V_{c1}(t) + L_c \frac{di_a}{dt};\quad (\pi/3 - u) \le \omega_r t \le \pi/3 \tag{14.51}$$

Finally, making use of Equations 14.45 through 14.51, we obtain the average inverter voltage $V_{Iav}$

$$V_{Iav} = \frac{3}{\pi}\int_0^{\pi/3} V_I(t)d(\omega_r t) = \frac{3}{\pi}(V_1\sqrt{6}\cos\gamma - L_c I_d \omega_r) \tag{14.52}$$

So the commutation process produces a reduction in the motor terminal voltage $V_1$ by $L_c I_d \omega_r$ for a given $V_{Iav}$.

Neglecting the power losses in the CSI and in the motor

$$I_{dc} V_{Iav} = 3V_1 I_1 \cos\phi_1 = T_{eav} \cdot \frac{\omega_r}{p} \tag{14.53}$$

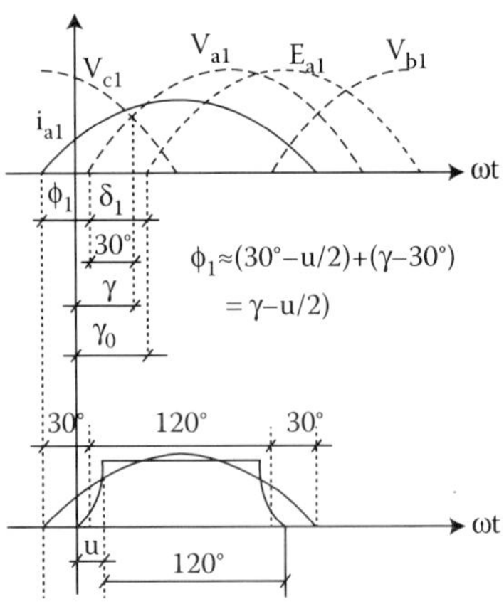


**FIGURE 14.31**
Fundamental phase current and voltage phasing.

As expected, the simultaneous torque $T_e(t)$ is

$$T_e(t) = \frac{I_{dc} V_I(t)}{\omega_r / p} \tag{14.54}$$

Consequently, the torque pulsates as does the inverter voltage (Figure 14.32).

The torque pulsations in Figure 14.32 are different from voltage pulsations during commutation periods, especially due to the filtering effect of the motor rotor damper winding.

### 14.9.2 Ideal No-Load Speed

To calculate the ideal no-load speed (zero $I_d$, zero torque), we have to use all equations in this section (from Equations 14.46 and 14.47) to obtain

$$\begin{aligned} &\delta_1 = 0, \ V_1 = E_1, \ \ u = 0, \gamma = \gamma_0 \\ &V_1 = L_{dm} I_F \omega_{r0} / \sqrt{2} \end{aligned} \tag{14.55}$$

Finally, from Equations 14.47 through 14.52

$$\omega_{r0} = \frac{V_{Iav0}\pi}{3\sqrt{3} L_{dm} I_F \cos\gamma_0} \tag{14.56}$$

Also, the rectifier voltage $V_r$ is related to inverter voltage $V_I$

$$V_{rav} = V_{Iav} + r_f I_d = V_{Iav}; \quad (I_d = 0) \tag{14.57}$$

### 14.9.3 Speed Control Options

To vary speed, we need to vary the ideal no-load speed $\omega_{r0}$. The available options are

- Inverter voltage variation through rectifier voltage variation up to maximum voltage available.
- Reducing the field current $i_F$ (flux weakening).

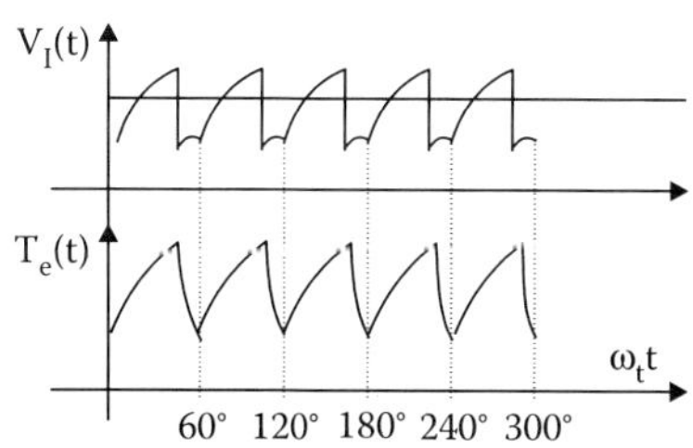


**FIGURE 14.32**
Inverter voltage $V_I(t)$ and motor torque pulsations during steady state.

- Modify the control angle $\gamma_0$. As $\gamma_0$ may lay in the 0° to 60° interval, this method is not expected to produce significant speed variations. However varying $\gamma$ (or $\gamma_0$) is used to keep the PF angle $\phi_1$ rather constant ($\phi_1 \approx \gamma - u/2$) or $\delta_{off}$ = const. for safe commutation.

**EXAMPLE 14.1**

A rectifier CSI-SM drive is fed from a $V_L = 4.8$ kV a.c. (line to line, rms), power source, the magnetization inductance $L_{dm} = 0.05$ H and rated field current (reduced to the stator) $i_{Fn} = 200$ A.

Determine

- The maximum average rectifier voltage available.
- The ideal no-load speed $\omega_{r0}$ for $\gamma_0 = 0$.
- For $L_c/L_{dm} = 0.3$, $\gamma = 45°$ and $I_d = 100$ A, $\omega_r/\omega_{r0} = 0.95$, calculate the fundamental voltage $V_1$ and the overlapping angle u.

**Solution**

The fully controlled rectifier produces an average voltage $V_r$ (Chapter 5, Equation 5.52).

$$V_r = \frac{3V_L\sqrt{2}}{\pi}\cos\alpha \tag{14.58}$$

So

$$V_{r\max} = \frac{3\cdot 4800\sqrt{2}}{\pi}\cdot 1 = 6466\,\text{V} \tag{14.59}$$

The ideal no-load speed (14.56) is

$$\omega_{r0} = \frac{6466\cdot\pi}{3\sqrt{3}0.05\cdot 200\cdot 1} = 390\,\text{rad/s} \tag{14.60}$$

The voltage fundamental $V_1$ becomes (14.53)

$$V_1 = \frac{6646\cdot\pi/3 + 0.95\cdot 390\cdot 0.3\cdot 0.05\cdot 100}{\sqrt{6}\cdot\cos 45^{\circ}} = 4336\,\text{V} \tag{14.61}$$

Now, from Equation 14.47 the overlapping angle u is found

$$\begin{gathered} 4336\sqrt{6}(\cos(\gamma - u) - \cos\gamma) = 2\cdot 0.3\cdot 0.05\cdot 390\cdot 0.95\cdot 100 = 1111.5 \\ \cos(\gamma - u) = 0.807;\quad u = 4.8^{\circ} \end{gathered} \tag{14.62}$$

The overlapping angle u is small, indicating an unusually strong damper winding. Higher values of u are practical.

### 14.9.4 Steady-State Torque Speed Curves

By now we know that the rectifier CSI-SM drives exhibit a finite ideal no-load speed much like d.c.-brush motors with separate excitation.

However, it is not indifferent if the speed increases or decreases with torque for given inverter voltage $V_{Iav}$. An increase of speed with torque means an unstable speed/torque characteristic, while speed decreasing with torque is a statically stable characteristic.

While successful commutation has to be provided up to the maximum load torque considered, the drive may work with $\gamma_0$ = const. (direct position sensor required) or with $\phi_1$ = const. = $\gamma$–u/2.

The above relationships suffice to calculate the ideal motor torque speed ($I_d$) curves and the variation of overlapping angle u with load ($I_d$).

For $\gamma_0$ = const. the torque speed curve proves to be stable (Figure 14.33a). This is not the case for $\gamma$ = const. (Figure 14.33b).

So far, the field current was considered constant with load ($I_d$). If the field current increases with load to keep the PF constant ($\phi_1$ = const. = $\gamma$–u/2), it means that we should increase $\gamma$ with load as (anyway) the overlapping angle u does so (Figure 14.33). This way the speed will decrease with load and thus, as a bonus, static stability is restored (Figure 14.34).

Results in Figure 14.34 lead naturally to the conclusion that, for high-performance variable speed control, *both the angle $\gamma$ and field current $i_F$ should increase with load, to secure constant PF and safe commutation*. These conclusions will be fully exploited in the dynamics and control section later in this chapter.

### 14.9.5 Line Commutation During Starting

As already mentioned, at low speeds (below 5% of rated speed), the e.m.f. is not high enough to produce a leading PF ($\phi_1$ < 0) and thus secure load commutation (the resistive

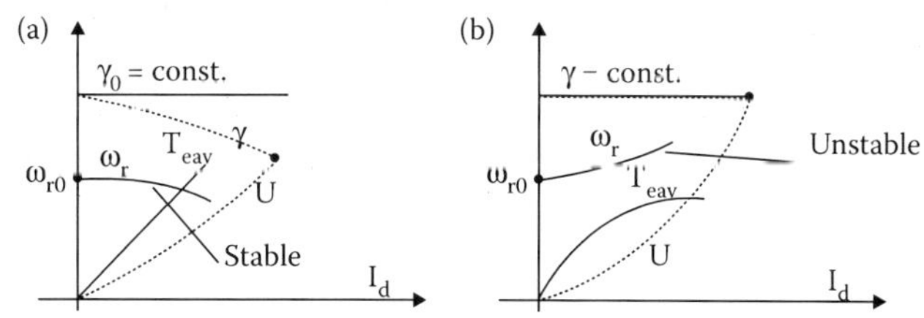


**FIGURE 14.33**
Steady-state curves ($V_{Iav}$ = const., $i_F$ = const.): (a) for constant $\gamma_0$ (position sensor); (b) for constant $\gamma$ (terminal voltage zero crossing angle).

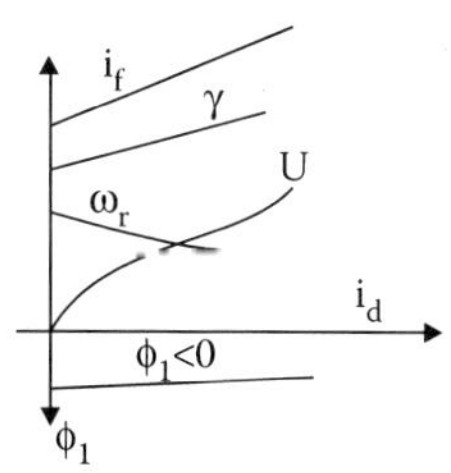


**FIGURE 14.34**
Steady-state curves ($V_{Iav}$ = const.), $i_F$ is a variable, $\phi_1$ = const. = $\gamma$–u/2.

voltage drop is relatively high). For starts, either line commutation or some forced commutation (with additional hardware) [4] is required.

To start, the angle $\gamma$ ($\gamma_0$) is chosen to be zero to obtain maximum torque per ampere. When the commutation of phases is initiated, both thyristors conducting previously are first turned off by applying a negative voltage at the machine terminals through increasing the rectifier delay angle to about 150°–160°. To speed up the d.c. link current attenuation to zero, the d.c. choke is short-circuited through the starting thyristor (Figure 14.24).

Current notches occur and thus the torque pulsations may also be slightly reduced, while the motor accelerates smoothly.

### 14.9.6 Drive Control Loops

While the basic scheme of the control system remains the same as in Figure 14.24, we will briefly discuss some practical solutions for speed, $\omega_r$, current, $I_d$, advance angle ($\gamma$), and field voltage control.

A typical speed controller capable of providing for safe load commutation during the transients is shown in Figure 14.35. The d.c. link current controller, the reference angle $\gamma^*$ and the field current–voltage $V_f^*$ are all included.

A PI speed controller is, in general, used

$$I_d^* = (\omega_r^* - \omega_r)G_\omega \tag{14.63}$$

with

$$G_\omega = K_{c\omega}\left[1 + 1/T_{i\omega}s\right] \tag{14.64}$$

where $K_{c\omega}$ is the gain and $T_{i\omega}$ is the integral time constant. Safe load commutation has to be provided, in general, up to 150% of rated current. A current limiter is required. The integral part of the speed controller is thus not desirable during the current limiter saturation period. This may be done inhibiting the integral part of the speed controller, as long as the current is higher than a limit value.

The current controller may also be of PI type

$$V_r^* = G_i(I_d^* - I_d) \tag{14.65}$$

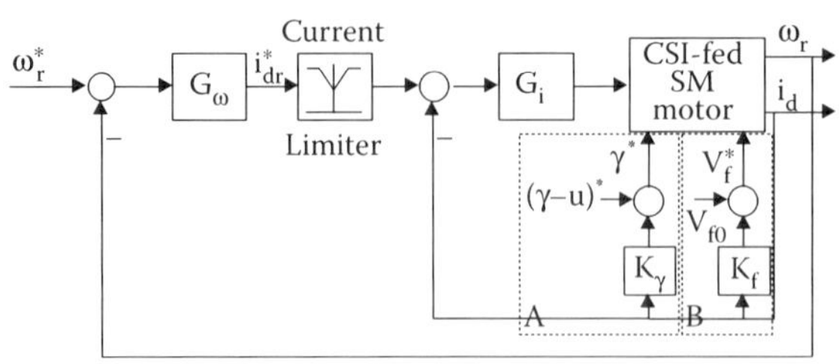


**FIGURE 14.35**
Speed control system (rectifier CSI-SM drive).

with

$$G_i = K_{ci}\left[1 + 1/T_{ii}s\right] \tag{14.66}$$

As expected, $T_{ii} \ll T_{i\omega}$ and, in general, $T_{i\omega} > 4T_{ii}$. The advance angle $\gamma^*$ will be increased with the load current as suggested in Section 14.6.5

$$\gamma^* = (\gamma - u)^* + K_\gamma I_d \tag{14.67}$$

The initial angle $(\gamma - u)^* \approx 15° - 25°$ provides for safety margin as required by slow, standard (low cost) thyristors

$$(\gamma - u)^* \geq \omega_{rmax} t_{off};\ \ t_{off} = 0.3 - 0.5\ \text{ms} \tag{14.68}$$

Though this is only a linear (intuitive) approximation, it has been proved practical. As in most CSI-SM drives, fast drive response is not a high priority, a field voltage referencer $V_f^*$ suffices instead of a field current controller

$$V_f^* = V_{f0} + K_f I_d \tag{14.69}$$

This way, a rather constant safety margin angle $(\gamma - u)^*$ with increasing loads and a constant PF (leading PF angle $\phi_1 = -(8°–10°)$) are obtained. For control numerical details related to a practical case, see Reference 10. Such drives are, by now, standard in industry for powers up to 30 MW (5500 rpm) per unit [11].

To improve (speed up) the current response at high speeds, an e.m.f. $V_c$ compensator may be added to the d.c. link current controller. The e.m.f. may be based on the flux $\lambda''$ behind the commutation inductance $L_c$

$$\bar{\lambda}'' = \int (\bar{V}_s - r_s \bar{i}_s) \cdot dt - L_c \bar{i}_s \tag{14.70}$$

$$V_c = \left|\bar{V}_c\right| = \left|j\omega_r \lambda''\right| \tag{14.71}$$

Above 5% of rated speed, the $\bar{\lambda}''$ calculator is good enough even if based only on the voltage model.

Feedforwarding the value of the $V_c$ to the d.c. link current controller produces notable increases in the current response quickness. The above control system—in fact, of indirect vector current type—looks rather simple but relies heavily on known machine data and parameters.

A more direct, robust approach, such as direct VC, has also been proposed [12].

As an alternative to this concept, the DTFC principle is introduced here.

### 14.9.7 DTFC of Rectifier CSI-SM Drives

As the stator current is rather trapezoidal, the stator flux is nonsinusoidal even under steady state. The subtransient flux linkage $\lambda''$, defined above, is however, very close to a traveling wave during steady state. For DTFC, we thus make use of $\lambda''$. A tentative DTFC configuration is shown in Figure 14.36.

The CSI provides six nonzero current space vectors $\bar{I}_{1,...,6}$ produced by two phases conducting at any time outside the commutation zones (Figure 14.37).

The subtransient flux $\lambda''$ position in one of the six sectors (Figure 14.37) is used to select the required current space vector, depending also on the torque error sign. As the machine is overexcited, the forthcoming current vector should be located at an angle greater than

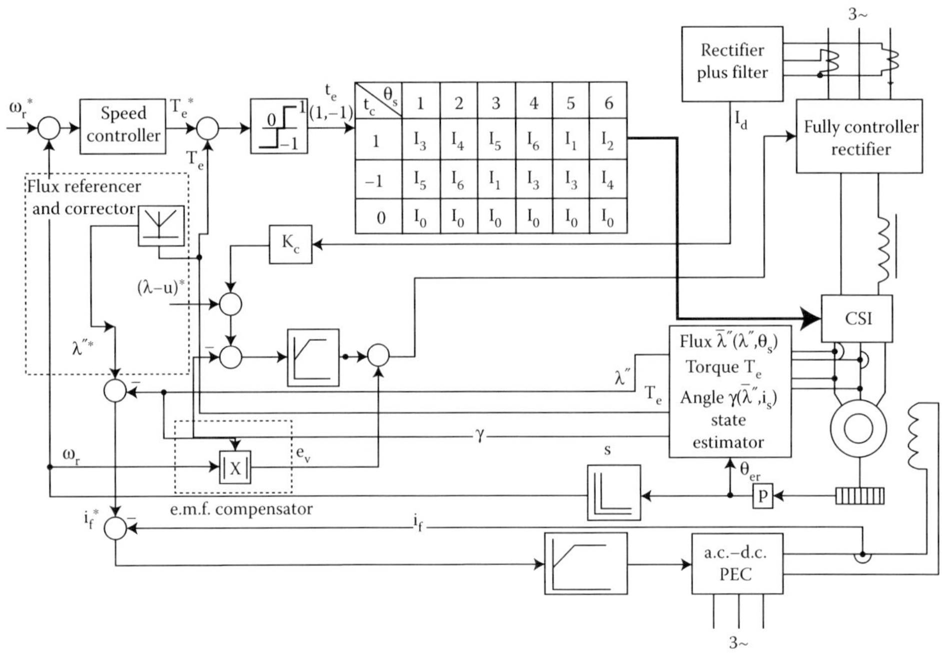


**FIGURE 14.36**
Tentative DTFC system for CSI-SM drives.

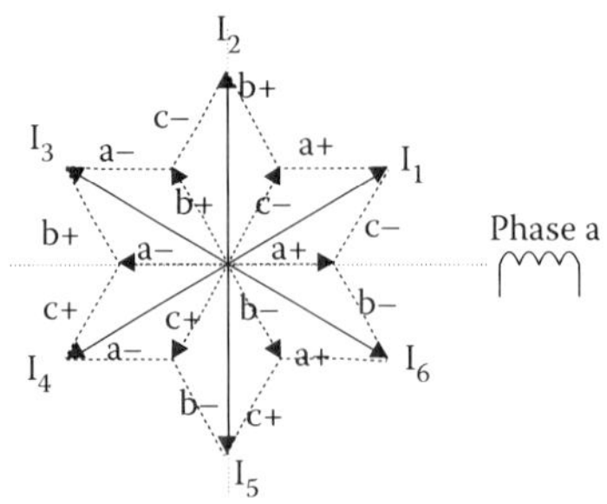


**FIGURE 14.37**
CSI-current space vectors.

90° with respect to the $\bar{\lambda}''$ position. So for $\bar{\lambda}''$ in the first sector (0–60°), $I_3$ should be applied for positive torque error and $I_5$ for negative torque error.

A hysteresis angle may be allowed at the boundary between sectors to reduce the switching frequency. The hysteresis band of the torque controller may be increased with speed to avoid PWM of current at high speeds.

For zero torque error, the thyristors of upper and lower leg of a phase are applied so as to minimize the switching frequency of the thyristors. In fact the current ($I_0$) in the machine is zero, and the a.c. choke limits the d.c. link current.

The subtransient flux estimator could be built in many ways. A combination of voltage (14.70) and current model is the obvious way to build a wide speed range subtransient flux estimator. Sensorless control may also be considered but a speed estimator is additionally required.

## 14.10 Sub- and Hyper-Synchronous IM Cascade Drives

### 14.10.1 Limited Speed Control Range for Lower PECs Ratings

Limited speed control range is required in many applications such as high-power pumps, fans, etc. Low motor speed range control, 20%–30% around rated speed, implies frequency control.

Stator frequency control, in either SMs or IMs, no matter the speed control range, requires full motor power PECs. So it is costly for the job done. It is very well known that the power balance in the wound rotor of IMs is characterized by the slip formula

$$SP_{elm} = P_{Co2} + P_r; \; S = \frac{\omega_1 - \omega_r}{\omega_1}; \; \omega_1 = 2\pi f \tag{14.72}$$

where $P_{elm}$ is the electromagnetic (active) power that "crosses the airgap," or is exchanged between rotor and stator in IMs; $P_{Co2}$ is the rotor winding loss and $P_r$ is the electric power extracted or introduced (injected) into the rotor.

Neglecting, for the time being, the rotor winding loss ($P_{Co2} = 0$), the electric power injected into the rotor at slip frequency $f_2 = Sf_1$ is

$$P_r \approx SP_{elm} = \frac{\omega_1 - \omega_r}{\omega_1} P_{elm} \tag{14.73}$$

The speed control range is defined through the minimum speed, $\omega_{rmin}$, or the maximum slip, $S_{max}$,

$$S_{max} = \frac{\omega_1 - \omega_{rmin}}{\omega_1} \tag{14.74}$$

So the maximum active electric power injected in the wound rotor is

$$P_{rmax} \approx S_{max} P_{elm} \tag{14.75}$$

For a 20% speed control range $S_{max} = 0.2$ and thus the PEC required to handle the rotor injected power $P_{rmax}$ is rated to $S_{max}$, that is 20%–30% of motor rated power.

In general, as the rotor voltage required at low slip frequency $f_{2max} = S_{max}f_1$ is low, a step-up transformer is mandatory to exchange energy with the power grid also used to feed the stator windings (Figure 14.38).

In essence, $P_r$ may be positive or negative and thus sub- or hyper-synchronous operation is feasible, but the rotor-side PEC has to be able to produce positive and negative sequence voltages as

$$f_1 = f_2 + \frac{\omega_r}{2\pi}; \quad f_2 = Sf_1 \gtrless 0 \tag{14.76}$$

For $S < 0$, $f_2 < 0$ and thus negative sequence rotor voltages are required. As the machine is reversible, both motoring and generating should be feasible, provided that a bidirectional power flow is allowed for through the rotor-side PECs.

Also, a smooth transition through $S = 0$ is required. For $S = 0$ (stator produced synchronism) $f_2 = 0$ and thus d.c. current (m.m.f.) should be injected into the rotor. These challenging constraints restrict the types of PECs to be used for direct a.c.–a.c. conversion.

Cyclo or matrix converters and two "back-to-back" voltage source PWM inverters are adequate for the scope.

**NOTE:** The so-called slip-recovery scheme is simple but, allowing only unidirectional power flow (from machine rotor to power grid), may work only as a subsynchronous motor (or as a hyper-synchronous generator). Moreover, the total PF of the machine is notably low. This is the reason why we do not detail it here.

Even with sub- and hyper-IM cascade the total PF is slightly lagging, though the rotor-side converter in special configurations such as back-to-back voltage source PWM inverters, may input reactive power in the rotor.

Let us note that the sub- and hyper-IM cascades may work both as a motor and a generator above and below $f_1/p$ and thus are used for pumped-storage power and wind energy plants. In this latter case, for motoring, the turbine will have, in general, to change the direction of motion to act as a pump. This explains the presence of a power switch 1–2 (Figure 14.38). Also, to protect the rotor-side converter from overvoltages or overcurrents "induced" by stator faults (short-circuits, etc.) and for starting, a controlled resistor is introduced at the rotor terminals (Figure 14.38).

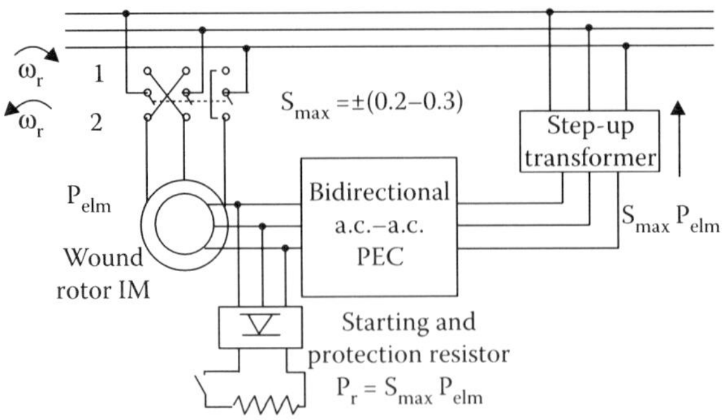


**FIGURE 14.38**
Sub- and hyper-synchronous IM cascade system.

### 14.10.2 Sub- and Hyper-Operation Modes

The induction machine equations in stator field (synchronous) coordinates are (Chapter 8, Equations 8.50 through 8.52)

$$\bar{V}_s = r_s \bar{i}_s + \frac{d\bar{\lambda}_s}{dt} + j\omega_1 \bar{\lambda}_s \tag{14.77}$$

$$\bar{V}_r = r_r \bar{i}_r + \frac{d\bar{\lambda}_r}{dt} + jS\omega_1 \bar{\lambda}_r \tag{14.78}$$

$$\bar{\lambda}_s = L_s \bar{i}_s + L_m \bar{i}_r; \quad \bar{\lambda}_r = L_r \bar{i}_r + L_m \bar{i}_s \tag{14.79}$$

$$T_e = \frac{3}{2} p \operatorname{Re}(j\bar{\lambda}_s \bar{i}_s^*) = -\frac{3}{2} p \operatorname{Re}(j\bar{\lambda}_r \bar{i}_r^*) \tag{14.80}$$

In the cascade configuration, the rotor circuits are fed with three-phase voltages (and currents) represented by the space vector $\bar{V}_r$.

Note that $\bar{V}_r$ and $\bar{i}_r$ in Equation 14.78 are referred to the stator and expressed in synchronous coordinates. During steady-state $\bar{V}_r$, $\bar{i}_r$ represent d.c. quantities, while in the real machine their frequency is $\omega_2 = S\omega_1$.

For steady-state ($d/dt = 0$) Equation 14.78 becomes

$$\bar{V}_r = r_r \bar{i}_r + jS\omega_1 \bar{\lambda}_r \tag{14.81}$$

The ideal no-load speed $\omega_{r0}$ is obtained for zero rotor current

$$\bar{V}_r = j(\omega_1 - \omega_{ro})\bar{\lambda}_r \tag{14.82}$$

From Equation 14.77 with $\bar{i}_r = 0$ and $d/dt = 0$

$$\bar{V}_s = r_s \bar{i}_{s0} + j\omega_1 L_s \bar{i}_{s0}; \quad \bar{\lambda}_{r0} = L_m \bar{i}_{s0} \tag{14.83}$$

Neglecting $r_s$ ($r_s = 0$)

$$\bar{\lambda}_{r0} = \bar{\lambda}_{s0} \cdot \frac{L_m}{L_s} = \frac{\bar{V}_s}{j\omega_1} \frac{L_m}{L_s} \tag{14.84}$$

From Equation 14.82 with 14.84

$$\bar{V}_{r0} = \bar{V}_{s0} \cdot S \frac{L_m}{L_s} \tag{14.85}$$

For ideal no-load, the stator and rotor voltages (in synchronous coordinates) are, for steady-state, d.c. quantities. They may be written as

$$V_{r0} = V_{s0}\left(\frac{\omega_1 - \omega_{r0}}{\omega_1}\right)\frac{L_m}{L_s} \tag{14.86}$$

$$\omega_{r0} = \omega_1\left(1 - \frac{V_{r0}}{V_{s0}}\frac{L_s}{L_m}\right) \tag{14.87}$$

For $V_{ro} > 0$ (zero phase shift) (and $V_{s0} > 0$), $\omega_{r0} < \omega_1$ and subsynchronous operation is obtained. In contrast, for $V_{r0} < 0$ (180° phase shift) (and $V_{s0} > 0$), $\omega_{r0} > \omega_1$, hyper-synchronous operation is obtained.

On the other hand, multiplying Equation 14.81 by $3/2i_r^*$ and extracting the real part, we obtain

$$P_{rotor} = \frac{3}{2}\mathrm{Re}(\overline{V}_r\overline{i}_r^*) = \frac{3}{2}r_r i_r^2 - S\frac{\omega_1}{p_1}T_e \tag{14.88}$$

As known, $\omega_1 T_e/p_1 = P_{elm}$ is the electromagnetic (airgap) power transferred through the airgap from stator to rotor

$$P_{rotor} = \frac{3}{2}r_r i_r^2 - SP_{elm} \tag{14.89}$$

It is now clear that we may have $P_{elm} > 0$ (motoring) or $P_{elm} < 0$ (generating) for both positive and negative slips, that is subsynchronous and hyper-synchronous operation, provided that the active electric power injected in the rotor $P_{rotor}$ may be either positive or negative.

The definition of slip frequency $\omega_2$

$$\omega_2 = \omega_1 - \omega_r = S\omega_1 \tag{14.90}$$

Equation 14.90 shows that for $S < 0$ ($\omega_r > \omega_1$), $\omega_2 < 0$, the phase sequence in the rotor is inverse with respect to the stator.

Also, multiplying Equation 14.81 by $3/2i_r^*$, but extracting the imaginary part, we obtain

$$Q_{rotor} = \frac{3}{2}\mathrm{Imag}(\overline{V}_r\overline{i}_r^*) = S\omega_1\frac{3}{2}\mathrm{Imag}(j\overline{\lambda}_r\overline{i}_r^*) \tag{14.91}$$

Neglecting the stator resistance and replacing $\overline{\lambda}_s$ with $\overline{\lambda}_r$ in Equation 14.77, we may obtain, after multiplication with $\overline{i}_s^*$ (and $\overline{i}_s = -\overline{i}_r$; $\overline{\lambda}_r = \overline{\lambda}_s - L_{sc}\overline{i}_s$; $\overline{V}_s = j\omega_1\overline{\lambda}_s$)

$$Q_1 = \frac{3}{2}\mathrm{Imag}(\overline{V}_s\overline{i}_s^*) = \frac{3}{2}\omega_1 L_{sc} i_s^2 - \frac{Q_{rotor}}{|S|} \tag{14.92}$$

So, if reactive power is injected in the rotor, it will be subtracted from the reactive power injected in the stator. Eventually, leading PF ($Q_1 < 0$) in the stator is obtained at the cost of lagging PF in the rotor. If the PEC on the rotor side itself is capable—through capacitors—of producing $Q_{rotor}$, then the total PF may be close to unity or even slightly leading. However, as the rotor voltage is rather low, this is hardly a practical way to produce reactive power through capacitors, unless the number of turns/phase in the rotor is larger than in the stator (Chapter 15).

On the other hand, the rotor-side PEC may be of a voltage source or current source type. It seems that rotor voltage control leads to motor-only stable operation in the sub-synchronous mode and to generator-only in the hyper-synchronous mode. For rotor current control, both motoring and generating are stable in sub- and hyper-synchronous modes [13].

### 14.10.3 Sub- and Hyper-IM Cascade Control

The power structure of the sub- and hyper-IM cascade drive is as shown in Figure 14.38. The control system is related to controlling the speed and, eventually, the stator reactive power for motoring and active and reactive stator power for generating.

As the electromagnetic power $P_{elm}$ is obtained at fixed stator frequency, the torque $T_e$ is

$$T_e = \frac{P_{elm}}{\omega_1} p \tag{14.93}$$

But the torque may be estimated from stator flux as

$$T_e = \frac{3}{2} p \operatorname{Re}(j\bar{\lambda}_s \bar{i}_s^*) = \frac{3}{2} p \lambda_s i_{sT} \tag{14.94}$$

Now we may define for the stator a "reactive torque"

$$T_{Q_1} = \frac{Q_1}{\omega_1} \tag{14.95}$$

$$T_{Q_1} = \frac{3}{2} \operatorname{Imag}(j\bar{\lambda}_s \bar{i}_s^*) = \frac{3}{2} \lambda_s i_{sQ} \tag{14.96}$$

$$(\bar{i}_s)_{ref} = i_{sT} - j i_{sQ} \tag{14.97}$$

As the value of $f_1(\omega_1)$ is rather large, the stator flux estimation may be performed through the voltage model

$$\bar{\lambda}_s = \int \left(\bar{V}_s - r_s \bar{i}_s\right) dt + \bar{\lambda}_{s0} \tag{14.98}$$

$$\bar{\lambda}_s \approx \left(\bar{V}_s - r_s \bar{i}_s\right) \frac{T_c}{1 + sT_c} + \bar{\lambda}_{s0} \tag{14.99}$$

In essence, the reference stator current $(\bar{i}_s)_{ref}$ has to be reproduced. What we need to control, in fact, is the rotor current $(\bar{i}_r)_{ref}$

$$\left(\bar{i}_r\right)_{ref} = -\left(\bar{i}_s\right)_{ref} - j\bar{I}_\mu = -i_{sT} - j\left(I_\mu - I_{sQ}\right) \tag{14.100}$$

where $\bar{I}_\mu$ is the main flux magnetizing current. Only approximately

$$\bar{I}_\mu = \frac{\bar{\lambda}_s}{L_s} \tag{14.101}$$

where $(\bar{i}_r)_{ref}$ from Equation 14.100—with $i_{ST}$ from Equation 14.94 and $i_{SQ}$ from Equation 14.96—is still in synchronous coordinates. It has to be transformed into rotor coordinates

$$\left(\bar{i}_r\right)^r_{ref} = e^{-j(\theta_{er} - \omega_1 t)} \left|\left(\bar{i}_r\right)_{ref}\right| \tag{14.102}$$

So we need a rotor position sensor. Once the reference rotor current is known, the a.c. current controller PWM may be used in the rotor-side PEC to produce sinusoidal currents at slip frequency $S\omega_1$.

NOTE: The rotor position angle $\theta_{er}$ and rotor speed considered as measured here, may be estimated and thus a motion-sensorless drive is obtained (as developed in Chapter 15).

Figure 14.39 summarizes the controller just described for both options—motoring and generating—with independent stator torque or reactive power (torque) control.

For generating, the stator active power input has to be commanded but, again (with $\omega_1$ = const.), the reference torque $T_e^*$ may be calculated, so the motor and generating control is similar. Regenerative braking is considered in the drive control mode.

The stator reference reactive power $Q_1^*$ is transformed into a reactive reference torque $T_Q^*$. In fact, the active (torque) $i_{sT}$ and reactive $i_{sQ}$ stator current components are proportional to $P_{elm}$ and $Q_1$ as the stator flux $\lambda_s$ is constant.

The control system requires a rotor position sensor (resolver, for example) or a rotor position observer if sensorless control is targeted.

For starting, the PEC in the rotor has to be separated from the rotor as high voltages and currents occur. Note that the PEC is designed for 20%–30% of machine full power (and voltage). Alternatively, the drive may be started from rotor side with short-circuited stator. Then the stator is opened and the conditions for self-synchronization are quickly provided at variable speed (within the slip speed range). After self-synchronization loading may be performed.

It also has been shown [4] that current faults (short-circuits) in the stator produce overvoltages in the rotor. Stator voltage variations induce rotor overcurrents. In all these situations, the starting resistors may be connected for protection purposes.

The sub- and hyper-IM cascade represents a rather generalized (unified) solution for limited variable speed motoring and generating applications.

While pumped-storage hydropower plants of various powers (up to 400 MW/unit, in principle) are the obvious applications, there are other industrial plants, such as wind generators (Chapter 15), which are performance-cost competitive.

Transient behavior for motor operation mode of such a doubly fed induction machine with VC in a 400 MW pump-storage power plant at Ohkawachi (Japan) [14] is shown in Figure 14.40.

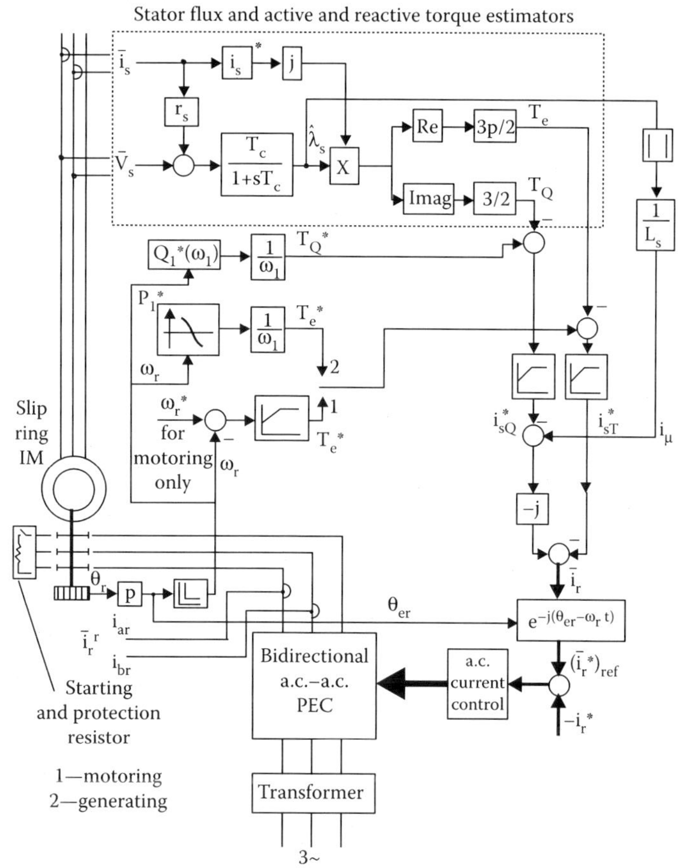


**FIGURE 14.39**
Fixed stator frequency motor-generating control with sub- and hyper-synchronous IM cascade.

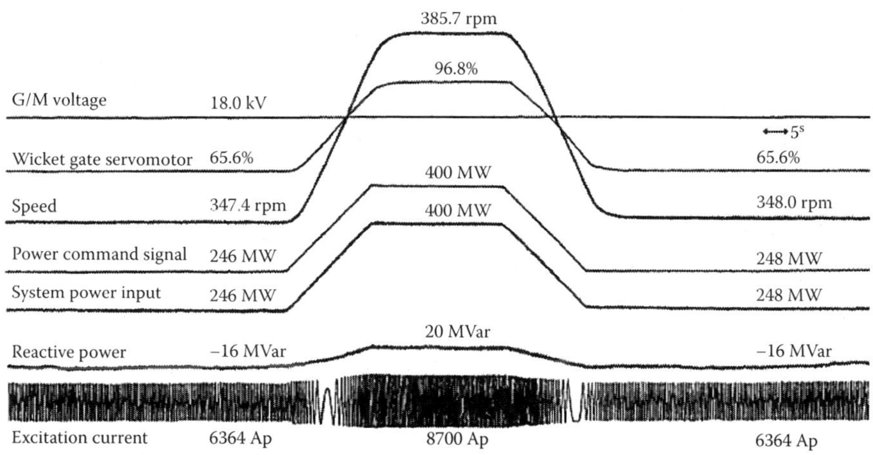


**FIGURE 14.40**
400 MW unit doubly fed induction machine motoring transient with smooth passage through conventional synchronous speed.

The smooth passage through the conventional synchronous speed ($n_1 = f_1/p$) is evident, together with fast active power (torque) response and small variation of reactive power. This particular case apparently refers to the largest-power electric motor drive so far.

## 14.11 Summary

- High power means power levels beyond the reach of two-level IGBT PWM converters. This limit goes up by the year and it is now about 4 MW.
- GTOs and thyristors are the standard SCRs used in large-power industrial drives. But IGCTs, GCTs, and IGBTs are catching on.
- Representative converters for large power may be of voltage source type or current source type.
- 3 (5)-level GTO |IGCT, two-stage PWM a.c.–a.c. voltage-source converters for high speeds and cycloconverters (for low speeds) represent the current-controlled voltage source types and are used with cageless SMs [15].
- Phase-delay rectifier CSIs are of the current source type and are used with cage-rotor SMs.
- Vector current control at unity PF—indirect or direct version—is used to control the current in the voltage source type a.c.–a.c. converter sensorless SM drives [8,16].
- E.m.f. compensation or voltage decoupler may be added for better response at high speeds.
- DTFC may be used for voltage source a.c.–a.c. PECs for a simpler and more robust system.
- Phase-delay rectifier CSI-SM drives have a low input PF at low speeds (due to the rectifier) though the CSI is load (e.m.f.) commutated as the SM works with the leading constant PF. Both vector current control and DTFC systems are feasible for such drives. There are methods to improve input (line) PF and reduce line current harmonics [17,18].
- For limited speed control range (±(20%–30%)) wound-rotor IMs are provided with a rotor-side PEC and, eventually, a step-up transformer to the power grid.
- The sub- and hyper-synchronous IM cascade, suitable for limited speed control (20%–30%), requires a starting and protection resistor, other than the transformer and a.c.–a.c. converter on the rotor side.
- The cyclo or matrix converters [19] can handle motoring and generating modes both sub- and hyper-synchronously (below and above conventional synchronous speed $\omega_1$ [stator frequency]).
- Though the rotor-side converter can handle bidirectional reactive power flow, the total PF of the system is, in general, slightly lagging (unless special capacitor power conditioners are added).
- As expected, vector current control or DTFC is to be applied to such systems for quick, precise, and robust speed (power) control.

- Dual stator winding stators with single-phase short-circuited (nested) winding or anisotropic rotors of adequate pole number combinations, with only one stator winding converter-fed, have been recently proposed for medium power but are still in the laboratories [20,21].

## 14.12 Problems

14.1. Unity PF SM drive: A nonsalient pole large SM motor is fed from a three-level PWM VSI at $\omega_r = 376.7$ rad/s, $p = 2$ with a $V_L = 4.6$ kV line to line voltage (rms). The motor rated phase current is $I_n = 1000$ A (rms), the synchronous inductance is $l_d = l_q = 0.75$ p.u. The drive works at the unity PF and the leakage inductance $l_{sl} = 0.17$ for star connection. Neglecting the stator resistance, determine

  a. The stator flux space vector amplitude.
  b. For a d-q flux angle $\delta = 30°$ and rated current calculate the $i_d$, $i_q$ current components.
  c. The field current $i_F$ and the corresponding torque $T_e$.

14.2. An IM cascade drive is designed for a ±20% speed control around synchronous speed. With $L_m/L_s = 0.93$ and stator voltage $V_{sn} = 5$ kV, (star connection line voltage [rms]) $I_{sn} = 1000$ A, $l_{sc} = 0.05V_{sn}/i_{sn}\omega_1$; $\omega_1 = 2\pi 60$ rad/s. Determine

  a. The rotor voltage $V_{r0}$ (value and sign)—in synchronous coordinates—for $\omega_{r0} = (1 \pm 0.2)\omega_1$ at no load.
  b. The required reactive power injection to the rotor side for $\omega_r = 0.8\omega_1$ ($S = 0.2$) to produce unity PF on stator side.
  c. The cycloconverter and step-up transformer—on rotor side—rating.

## References


1. P.H. Hammond, A new approach to enhance power quality for medium voltage a.c. drives, *IEEE Trans.*, vol. IA-33, no. 1, 1997, pp. 202–208.
2. R.A. Errath, 15000HP gearless ball mill drive in cement—why not, *IEEE Trans.*, vol. IA-32, no. 3, 1996, pp. 663–669.
3. H. Okayama, Large capacity high performance 3 level GTO inverter system for steel main rolling mill drives, *Record of IEEE-IAS, Annual Meeting*, vol. I, 1996, pp. 174–179 (IEEExplore).
4. H. Stemmler, High power industrial drives, *Proc. IEEE*, vol. 82, no. 8, 1994, pp. 1266–1286.
5. B. Wu, *High Power Converters and AC Drives*, IEEE Press, New Jersey and Wiley Interscience, New Jersey, 2006.
6. D. Andler, R. Alvarez, S. Bernet, J. Rodriguez, Experimental investigation of the commutations of a 3L-ANPC phase leg using 4.5 kV—5.5 kA IGCTs, *IEEE Trans.*, vol. IE-60, no. 11, 2013, pp. 4820–4830.
7. T. Nakama, H. Ohsawa, K. Endoh, A high performance cycloconverter-fed synchronous machine drive system, *IEEE Trans.*, vol. IA-20, no. 5, 1984, pp. 1278–1284.

8. I. Boldea, G. Andreescu, A. Pillati, D. Casadei, C. Rossi, Active flux based sensorless vector control of a DC excited SM drive, *Record of IEEE-ECCE*, 2009, pp. 2496–2503.
9. J. Rosa, Utilization and rating of machine commutated inverter-synchronous motor drives, *IEEE Trans.*, vol. IA-15, no. 2, 1979, pp. 395–404.
10. S. Nishikata, T. Kataoka, Dynamic control of a selfcontrolled synchronous motor drive system, *IEEE Trans.*, vol. IA-20, no. 3, 1984, pp. 598–604.
11. M. Miyazaki et al., New application of thyristor motor drive system, *Record of IEEE-IAS, Annual Meeting*, 1984, pp. 649–654 (IEEExplore).
12. L. Bojtor, L. Schmidt, Microprocessor controlled converter fed synchronous motor using subtransient flux model, *Record of IEEE Electric Machines and Power Systems*, vol. 20, no. 3, 1992, MIT, Cambridge, Part 2, pp. 297–308.
13. A. Masmoudi, M.B.H. Kamoun, M. Poloujadoff, A comparison between voltage and current controls of doubly fed synchronous machine with emphasis on steady-state characteristics, *Record of ICEM-1996*, vol. 3, 1996, Vigo, Spain, pp. 212–217.
14. T. Kawabara, A. Shibuya, H. Furata, Design and dynamic response characteristics of 400 MW adjustable speed pump storage unit for Okawachi power station, *IEEE Trans.*, vol. EC-11, no. 2, 1996, pp. 376–394.
15. S. Nonaka, Y. Ueba, A PWM GTO current source converter—inverter system with sinusoidal input and output, *Record of IEEE-IAS, Annual Meeeting*, vol. I, 1987, pp. 247–252.
16. Y. Zhou, S. Long, Sensorless direct torque control for electrically excited synchronous motor based on injecting high-frequency ripple current info rotor winding, *IEEE Trans.*, vol. EC—30, no. 1, 2015, pp. 246–253.
17. Z. Wang, B. Wu, D.I. Xu, N.R. Zargari, A current–source–converter-based high-power high-speed PMSM drive with 420 Hz switching frequency, *IEEE Trans.*, vol. IE-59, no. 7, 2012, pp. 2970–2981;
18. A. Tessarolo, C. Bassi, G. Ferrari, D. Giulivo, R. Macuglia, R. Menis, Investigation into the high-frequency limits and performance of load commutated inverters for high speed synchronous motor drives, *IEEE Trans.*, vol. IE-60, no.6, 2013, pp. 2147–2157.
19. L. Empringham, J.W. Kolar, J. Rodriguez, P.W. Wheeler, J.C. Clare, Technological issues and industrial application of matrix converters: A review, *IEEE Trans.*, vol. IE-60, no. 10, 2013, pp. 4260–4271.
20. W.R. Brassfield, R. Spee, T.G. Habetler, Direct torque control of brushless doubly-fed machines, *IEEE Trans.*, vol. IA-32, no. 5, 1996, pp. 1098–1104.
21. Y. Liao, L. Xu, Li Zhen, Design of a doubly fed reluctance motor for adjustable speed drives, *IEEE Trans.*, vol. IA-32, no. 5, 1996, pp. 1195–1203.

# 15

# *Control of Electric Generators*

## 15.1 Introduction

Practically all electric energy is produced through electric generators driven by prime movers. The solar electric (photovoltaic) power panels are the notable exception.

The prime movers extract their mechanical energy from the water or wind energy, from steam (by burning oil or coal or nuclear fuel) or gas—burning thermal energy, gasoline, or diesel fuel in internal combustion engines (ICEs).

The prime movers are basically wind, hydro, steam turbines, or ICEs.

The standard is that electric energy is produced in power plants with the prime movers (turbines) operated at controlled, almost constant, speed, through a speed governor that acts on the fuel input rate of the prime movers.

Synchronous generators (SGs) (up to 1400 MVA/unit) are used in standard power systems. They provide controlled constant voltage (through their d.c. rotor excitation current control) and constant frequency ($f_n = n_n p$) through prime-mover speed control.

Regional, national, and continental electric power systems have been built by adding SGs in parallel.

The rather weak coupling between the frequency (active power) and voltage (reactive power) control in SGs works well in stiff power systems with sizeable spinning power reserves.

The recent opening of electrical energy markets leads to the separation of electric energy production, transmission, and distribution to customers. Also, decentralization in energy production is being accentuated and results in more distributed ("smart") electric power systems.

The tendency to produce the daily peaks of electric energy closest to their consumption has led to smaller generator units, with variable output.

Tighter environmental requirements have prompted more renewable energy conversion systems. Wind (up to 8 MW/unit) and hydro energy (up to 770 MW/unit) are typical renewable energy sources with a softer effect on the environment.

In distributed electric power systems, the coproduction of heat and electricity is proposed through super-high-speed gas turbine-electric generators up to 3 MVA/unit at 15,000 rpm or 100 kVA/unit at 70,000 rpm. Small companies, villages, or town sections may profit from such highly efficient decentralized energy production solutions.

Autonomous generators sets, driven mainly by gas turbines or ICEs, even by small hydro or wind turbines, are also used for emergency (or standby) power in telecom, hospitals, banks, etc. with power/unit up to a few hundred kW.

Finally, more and more electric energy in vehicles is needed for better comfort, better gas-mileage in standard ICE vehicles, and also for the newly developed hybrid electric vehicles (HEV). Electric machines on HEV for powers up to 100 kW and more should work both as motors, for vehicle propulsion assistance, and as electric generators for battery recharging and electrical vehicle braking.

Large- and medium-power SGs in power systems are controlled in terms of voltage (reactive power), while the frequency (speed) or active power is controlled through the mechanical prime mover's speed governor.

In the distributed power systems of the future, fast active power and voltage control may be harnessed better when variable speed is allowed for.

Variable speed electric generators imply PEC interfaces to produce constant voltage and frequency output.

Wind and small hydro turbine electric generators are typical for almost mandatory variable speed control.

So full rating (100%) PECs with electrically excited, or PM, SGs, or cage-rotor induction generators up to 4–10 MW are to be expected for wind and small hydro applications.

Partial rating (20%–30%) PECs connected to the wound rotor of induction generators have already been introduced for limited (±(20–30) %) speed control range up to 400 MVA in pump-storage hydropower plants.

Switched reluctance generators (SRGs) with full rating power electronics control have been proposed for aircraft (up to 250 kW) or for HEVs (up to 100 kW).

Also, full rating (100%) power electronics control PM brushless generators/motors are introduced as starter—alternators up to 100 kW and more, on board of hybrid (or electric) vehicles.

In what follows we will concentrate on a few essential control aspects of electric generators, recommending the interested reader to go to Reference 1 and then take it from there.

The main issues treated here are

- Rather constant frequency (active power) and voltage (reactive power) control of SGs connected to electric power systems
- Limited variable speed wound-rotor induction generator (WRIG) control
- Full-power electronics control of variable speed PM SGs.
- Full-power electronics control of variable speed cage-rotor induction generators
- The control of claw-pole rotor alternators on automobiles
- IPM synchronous starter-alternator control on HEV
- Switched reluctance generator (SRG) output control at variable speed

## 15.2 Control of SGs in Power Systems

The a.c. power systems operate satisfactorily when their frequency and voltage remain nearly constant and thus vary in a limited and controllable manner, when active and reactive power loads they serve are modified.

Active power flow is controlled by modifying the prime-mover fuel flow rate such that to control the speed of the generator with a few percent variation range.

On the other hand, reactive power flow is related to generator terminal voltage control, within a few percent change range, by varying the d.c. field rotor winding current.

When too a large electric power load occurs, the speed may collapse. Similarly, too a large reactive power load leads to voltage collapse.

When a SG acts alone on its loads such as in generator sets or it is the largest by far in the respective power system, the SG is controlled for constant (isochronous) speed.

In contrast, when an SG operates in a large-power system, the load power is shared by many SGs. In this case it is mandatory to control the generator with a speed droop.

A voltage droop is also required for voltage control, or to provide for reactive power load sharing between SGs.

Automatic generation control (AGC) divides the electric power generation between SGs, while the automatic reactive power control (AQC) distributes the reactive power contribution between same SGs.

As there is some coupling between P (active power) and Q (reactive power) controls, especially for weak-power systems, power system stabilizers (PSS) are added for decoupling.

All these principles are visible in the general SG control system in Figure 15.1.

The speed governor and its controller and the frequency (speed)/power droop curve serve to control the speed to produce the required active power P*.

The speed droop curve may be raised or lowered to decrease or increase the power share of the respective generator.

On the generator side, the voltage controller and limiters act on the voltage error $V_c^* - V_c$, corroborated with the PSS output. A voltage load compensator is added. It is beyond our scope here to explain the whole, multilevel, control system of an SG into a large-power system ([1], Chapter 2).

We will concentrate a little only on SG exciters and on the automatic voltage regulator (AVR).

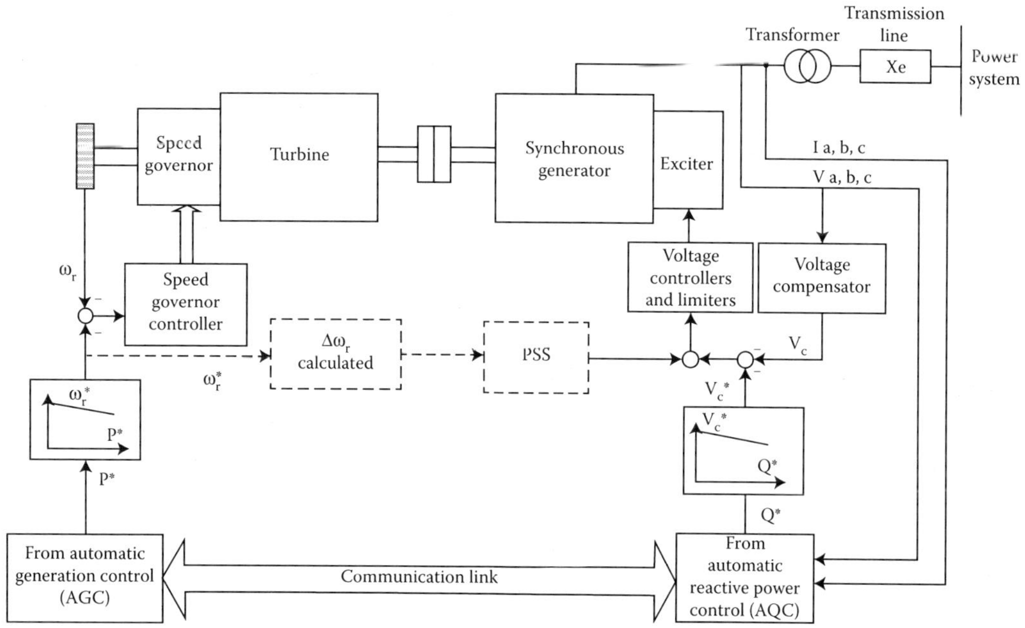


**FIGURE 15.1**
General SG control system.

### 15.2.1 Exciters for SG

The exciter is an electric power source that is capable to d.c. supply in a controllable manner the field winding of the SG such that the SG terminal voltage is satisfactorily controlled under designed active and reactive power load variation range.

There are two main types of exciters in fabrication today:

- a.c. brushless exciters
- Static exciters

### 15.2.2 A.C. Brushless Exciter

The standard a.c. brushless exciter (Figure 15.2) is made of an inside-out synchronous machine whose stator holds a d.c. fed field winding while its rotor contains the three-phase (armature) winding and a diode rectifier which supplies directly the SG field winding. The stator-placed exciter d.c. field winding current is controlled through a small-power static power converter, to provide controlled voltage at SG terminals.

It is evident that the time constant and transfer functions of both, SG and the a.c. brushless exciter (basically an inverted, smaller, 3%, rating SG), stand between the command voltage $V_{con}$ at the output of the voltage regulator and the regulated terminal voltage of SG.

There is no easy way to measure any variable such as the SG field current or voltage and thus a very robust AVR is needed.

The time allowed for the field winding voltage $V_f$ to be increased from 0% to 95% rated value is limited by standards to a few tens of milliseconds. On the other way, forcing the voltage $V_f$ to 1.5–3 times rated value $V_{fn}$ is needed to produce acceptably fast terminal voltage response to perturbations.

The a.c. brushless exciter has the inherent advantage that it is rather immune to SG faults, when the SG terminal voltage decreases notably, if the small-power static converter (Figure 15.2) has a back-up source during SG faults or is fed from a rechargeable battery.

An approximate model of the a.c. exciter (alternator without a rotor cage) is shown in Figure 15.3.

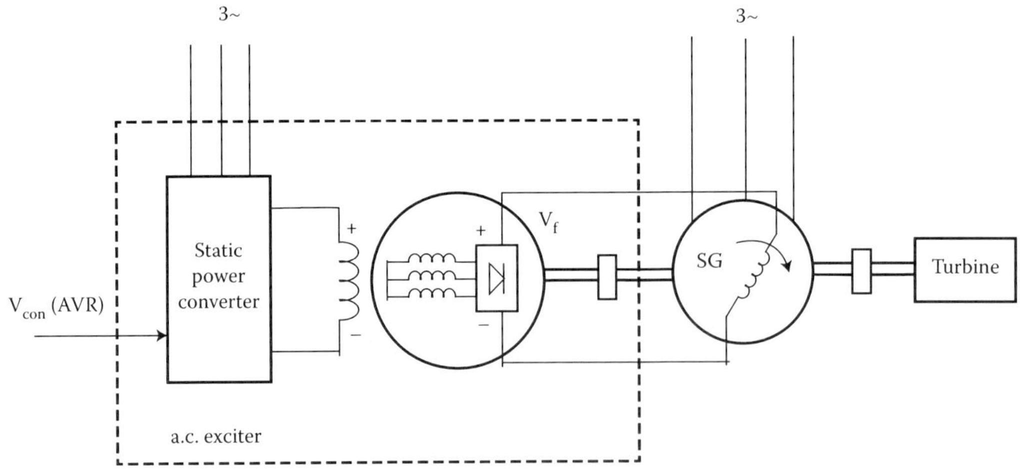


**FIGURE 15.2**
The a.c. brushless exciter.

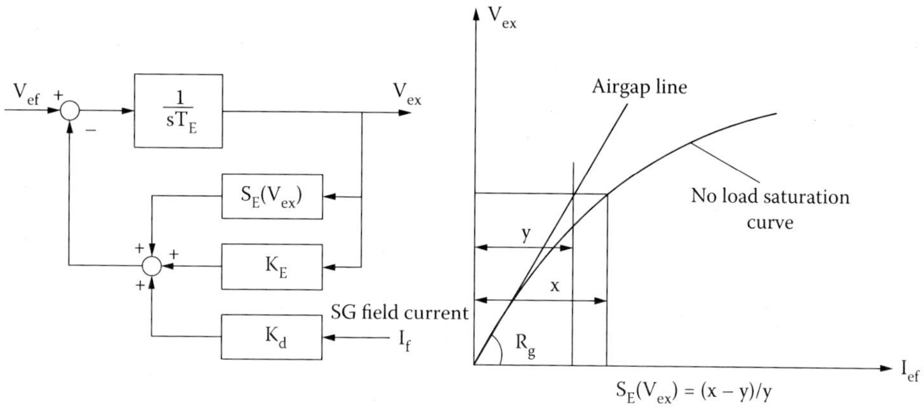


**FIGURE 15.3**
The a.c. exciter alternator structural diagram for control.

The load of the a.c. exciter alternator comes into play only through the SG field current $I_f$ produced armature reaction in the a.c. exciter. As the diode rectifier provides for almost unity power factor, $I_f$ influence on $V_{ex}$ regulation is rather straightforward (Figure 15.3).

$S_E$ ($V_{ex}$) introduces the influence of the magnetic saturation in the a.c. exciter on the $V_{ex}$ limitation. $T_E$ is the a.c. exciter alternator excitation time constant.

The diode rectifier (Figure 15.4) introduces an additional voltage regulation and its model is needed also.

The $V_f$ ($I_f$) output characteristic is rather nonlinear and depends heavily on diode commutation overlapping.

The a.c. exciter reactance ($X_{ex}$) plays a major role in the diode commutation process.

Three main operation modes may be identified:

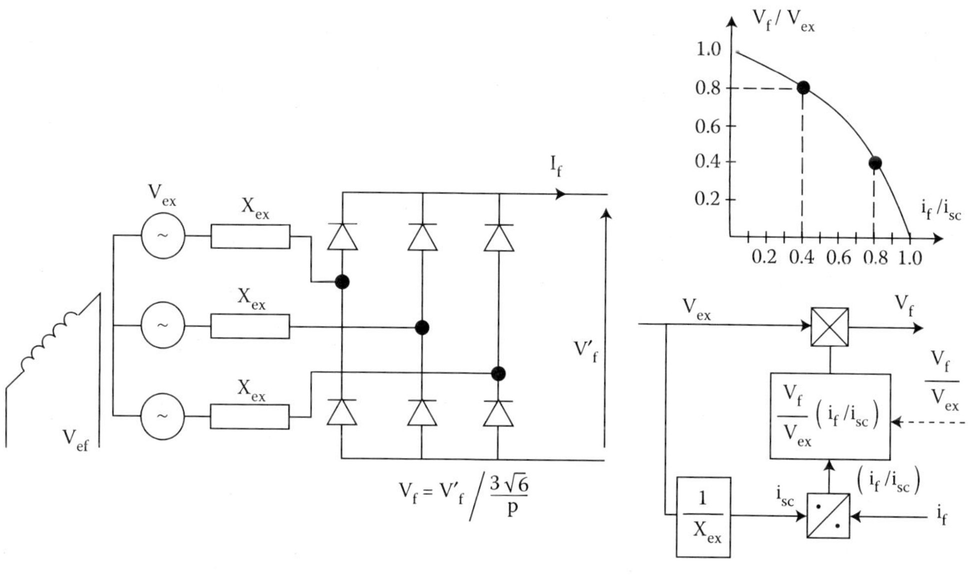


**FIGURE 15.4**
The a.c. exciter alternator + diode rectifier.

*Stage 1*: Two diodes conducting (low load):

$$\frac{V_f}{V_{ex}} \approx 1 - \frac{1}{\sqrt{3}}\frac{I_f}{I_{sc}} \quad \text{for } \frac{I_f}{I_{sc}} < \left(1 - \frac{1}{\sqrt{3}}\right) \tag{15.1}$$

with

$$I_{sc} = \frac{V_{ex}\sqrt{2}}{X_{ex}} \tag{15.2}$$

*Stage 2*: Each diode conducts only when the other one on the same leg ended conduction (medium load):

$$\frac{V_f}{V_{ex}} = \sqrt{\frac{3}{4} - \left(\frac{I_f}{I_{sc}}\right)^2} \quad \text{for } \left(1 - \frac{1}{\sqrt{3}}\right) \le \frac{I_f}{I_{sc}} \le \frac{3}{4} \tag{15.3}$$

*Stage 3*: Four diodes are conducting at any time:

$$\frac{V_f}{V_{ex}} = \sqrt{3}\left(1 - \frac{I_f}{I_{sc}}\right) \quad \text{for } \frac{3}{4} \le \frac{I_f}{I_{sc}} \le 1 \tag{15.4}$$

Figure 15.4 illustrates Equations 15.1 through 15.4 and introduces the nonlinear structural diagram that represents the diode rectifier.

### 15.2.3 Static Exciter

The standard static exciter for SGs is the fully controlled phase rectifier (Chapter 4)—Figure 15.5 where $V_{ex}$, $X_{ex}$ represent the voltage and internal (transient) reactance of the power source (a transformer in general) that supplies the static exciter. The static exciter is placed on ground and thus it transmits the power to the SG field winding through brushes and copper slip rings.

As shown in Chapter 5, the controlled rectifier output voltage $V_f$ is

$$V_f' = \frac{3\sqrt{2}V_{ex}\sqrt{3}\cos\alpha}{\pi} - \frac{3}{\pi}X_{ex}I_f; \quad I_{sc} = \frac{V_{ex}\sqrt{2}}{X_{ex}} \tag{15.5}$$

Consequently

$$\frac{V_f}{V_{ex}} = \cos\alpha - \frac{1}{\sqrt{3}}\frac{I_f}{I_{sc}}; \quad V_f = \frac{V_f'\pi}{3\sqrt{6}} \tag{15.6}$$

For $\alpha = 0$—zero-phase delay angle—Equation 15.6 degenerates into Equation 15.1, the diode rectifier.

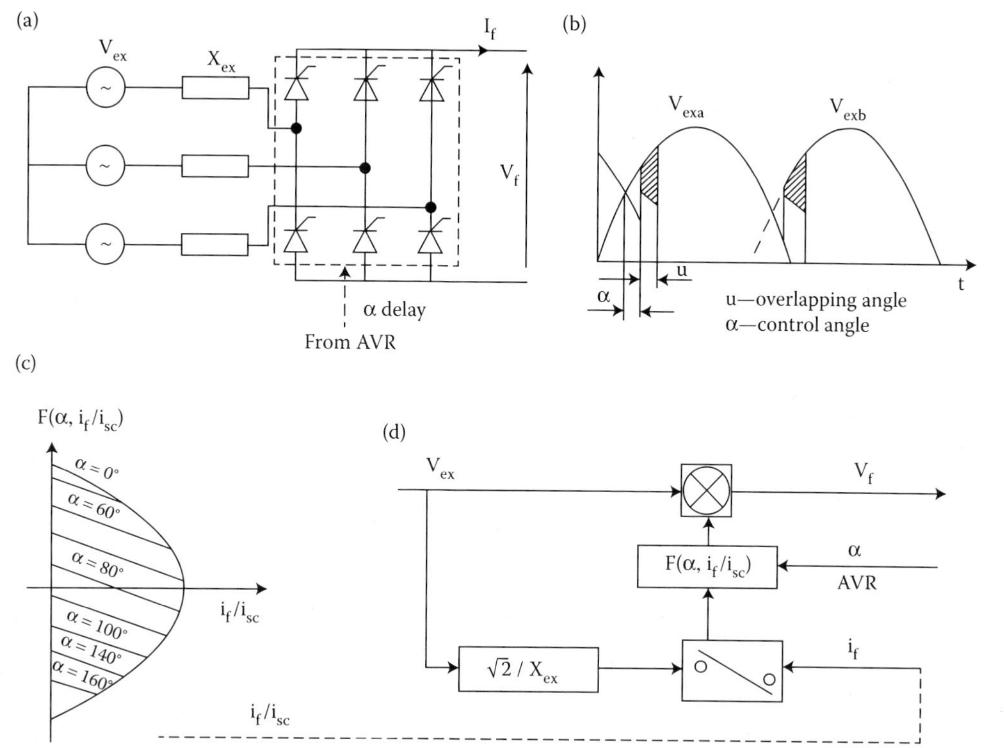


**FIGURE 15.5**
(a) The controlled rectifier as static exciter, (b) its output voltage waveform, (c) voltage/current curve, (d) nonlinear structural diagram.

With the exciter main parts already modeled here, only the automatic voltage regulator (AVR) is still needed to complete the system [2].

Apart from standardized excitation, AVR systems such as IEEE 1992 AC1A (with a.c. brushless exciter), and IEEE 1992 ST1A (with static exciter) which are basically analog systems, let us consider here a digital PID AVR system.

### 15.2.4 A Digital PID AVR System

The a.c. brushless exciter is simplified (armature reaction is neglected) and the diode rectifier model is a gain, as in Figure 15.6. A proportional–derivative–integral (PID) voltage regulator is used.

With a sampling frequency more than 20 times the damped frequency of the closed-loop system, we may design the PID controller as if it were continuous, with

$$G_c(s) = K_P + \frac{K_I}{s} + sK_D \tag{15.7}$$

where $K_P$ is the proportional gain, $K_I$ is the integral gain, and $K_D$ is the derivative gain.

As the exciter was considered here as a first-order model (only one time constant, $T_E$), the SG may be modeled also as a first-order system with the excitation transient time constant $T'_{d0}$, at constant speed and for small deviation transients.

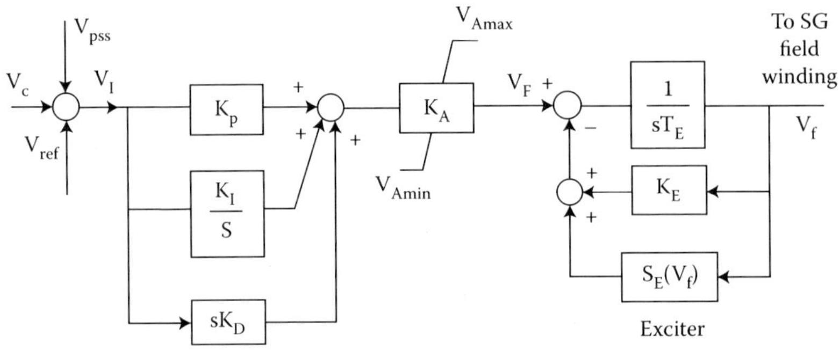


**FIGURE 15.6**
PID voltage regulator for SG.

Consequently, the a.c. exciter plus the SG exhibits a simplified second-order transfer function G(s):

$$G(s) = \frac{l_{dm}/r_f(S_E + K_E)}{(1 + sT'_{d0}) \cdot (1 + sT_e)}; \quad T_e = \frac{T_E}{S_E + K_E}; \quad T'_{d0} = \frac{l_{dm} + l_{fl}}{\omega_b \cdot r_f} \tag{15.8}$$

where $l_{dm}$ is the SG magnetization inductance in p.u. and $r_f$ is the SG field winding resistance in p.u; $l_{fl}$ is the SG field winding leakage inductance in p.u; and $\omega_b$ is the base (rated) angular frequency of the SG.

The closed-loop system, which includes the PID controller, has the known characteristic equation:

$$G(s) \cdot G_c(s) + 1 = 0 \tag{15.9}$$

Let us consider $l_{dm}/(r_f(S_E + K_E)) = 1$ for simplicity.

Making use now of Equations 15.7 and 15.8 in Equation 15.9 we obtain

$$K_D s^2 + K_P s + K_I = -s(1 + sT'_{d0})(1 + sT_e) \tag{15.10}$$

As it would be desirable to deal with a second-order system, we select from start a negative pole $s_3 = c$ in the left half-plane. The other two poles are chosen as complex and conjugate: $s_{1,2} = a \pm jb$.

The peak overshoot and settling time, or other method of pole placement, may be used to find the two remaining unknowns $s_{1,2}$ in Equation 15.10 and thus find the controller gains $K_P$, $K_I$, $K_D$.

The controller gains give rise to two zeros that might affect the transients unfavorably and thus, by trial and error, the final values of $K_P$, $K_I$, $K_D$ are settled.

As an example, for $T'_{d0} = 1.5$ s, $T_e = 0.3$ s, $f_1 = 60$ Hz, settling time = 1.5 s, peak overshoot = 10%, a satisfactory analog PID controller gain set would be $K_P = 39.3$, $K_I = 76.5$, $K_D = 5.4$ [3].

The conversion into discrete form may use the trapezoidal integration method

$$s \to \frac{1 - z^{-1}}{T}; \quad \frac{1}{s} \to \frac{T}{2} \frac{1 + z^{-1}}{1 - z^{-1}} \tag{15.11}$$

where $z^{-1}$ is the unit delay.

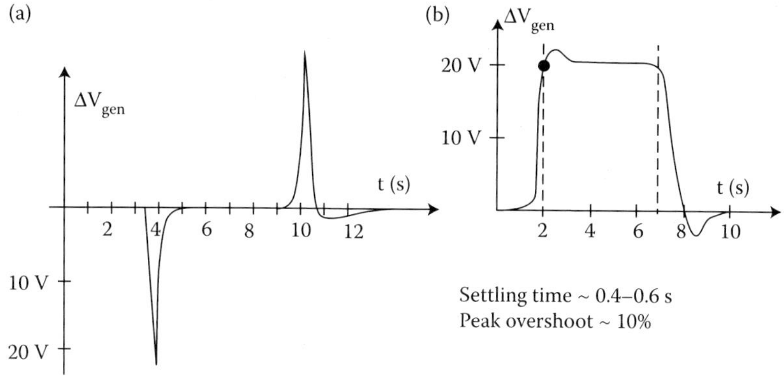


**FIGURE 15.7**
SG voltage response with PID AVR: (a) Step reactive power introduction and rejection, (b) step-up and down voltage response. (After Godhwani, A., Bassler, M.J., *IEEE Trans.*, EC-11(3), 616–620, 1996.)

The PID controller discrete transfer function $G_c(z)$ is thus

$$G_c(z) = \left[K_{PD} + \frac{K_{ID}}{1 - z^{-1}} + K_{DD}(1 - z^{-1})\right] \cdot K_{AA} = \frac{\Delta V_F(z)}{\Delta V_I(z)} \tag{15.12}$$

with $K_{PD} = K_P - K_I(T/2)$, $K_{ID} = K_I T$, $K_{DD} = K_D/T$.

The $K_{AA}$ gain was added in Equation 15.12.

Making use of the property: $z^{-1}X(k) = X(k - 1)$ leads to the discrete form of the PID voltage controller output $\Delta F(k)$

$$\Delta F(k) = \Delta F(k-1) + (K_{PD} + K_{ID} + K_{DD})\Delta V_I(k) - (K_{PD} + 2K_{DD})\Delta V_I(k-1) + K_{DD}\Delta V_I(k-2) \tag{15.13}$$

with $\Delta V_I$ as the SG voltage error variation.

For a 75 kVA, 208 V, 0.8 PF (lagging) SG with $T = 12.5$ ms, $K_{PD} = 777$, $K_{ID} = 19$, $K_{DD} = 8640$, $K_{AA} = 7.00$, a 50-KVAR reactive load application and rejection response is shown in Figure 15.7a. A step reference voltage response (up and down) is shown in Figure 15.7b.

Stable voltage response under severe reactive power surge is evident.

NOTE: The PID voltage regulator is a mere example of AVR and many other more robust configurations have been introduced. Among them the ones based on sliding mode and fuzzy logic stand out as apparently more practical.

## 15.3 Control of WRIGs with Limited Speed Range

Wound rotor induction generators (WRIGs) have been treated in Chapter 14 for large-power drive applications with limited speed control range.

The stator and rotor active power balance for steady state still holds (from Chapter 14):

$$P_s \approx -P_{Cos} + P_{elm}; \quad P_{elm} = \frac{T_e\omega_1}{p} \tag{15.14}$$

$$P_r \approx -p_{Cor} - SP_{elm}; \quad S = \frac{\omega_1 - \omega_r}{\omega_r} \tag{15.15}$$

$P_{Cos}$ and $P_{Cor}$ are the stator and rotor winding losses.

$P_s$, $P_r$ positive means generated (delivered) electric powers; S is slip and is positive in subsynchronous ($\omega_r < \omega_1$) and negative in supersynchronous ($\omega_r > \omega_1$) operation.

$P_{elm}$ is the electromagnetic power, which again is positive for the generator operation mode. The torque $T_e$ is considered positive for the generator mode ($\omega_1 = 2\pi f_1$, $f_1$—stator frequency, p—pole pairs).

For generator mode in subsynchronous operation, $P_s > 0$, $P_r < 0$, while for supersynchronous operation, $P_s > 0$, $P_r > 0$ (Figure 15.8).

The core and mechanical losses have been neglected here, for more clarity.

It should be noted that the stator power $P_s$ comes at frequency $f_1$ and the rotor electric power $P_r$ at frequency $f_2 = Sf_1$ with $|f_2| < 0.3 \cdot f_1$, in general.

Also $f_1$ = constant and $f_2$ is variable with speed:

$$f_1 = f_2 + np; \quad n - \text{speed}\,(\text{rps}) \tag{15.16}$$

For supersynchronous operation ($n > f_1/p$), $f_2 < 0$, which means that the sequence of rotor phases (supplied through a static power converter) is changed from abc to acb.

For supersynchronous operation, say at $S_{max} = -0.25$, with copper losses neglected, $P_r > 0$ and thus the total power of WRIG is

$$P_{max} = P_{elm} + |S_{max}| \cdot P_{elm} \tag{15.17}$$

The machine is electromagnetically designed for $P_{elm}$, at synchronous speed $n_1 = f_1/p$ but, because it runs at larger speed $n_{max} = n(1 + |S_{max}|)$ and produces additional electric power $P_r$ through the rotor $P_r = |S_{max}|\ P_{elm}$. This is considered a notable cost advantage of (WRIG).

The reactive power flow in WRIGs (wound-rotor induction generators) is not dependent on slip S sign, as expected, and we may magnetize the machine from the rotor side (through the static power converter) or from the stator. In addition, it may be feasible to

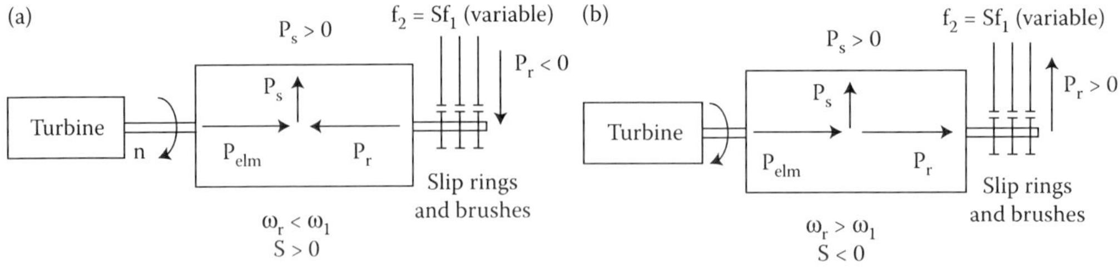


**FIGURE 15.8**
WRIG power balance: (a) subsynchronous generator mode, (b) supersynchronous generator mode.

work with reactive power delivery through the stator or the rotor-side converter if the latter is capable of providing it. Fortunately the ratio of the rotor reactive power $Q_r$ (at $f_2$), necessary to produce a reactive power $Q_s$ (at $f_1$), is

$$\frac{Q_r}{Q_s} = \frac{f_2}{f_1} = S \tag{15.18}$$

This is so because the magnetic energy, which is conserved, does not depend on frequency.

To limit the costs of the rotor-side converter, WRIGs in general are operated at best at unity power factor in the stator. In this case the current oversizing of the rotor static power converter over $|S_{max}|\ P_{elm}$ is not larger than 10%–15%, to provide for the reactive power required to magnetize the machine.

Typical static power converters used for the scope are

- Back to back voltage source two (or multi) level PWM converters
- Cycloconverters
- Matrix converters

They all provide four quadrant operations (Figure 15.9) but only back to back PWM converters and matrix converters have enough secondary frequency range to provide the starting of the machine as a motor (for pump-storage applications) from the rotor side, with short-circuited stator.

The two converters in Figure 15.9 are practically identical and constitute the typical two (or multi)-level voltage source PWM converters used for a.c. drives.

There is a filter $L_f$ at the source side but the voltage-matching transformer to the a.c. source is missing. It is feasible to leave it out if we design the WRIG such at maximum slip ($|S_{max}|$) the required rotor voltage $V_{rmax}$ is equal to the constant stator voltage $V_s$

$$V_{rmax} \approx V_s \tag{15.19}$$

where $E_s$ and $E_r$ are the stator and rotor e.m.f.s

$$S_{max} E_s \frac{W_r K_{Wr}}{W_s K_{Ws}} = E_{rmax} = V_{rmax} \tag{15.20}$$

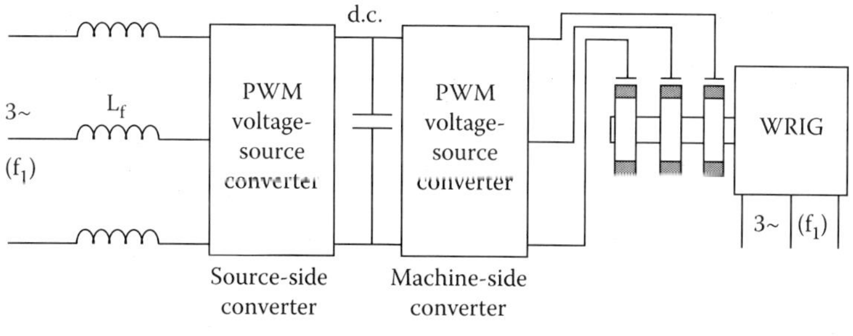


**FIGURE 15.9**
Bidirectional back to back PWM converter for the rotor of WRIGs.

where $W_s$, $W_r$ are the numbers of turns per current path in the stator and, respectively, in the rotor, while $K_{Ws}$, $K_{Wr}$ are the corresponding winding factors. For example, with $|S_{max}| = 0.25$ and identical winding factors ($K_{Ws} = K_{Wr}$), $W_r/W_s = 4/1$. This situation leads to a corresponding reduction of rotor rated current $I_r$: $I_r/I_s \cong 1/4$, as expected.

### 15.3.1 Space Phasor Model of WRIG

WRIG is a typical IM and thus the space-phasor model developed in Chapter 8 stands valid here

$$\begin{aligned} &\bar{I}_s R_s + \bar{V}_s = -\frac{d\bar{\lambda}_s}{dt} - j\omega_b \bar{\lambda}_s; \quad \bar{\lambda}_s = L_{st}\bar{I}_s + L_m \bar{I}_m; \quad \bar{I}_m = \bar{I}_s + \bar{I}_r \\ &\bar{I}_r R_r + \bar{V}_r = -\frac{d\bar{\lambda}_r}{dt} - j(\omega_b - \omega_r)\bar{\lambda}_r; \quad \bar{\lambda}_r = L_{rl}\bar{I}_r + L_m \bar{I}_m \end{aligned} \tag{15.21}$$

The electromagnetic torque $T_e$ ($T_e > 0$ for generating) is

$$T_e = \frac{3}{2} p I_{mag}\left[\bar{\lambda}_s \bar{I}_s^*\right] = \frac{3}{2} p(\lambda_d I_q - \lambda_q I_d) \tag{15.22}$$

With $d/dt = s$, Equations 15.21 become

$$\begin{aligned} &(R_s + (s + j\omega_b)L_{sl}) \cdot \bar{I}_s + \bar{V}_s = -L_{mt}s(\bar{I}_s + \bar{I}_r) - j\omega_b L_m(\bar{I}_s + \bar{I}_r) \\ &(R_r + (s + j(\omega_b - \omega_r))L_{rl}) \cdot \bar{I}_r + \bar{V}_r = -L_{mt}s(\bar{I}_s + \bar{I}_r) - j(\omega_b - \omega_r)L_m(\bar{I}_s + \bar{I}_r) \end{aligned} \tag{15.23}$$

These equations lead to the equivalent circuit of Figure 15.10.

$L_{mt}$ is the transient magnetization inductance while $L_m$ is the steady state one. Both depend on magnetization current $I_m$, due to magnetic saturation.

The speed of the reference system $\omega_b$ is free to choose but stator coordinates ($\omega_b = 0$), rotor coordinates ($\omega_b = \omega_r$), and synchronous coordinates ($\omega_s = \omega_1$) are favored, depending on the application.

For steady-state $s \to j(\omega_1 - \omega_b)$ with $\omega_1$ the frequency of actual stator variables (currents and voltages).

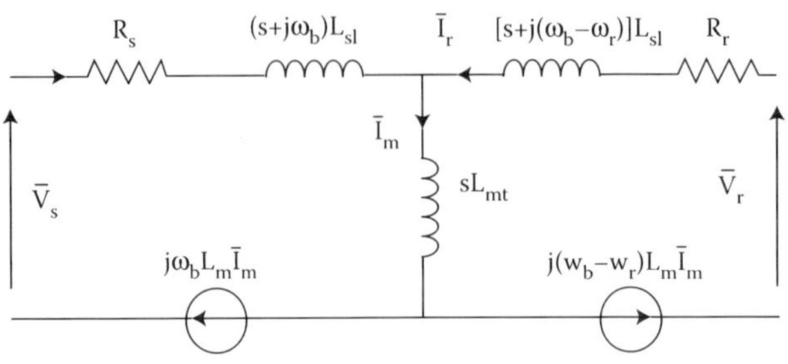


**FIGURE 15.10**
The space-phasor equivalent circuit of WRIG.

We may add two equations for the stator and rotor homopolar components, which are however independent of the space phasor components (Chapter 8):

$$I_{s0}R_s + V_{s0} \approx -L_{sl}\frac{dI_{s0}}{dt}$$
$$I_{r0}R_r + V_{r0} \approx -L_{rl}\frac{dI_{r0}}{dt} \tag{15.24}$$

NOTE: In all equations above, the rotor variables are reduced to the stator ones.

For steady state, the stator and rotor voltages in their phase coordinates, when symmetric, are

$$V_{abc}(t) = V_s\sqrt{2}\cos\left(\omega_1 t - (i-1)\frac{2\pi}{3}\right)$$
$$V_{ar\ br\ cr}(t) = V_r\sqrt{2}\cos\left((\omega_1 - \omega_r)t + \gamma_v - (i-1)\frac{2\pi}{3}\right) \tag{15.25}$$

Also for steady state a simplified vector diagram is obtained by using Equation 15.23 with $s \to j(\omega_1 - \omega_b)$.

In Figure 15.11 such a vector diagram is presented for unity power factor in the stator and subsynchronous operation.

A few remarks are in order here.

- As the stator power factor is unity, the machine magnetization comes from the rotor and thus the rotor flux amplitude $\lambda_r$ is larger than the stator flux amplitude $\lambda_s$.

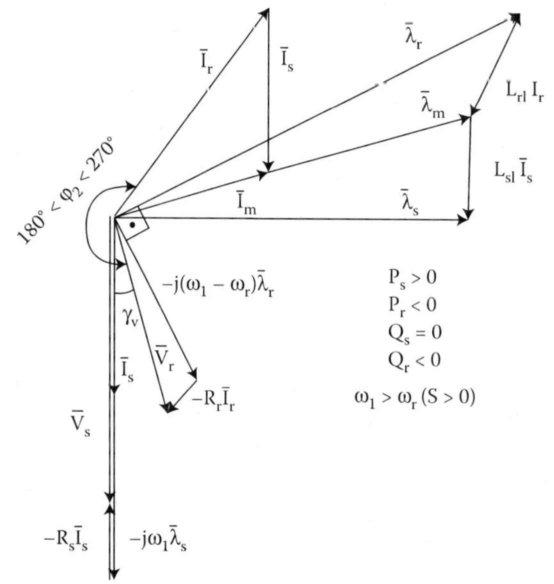


**FIGURE 15.11**
WRIG steady-state vector diagram for subsynchronous operation $((\omega_1 - \omega_b) > 0)$ at unity stator power factor.

- Also, the rotor power factor angle $\varphi_2$ is between 180° and 270° to provide for negative (drained) active and reactive rotor electric power.
- At zero slip (conventional synchronous speed, $\omega_1 = \omega_r$) the vector diagram still holds but then $\overline{V}_r \equiv \overline{I}_r R_r$ and the rotor flux $\lambda_r$ loses its paramount role in explaining the machine behavior.

As the stator flux voltage and frequency are kept rather constant, the reference system may be tied to stator voltage $\overline{V}_s$ or stator flux $\overline{\lambda}_s$, for VC.

### 15.3.2 Vector Control Principles

Let us align the reference system to stator flux $\overline{\lambda}_s$:

$$\overline{\lambda}_s = \lambda_s = \lambda_d; \quad \lambda_q = 0; \quad \frac{d\lambda_q}{dt} = 0 \tag{15.26}$$

In dq coordinates, the stator equation in Equation 15.19, with $\lambda_s$ ~ constant and $R_s$ ~ 0, becomes

$$\begin{aligned} &V_d = 0; \quad \lambda_q = L_s I_q + L_m I_{qr} = 0 \\ &V_q = -\omega_r \lambda_d; \quad \lambda_d = L_s I_d + L_m I_{dr} \end{aligned} \tag{15.27}$$

Consequently, the stator active and reactive powers $P_s$, $Q_s$ are

$$\begin{aligned} P_s &= \frac{3}{2}(V_d I_d + V_q I_q) = \frac{3}{2} V_q I_q \approx \frac{3}{2}\omega_1 \lambda_d \frac{L_m I_{qr}}{L_s} \\ Q_s &= \frac{3}{2}(V_d I_q - V_q I_d) = \frac{3}{2}\omega_1 \lambda_d I_d = \frac{3}{2}\omega_1 \frac{\lambda_d}{L_s}(\lambda_d - L_m I_{dr}) \end{aligned} \tag{15.28}$$

If is now evident that, with constant stator flux amplitude ($\lambda_d$), the stator active power, $P_s$, may be controlled by $I_{qr}$ (rotor q axis) current control, while stator reactive power, $Q_s$, control is performed through $I_{dr}$ (rotor d axis) current control. This is in fact the principle of current VC.

The rotor voltage equations, for steady state, in these conditions are simply

$$V_{dr} = -R_r I_{dr} + L_{sc} S\omega_1 I_{qr} \tag{15.29}$$

$$V_{qr} = -R_r I_{qr} - S\omega_1 \left( \frac{L_m}{L_r} \lambda_d + L_{sc} I_{dr} \right) \tag{15.30}$$

Now Equations 15.29 and 15.30 pave the way for combined voltage/current VC as they represent the voltage decoupling conditions.

### 15.3.3 Vector Control of the Machine-Side Converter

Combined voltage/current VC is straightforward by using Equations 15.28 and 15.30. Such a scheme is illustrated in Figure 15.12 where $P_s$ and $Q_s$ power controllers are added, and $\theta_s$ is the stator flux position angle with respects to phase a in the stator and $\theta_{er}$ is the rotor phase $a_r$ axis position with respect to stator phase a in electrical degrees.

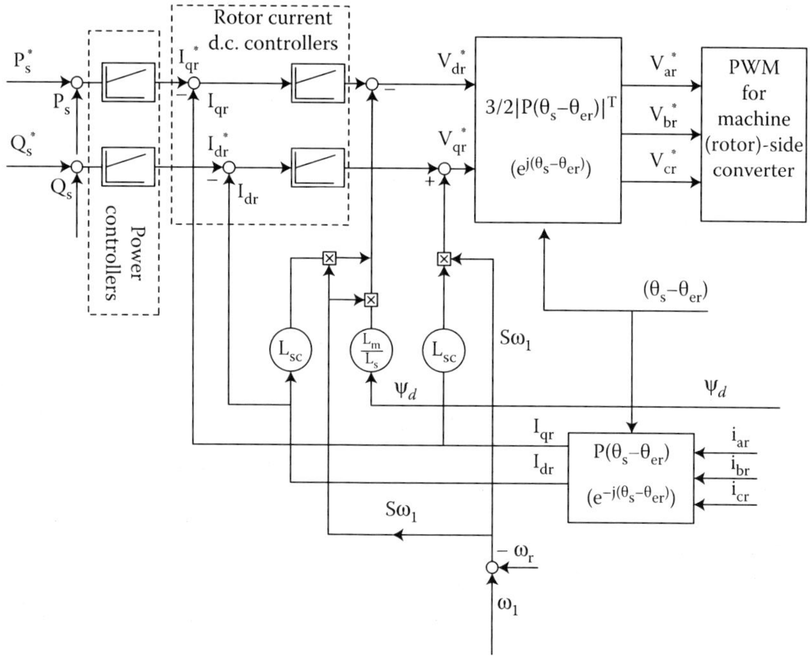


**FIGURE 15.12**
$P_s$, $Q_s$ VC of WRIG-side converter.

Dual Park transformation is required first to bring the rotor currents into stator flux coordinates and then, once again, to "translate" the reference rotor voltages $V_{dr}$, $V_{qr}$ into rotor coordinates.

It is evident that the stator voltages, stator currents, and rotor currents have to be measured. The stator flux estimator is straightforward

$$\bar{\lambda}_s - \bar{\lambda}_{s0} \int (V_s + R_s I_s) \cdot dt \tag{15.31}$$

The offset in the integrator has to be taken care of. A first-order delay replacement of the integrator might do it. This way the stator flux amplitude $\lambda_s = \lambda_d$ and angle $\theta_s$ are obtained.

The rotor position $\theta_{er}$ has to be either measured by a robust and precise encoder or may be estimated to provide for motion-sensorless control.

### 15.3.4 Rotor Position Estimation

For rotor position ($\theta_{er}$) estimation we first have to investigate the angle relationships between rotor axis angle $\theta_{er}$, stator flux angle $\theta_s$, and the rotor current vector (Figure 15.13).

Let us consider that we know the magnetization curve of the machine, say $I_{ms}(\lambda_s)$ and start with a given value of $I_{ms}$.

The αβ components of $I_{ms}$ are evidently (Figure 15.13)

$$\begin{aligned} I_{ms\alpha} &= I_{ms} \cos\theta_s \\ I_{ms\beta} &= I_{ms} \sin\theta_s \end{aligned} \tag{15.32}$$

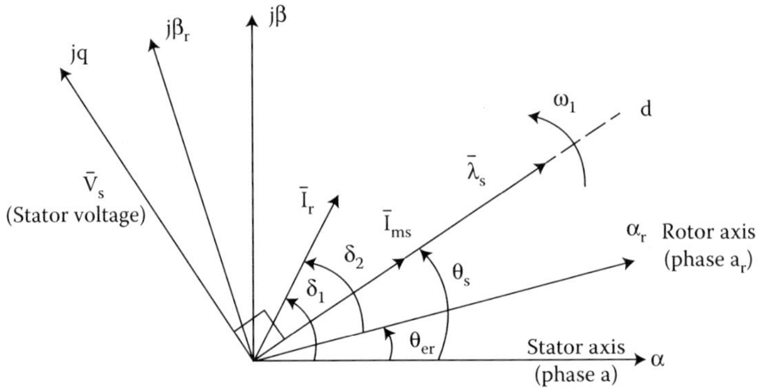


**FIGURE 15.13**
Location of rotor current vector $\bar{I}_r$.

Then, in stator coordinates, the rotor current components are (from $\bar{\lambda}_s = L_s\bar{I}_s + L_m\bar{I}_r$)

$$
\begin{aligned}
I_{r\alpha} &= (I_{ms\alpha} - I_{s\alpha})\frac{L_s}{L_m}; \quad I_{s\alpha} = I_a \\
I_{r\beta} &= (I_{ms\beta} - I_{s\beta})\frac{L_s}{L_m}; \quad I_{s\beta} = \frac{1}{\sqrt{3}}(2I_b + I_a) \\
I_r &= \sqrt{I_{r\alpha}^2 + I_{r\beta}^2}
\end{aligned}
\tag{15.33}
$$

The rotor currents are, however, measured in rotor coordinates, $I_{r\alpha r}$, $I_{r\beta r}$, and

$$
\begin{aligned}
\cos\delta_2 &= \frac{I_{r\alpha r}}{I_r}, \quad \sin\delta_2 = \frac{I_{r\beta r}}{I_r} \\
I_{r\alpha r} &= I_{ar}; \quad I_{r\beta r} = \frac{1}{\sqrt{3}}(2I_{br} + I_{ar})
\end{aligned}
\tag{15.34}
$$

But $\hat{\theta}_{er} = \delta_1 - \delta_2$ and thus

$$
\begin{aligned}
\sin\hat{\theta}_{er} &= \sin(\delta_1 - \delta_2) = \sin\delta_1\cos\delta_2 - \cos\delta_1\sin\delta_2 \\
&= \frac{I_{r\beta}I_{r\alpha r} - I_{r\alpha}I_{r\beta r}}{I_r^2} \\
\cos\hat{\theta}_{er} &= \cos(\delta_1 - \delta_2) = \frac{I_{r\alpha}I_{r\alpha r} + I_{r\beta}I_{r\beta r}}{I_r^2}
\end{aligned}
\tag{15.35}
$$

Knowing $\cos\hat{\theta}_{er}$ and $\sin\hat{\theta}_{er}$ the rotor speed $\hat{\omega}_r$ is estimated by digital filtering because

$$
\frac{d\hat{\theta}_{er}}{dt} = \hat{\omega}_r = -\sin\theta_{er}\frac{d}{dt}(\cos\theta_{er}) + \cos\theta_{er}\frac{d}{dt}(\sin\theta_{er})
\tag{15.36}
$$

To account even for magnetic saturation state change (especially during faults when the stator voltage and flux vary notably), after starting the online computation cycle by an initial value of $I_{ms}$, we will recalculate it as $I'_{ms\alpha}(k)$, $I'_{ms\beta}(k)$, after every computation cycle, in two stages [4]. The magnetization current computation is one step behind but this is acceptable.

$$\begin{aligned} I'_{ms\alpha}(k) &= I_{s\alpha}(k) + \frac{L_m}{L_s} I'_{r\alpha}(k) \\ I'_{ms\beta}(k) &= I_{s\beta}(k) + \frac{L_m}{L_s} I'_{r\beta}(k) \end{aligned} \tag{15.37}$$

With

$$\begin{aligned} I'_{r\alpha}(k) &= I_{r\alpha r}(k)\cos\hat{\theta}_{er}(k-1) - I_{r\beta r}(k)\sin\hat{\theta}_{er}(k-1) \\ I'_{r\beta}(k) &= I_{r\beta r}(k)\cos\hat{\theta}_{er}(k-1) + I_{r\alpha r}(k)\sin\hat{\theta}_{er}(k-1) \end{aligned} \tag{15.38}$$

Other rotor position and (or) slip frequency observers keep being proposed.

### 15.3.5 Control of the Source-Side Converter

The source-side PWM converter (Figure 15.14) includes at least an LR filter, in order to reduce the source-side current harmonics.

We start with the LR filter equations

$$\begin{vmatrix} V_a \\ V_b \\ V_c \end{vmatrix} = R \begin{vmatrix} I_{as} \\ I_{bs} \\ I_{cs} \end{vmatrix} + L\frac{d}{dt}\begin{vmatrix} I_{as} \\ I_{bs} \\ I_{cs} \end{vmatrix} + \begin{vmatrix} V_{as} \\ V_{bs} \\ V_{cs} \end{vmatrix} \tag{15.39}$$

After translation into dq synchronous coordinates, aligned with d axis voltage ($V_d = V_s$, $V_q = 0$), Equation 15.39 becomes

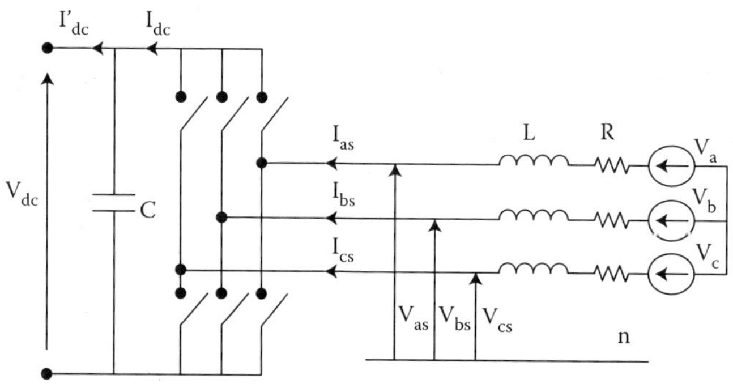


**FIGURE 15.14**
The source-side converter with LR filter.

$$V_d = RI_{ds} + L\frac{dI_{ds}}{dt} - \omega_1 LI_{qs} + V_{ds}$$
$$V_q = RI_{qs} + L\frac{dI_{qs}}{dt} + \omega_1 LI_{ds} + V_{qs} \tag{15.40}$$

where $\omega_1$ is the frequency of the a.c. supply voltages.

Neglecting all losses in the machine and in the converter

$$V_{dc}I_{dc} = \frac{3}{2}V_dI_d = P_r; \quad V_q = 0 \tag{15.41}$$

With the PWM depth $m_1$

$$V_d = \frac{m_1}{2\sqrt{2}}V_{dc} \tag{15.42}$$

Making use of Equation 15.41 in Equation 15.42

$$I_{dc} = \frac{3m_1I_d}{4\sqrt{2}} \tag{15.43}$$

The d.c. link equation can be written as

$$C\frac{dV_{dc}}{dt} = I_{dc} - I'_{dc} = \frac{3m_1I_d}{4\sqrt{2}} - I'_{dc} \tag{15.44}$$

It is thus evident that the d.c. link voltage may be controlled by controlling the current $I_d$.

The reactive power $Q_r$ at rotor terminals is

$$Q_r = \frac{3}{2}(V_dI_q - V_qI_d) = V_dI_q; \quad V_q = 0 \tag{15.45}$$

Consequently, the reactive rotor power flow may be controlled by controlling the current $I_q$. We may add the voltage decoupler

$$V'_{ds} = V_d + \omega_1 LI_q$$
$$V'_{ds} = -\omega_1 LI_q \tag{15.46}$$

This way a combined voltage/current VC system ([1], Chapter 9) is born (Figure 15.15).

For operation at power grid the d.c. voltage $V_{dc}$ is controlled to remain constant. The Park transformation is operated into two stages (abc to αβ, αβ to dq) and $I_d$, $I_q$, $V_d$, and the voltage vector angle $\theta_e$ are calculated from measured $I_a$, $I_b$, $V_a$, $V_b$. A rather complete design of such a VC system is given in Reference 5.

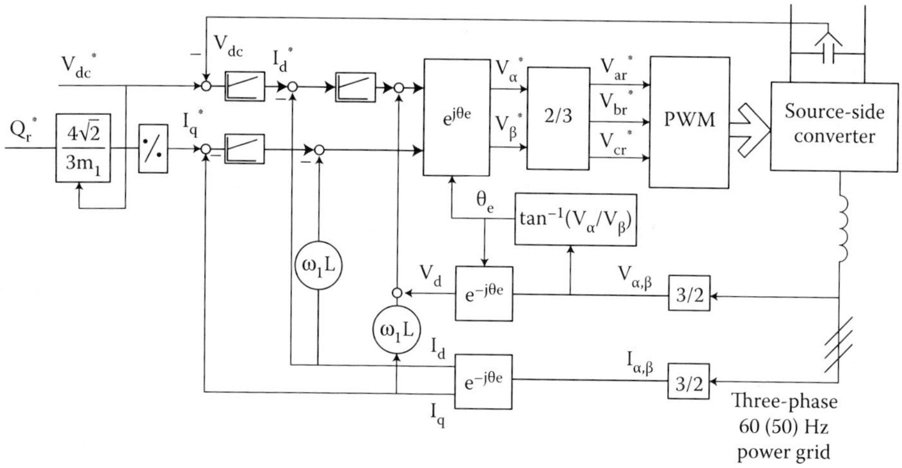


**FIGURE 15.15**
VC of the source-side converter.

### 15.3.6 Wrig Control: A Case Study in MATLAB/Simulink (Available Online)

The case of a wind turbine driving a 2.0 MW WRIG with the following parameters is considered:

$V_{SN}$ (line, RMS) = 690 V, 2p = 4, $f_N$ = 50 Hz, $l_m$ = 3.658 p.u., $l_{sl}$ = 0.0634 p.u., $l_{rl}$ = 0.08466 p.u., $r_s = 4.694 \times 10^{-3}$ p.u., $r_r = 4.86 \times 10^{-3}$ p.u., H = 3.611 s.

A ±25% maximum slip is considered and, to eliminate the voltage-matching transformer from the rotor circuit, the rotor/stator turns ratio $W_r/W_s = 1/|S_{max}| = 4/1$.

The wind turbine is simulated by a two-dimensional look-up table that produces the mechanical torque $T_M$ as a function of wind speed U (m/s) and generator speed ($\omega_r$) such that to extract maximum power, up to the speed limit.

#### *15.3.6.1 Control of the Source-Side Converter*

The q axis current controller design is obtained from the system's transfer function in z-domain with a sampling time T = 0.45 ms. The closed-loop natural frequency is 125 Hz and the damping factor is 0.8. In these conditions the d.c. loop PI controller transfer function is 1.8(z – 0.9638)/(z – 1).

The current controllers' loop are much faster than the d.c. voltage ($V_{dc}$) control loop and their transfer function is 0.5 (z – 0.8943)/(z – 1).

#### *15.3.6.2 Generator-Side Converter Control*

Again, by imposing the natural closed-loop frequency and the damping factor of the plant, the dq PI current controllers' transfer functions are 12(z – 0.995)/(z – 1). The power control loops are much slower, with their transfer functions as 0.00009 (z – 0.9)/(z – 1).

Space vector PWM is used in both converters to generate their reference a.c. voltages.

Simulation results are obtained through a MATLAB/Simulink dedicated program (WRIG.sim: ex9.mdl—see online case studies).

At 1.5-s in time, the wind speed is increased from 7 to 11 m/s, then, at 6 s, the reference value of active power is increased from 0 to 1.2 MW for constant 0.3 MVAR reactive power.

The rotor currents waveforms (Figure 15.16a) and the generator speed (Figure 15.16b) show clearly the smooth passing of the machine through the synchronous speed around 2.1 s in time.

The active and reactive power fast transients are shown in Figure 15.16c and d.

A three-phase short-circuit at the power grid for same WRIG is illustrated in Figure 15.17. There are two main current peaks, one at the initiation of the short-circuit at half length of a symmetric power line (Figure 15.16e) and the other after the short-circuit.

Decreasing these peaks may be accomplished by simply limiting the output of the current controllers at the machine-side converter at, say 150% rated value.

This way the WRIG may remain connected to the power grid during the short-circuit, to be ready to contribute to the voltage restoration once the short-circuit is cleared.

## 15.4 Autonomous D.C.-Excited SG Control at Variable Speed

The d.c.-excited SG may perform as an autonomous power source at constant speed (and voltage) such as in standard standby emergency or gensets, many driven by diesel engines. In this case, however, the speed (frequency) and voltage are controlled separately such as for power grid operation.

The speed is controlled to remain constant by the speed governor of the prime mover (diesel engine) while the voltage is controlled, to stay constant by the field current control.

As this case has been treated in Section 15.2, we will dwell here only on variable speed control of d.c.-excited autonomous SG.

Even here there are two main situations:

- With battery back-up and d.c. controlled output (as for vehicle alternators)
- With a.c. constant voltage and frequency controlled output

We will treat both cases in some detail.

### 15.4.1 Control of the Automotive Alternator

The automotive industry's alternator is of a single topology: the Lundell–Rice configuration. It has a single ring-shape rotor coil which is d.c. fed through slip rings and brushes to produce a multipolar (2p = 10, 12, 14, 16, 18) magnetic field in the airgap. The ring-shape rotor coil is surrounded by solid-iron claws to help configure the multipolar magnetic field in the airgap.

The stator is made of standard rotary machine silicon iron laminations with uniform slots that hold a q = 1 slot/pole/phase single-layer three-phase a.c. winding (Figure 15.18).

The rotor claw poles are made of solid iron; so they act as a rather weak damper cage, to be neglected, to a first approximation.

The dq axis reactances $X_d$ and $X_q$ differ from each other ($X_d > X_q$).

The claw-pole alternator is to deliver d.c. power to the battery backup d.c. loads on automobiles.

To do so, it uses a full-power diode rectifier connected between the stator terminals and the battery and a d.c. field current controller.

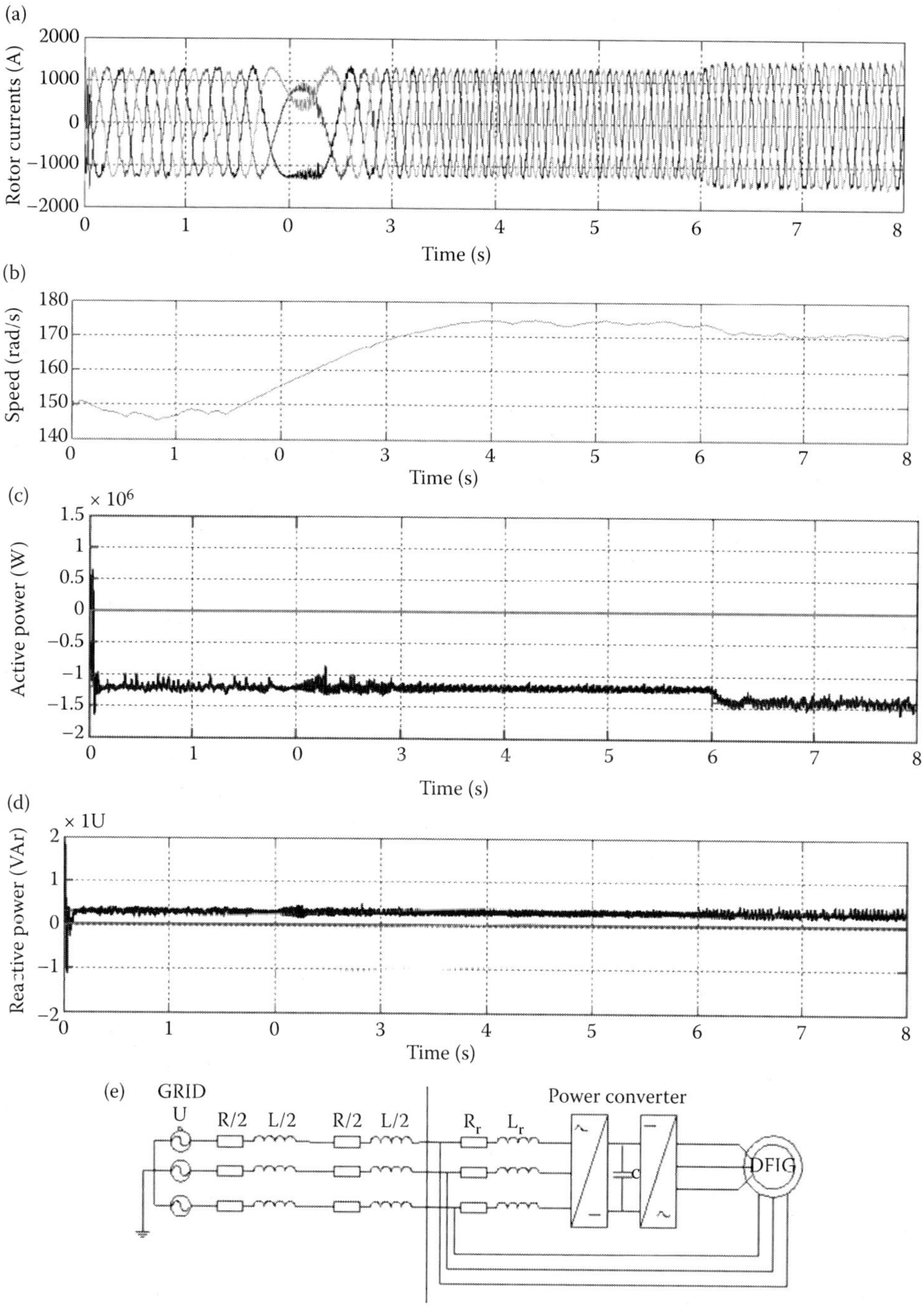


**FIGURE 15.16**
2 MW WRIG transients: (a) rotor currents, (b) generator speed, (c) active power, (d) reactive power, (e) the system structure. (After Serban, I. et al., A study of the double fed wind power generator under power system faults, *Record of EPE*, 2003, Toulouse, France.)

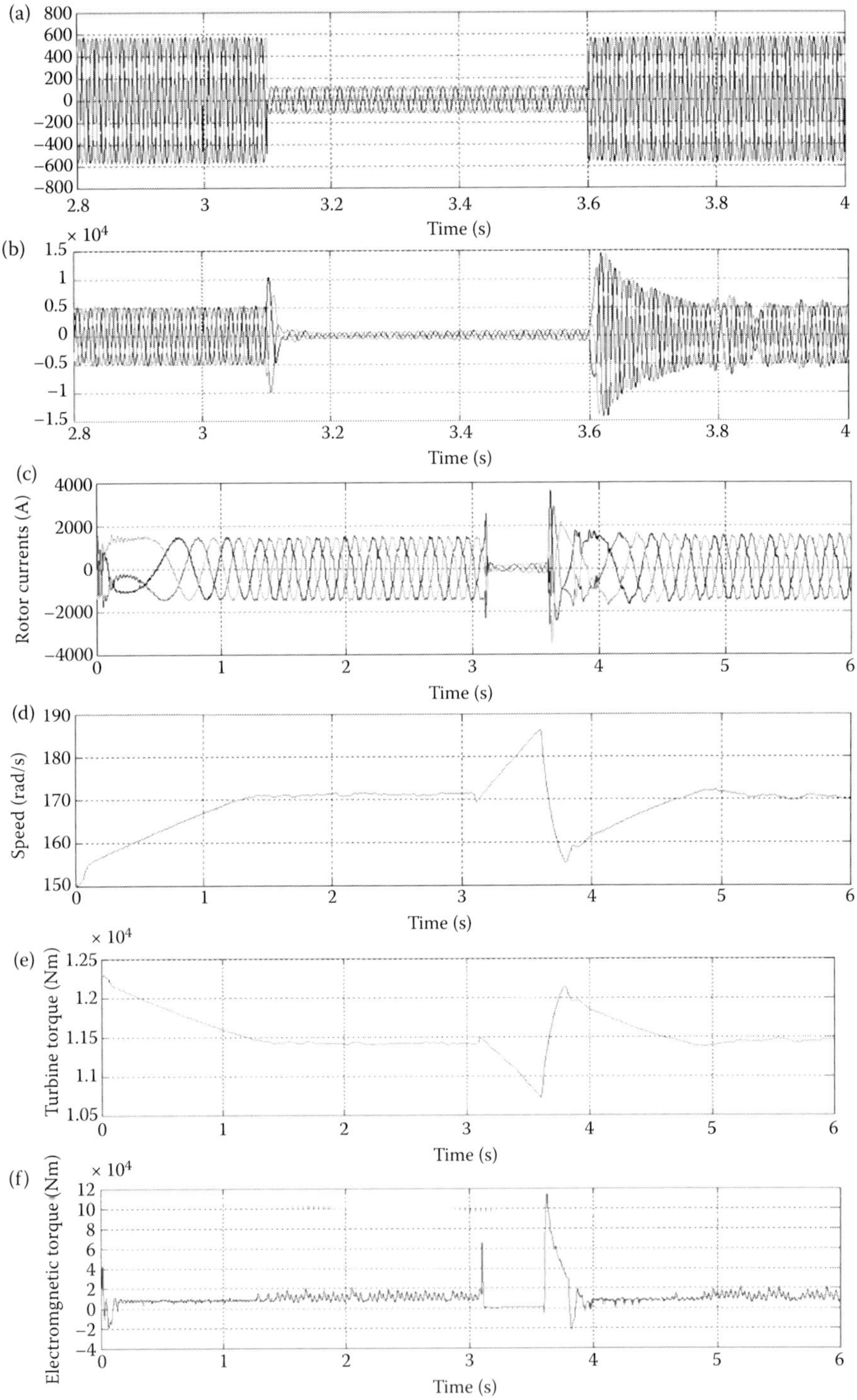


**FIGURE 15.17**
Three phase short-circuit at power grid: (a) stator voltage, (b) stator currents, (c) rotor currents, (d) speed, (e) turbine torque, (f) electromagnetic torque. (After Serban, I. et al., A study of the double fed wind power generator under power system faults, *Record of EPE*, 2003, Toulouse, France.)

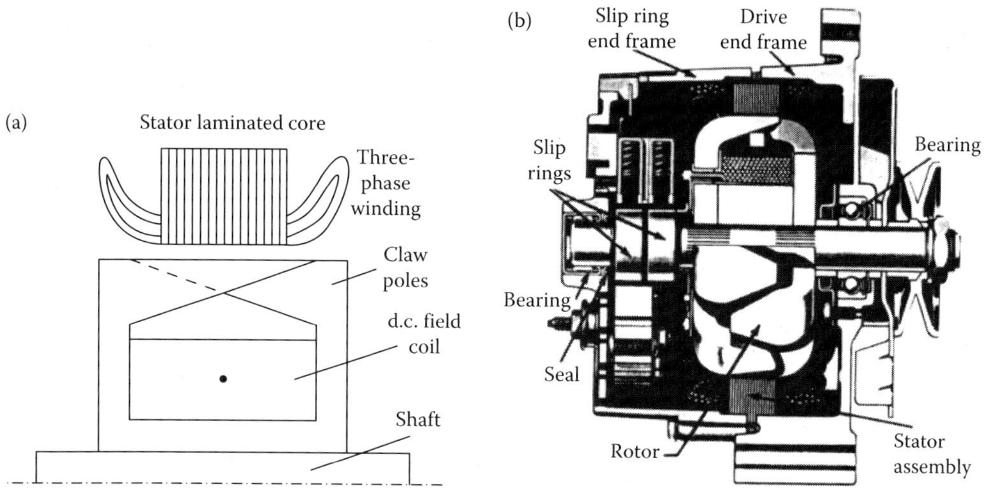


**FIGURE 15.18**
Lundell–Rice automotive alternator: (a) cross section, (b) general view.

The field winding is fed from the alternator terminals through a three-diode (half) rectifier and a d.c.–d.c. converter (Figure 15.19).

Though Figure 15.19 depicts a control system for automotive alternators, any variable speed alternator with battery backup d.c. output will fit into the category.

A voltage closed-loop controller is included. For maximum field current, the d.c. output current $I_{dc}$ increases with speed as in Figure 15.19. Slightly below the engine idle speed the

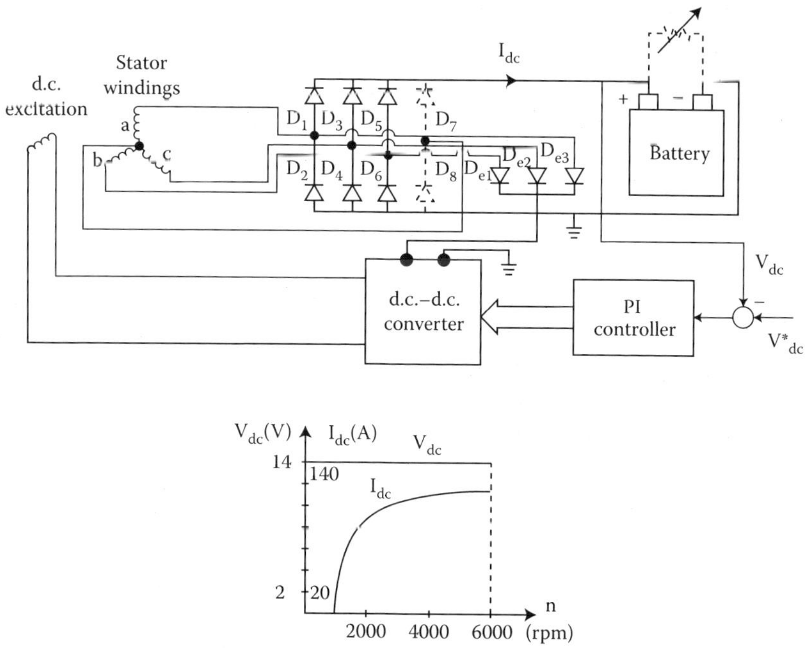


**FIGURE 15.19**
Typical automotive alternator control system.

d.c. current is zero because the alternator e.m.f. is insufficient so that the diode rectifier does not open up the energy flow.

The rather large machine transient inductances $L_d', L_q'$, due to the weak solid-iron rotor claw-poles damping effect, make the response in d.c. current, to a step load power increase, not very fast.

As the diode rectifier provides for almost unity displacement power factor, the phasor diagram of the alternator with $L_d \approx L_q$ gets simplified (Figure 15.20).

$$\underline{I}_1 R_s + \underline{V}_1 = \underline{E}_1 - jX_s \underline{I}_1 \tag{15.47}$$

$$\underline{E}_1 = -j\omega_1 X_{dm} \underline{I}_F \tag{15.48}$$

Equations 15.47 and 15.48 have been documented in Chapter 10 but they are straightforward as $\underline{E}_1$ is the e.m.f. and $R_s$, $X_s$ resistance and cyclic synchronous reactance. $\underline{I}_F$ is the field current equivalent phasor seen from the stator as Equation 15.47 is written in stator coordinates.

From the power balance:

$$V_1 \approx \frac{V_{dc}I_{dc} + p_{diode}}{3I_1 \cos\phi_1}; \quad K_i = \frac{I_{dc}}{I_1} \approx \frac{3\sqrt{3}}{\pi}\cos\phi_1 \tag{15.49}$$

As the speed increases, $E_1$ (for a given field current value) and $X_s = \omega_1 L_s$ increase and thus the current $I_1$ (and $I_{dc}$) tends to be limited.

From the phasor diagram (with $R_s \sim 0$) the current $I_1$ is approximately

$$I_1 = \frac{\sqrt{E_1^2 - V_1^2}}{\omega_1 L_s} \tag{15.50}$$

But, for constant field current $I_F$

$$\begin{aligned} E_1 &= K_E(I_F)\omega_1 W_1 \\ X_s &= K_x \omega_1 W_1^2 \end{aligned} \tag{15.51}$$

where $W_1$ is the number of turns/phase and $\omega_1$ is the stator angular frequency.

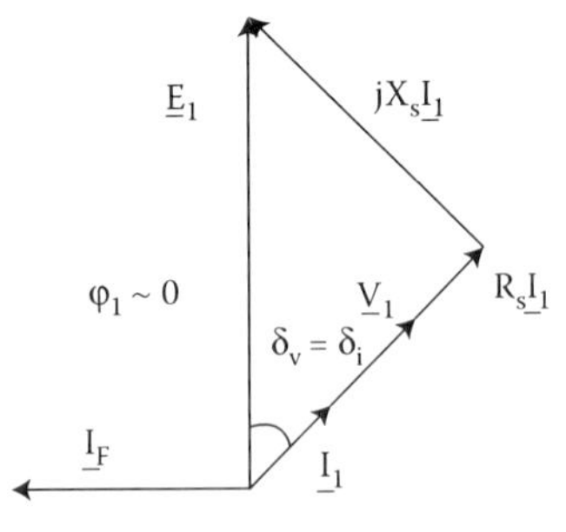


**FIGURE 15.20**
The simplified phasor diagram of claw-pole alternator with diode rectifier and $L_d \approx L_q$.

Making use of Equation 15.51 in Equation 15.50, the maximum current $I_1$ is obtained for $W_{1opt}$

$$W_{1opt} = \frac{V_1\sqrt{2}}{K_e\omega_1} \text{ or } E_{1opt} = V_1\sqrt{2} \tag{15.52}$$

Consequently, $X_sI_{1opt} = V_1$ for maximum current ($\delta_v = \delta_i = \pi/4$ in Figure 15.20).

It is evident that $W_{1opt}$ should decrease with speed ($\omega_1$) to maintain the maximum optimum current with increasing speed.

Winding tapping or operating at larger voltage are thus practical solutions to increase alternator output current at high speeds. Alternatively, an alternator designed for given power $P_n$ at 14 V d.c. with a switched-mode diode rectifier (as voltage booster) can produce almost twice as much power at high speeds [7] (Figure 15.21) at 42 V d.c.

By operating the automotive alternator at a larger d.c. voltage, not only was the output doubled, but implicitly, the efficiency was improved from less than 50% to 70% [7].

When the load is dumped the maximum voltage has to be limited to 80 V d.c. by adequate field current control (protection) provisions, to meet the current standards in the automotive industry. Results in Figure 15.22 prove such good performance with the switched-mode rectifier (Figure 15.21).

As the required electric energy level in vehicles is on the rise, the adopting of a 42 V d.c. bus becomes necessary and beneficial in the sense that today's alternators, designed for 14 V d.c. at maximum power $P_n$, may be used with a switched-mode rectifier and produce $2P_n$ maximum power at 42 V d.c., for increased efficiency.

While efforts to improve the output and efficiency of claw-pole alternators continue, other d.c.-excited alternators such as those with smaller voltage regulation (smaller $L_q$, with PMs in rotor q axis) have been proposed [8].

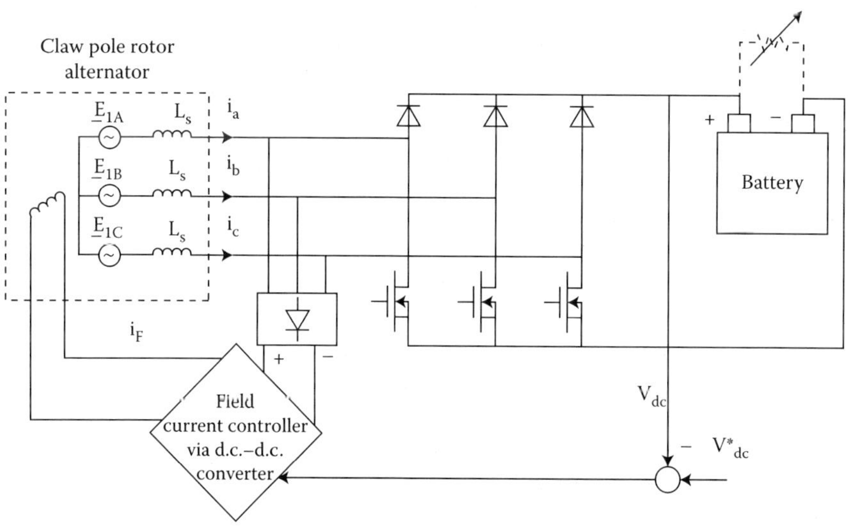


**FIGURE 15.21**
Alternator with switched-mode diode rectifier and voltage controller.

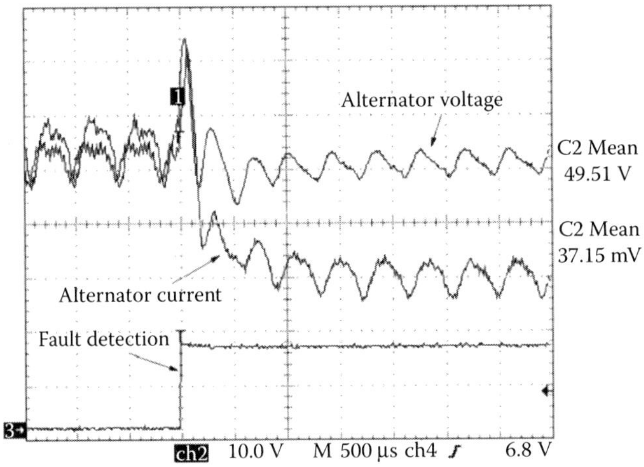


**FIGURE 15.22**
Limited load dump voltage transients with the switched-mode diode rectifier.

### 15.4.2 A.C. Output Autonomous Alternator Control at Variable Speed

The autonomous alternator may also provide for constant voltage and frequency output at variable speed (Figure 15.23). A full-power PWM inverter is required.

The field circuit is designed and controlled to provide for constant voltage amplitude $V_1$ (at variable frequency) at alternator terminals for the whole speed range.

Further on, a standard full-power diode rectifier with a PWM voltage source inverter will provide for constant frequency and voltage a.c. output at the power grid or for autonomous (separate) loads.

Only unidirectional power flow—from alternator to power grid or to loads—is possible, but, at moderate costs, as the standard PWM inverter at the power grid is similar to the active front end rectifier [9].

V/f standard control of the latter is adaptable for autonomous loads when V and f have to be controlled to remain constant with load, up to a point.

It is feasible to use a single d.c. voltage link for a cluster of alternators and thus a single PWM inverter is required for all of them.

The desired power sharing of each alternator is tailored through its field current controller.

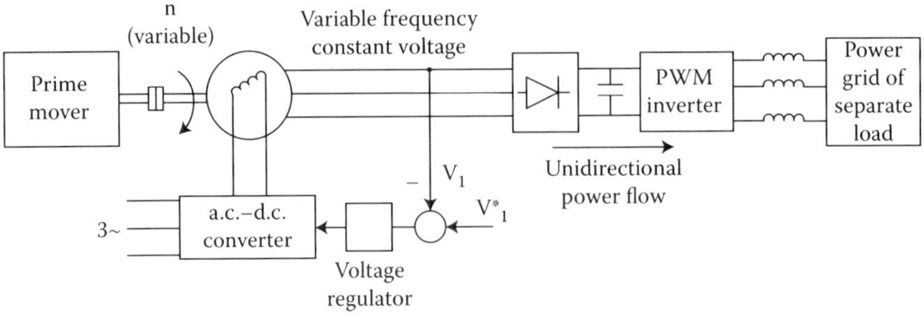


**FIGURE 15.23**
Alternator with full-power PWM converter control at variable speed.

## 15.5 Cage-Rotor Induction Generator Control

The cage-rotor induction machine is known for its ruggedness, moderate costs, and rather good performance.

In some applications, such as wind or small hydro energy conversion (up to 1–5 MW/unit), variable speed is required to track most of the available wind and hydro energy when the wind speed, and respectively, the water head vary. Also, variable speed with power increasing with generator speed, results in better system efficiency.

One drawback of cage-rotor induction machines is the need for an external source to magnetize them, corroborated with reactive power "production" for voltage control, at constant frequency.

The cage-rotor induction generator (called simply IG) has to be associated with a bidirectional a.c.–a.c. PEC in the stator in order to produce active and reactive power control, both at power grid and in stand-alone operation. For ±100% $P_1$ and $Q_1$ control, a back to back PWM voltage-source converter (Figure 15.24) is required.

The bidirectional back to back PWM inverter contains the grid-side converter and the IG-side converter. The source-side converter control is the same as for WRIG (see Section 15.3) where this converter was connected to the rotor. In essence, for power grid operation, the VC in grid voltage coordinates is used such that the active power flow is controlled by controlling the d.c. link voltage along axis d and the reactive power $Q_1$ along axis q. But in this case the d.c. link capacitor has to be designed for ±100% reactive power control. This way the IG behaves toward the loads as a SG, with faster ±100% active and reactive power control.

The IG-side converter may be either vector or DTFC type, for motoring and generating, as widely discussed in Chapter 9 on induction motor drives.

When the IG works in autonomous load mode, the load-side converter may be either vector or V/f controlled, for constant or drooped voltage amplitude control and constant frequency output control (for more details see Reference 1, Chapter 13). Typical a.c. grid-side voltage and current waveforms for 100% active power and zero reactive power control of a 10 kW IG (connected to an off the shelf bidirectional back to back PWM converter designed for fast braking electric drives) is shown in Figure 15.25 ([1], Chapter 13).

More filtering of the current is required to meet the today's strict harmonics content limitation standards.

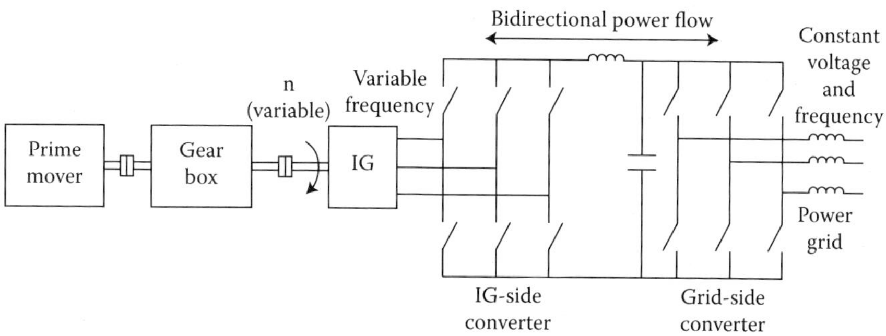


**FIGURE 15.24**
IG with back to back PWM converter to power grid: ±100% $P_1$ and $Q_1$ control.

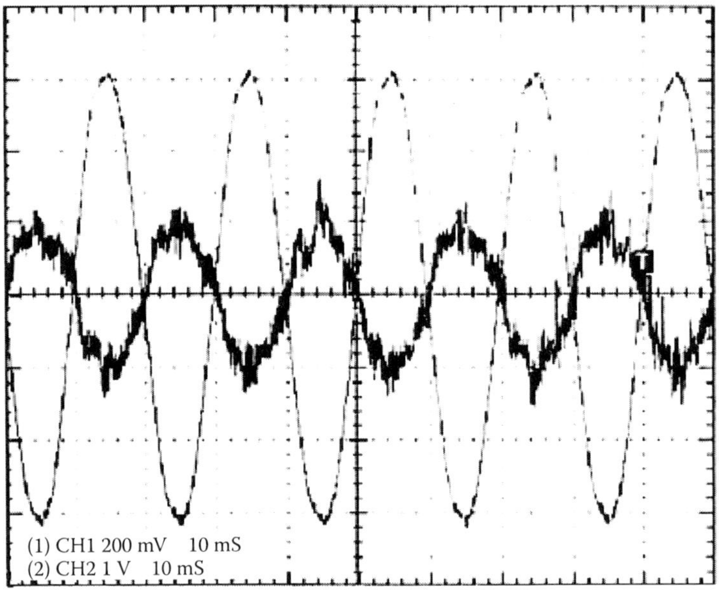


**FIGURE 15.25**
Grid voltage and current at 1500 rpm for generator mode at 100% active (10 kW) and zero reactive power control.

## 15.6 PM Synchronous Generator Control for Variable Speed

PM synchronous generators (PMSG) are built either with surface-pole PM or interior PM rotor, with radial or axial airgap, with distributed or nonoverlapping windings, just as are PMSMs (Chapter 10).

For convenience we will show here only the radial-airgap, distributed stator winding configuration with surface-pole-PM and, respectively, inset-pole-PM rotor (Figure 15.26).

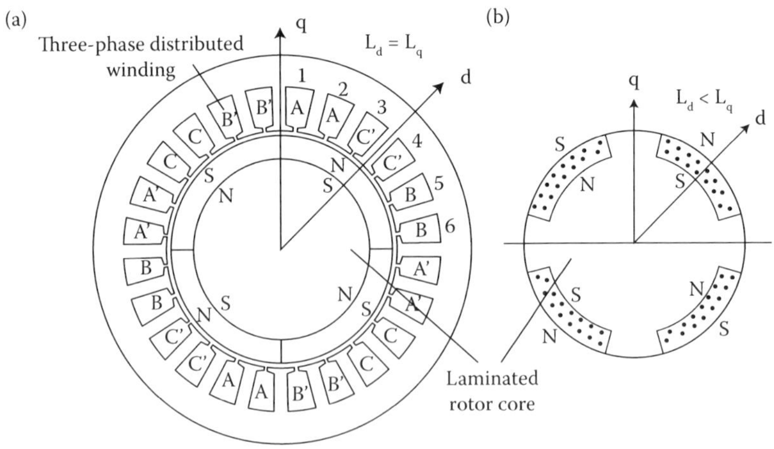


**FIGURE 15.26**
PMSG with distributed single layer stator winding ($2p_1 = 4$, $N_c = 24$ slots): (a) surface PM-pole rotor, (b) inset-PM-pole rotor.

In essence, the distributed three-phase winding in Figure 15.26 is characterized by rather sinusoidal stator phase e.m.f.s. However, as a single-layer winding is chosen, the stator current m.m.f. produces notable 5 and 7 space harmonics which, in turn, may produce notable eddy current losses in the rotor PMs.

Surface PM pole rotors are characterized by the zero magnetic saliency ($L_d = L_q$) while inset-PMs may lead to a special design when, for rated resistive load, the rated PMSG voltage $V_{1N}$ is equal to no-load voltage $E_1$, for given speed (Figure 15.27).

The phasor diagram for the two rotors (Figure 15.27) is based on the phase equations

$$\underline{I}_1 R_s + \underline{V}_1 = \underline{E}_1 - jX_d \underline{I}_d - jX_q \underline{I}_q; \quad \underline{I}_1 = \underline{I}_d + \underline{I}_q$$
$$\underline{E}_1 = -j\omega_1 \lambda_{PM} \tag{15.53}$$

The dq model of the PMSG is practically the same as for PMSM (Chapter 10)

$$I_d R_s + V_d = -L_d \frac{dI_d}{dt} + \omega_r \lambda_q; \quad \lambda_d = L_d I_d + \lambda_{PM}$$
$$I_q R_s + V_q = -L_q \frac{dI_q}{dt} - \omega_r \lambda_d; \quad \lambda_q = L_q I_q \tag{15.54}$$

$$V_d = \frac{2}{3}\left(V_a \cos(-\theta_{er}) + V_b \cos\left(-\theta_{er} + \frac{2\pi}{3}\right) + V_c \cos\left(-\theta_{er} - \frac{2\pi}{3}\right)\right)$$
$$V_q = \frac{2}{3}\left(V_a \sin(-\theta_{er}) + V_b \sin\left(-\theta_{er} + \frac{2\pi}{3}\right) + V_c \sin\left(-\theta_{er} - \frac{2\pi}{3}\right)\right) \tag{15.55}$$

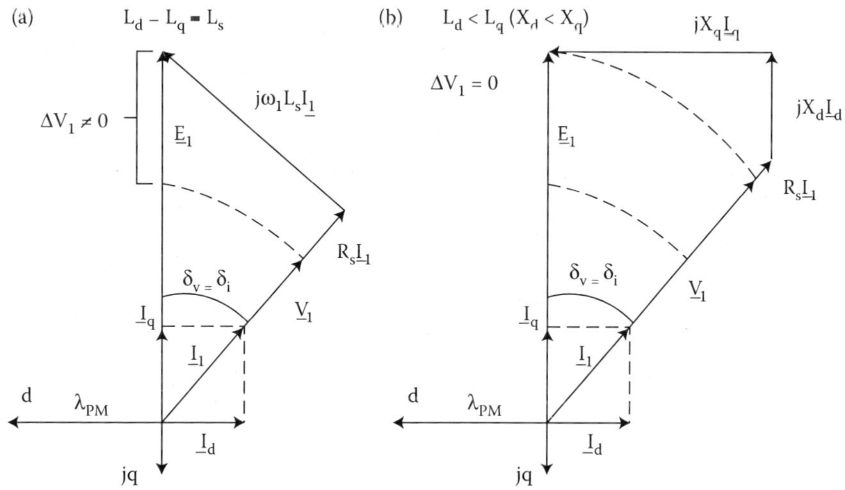


**FIGURE 15.27**
The phasor diagram of PMSG on resistive (or diode rectifier) load: (a) with surface-PM-pole rotor, (b) with inset-PM-pole rotor.

The electromagnetic torque can be written as (Chapter 10)

$$T_e = -\frac{3}{2}p(\lambda_d I_q - \lambda_q I_d) = -\frac{3}{2}p(\lambda_{PM} + (L_d - L_q)I_d)\cdot I_q \quad (15.56)$$

As the equations are written for generator association of signs $T_e > 0$ for generator. Negative torque (motoring) is obtained with positive $I_q$.

The motion equations have to be added:

$$\frac{J}{p}\frac{d\omega_r}{dt} = T_{mec} - T_e; \quad \frac{d\theta_{er}}{dt} = \omega_r \quad (15.57)$$

The prime-mover torque $T_{mec} > 0$ for PMSG generator operation mode and negative for PMSG motoring operation.

A numerical example to calculate PMSG steady-state performance follows.

**EXAMPLE 15.1**

Let us consider a surface-PM-pole rotor PMSG with the data: $R_s = 0.1\ \Omega$, $L_s = L_d = L_q = 0.005$ H, $\lambda_{PM} = 0.5$ Wb, $p_1 = 2$ pole pairs, that has to deliver power into a three-phase resistance $R_L = 3\ \Omega$/phase at $n_1 = 1800$ rpm:

a. Calculate the phase current, voltage and power.
b. For same phase current, calculate the load resistance $R_L'$ and power and $L_q > L_d$ (inset-PM-pole rotor), when the load voltage $V_{1r} = E_1$ (e.m.f.).

**Solution**

a. Equations 15.54 apply directly for $V_d = R_L I_d$, $V_q = R_L I_q$ and $d/dt = 0$:

$$(R_s + R_L)I_d = \omega_r L_q I_q$$

$$(R_s + R_L)I_q = -\omega_r(\lambda_{PM} + L_d I_d)$$

or

$$\begin{matrix}(0.1+3)I_d = 2\pi\cdot 30\cdot 2\cdot 0.005\cdot I_q \\ (0.1+3)I_q = -2\pi\cdot 30\cdot 2(0.5+0.005\cdot I_d)\end{matrix} \rightarrow \begin{matrix} I_d = -19.5766\ \text{A} \\ I_q = -53.664\ \text{A}\end{matrix}$$

The phase current $I_1$ is

$$I_1 = \sqrt{\frac{1}{2}\left(I_d^2 + I_q^2\right)} = 40.3922\ \text{A(rms)}$$

Consequently, the delivered power $P_1$ can be written as

$$P_1 = 3R_L I_1^2 = 14.683\ \text{kW}$$

The phase voltage $V_1$ is

$$V_1 = R_L I_1 = 121.1766\ \text{V(rms/phase)}$$

b. For zero voltage regulation:

$$V_1 = R_L I_1 = 121.1766\ V(rms/phase)$$

If the current $I_1$ is kept the same, the power is delivered on a new load resistance $R_L'$

$$R_L' = \frac{V_{1r}}{I_1} = 4.6667\ \Omega/phase$$

We return to the above equations with known $R_s$, $R_L'$, $L_d$, and $I_1$, and need to find out $L_q$ (and $I_d$)

$$(0.1 + 4.6667)I_q = -377(0.5 + 0.005I_d)$$

$$4.7667I_d = 377L_qI_q$$

From the first equation with $I_q = \sqrt{3264 - I_d^2}$, we finally calculate

$$I_d = -31.128\ A$$

Consequently

$$L_q = \frac{4.7667 \cdot (-31.128)}{377 \cdot (-47.906)} = 0.00821\ H$$

$L_q/L_d = 0.00821/0.005 = 1.643$, a ratio which may be reached with inset-PMs on the rotor if only with a slightly smaller mechanical airgap.

The power delivered now is

$$P_1' = 3R_L' I_1^2 = 23.153\ kW$$

Consequently, 57.69% more power is gained from basically the same machine and stator windings losses, just by using inset magnets on the rotor poles. Definitely the inverse saliency seems to pay off. However the price to pay is notably less than 1 p.u. machine reactances (thicker PMs).

### 15.6.1 Control Schemes for PMSG

The PMSG may work alone or at the power grid. Let us consider the power grid operation, at variable speed, by two essential power electronics control configurations (Figure 15.28).

The voltage booster active switch $T_1$ (Figure 15.28a) makes use of machine inductances and the additional $L_{dc}$ to boost the voltage of the d.c. link in a controllable manner.

A constant d.c. voltage link may be secured from $\omega_{rmin}$ to $\omega_{rmax}$, provided there is still room at $\omega_{rmax}$ to boost a little the d.c. link voltage, to keep $T_1$ active.

The grid-side PWM inverter is controlled as for the WRIG (Section 15.3).

On the other hand, if motoring (for starting the prime mover) is required, an active front end rectifier is required on the machine side (Figure 15.28b). This PWM converter is to

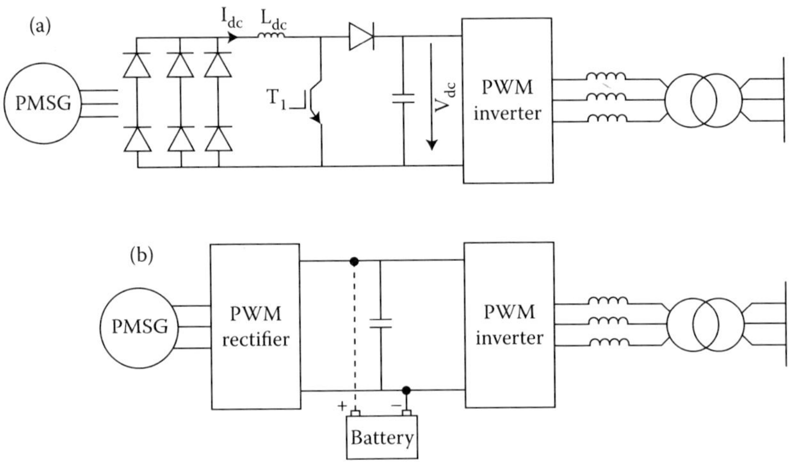


**FIGURE 15.28**
Variable speed PMSG with constant voltage and frequency a.c. output: (a) with hard—switched diode—rectifier and PWM inverter; (b) with active front rectifier and PWM inverter.

be controlled as for motoring and regenerative braking of PMSM. Vector or direct torque (or power) and flux control (DTFC), as in Chapter 11 for PMSM drives, is applicable with or without motion sensors.

On the other hand, the PMSG may work as a stand-alone generator driven by a medium speed diesel engine or by a high-speed gas turbine. Standby emergency or remote-area electric energy generation require such solutions. Again, the load-side PWM converter may be V/f or vector controlled with an output harmonics power filter as for the IG.

An up to date diesel engine PMSG system with multiple output frequency, single-phase, and three-phase a.c. output, for variable speed, with fuel saving, is presented in Reference 10.

### 15.6.2 PMSG Sensorless Control: A Case Study

PMSG control may be approached by scalar, field oriented, or direct power control methods, preferably without a rotor position sensor (after [17]). In here we detail on sensorless VC of PMSG with stand-alone to grid and vice versa transitions, with asymmetric voltage sags and with harmonic filtering, to offer a better illustration of today's requirements from PMSG systems in distributed power systems [17]. A general block diagram of such a complex control system is shown in Figure 15.29.

#### *15.6.2.1 Grid-Side Converter Control for Grid Connection*

As shown in Figure 15.30 the grid-side converter in Figure 15.29 for grid-connection control mode has the following components:

- PI d.c. link voltage controller
- BPF based or D module to extract safely the positive sequence of grid voltage $\underline{V}^+$ even if slightly nonsinusoidal
- Independent active and reactive power control is targeted (through $i_d^*$, $i_q^*$); for the small PMSG in investigation $K_{p\_Vdc} = 0.02$ A/V, $K_{i\_Vdc} = 5$ s$^{-1}$, $V_{dc}^* = 600$ V

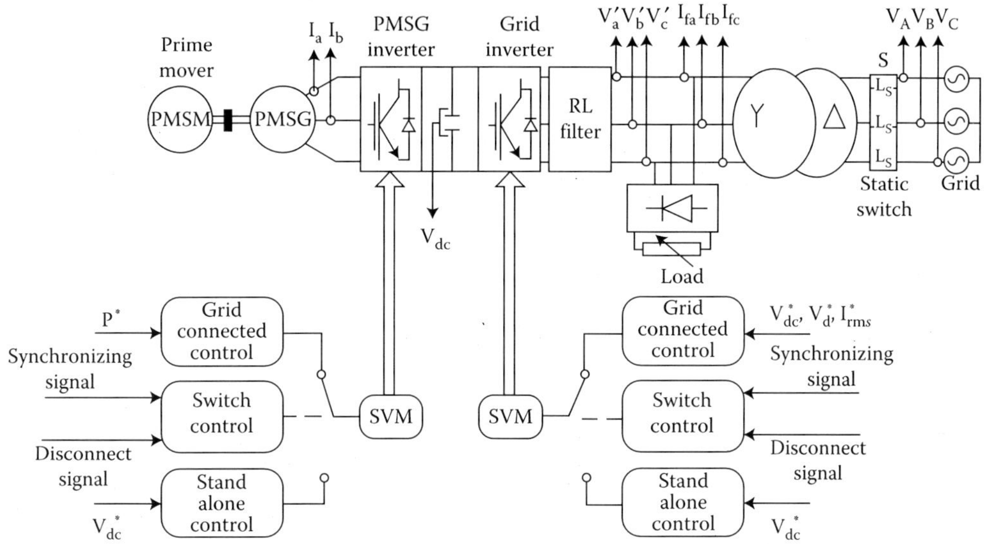


**FIGURE 15.29**
Block diagram for transfer from stand-alone to grid-connected (and back) PMSG. (After M. Fatu, F. Blaabjerg, I. Boldea, *IEEE Trans. Power Electron.*, vol. 29, no. 7, 2014, pp. 3463–3472.)

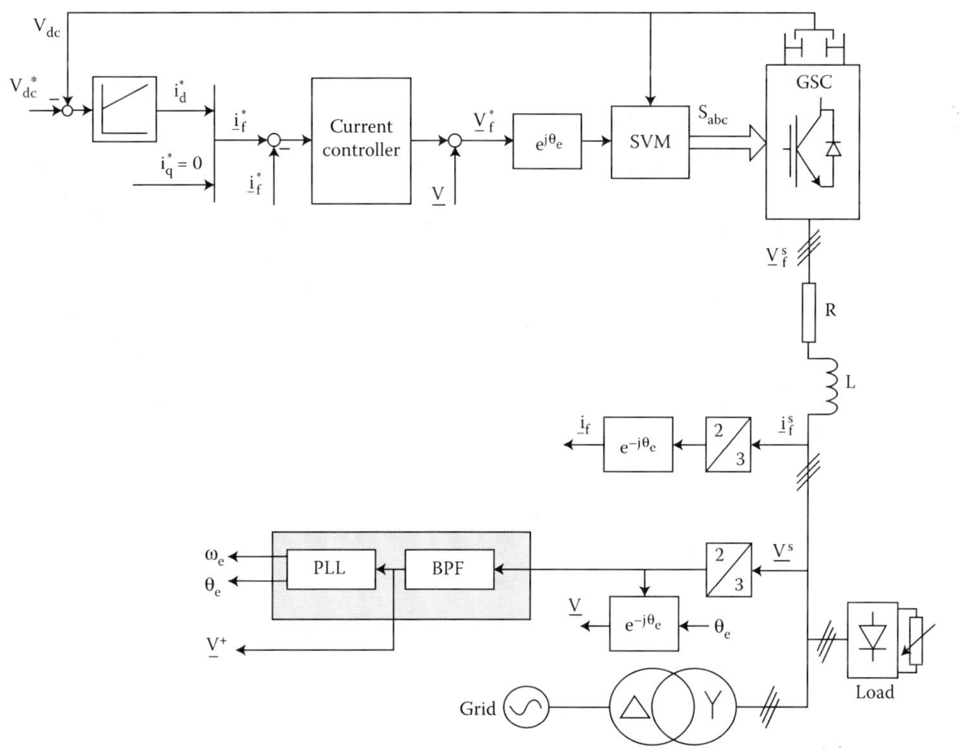


**FIGURE 15.30**
Grid-connected control of grid-side converter. (From M. Fatu, F. Blaabjerg, I. Boldea, *IEEE Trans. Power Electron.*, vol. 29, no. 7, 2014, pp. 3463–3472.)

- The complex current control (Figure 15.31) includes decoupling and feedforward line voltage compensation to avoid load heavy influence.
- The line voltage positive sequence phase and frequency $\theta_e$, $\omega_e$ are obtained through a PLL observer, that is used in control to maintain synchronism even for asymmetric voltage sags (Figure 15.32) with $\Delta\theta_e \approx (V_\beta^+ \cos\theta - V_\alpha^+ \sin\theta)/V^+$ ($V^+ = 310$ V, h = 110 μs, $K_p = 6$, $K_i = 300\ s^{-1}$).

### 15.6.2.2 Grid-Side Converter in Stand-Alone Control

When the grid-side converter operates as stand-alone, the control changes as shown in Figure 15.33.

In stand-alone mode, harmonics pollution caused by nonlinear loads should be reduced. Harmonics up to 23 order are reduced in Figure 15.33a and b with proportional–resonant PR individual controllers. In order to maintain voltage symmetry for a.c. loads, a negative

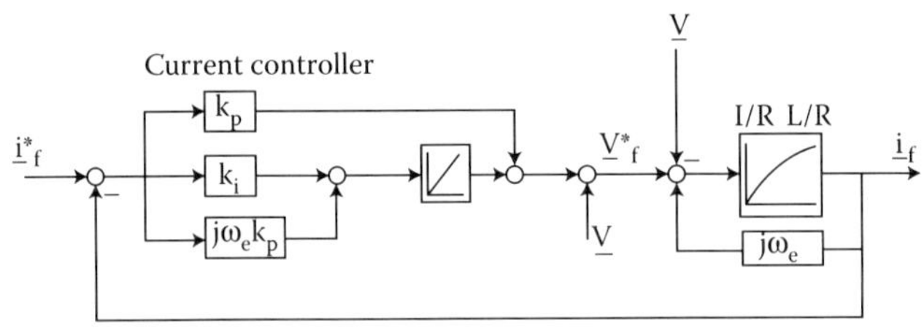


**FIGURE 15.31**
Complex (d-q) current control in synchronous/grid voltage synchronous reference frame. (From M. Fatu, F. Blaabjerg, I. Boldea, *IEEE Trans. Power Electron.*, vol. 29, no. 7, 2014, pp. 3463–3472.)

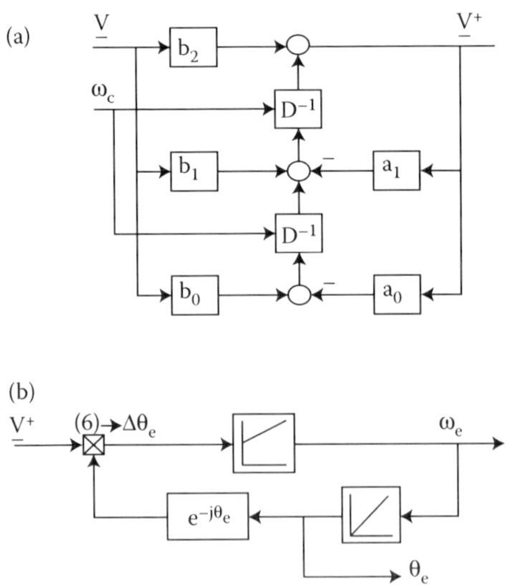


**FIGURE 15.32**
(a) Second-order BPF (D(s, $\omega_c$)) with D-module filter, (b) PLL observer. (After Fatu, M. Fatu, F. Blaabjerg, I. Boldea, *IEEE Trans. Power Electron.*, vol. 29, no. 7, 2014, pp. 3463–3472.)

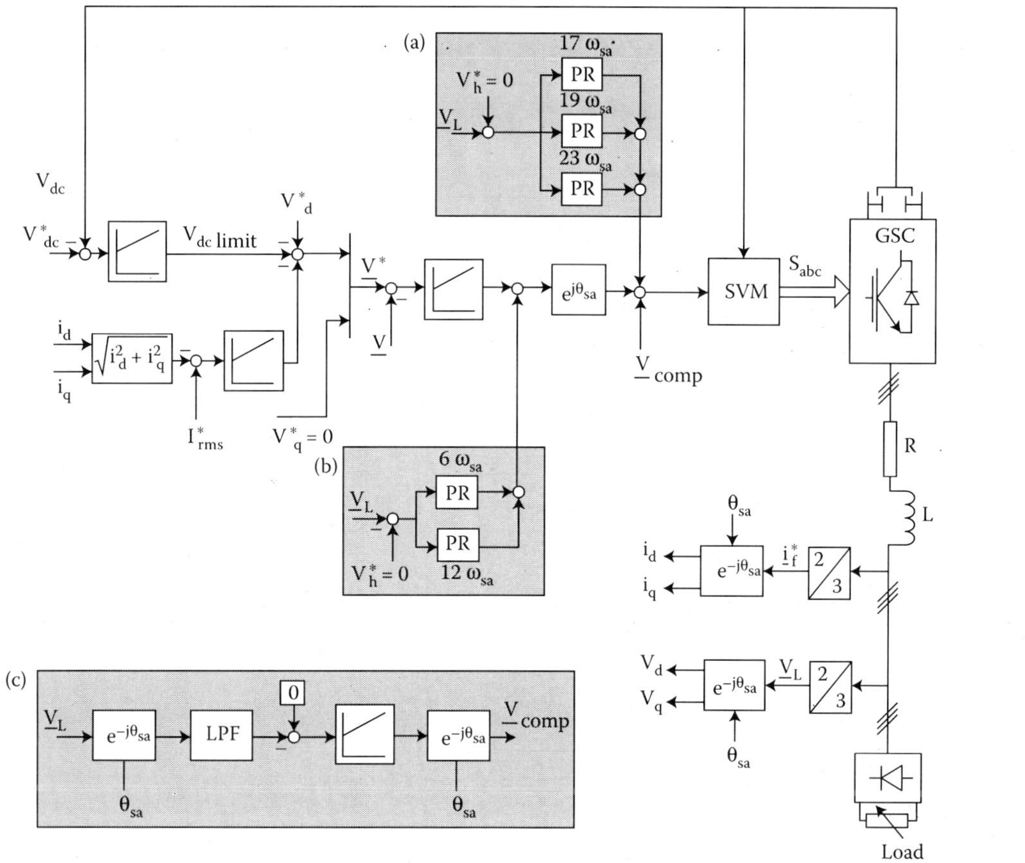


**FIGURE 15.33**
Grid-side converter in stand-alone mode: (a, b) with harmonic voltage compensation and negative sequence voltage compensator (c). (After Fatu, M., Blaabjerg, F., Boldea, I., *Power Electronics, IEEE Transactions on*, 29 (7), 3463–3472, 2014.)

voltage sequence compensation is required (Figure 15.33c) when unbalanced loads are present.

Fundamental voltage control is obtained through PI control: $V_d^* = 220\sqrt{2}$ V and $V_q^* = 0$ controllers.

#### 15.6.2.3 Transition Controls

The seemless transition from stand-alone to grid mode is illustrated in Figure 15.34.

A similar controller is used for grid to stand-alone transition (Figure 15.35).

#### 15.6.2.4 PMSG-Side Converter Sensorless Control

The space vector model of a PMSG in rotor coordinates can be written as

$$-\underline{V}_s = \underline{i}_s R_s + d\underline{\Psi}_s / dt + j\omega_r \underline{\Psi}_s, \quad \underline{\Psi}_s = \lambda_{PM} + L_d i_d + jL_q i_q \tag{15.58}$$

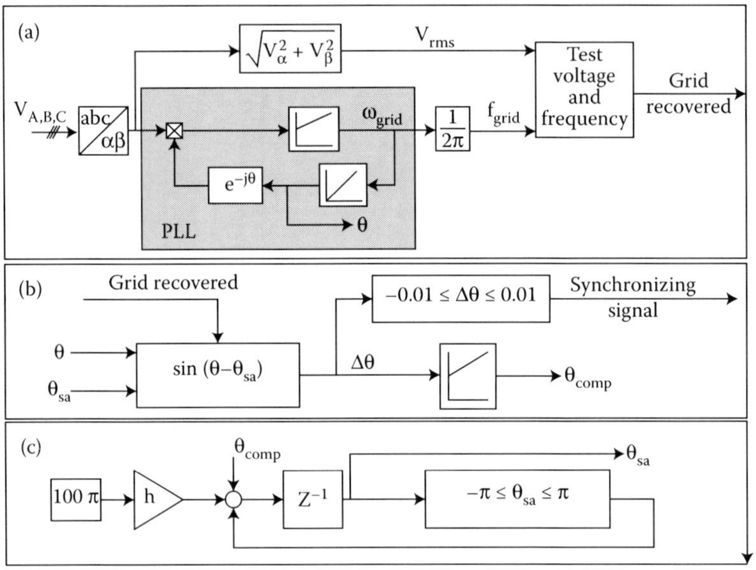


**FIGURE 15.34**
Stand-alone to grid operation transition control: (a) grid parameters are checked for validity; (b) if the grid recovered synchronization starts; (c) stand-alone voltage angle estimation. (After M. Fatu, F. Blaabjerg, I. Boldea, *IEEE Trans. Power Electron.*, vol. 29, no. 7, 2014, pp. 3463–3472.)

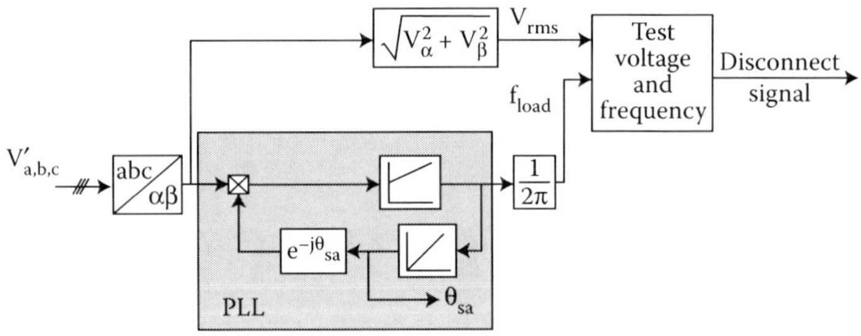


**FIGURE 15.35**
Grid to stand-alone control transition. (After M. Fatu, F. Blaabjerg, I. Boldea, *IEEE Trans. Power Electron.*, vol. 29, no. 7, 2014, pp. 3463–3472.)

Basically active power and current $(i_d^*, i_q^*)$ control is operated slightly differently for grid and stand-alone operation modes. The reference active power P* current $i_q^*$ is

$$i_q^* = k_{p\_pow}\left(1+\frac{k_{i\_pow}}{s}\right)(P^* - V_{dc}i_{dc}) \tag{15.59}$$

for grid mode and

$$i_q^* = k_{P\_V_{dc}}\left(1+\frac{k_{i\_V_{dc}}}{s}\right)\left(V_{dc}^* - V_{dc}\right) \tag{15.60}$$

for stand-alone mode with $i_d^* = 0$ in both cases here, for simplification (Figure 15.36a).

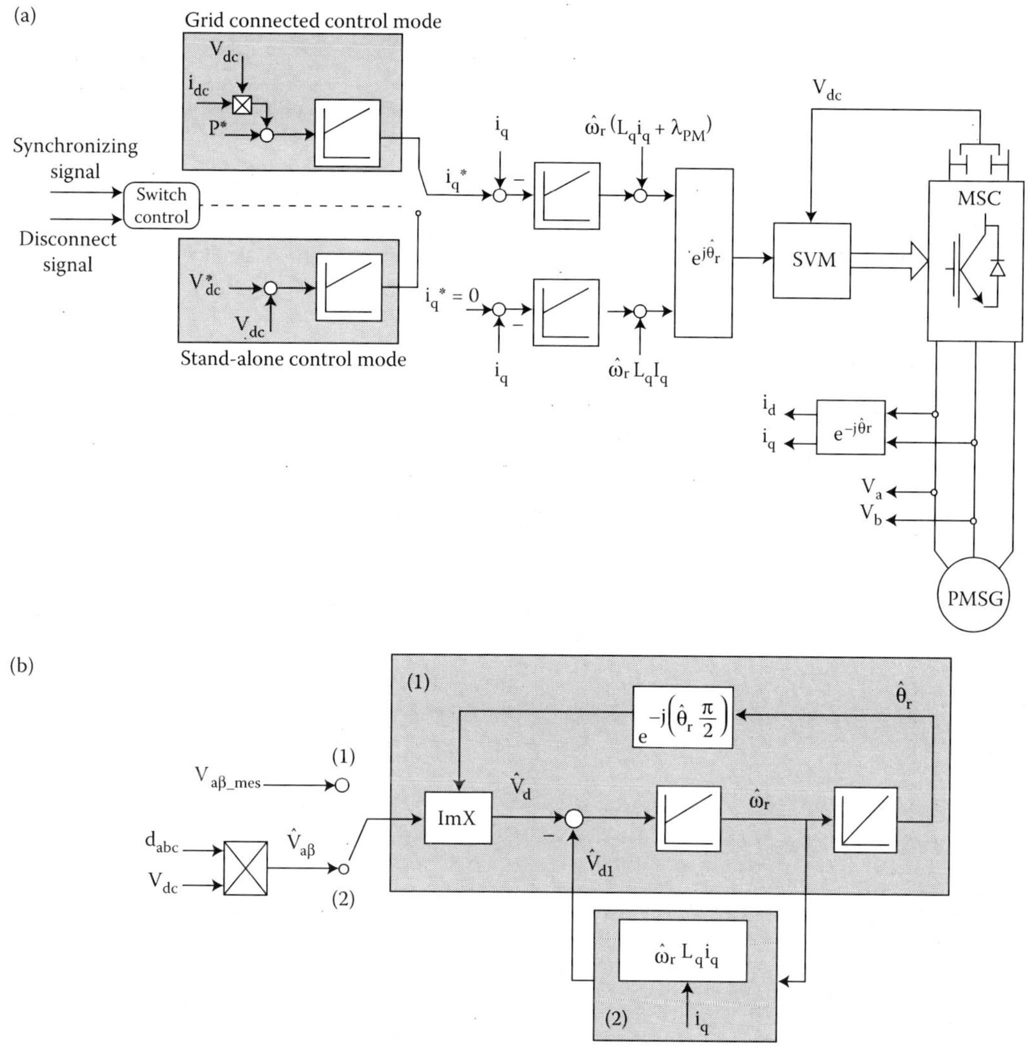


**FIGURE 15.36**
(a) PMSG-side converter control for grid and stand-alone operation. (b) Rotor speed and position observer. (After M. Fatu, F. Blaabjerg, I. Boldea, *IEEE Trans. Power Electron.*, vol. 29, no. 7, 2014, pp. 3463–3472.)

The rotor speed and position observer [17] is shown in Figure 15.36b.

For our case study $K_{p\_pow} = 0.002$ A/W, $K_{p\_Vdc} = 0.002$ V/W, $K_{ipow} = K_{i\_Vdc} = 25\ s^{-1}$.

### *15.6.2.5 Test Platform and Sample Results*

The test platform includes a 12 Nm PMSG connected to the grid by two back-to-back Danfoss VLT 5005 voltage source converters driven by a Simovert Master drive. The grid is replaced by a programmable three-phase a.c. power source (California Instruments 5005) capable to create asymmetric voltage faults (sags).

As the a.c. power source cannot receive average power, a resistive nonlinear (bump) load has been connected to the power line.

#### 15.6.2.6 Voltage Ride Through

A single-phase voltage sag ride through is illustrated in experiments of Figure 15.37a–c.

As expected the system performs well though the increase and oscillation of $i_q^*$ leads to limited oscillations in the d.c. link voltage too, as seen in Figure 15.38 for a symmetric (three-phase) voltage sag.

The beneficial effect of voltage harmonics compensation is shown in Figure 15.39a and b.

Transitions from stand-alone to grid (Figure 15.40) and vice versa (Figure 15.41) lead to remarks such as

- At $t = 10.8$ s the grid is connected by the switch S (Figure 15.29).
- A synchronization signal (Figure 15.40a) is automatically initiated to match the phases between load and grid voltages.
- When the error $\Delta\theta < 0.01$ rad the system switches from stand-alone to grid operation.
- The transition is rather smooth as illustrated in Figure 15.41b–d.
- For the grid to stand-alone operation triggering, a fault event is simulated by opening the switch S (Figure 15.29) at $t = 15.8$ s. As we still have current control in the grid-side converter there is a temporary increase in load voltage (Figure 15.41c); when this is too large or the grid frequency is out of range, the system automatically switches to stand-alone mode (Figure 15.41b). Load currents and PMSG currents (Figure 15.41d and e) show smooth transition while the speed observer operates satisfactorily during the transition (Figure 15.41f and g).

## 15.7 Switched Reluctance Generator Control

The switched reluctance machine (SRM), which can be built in single-phase or multiphase configurations, may work as a motor or as a generator.

Motoring is required if prime-mover starting or assistance is required as in HEV [1,11,12].

If only generator operation is necessary, single-phase (for low power) and three-phase configurations seem favored.

The circuit mathematical model of SRM developed in Chapter 12 holds valid here and thus will not be derived again.

We will rather concentrate on typical current waveforms for generator mode, typical PWM converters, and a voltage controller for d.c. output at variable speed.

The turn on $\theta_{on}$ and $\theta_{off}$ angles for generator mode occur at the beginning, and respectively, long before the end of negative inductance slope of the conducting phase (Figure 15.42a).

The phase i voltage equation is

$$V_i = R_s I_i + L_{ti}(I_i, \theta_r)\frac{dI_i}{dt} + K_E(I_i, \theta_r)\cdot 2\pi n \tag{15.61}$$

where $K_E$ is the pseudo e.m.f. coefficient.

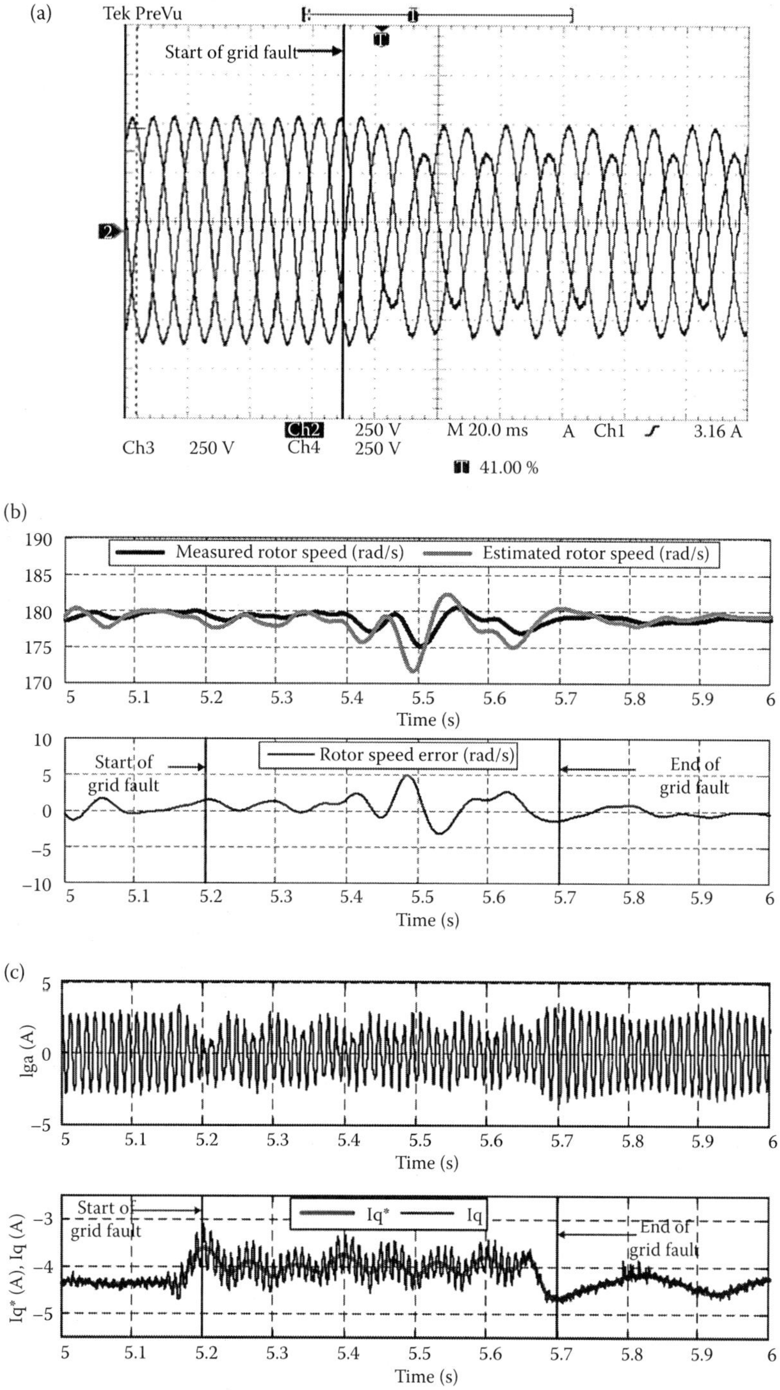


**FIGURE 15.37**
Single-phase voltage sag operation of PMSG: (a) Grid phase voltages after transformer (Figure 15.29), (b) Estimated, measured, and error of rotor speed, (c) PMSG generator current; $i_q^*$; and $i_q$ variation during single-phase voltage sag. (After M. Fatu, F. Blaabjerg, I. Boldea, *IEEE Trans. Power Electron.*, vol. 29, no. 7, 2014, pp. 3463–3472.)

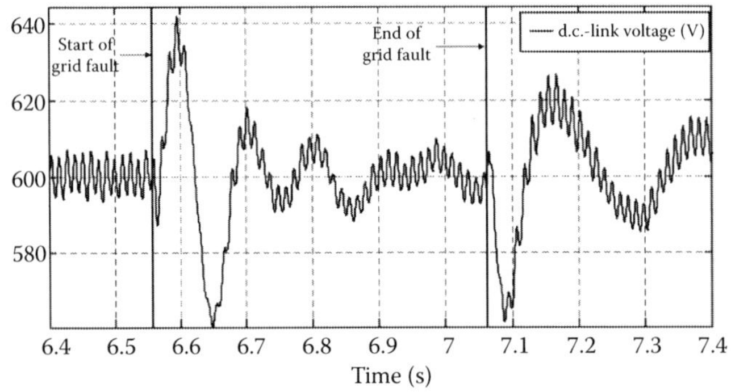


**FIGURE 15.38**
The d.c. link voltage during three-phase voltage sags.

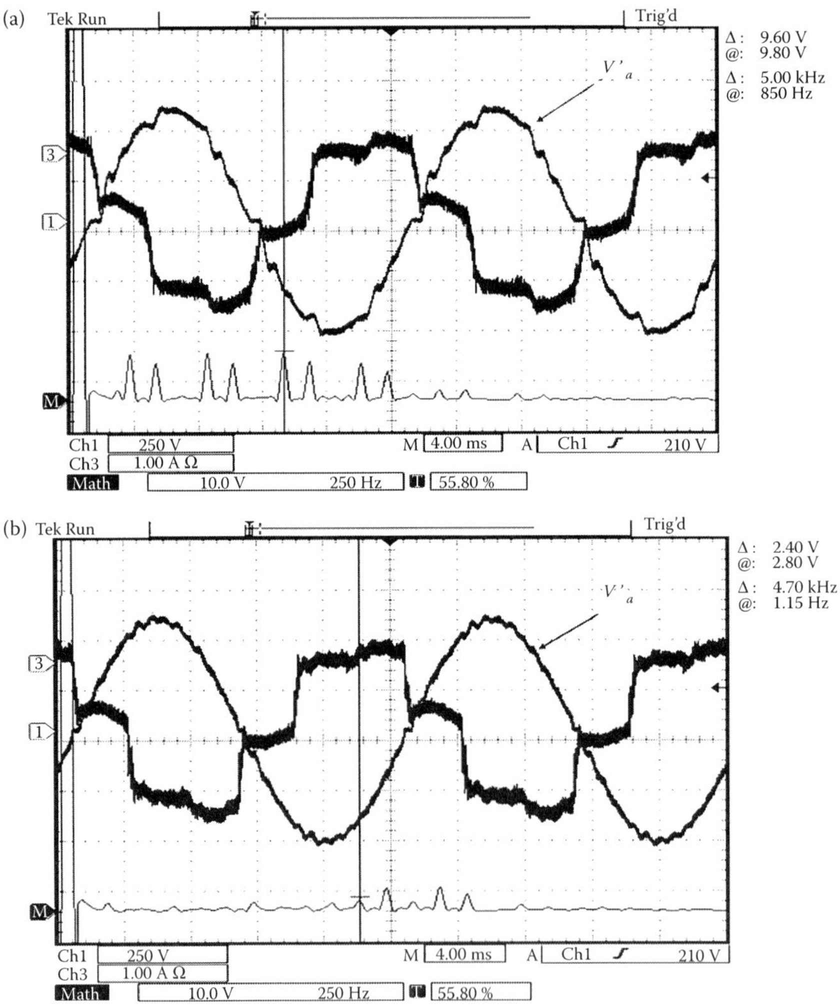


**FIGURE 15.39**
Nonlinear load current and voltage (a) without harmonics compensation (b) with harmonics compensation. (After M. Fatu, F. Blaabjerg, I. Boldea, *IEEE Trans. Power Electron.*, vol. 29, no. 7, 2014, pp. 3463–3472.)

$$E = K_E 2\pi n; \; K_E(I_i, \theta_r) = I_i \frac{\partial L_i}{\partial \theta_r} < 0, \quad \text{for generating} \tag{15.62}$$

We may consider $L_{ti} \cong L_u$ (unaligned) = constant for the heavily saturated core machine.

Voltage $V_i = +V_{dc}$ for phase energization and $V_i = -V_{dc}$ during power delivery (phase de-energization).

There are three generator operation modes as illustrated in Figure 15.42a and b.

As during generator mode $V_i = -V_{dc} < 0$ and $E < 0$, the current, which remains positive, increases as long as $|E| > |V_{dc}|$. This case is typical for high speeds when the torque (energy cycle area) is smaller (Figure 15.42b, cycle 1).

For case a2 in Figure 15.42a, at the turn off angle $\theta_c$, $|\underline{E}| = V_{dc}$ and thus, with $R_s \sim 0$, $di/dt = 0$, the current stays constant until the phase inductance reaches its minimum at angle $\theta_d$ (energy cycle 2 in Figure 15.28b).

In case a3 in Figure 15.42a the maximum current is reached at $\theta_c$ and, after that, the current decreases monotonously because $|E| < V_{dc}$, which corresponds to low speeds (energy cycle 3 in Figure 15.42b).

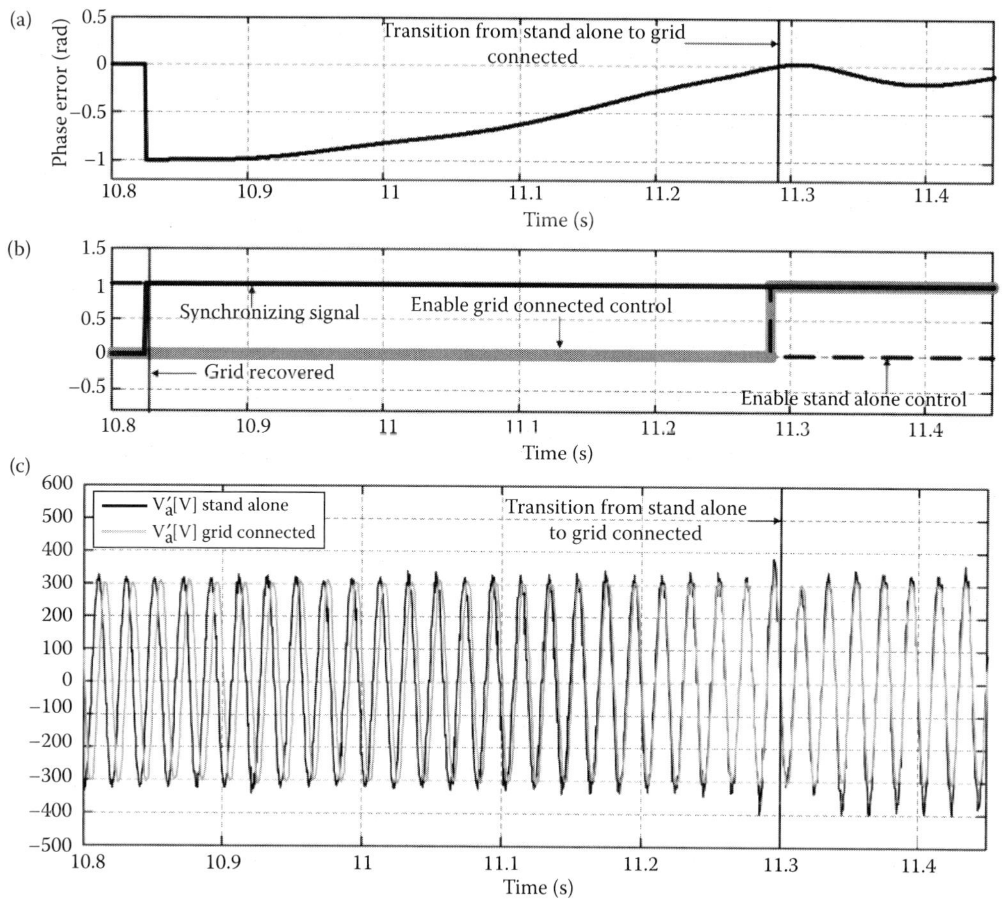


**FIGURE 15.40**
Measured stand-alone to grid-connection transition: (a) phase error; (b) enable grid control/stand-alone signals, synchronizing signal; (c) phase A load voltage ($V_a'$), phase A grid voltage ($V_a$). *(Continued)*

(d)
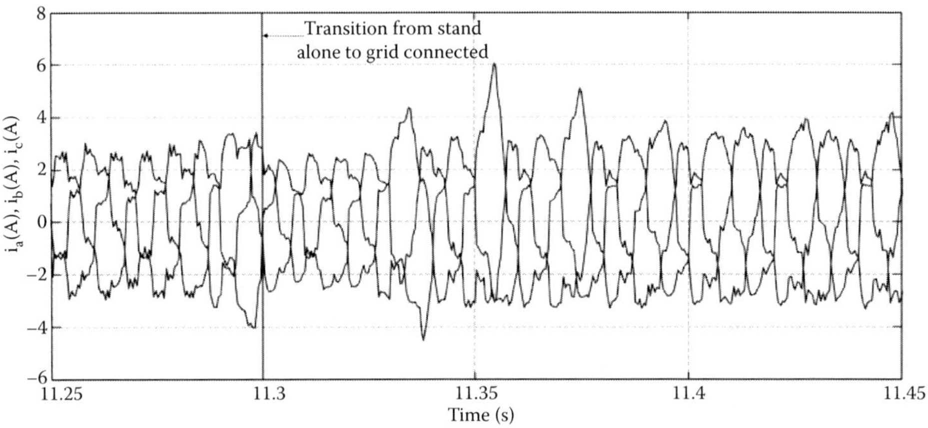


(e)
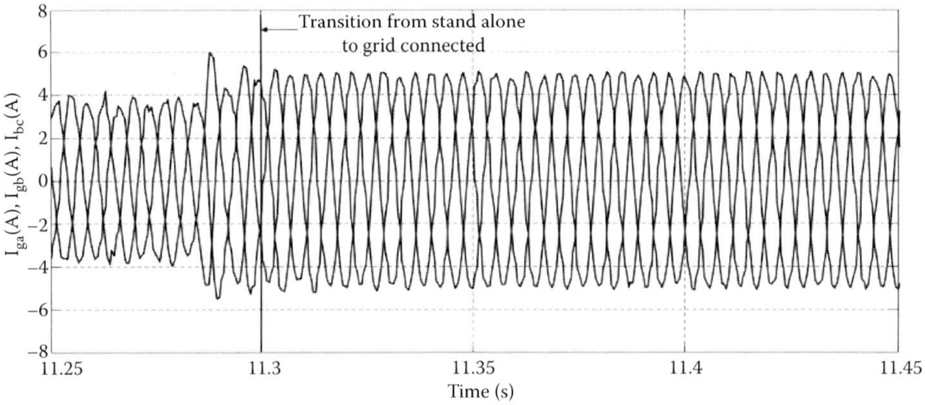


**FIGURE 15.40 (*Continued*)**
Measured stand-alone to grid-connection transition: (d) load currents; (e) PMSG currents. (After M. Fatu, F. Blaabjerg, I. Boldea, *IEEE Trans. Power Electron.*, vol. 29, no. 7, 2014, pp. 3463–3472.)

For constant d.c. voltage power delivery, the excitation (energization) energy per cycle $W_{exc}$ is

$$W_{exc} \approx \frac{V_{dc}}{2\pi n} \int_{\theta on}^{\theta c} i d\theta_r \tag{15.63}$$

while the energy delivered per cycle $W_{out}$ is

$$W_{out} \approx \frac{V_{dc}}{2\pi n} \int_{\theta c}^{\theta off} i d\theta_r \tag{15.64}$$

We may define the excitation penalty ε as

$$\varepsilon = \frac{W_{exc}}{W_{out}} \tag{15.65}$$

$|E| > V_{dc}$ leads to a smaller excitation penalty but not necessarily to the highest energy conversion ratio, which seems to correspond to $|E| \sim V_{dc}$.

Maintaining this latter condition, however, means to increase $V_{dc}$ with speed because E increases with speed. A step-down d.c.–d.c. converter has to be added as the d.c. output voltage has to stay constant for $\omega_{rmin}$ to $\omega_{rmax}$.

To control the power output, the turn on and turn off angles $\theta_{on}$ and $\theta_c$ should be changed in a certain manner.

If such a step-down d.c.–d.c. converter is not available, then the generator mode will evolve from case a3 ($|E| < V_{dc}$), below base speed, to case a2 ($|E| = V_{dc}$) and a1 ($|E| > V_{dc}$)—Figure 15.42—as the speed increases. A rather moderate speed range constant output voltage control is provided this way.

An average excitation penalty $\varepsilon = 0.3–0.4$ should be considered acceptable for good energy conversion ratio, though a notable energy reserve is lost.

Typical static power converters for $|E| = V_{dc}$ and, respectively, variable $|E|/V_{dc}$ ratio (with speed) control are shown in Figure 15.43.

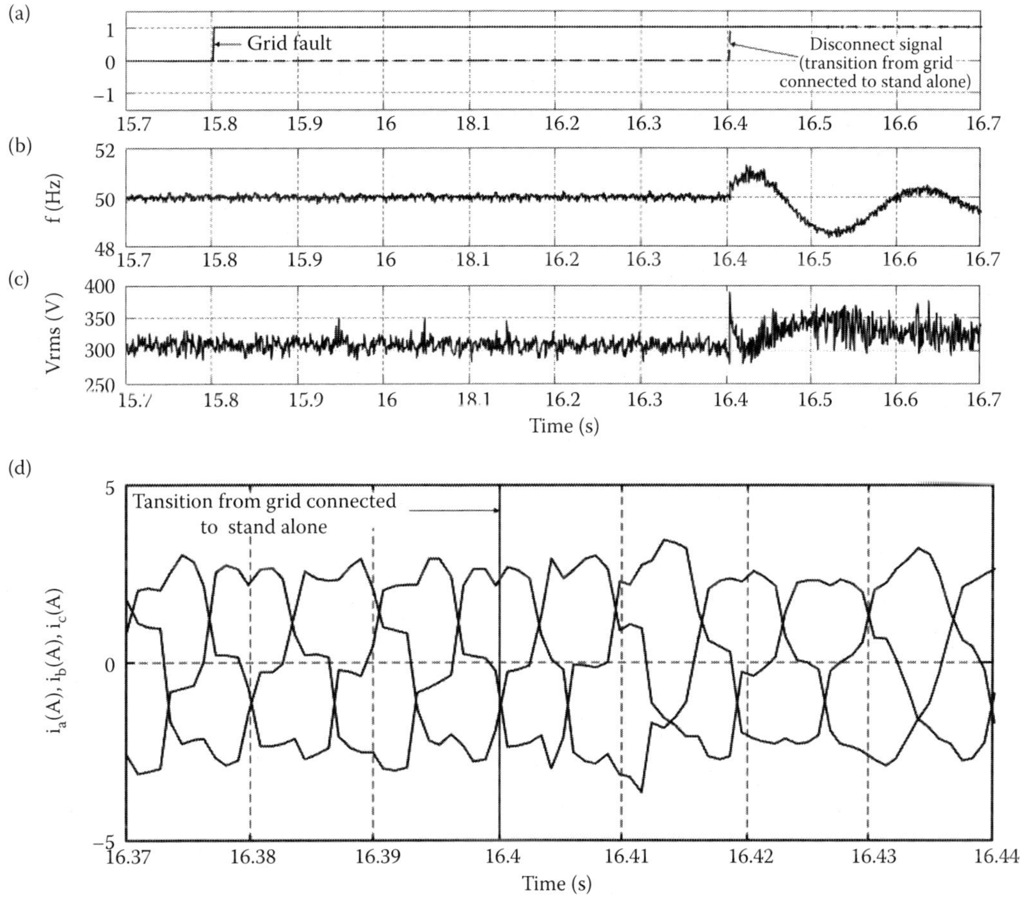


**FIGURE 15.41**
Measured transition from grid-connection to stand-alone control mode: (a) disconnect signal, (b) $f_{grid}$, (c) $V_{rms}$, (d) grid currents. *(Continued)*

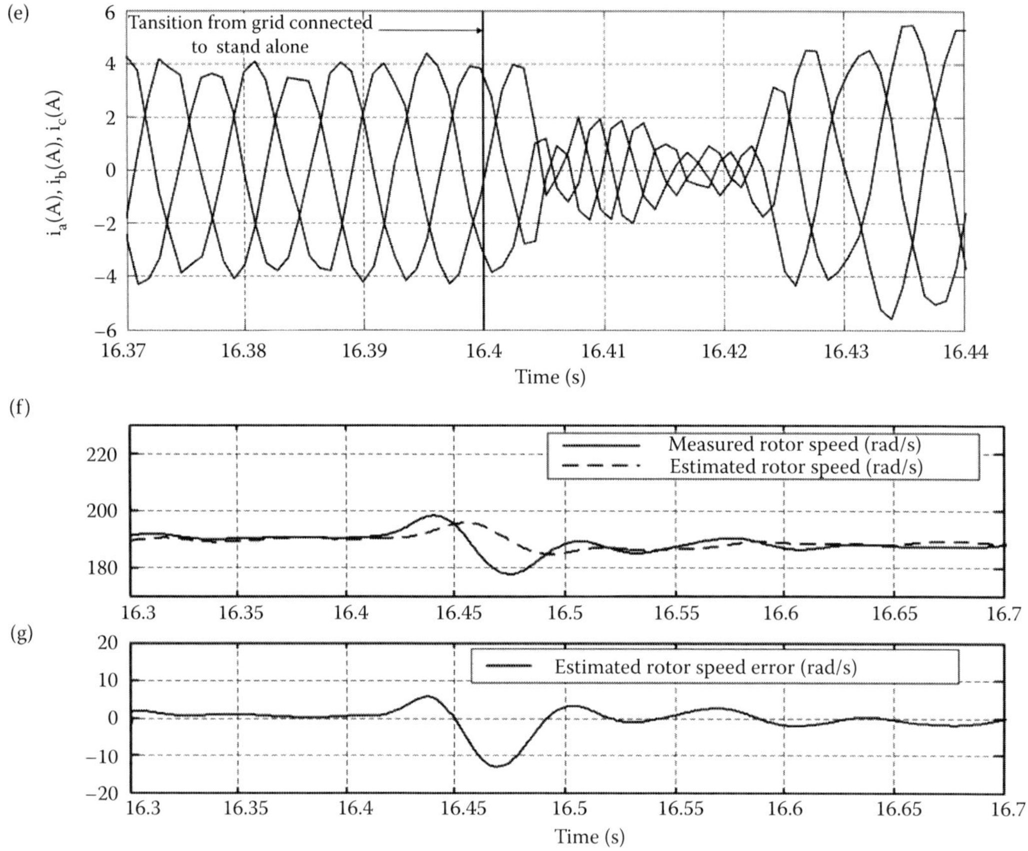


**FIGURE 15.41 (*Continued*)**
Measured transition from grid-connection to stand-alone control mode: (e) PMSG currents, (f) estimated and measured rotor speed, (g) rotor speed estimation error. (After M. Fatu, F. Blaabjerg, I. Boldea, *IEEE Trans. Power Electron.*, vol. 29, no. 7, 2014, pp. 3463–3472.)

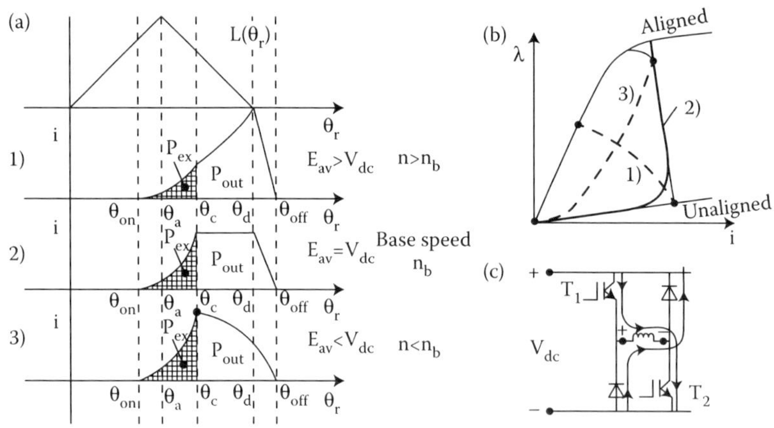


**FIGURE 15.42**
SRG: (a) inductance/position and current waveforms, (b) corresponding energy cycles, (c) typical phase PWM converter.

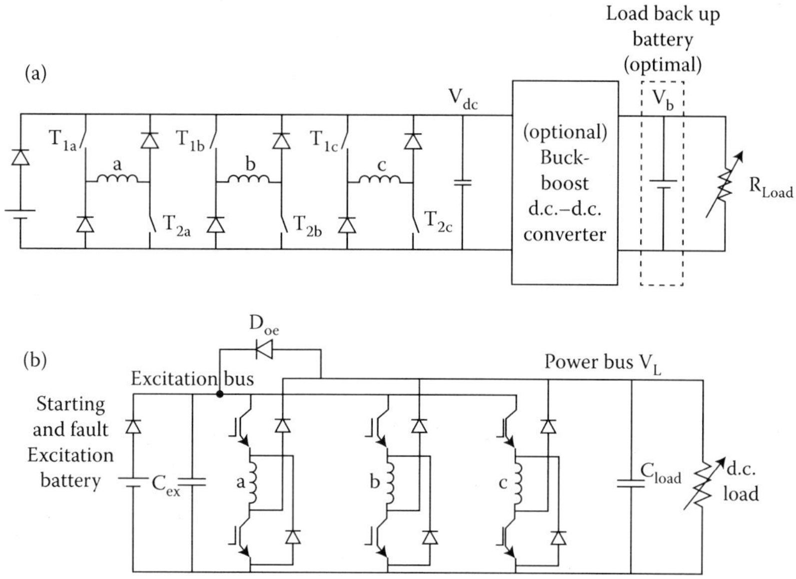


**FIGURE 15.43**
Three SRG: (a) with asymmetrical PWM converter, selfexcitation (from battery) or load back-up battery and d.c.–d.c. converter; (b) with separate excitation power bus and fault clearing capability.

The additional boost capability to the buck converter is required when motoring is needed, as for starter-alternators on hybrid vehicles (Figure 15.43a).

To stabilize the output voltage of stand-alone loads it may be suitable to supply the excitation power from a separate battery (Figure 15.43b, left). The diode $D_{oe}$ allows for battery assistance during generating mode and thus the battery capacitor cost is rather small.

The excitation battery also allows SRG to operate during load faults and clear them quickly (Figure 15.44) [13].

Intricate SRG control at variable speed for electric vehicles is demonstrated in Reference 14 with feedforward torque control and in References 15–17 with direct torque control.

The ruggedness of SRG and its wide speed range constant power capability at constant d.c. voltage are strong assets in its favor, especially in battery back-up d.c. loads, typical for automobiles, ships, and aircraft.

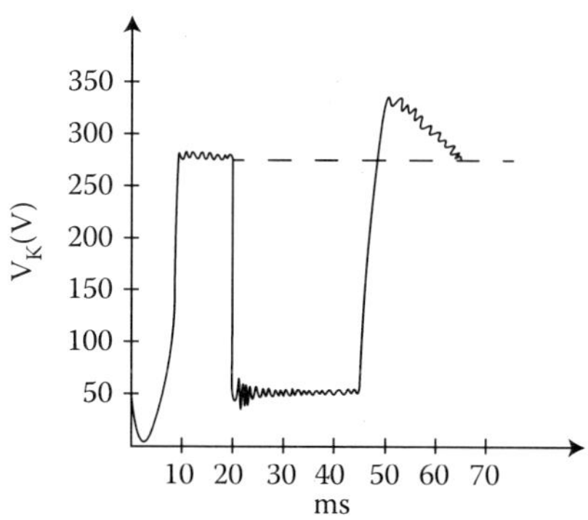


**FIGURE 15.44**
SRG voltage rise-up, experiencing a fault (by sharp load resistance reduction) and voltage swift recovery.

## 15.8 Summary

- Practically all electric energy is produced (from primary fuel energy) through electric generators driven by prime movers (turbines).
- In today's standard electric power systems many synchronous electric generators are connected in parallel, to provide rather constant voltage and frequency.
- Their voltage (reactive power) and speed (active power) is controlled with a small droop to allow for desired power sharing between paralleled SGs.
- The voltage (reactive power) control of SG is performed through the SG d.c. excitation current; an exciter source is required; it may be a brushless a.c. exciter or a static exciter. Fast voltage recovery response is required; various automatic voltage regulators (AVRs) have been proposed, and a PID digital version is illustrated in this chapter.
- While SGs in power systems are controlled with a small speed droop, when they act alone on a few loads their speed is controlled either to stay constant or is left to vary (to reduce speed with power decrease) to save fuel in the prime mover, when a full-power converter is needed to yield constant voltage and frequency output over the entire speed range.
- SGs are present on automobiles as alternators with full-power diode rectifier output to a d.c. battery. The alternator provides d.c. current, in the battery and loads, increasing with speed through proper d.c. excitation current control via voltage closed-loop regulator and a low-power electronics converter.
- For wind and small hydro energy conversion in the multi MW/unit range variable speed, for best primary energy tapping and better stability, the doubly fed WRIG is frequently used.
- WRIG contains a dual (a.c.–d.c.–a.c.) PWM converter connected to the wound rotor through brushes and slip rings. This converter is feeding the rotor with a voltage $V_r \le V_s$ at a frequency $f_2 = f_1 - np >< 0$; ($V_r$, $V_s$, rotor and stator voltages, $f_2$, $f_1$ is the rotor and stator frequency, n the speed rps, and p is pole pairs). The speed range is defined by the maximum slip $S_{max} = \pm(5\text{–}30)\ \% = f_2/f_1$; smaller values are typical for larger powers.
- The PWM bidirectional converter is practically rated at $|S_{max}|P_n < 30\%P_n$ and thus lower system costs are obtained.
- WRIG has been implemented up to 400 MW/unit in pump-storage hydro-generators. The fast active and reactive power control within the speed range from 70% to 130% of rated speed is a flexibility asset that fixed speed SGs are lacking. WRIG is practically an SG at all speeds within the speed range and its synchronization to the grid is very quick and safe.
- Cage-rotor induction generators with full-power bidirectional PWM converters connected to the stators seem the favorite solution in wind, small hydro, and even gensets up to 3–5 MW/unit, as they provide fast and smooth connection to the power grid. From the grid point of view they behave like very fast controlled SGs, but at variable speed.
- Permanent magnet SGs are characterized by low losses and, for inset (or interior) PM-rotor, they may exhibit even zero voltage regulation on resistive rated load.

- PMSG may be controlled for variable speed with constant d.c. voltage output or with constant frequency and voltage a.c. output, when a full-power electronic converter is required. PMSGs may work in stand-alone applications (starter/alternators on cars) or in power grid applications, driven by high speed gas turbines to produce cogeneration or emergency, or peak power in various applications. Better direct power PMSG dual converter control solutions keep surfacing [18].
- SRGs have been proposed especially for vehicle applications (automobile and aircraft starter alternators) at variable speed, when they deliver power on d.c. battery back-up loads.
- Electric generator control is needed both in power systems, now still standard but more distributed in the future, and also in stand-alone vehicular or on ground applications. Decentralized, controlled, electric energy production requires variable speed in general and thus power electronics control of active and reactive power comes into play.
- Control of electric generators at variable speed can draw on the immense heritage of electric drives control techniques, and should constitute a major technology of the future.
- The present chapter should be considered only a tiny introduction to EG control; if a newcomer, see Reference 1 to start your own journey into this field.

## References

1. I. Boldea, *Electric Generators Handbook*, Taylor & Francis Group, New York, 2005, second edition, 2015.
2. P. Kundur, *Power Systems Stability and Control*, McGraw-Hill, New York, 1994.
3. A. Godhwani, M.J. Bassler, A digital excitation control system for use on brushless excited synchronous generators, *IEEE Trans.*, vol. EC-11, no. 3, 1996, pp. 616–620.
4. R. Datta, V.T. Ranganathan, A simple position—Sensorless algorithm for rotor-side field oriented control of wound rotor induction machine, *IEEE Trans.*, vol. IE-48, no. 4, 2001, pp. 786–793.
5. R. Rena, J.C. Clare, G.M. Asher, Doubly fed induction generator using back to back PWM converters and its application to variable speed wind-energy generation, *Proc. IEE.*, vol. EPA-143, no. 3, 1996, pp. 231–241.
6. I. Serban, F. Blaabjerg, I. Boldea, Z. Chen, A study of the double fed wind power generator under power system faults, *Record of EPE*, 2003, Toulouse, France (IEEExplore).
7. D.J. Perreault, V. Caliskan, Automotive power generation and control, *IEEE Trans.*, vol. PE-19, no. 3, 2004, pp. 618–630.
8. S. Scridon, I. Boldea, L. Tutelea, F. Blaabjerg, E. Ritchie, BEGA—A biaxial excitation generator for automobiles: Comprehensive characterization and test results, *Record of IEEE-IAS*, 2004, Seattle, Washington (IEEExplore).
9. M. Kazmierkowski, R. Krishnan, F. Blaabjerg (eds.), *Control in Power Electronics*, Academic Press, Amsterdam, 2002.
10. L.M. Tolbert, W.A. Peterson, T.J. Theiss, M.B. Sardiere, Gen-sets, *IEEE-IA Mag.*, vol. 9, no. 2, 2003, pp. 48–54.
11. J. Hussain, *Electric and Hybrid Vehicles*, CRC Press, Boca Raton, Florida, 2003.
12. A. Emadi, M. Ehsani, J.M. Miller, *Vehicular Electric Power Systems*, Marcel Dekker, New York, 2004.

13. A.V. Radun, C.A. Ferreira, E. Richter, Two channel switched reluctance starter-generator results, *IEEE Trans.*, vol. IA-34, no. 5, 1998, pp. 1106–1109.
14. H. Bausch, A. Grief, K. Kanelis, A. Mickel, Torque control battery supplied switched reluctance drives for electrical vehicles, *Record of ICEM*, 1998, vol. 1, pp. 229–234, Istanbul, Turkey.
15. R.B. Inderka, R.W. De Doncker, Simple average torque estimation for control of switched reluctance machines, *Record of EPE-PEMC*, 2000, Kosice, vol. 5, pp. 176–181.
16. S. Dixon, B. Fahimi, Enhancement of output electric power in SRGs, *Record of IEEE-IEMDC*, 2003, Madison, Wiscosin, vol. 2, pp. 849–856.
17. M. Fatu, F. Blaabjerg, I. Boldea, Grid to standalone transition motion-sensorless dual-inverter control of PMSG with asymmetrical grid voltage sags and harmonics filtering, *IEEE Trans. Power Electron.*, vol. 29, no. 7, 2014, pp. 3463–3472.
18. Y. Zhang, I. Fu, I. Hu, I. Zhu, Three vectors based predictive direct power control of the direct fed induction generator for wind energy applications, *IEEE Trans.*, vol. PE-29, no. 7, 2014, pp. 3485–3500.

# 16

# *Scalar V/f and I–f Control of A.C. Motor Drives: An Overview*

## 16.1 Introduction

Motion (position, speed, torque) control of electric motors and active and reactive power control in electric generators are the first means for increased productivity and energy savings in all industries with variable-in-time output. When torque fast response and control precision are required, the so-called field oriented control (FOC) and DTFC are preferred for a.c. motors/generators control. As expected, FOC and DTFC imply rather complex control.

However, there are many applications (the majority in terms of total installed power), such as pumps, compressors, ventilators, very-high-speed small-power machines, where the simplicity and low online computation effort for control decisions is imperative, either based on total system cost or on small online available computation time cycle for control (in super-high-speed drives). For such applications, scalar control, initially implemented as open-loop voltage/frequency (V/f) and current and frequency (I–f) control, have been introduced, first for cage-rotor IM general drives and later for PMSM drives.

Considerable efforts in the last few decades have been put into V/f and I–f control of a.c. motor drives:

- To increase the speed control range to more than 10 (20)/1 without a speed control loop, by thorough calculations of command voltage to frequency relationships, all through flux weakening
- For faster torque dynamics by stabilizing loops (one or two)
- For implicit reduction of losses for lower torque
- To provide for self-starting of SM drives

The main purpose of these efforts is to introduce new attributes to the simplified (lower cost) scalar control in variable speed drives, to bridge the gap to FOC and DTFC and extend the area of their applications. Such advanced scalar control methods may be used also for redundancy in safety critical drives (on HEV, EV, aircraft, vessel, etc.).

While it is true that the price of NdFeB magnets has increased lately above 100 USD/kg, their use in small quantities in small-power electric drives or even in medium- and large-power drives at high speeds (when the weight of NdFeB in kg/kW is still economic in terms of performance/cost) makes the investigations into PMSM drives, besides IM drives, still necessary.

Also, the use of less NdFeB magnets (or of ferrite magnets) in PM-assisted (low e.m.f.) RSMs, in wide CPSR applications, justifies the interest in simpler (lower cost) control methods for them, in general-purpose variable speed drives.

Even in the range of hundreds of kW variable low-speed drive (around maximum 500 rpm) applications, the PMSM additional cost may be offset by the much higher power factor (above 0.9; only 0.6 for IMs with 12 poles), which reduces drastically the peak kVA (and thus the costs) of the inverter supply (ultimately the total initial cost of the drive may be reduced for lower capitalized cost of losses).

This chapter attempts an overview, with representative results, of scalar (V/f and I–f) control with open-loop and closed-loop stabilization (and loss reduction) for wide speed range control (with flux weakening), of induction and synchronous motor drives.

## 16.2 Induction Machines V/f and I–f Open and Closed-Loop Control

What characterizes all scalar controls of a.c. motor drives is the absence of a motion sensor and of closed-loop speed control, which implies the elimination of the speed observer, typical to sensorless FOC and DTFC control. The simplest (and historically the first) control system for IM drives was the open-loop $V_1/f_1$ control (Figure 16.1).

Very soon the main liabilities of this simple scheme became apparent:

- The nonlinear model of IM revealed instabilities at/around certain small frequency $\omega_{1min}$, dependent on machine parameters and on the number of pole pairs [1]:

$$\omega_{1\min} > 2/\tau_n \tag{16.1}$$

where $\tau_n$ is the rotor time constant; also instabilities occur [1] if the figure of merit $f_m$ is

$$f_m = \tau_m/\tau_r < 0.5 \tag{16.2}$$

where $\tau_m$ is the mechanical time constant: $\tau_m = J/p \cdot H$ ($H \approx T_e/(\omega_1 - \omega_r)$), p is the pole pairs count, and $T_e$ is the torque, with $\omega_1 - \omega_r = S \cdot \omega_1$ as slip frequency.

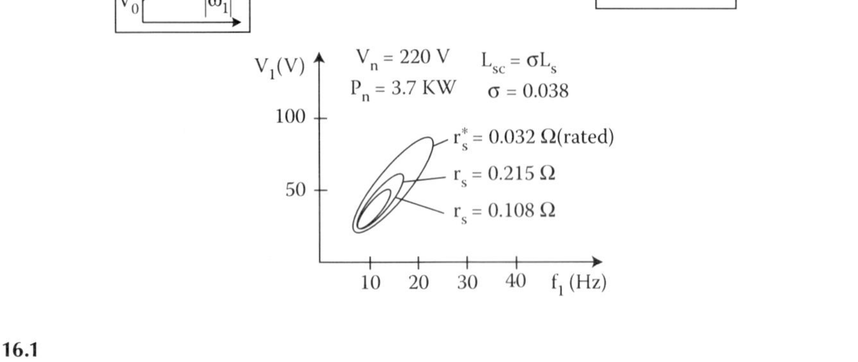


**FIGURE 16.1**
"Primitive" $V_1/f_1$ open-loop control of IM drives.

Two-pole IMs tend to be unstable ($f_m < 0.5$), while 4-, 6-, 8-pole IMs are less susceptible to instabilities in open-loop $V_1/f_1$ control.

- Open-loop $V_1/f_1$ control is vulnerable to fast ramping of reference speed and to large torque perturbations.
- The instabilities limit the lowest speed to $\omega_{r\,min}$ and also, as the load implies an 1–3 Hz slip frequency under full load torque, this means an up to 100% reduction of speed under load, from $\omega_{r\,min}$ to $\omega_r = 0$. Compensating for the slip frequency in a feedforward manner becomes thus necessary [2] (Figure 16.2).

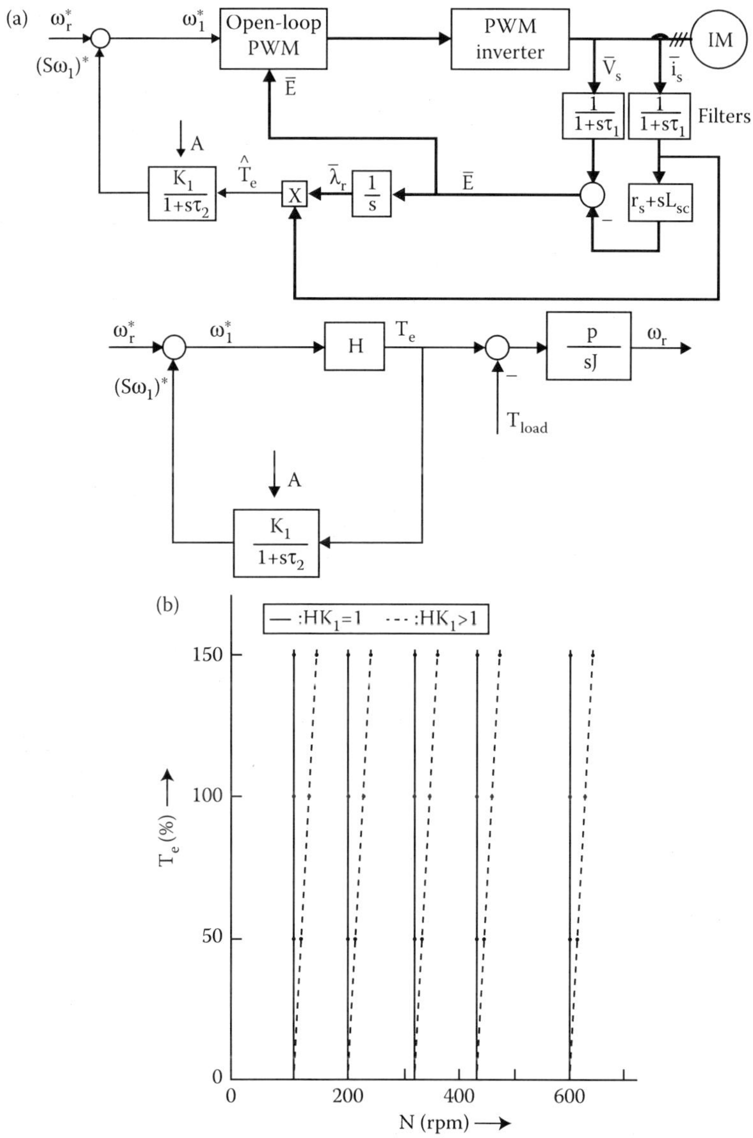


**FIGURE 16.2**
$V_1/f_1$ scalar control with feedforward slip frequency compensation: (a) generic scheme; (b) slip frequency compensator. *(Continued)*

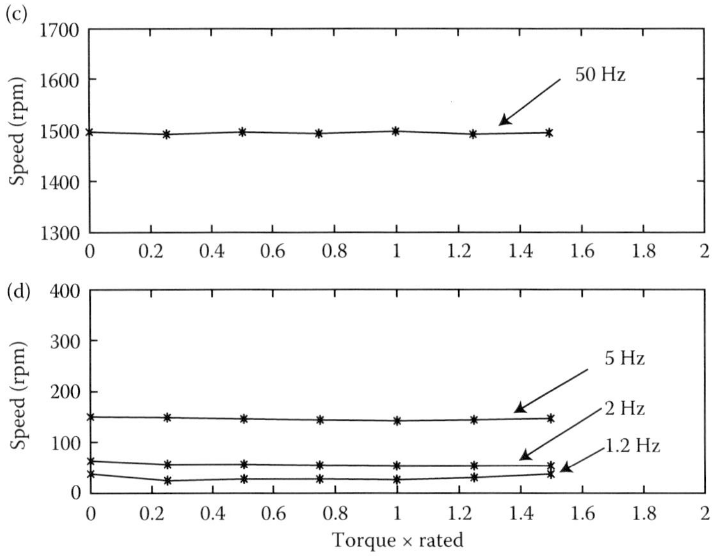


**FIGURE 16.2 (*Continued*)**
$V_1/f_1$ scalar control with feedforward slip frequency compensation: (c) extension of speed–torque range (After K. Koga, R. Ueda, T. Sonoda, *IEEE Trans.*, IA-28(2), 463–471, 1992; (d) nonlinear slip compensation. (After A.M. Garcia, T.A. Lipo, D.W. Novotny, *IEEE Trans.*, IA-34(4), 813–821, 1998.)

The generic scheme in Figure 16.2 assumes implicitly that the rotor flux $|\bar{\Psi}_r|$ $(\bar{E} = +j\omega_1 \cdot \bar{\Psi}_r \cdot L_m/L_r)$ is given (constant) and leads to stable speed dynamics if $HK_1 \le 1$ [2].

A similar solution for slip compensation is applied in Reference 3, but with voltage amplitude correction. However, if energy savings operation is intended for all torque levels, then the control scheme with only $V_1 = V_0 + k_f \cdot \omega_1$ (Figure 16.2) is not sufficient and the rotor flux level $\Psi_r$ (E, implicitly) varies. Consequently, H and K should be made adaptable. A V/f with nonlinear slip compensation scheme with more extended online calculation effort, at constant stator flux, has produced safe operation down to 1.2 Hz and 150% torque, with reasonable speed errors (Figure 16.2d) [4]. Voltage amplitude correction may be added to provide energy savings operation. Such a solution may come as a variable voltage boost $V_0$, calculated from loss minimization with steady-state equations [4–6], but it is heavily dependent on machine parameters. Operation at the lowest frequency of 1.2 Hz was performed [4] safely, but still no fast torque response was demonstrated.

The rather simple robust (but slow) open-loop energy savings, by modifying stepwise the voltage correction ΔV as long as the d.c.-link current decreases, has been proposed as early as 1984 [7].

Alternatively, a rather simplified feedforward voltage correction ΔV could be obtained, for example, if the machine is planned to operate around maximum power factor (which is good for energy savings control) as long as the reference voltage $V_S^*$ is smaller than the maximum inverter voltage $V_{S\,max}$, and at maximum torque/flux otherwise. By using the concept of "active flux" $\bar{\Psi}_d^a$ [8]

$$\left(\bar{\Psi}_d^a\right)^{\theta_{\Psi r}} = (L_S - L_{Sc}) \cdot i_d; \quad T_e = \frac{3}{2} p \cdot \Psi_d^a \cdot i_q \tag{16.3}$$

where $L_S$, $L_{Sc}$ are the no-load and short-circuit inductances of IM.

The maximum torque/power factor is obtained [9] for

$$\frac{i_d}{i_q} = \sqrt{\frac{L_{Sc}}{L_S}} \tag{16.4}$$

As the stator flux $\Psi_S$ (in rotor flux coordinates) is

$$\Psi_S = L_S \cdot i_d + j \cdot L_{sc} \cdot i_q \quad \text{and} \quad \bar{\Psi}_r = \bar{\Psi}_d^a \cdot \frac{L_m}{L_S - L_{Sc}} \tag{16.5}$$

where $L_m$ is the magnetization inductance, the magnetic energy in the machine $W_m$ in "active (rotor) flux" coordinates, with Equation 16.4, is

$$(W_m)_{(\cos\varphi)_{max}} = \frac{3}{2}\left(L_S \cdot i_d^2 + L_{Sc} \cdot i_q^2\right) = 3 \cdot L_{Sc} \cdot i_q^2 \tag{16.6}$$

But with $\Psi_S$ estimated in stator coordinates (above 1 Hz)

$$\Psi_S \approx \int \left(\bar{V}_S - R_S \cdot \bar{i}_S\right) dt = \frac{T}{1+sT}\left(\bar{V}_S - R_S \cdot \bar{i}_S\right) \tag{16.7}$$

$\bar{\Psi}_d^a$ from Equation 16.5, and $T_e$ (torque) as

$$T_e = \frac{3}{2} \cdot p \cdot \mathrm{Real}\left(j\bar{\Psi}_s \cdot \bar{i}_s^*\right) = \frac{3}{2} \cdot p \cdot \left|\bar{\Psi}_d^a\right| \cdot \hat{i}_q \tag{16.8}$$

the value of $\hat{i}_q$ (torque current) may be estimated now from Equation 16.8. The magnetic energy for maximum power factor is obtained from Equation 16.6. Finally, a PI loop to correct the magnetic energy in stator coordinates $W_{m\alpha\beta}$

$$W_{m\alpha\beta} = \frac{3}{2} \cdot \mathrm{Real}(\bar{\Psi}_S \cdot \bar{i}_S^*) \tag{16.9}$$

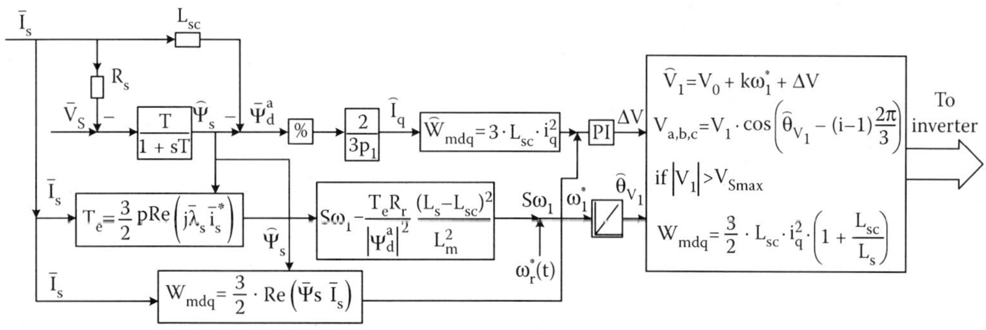


**FIGURE 16.3**
Active flux-based V/f IM drive with maximum power factor control.

is used to produce the stator voltage correction ΔV ($W_{mdq}$ from Equation 16.6)

$$\Delta V = PI(W_{m\alpha\beta} - W_{mdq}) \tag{16.10}$$

The hereby introduced V/f control with maximum power factor operation and slip frequency compensation obtained this way is shown in Figure 16.3.

When the voltage command limit of inverter is met ($V_s^* > V_{Smax}$), the maximum torque for flux condition applies ($i_d/i_q = L_{sc}/L_s$) and thus the dq magnetic energy $W_{mdq}$ becomes

$$(W_{mdq})_{max\_torque/flux} = \frac{3}{2} L_{Sc} \cdot i_q^2 \cdot \left(1 + \frac{L_{Sc}}{L_S}\right) \tag{16.11}$$

This option is already introduced in Figure 16.3, which depicts a more (though not completely) robust wide speed range V/f IM drive with maximum power factor operation, unless the voltage limit of the inverter is surpassed.

As the current $\bar{i}_S(i_a, i_b)$ is measured, fast current protection through inverter is secured; the stator voltages may be estimated from the d.c.-link voltage and from inverter state knowledge (or from reference voltages), to eliminate two wide-frequency-band (expensive) voltage sensors. Making use of current simplified reconstruction from d.c.-link current $i_{dc}$ [10] may eliminate the direct measurement of $i_a$, $i_b$ and thus reduce drastically the cost of current sensing.

Additional output voltage correction for dead time and of power switch saturation voltage, based on a disturbance observer, has been recently proposed in order to reduce THD of current in induction motor V/f drives.

In essence, voltage correction and boost are producing $V_q^*$ (aligned with voltage phase, in its coordinates) while the $i_d$ current only regulator is used to yield $V_d^*$. Good results in THD reduction are reported, but operation with energy savings or in flux weakening is not attempted [11].

The absence of speed closed-loop control in V/f advanced control of IMs with one stabilizing loop around magnetic energy (as above) or magnetization current [11] with slip frequency compensation has not been yet proven to offer wide speed range with fast torque perturbation rejection.

Fast torque response (in the millisecond range, as in sensorless FOC), proven by very fast acceleration and regenerative braking, with incorporated energy savings operation and full (maximum) flux weakening, are still due for V/f IM drives.

## 16.3 V/f Advanced Control of PMSMs

### 16.3.1 Open-Loop V/f Control

V/f open-loop control of high-speed SPMSM has been investigated up to very high speeds, 100 krpm [12], by carefully "programming" the reference speed versus time (Figure 16.4a) with the voltage boost $V_0$ additional increasing with stator current (load torque [Figure 16.4b]) [11].

The tedious calibration process of $\omega_r^*(t)$ and of $V_s^*(\omega_r^*)$ and the lack of guarantee of stable response to even moderately fast torque perturbations led to the search for more complicated solutions.

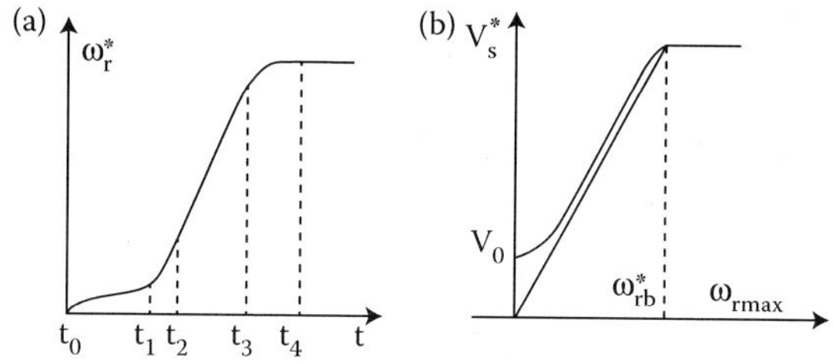


**FIGURE 16.4**
Reference speed ($\omega_r^*$) versus time (a) and reference voltage amplitude ($V_S^*$) versus reference speed ($\omega_r^*$) (b) in open-loop SPMSM drives.

### 16.3.2 V/f Control with Frequency Modulation Stabilizer

A stable $V_1/f_1$ control was introduced in Reference 13 (Figure 16.5).

In essence, for given power factor angle $\varphi_0$, the stator voltage command is calculated in steady state (Figure 16.5a) from the equivalent circuit; then, after proof of instabilities occurrence for open-loop V/f control, the command frequency (speed) modulation signal $\Delta\omega_e$ is provided for motor stabilization (Figure 16.5b), which is based on high-pass filtering of the instantaneous active power of the motor $P_e$:

$$\Delta\omega_e = -k_p \cdot \Delta p_e \tag{16.12}$$

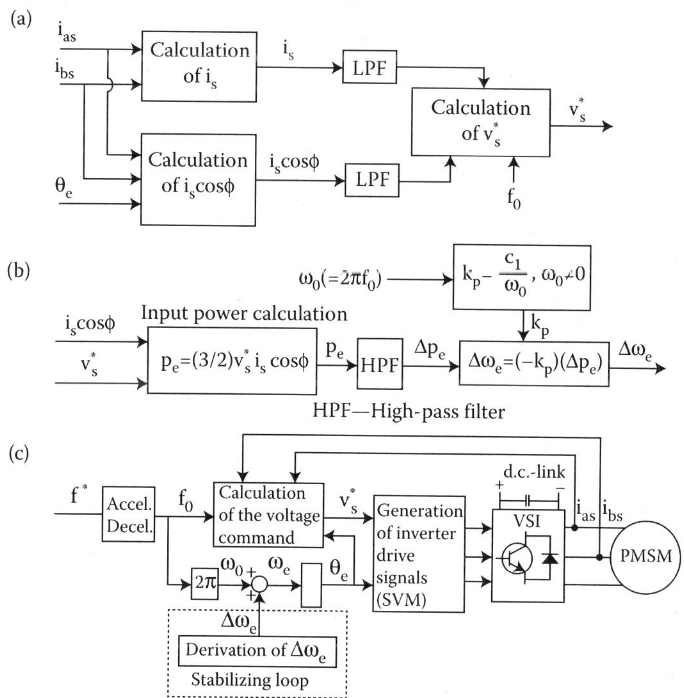


**FIGURE 16.5**
Voltage command calculation (a), derivation of frequency modulation signal $\Delta\omega_e$ (b), and the complete drive with the stabilizing loop (c). (Adapted after P.D.C. Perera et al., *IEEE Trans. Ind. Appl.*, 39(3), 783–791, 2003, ISSN: 0093-9994.)

The complete control system, given in Figure 16.5c [13], has been proven to operate stably from 5 to 87.5 Hz against rather fast torque perturbations on a 2.2 kW low saliency PMSM.

Given the complexity of online calculations and the absence of energy savings operation (with variable load), other attempts have been made to provide stable V/f control of PMSM by using one or two stabilizing loops. However, the frequency modulation signal approach (Figure 16.5b) has been kept in a few such efforts, as revealed in what follows.

### 16.3.3 V/f Control with Stabilizing Loops and Energy Savings

Energy savings with variable load is essential to variable speed drives. As early as 1988 [14] a $V_1/f_1$ control system for PMSMs with the d.c.-link current variation $\Delta i_{dc}$ used for frequency modulation ($\Delta\omega_e$) has been proposed:

$$\Delta\omega_e = -k_C \cdot \Delta i_{dc} \tag{16.13}$$

The entire scheme is shown in Figure 16.6, where the fast attenuation of speed oscillations for 10% torque perturbation and notable energy savings at lower loads are visible.

The d.c.-link power (or average d.c.-link current) is measured and, as long as it decreases, the voltage correction $\Delta V$ (in given steps) is applied.

This is a still slow torque response, but practical, approach, as no parameters of the motor are needed, though a notable calibration time (effort) is required.

An elaborated energy-saving V/f control for IPMSM, which calculates the power factor angle $\varphi$, the voltage load angle $\delta$ and current load angle $\Psi$, to correct the voltage amplitude (only), is given in Reference 15 (Figure 16.7).

Still, the dependence of $\varphi_{MTPA}$ not only on torque but also on speed makes the calculation efforts strenuous while the damping of speed oscillations is not considered in Reference 15.

The simplification of solution 15,16 calculates the reactive power $Q_1$ in dq coordinates at steady state for maximum torque/current (MTPA) conditions for IPMSMs [5].

$$(Q_1)_{MTPA} = \frac{3}{2} \cdot \omega_{re}^* \left(L_q \cdot i_d^2 + L_d \cdot i_q^2\right) \tag{16.14}$$

However, a fourth-order equation in torque $T_e$ is to be solved online in order to calculate $i_q$ and then, from measured $i_s$, $|i_d| = \sqrt{i_S^2 - i_q^2}$ and finally $(Q_1)_{MTPA}$ in Equation 16.14 are obtained.

For surface PMSMs, MTPA conditions are met with $i_d = 0$ when $Q_1$ becomes

$$(Q_r)_{i_d=0} = \frac{3}{2} \cdot i_S^2 \cdot \omega_{re}^* \cdot L_S; \quad L_d = L_q = L_S \tag{16.15}$$

Making use of $\Delta\omega_e$ frequency modulation [13] and adding a reactive power closed loop with the reactive power $Q_S$, in stator coordinates

$$Q_S = \frac{3}{2}(V_\beta \cdot i_\alpha - V_\alpha \cdot i_\beta) \tag{16.16}$$

leads to a voltage correction $\Delta V$

$$\Delta V = PI\left(Q_S - Q_{r(i_d=0)}\right) \tag{16.17}$$

This is used for V/f control of a SPMSM [16] (Figure 16.8).

In a further simplification, a V/f control system with reactive power-based PI voltage amplitude and phase angle correction (ΔV, Δθ) led to a good performance SPMSM drive for $i_d = 0$ conditions (MTPA) [17] (Figure 16.9a and b).

The reduced current at no load is a clear sign of $i_d = 0$ (MTPA) operation for the SPMSM in Figure 16.9c.

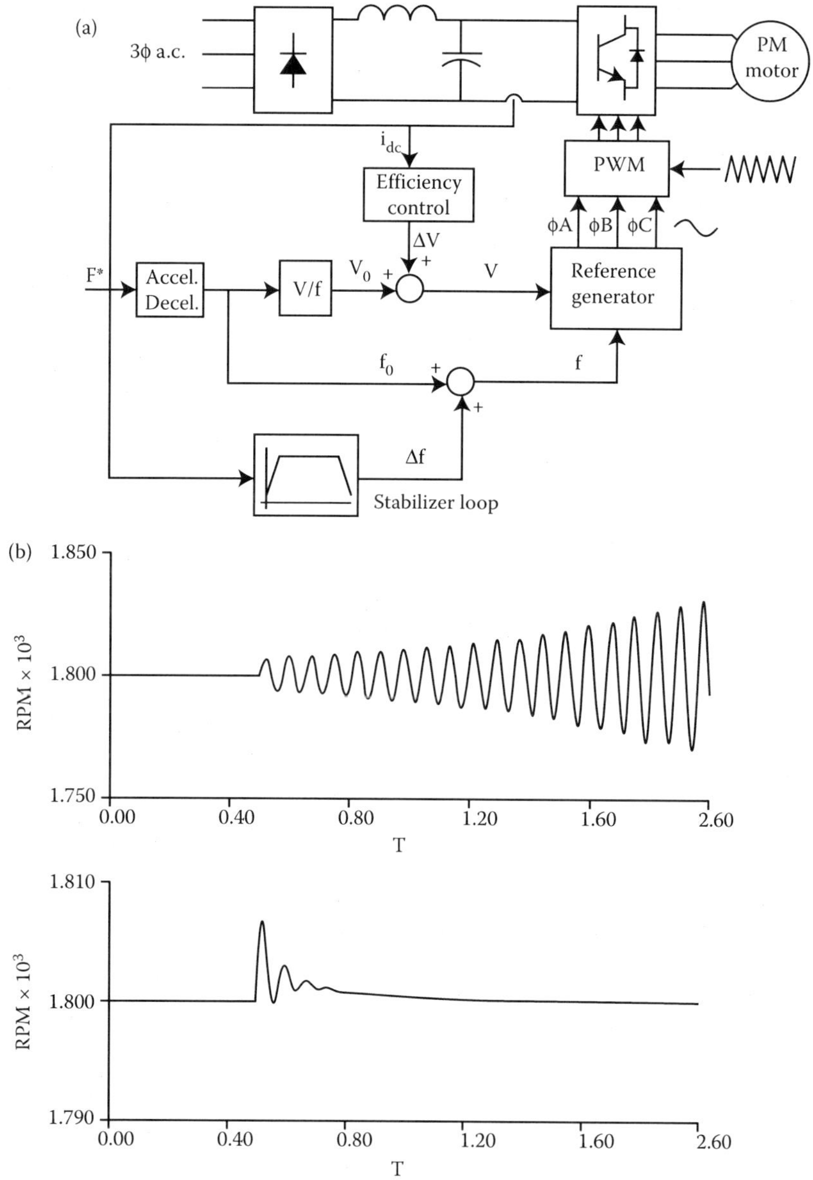


**FIGURE 16.6**
Efficiency optimizing PMSM V/f control: (a) the generic scheme, (b) stability effects of frequency modulation feedback ($\Delta\omega_c$) for 10% torque perturbation. (Adapted after R.S. Colby, D.W. Novotny, *IEEE Trans. Ind. Appl.*, 24(3), 462–469, 1988.)

*(Continued)*

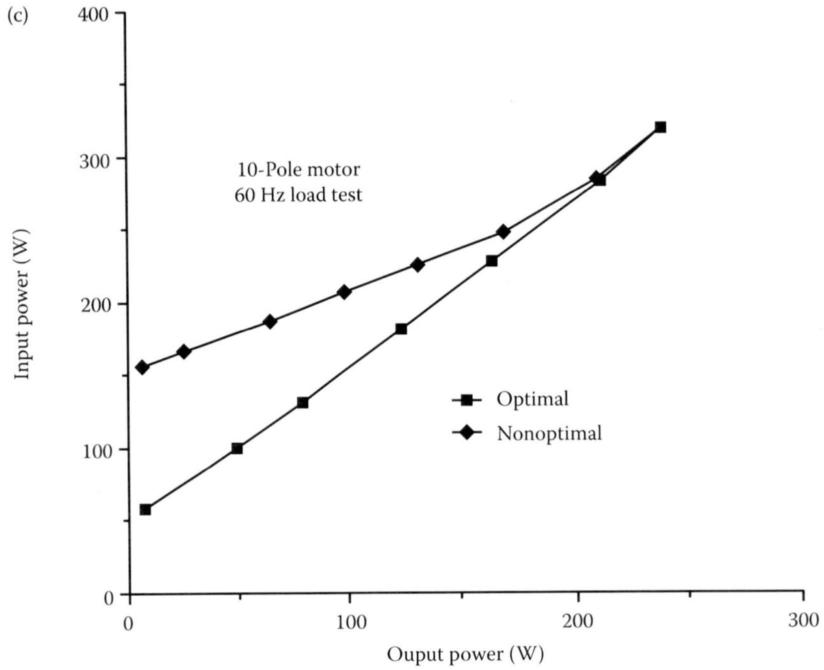


**FIGURE 16.6 (*Continued*)**
Efficiency optimizing PMSM V/f control: (c) input power versus output power. (Adapted after R.S. Colby, D.W. Novotny, *IEEE Trans. Ind. Appl.*, 24(3), 462–469, 1988.)

Also, the fast (150 ms) acceleration time from 0 to 10,000 rpm in this 0.4 Nm drive is to be noted.

Quite a few V/f schemes with various dual stabilizing loops that provide MTPA in IPMSMs (and SPMSMs) have been introduced in References 18 and 19, based on the novel concept of "active flux" [8], which allows notable simplifications of online calculations. Comparable speed dynamics of V/f control versus FOC have been proven down to 5 rpm (0.25 Hz!) for an IPMSM drive (Figure 16.10 [19]).

In yet another approach, direct power factor angle error ($\varphi^* - \varphi$) closed-loop control [20] for voltage amplitude correction with frequency modulation (as in Reference 13) has been applied to an SPMSM. One more loop is added to input robustly the power factor command angle $\varphi^*$ as the output of the ($i_d^* - \hat{i}_d$) PI loop with

$$\hat{i}_d = \frac{2 \cdot Q_S}{3 \cdot \omega_r^* \cdot \Psi_{PM}} - \frac{L_S \cdot i_S^2}{\Psi_{PM}}; \quad \varphi^* = PI(i_d^* - \hat{i}_d) \quad \text{for SPMSM} \tag{16.18}$$

and

$$\hat{i}_d = \frac{(2/3) \cdot |Q_S/\omega_r^*| - L_S \cdot i_S^2}{|\hat{\Psi}_d^a|}; \quad \hat{\Psi}_d^a = \frac{T}{1+sT}(\bar{V}_S - R_S \cdot \bar{i}_S) - L_q \cdot \bar{i}_S \Rightarrow$$

$$\hat{\Psi}_d^a = \Psi_{S\alpha} + j \cdot \Psi_{S\beta} - L_q \cdot \bar{i}_S, \quad \text{for IPMSMs} \tag{16.19}$$

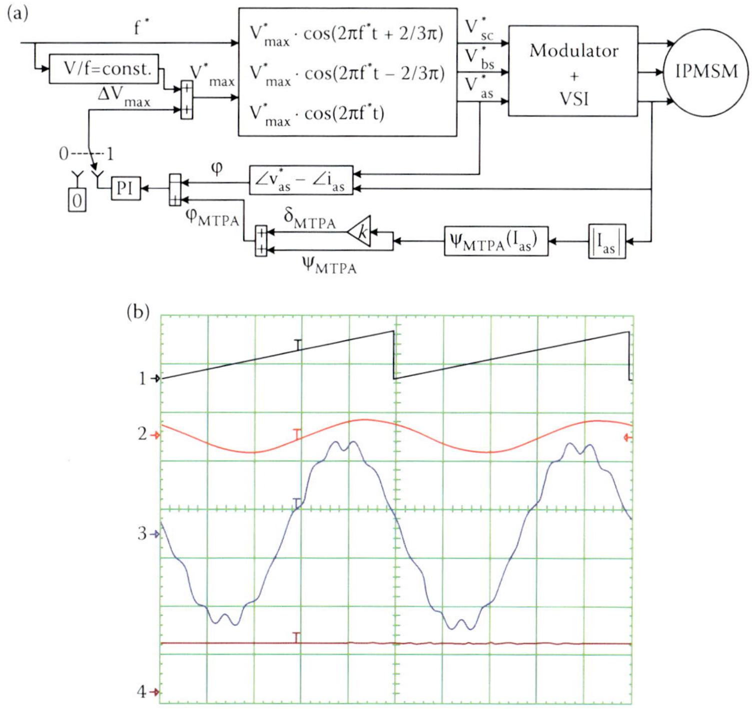


**FIGURE 16.7**
Scalar $V_1/f_1$ control with voltage correction ΔV based on power factor angle φ correction with respect to maximum torque/current ($\varphi_{MTPA}$) (a), constant V/f control: (1) rotor position (2π/div); (2) $V_{as}$ (160 V/div); (3) $I_{as}$ (2 A/div); (4) $\omega_e$ (31.4 rad/div); time (20 ms/div) (b). (Adapted after M. Cacciato et al., A novel efficiency optimization scalar control technique for industrial IPMSM drives, *IEEE International Symposium on Industrial Electronics (ISIE)*, 2010, pp. 1181–1186.)

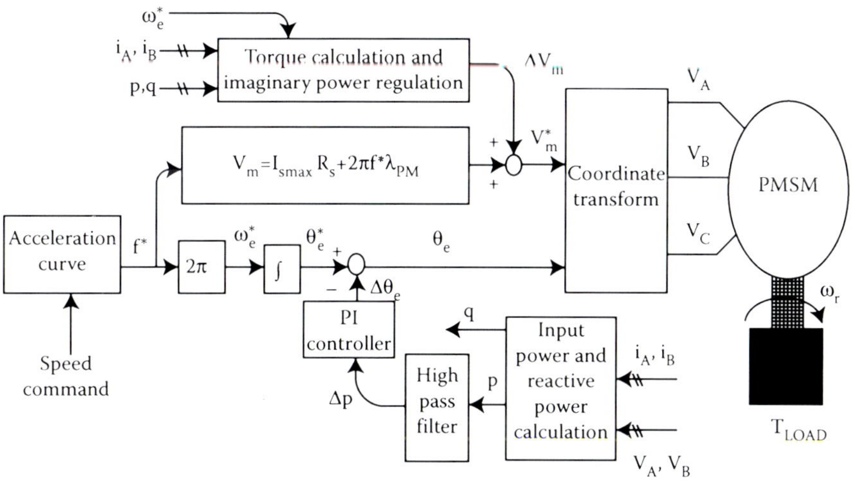


**FIGURE 16.8**
V/f with frequency (power) modulation and reactive power control loop for voltage correction (ΔV) for $i_d = 0$ in SPMSM drives. (Adapted after S.M. Sue et al., A new MTPA control strategy for sensorless V/f controlled PMSM drives, *Industrial Electronics and Applications (ICIEA), 6th IEEE Conference on*, 2011, pp. 1840–1844.)

(Now for MTPA: $i_d^* = 0$, for SPMSM)

$$\text{Alternatively}\quad i_d^* = -\left|\bar{i}_s\right|\sqrt{\frac{\hat{\Psi}_d^a - \Psi_{PM}}{2\hat{\Psi}_d^a - \Psi_{PM}}};\quad i_q = \frac{\Psi_{S\alpha}\cdot i_\beta - \Psi_{S\beta}\cdot i_\alpha}{\left|\bar{\Psi}_d^a\right|} \tag{16.20}$$

for IPMSMs with $\Psi_d^{*a}$ as "active flux" [8].

Sample test results for a SPMSM with such a method are exhibited in Figure 16.11a; they show stable fast dynamics in response to quick torque variations.

Figure 16.11b illustrates even a regenerative braking episode as a proof of servodrive attributes of such an advanced V/f SPMSM control strategy.

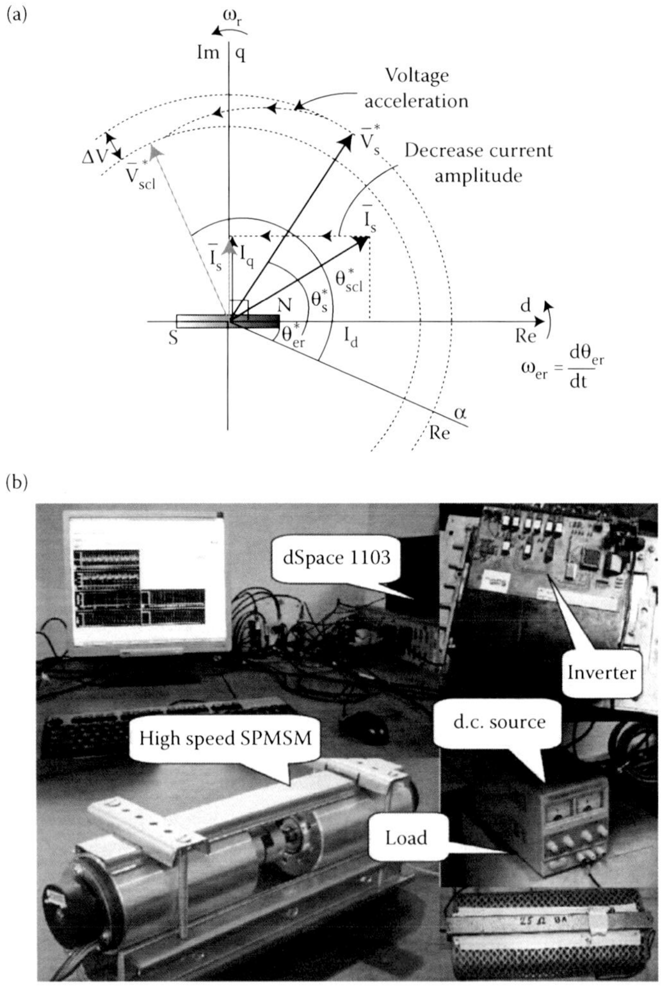


**FIGURE 16.9**
V/f with two stabilizing loops both based on reactive power error for $i_d = 0$ in SPMSM: (a) the phasor diagram, (b) test rig. (Adapted after R. Ancuti, I. Boldea, G.D. Andreescu, *Electr. Power Appl., IET,* 4(3), 40–57, 2010.)
*(Continued)*

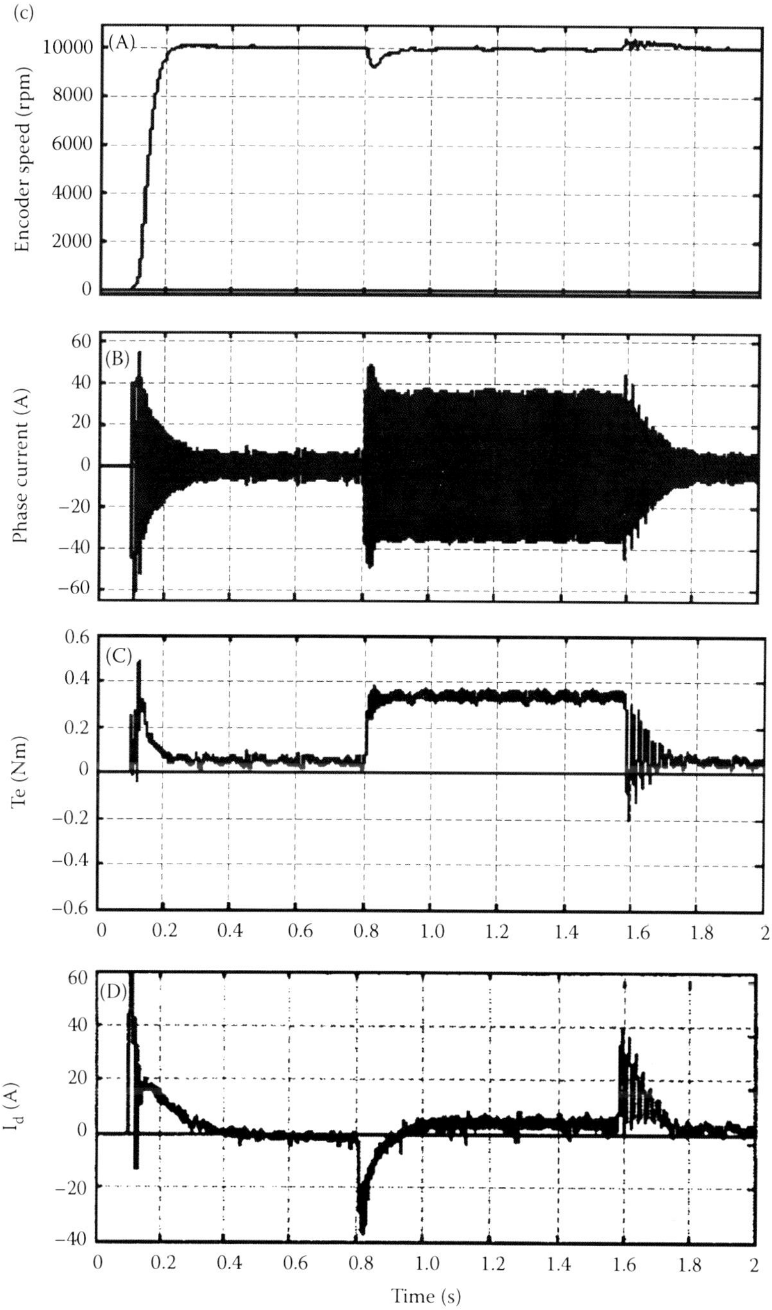


**FIGURE 16.9** (*Continued*)
V/f with two stabilizing loops both based on reactive power error for $i_d = 0$ in SPMSM: (c) test results—(A) encoder speed, (B) phase current, (C) torque, (D) d-axis current. (Adapted after R. Ancuti, I. Boldea, G.D. Andreescu, *Electr. Power Appl., IET*, 4(3), 40–57, 2010.)

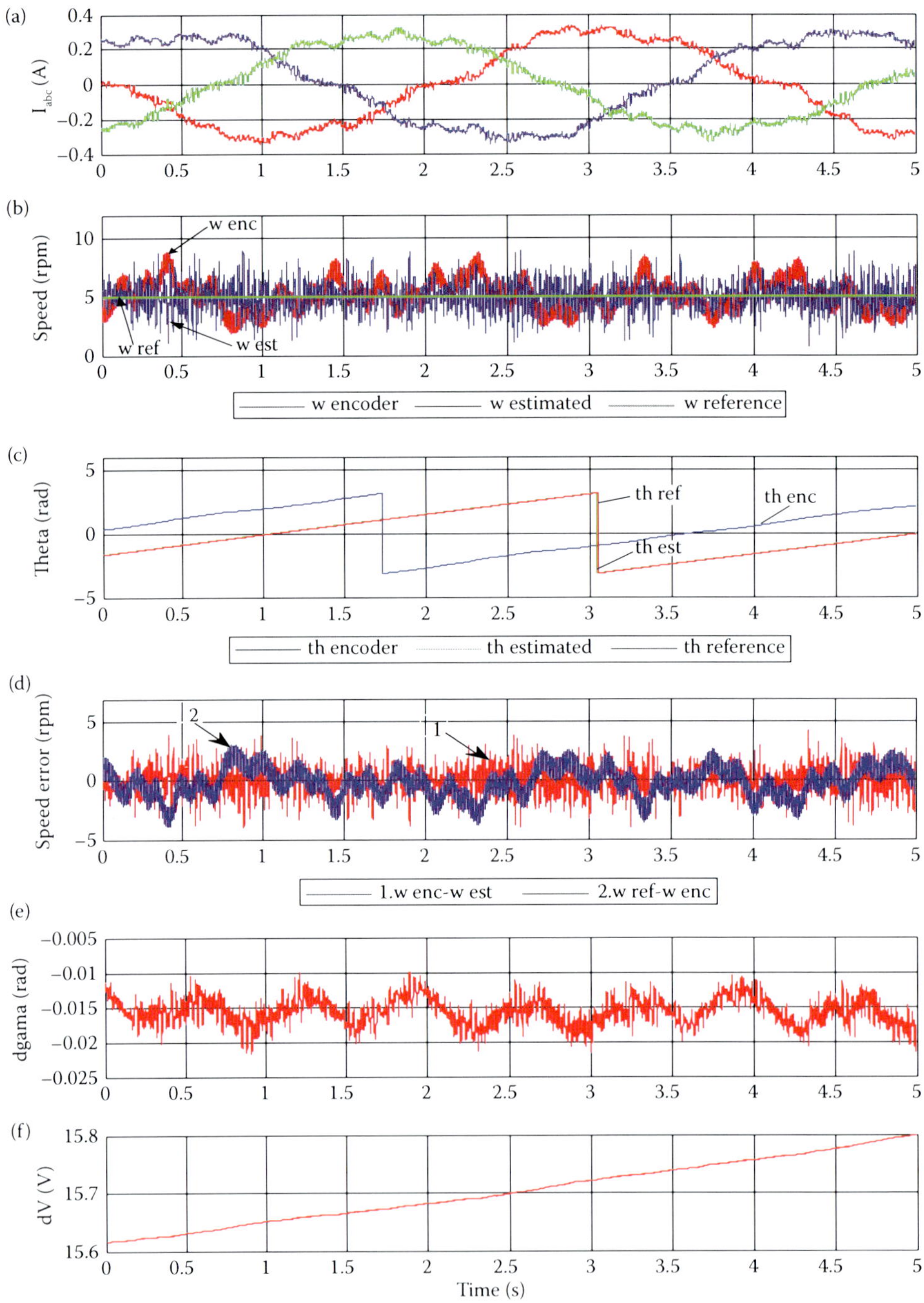


**FIGURE 16.10**
PMSM V/f control with stabilizing loops—experimental results for 5 rpm at no load, during steady state: (a) phase currents, (b) speed, (c) rotor position angle, (d) speed error, (e) rotor position error, (f) voltage correction. (Adapted after A. Moldovan, F. Blaabjerg, I. Boldea, Active flux based V/f with stabilizing loops versus sensorless vector control of IPMSM drives. *IEEE International Symposium on Industrial Electronics (ISIE)*, 2011, pp. 514–519.)

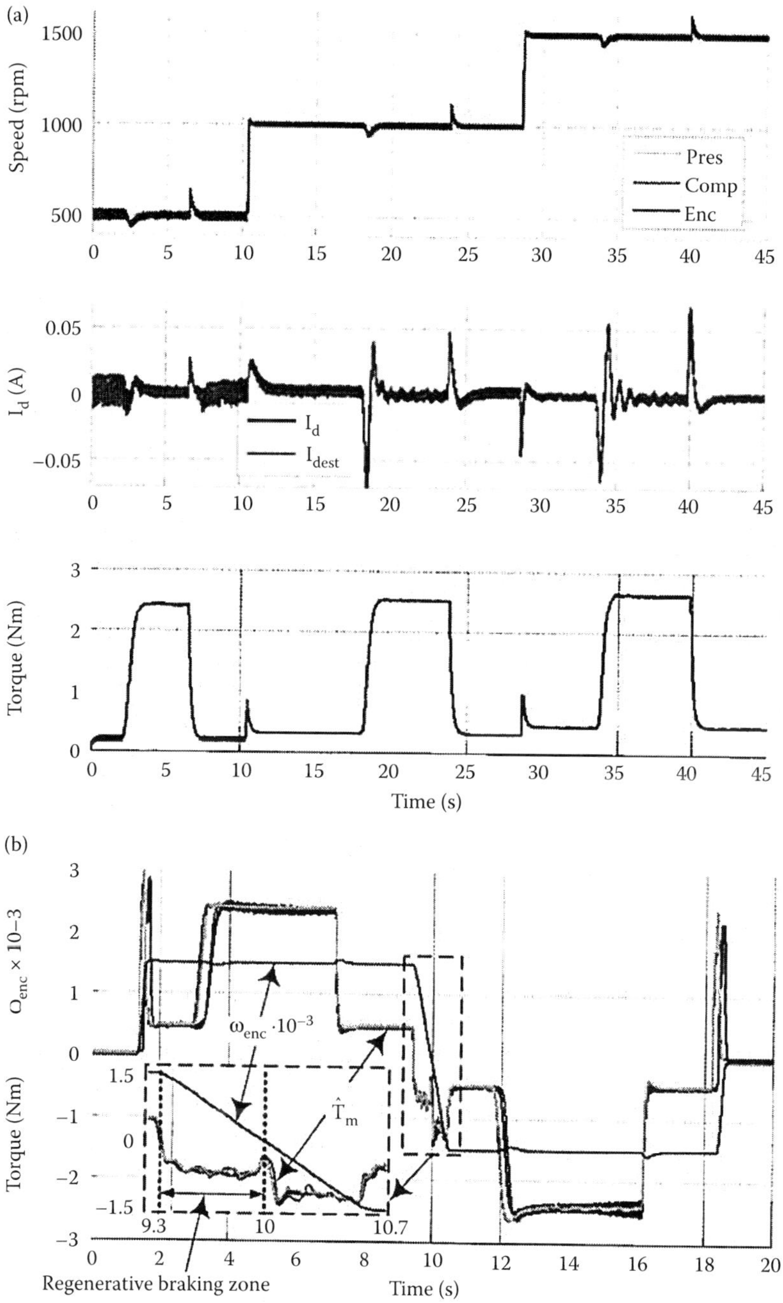


**FIGURE 16.11**
SPMSM speed and torque measured dynamics in $V_1/f_1$ control with frequency modulation and cascaded, $i_d$ and power factor angle φ, stabilizing loops (a) and regenerative braking episode (b). (Adapted after S. Agarlita et al., *Proc. IET*, EPA, no. 6, 2013.)

Though two (cascaded) $i_d$ and φ closed loops are used to correct voltage amplitude (Figure 16.12), it has been proven that the power factor angle φ loop may be eliminated without notable consequences.

This is a remarkable simplification for the scheme in Figure 16.12.

### 16.3.4 I–f Control for High-Speed PMSMs

A rather elaborate industrial solution for high-speed (frequency) SPMSMs with V/f control (and I–f starting) and only d.c. voltage boost, but with torque dynamic damping in frequency command (Figure 16.13), is presented in detail in Reference 21, for asynchronous, synchronous, and 6 pulse PWM, for fundamental frequencies of up to 2.343 kHz ($f_{SM}$ = 11.665 kHz) with acceptable voltage THD.

Though the V/f control with frequency modulation (as in Reference 13) is proven stable, I–f control with the same stabilizer is introduced [22] in order to simplify the online computation (Figure 16.14) in a drive with load proportional with speed, when the reference torque current $i_\delta^*$ may be written as

$$i_\delta^* = \frac{2}{3 \cdot p_1 \cdot \Psi_{PM}}\left(k_T \cdot \omega_r^* + T_{l0}\right) \cdot 1.1; \quad T_{load} = k_T \cdot \omega_r^* + T_{l0} \tag{16.21}$$

The flux–current $i_\gamma^*$ may be calculated again to correspond to MTPA conditions, as required.

The coefficient of 1.1 in Equation 16.21 accounts empirically for errors in MTPA condition calculations and allows for a "dynamic" torque reserve for motor acceleration during starting, or when otherwise needed. The measured current waveform at 100 krpm in

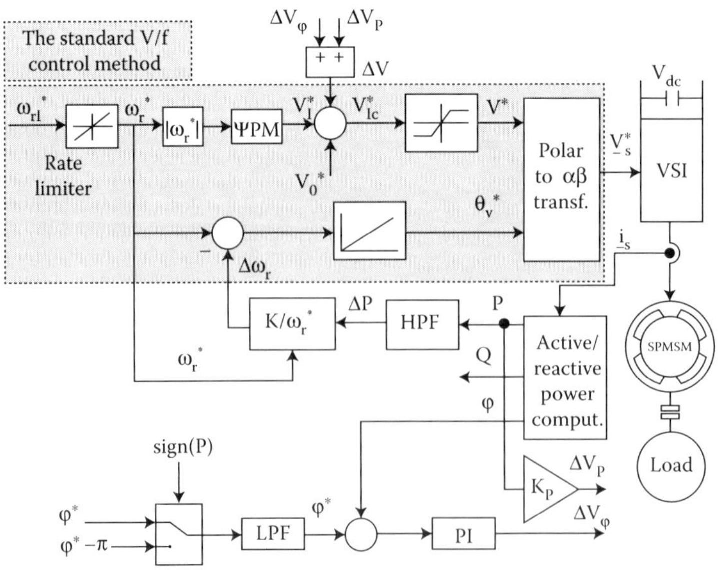


**FIGURE 16.12**
$V_1/f_1$ control with two stabilizing loops. (Adapted after S. Agarlita et al., *Proc. IET*, EPA, no. 6, 2013.)

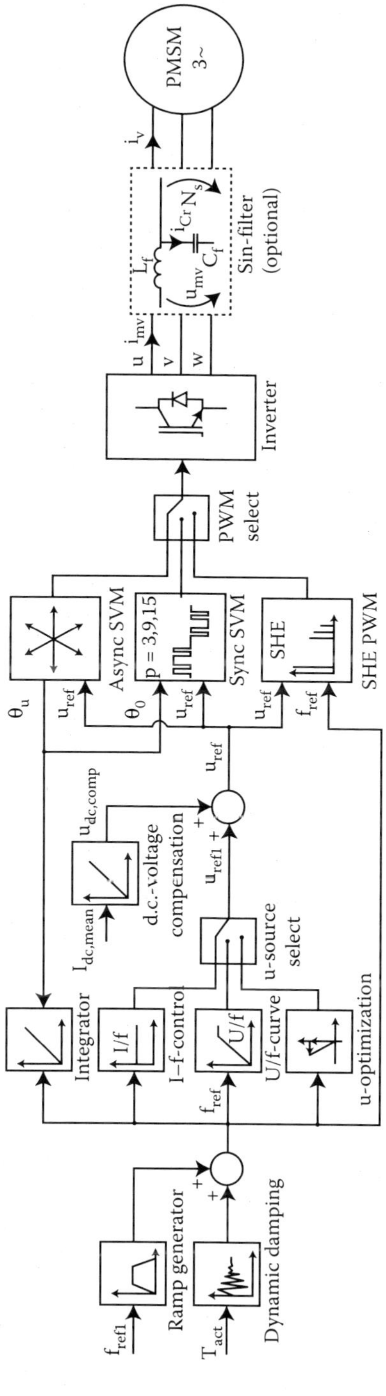


**FIGURE 16.13**
Scalar control of high-speed SPMSM with 3 PWM modes. (Adapted after T. Halkosaari, Optimal U/f-control of high speed permanent magnet motors, *Proceedings of IEEE-ISIE*, 2006, pp. 2303–2308.)

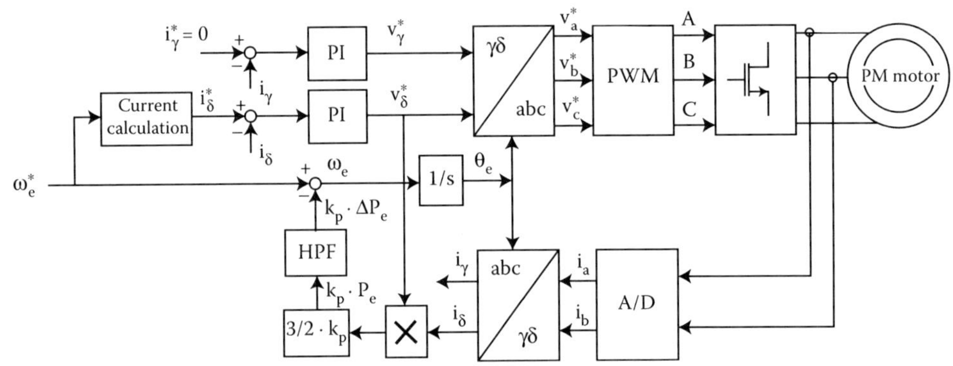


**FIGURE 16.14**
Block diagram of I–f control of PMSM with speed-proportional load.

Figure 16.15 [22] shows remarkable quality and the absence of notable amplitude modulation, which means reasonably low-speed pulsations around target speed.

### 16.3.5 I–f Starting in e.m.f.-Based Sensorless FOC of RELSYN Motor Drives

A starting strategy is needed in (most!) e.m.f.-based sensorless SM drives as torque production is related to initial rotor d axis position. In the absence of signal injection, the rotor is brought first to a known initial position if the load accepts it. Alternatively, a V/f of I–f self-starting strategy of SMs may be blended into an e.m.f.-based sensorless FOC (or DTFC) system.

As less dependent on machine parameters here, an I–f starting strategy is added to an "active flux"-based sensorless vector control of a RSM [23].

The entire control system is shown in Figure 16.16.

Figure 16.17 illustrates the transition strategies back and forth to I–f control from/to active flux-based sensorless control of an RSM. Sample experimental results are visible in Figure 16.18 for a typically 1 Nm 4-pole machine with a saliency above 4.5/1. Smooth transitions (Figure 16.17) and wide speed range (Figure 16.18) are both visible.

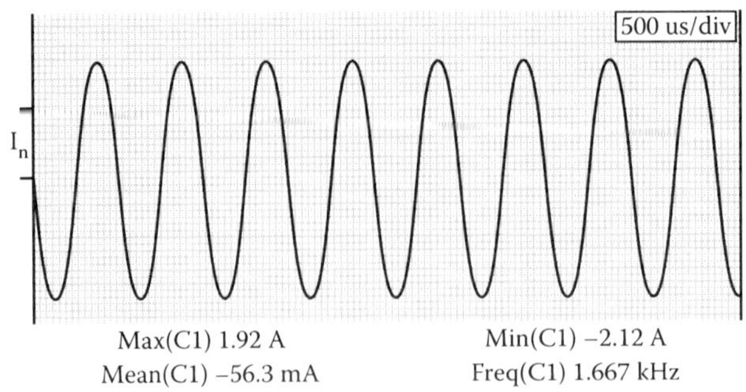


**FIGURE 16.15**
Current waveforms 100 krpm with I–f plus frequency modulation control. (Adapted after A. Borisavljevic, H. Polinder, J.A. Ferreira, Realization of the I/f control method for a high-speed permanent magnet motor, *Record of ICEM*, 2010, Rome, Italy, pp. 1–6.)

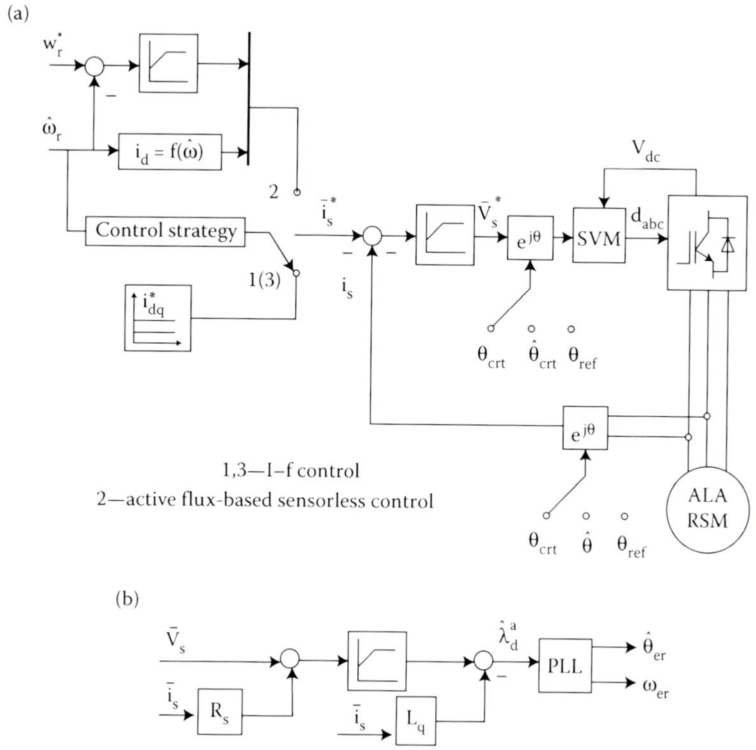


**FIGURE 16.16**
Motion-sensorless control of RSM: (a) current controllers, (b) rotor speed and position estimator. (Adapted after S. C. Agarlita et al., I–f starting and active flux based sensorless vector control of RSM with experiments, *Record of OPTIM-2010* [IEEExplore].)

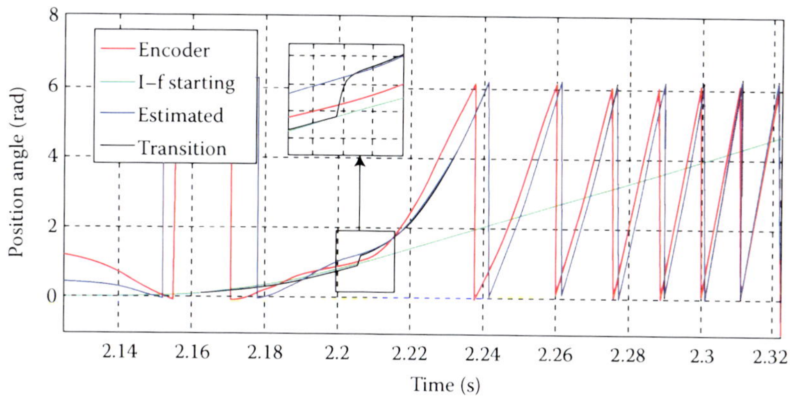


**FIGURE 16.17**
Transition from I–f starting to active flux sensorless control. (Adapted after S. C. Agarlita et al., I–f starting and active flux based sensorless vector control of RSM with experiments, *Record of OPTIM-2010* [IEEExplore].)

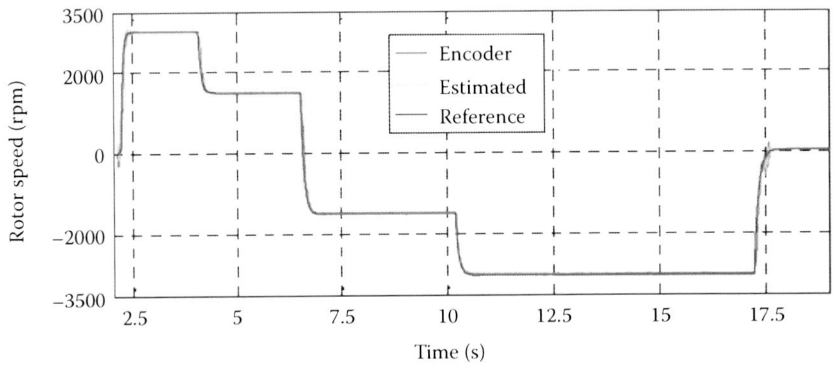


**FIGURE 16.18**
Reference, estimated, and measured rotor speed. (Adapted after S.C. Agarlita et al., I–f starting and active flux based sensorless vector control of RSM with experiments, *Record of OPTIM*, 2010 [IEEExplore].)

## 16.4 One-Phase PMSM I–f Starting and e.m.f.-based Sensorless Control

The one-phase PMSM, with preferred self-starting position (obtained by tapered airgap, etc.), when e.m.f.-based, sensorlessly controlled, still needs a starting strategy. For soft-starting (low torque at start), when no signal injection is available for safe operation at very low speeds, an I–f starting strategy may be added to the control system (Figure 16.19). Typical experimental results for a 12 V d.c., 40 W, 0.02 Nm, one-phase PMSM are shown in Figure 16.20 where the transition from I–f to e.m.f.-based sensorless control is evident [24].

Moreover, the same sensorless drive is shown in Figure 16.21 capable of regenerative braking [24].

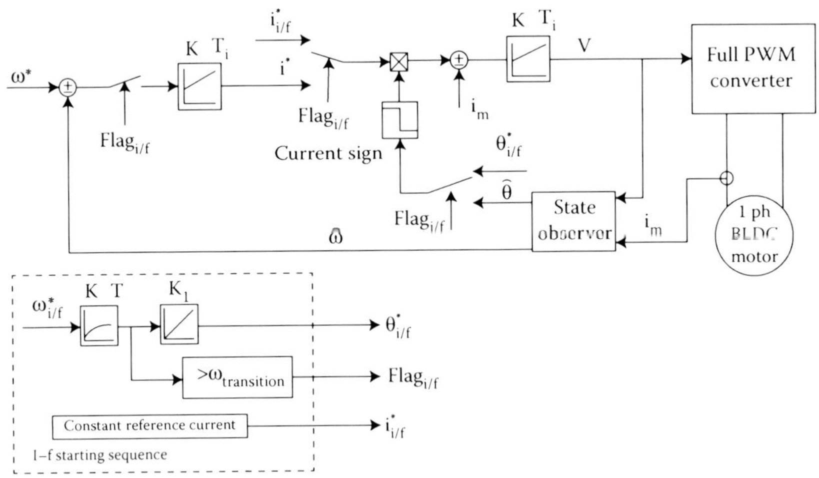


**FIGURE 16.19**
One-phase PMSM: sensorless control block diagram. (Adapted after L.I. Iepure, F. Blaabjerg, I. Boldea, *IEEE Trans.*, IE-59(9), 3436–3444, 2012.)

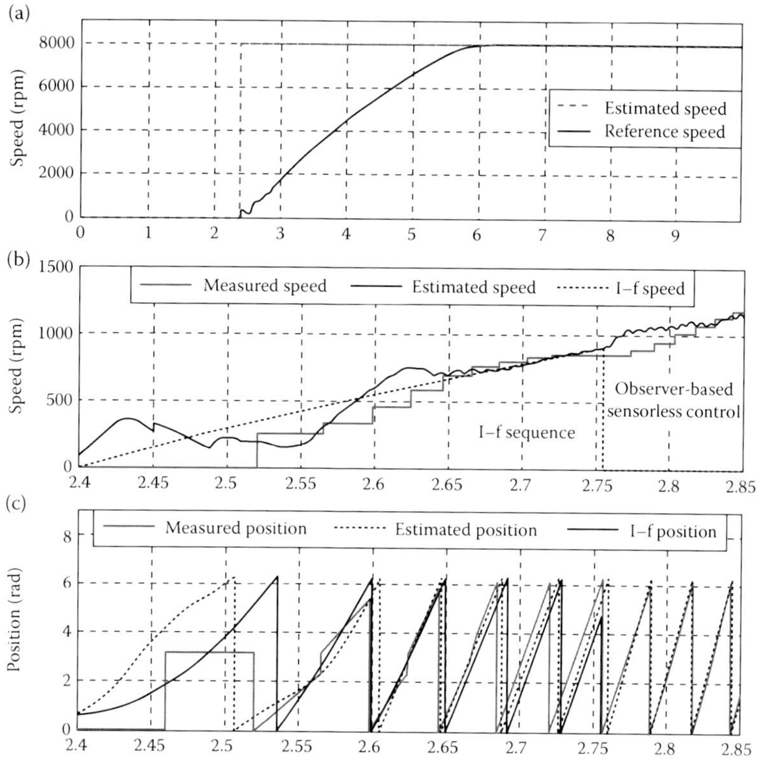


**FIGURE 16.20**
One-phase PMSM: experimental results for starting. (a) Estimated speed, (b) speed comparison, (c) position. (Adapted after L.I. Iepure, F. Blaabjerg, I. Boldea, *IEEE Trans.*, IE-59(9), 3436–3444, 2012.)

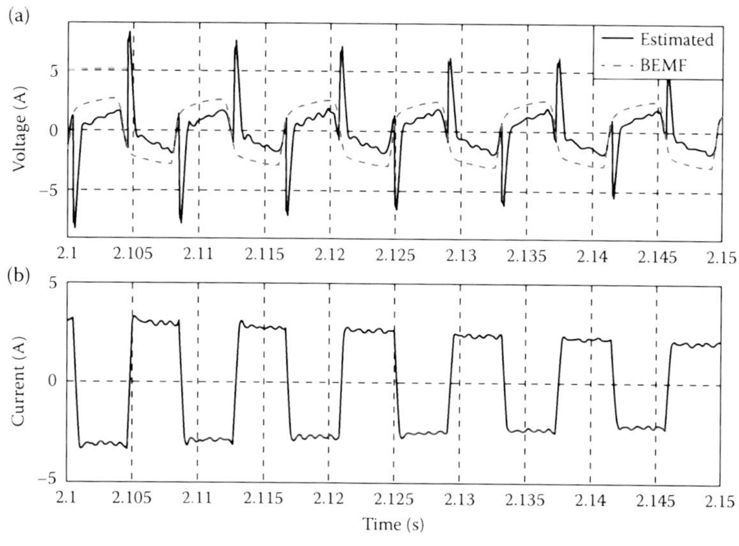


**FIGURE 16.21**
One-phase PMSM: representative signals during regenerative braking. (a) Estimated voltage versus BEMF, (b) phase current. (Adapted after L.I. Iepure, F. Blaabjerg, I. Boldea, *IEEE Trans.*, IE-59(9), 3436–3444, 2012.)

## 16.5 Summary

The overview of a.c. motor drives V/f advanced control methods recent progress synthesized in this chapter leads to remarks such as

- Open-loop IM V/f drives show instabilities in some V/f region, which may be removed with slip frequency compensation and voltage amplitude correction, based, for example, on boost voltage increasing with load current.
- A rather robust steady-state (dynamically slow) energy savings operation of V/f induction motor drives by a feedforward step variation of voltage ΔV as long as the d.c.-link current decreases, has been demonstrated.
- A potentially more coherent and practical V/f IM drive with energy savings slip compensation and ΔV correction at maximum power factor conditions based on "active flux" concept is introduced.
- As open-loop $V_1/f_1$ of PMSMs are inherently unstable (no notable rotor currents damping effect), stabilizing means have been proposed. Among them two stand out: one via frequency modulation by input active power high-pass filtering and the second one (first in historical order) by d.c.-link current high-pass filtering. In addition, MTPA online conditions control by analytical calculations, or by using magnetic energy PI loops, or power factor angle or "$i_d$" PI loops for ΔV correction, have been proven to produce energy savings and stable response.
- Even regenerative braking and fast torque response has been proven [20], with a SPMSM, to bridge further the gap to servodrive performance.
- The methods investigated in this chapter may be extended rather straightforwardly to RSMs and to one-phase PMSMs; also, to d.c.-excited SMs (operating on pure $i_q$ control or on unity power factor control).
- Finally, a similar investigation may be launched for scalar advanced control of a.c. generators.

## References


1. R. Ueda, T. Sonoda, K. Koga, M. Ichikawa, Stability analysis in induction motor driven by V/f controlled general-purpose inverter, *IEEE Trans.*, vol. IA-28, no. 2, 1992, pp. 472–481.
2. K. Koga, R. Ueda, T. Sonoda, Constitution of V/f control for reducing steady state error to zero in induction machine drive to zero in induction machine drive system, *IEEE Trans.*, vol. IA-28, no. 2, 1992, pp. 463–471.
3. M. Tsuji, S. Chen, S.I. Hamasachi, X. Zhao, Y. Yamada, A novel V/f control of induction motors for wide and precise speed operation, *Record of IEEE-SPEEDAM*, 2008, pp. 1130–1135 (IEEExplore).
4. A.M. Garcia, T.A. Lipo, D.W. Novotny, A new induction motor V/f control method capable of high-performance regulation at low speeds, *IEEE Trans.*, vol. IA-34, no. 4, 1998, pp. 813–821.

5. A.V. Ravi Teja, C. Chakraborty, A novel model reference adaptive controller for estimation of speed and stator resistance for vector controlled induction motor drives, *IEEE International Symposium on Industrial Electronics*, ISIE, July 4–7, 2010, pp. 1187–1192 (IEEExplore).
6. M.A. Garcia, T.A. Lipo, D.A. Novotny, A new induction motor V/f control method capable of high performance regulation at low speeds, *IEEE Trans. Ind. Appl.*, vol. 34, 1998, pp. 813–821.
7. D.S. Kirschen, D.W. Novotny, T.A. Lipo, On-line efficiency optimization of a variable frequency induction motor drive, *IEEE Trans. Ind. Appl.*, vol. IA-21, no. 3, 1985, pp. 610–616.
8. I. Boldea, M.C. Paicu, G.D. Andreescu, Active flux concept for motion-sensorless unified ac. drives, *IEEE Trans. Power Electron.*, vol. 23, no. 5, 2008, pp. 2612–2618.
9. I. Boldea, S.A. Nasar, *Electric Drives*, second edition, Taylor & Francis Group, New York, 2006.
10. H. Kim, T.M. Jahns, Phase current reconstruction for AC motor drives using a DC link single current sensor and measurement voltage vectors, *IEEE Trans. Power Electron.*, vol. 21, no. 5, 2006, pp. 1413–1419.
11. T. Hoshino, J.-I. Itoh, Output voltage correction for a voltage source type inverter of an induction motor drive, *IEEE Trans. Power Electron.*, vol. 25, no. 9, 2010, pp. 2440–2449.
12. L. Xu, C. Wang, Implementation and experimental investigation of sensorless control schemes for PMSM in super-high variable speed operation, *Industry Applications Conference, 1998. Thirty-Third IAS Annual Meeting. The 1998 IEEE*, vol. 1, 1998, pp. 483–489 (IEEExplore).
13. P.D.C. Perera, F. Blaabjerg, J.K. Pedersen, P.T. Thogersen, A sensorless stable V/f control method for PMSM drives, *IEEE Trans. Ind. Appl.*, vol. 39, no. 3, 2003, pp. 783–791, ISSN: 0093-9994.
14. R.S. Colby, D.W. Novotny, An efficiency-optimizing permanent-magnet synchronous motor drive, *IEEE Trans. Ind. Appl.*, vol. 24, no. 3, 1988, pp. 462–469.
15. M. Cacciato, A. Consoli, G. Scarcella, G. Scelba, A novel efficiency optimization scalar control technique for industrial IPMSM drives, *IEEE International Symposium on Industrial Electronics (ISIE)*, 2010, pp. 1181–1186 (IEEExplore).
16. S.M. Sue, T.-W. Hung, J.-H. Liaw, Y.-F. Li, C.-Y. Sun, A new MTPA control strategy for sensorless V/f controlled PMSM drives, *Industrial Electronics and Applications (ICIEA), 6th IEEE Conference on*, 2011, pp. 1840–1844 (IEEExplore).
17. R. Ancuti, I. Boldea, G.D. Andreescu, Sensorless V/f control of high-speed surface permanent magnet synchronous motor drives with two novel stabilizing loops for fast dynamics and robustness, *Electr. Power Appl., IET*, vol. 4, no. 3, 2010, pp. 40–57.
18. I. Boldea, A. Moldovan, V.C. Schramel, G.D. Andreescu, L. Tutelea, A class of fast dynamics V/f sensorless AC general drives with PM-RSM as a case study, *Optimization of Electrical and Electronic Equipment (OPTIM), 2010 12th International Conference on*, Brasov, Romania, 2010, pp. 453–459 (IEEExplore).
19. A. Moldovan, F. Blaabjerg, I. Boldea, Active-flux-based, V/f-with-stabilizing-loops versus sensorless vector control of IPMSM drives, *IEEE International Symposium on Industrial Electronics (ISIE)*, 2011, pp. 514–519 (IEEExplore).
20. S. Agarlita, C. Coman, G.D. Andreescu, I. Boldea, Stable V/f control system with controlled power factor angle for SPMSM drives, *Proc. IET*, vol. EPA, no. 6, 2013.
21. T. Halkosaari, Optimal U/f-control of high speed permanent magnet motors, *Proceedings of IEEE-ISIE*, 2006, pp. 2303–2308 (IEEExplore).
22. A. Borisavljevic, H. Polinder, J.A. Ferreira, Realization of the I/f control method for a high-speed permanent magnet motor, *Record of ICEM-2010*, Rome, Italy, pp. 1–6.
23. S.C. Agarlita, M. Fatu, L.N. Tutelea, F. Blaabjerg, I. Boldea, I–f starting and active flux based sensorless vector control of RSM with experiments, *Record of OPTIM*, Brasov, Romania, 2010 (IEEExplore).
24. L.I. Iepure, F. Blaabjerg, I. Boldea, Hybrid I–f starting and observer-based sensorless control of single phase BLDC-PM motor drives, *IEEE Trans.*, vol. IE-59, no. 9, 2012, pp. 3436–3444.

# 17

# *Multiphase Electric Machine Drives: An Overview*

## 17.1 Introduction

It was in the 1960s that some sporadic large-power multiphase ($m > 3$) a.c. variable speed drives surfaced. But only after the 1990s, with the advent of multiphase PWM inverters, did multiphase electric machine drive technology gain considerable academic and industrial heavy attention worldwide, mainly (so far) for electric locomotives traction, HEV, EV, and ship propulsion, more electric aircraft motor/generator drives, electric power steering, high-power compressors, large pumps, PMSG wind generators, etc.

Multiphase electric machine drives may also be classified as of induction, PM synchronous, d.c.-excited synchronous, reluctance synchronous, etc. types (switched reluctance machines, typically multiphase, have been treated in Chapter 12).

They all, however, share quite a few common attributes:

- Owing to a larger number of phases, the maximum switching frequency in the PWM converter supply, for the same torque and stator flux pulsations, is lower; so the inverter commutation losses are lower than in three-phase inverter-fed drives, of the same power/speed envelope
- The total power is split in more phases to use smaller controlled switching power devices and thus secure lower inverter cost
- Fault tolerance with $m > 3$ is much better even in machines with nonzero mutual inductance between phases (all quasi-traveling field a.c. machines); the control may be reconfigured online after fault detection to reduce torque pulsations or minimize machine losses during faults (open phase(s), in general)
- Fundamental stator currents in machines with quasi-sinusoidal (distributed, $q \geq 2$) stator windings produce an airgap flux density with lower harmonics content. The lowest frequency of torque ripple is larger (2 m)
- As only two orthogonal currents (d, q) are needed to produce FOC or DTFC, the remaining degrees of freedom may be used for
    - Independent control of (regularly) two motors (in series) from a single inverter, for sinusoidal m.m.f. stator winding machines ($q = 1$ diametrical or tooth wound).

- For concentrated stator winding machines, the third (in 5-phase machines) stator m.m.f. harmonic may be used to inject third harmonic currents and produce more torque density.
- During faulty operation, the third current harmonics is inhibited.
- Multiphase inverters may be two level (with m (m + 1) legs), multilevel (with m (m + 1) legs), and modular (m H bridge) (Figure 17.1).
- Multiphase machine modeling by orthogonal axis transformation for three phases has been investigated since early in the twentieth century; for odd prime number of phases m = (5, 7, 11, 13) the phase windings are dephased by $\alpha_{ph} = 2\pi/m$ while for an even number or phases (or a multiple of 3) we may have symmetric ($\alpha_{ph} = 2\pi/m$) or asymmetric ($\alpha_{ph} = \pi/m$) windings, for m = even. We may divide the windings in groups of three phases, each with their own null point, but no third harmonic current will flow in such cases; so only sinusoidal windings ($q \geq 2$) qualify for the situation. A detailed discussion about this issue is to be found in Reference 1 (with 229 citations!). We will come back in the chapter with IM, PMSM, BLDC-MRM (multiphase reluctance machine) multiphase case studies to grasp some quick fundamental knowledge of this involved field.
- Also, open multiphase windings allow for dual two-level inverters that behave like multilevel inverters. Multilevel inverters are needed when the ratio of inverter maximum switching frequency $f_{switch} < (15 - 20)f_{1max}$, where $f_{1max}$ is the fundamental

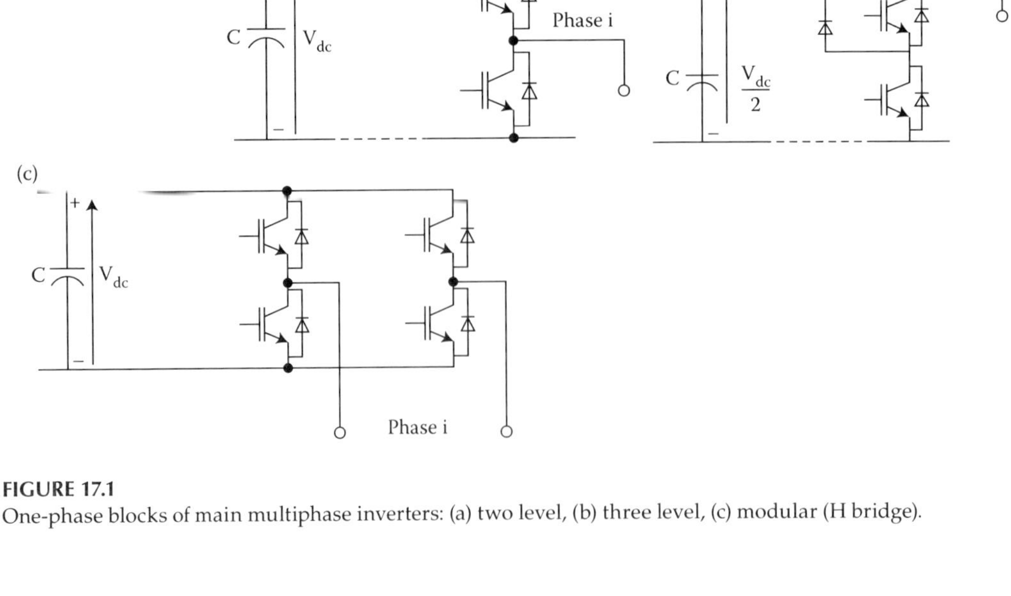


**FIGURE 17.1**
One-phase blocks of main multiphase inverters: (a) two level, (b) three level, (c) modular (H bridge).

maximum frequency (speed). This operation is typical to both high-torque low-speed multiphase machines and to smaller-torque super-high-speed machines.

## 17.2 Multiphase IM Modeling and Parameter Estimation

We will start here with the simplest case of a 5-phase IM with a quasi-distributed stator winding (q = 2), 2-pole configuration [2] (Figure 17.2).

Based on the fact that adding a third harmonic to the flux density in the machine, by a third space m.m.f. harmonic, the fundamental of flux density in the core is increased without too heavy oversaturation of core; a 10% additional average torque is obtained from same stator with three-phase winding.

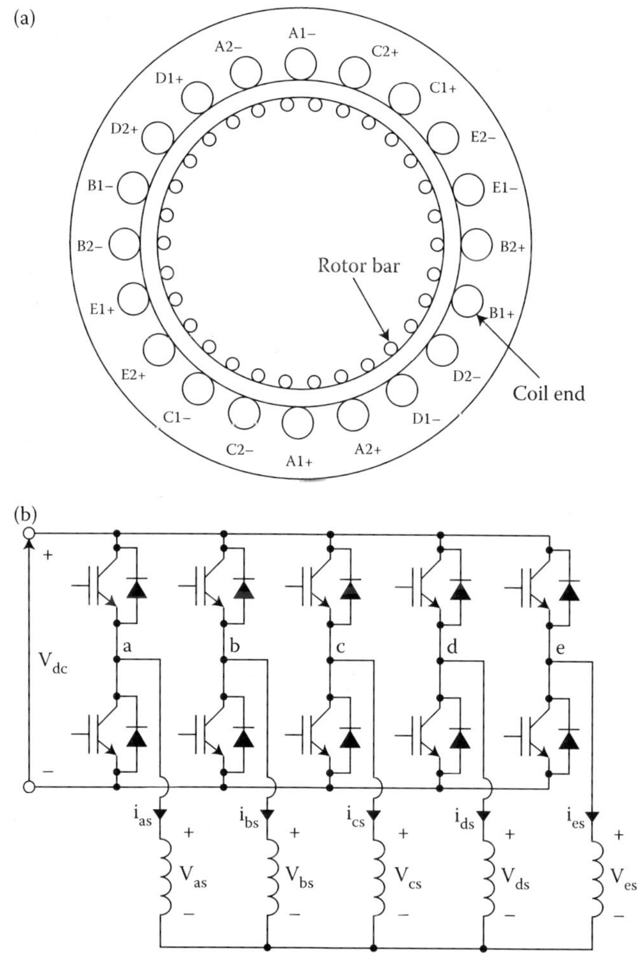


**FIGURE 17.2**
Five-phase 2-pole IM (a) and its inverter (b).

So the transformation matrix of coordinates $[T_s]$ for both the fundamental $d_1$, $q_1$ and for the third harmonic $d_3$, $q_3$ is required (17.1):

$$[T_s] = \sqrt{\frac{2}{5}} \begin{bmatrix} \cos(\theta_{e1}) & \cos(\theta_{e1}-\alpha) & \cos(\theta_{e1}-2\alpha) & \cos(\theta_{e1}-3\alpha) & \cos(\theta_{e1}-4\alpha) \\ -\sin(\theta_{e1}) & -\sin(\theta_{e1}-\alpha) & -\sin(\theta_{e1}-2\alpha) & -\sin(\theta_{e1}-3\alpha) & -\sin(\theta_{e1}-4\alpha) \\ \cos(3\theta_{e3}) & \cos(3\theta_{e3}-3\alpha) & \cos(3\theta_{e3}-\alpha) & \cos(3\theta_{e3}-4\alpha) & \cos(3\theta_{e3}-2\alpha) \\ -\sin(3\theta_{e3}) & -\sin(3\theta_{e3}-3\alpha) & -\sin(3\theta_{e3}-\alpha) & -\sin(3\theta_{e3}-4\alpha) & -\sin(3\theta_{e3}-2\alpha) \\ \frac{1}{\sqrt{2}} & \frac{1}{\sqrt{2}} & \frac{1}{\sqrt{2}} & \frac{1}{\sqrt{2}} & \frac{1}{\sqrt{2}} \end{bmatrix}$$

$$\alpha = \frac{2\pi}{5} \tag{17.1}$$

We may write the $d_1$, $q_1$, $d_3$, $q_3$ IM machine equations in space phasors as

$$\begin{aligned} &\bar{V}_{s1} = r_s \bar{i}_{s1} + \frac{d}{dt}\bar{\lambda}_{s1} + j\omega_{e1}\bar{\lambda}_{s1}; \quad \bar{\lambda}_{s1} = L_{s1}\bar{i}_{s1} + L_{m1}\bar{i}_{r1} \\ &\bar{V}_{r1} = r_{r1}\bar{i}_{r1} + \frac{d}{dt}\bar{\lambda}_{r1} + j(\omega_{e1}-\omega_r)\bar{\lambda}_{r1}; \quad \bar{\lambda}_{r1} = L_{r1}\bar{i}_{r1} + L_{m1}\bar{i}_{s1} \\ &\bar{V}_{s3} = r_s \bar{i}_{s3} + \frac{d}{dt}\bar{\lambda}_{s3} + j3\omega_{e3}\bar{\lambda}_{s3}; \quad \bar{\lambda}_{s3} = L_{s3}\bar{i}_{s3} + L_{m3}\bar{i}_{r3} \\ &\bar{V}_{r3} = r_{r3}\bar{i}_{r3} + \frac{d}{dt}\bar{\lambda}_{r3} + j3(\omega_{e3}-\omega_r)\bar{\lambda}_{r3}; \quad \bar{\lambda}_{r3} = L_{r3}\bar{i}_{r3} + L_{m3}\bar{i}_{s3} \end{aligned} \tag{17.2}$$

with

$$L_{s1} = L_{ls} + L_{m1}; \quad L_{s3} = L_{ls} + L_{m3}, \quad L_{r1} = L_{lr1} + L_{m1}; \quad L_{r3} = L_{lr3} + L_{m3} \tag{17.3}$$

As the inverter has a null point, the sum of stator currents is always forced to zero and the zero sequence of transformation is missing (see $T_s$ in Equation 17.1, the last line).

The torque expression is simply

$$T_e = p\cdot[L_{m1}(i_{sq1}\cdot i_{rd1} - i_{sd1}\cdot i_{rq1}) + 3L_{m3}(i_{sq3}\cdot i_{rd3} - i_{sd3}\cdot i_{rq3})]$$

where

$$\bar{i}_{s1} = i_{sd1} + ji_{sq1}; \quad \bar{i}_{s3} = i_{sd3} + ji_{sq3}; \quad \bar{i}_{r1} = i_{rd1} + ji_{rq1}; \quad \bar{i}_{r3} = i_{rd3} + ji_{rq3} \tag{17.4}$$

The stator magnetization inductances ($L_{m1}$, $L_{m3}$) and rotor resistances and leakage inductances ($r_{r1}$, $r_{r3}$, $L_{lr1}$, $L_{lr3}$) for the fundamental and for the third harmonic are considered different.

Their expressions, without considering magnetic saturation are [2,3]

$$L_{m1} = \frac{5}{2}\cdot\frac{\mu_0\cdot l_{stack}\cdot r_r\cdot\pi\cdot N_{s1}^2}{g}; \quad L_{m3} = \frac{5}{2}\cdot\frac{\mu_0\cdot l_{stack}\cdot r_r\cdot\pi\cdot N_{s3}^2}{g} \tag{17.5}$$

$$r_{r1} = \left(\frac{5\pi^2}{2}\right)\cdot\frac{N_{s1}^2}{N_r \sin^2(\alpha_r/2)}[R_b(1-\cos\alpha_r)+R_e] \quad (17.6)$$

$$L_{lr1} = \left(\frac{5\pi^2}{2}\right)\cdot\frac{N_{s1}^2}{N_r \sin^2(\alpha_r/2)}[L_b(1-\cos\alpha_r)+L_e] \quad (17.7)$$

$$r_{r3} = \left(\frac{45\pi^2}{2}\right)\cdot\frac{N_{s3}^2}{N_r \sin^2(3\alpha_r/2)}[R_b(1-\cos 3\alpha_r)+R_e] \quad (17.8)$$

$$L_{lr3} = \left(\frac{45\pi^2}{2}\right)\cdot\frac{N_{s3}^2}{N_r \sin^2(3\alpha_r/2)}[L_b(1-\cos 3\alpha_r)+L_e] \quad (17.9)$$

with

$$N_{s1} = \frac{4}{\pi}N_s\cos\frac{\pi}{20}, \quad N_{s3} = \frac{4}{3}\frac{N_s}{3}\cos\frac{3\pi}{20} \quad (17.10)$$

where $l_{stack}$ is the stack length, g is the airgap, $N_{s1}$, $N_{s3}$ are the amplitude of fundamental and third harmonics winding functions, $N_r$ is the rotor bar number, $R_b$, $R_e$ are rotor bar and end-ring segment resistance, $L_b$, $L_e$ are rotor bar and end-ring leakage inductances, $\alpha_r = 2\pi/N_r$ is the rotor bar angle, $N_s$ is the number of conductors/slot, g is the airgap, and $r_r$ is the rotor outer radius.

## 17.3 Multiphase IM Drives Control Strategies

The essential control problem is to "synchronize" the fundamental and third harmonic fluxes, speeds or slip speeds ($\theta_{e1} = \theta_{e3}$ in Equation 17.1). So there is a strict relationship between $i_{sq3}$ and $i_{sq1}$ [2]:

$$i_{sq3} = k_c \cdot i_{sq1}; \quad k_c = 3\cdot\frac{T_{r3}}{T_{r1}}\cdot\frac{i_{sd3}}{i_{sd1}}; \quad T_{r3} = L_{r3}/r_{r3}; \quad T_{r1} = L_{r1}/r_{r1} \quad (17.11)$$

This automatically means a rigid ratio between the fundamental ($T_{r1}$) and third harmonic torque ($T_{e3}$) component:

$$k_t = \frac{T_{e1}}{T_{e1}+T_{e3}}; \quad k_t = \frac{1}{1+\alpha}; \quad \alpha = 9\cdot\frac{r_{r1}}{r_{r3}}\cdot\left(\frac{L_{m3}}{L_{m1}}\right)^2\cdot\left(\frac{i_{sd3}}{i_{sd1}}\right)^2 \quad (17.12)$$

Though Equation 17.11 is ideally enough for maintaining a trapezoidal flux pattern, in practice, a correction is added:

$$i_{sq3}^* = k_c \cdot i_{sq1}^* + k_p\cdot(\theta_{e1}-\theta_{e3}) \quad (17.13)$$

The generic control system is shown in Figure 17.3.

To verify if the control is feasible, first, we inject only either the fundamental or the third harmonic.

Results as in Figure 17.4 [2] have been obtained.

It should also be noted that the order of phases is natural (a, b, c, d, e) for the fundamental but it is (a, c, e, b, d) for the third harmonic, and that the attained speed ratio is 3/1, for the same target frequency.

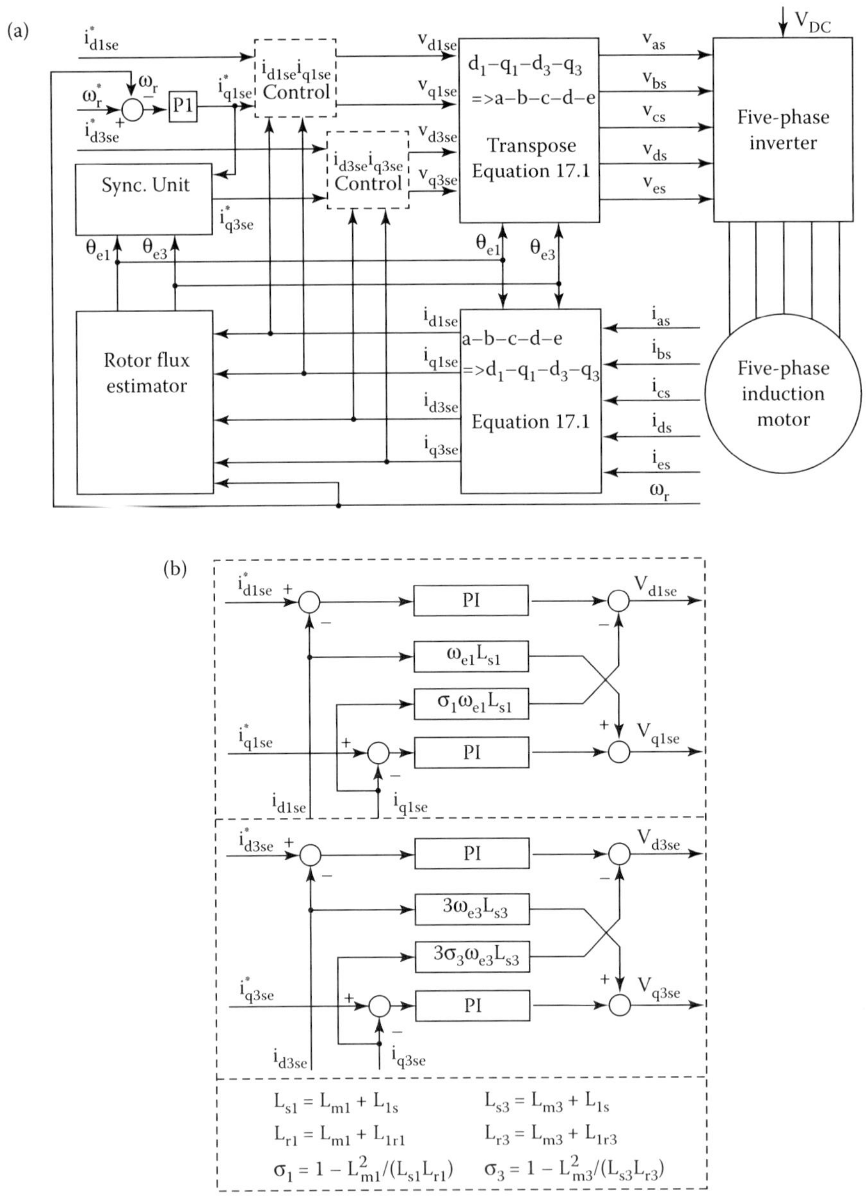


**FIGURE 17.3**
Five-phase IM control with quasi-trapezoidal flux waveform (a) and $d_1 - q_1$, $d_3 - q_3$ current controller plus e.m.f. compensation to $V_{sd1}$, $V_{sq1}$, $V_{sd3}$, $V_{sq3}$ (b). (After L. Zheng et al., *IEEE-Trans.*, IA-55(5), 1996–2005, 2008.)

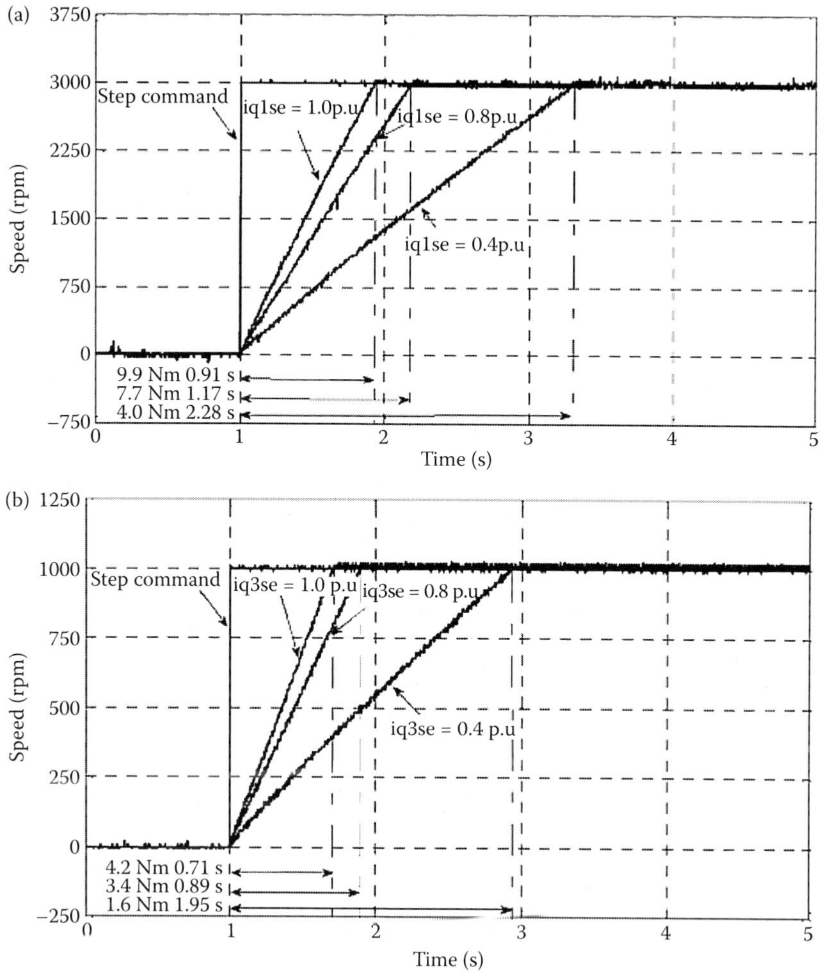


**FIGURE 17.4**
Acceleration to speed: (a) first harmonic only, (b) third harmonic only. (After L. Zheng et al., *IEEE-Trans.*, IA-55(5), 1996–2005, 2008.)

When both the fundamental and the third harmonic are considered, the steady-state operation at 1200 rpm is as illustrated in Figure 17.5 [2].

The results in Figure 17.5 may be characterized as

- The rotor flux components position estimation (standard) shows small errors between $\theta_{e1}$ and $\theta_{e3}$.
- The current dq components hodographs are well checked around a circle.
- The filtered phase voltage has a quasi-flat top as it should (the maximum modulation index in the inverter is raised from 0.526 to 0.602 [a 14% increase]).
- The speed is rather tightly controlled.

The investigated motor parameters are [2]: $r_s = 0.67\ \Omega$, $r_{r1} = 0.52\ \Omega$, $r_{r3} = 0.32\ \Omega$, $L_{ls} = 2.9$ mH, $L_{lr1} = 2.9$ mH, $L_{lr3} = 1.6$ mH, $L_{m1} = 104$ mH, $L_{m3} = 11.1$ mH.

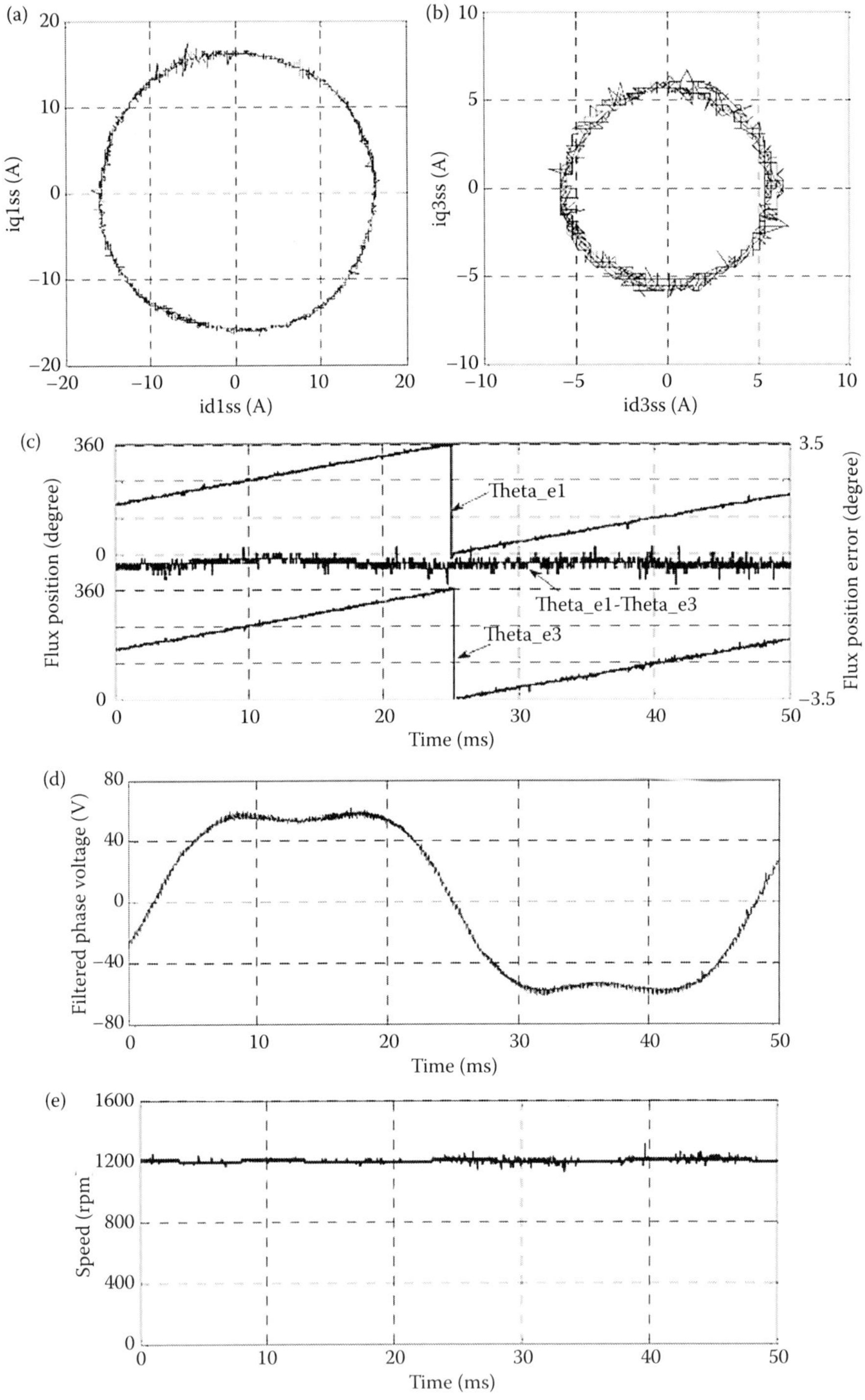


**FIGURE 17.5**
Steady-state operation at 1200 rpm: (a), (b) fundamental and third harmonic current hodographs, (c) estimated flux positions $\theta_{e1}$, $\theta_{e3}$, (d) filtered phase voltage, (e) speed versus time. (After L. Zheng et al., *IEEE-Trans.*, IA-55(5), 1996–2005, 2008.)

*Note on multiphase IM drives*

Other number of phases (6, e.g., asymmetric (dual star [4]) or asymmetric [5]) may be investigated for fault tolerance, especially.

PWM in the inverters for multiple phases has led to numerous ingenious solutions ([6], for 7 phases). Control over wide speed range of IMs is treated in Reference 7. Also, the open-end IM winding with 6 poles is analyzed in Reference 8 in an equivalent four-level inverter for improved d.c.-link utilization, while reduced switch count converters are investigated thoroughly in Reference 9. Finally, fuzzy adaptive control is investigated in Reference 10.

Though the efficiency of multiphase IM is not higher than for three phases [11], other attributes, related to additional torque and inverter savings with better fault tolerance, keep it as a strong competitor in safety critical applications.

Pentagon connection of stator phases is proven better under open-phase faults, while star connection is superior in healthy conditions, as shown in Reference 12.

## 17.4 Multiphase PMSM Drives Control under Open-Phase Faults

Multiphase PMSM, as IMs, may be used with

- Even (or 3k) number of phases ($2 \times 3$, $3 \times 3$, etc. [13–15])
- Odd (prime) number of phases (5, 7) [16–18]

To shorten the presentation but integrate most features of multiphase PMSM drives, the optimal control for minimum copper losses and low torque pulsations under open-phase faults is treated here [16]. The phase currents are still controlled according to rotor position, but an open-phase fault is present in phase j. A few additional variables are defined [16]:

- The fault column matrix F with 1 for healthy and 0 for open phase is

$$F = [1, 1, 0, 1, 1]^T \tag{17.14}$$

where Q is a diagonal terms matrix with F as its only nonzero diagonal element:

$$Q = \begin{bmatrix} 1 & 0 & 0 & 0 & 0 \\ 0 & 1 & 0 & 0 & 0 \\ 0 & 0 & 0 & 0 & 0 \\ 0 & 0 & 0 & 1 & 0 \\ 0 & 0 & 0 & 0 & 1 \end{bmatrix} \tag{17.15}$$

And e, i as total e.m.f. and phase currents column matrixes

$$e = \begin{bmatrix} e_1 = k_1\omega \\ e_2 = k_2\omega \\ e_3 = k_3\omega \\ e_4 = k_4\omega \\ e_5 = k_5\omega \end{bmatrix} = |k|\omega; \quad i = \begin{bmatrix} i_1 \\ i_2 \\ i_3 \\ i_4 \\ i_5 \end{bmatrix} \tag{17.16}$$

The total number of healthy phases is $n_{healthy} = F^T \cdot F_0$, and $\omega$ is the stator frequency. The instantaneous electromagnetic torque is then

$$T_e = k^T Qi \tag{17.17}$$

To minimize the torque ripple and the copper losses, we have first to build the corresponding error functions:

$$T^* - T_e = T^* - k^T Qi = g_1(i) \rightarrow 0 \tag{17.18}$$

For no null current path, the sum of currents is zero and thus

$$F^T i = g_2(i) \rightarrow 0 \tag{17.19}$$

As there is not an unique solution for Equations 17.18 and 17.19 functions $g_1(i)$, $g_2(i)$ minimization, we may add the condition of minimizing copper loss. For equal stator phase resistances

$$\frac{1}{2} i^T i = h(i) \rightarrow \text{minimum} \tag{17.20}$$

The optimization objective function $f(i,p_1,p_2)$ may be then built as [16]

$$f(i, p_1, p_2) = \frac{1}{2} i^T i + p_1(T^* - k^T Qi) + p_2 F^T i \tag{17.21}$$

with $p_1$, $p_2$ Lagrangian multipliers.

Now by using partial derivatives of function f to make them zero, the optimum current matrix fulfills the condition

$$i = p_1 \cdot Q \cdot k - p_2 \cdot F \tag{17.22}$$

Finally, $p_1$ and $p_2$ are calculated:

$$\begin{aligned} p_1 &= \frac{n_h}{G(k)}; \quad G(k) = n_h \cdot k^T \cdot Q \cdot k - F^T \cdot Q \cdot k \cdot k^T \cdot F \\ p_2 &= \frac{F^T \cdot Q \cdot k}{G(k)} \end{aligned} \tag{17.23}$$

Substituting $p_1$ and $p_2$ in Equation 17.22, the phase currents for open-phase fault conditions are obtained (torque $T_e \approx T^*$).

The matrix k (e.m.f. = $k\omega$) may be obtained either by experiments or by FEM. Note that k has terms dependent on rotor position.

The maximum phase current condition may be added during open-phase fault operation, for safety.

A 4-pole 15-slot 5-phase, 3-kW, 3-krpm PMSM is considered here [16]. Typical FEA stator flux linkages, e.m.f.s, and optimal current waveforms as a function of rotor position are shown in Figure 17.6 for the healthy machine and in Figure 17.7 for phase 1 open, phases 1 + 3 open, and adjacent phases 1 + 2 open, respectively.

The corresponding torque for the four cases in Figures 17.6 and 17.7, shown in Figure 17.8, warns remarks such as

- In all cases, with limited maximum phase current peak value, the PMSM with 1, 1 + 2, 1 + 3 phases open can produce sufficient torque with reasonable torque pulsations, for minimum copper losses.
- As expected, for some phases open, less torque is obtained, with the worst situation experienced for two adjacent phases open.
- For the optimization of current patterns, the currents are sampled in $N_s$ points over a cycle.
- So far in this paragraph the voltage ceiling in the inverter was not considered in the provision of the desired current patterns. By the fast raises and decreases of current patterns for two adjacent open phases it seems that here lies the design bottleneck for maximum desired speed and torque for the drive [17].
- Hysteresis current a.c. controllers may be used to realize the current shapes after the fault detection was done.
- While $k(\theta_{er})$ matrix reflects total (stator flux) e.m.f. components, the solution here, with FEM extracted data, seems most adequate for high number custom-made critical safety drives.

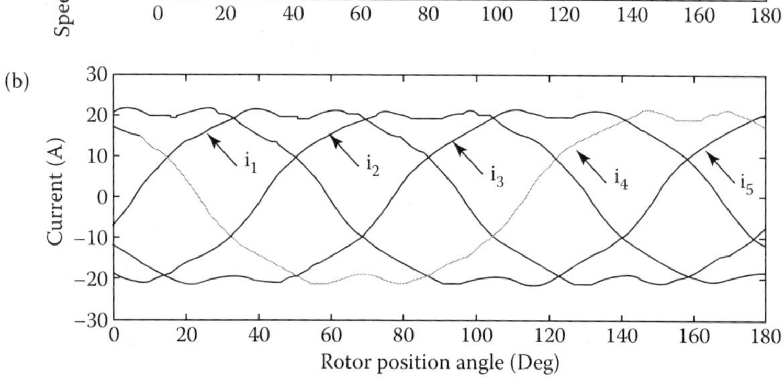


**FIGURE 17.6**
Phase flux linkages (a), e.m.f.s and optimal current waveforms (b) for the healthy PMSM. (After S. Dwari, L. Parsa, *IEEE-Trans.*, IE-55(5), 1988–1995, 2008.)

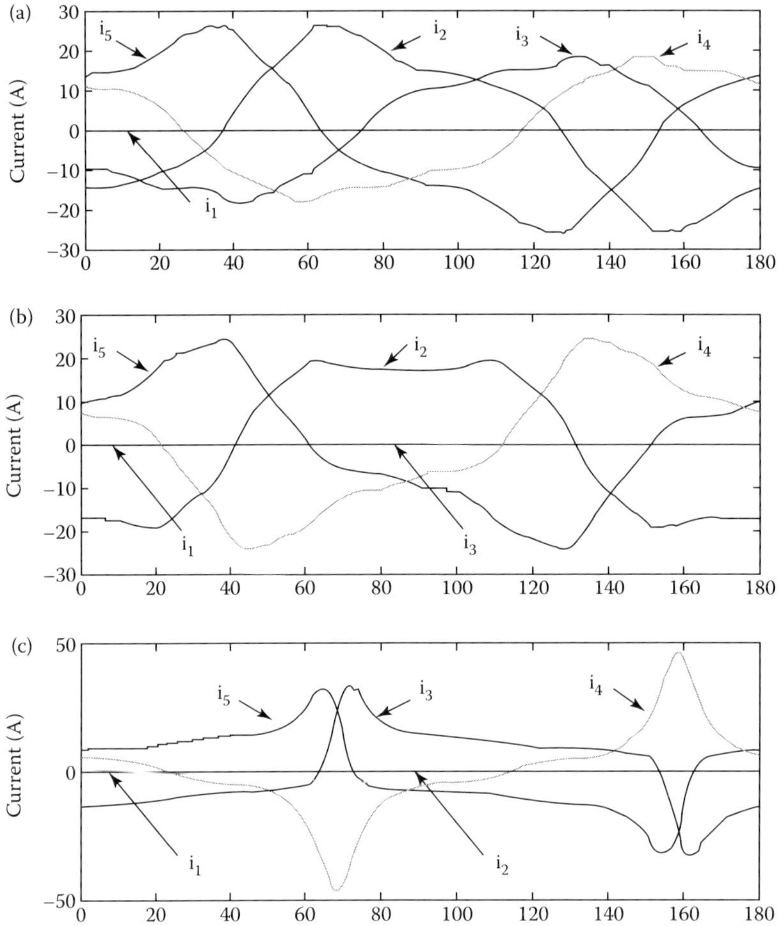


**FIGURE 17.7**
The optimal phase currents for: (a) phase 1 open, (b) phases 1 + 3 open, (c) phases 1 + 2 open. (After S. Dwari, L. Parsa, *IEEE-Trans.*, IE-55(5), 1988–1995, 2008.)

- For less critical safety drives, 2 × 3 or 3 × 3 phase PMSM drives may be used to be controlled on 1 × 3 respectively 1(2) × 3 phase components in case of open fault by simply disconnecting the faulty three-phase component [13,18].

## 17.5 BLDC Multiphase Reluctance Machines: Topology, Modeling, and Control—A Case Study

### 17.5.1 Introduction

Eliminating the high-energy product permanent magnets (or replacing them by low-cost ferrite PMs), but maintaining good performance in electric machine drives, has lately become the main way to go. The high saliency rotor (with multiple flux barriers, or

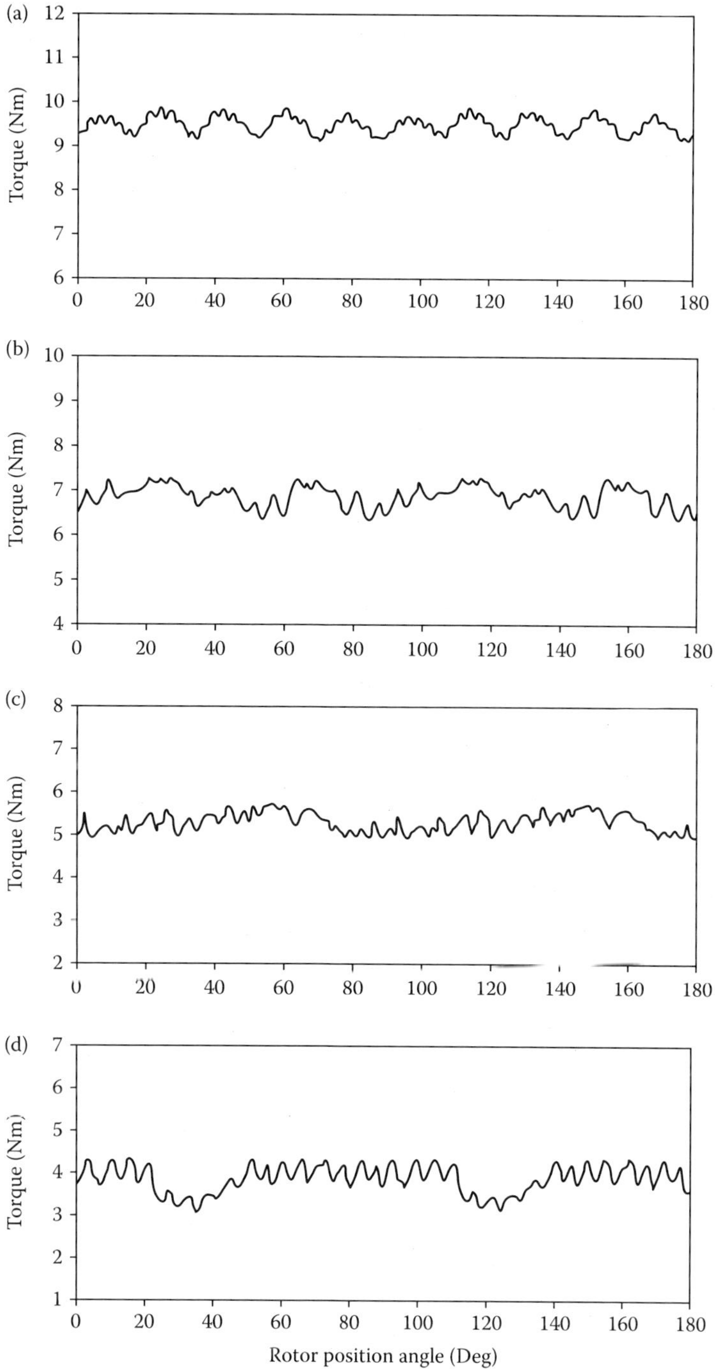


**FIGURE 17.8**
FEA torque versus rotor position: (a) healthy machine, (b) phase 1 open, (c) phases 1 + 3 open, (d) phases 1 + 2 open. (After S. Dwari, L. Parsa, *IEEE-Trans.*, IE-55(5), 1988–1995, 2008.)

ALA rotor) reluctance synchronous machine—without or with ferrite PMs assistance (in q axis)—with a three-phase inverter with field oriented or direct torque sinusoidal current control seems the obvious way to good performance without high-energy product PMs [19]. However, the reluctance three-phase synchronous machine with sinusoidal current control does not make full use of inverter kVA (as known from the comparison with brushless d.c. (BLDC) PM motors—trapezoidal current control). The three-, four-, five-phase 400-Nm switched reluctance machine drives recently proved to produce 45 Nm/L and 95% efficiency at 5000 rpm, with special magnetic core [20], but still suffer from rather low utilization of inverter voltage and limited torque density production typical to the variable reluctance machine principle. Recently, stator flux switching (reversal) machines with 2k d.c. heteropolar stator excitation coils interleaved with a 2k three-phase stator a.c. coils and 4k-salient pole rotor have been proposed for hybrid electric vehicles [21–23]. The torque density is about 75% of that interior-PM synchronous machine used for Toyota Prius 3, 2005 (for the same geometry). Unfortunately, this is obtained for a very low (0.4 ÷ 0.6) power factor [22], which means strong overrating of the PWM inverter used for the scope. As the inverter is more expensive than the electric machine (even at 500 V d.c., 123 kW range), the search for better solutions should go on. In the hunt for higher torque density without PMs, References 24 through 26 have investigated decades ago the BLDC multiphase reluctance machine (BLDC-MRM) principle with promising preliminary results. However, the spectacular advance of PM synchronous motors had led to their practical dormancy for 20 years. There is a notable exception, that is, [27], which pertinently introduces a two-dimensional (2-D) finite-element method (FEM)-based optimal design for BLDC-MRMs (multiphase reluctance machines) with given outer diameter, stack length, and a small airgap. In an effort to reduce armature reaction, Reference 27 compares standard with slit and chamfered salient rotor poles showing small progress for a six-phase q = 2 (slots/pole/phase) stator winding, mainly due (in our view) to a still small magnetic reluctance in axis q. Also, they investigated the independent trapezoidal current shape beneficial influence of field and torque currents on the torque ripple. Finally, Reference 27 shows comprehensive experimental results about the torque pulsations and torque versus speed, noticing openly/fairly the "poor torque performance of the machines in the field weakening" mainly because (in our view) the airgap and the q axis magnetic reluctance were too small.

A survey of multiphase machines and drives is given in Reference 1, but the closest work related to this chapter, published in the past few years, is in References 11 and 12. The separately excited d.c. brush machine principle has been successfully applied for a decoupled field and torque control in a high phase order IM, without the use of coordinate transformation and flux linkage estimators, and simplifying the control algorithm [28,29]. Two equivalent windings essentially supplied by two trapezoidal currents (field and torque) were used, thus making full use of machine iron, through an even distribution of iron core saturation.

The key ingredients of wide constant power speed range of BLDC-MRM is constituted, in our view, by an optimal (not minimum) airgap, which will mitigate excitation losses and commutating phase inductance, on a moderate d axis inductance (to use best inverter voltage) and a high magnetic reluctance in axis q, to decrease armature reaction and thus allow higher torque density (by avoiding magnetic stator teeth local saturation) and, again, to allow for better use of d.c.-link (input) inverter voltage. This paragraph aims to use ALA and (or) multiple flux barrier rotors and q = 1 windings that work best (as known from BLDC PM motors) with two-level flat-top single trapezoidal phase currents, to further reduce inverter commutation kVA and copper additional losses.

The main items treated in what follows are:

- A deduction of the BLDC-MRM principles of operation from an exciter-less d.c. brush machine with brushes off the neutral axis and high stator saliency or higher torque density
- For a given ALA-rotor six-phase, six-pole, and 36-slot stator prototype, parameter identification from standstill tests and thorough 2D FEA investigation, considering magnetic saturation and including core loss calculation from FEM imported flux density local variation are performed
- 2D FEA computation and curve fitting the machine inductances to account for magnetic saturation and rotor position and their use in a nonlinear circuit model of BLDC-MRM
- Development of a four-quadrant control method (with flux weakening), for a 6 + 1 leg inverter for which the machine null-current is indirectly controlled through the null-point voltage pulsation control, to provide symmetric phase currents, and its implementation on a test platform with promising results (direct comparison of current measured and calculated waveforms)

### 17.5.2 Principle of BLDC-MRM

The principle of operation, as first presented in References 30 and 31, will be briefly discussed next. Let us consider a six-coil rotor exciter-less d.c. brush machine, with brushes moved away from neutral axis (q axis) to the corners of the rotor poles, and with enlarged magnetic saliency (reluctance) in the stator (see Figure 17.9a). The two coils (E–E′, F–F′) in the q axis produce field flux ($\Psi_d$) along the d axis, whereas four coils (A–A′, B–B′, C–C′, D–D′) produce torque, with low armature reaction. A large saliency ALA rotor is adopted to the BLDC-MRM in Figure 17.9b (a multiple flux barrier salient rotor will also do), which is equivalent to the d.c. brush motor in Figure 17.9a, but with power electronic-assisted commutation of phase currents (see Figure 17.10); each phase current has a "field current" time interval and a "torque current" time interval and is bipolar. The "field phase current" sequence directly goes to torque sequence without going to zero, in order to reduce copper losses in the machine and commutation losses in the inverter. Now, it is evident that phases E and F play the role of field phases in Figure 17.9b.

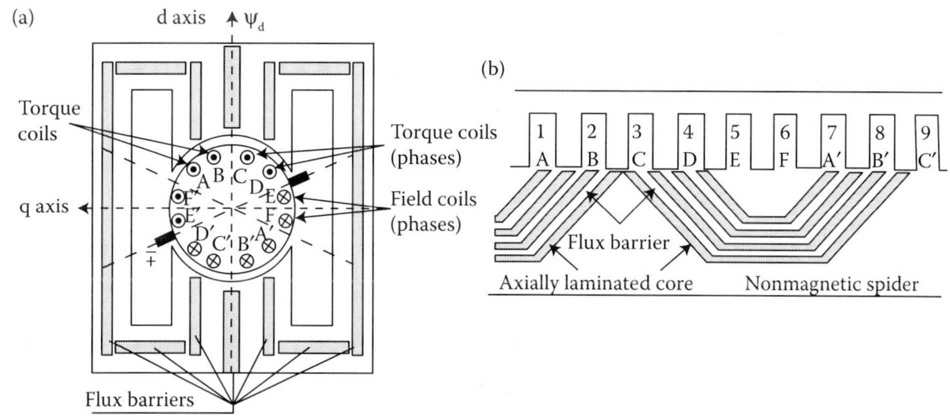


**FIGURE 17.9**
(a) Equivalent six-phase exciter-less d.c. machine. (b) Linear (rotary) BLDC-MRM simplified geometry.

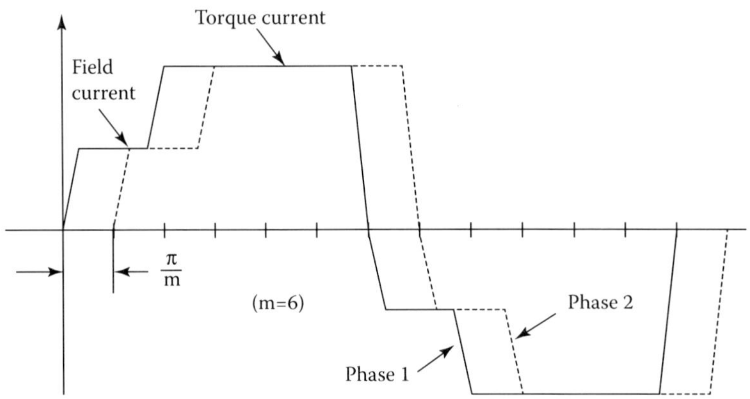


**FIGURE 17.10**
Typical flat-top bipolar level current waveform.

The interaction between field and torque phase fields (m.m.f.s) takes place close (within ± 15°) to 90° (not between 60° and 120° as in standard three-phase BLDC PM motors). The high magnetic saliency (by ALA rotor or by multiple flux barriers rotor) serves to reduce further "armature" reaction field. Hence, in fact, this machine goes back and mimics the d.c. brush machine, but we took out the stator excitation and moved the brushes from the neutral axis.

The price for this will be paid in the inverter because we need rotor position triggered electronic commutation of phases, one after the other. The field current control is not fully independent of armature (q axis) reaction field, but not far from that, as the field phase's field axis stays close to the d axis. The machine may also be used as a generator, for example, for wind or micro-hydro energy conversion [30,31].

The airgap needs not be very small as it will impede on phase current commutation and, thus, on wide constant power speed range. An additional benefit is that a larger airgap will allow a mechanically rugged rotor and lower mechanical losses.

### 17.5.3 BLDC-MRM Parameter Identification

The machine electrical and mechanical parameters (inductances and moment of inertia) have been determined by the flux decay and free-deceleration tests [15]. These are mandatory for as close to reality as possible electric drive simulation.

An FE model (see Figure 17.11) of the BLDC-MRM (script base built in Cobham/VectorFields Opera 2-D) has also been checked by direct comparison between experiment and simulation results.

A 6-phase, 6-pole, 35-Nm, 250-rpm, and one-slot-per pole-per-phase (q = 1) ALA rotor machine was investigated and built (see Figures 17.12 and 17.13) for experimental purpose (the q = 1 winding "collaborates" best with trapezoidal current to reduce torque pulsations). The main dimensions are as follows: tooth width—2.5 mm, average slot width—6.5 mm, airgap length—0.3 mm (still small), slot area—130 mm$^2$, stator inner diameter—104 mm, stator outer diameter—170 mm, and core length—145 mm.

#### *17.5.3.1 Experimental Setup*

#### *17.5.3.2 Flux Decay Tests*

The flux decay test involves reading the decaying current and voltage in a circuit whose d.c. source has been interrupted (see Figure 17.14). One inverter leg (upper switch and lower

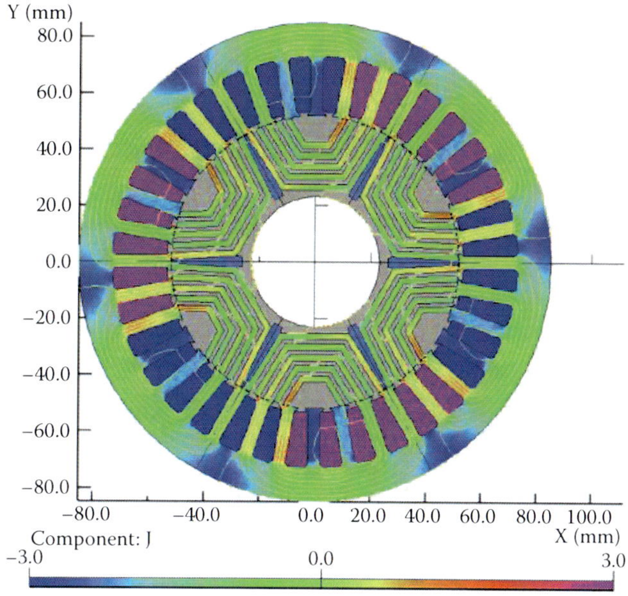


**FIGURE 17.11**
Six-phase/six-pole BLDC-MRM FEM 2-D model (nominal load conditions—I phase = 12 A, 28 turns/coil, $T_e$ = 35 Nm).

**FIGURE 17.12**
Experimental lab model of the six-phase BLDC-MRM.

diode switch) has been used for this purpose. By summing up the area of the decaying current multiplied by the circuit resistance (R) and the diode voltage area, and dividing the result to the initial conduction current ($I_{cond}$), one can find out the total inductance of the circuit, L, expressed by Equation 17.24

$$L = \frac{R \cdot \int_{t_{off}}^{t,i=0} i(t)dt + \int_{t_{off}}^{t,Vd=0} V_D(t)dt}{I_{cond}} \tag{17.24}$$

where the circuit resistance R is measured at the beginning of every test, i(t) represents the instantaneous value of current, and $V_D(t)$ is the instantaneous value of the diode voltage.

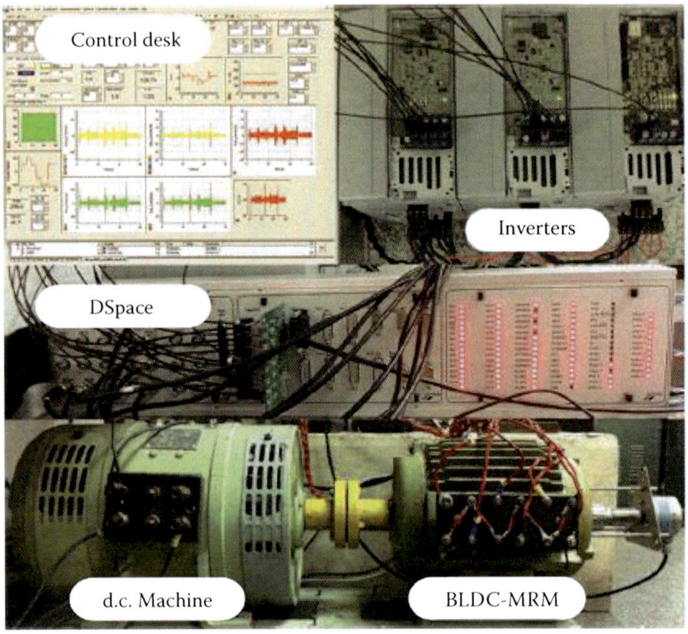


**FIGURE 17.13**
Test rig for the six-phase BLDC-MRM.

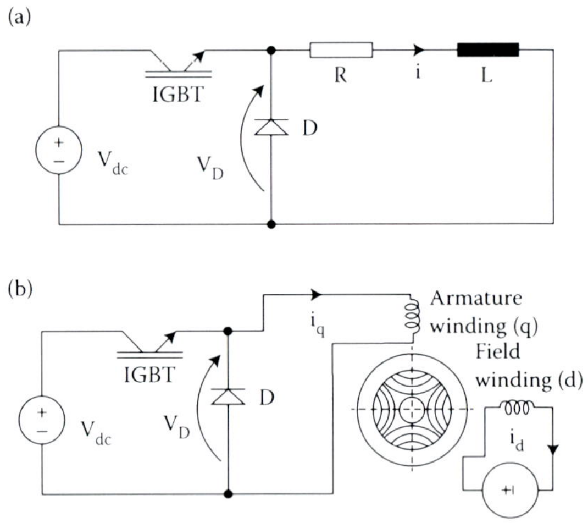


**FIGURE 17.14**
Electrical diagram. (a) One-phase current decay test. (b) Equivalent two-phase (field/torque) BLDC-MRM.

The integral is done from the moment the transistor is blocked until the voltage and the current become 2% of their initial values (to reduce integration errors). Figure 17.14a presents the electrical diagram for one-phase current decay test (R and L represent the parameters of one phase). This will yield the self-inductance $L_{aa}$ and, by reading the induced e.m.f.s in the remaining five phases, the mutual inductances $L_{ba}$, $L_{ca}$, $L_{da}$, $L_{ea}$, $L_{fa}$ versus rotor position, Figure 17.15, which also shows acceptable agreement with FEA results.

The results in Figure 17.16 were obtained by current (flux) decay tests done at a maximum torque position to see the influence of one magnetic axis to the other, using the

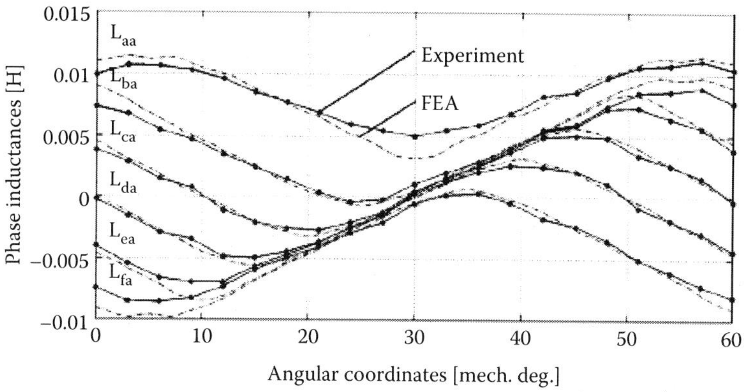


**FIGURE 17.15**
Self- and mutual inductances of the BLDC-MRM at rated current (I = 12 A). (Continuous line) Experiment (dotted line) FEA. (After D. Ursu et al., *IEEE-Trans.*, IA-51(3), 2105–2115, 2015.)

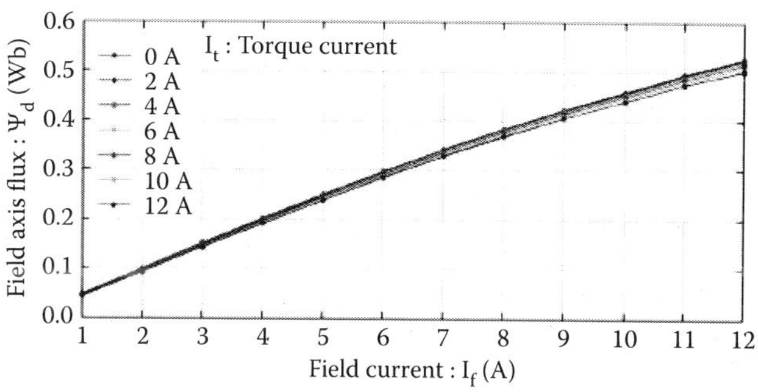


**FIGURE 17.16**
Experimental—BLDC-MRM d axis (field axis) flux versus d axis (field) current (in two phases) at different torque current $I_t$ values. (After D. Ursu et al., *IEEE-Trans.*, IA-51(3), 2105–2115, 2015.)

electrical diagram presented in Figure 17.14b (torque winding represents four adjacent phases in series, whereas field winding represents the remaining two adjacent phases in series). A good agreement between experiments in Figure 17.16 and FEA results in Figure 17.17 is apparent and shows only a small cross-coupling saturation effect (expected in an ALA rotor).

### *17.5.3.3 Mechanical Losses and the Moment of Inertia*

The sum of d.c. machine iron and total (drive) mechanical losses (see Figure 17.18) was computed by extracting the copper losses from input power in a no-load motor test. The speed is held constant by decreasing the field flux together with armature voltage. The lower speed limit is imposed when the armature current increases in the process. Two linear algebraic equations $y_1(x)$ and $y_2(x)$ are obtained (by curve fitting) function of d.c. machine field winding voltage, for 250 and 500 r/min. It is obvious that the mechanical losses do not depend on field flux; thus, they are 17.64 W at 500 r/min and 8.05 W at 250 r/min.

Drive transient behavior is strongly connected with the moment of inertia. The run-out test implies acceleration of the drive at rated speed when the drive power is switched

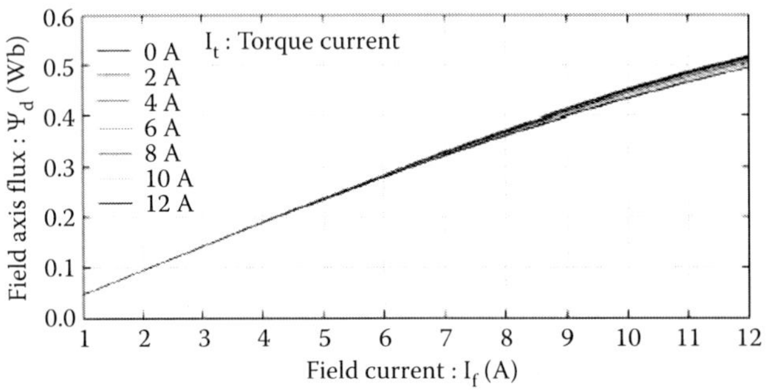


**FIGURE 17.17**
FEA—BLDC-MRM d axis (field axis) flux versus d axis (field) current (in two phases) at different torque current $I_t$ values. (After D. Ursu et al., *IEEE-Trans.*, IA-51(3), 2105–2115, 2015.)

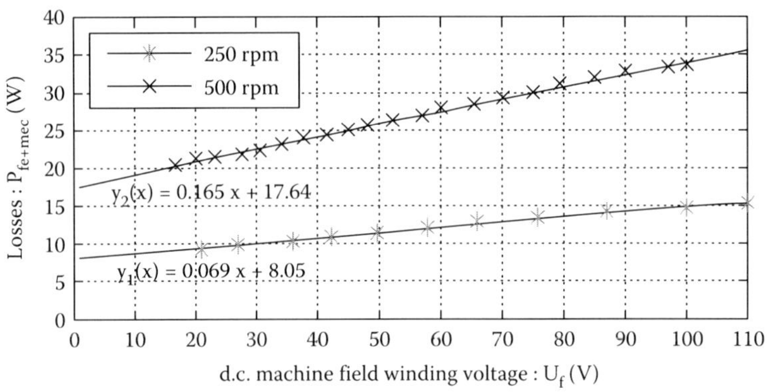


**FIGURE 17.18**
Experimental computation of mechanical losses.

off (t = 1 s in Figure 17.19). The drive inertia can be determined from the slope of coasting curve, if all the losses during the decelerating time period are known. The BLDC-MRM is mechanically coupled to a 2.7-kW 1500-r/min separately excited d.c. motor with the following data: $V_{field} = 110$ V, $I_{armature} = 30$ A, $V_{armature} = 110$ V, $R_{armature} = 0.33\ \Omega$, poles = 4 J was computed at base speed and twice of it, the average value being used. The tests were

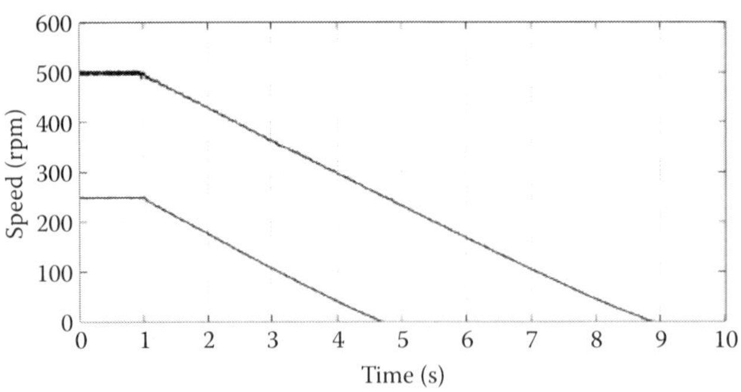


**FIGURE 17.19**
Run-out test at 250 and 500 r/min.

conducted after bringing the temperature of the bearings at rated temperature by 15 min of running at 1500 r/min.

Studying Figure 17.19, we can conclude that, at these speeds, the B term from the motion equation is negligible since the speed derivative value does not change during deceleration. The average value of J from the two tests is found to be 0.049 kg $m^2$.

#### *17.5.3.4 Standstill Torque Measurements*

At the maximum torque position, the developed torque of the six-phase BLDC-MRM was painstakingly measured by a load cell (AEP transducer, type TCS-C2, 100 kg) connected to the rotor through a steel bar. The torque was computed for different pairs of torque/field currents, up to rated current (12 A), by multiplying the exerted force on the sensor with the arm length. These results were also compared with FE results as can be seen in Figure 17.20. The agreements of both inductances and torque values between tests and FEA are considered satisfactory, given the prototype manufacturing tolerances and the heavy local magnetic saturation at high currents (stator slot width ≫ tooth width).

### 17.5.4 Extended 2-D FEA

Now that the 2-D FEM model of the machine has been validated in Section 17.5.3, extensive simulations with various excitation ($I_FW_1$) and torque ($I_TW_1$) coil m.m.f. values have been run to expose the torque capability of the prototype in Figure 17.12, for a wider range of currents than in the previous paragraph. For a more general analysis of the machine, total current density in field/torque phases is given as a parameter. The rotor is "fixed" in a maximum torque position, and the current densities are varied. Field current density $j_F$ would give rise to a field phase current equivalent to d axis current, and torque current density $j_T$ to a q axis current, responsible for torque production.

Studying Figure 17.21, we can deduce that, at $j_F = 12$ A/mm$^2$ (flat top) and $j_T = 24$ A/mm$^2$ (flat top), an almost 120 Nm torque (average value) is obtained at 6.7 Nm/kg (36.5 Nm/L) of active materials, which is quite competitive.

$j_F = j_T$ is an ideal optimum [26], but Reference 27 found pertinently by optimizations, a $j_F = (0.83 \div 0.85)j_T$ optimum for their small airgap (and particular rotor topology). Torque pulsations can be reduced by 50% if half of the rotor is shifted with half a slot pitch, while

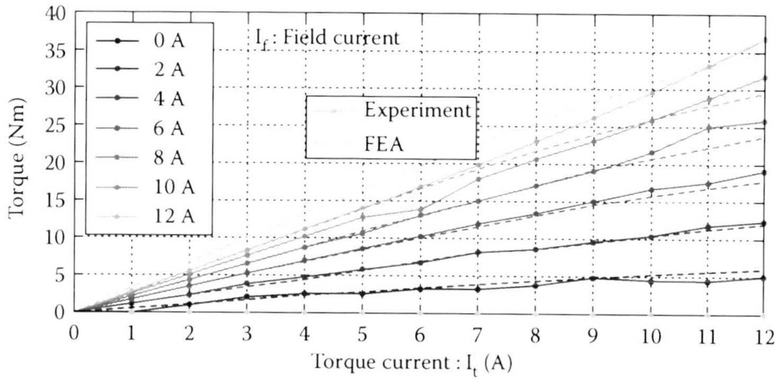


**FIGURE 17.20**
BLDC-MRM torque versus torque current at different field current values. (After D. Ursu et al., *IEEE-Trans.*, IA-51(3), 2105–2115, 2015.)

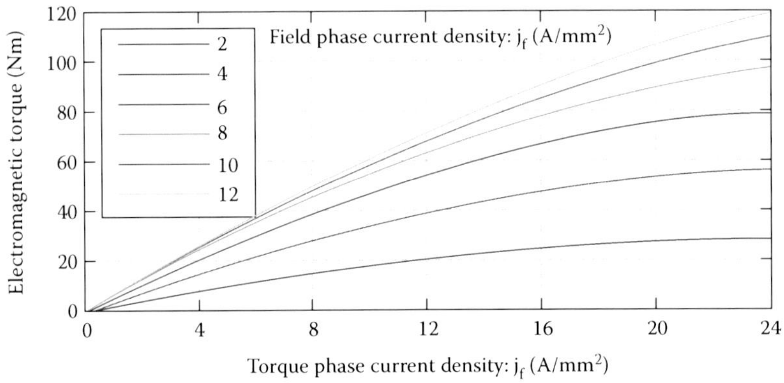


**FIGURE 17.21**
BLDC-MRM electromagnetic torque for different field/torque current levels, by 2-D FEM.

maintaining the 35 Nm average torque value; for j = 20 A/mm$^2$, the torque ripple decreased via skewing by 40% and the average torque increased by 10% (see Figure 17.22).

Ferrite PMs were added in the bottom of rotor flux barriers (in the q axis), thus further reducing the armature reaction and bringing 10% more torque to the motor (see Figure 17.23). Adding the same volume of NdFeB magnets (as for ferrites) will actually make the situation worse since too much flux is produced in the q axis, which leads to deeper local magnetic saturation; more than is needed to cancel the q axis armature reaction, resulting in a decrease of average torque by 15% (in our case) and an increase of torque pulsations by 75% (see Figure 17.23).

### 17.5.5 Iron Loss Computation

When the machine topology implies utilization of an ALA rotor and a nonsinusoidal airgap m.m.f., one has to have in mind the rotor iron losses. Nowadays, it is usual to compute iron losses based on Equation 17.25 [33]. This three-term equation started from Steinmetz, followed 30 years later by a contribution of Jordan who defined the two components: static (hysteresis) and dynamic (eddy current) losses. Bertotti introduced the third term for

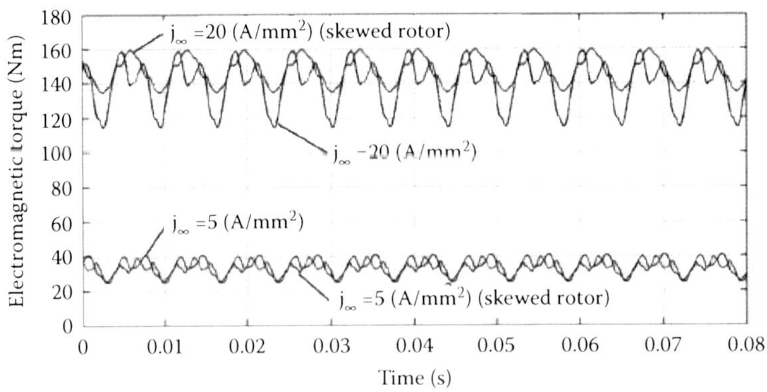


**FIGURE 17.22**
Torque pulsations of the BLDC-MRM at rated ($j_{Co} = j_F = j_T = 5$ A/mm$^2$) and peak conditions ($j_{Co} = 20$ A/mm$^2$), by 2-D FEM. (After D. Ursu et al., *IEEE-Trans.*, IA-51(3), 2105–2115, 2015.)

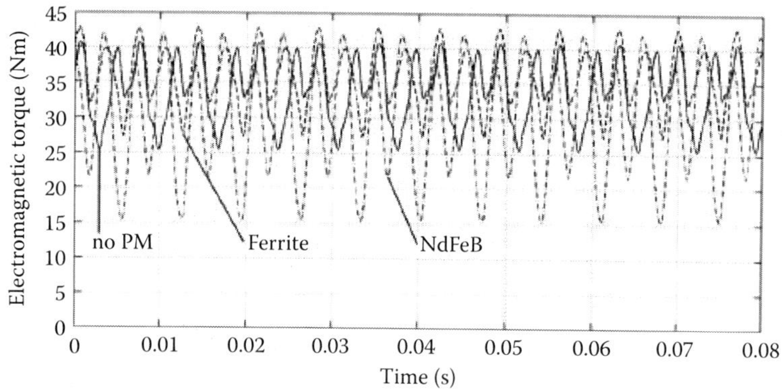


**FIGURE 17.23**
Torque pulsations at rated conditions j = 5 A/mm$^2$ for normal rotor (no q axis PMs) and modified rotor (q axis Ferrite/NdFeB PM), by 2-D FEM. (After D. Ursu et al., *IEEE-Trans.*, IA-51(3), 2105–2115, 2015.)

excess-anomalous losses [33]; $k_h$ (hysteresis iron loss coefficient) and $k_a$ (anomalous iron loss coefficient) cannot be analytically estimated, both of them having a strong dependence on magnetic flux density B (amplitude and frequency) and material properties, that is,

$$P_{fe} = k_h \cdot f \cdot B^{\alpha} + k_e \cdot f^2 \cdot B^2 + k_a \cdot f^{1.5} \cdot B^{1.5}\ W/m^3 \tag{17.25}$$

$k_e$ (eddy current iron loss coefficient) has been proven to be (from Maxwell equations)

$$k_e = \frac{\pi^2 \cdot \sigma \cdot d^2}{6} \tag{17.26}$$

where σ is the material electric conductivity and d is the lamination thickness. Boglietti studied eight different materials at frequencies between 10 and 150 Hz for magnetic flux densities in the range 0.6–1.7 T [33]. His results show that the excess losses coefficient ($k_a$) is zero, if the experimental data from an Epstein frame are interpolated using the third term in the Steinmetz–Jordan–Bertotti equation. We conclude that the individual contribution of eddy current losses and excess losses cannot be separated by measurements on an Epstein test bed. The two terms Steinmetz formula (17.27) is used here for iron loss estimation in the BLDC-MRM, that is,

$$P_{fe} = k_h \cdot f \cdot B^{\alpha} + k_e \cdot f^2 \cdot B^2 \tag{17.27}$$

It should be kept in mind that separation of iron losses in different terms is an empirical approach, in the sense of trying to separate the influence due to frequency and magnetic flux density, rather than explaining the physical phenomena directly. Steinmetz-based equations are best suited for fast and rough iron loss determination and can be easily integrated in an FE simulation. A script was written in Opera 2-D in order to compute the iron losses for each harmonic (from 1 to 30) of the magnetic field density components (radial and tangential) for every element of the mesh [31].

Samples of radial (Br) and tangential (Bt) magnetic flux density components versus time were plotted for the selected points in Figure 17.24. Figure 17.25a–f presents the variation

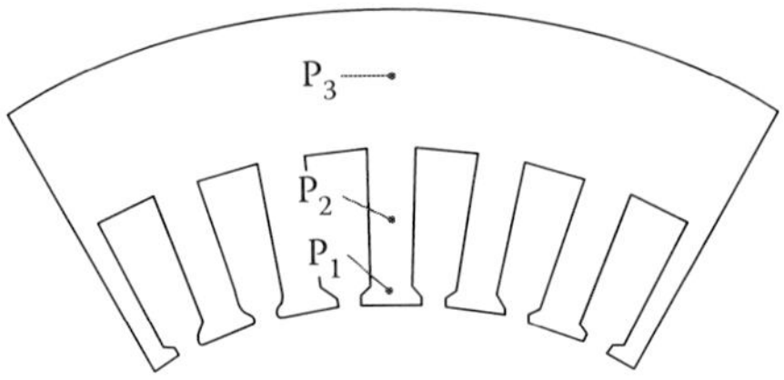


**FIGURE 17.24**
Selected stator points for magnetic flux density analysis.

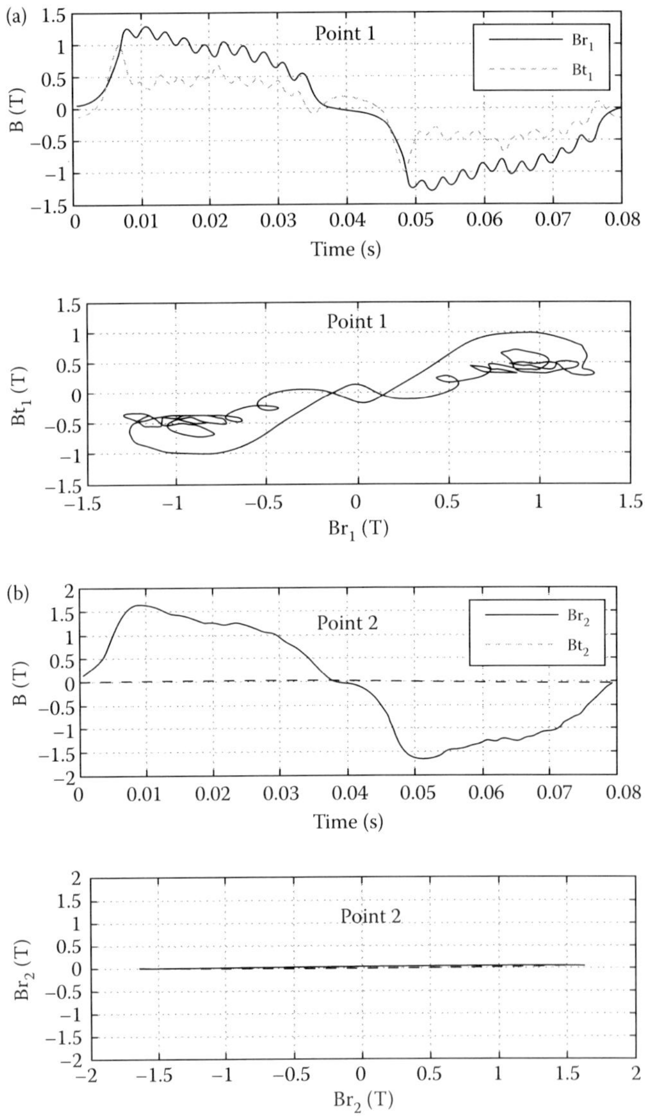


**FIGURE 17.25**
Magnetic flux density components variation in three different locations at 250 r/min: (a) in Point 1, (b) in Point 2,—see Figure 17.24. (After D. Ursu et al., *IEEE-Trans.*, IA-51(3), 2105–2115, 2015.) *(Continued)*

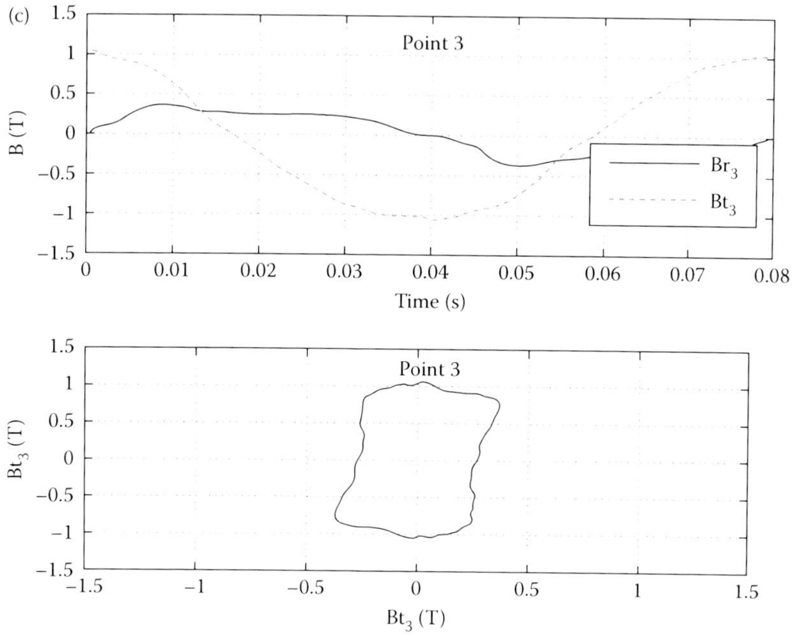


**FIGURE 17.25 (*Continued*)**
Magnetic flux density components variation in three different locations at 250 r/min: (c) in Point 3—see Figure 17.24. (After D. Ursu et al., *IEEE-Trans.*, IA-51(3), 2105–2115, 2015.)

of these components as functions of time and of each other. It is clear that at steady-state operation, a Fourier harmonic analysis (under the assumption that the contribution of the fundamental frequency is largely dominant) is mandatory, because the Steinmetz formula only holds up for a sinusoidal waveform of B. Hence, Equation 17.27 is applied for each of the studied harmonics (in our case, 30 terms), the total core losses being obtained by summing up the core losses for each element according to the variation of B in that element.

Since we did not have an Epstein test bed, but knew the lamination material, we used the data for M19 material from Reference 33. The computations are based on the following data: $k_h = 139.95$ W/m$^3$, $\alpha = 1.92$; $k_e$ computed by Equation 17.26 is 0.589 W/m$^3$ (d = 0.35 mm, $\rho = 7650$ kg/m$^3$, $\sigma = 2.17 \cdot 10^6$ S/m).

Total iron losses (11 W in stator iron losses plus rotor iron losses), for the first 30s harmonics of B, are shown in Figure 17.26. These full-load results were somehow expected, since the machine was designed for a low-speed operation (250 r/min, e.g., 12.5 Hz), with the possibility of being used as a small wind generator. Although the total rotor iron lamination volume is 35% of the stator volume, 50% of the total iron losses (5.4 W out of 11 W) occur in the rotor due to a high number of magnetic flux harmonics, as expected due to m.m.f. waveform. First, the third, fifth, and seventh harmonics are responsible for stator iron losses, whereas all the other harmonics produce mainly rotor iron losses. All these computations are a rough estimation since the magnetic material is not perfect, and the lamination cutting process can notably increase the iron losses (particularly in the tooth region where they can be doubled [34]).

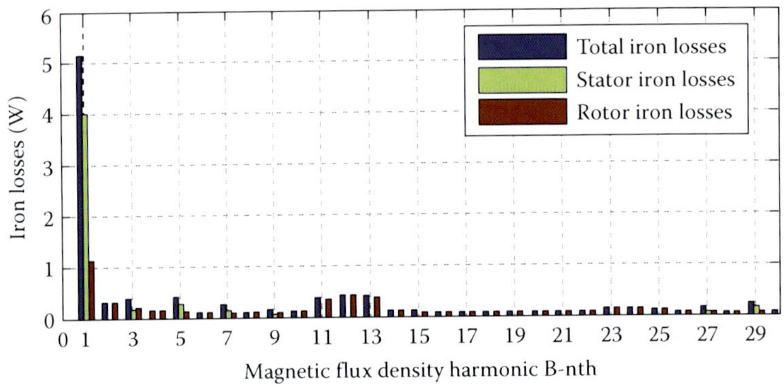


**FIGURE 17.26**
Computed iron loss for each harmonic of magnetic flux density.

## 17.5.6 Control Strategy

### *17.5.6.1 Circuit Model*

The circuit model of the BLDC-MRM for control design is based on the phase coordinate model, which makes use (for a six-phase machine) of a 6 × 6 inductance matrix $L(\vartheta)$ with all terms dependent only on rotor position, although in principle, they also depend on $I_F$ and $I_T$ (experimental measurements on our machine have shown no notable saturation up to rated current).

Taking advantage of the machine symmetry, only phase a inductances must be curve fitted using a fast Fourier transform (FFT) routine and cosine functions (17.28), whereas the other phases are reconstructed from the first one.

The curve fitting process (already performed for a special configuration six-phase BLDC-MRM [35]) was employed on experimental data (see Figure 17.27). FFT analysis is used to give for each (self and mutual) phase inductance the average ($L_{xx\_0}$) and amplitude value ($L_{xx\_2}$) for the first harmonic of the inductance waveform (function of twice the electrical rotor angular displacement $2 \cdot \theta_{er}$). Following this process, they were found to have these values: $L_{aa\_0} = 0.0077$ H, $L_{ba\_0} = 0.0041$ H, $L_{ca\_0} = 0.0021$ H, $L_{da\_0} = 0$ H,

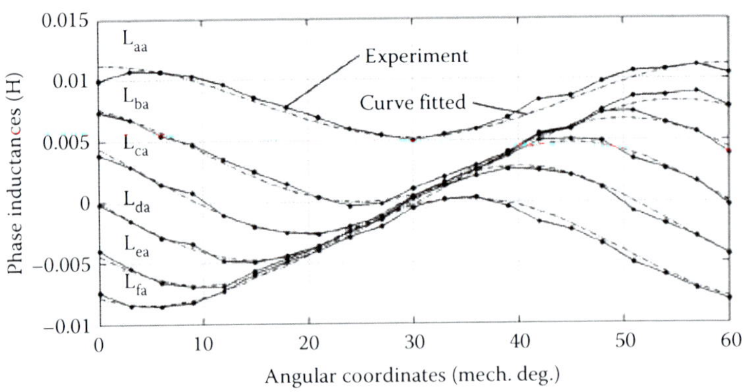


**FIGURE 17.27**
Curve fitting phase inductances (dotted line).

$L_{ea\_0} = -0.0021$ H, $L_{fa\_0} = -0.0041$ H, $L_{aa\_2} = 0.025$ H, $L_{ba\_2} = 0.035$ H, $L_{ca\_2} = 0.045$ H, $L_{da\_2} = L_{ea\_2} = L_{fa\_2} = 0.039$ H.

BLDC-MRM model is based on electrical equations (17.29), (17.30) and mechanical equation 17.31. The electromagnetic torque $T_e$ is computed from Equation 17.32.

The resulting system of nonlinear equations is not an issue for today's computing power; at the same time, the phase model is closer to reality than a d-q model is:

$$\begin{aligned} L_{aa} &= L_{aa\_0} + L_{aa\_2} \cdot \cos(2 \cdot \theta_{er}) \\ L_{ba} &= L_{ba\_0} + L_{ba\_2} \cdot \cos(2 \cdot \theta_{er} - \pi/6) \\ L_{ca} &= L_{ca\_0} + L_{ca\_2} \cdot \cos(2 \cdot \theta_{er} - 2 \cdot \pi/6) \\ L_{da} &= L_{da\_0} + L_{da\_2} \cdot \cos(2 \cdot \theta_{er} - 3 \cdot \pi/6) \\ L_{ea} &= L_{ea\_0} + L_{ea\_2} \cdot \cos(2 \cdot \theta_{er} - 4 \cdot \pi/6) \\ L_{fa} &= L_{fa\_0} + L_{fa\_2} \cdot \cos(2 \cdot \theta_{er} - 5 \cdot \pi/6) \end{aligned} \tag{17.28}$$

$$\frac{d[\psi]}{dt} = [u] - [R] \cdot [i] - [i] \cdot \frac{\partial [L(\theta_{er})]}{\partial \theta} \tag{17.29}$$

$$[L(\theta_{er})] = \begin{bmatrix} L_{aa} & L_{ab} & L_{ac} & L_{ad} & L_{ae} & L_{af} \\ L_{ba} & L_{bb} & L_{bc} & L_{bd} & L_{be} & L_{bf} \\ L_{ca} & L_{cb} & L_{cc} & L_{cd} & L_{ce} & L_{cf} \\ L_{da} & L_{db} & L_{dc} & L_{dd} & L_{de} & L_{df} \\ L_{ea} & L_{eb} & L_{ec} & L_{ed} & L_{ee} & L_{ef} \\ L_{fa} & L_{fb} & L_{fc} & L_{fd} & L_{fe} & L_{ff} \end{bmatrix}$$

$$[\psi] = [L(\theta_{er})] \cdot [i] \tag{17.30}$$

$$[i] = [i_a \quad i_b \quad i_c \quad i_d \quad i_e \quad i_f]^t; \quad [u] = [u_a \quad u_b \quad u_c \quad u_d \quad u_e \quad u_f]^t$$

$$[\psi] = [\psi_a \quad \psi_b \quad \psi_c \quad \psi_d \quad \psi_e \quad \psi_f]^t$$

$$J \frac{d\Omega}{dt} = T_e - T_{load} - B \cdot \Omega \tag{17.31}$$

$$T_e = \frac{1}{2} [i]^t \cdot \frac{\partial [L(\theta_{er})]}{\partial \theta} \cdot [i] \tag{17.32}$$

#### *17.5.6.2 Control System*

Controlling the BLDC-MRM with an inverter with a reduced number of switches, $2 \cdot (m + 1)$ compared with 4 m (full bridge), could become a problem since the null current varies between $\pm m \cdot I_{phase}$ (in our case $\pm 6 \cdot 12$ A = $\pm 72$ A) and the null inverter leg needs to be oversized. This situation drastically improves if each even phase is connected backwards to the inverter, while prescribing inverse current through, so that from the machine point of view, the same m.m.f. is created [36]. The result is a null current variation in the range $\pm 2 \cdot I_{phase}$ (for an even number of phases) or $\pm I_{phase}$ (for an odd number of phases). The even number of phases, though, allows for simplified fault tolerant operation of three out of six phases in a RelSyn (sinusoidal current) mode. The inverter null leg should be designed with the above observation into consideration. Since, for convenience, we used three industrial voltage source inverters (5.5 kVA, Danfoss FC-302P7K5), for protection reasoning, the null point is divided in three and connected to the third inverter (see Figure 17.28).

The control scheme for the BLDC-MRM is presented in Figure 17.29 and involves a speed PI controller with limitation (±75, $k_p = 3$, $k_i = 300$). The speed controller output is fed to the current generation block that will generate the reference currents based on rotor position (by the encoder in Figure 17.13) and reference speed. This block assures a complete control of field/torque current levels and the commutation period (Figure 17.10).

Current error is regulated by PI controllers ($k_{p\_i} = 5$, $k_{i\_i} = 5000$) similar to the null point voltage error (kp_n = 0.001, ki_n = 0.002), both of them being presented in Figures 17.30 and 17.31.

The control scheme was implemented as a discrete model in MATLAB/Simulink, simplifying the control algorithm implementation on the dSpace 1103 platform (see Figure 17.13). Initial simulation results [37] used bang-bang current controllers and a full-bridge inverter, but practical implementation has forced us to make use of PI controllers for regulating current and null point voltage, since a hysteresis controller, though simpler, has the same

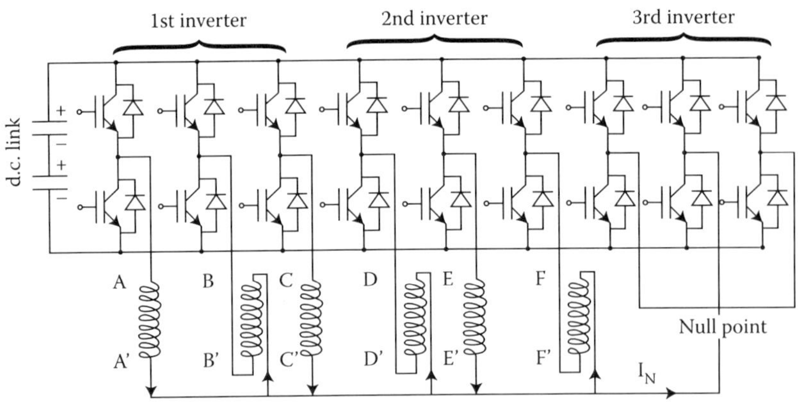


**FIGURE 17.28**
Phase connection for control with reduced number of switches.

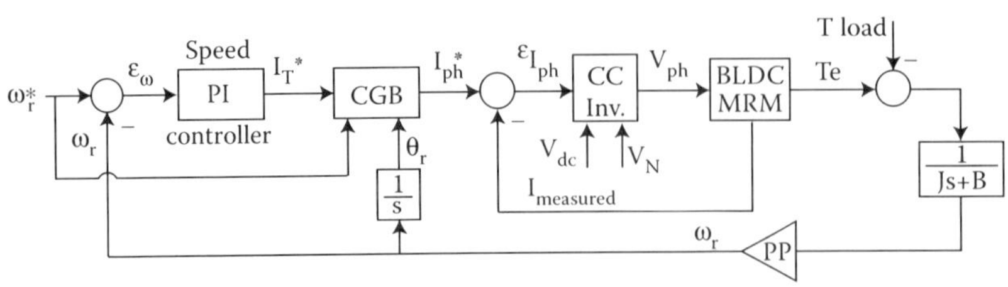


**FIGURE 17.29**
Speed control scheme of the BLDC-MRM. (After D. Ursu et al., *IEEE-Trans.*, IA-51(3), 2105–2115, 2015.)

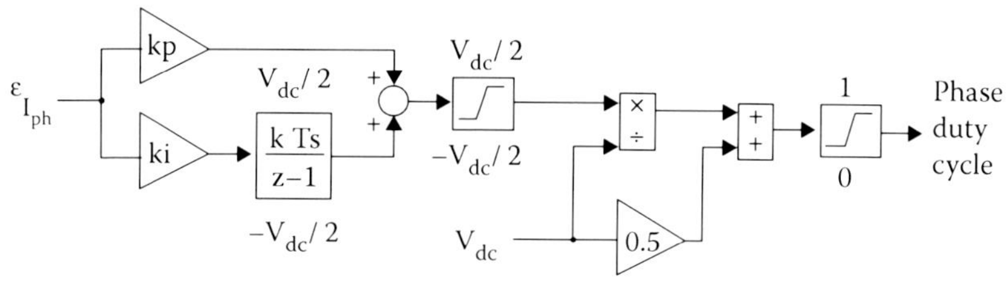


**FIGURE 17.30**
Phase current controller ($V_{dc}$ = 250 V). (After D. Ursu et al., *IEEE-Trans.*, IA-51(3), 2105–2115, 2015.)

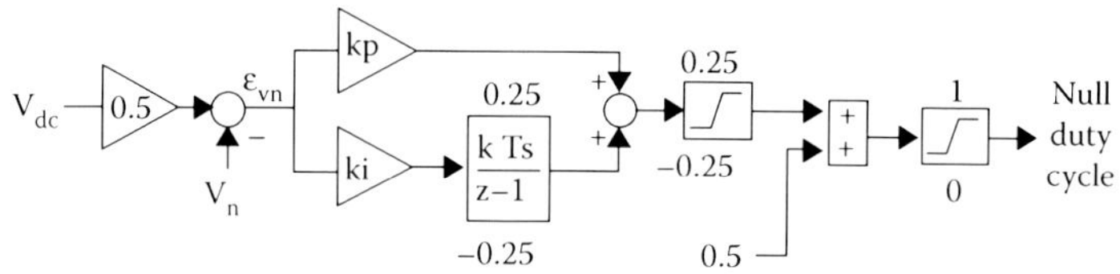


**FIGURE 17.31**
Null point voltage regulation ($V_{dc}$ = 250 V). (After D. Ursu et al., *IEEE-Trans.*, IA-51(3), 2105–2115, 2015.)

problem as in DTFC control applied on cage-less machines: high current (torque) ripple, high acoustical noise, and vibrations [38]. Three particular cases have been studied and presented next: motoring mode at 250 r/min (base speed) and 500 r/min, generating mode at base speed, and speed reversal with flux-weakening operation.

### *17.5.6.3 Motoring Mode*

The BLDC-MRM is started at no load (t = 1.2 s), then it is loaded through the d.c. machine, which, at t = 4.7 s, becomes a generator. Since the d.c. machines induced voltage is small at 250 rpm, ($V_a$ = 14 V), its armature circuit is short-circuited at base speed giving an armature current of $I_a$ = 37 A. Knowing all the losses of the d.c. machine, the computed shaft torque is 20 Nm. This result agrees with static torque measurements (at $I_f = I_t$ = 9 A, T = 22 Nm) since in running conditions, torque pulsations due to magnetic, mechanical, and geometrical causes make the average torque smaller. Measured phase currents are plotted in Figure 17.32, whereas in Figure 17.33, for clarity of representation, only one measured phase current is compared with a simulated phase current in identical loading conditions. The null current waveform is almost as predicted by theory [36], thus ensuring that the PI null point voltage controller is the right solution to implement for this control system (see Figure 17.34). The measured speed is plotted in Figure 17.35, and it shows a speed ripple of 0.8% at no load and 3.2% in load conditions.

At 500 r/min, the induced voltage in the d.c. machine becomes too high ($V_a$ = 28 V), and an external resistance is needed in order to load the BLDC-MRM (at t = 3.9 s, Figure 17.36) Now, the d.c. machine armature current is $I_a$ = 39 A, which means that the shaft torque is 21 Nm. This represents the upper limit at this speed, slightly higher than the expected 35 Nm/2 = 17.5 Nm, needed for constant power speed range. Field weakening can be seen in phase current waveforms (see Figures 17.36 and 17.37) by a decreased value of the field current over the span of two slots (see Figure 17.37). Measured null current appears in Figure 17.38 for the transients at 500 rpm (Figure 17.35). Speed pulsations at no load are

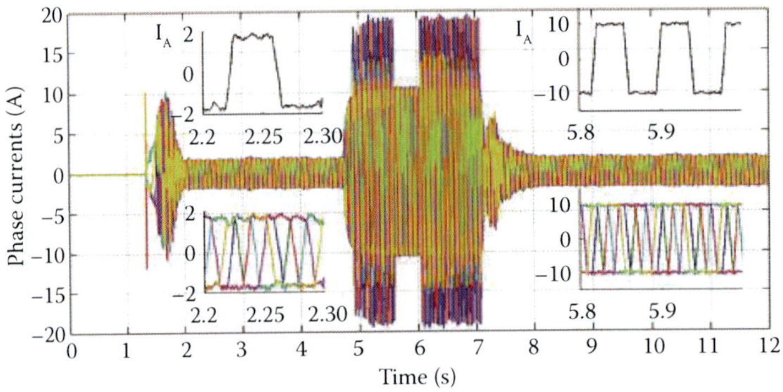


**FIGURE 17.32**
Measured phase currents at 250 r/min with zooming boxes at no load and load. (After D. Ursu et al., *IEEE-Trans.*, IA-51(3), 2105–2115, 2015.)

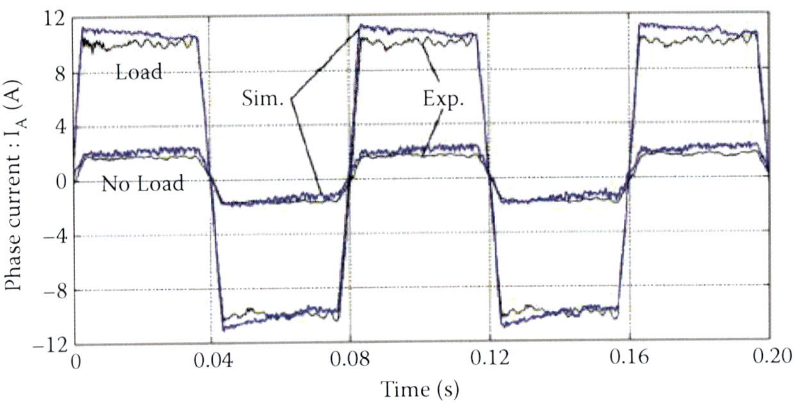


**FIGURE 17.33**
Simulated versus measured phase current at base speed (no load and load). (After D. Ursu et al., *IEEE-Trans.*, IA-51(3), 2105–2115, 2015.)

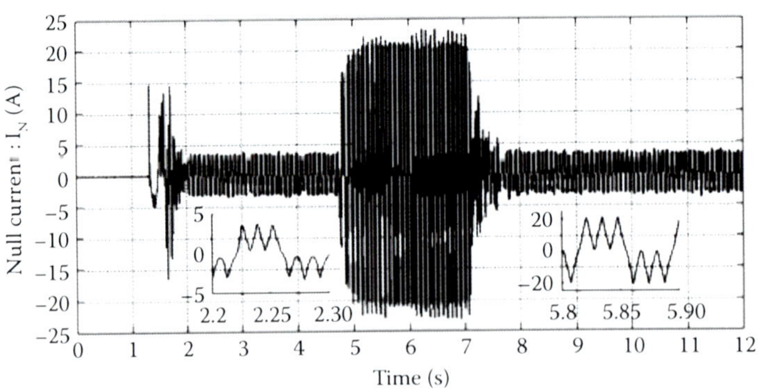


**FIGURE 17.34**
Measured null conductor current at base speed with zooming boxes at no load and load. (After D. Ursu et al., *IEEE-Trans.*, IA-51(3), 2105–2115, 2015.)

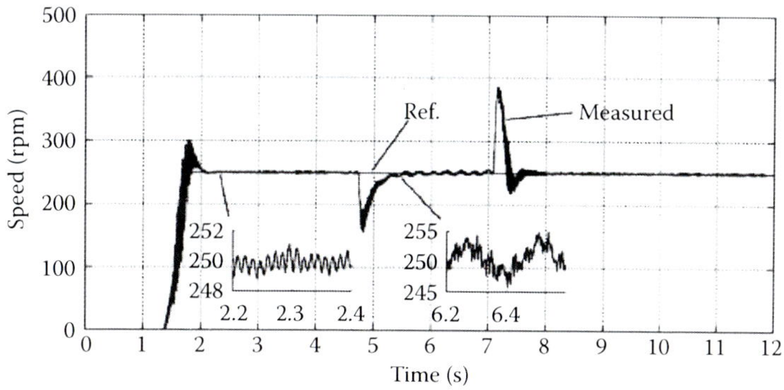


**FIGURE 17.35**
BLDC-MRM measured speed during transients. (After D. Ursu et al., *IEEE-Trans.*, IA-51(3), 2105–2115, 2015.)

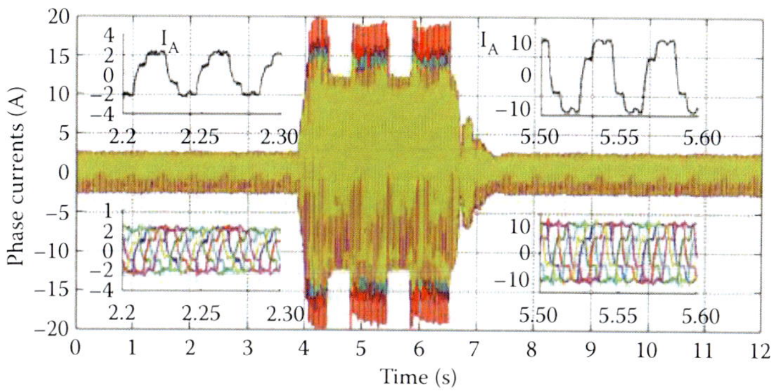


**FIGURE 17.36**
Measured phase currents at 500 r/min (motoring). (After D. Ursu et al., *IEEE-Trans.*, IA-51(3), 2105–2115, 2015.)

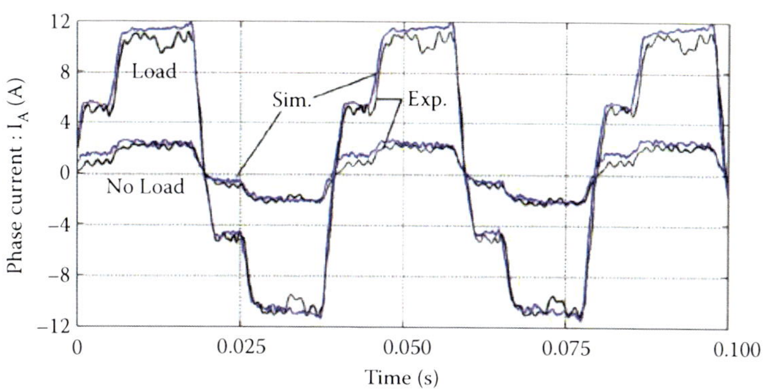


**FIGURE 17.37**
Simulated versus measured phase current at 500 r/min (no load and load). (After D. Ursu et al., *IEEE-Trans.*, IA-51(3), 2105–2115, 2015.)

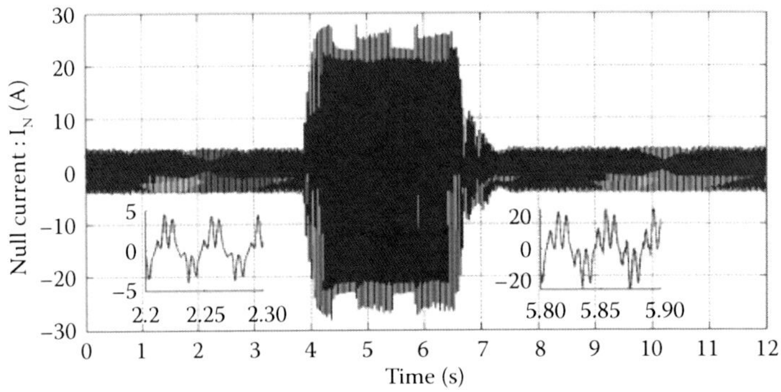


**FIGURE 17.38**
Measured null current at 500 r/min with zooming boxes at no load and load (monitoring). (After D. Ursu et al., *IEEE-Trans.*, IA-51(3), 2105–2115, 2015.)

0.3%, whereas, at load, they are 1.2% (see Figure 17.39). In the absence of a torque-meter, we can conclude by studying speed ripple that torque ripple is also small.

#### 17.5.6.4 Generating Mode

Generator mode is tested by driving the d.c. machine as a motor at t = 1.9 s, with a d.c. motor armature applied voltage $V_a = 32$ V (which corresponds to a speed of 600 r/min). Since our drive has a speed loop with a prescribed reference speed equal to base speed, the BLDC-MRM becomes a generator and receives mechanical power from the d.c. motor (see Figures 17.40 through 17.42). The d.c. machine armature current in this case is equal to $I_a = 37$ A. Figure 17.43 shows the prescribed BLDC-MRM torque current from the PI speed controller, which clearly changes sign.

#### 17.5.6.5 Flux Weakening and Speed Reversal

Flux-weakening operation is presented in Figure 17.44a through c. The ±500 r/min bound (2:1 speed range) is related to machine limit (small airgap: 0.3 mm, kept so small for eventual testing as a reluctance synchronous motor, for comparisons) and not to control system

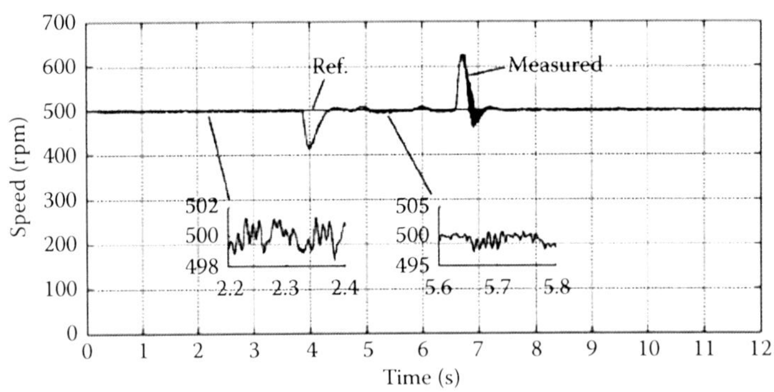


**FIGURE 17.39**
BLDC-MRM measured speed for sudden loading and unloading at 500 r/min (motoring). (After D. Ursu et al., *IEEE-Trans.*, IA-51(3), 2105–2115, 2015.)

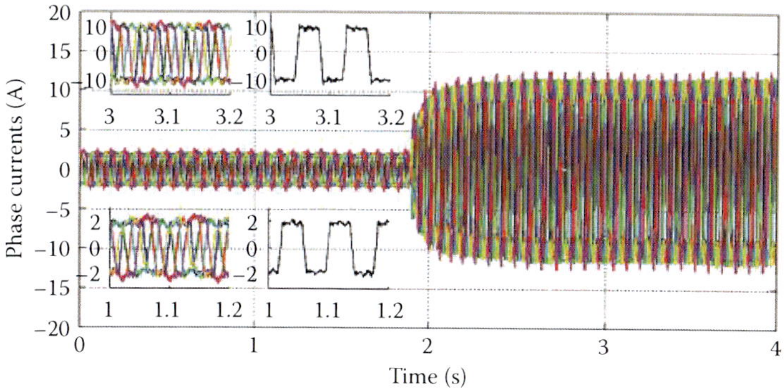


**FIGURE 17.40**
Measured phase currents at 250 r/min with zooming boxes at no-load motor and then loaded generator. (After D. Ursu et al., *IEEE-Trans.*, IA-51(3), 2105–2115, 2015.)

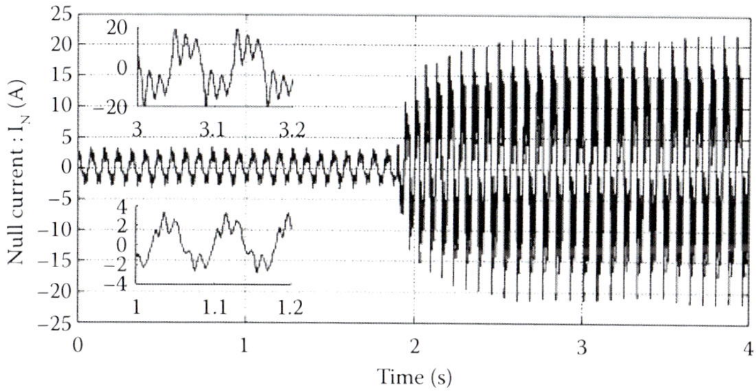


**FIGURE 17.41**
Measured null current at base speed with zooming boxes at no load and load at 250 r/min (as in Figure 17.32). (After D. Ursu et al., *IEEE-Trans.*, IA-51(3), 2105–2115, 2015.)

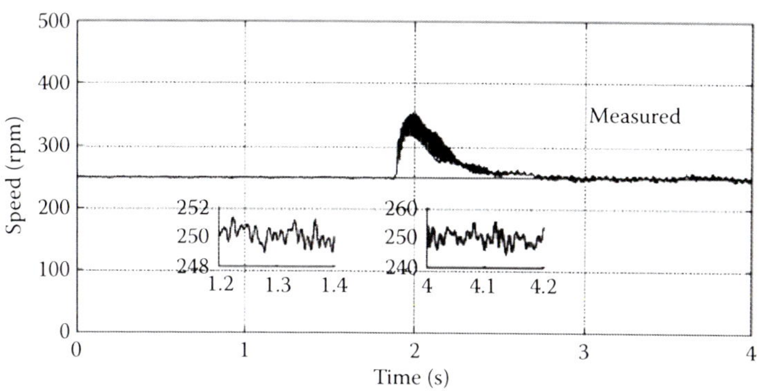


**FIGURE 17.42**
BLDC-MRM measured speed at no-load motor and loaded generator. (After D. Ursu et al., *IEEE-Trans.*, IA-51(3), 2105–2115, 2015.)

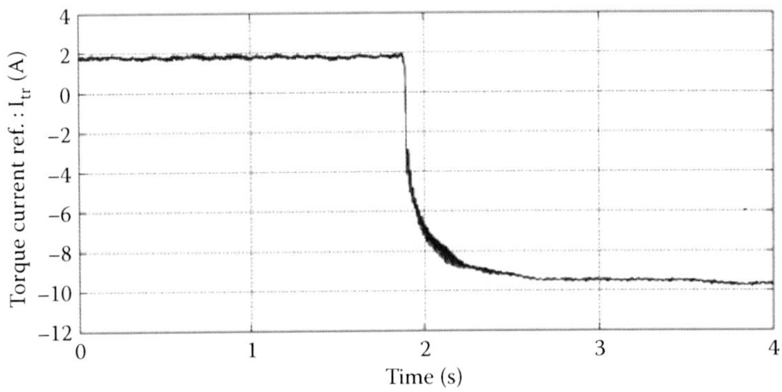


**FIGURE 17.43**
Prescribed torque current reference value (from PI speed controller). (After D. Ursu et al., *IEEE-Trans.*, IA-51(3), 2105–2115, 2015.)

or physical principles, since Reference 39 shows a 4:1 constant power speed range on a new (larger airgap, multiple flux barriers) rotor design, implemented to a five-phase BLDC-MRM.

Based on the computed mechanical and iron losses and on the measured phase resistance (at 70°C) $R_p = 0.4\ \Omega$, knowing the root-mean-square value of the phase current ($I_{ph} = 9$ A) at which we obtained a 20-Nm shaft torque (see Section 17.5.6.3), an efficiency of $\eta = 71\%$ at base speed $n_b = 250$ r/min was obtained for the six-phase BLDC-MRM. Since the machine was not "born" from an optimization process, and the objective was the testing of control solutions, the result is considered acceptable (much higher efficiency [above 90%] in a 5 phase similar torque/speed BLDC-MRM [39–41] was measured).

## 17.6 Summary

- This review chapter refers to a vibrant new direction for electric drives (for motoring and generating purposes) which seems now very close to reaching wide markets.
- The multiphase electric machines, controlled by multiphase PWM converters for variable speed operation with controlled speed (or torque) for motoring and active and reactive power for generating, are dedicated to safety critical applications such as transportation (by air, road, or railway) and to larger power, for lower-power electronics costs.
- With multiphase induction motors, but probably with PMSMs and BLDC-MRMs, and other types, the efficiency will not be dramatically improved over three-phase drives but the torque density may be improved by adding a third (and fifth) current harmonic in 5(7) phase motors with concentrated ($q \leq 1$) stator windings. This addition also improves the stability, but it worsens the performance during faulty operation, when the harmonics should be inhibited.
- The PWM in the converter and the control has more options with multiphase electric machine drives, but also more current sensors are required, in general, for closed-loop control.

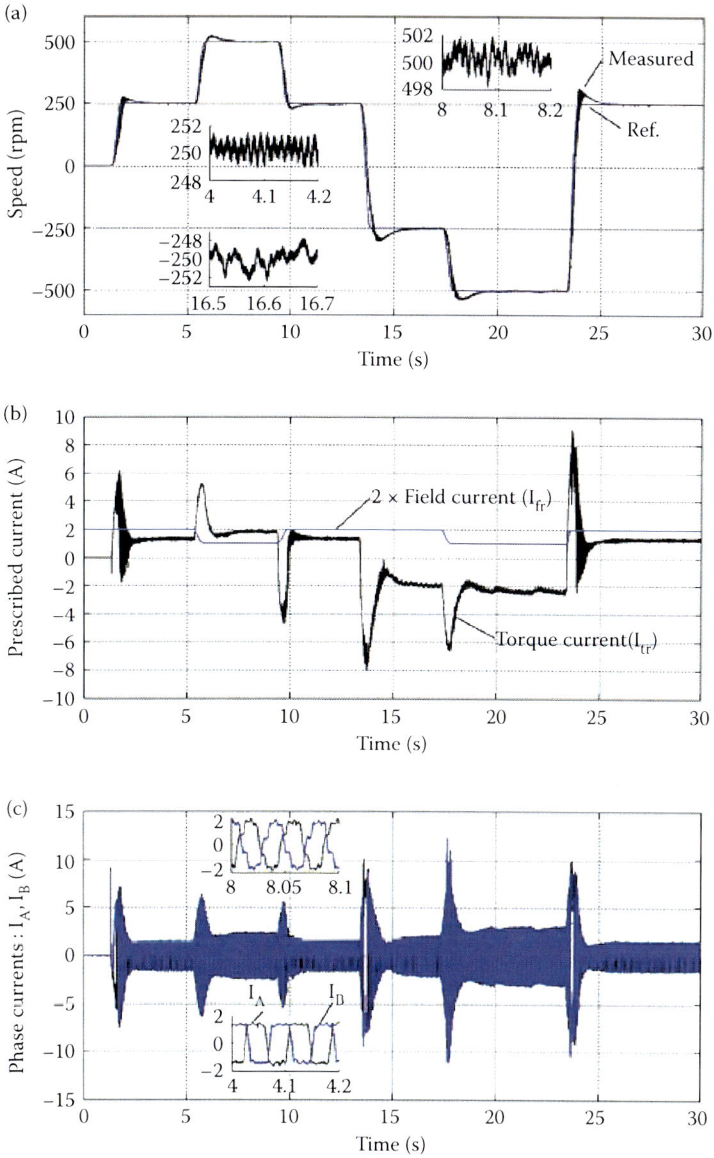


**FIGURE 17.44**
Speed reversal tests of the BLDC-MRM. (a) Measured speed. (b) Prescribed torque and field current reference value. (c) Measured phase currents (only two shown: $I_A$ and $I_B$). (After D. Ursu et al., *IEEE-Trans.*, IA-51(3), 2105–2115, 2015.)

- The reduction of the number of current sensors and motion-sensorless control (under healthy and open-phases faults) are still due to be accomplished.
- With two series connected multiple phase machines supplied from a single multiphase inverter, with independent speed control, new multimotor/generator applications (such as propulsion systems for HEVs or dual motors in one in HVAC) may also be targeted [1].

## References


1. E. Levi, Multiphase electric machines for variable-speed applications, *IEEE-Trans. Ind. Electron.*, vol. IA-55, no. 5, 2008, pp. 1893–1909.
2. L. Zheng, J.E. Fletcher, B.W. Williams, X. He, Dual-plane vector control of a five-phase induction machine for an improved flux pattern, *IEEE-Trans.*, vol. IA-55, no. 5, 2008, pp. 1996–2005.
3. A.R. Munoz, T.A. Lipo, Complex vector model of the squirrel cage induction machine including rotor bar currents, *IEEE-Trans.*, vol. IA-35, no. 6, 1999, pp. 1330–1340.
4. K. Marouani, L. Baghli, D. Hadiouche, A. Kheloui, A. Rezzoug, A new PWM strategy based on a 24-sector vector space decomposition for a six-phase VSI fed dual stator induction motor, *IEEE-Trans.*, vol. IA-55, no. 5, 2008, pp. 1910–1920.
5. R. Kianinezhad, B. Nahid-Mobarakeh, L. Baghli, F. Betin, G.A. Capolino, Modeling and control of six-phase symmetric induction machine under fault condition due to open phases, *IEEE-Trans.*, vol. IA-55, no. 5, 2008, pp. 1966–1977.
6. D. Casadei, D. Dujic, E. Levi, G. Serra, A. Tani, L. Zarri, General modulation strategy for seven-phase inverters with independent control of multiphase voltage space vectors, *IEEE-Trans.*, vol. IA-55, no. 5, 2008, pp. 1921–1932.
7. M. Mengoni, L. Zarri, A. Tani, L. Parsa, G. Serra, D. Casadei, High torque-density control of multiphase induction motor drives operating over a large speed range, *IEEE-Trans.*, vol. IE-62, no. 2, 2015, pp. 814–825.
8. V.T. Somasekhar, V.B. Reddy, K. Sivakumar, A four level inversion scheme for a 6n-pole open-end winding IM drive for an improved dc. link utilization, *IEEE-Trans.*, vol. IE-61, no. 9, 2014, pp. 4565–4572.
9. C.B. Jacobina, I. Soares de Freitas, C. Ricarte da Silva, M. Beltrao de Rossiter, E.R. Cabral da Silva, Reduced switch count six phase ac motor drive systems without input reactor, *IEEE-Trans.*, vol. IE-55, no. 5, 2008, pp. 2024–2032.
10. T. Laamayad, F. Naceri, R. Abdessemed, Fuzzy adaptive controller based on Lyapunov theory for control of the dual star induction machine, *J Electr. Eng.*, vol. 13, no. 1, 2012.
11. A. Boglietti, R. Bojoi, A. Cavagnino, A. Tenconi, Efficiency analysis of PWM inverter fed three-phase and dual three-phase high frequency induction machines for low/medium power applications, *IEEE-Trans.*, vol. IE-55, no. 5, 2008, pp. 2015–2023.
12. A.S. Abdel-Khalik, A.S. Morsy, S. Ahmed, A.M. Massoud, Effect of stator winding connection on performance of five-phase induction machines, *IEEE-Trans.*, vol. IE-61, no. 1, 2014, pp. 3–19.
13. M.A. Shamsi-Nejad, B. Nahid-Mobarakeh, S. Pierfederici, F. Meibody-Tabar, Fault tolerant and minimum loss control of double star synchronous machines under open phase conditions, *IEEE-Trans.*, vol. IE-55, no. 5, 2008, pp. 1956–1965.
14. N. Bianchi, S. Bolognani, M. Dai-Pre, Impact of stator winding of a five-phase PM motor on postfault operation, *IEEE-Trans.*, vol. IE-55, no. 5, 2008, pp. 1978–1987.
15. S. Kallio, J. Karttunen, P. Peltoniemi, P. Silventoinen, Online estimation of double-star IPM machine parameters using RLS algorithm, *IEEE-Trans.*, vol. IE-61, no. 9, 2014, pp. 4519–4530.
16. S. Dwari, L. Parsa, An optimal control technique for multiphase PM machines under open-circuit faults, *IEEE-Trans.*, vol. IE-55, no. 5, 2008, pp. 1988–1995.
17. A. Mohammadpour, S. Sadeghi, L. Parsa, A generalized fault-tolerant control strategy for 5-phase PM motor drives considering star, Pentagon and Pentacle connections of stator windings, *IEEE-Trans.*, vol. IE-61, no. 1, 2014, pp. 63–75.
18. M. Ruba, D. Fodorean, Analysis of fault-tolerant multiphase power converter for a nine-phase permanent magnet synchronous machine, *IEEE-Trans.*, vol. IA-48, no. 6, 2012, pp. 2092–2101.
19. S. Oai, S. Morimoto, M. Samada, and Y. Inoue, Performance evaluation of high power density PMs SyncRM with Ferrite magnets, *Proc. IEEE ECCE*, 2011, pp. 4195–4200 (ECCExplore).
20. M. Takeno et al., Test results and torque Improvement of the 50-kW switched reluctance motor designed for hybrid electric vehicles, *IEEE Trans. Ind. Appl.*, vol. 48, no. 4, 2012, pp. 1327–1334.

21. E. Sulaiman, T. Kosaka, M. Matsui, High power density design of a 6-slot, 8 pole hybrid excitation flux switching machine for HEV, *IEEE Trans. Mag.*, vol. MAG-48, no. 10, 2011, pp. 4453–4456.
22. E. Sulaiman, T. Kosaka, M. Matsui, A new structure of 12 slots-10 pole field-excitation flux switching synchronous machine for HEV, *Proc. EPE Appl.*, Birmingham, UK, 2011, pp. 1–10.
23. E.B. Sulaiman, T. Kosaka, N. Matsui, Design study and experimental analysis of wound field flux switching motor for HEV applications, *Proc. ICEM*, 2012, pp. 1269–1275 (IEEExplore).
24. R. Mayer, H. Mosebach, U. Scroder, H. Weh, Inverter fed multiphase reluctance machine with reduced armature reaction and improved Power density, *Proc. ICEM*, Munich, Germany, Part 3, 1986, pp. 1138–1141.
25. I. Boldea, G. Papusoiu, S.A. Nasar, Z. Fu, A novel series connected switched reluctance motor, *Proc. ICEM*, Part 3, 1990, pp. 1212–1217, MIT, USA.
26. J.D. Law, A. Chertok, T.A. Lipo, Design and performance of field regulated reluctance machine, *Conference Record of IEEE IAS Annual Meeting*, 1992, vol. 1, pp. 234–241.
27. E.T. Rakgati, M.J. Kamper, A.D. Le Roux, Torque performance of optimally designed six-phase reluctance DC machine, *Conference Record of IEEE IAS Annual Meeting*, 2006, vol. 3, pp. 1186–1192 (IEEExplore).
28. Y.-L. Ai, M.J. Kamper, A.D. Le Roux, Novel direct flux and direct torque control of six-phase induction machine with nearly square airgap flux density, *IEEE Trans. Ind. Appl.*, vol. 43, no. 6, 2007, pp. 1534–1544.
29. N. Gule, M.J. Kamper, Multiphase cage-rotor induction machine drive with direct implementation of brush DC operation, *IEEE Trans. Ind. Appl.*, vol. 48, no. 6, 2012, pp. 2012–2020.
30. I. Boldea, L.-N. Tutelea, D. Ursu, BLDC multiphase reluctance machines for wide range applications: A revival attempt, *Proc. EPE/PEMC*, 2012, pp. LS1b.1-1–LS1b.1-6 (IEEExplore).
31. D. Ursu, L. Tutelea, I. Boldea, Proposal with 2-D FEM analysis of a six phase, 12 poles, 3 kW, 200 rpm BLDC multiphase reluctance machine wind generator, *Proceedings of the EPE Applications Conference*, 2013, pp. 1–9.
32. D. Ursu, V. Gradinaru, B. Fahimi, I. Boldea, Six-phase BLDC reluctance machines: FEM-based characterization and four-quadrant control, *IEEE-Trans.*, vol. IA-51, no. 3, 2015, pp. 2105–2115.
33. A. Boglietti, A. Cavagnino, M. Lazzari, M. Pastorelli, Predicting iron losses in soft magnetic materials with arbitrary voltage supply: An engineering approach, *IEEE Trans. Mag.*, vol. 39, no. 2, 2003, pp. 981–989.
34. A. Kedous-Lebouc, B. Cornut, J.C. Perrier, P.F. Manfe, T. Chevalier, Punching influence on magnetic properties of the stator teeth of an induction motor, *J. Magn. Magn. Mater.*, vol. 254/255, 2003, pp. 124–126.
35. E.T. Rakgati, M.J. Kamper, Equivalent circuit parameters identification for 6 phase reluctance DC machine, *Proc. IEEE IEMDC*, 2007, vol. 2, pp. 910–915 (IEEExplore).
36. H. Weh, U. Schroder, Static Inverter concepts for multiphase machines with square-wave current-field distribution, *Proc. EPE*, Brussels, Belgium, 1985, pp. 1147–1152.
37. S. Agarlita, D. Ursu, L. Tutelea, I. Boldea, B. Fahimi, BLDC multiphase reluctance machines: A revival attempt with 2D FEM investigation and standstill tests, *Proc. IEEE ECCE*, 2013, pp. 1850–1857 (IEEExplore).
38. C. Lascu, I. Boldea, F. Blaabjerg, Super-twisting sliding mode control of torque and flux in permanent magnet synchronous machine drives, *Proc. IEEE IECON*, 2013, pp. 3171–3176.
39. D. Ursu, P. Shamsi, B. Fahimi, I. Boldea, 5 phase BLDC-MRM: Design, control, FEA and steady-state operation experiments, *Proceedings of the International Conference on OPTIM Electrical and Electronic Equipments*, 2014, Brasov, Romania, pp. 354–361 (IEEExplore).
40. R. Babau, I. Boldea, T.J.E. Miller, and N. Muntean, Complete parameter identification of large induction machines from no load acceleration–deceleration tests, *IEEE Trans. Ind. Electron.*, vol. 54, no. 4, 2007, pp. 1962–1972.
41. G. Bertotti et al., An improvement estimation of iron losses in rotating electrical machine, *IEEE Trans. Magn.*, vol. 27, no. 6, 1991, pp. 5007–5009.

# Index

## B



## C

## D

## K



## L